This volume is dedicated to all those miners who have lost their lives in rockbursts, and to the memory of my daughter Emma.

PROCEEDINGS OF THE 3RD INTERNATIONAL SYMPOSIUM ON ROCKBURSTS AND SEISMICITY IN MINES / KINGSTON / ONTARIO / CANADA / 16 - 18 AUGUST 1993

Rockbursts and Seismicity in Mines 93

Edited by
R. PAUL YOUNG
Queen's University, Kingston, Ontario, Canada

A.A. BALKEMA / ROTTERDAM / BROOKFIELD / 1993

Cover: 359 source located Acoustic Emission/Microseismic (AE/MS) events (M < – 2) following the excavation of a one metre section of the Mine-by tunnel, at 420 m depth in granite, at Atomic Energy of Canada's Underground Research Laboratory, Manitoba, Canada. Photograph (courtesy of Atomic Energy of Canada Ltd.) shows the Mine-by tunnel after excavation and the breakout notches in the roof and floor. Note the coincidence of AE/MS events in the notch regions and the measured orientation of σ1 (after Feignier and Young and Martin and Young, this volume).

The texts of the various papers in this volume were set individually by typists under the supervision of each of the authors concerned.

Authorization to photocopy items for internal or personal use, or the internal or personal use of specific clients, is granted by A.A. Balkema, Rotterdam, provided that the base fee of US$1.00 per copy, plus US$0.10 per page is paid directly to Copyright Clearance Center, 27 Congress Street, Salem, MA 01970, USA. For those organizations that have been granted a photocopy license by CCC, a separate system of payment has been arranged. The fee code for users of the Transactional Reporting Service is: 90 5410 320 5/93 US$1.00 + US$0.10.

Published by
A.A. Balkema, P.O. Box 1675, 3000 BR Rotterdam, Netherlands
A.A. Balkema Publishers, Old Post Road, Brookfield, VT 05036, USA

ISBN 90 5410 320 5
© 1993 A.A. Balkema, Rotterdam
Printed in the Netherlands

Rockbursts and Seismicity in Mines, Young (ed.) © 1993 Balkema, Rotterdam, ISBN 90 5410 320 5

Table of contents

3 Monitoring of seismicity and geomechanical modelling

Introduction

R. Paul Young
Queen's University, Kingston, Ont., Canada

This International Symposium on Rockbursts and Seismicity in Mines is the third in a series that began in Johannesburg[1] in September 1982 and continued in Minneapolis[2] in June 1988. One of the outstanding achievements of the past symposia was the bringing together of experts from many disciplines to contribute to a solution to the rockburst problem. This third symposium continues this tradition and I am pleased that contributions from rock mechanics, seismology, mining, geology and electronic and computer engineering provide a significant enhancement to the state of our knowledge in the field. The papers in this symposium volume explore international developments in both fundamental and practical aspects of the subject.

This volume contains six invited keynote lectures and sixty five written contributions. The publication date of this proceedings was organized to coincide with the symposium meeting held in Kingston, Canada, August 16-18, 1993. Approximately one hundred and twenty extended abstracts were received for consideration of which ninety were approved for submission as full papers. Each of the abstracts was evaluated by two reviewers and the submitted manuscript was considered for publication by the same reviewers. Manuscripts were accepted or rejected on an as submitted basis. Special allowance has been made for contributions from non-English speaking countries. Several manuscripts were completely retyped by my editorial staff in an attempt to improve the English when it was felt that these papers were essential to the symposium. This was deemed necessary in order to try and maintain the quality of publication achieved by Drs Gay and Wainwright following the first symposium and Dr Fairhurst following the second symposium. The result is five hundred pages of current international research and practice in the field.

The symposium program was developed to specifically address critical questions regarding mining-induced seismicity such as: What are the basic mechanisms of rockbursting? What is the present state-of-the-art in monitoring seismicity and prediction of rockbursts? What is the status of prevention and ground control methods? Can analytical or numerical methods be used to improve mine design for rockburst reduction? Papers in this volume have been grouped in three sections, each highlighted by two keynote lectures. The section on strong ground motion and rockburst hazard is introduced by Drs McGarr and Kaiser. The section on the mechanics of seismic events and stochastic methods is introduced by Drs Gibowicz and Rudajev. Finally, the section on monitoring of seismicity and geomechanical modelling is introduced by Drs Mendecki and Salamon. I chose keynote speakers for each section to address both the rock mechanics and seismological aspects of the problem, as well as choosing individuals who have made significant contributions to their discipline. I was delighted that they each agreed to contribute their insight and experience in the form of invited papers on the three key symposium themes.

It is not possible to organize an international symposium of this type without the cooperation of numerous individuals. I am grateful to all the authors who submitted papers to this symposium and volume. I thank the keynote authors and Dr Fairhurst for his very insightful preface to this volume. I am grateful to my international advisory committee, W. Blake, B.H.G. Brady, N.G.W. Cook, R. Dmowska, C. Fairhurst, B. Feignier, N.C. Gay, S.J. Gibowicz, B. Hobbs, A. McGarr, D. Morrison, P. Mottahed, W.D. Ortlepp, M.D.G. Salamon, S. Talebi, J. Udd, and R. Wetmiller, who provided me with input at strategic points during the organization of this symposium. I am deeply indebted to my reviewers and local organizing committee, S. Carlson, L. Farrell, B. Feignier, S. Maxwell, V. Putnam, C. Trifu, T. Urbancic and C. Yalin, who gave unsparingly of their time to make this proceedings and the Kingston meeting a reality. I would also like to thank Queen's University and the Mining Research Directorate of Canada for providing an interest free loan to start the organization of the symposium. I am also personally grateful to the Natural and Engineering Research Council of Canada, the Canadian mining industry and Queen's University who have sponsored my research work in the field since coming to Canada in 1984. I also thank all my staff and

1. Proceedings, Rockbursts and Seismicity in Mines, N.C. Gay and E.H. Wainwright (Editors), South African Institute of Mining and Metallurgy, Kelvin House, 2 Holland Street, Johannesburg, South Africa (1984), 363 p.
2. Proceedings, Rockbursts and Seismicity in Mines, C. Fairhurst (Editor), A.A. Balkema, P.O. Box 1675, 3000 BR Rotterdam, Netherlands (1990), 439 p.

students at the Engineering Seismology Laboratory, Department of Geological Sciences, Queen's University, without whose collective efforts we would not have been invited to host this very important symposium. Last but not least, I am indebted to the continuous support provided by my wife Teresa and the patience of my children Christopher and Adrian while I was burning that midnight oil! The symposium marks the end of a very rewarding and enjoyable time for me at Queen's University and the start of a new and exciting challenge at the University of Keele in the UK[3], where I will be establishing a Centre for Induced Seismicity Research. I hope that all who attend the symposium at Queen's University, August 16-18, 1993 will find it beneficial and see the progress which has been made in the field in recent years. Hopefully we will have the opportunity to meet again in Poland[4], which is the proposed location of the fourth International Symposium on Rockbursts and Seismicity in Mines.

Paul Young
Editor
Queen's University, Kingston, Canada

3. Professor R. Paul Young, Chair: Applied Seismology and Rock Physics, Department of Geology, University of Keele, Staffordshire, United Kingdom ST5 5BG.
4. For further details, contact Dr S.J. Gibowicz, Institute of Geophysics, Polish Academy of Sciences, Ks. Janusza 64, 01-452 Warsaw, Poland, and/or Dr S. Lasocki, University of Mining and Metallurgy, Institute of Geophysics, Al. Mickiewicza 30, 30-059 Cracow, Poland.

Rockbursts and Seismicity in Mines, Young (ed.) © 1993 Balkema, Rotterdam, ISBN 90 5410 320 5

Preface

Charles Fairhurst
University of Minnesota, Minneapolis, Minn., USA

One of the main attractions of the photoelastic modelling technique of stress analysis – used extensively before the advent of numerical modelling – was the vivid graphical portrayal of the global stress-distribution within a two-dimensional elastically stressed body, and the regions of stress concentrations, immediately recognizable through the "bunching together" of the photoelastic fringes or stress contours. The three-dimensional region affected by extraction in a rockburst-prone mine is much more complicated structurally than the photoelastic plate, but the benefits of rendering the solid opaque earth more transparent with respect to the distribution of in-situ stresses and regions of potential instabilities due to mining are significant enough that the search is worth pursuing vigorously

Impressive advances are being made in seismic (and radar) tomography, whole waveform analysis of seismic and microseismic releases, with associated insight into the source mechanisms of fault slip and energy release. This research is valuable in a variety of geotechnical engineering applications, e.g., the study of rock mass behaviour at potential underground sites for high-level radioactive waste repositories, but it assumes special significance in rockburst and induced seismicity studies, where a better understanding of the mechanics of strain energy accumulation and its violent release in the vicinity of excavations – or in surface vibrations – is critical to the development of safer mining and engineering procedures.

The difficulties of extrapolating from a few "hard-won" point determinations of in-situ stress to a global stress distribution are severe, so it is hard not to be excited when one sees global contours of "apparent stress" for a mining region, derived on the basis of seismic energy release observations. Clearly, there are questions and uncertainties as to the validity of the interpretations of the seismic signals, but it seems that a promising start has been made on a potentially very rewarding line of investigation. The parallel development of three-dimensional discontinuum numerical codes that will allow simulation of slip on faults and similar discontinuities in the rock mass, and calculation of the partition of the energy released dynamically by slip, gives added impetus to the seismic studies. The gain in understanding of the mechanics of rockbursts and induced seismicity through the simultaneous application of both the numerical modelling and the seismic/microseismic monitoring and interpretation could be major.

Complementary to such studies of unstable movements on existing discontinuities, recent research into the mechanics of the development of new discontinuities, through localization of deformation in intact rock to form shear bands and extensile slabbing, is making significant progress. Here again, numerical models of possible localization processes are starting to provide new insights into instabilities that may develop in rock in this manner.

Much of rockburst research to date has been concerned primarily with mining in tabular (i.e. essentially two-dimensional) ore bodies. The appearance of severe rock bursting in association with large-scale caving and other three-dimensional mining operations, and the growing number of reports of seismicity induced by fluid withdrawal and/or injection in oil and gas exploration, adds emphasis to the need to develop a better fundamental understanding of the mechanics of unstable energy releases induced in the earth's crust. The practical rules for planar mining operations do not apply.

How the severity of rockbursting is affected by increase in mining depth is a question of extreme importance in assessing the available resources and economics of ore bodies. The simple suggestion that, since the strain energy density in the rock (or at least the gravitational component) increases as the square of the depth, the severity of rockbursting will increase correspondingly, has been challenged – with arguments for both a more rapid and a less rapid increase. The answer may depend also on the development of improving mining technologies and techniques such as de-stressing. Again, understanding the mechanics of in-situ rock mass behaviour is a key.

Thus, the benefits of closer collaboration between seismologists, geologists, rock mechanics, mining and petroleum engineers in the common goal of better understanding and 'real-time' visualization of the stressing, deformation and energy release within an active mining region, are many. Recognition of these benefits was a prime reason for starting these international symposia. But the problem of making mines safer, and controlling the surface tremors induced by fluid extraction/injection, are immediate and urgent. We must consider what can be recommended in the light of what is known now – some answers may not need to await the sometimes painfully slow advance through basic research.

The Third International Symposium on Rockbursts and Seismicity in Mines takes place five years after the Second Symposium and promises to be an exciting and rewarding event, as scientists and engineers meet again, some with possible answers, some with urgent questions, to review developments worldwide and collectively seek to advance understanding in this serious and challenging geotechnical problem.

The effort in organizing an international meeting of this magnitude is considerable. It is the culmination of five years of labor starting with the end of the previous meeting. Everyone involved in or affected by the rockburst and induced seismicity problem owes a deep debt of gratitude to Professor R. Paul Young, Symposium Chairman, and his associates, for their tireless work on our behalf. We look forward to a stimulating and rewarding meeting.

Charles Fairhurst
University of Minnesota

1 Strong ground motion and rockburst hazard

Rockbursts and Seismicity in Mines, Young (ed.) © 1993 Balkema, Rotterdam, ISBN 90 5410 320 5

Keynote address: Factors influencing the strong ground motion from mining-induced tremors

A. McGarr
US Geological Survey, Menlo Park, Calif., USA

ABSTRACT: Mining in the Witwatersrand gold fields, South Africa, at depths between 2000 and 3300 m induces tremors with magnitudes M ranging up to 5.2. Although most of the energy released by these events is consumed in overcoming a stress that resists fault slip of approximately 45 MPa, the remaining energy, typically about 0.01 of that released, is often sufficient to cause substantial damage underground for tremors of $M > 2$, or so. The potentially-damaging ground motion is largely restricted to the seismic source region, adjacent to the causative fault slip, where peak particle velocities of as much as 4 m/s may occur. Whereas the peak ground velocities appear to be independent of magnitude, the corresponding fault slips increase systematically with M up to a maximum somewhat in excess of 400 mm. The size of the source region increases with M similarly. Broad-band, wide-dynamic-range seismic data recorded underground, in two of the most seismically active Witwatersrand gold fields, were inverted to determine the complete moment tensors for 16 events in the magnitude range 1.7 to 3.3. The events fall into two categories. In the first, the deformation is exclusively deviatoric involving normal fault slip. In the second, the moment tensor includes normal fault slip plus coseismic volume reduction comparable in magnitude to the shear deformation as measured by the product of slip and fault area. For events in the second category, high-frequency seismic ground motion is markedly suppressed compared to corresponding ground motion associated with purely deviatoric tremors. This suppression, which increases with decreasing magnitude below 3, is attributed to coupling between the fault slip and the nearby mine stope, which undergoes coseismic closure.

1 INTRODUCTION

One of the conclusions in Ortlepp (1984) serves as a good introduction to this report. "The design and selection of materials for various tunnel support components, such as rockbolts and grouted steel tendons, and for hydraulic props in stopes, would be improved if quantitative measurements were made of the velocity and acceleration of the damaged rock surfaces, particularly in the vicinity of large seismic sources." In this concluding statement Ortlepp defined very nicely the essential goal of strong motion seismology in deep mines. Unfortunately, even today, almost no ground motion data exists from the immediate vicinity of a large seismic source. Accordingly, this report describes how we attempt to infer the ground motion within the large seismic sources in the absence of actual ground motion data there.

Following Wagner (1984), the measure of damaging ground motion emphasized here is peak ground velocity as this parameter can be most easily related to underground support requirements. Moreover, both Wagner (1984) and Ortlepp (1984) reviewed considerable evidence to the effect that underground damage is primarily restricted to the immediate environs of the seismic source. Thus, the primary focus of this report is the maximum credible level of ground velocity at the source. What are the important factors to consider when estimating this level?

Although measurements of peak acceleration have been widely used in earthquake engineering studies (e.g., Joyner and Boore, 1981), peak acceleration measured underground is generally undiagnostic of damage potential. Figure 5 of McGarr et al. (1981) illustrates this point. That figure shows a peak acceleration of 7.7 g recorded in the sidewall of an experimental tunnel in the ERPM gold mine near Johannesburg. In the earthquake engineering context, such a huge acceleration would be completely catastrophic, whereas this ground motion, in fact, caused no damage whatsoever to the tunnel. Integration of the accelerogram to velocity (Figure 5 of Mc-

Garr et al., 1981) revealed a peak velocity of 4.6 cm/s, which is more than an order of magnitude below even a conservative damage threshold in a well-supported area.

In the deep gold mines of South Africa, the most commonly used support behind the stope faces, the rapid-yielding hydraulic props (Tyser and Wagner, 1977; van Antwerpen and Spengler, 1984) can accommodate ground velocities in excess of 1 m/s but less than 2 m/s according to Wagner (1984). Because the extensive stoping leads directly and predictably (e.g., Cook, 1963; McGarr et al., 1975; McGarr, 1976) to induced earthquakes these production areas tend to be the loci of both the tremors and the intense damage often associated with them. The problem, then, is to figure out whether or not these props are capable of sustaining the maximum credible ground velocity at the source. In the absence of ground velocity data at the source, the strategy, reviewed here, is to model the source process so as to be able to interpret seismic data measured well outside the source in terms of the time history of the causative fault slip. In this way we try to predict the ground velocity at the source as a function of seismic source parameters that can readily be measured. One of the key issues, therefore, is how these source parameters scale. That is, how does the ground motion of small tremors, which are frequently observed and analyzed, relate to that of the comparatively rare, large and damaging events?

2 SOURCE PROCESSES

I review here some recent developments in the measurement and interpretation of various source parameters. First, measuring the size of an earthquake, or mining-induced tremor, has progressed from the traditional magnitude scale (e.g., Richter, 1958) to the scalar seismic moment (Aki, 1966; Brune, 1970; Spottiswoode and McGarr, 1975) and more recently to moment tensors (e.g., Stump and Johnston, 1977; Spottiswoode, 1980; Gibowicz, 1990; Feignier and Young, 1992;

McGarr, 1992, 1993). Whereas the magnitude scale was developed simply to compare one earthquake to another and has no intended significance regarding earthquake source processes, the scalar seismic moment, M_0, has the specific source definition

$$M_0 = \mu A D_0 \qquad (1)$$

where μ is the modulus of rigidity, A is the fault area and D_0 is the average slip across the fault.

The moment tensor, a 3×3 symmetrical tensor, is the most complete description of a compact seismic source. As described in detail by McGarr (1992, 1993), broad-band, wide-dynamic-range ground motion can be inverted to determine the six independent components of the moment tensor. It is then straightforward to decompose this tensor into deviatoric and volumetric components. The deviatoric component can be interpreted as slip across one or more fault planes

$$M_0(\text{dev}) = \mu \Sigma A_i D_i \qquad (2)$$

which can be summed to yield an equivalent scalar moment, as in (1). The trace of the moment tensor is the volumetric component, which is related to the coseismic volume change ΔV according to (Aki and Richards, 1980, equation 3.34)

$$M_0(\text{vol}) = (3\lambda + 2\mu)\Delta V \qquad (3)$$

where λ is one of Lamé's elastic parameters.

To many readers, it might seem inappropriate to be concerned with complete moment tensor descriptions of seismic sources in a report about the practical issue of damaging ground motion. Surprisingly, it turns out, as will be seen, that the nature of the moment tensor, as decomposed into deviatoric and volumetric components, is an important factor influencing the strong ground motion. Whereas natural tectonic earthquakes are thought to involve exclusively deviatoric moment tensors, a substantial fraction of mining-induced tremors, at least in the deep mines of South Africa, show a significant volumetric component which has a first-order effect on the high-frequency ground motion.

Before describing these results, though, it is necessary to review some measures of the source that are related to the strong ground motion. The three measures considered here are radiated seismic energy, E_s, and two peak ground motion parameters $R\underline{v}$ and $\rho R\underline{a}$, where R is hypocentral distance, $\underline{v}$ and $\underline{a}$ are peak ground velocity and acceleration, respectively, as recorded in a whole space (underground) and ρ is density. R is a factor in the two ground motion parameters so as to compensate for the geometrical spreading of the far-field shear wave amplitudes according to $1/R$; other sources of wave attenuation, scattering for example, are neglected in these parameters. The factor ρ in the peak acceleration parameter gives it units of stress that can be compared to stress changes thought to take place at the seismic source (Hanks and Johnson, 1976; McGarr, 1991).

To interpret the radiated seismic energy, it is useful to consider the fundamental model of a fault of area A loaded to failure by a shear stress τ_1. As slip increases to its final value D, the loading stress decreases linearly from τ_1 to τ_2 (Figure 1). At the same time, the stress resisting slip τ_r decreases abruptly from τ_1 to a lower level about which it fluctuates for the remainder of the slip event; the average level of τ_r during this process is denoted $\overline{\tau}_r$. The energy released W due to the slip event is the area under the line joining τ_1 and τ_2 (Figure 1). That is

$$W = \left(\frac{\tau_1 + \tau_2}{2} \right) DA = \overline{\tau} DA \qquad (4)$$

where $\overline{\tau} = (\tau_1 + \tau_2)/2$. Similarly, the energy consumed E_r because of frictional resistance is

$$E_r = \overline{\tau}_r DA \qquad (5)$$

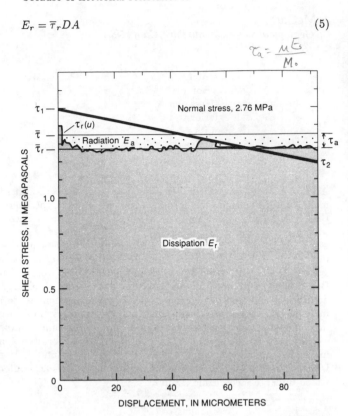

Figure 1. Relation between resisting stress and displacement in an unloading elastic medium (inclined line) during an earthquake. As slip increases, stress in the rock diminishes linearly from τ_1 to τ_2, with average value $\overline{\tau}$. The area under this line is the total work expended per unit fault area; the area below the curve of resisting stress τ_r is the energy dissipated per unit fault area. The difference between the total work expended and the dissipated energy (shaded area) is the work done by the apparent stress (stippled area) τ_a that is available for seismic ground motion. (Adapted from Figure 10.3 of Lachenbruch and McGarr (1990) and based on a laboratory experiment on a large granite sample reported by Lockner and Okubo (1983)).

As seen in Figure 1, the apparent stress τ_a is the difference between the average loading stress and the average resisting stress. That is,

$$\tau_a = \overline{\tau} - \overline{\tau}_r = \eta \overline{\tau} \qquad (6)$$

where η, the seismic efficiency, is the fraction of the average loading stress available for the seismic radiation. Thus, the seismic energy is given by

$$E_s = \tau_a AD \qquad (7)$$

Combining (1) and (7) yields a means of measuring the apparent stress

$$\tau_a = \frac{\mu E_s}{M_0} \qquad (8)$$

where, in general, M_0 is the deviatoric component of the seismic moment given by (3).

The seismic energy, E_s, is clearly relevant in terms of damage potential inasmuch as this is a robust measure of the high-frequency ground motion. Similarly, τ_a represents the stress available to cause damaging ground motion.

Interpretation of the peak ground motion parameters $R\underline{v}$ and $\rho R\underline{a}$ (McGarr et al., 1981) in terms of source processes is model dependent (McGarr, 1991), but underground observations at the source constrain the model. In essence, McGarr and Bicknell (1990) and McGarr (1991) calculated the ground motion due to the failure of a circular patch of fault loaded to failure by the ambient stress plus the stress due to pre-existing slip across the fault exterior to the circular patch, or asperity. This asperity fails from its periphery inward resulting in slip D localized within a broader source region of average slip D_0 (McGarr, 1991). This maximum slip is related to the peak velocity of the far-field S wave according to

$$R\underline{v} = 0.124 \, \beta D \qquad (9)$$

where β is the shear wave speed. Similarly, the maximum particle velocity $\dot{D}/2$ adjacent to the slippery fault is linearly related to the far-field S wave peak acceleration as

$$\rho R\underline{a} = \frac{0.78 \, \mu}{\beta} \, \dot{D}/2 \qquad (10)$$

where $\dot{D}$ is the maximum slip rate. $\dot{D}/2$ is, thus, the maximum particle velocity adjacent to the causative fault that undergoes total slip D.

The two principal assumptions in the model underlying (9) and (10) are (1) that the far-field velocity waveform can be represented as one or more single cycles of sine wave and (2) that the radius r of the asperity is related to the duration $T(= 1/f)$ of the corresponding far-field velocity pulse as $r = 2.34 \, \beta/(2\pi f)$ (Brune, 1970). McGarr (1991) presented various observations supporting these assumptions.

Figures 2 and 3 illustrate how (9) and (10) are used to relate the far-field ground motion parameters to the near-fault ground motion. In Figure 2 we see that underground observations of slip D associated with tremors in the magnitude range 3.6 to 4.6 are consistent with corresponding seismic observations of $R\underline{v}$ in terms of (9). Moreover, $R\underline{v}$ increases with M_0 approximately as $M_0^{1/3}$ for the magnitude range greater than 2. Figure 3, which shows peak far-field acceleration as a function of M_0, is especially relevant with regard to underground damage potential in that for magnitudes greater than 2, peak particle velocity $\dot{D}/2$ (right-hand scale) at the causative fault shows no tendency to increase or decrease with magnitude or moment, but is generally high enough to cause prop failure ($\dot{D}/2 > 1.5$ m/s) if (10) is valid. The fact that prop failure sometimes occurs suggests that (10) is reasonable. According to Wagner (1984) "Judging by the performance of rapid yielding hydraulic props, the rate of stope closure caused by movement along these [fault] planes can be several meters per second . . . ." From the evidence in hand, then, (9) and (10) appear to provide an effective means of relating far-field seismic observations of peak ground motion to the causative fault slip and near-fault particle velocity.

3   MOMENT TENSORS AND SEISMIC GROUND MOTION

An exceptionally fine data set for relating strong ground motion to source processes was obtained during three special field experiments, of about two weeks duration each, embedded in a three-year project, from early 1986 to early 1989, during which the U.S. Geological Survey operated a seven-station seismic network (McGarr et al., 1989) in and around the Witwatersrand gold fields (Figure 4). During the short-term experiments, GEOS (Borcherdt et al., 1985) recorders

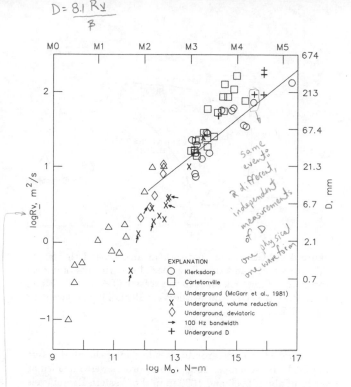

Figure 2. Peak velocity parameter $R\underline{v}$ and corresponding fault slip D as functions of M_0 and moment magnitude M (Hanks and Kanamori, 1979). The underground observations of D are from Brummer and Rorke (1990) and van Aswegen (1990). The solid line of slope $\frac{1}{3}$ is from equation 16a of McGarr (1984) and is for hypocentral depths of 2 km. The circles and squares are for data recorded at the surface with a 50 Hz bandwidth. The diamonds and x's denote data recorded underground at a bandwidth of 200 Hz or 100 Hz (arrow). The triangles indicate a bandwidth of 400 Hz. The datum in the upper right-hand corner at $M > 5$ was reported by Fernandez and van der Heever (1984). (Modified from Fig. 3 of McGarr (1991) by the addition of the data listed in Table 1.)

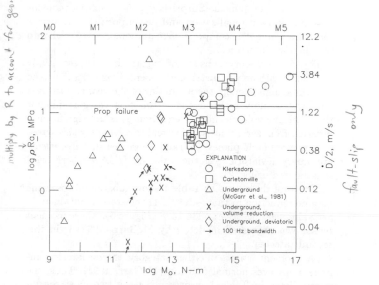

Figure 3. Peak acceleration parameter $\rho R\underline{a}$ and corresponding estimates of near-fault particle velocity $\dot{D}/2$ as functions of M_0 and M. (Modified from Figure 5 of McGarr (1991) by the addition of data listed in Table 1.) See caption of Figure 2 for more details.

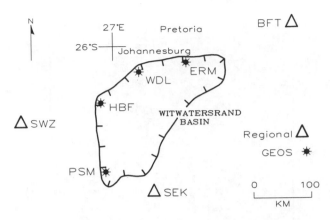

Figure 4. USGS seismic network in and around the major Witwatersrand gold fields. The special experiments of underground recording took place in the Western Deep Levels Mine, below station WDL, and in the Vaal Reefs Mine adjacent to station HBF.

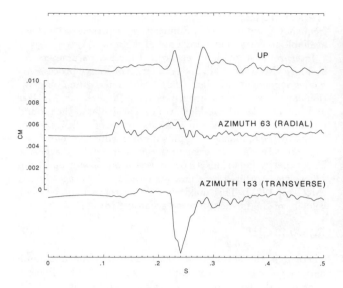

Figure 5a. Ground displacement from event 323 1839 recorded at station W10, sited at a depth of 2866 m in the Two Shaft Pillar of Western Deep Levels mine.

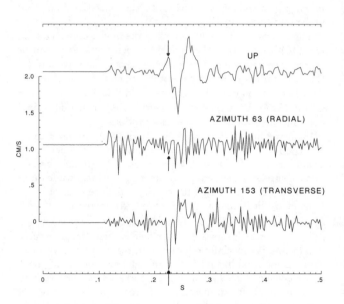

Figure 5b. Ground velocity from event 323 1839 recorded at W10. Arrows indicate where measurements were made on the S wave pulse to estimate the ground motion parameter $R\underline{v}$.

were installed at underground sites in two of the most active gold fields. In November 1986, two recorders were installed at depths of about 2900 and 1600 m in the Western Deep Levels gold mine (WDL in Figure 4). The other two experiments, in early 1988 and 1989, involved one GEOS unit installed at a depth of 2065 m in the Vaal Reefs Mine to record data in conjunction with the nearby surface recorder HBF (Figure 4). Each underground GEOS unit recorded in one of three modes: three components of ground acceleration at 400 samples/sec per component, three components of velocity at the same sampling rate, or three components of both acceleration and velocity at 200 samples/sec/channel. Of the 16 tremors presented here, 10 were recorded at the higher sampling rate.

Figures 5 and 6 illustrate some of the data obtained during the special experiments. As described in detail by McGarr (1993), complete moment tensors were determined for each event by linearly inverting ground displacement (Figures 5a and 6a) using synthetic seismograms calculated for each moment tensor component. Surprisingly, the resulting moment tensors, after diagonalization and decomposition into deviatoric and volumetric components (equations 2 and 3), fall into two distinct categories.

The first category consists of events that are essentially deviatoric with no volumetric component. Event 323 1839 (the first three digits denote the julian day of the year and the last four are the universal time) is in this first category in that the coseismic volume change (equation 3) is not significantly different from zero; that is, this event involves only slip across one or more fault planes. In Table 1, we see that five events are purely deviatoric from the ratio $-\Delta V/\Sigma AD$, where ΣAD is estimated using (2).

Eleven of the events in Table 1 show substantial coseismic volume reduction in that $-\Delta V/\Sigma AD$ clusters about an average value near 0.7. Event 034 1528 (Figure 6), whose moment tensor was described in detail by McGarr (1992) is in this second category.

For moment tensors in either category, the deviatoric component involves normal faulting (McGarr, 1993). Thus, the events listed in Table 1, representing the two most seismically active mining districts in South Africa (Figure 4) involve either normal slip across one or more fault planes or normal shear deformation plus a comparable amount of coseismic volume reduction. As stated by McGarr (1993), the marked separation of the two categories of moment tensor, in terms of $-\Delta V/\Sigma AD$, is not understood.

The issue explored in Table 1 is the question of whether the moment tensor type, with or without coseismic volume

reduction, influences the high-frequency ground motion. In addressing this question events 323 1839 and 034 1528 are of substantial interest because they represent the first and second categories of source but have nearly the same deviatoric deformation (M_0 (dev)) and, therefore, the same moment-magnitudes M. In the case of event 323 1839 (Figure 5) the ray path from the source to station W10 is horizontal and 1131 m in length. The vectorially-summed peak velocity (Figure 5b) is 0.72 cm/s yielding an estimate of $R\underline{v}$ for this site of 8.2 m²/s. For event 034 1528 (Figure 6) the ray path to station HMN is also horizontal and 1423 m in length. From the peak velocities indicated in Figure 6b, $R\underline{v} = 2.0$ m²/s, about 25 per cent that of event 323 1839. The contrast in peak acceleration (Figures 5c and 6c) is even greater with $\rho R\underline{a}$ for event 323 1839 exceeding that for 034 1528 by a factor in excess of eight (Table 1). From these two examples, then, it seems

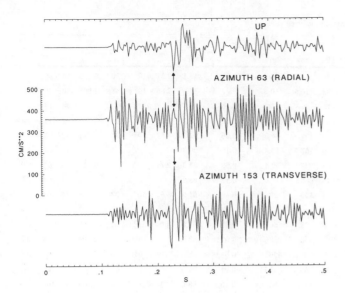

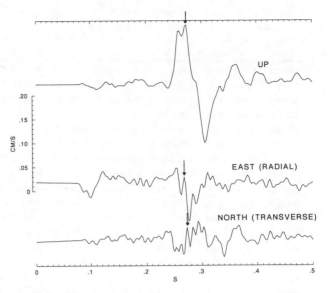

Figure 5c. Ground acceleration from event 323 1839 recorded at W10. Arrows indicate where measurements were made to estimate the ground motion parameter $\rho R \underline{a}$.

Figure 6b. Ground velocity from event 034 1528 at HMN. Arrows indicate where measurements were made of the S wave to estimate the ground motion parameter $R\underline{v}$.

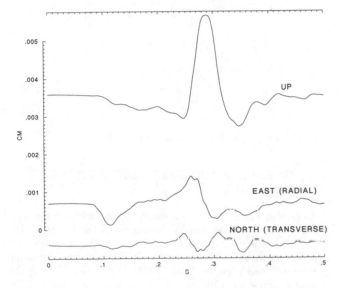

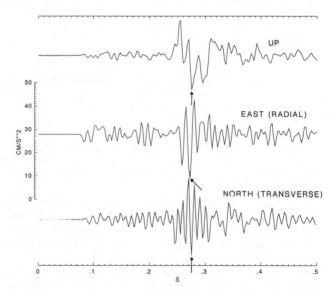

Figure 6a. Ground displacement for event 034 1528 recorded at station HMN sited at a depth of 2065 m in the Vaal Reefs Mine.

Figure 6c. Ground acceleration for event 034 1528 at HMN. Arrows show where measurements were made to estimate $\rho R \underline{a}$.

that if the moment tensor includes a substantial component of coseismic volumetric reduction (e.g., event 034 1528) the high-frequency ground motion is markedly reduced in amplitude, as recorded at a fixed distance, compared to that for purely deviatoric moment tensors (event 323 1879) of similar deviatoric moment or magnitude.

In Figures 7 and 8, we see that this effect is quite consistent in that for both the peak velocity and peak acceleration parameters, the data for purely deviatoric sources are situated distinctly above those whose sources include a significant implosive component.

The radiated seismic energy E_s can be calculated from the ground velocity using (e.g., Cook, 1963; Spottiswoode and McGarr, 1975)

$$E(C) = 4\pi\rho C R^2 \int v^2 dt \qquad (11)$$

where C is either the P wave speed α or the S wave speed β and the integral of squared ground velocity is taken within a window that includes the P or S wave. Then the total radiated energy is

$$E_s = E(\alpha) + E(\beta) \qquad (12)$$

The influence of moment tensor type on E_s is best seen by first calculating the apparent stress τ_a, using (8) and then comparing results for the two source types. As seen in Figure 9, τ_a is sharply reduced for tensors having a volume reduction component compared to what it would presumably be

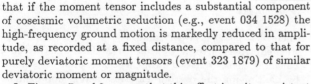

7

Table 1. Seismic data and inferred results.

Event	Mine[a]	Depth m	$M_0(\text{dev})$[b] N-m	$M_0(\text{vol})$[c] N-m	E_s[d] J	$R\underline{v}$ m²/s	$\rho R\underline{a}$ MPa	τ_a[e] MPa	ΣAD[f] m³	$-\Delta V$[g] m³	$-\Delta V/\Sigma AD$	σ_v MPa	W[h] J	$\bar{\tau}$ MPa	η	f[i] Hz	r_0[j] m	M[k]
3102038	WDL	2044	3.65×10^{11}	-8.22×10^{11}	6.39×10^5	0.44	0.22	0.066	9.7	5.	0.52	58.1	2.91×10^8	30	0.002	14	96	1.7
3121332	WDL	2683	6.14×10^{11}	-1.69×10^{12}	5.25×10^6	1.34	0.96	0.32	16.6	10.3	0.62	76.3	7.86×10^8	48	0.007	30	45	1.9
3141709	WDL	3081	1.50×10^{12}	1.03×10^{12}	2.78×10^7	3.24	1.97	0.70	40.	$-$6.3	-0.16	87.7				18	74	2.1
3151552	WDL	2042	1.98×10^{12}	-8.94×10^{12}	2.26×10^7	1.97	1.23	0.43	52.7	55.	1.04	58.1	3.20×10^9	61	0.007	12	112	2.2
3151554	WDL	2205	4.85×10^{12}	-1.21×10^{13}	6.41×10^7	3.13	1.44	0.50	129.	74.5	0.58	62.7	4.67×10^9	36	0.014	13	103	2.5
3231839	WDL	2822	4.06×10^{12}	1.64×10^{12}	2.47×10^8	9.0	8.65	2.29	107.9	-10.4	-0.10	80.3				18	74	2.4
3241523	WDL	3243	1.75×10^{12}	-4.82×10^{12}	3.09×10^7	2.90	1.92	0.66	47.1	29.4	0.62	92.3	2.71×10^9	58	0.011	25	54	2.2
3241529	WDL	3100	4.18×10^{12}	1.95×10^{12}	2.39×10^8	11.1	9.2	2.15	111.3	-12.0	-0.11	88.2				13	103	2.5
3241624	WDL	2175	5.64×10^{12}	-1.47×10^{13}	9.15×10^7	3.89	3.60	0.61	150.	90.	0.60	61.9	5.57×10^9	37	0.016	16	84	2.5
3251544	WDL	2443	7.37×10^{11}	~0	1.14×10^7	2.28	2.6	0.58	19.6	~0	~0	69.5				29	46	1.9
3251605	WDL	2166	2.95×10^{12}	-1.07×10^{13}	3.31×10^7	2.28	1.42	0.36	78.	66.	0.84	61.6	4.07×10^9	52	0.008	15	89	2.3
0301411	HBF	2125	8.1×10^{13}	-3.20×10^{14}	2.78×10^9	25.9	15.3	1.29	2160.	1980.	0.92	60.5	1.20×10^{11}	56	0.023	6	223	3.3
0301411a[l]	HBF	2055	2.75×10^{13}	-6.93×10^{13}	7.04×10^8	10.5	9.09	0.96	731.	426.	0.58	58.5	2.49×10^{10}	34	0.028	8	168	3.0
0331352	HBF	2145	2.10×10^{12}	1.15×10^{11}	7.51×10^7	4.44	3.89	1.34	55.9	$-$ 0.71	-0.01	61.0				22	61	2.2
0341528	VR	2060	4.40×10^{12}	-1.43×10^{13}	1.88×10^7	2.03	1.09	0.16	117.	88.	0.75	58.6	5.16×10^9	44	0.004	13	103	2.4
0271046	HBF	1918	6.19×10^{12}	-2.45×10^{13}	1.51×10^8	4.	1.98	0.92	165.	150.	0.91	54.6	8.19×10^9	50	0.018	13	103	2.5

[a] WDL: Western Deep Levels; HBF: Hartebeestfontein; VR: Vaal Reefs.

[b] Obtained by subtracting isotropic component of moment tensor (McGarr, 1993).

[c] Trace of moment tensor.

[d] From equation (12).

[e] From equation (8), with $M_0 = M_0(\text{dev})$.

[f] From equation (2).

[g] From equation (3).

[h] From equation (13).

[i] Inverse of the duration of the clearest S wave pulse recorded underground.

[j] $r_0 = 2.34\beta/(2\pi f)$ (Brune, 1970).

[k] $M = (\log M_0 - 9)/1.5$ (Hanks and Kanamori, 1979), where $M_0 = M_0(\text{dev})$.

[l] Occurred less than one minute after event 0301411.

otherwise. Thus, referring to Figure 1, the effect of a coseismic volume reduction is to reduce the difference between the average loading stress and the average resisting stress during an event.

It is worth noting that for events with a coseismic volume reduction, τ_a is overestimated somewhat inasmuch as the seismic energy, estimated from (12), contains contributions from both the fault slip and volume components of the source. For a proper estimate of τ_a, using (8), E_s due to any process other than fault slip should be excluded. Accordingly, if τ_a had been calculated properly for each event the contrast in τ_a between the two types of events represented in Figure 9 would be greater than shown.

The demonstration (Table 1, Figures 7–9) that events with significant coseismic volume reduction produce much less high-frequency ground motion compared to purely deviatoric events raises the question of the mechanism by which the ground motion is suppressed. Nearby stopes, of course, provide a source of coseismic volume reduction. If the fault plane actually intersects the stope (Figure 10) one expects coseismic volume reduction (e.g., McGarr, 1971) such that $-\Delta V$ is of the order of ΣAD. Thus, Figure 10 illustrates conjecture to the effect that events for which the slipping fault intersects the nearby stope entail coseismic volume reduction whereas those whose fault planes are further away from the mining have purely deviatoric mechanisms.

Accordingly, the volumetric component of the moment tensor may indicate interaction between the fault slip and the nearby mine stope. This interaction, because of the extent of typical Witwatersrand mine stopes, reduces the rate of fault slip, $\dot{D}$, compared to what it would be for faults remote from mining (e.g., McGarr et al., 1979). Thus, the slip rate is controlled by the size of the interacting stope more than the

dimension of the slipping fault plane itself.

4 AVERAGE LOADING STRESS AND SEISMIC EFFICIENCY

As seen in Figure 1, which involves measurements made during a laboratory earthquake (Lockner and Okubo, 1983; Dieterich, 1979) the apparent stress τ_a is only a tiny fraction of $\bar{\tau}$, the average loading stress. Is this also true for mine tremors? According to Cook (1963) and McGarr (1976) the seismic efficiency η (or $\tau_a/\bar{\tau}$, equation 6) is very small, of the order of 0.01. These findings, however, were based on comparisons between energies estimated for suites of earthquakes recorded over extended periods of time and estimates of the energy released by mining over the same periods. The data listed in Table 1 provide the means of estimating the average loading stress and seismic efficiency for individual events.

This is done by first calculating the coseismic energy release

$$W = -\sigma_v \Delta V \tag{13}$$

where $\sigma_v = \rho g z$, g is gravity, and z is depth; σ_v is the vertically-oriented stress due to the overburden (McGarr and Gay, 1978). Thus, the energy released during a seismic event is estimated by multiplying the overburden stress acting approximately normal to the tabular stopes (Figure 10) by the measured volume reduction $-\Delta V$ (Table 1).

An estimate of the seismic efficiency η for each event is directly obtained by

$$\eta = E_s/W \tag{14}$$

8

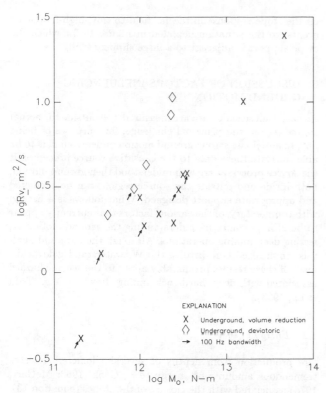

Figure 7. log $R\underline{v}$ as a function of log M_0 for the events listed in Table 1. 10 of the events were recorded with a bandwidth of 200 Hz and the other 6 (arrowed) at 100 Hz.

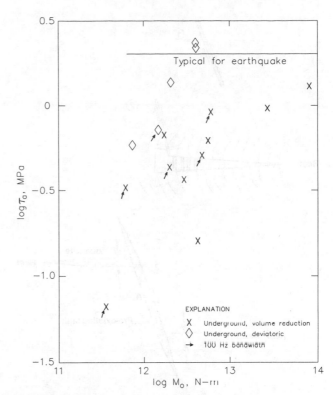

Figure 9. log τ_a as a function of log M_0 for the events in Table 1. Arrows indicate data recorded at 100 Hz bandwidth. A typical level of τ_a for crustal earthquakes of 2 MPa (Lachenbruch and McGarr, 1990) is shown for comparison.

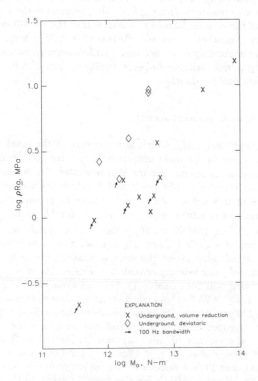

Figure 8. log $\rho R\underline{a}$ as a function of log M_0 for the events listed in Table 1. Arrows indicate 100 Hz recording bandwidth. Otherwise, bandwidth was 200 Hz.

and as seen in Table 1, the measured efficiencies are of the order of 0.01.

Similarly, the average loading stress $\bar{\tau}$ can be estimated by combining (1) and (4)

$$\bar{\tau} = \frac{\mu W}{M_0} \tag{15}$$

and the results listed in Table 1 range from 30 to 61 MPa with an average for $\bar{\tau}$ of about 45 MPa. $\bar{\tau}$ is remarkably constant for events spanning more than two orders of magnitude in seismic moment. Moreover as for the laboratory earthquake of Figure 1, the ratios $\tau_a/\bar{\tau}(=\eta)$ for the induced earthquakes with coseismic volume reduction are tiny. Thus, nearly all of the loading stress, or released energy, is expended in overcoming sliding friction, with only a small fraction left over to produce seismic ground motion.

Interestingly, the values of $\bar{\tau}$ listed in Table 1 are in excellent agreement with estimates based in the analysis of fault gouge surface area (Spottiswoode, 1980; McGarr et al., 1979). Gouge samples were taken from a well-studied shear zone (Ortlepp, 1978; Gay and Ortlepp, 1979) and compared to gouges produced in the laboratory in stress states similar to those in situ, in the ERPM mine at 2 km depth, to conclude that $\bar{\tau} \simeq 40$MPa.

With brief regard to the five events (Table 1) with no significant coseismic volume change, it seems likely that in the long term, these events bear much the same relationship to the causative mining, as the other 11 tremors. That is, for these events, also,

$$M_0 \simeq \mu \Delta V \tag{16}$$

9

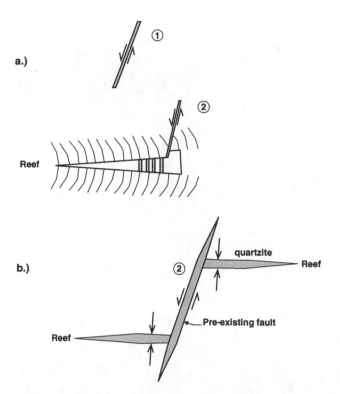

Figure 10. a.) Schematic cross section of mine stope with hydraulic yielding props. A fracture zone forms continuously ahead of the advancing face (rightward). Stoping widths are typically of the order of 1.2 m. Normal fault, labeled "2", has intersected stope resulting in strong coupling between fault slip and stope closure. Normal faulting labeled "1" is sufficiently far from the stope that coupling is presumably weak. b.) In the mining district that includes station HBF (Figure 4), the gold-bearing reef is offset by major throughgoing faults, typically by several hundred meters. This leads to the cross section geometry shown here that is conducive to renewed seismic slip on the offsetting faults resulting in strong coupling between the stopes and normal fault slip.

as analyzed by McGarr (1976). Here M_0 represents the total seismic deformation of a suite of events occurring during a given period and ΔV is estimated from ore production over the same period. Thus, I developed (16) on the assumed basis that the volumetric reduction, leading to the induced seismic deformation, took place on time scales that were much longer than seismic. Presumably, the purely deviatoric events are coupled to the mining approximately according to (16), but the coupling is not so tight as to result in coseismic volume reduction, perhaps due to position relative to the nearest stope (Figure 10). (For reasons mentioned by McGarr (1993) the hypocentral locations do not provide a good test of this hypothesis.)

5 STOPE RESPONSE

From evidence presented by Spottiswoode and Churcher (1988), the relative velocity of the hangingwall and footwall, toward or away from each other, of a mine stope can exceed the average or bulk, stope velocity by a factor ranging from two to three. Accordingly, it may be that the particle velocities $\dot{D}/2$ at the source (Figure 3) result in amplified differential motion affecting the stope support. Currently, it is difficult to relate the Spottiswoode and Churcher (1988) result based

on the stope response in the far field to tiny, high-frequency events, to the situation depicted in Figure 10 for which the stope support is adjacent to a large slipping fault.

6 DISCUSSION OF FACTORS INFLUENCING GROUND MOTION.

As nearly all seismic data are measured well outside the actual source region, the principal challenge, for purposes of being able to anticipate strong ground motion underground, is to be able to relate these data to the causative source processes. If the source processes are well understood then ground motion, both inside and outside the source region, can be predicted and appropriate support designed. What follows is a listing, with commentary, of the various factors that currently appear to be of importance in understanding the ground motion affecting deep mining operations. Although the data and analysis emphasized here involve the Witwatersrand gold fields, most of these results presumably apply to the seismic hazard associated with deep, hard-rock mining elsewhere (e.g., Trifu et al., 1993).

6.1 Energy budget

Ore production from depths of several kilometers releases tremendous amounts of energy (e.g., Cook, 1963; McGarr, 1976) associated with the closure of the stopes (equation 13). From the viewpoint of seismic hazard, it is fortunate that only about 0.01 of energy released by a seismic event, as measured by the coseismic volume reduction (Table 1) is available to produce ground motion. That is, nearly all of the available energy is spent in overcoming sliding friction across the fault surface and comparatively little is left over to accelerate the slip. That is, τ_a (Figure 1), which is only a tiny fraction of $\overline{\tau}$, the average stress causing fault slip, represents the magnitude of the seismic hazard problem within the overall energy budget. Nonetheless, the observation that $\tau_a/\overline{\tau}$ is very small is comforting only in an academic sense inasmuch as tremors in deep mines continue to be an enormous problem in terms of safety and production.

6.2 Seismic moment scaling

Not surprisingly, M_0, which is a measure of the total shear deformation, is the most important single factor to consider when assessing seismic hazard. Appropriately M_0 is the independent variable in Figures 2 and 3, which are the two key figures of this report. First, with regard to ground velocity external to the source region, we see that for M_0 greater than about 10^{12} or 10^{13} N-m, $R\underline{v}$ scales approximately as $M_0^{1/3}$. In terms of seismic hazard, $R\underline{v}$ is much more important for what it indicates about the seismic source than as a direct measure of peak velocity outside the source. To see why, consider event 030 1411 (Table 1). For this event, the largest in Table 1, $R\underline{v} = 25.9 \text{ m}^2/\text{s}$ and the source radius r_0 is estimated to be 223 m. Clearly $R\underline{v}$, measured from the far-field S pulse, is not a proper measure of ground motion within the source and so the validity of this parameter is limited to $R > 223$ m. At these distances, $\underline{v} < 12\text{cm/s}$, a level of ground velocity that is of little consequence for underground seismic hazard (e.g., Wagner, 1984). Similar remarks apply to the other events listed in Table 1. Accordingly, $R\underline{v}$ itself is not of much importance as a measure of potentially-damaging ground velocity because exterior to the source region $\underline{v}$ is nearly always too small to be very damaging. $R\underline{v}$ is, however, important in terms of source implications because, as seen in Figure 2, underground observations of fault slip D can be compared to $R\underline{v}$

as a test of the validity of equation (9) and the corresponding source model (McGarr, 1991). Although there are few published values of D for specific events of known magnitude or moment, the five values of D shown in Figure 2 are compatible with (9) in that they coincide with corresponding $R\underline{v}$ data.

The maximum fault slip D, which presumably also scales approximately as $M_0^{1/3}$ for $M_0 \geq 10^{12}$ N–m, can be of importance if the corresponding fault offsets underground workings (e.g., a shaft or tunnel). As seen in Figure 2 the largest magnitude events are associated with offsets of as much as 0.41 m.

The peak acceleration parameter $\rho R\underline{a}$ (Figure 3) provides the most critical information about peak ground motion within the source (equation 10) but there are some complications to be taken into account, such as the effect of band-limited recording. The complications notwithstanding, however, the first-order feature of Figure 3 is the lack of any systematic M_0 dependence for maximum values of $\rho R\underline{a}$, and presumably $\dot{D}/2$ (equation 10) for M_0 in the range 10^{12} to 10^{17} N–m. In this size range $\rho R\underline{a}$ has an upper bound of about 30 MPa and the corresponding bound for the near-fault particle velocity $\dot{D}/2$ is close to 3.8 m/s. Moreover, as will be described, it seems likely that the reduction in $\rho R\underline{a}$ (Figure 3) with decreasing magnitude below M 2 may be due to factors external to the source, as will be described. If so then $\dot{D}/2$ might be as high as 3.8 m/s even for events of $M < 2$.

As events of $M < 1$ are rarely of any consequence in terms of damaging ground motion, factors in addition to $\dot{D}/2$ must play a role in the damage potential of a tremor. The two obvious factors are size r_0 and fault slip D. For an event of $M = 1$ r_0 is likely to be less than 20 m (McGarr et al., 1981) and thus the dimension of the source is small. More importantly, the maximum fault slip (Figure 2) is unlikely to exceed 3 mm for $M = 1$; displacements this small could easily be accommodated elastically by the hydraulic props regardless of the slip velocity. In contrast, consider event 030 1411 with M 3.3. From Table 1 and equations (9) and (10) we see that $\dot{D}/2 = 1.9$ m/s, $D = 58$ mm, and $r_0 = 223$ m. This event is potentially damaging because of its high inferred particle velocity affecting a region several hundred meters in extent together with maximum slip that certainly exceeds the elastic limit of the hydraulic yielding props.

6.3 Band-limited recording.

Trifu et al. (1993) argued that whereas the peak velocity parameter $R\underline{v}$ is not very sensitive to the recording bandwidth, the parameter for peak acceleration can depend strongly on bandwidth. The different data sets plotted in Figures 2 and 3 support this conclusion. With the exception of the events involving a coseismic volume reduction in the magnitude range less than 3, values of $R\underline{v}$ (Figure 2) recorded over a variety of bandwidths, ranging from 25 Hz to 400 Hz, are reasonably consistent. For instance, near M 3 the surface data, with bandwidth 50 Hz, are compatible with the underground data recorded over shorter paths and 200 Hz bandwidth. As will be discussed, the data for events involving coseismic volume reduction must be suppressed due to factors other than recording bandwidth.

In contrast, the effects of band-limited recording on $\rho R\underline{a}$ are abundantly clear (Figure 3). For the bandwidth of 400 Hz the data diminish as magnitude decreases below 2. For the events with no coseismic volume change (Table 1) recorded underground with bandwidth 200 Hz, events of M less than about 2.5 are similarly suppressed. The surface data, recorded over a 50 Hz band diminish below M 3.5. Generally, then, for a given bandwidth, there is a magnitude or moment threshold above which the value of $\rho R\underline{a}$ apparently reflects source

processes but, below which the parameter is suppressed due to limited bandwidth. In some cases the bandwidth may be limited by attenuation during wave propagation instead of the recording system; for the Witwatersrand quartzites Q is sufficiently high (e.g., Churcher, 1990) and the ray paths sufficiently short that few, if any, of the data in Figure 3 are affected significantly by wave propagation effects.

6.4 Coseismic volume reduction

As indicated before and as seen in Figures 2 and 3, events for which the moment tensor includes a coseismic volume reduction (Table 1) have ground motion parameters that are suppressed relative to those from exclusively deviatoric events. If, as suggested before, this ground motion suppression is associated with substantial coupling between the nearby stope and fault slip (Figure 10), then, as suggested by McGarr et al. (1979) the source time history is extended if the dimension of the nearby stope exceeds that of the slipping fault. For example, comparison of Figures 5b and 6b indicates that the sinusoidal pulse of peak amplitude velocity for event 034 1528 (top trace, Figure 6b) has a substantially greater duration, of about 0.09 sec, than that of event 323 1839 (bottom trace, Figure 5b), for which the corresponding pulse lasts approximately 0.04 sec. Generally, the time history of the purely deviatoric event 323 1839 (Figure 5) appears significantly compressed compared to that of the event with coseismic volume reduction, 034 1528 (Figure 6). Perhaps fortuitously, McGarr et al. (1979) estimated a stope response time of $T = 0.13$ sec, corresponding to a frequency of about 8 Hz. Thus, events for which f_0 (Table 1) is much greater than 8 Hz would presumably be affected by the stope such that the radiated ground motion is suppressed. For larger events (e.g., 030 1411 and 030 1411a), the response of the nearby stope is sufficiently rapid so as not to cause reduction in the seismic ground motion.

7 CONCLUDING REMARKS

From a practical viewpoint, the primary conclusion of this study is that the rapid yielding hydraulic props used in the Witwatersrand gold fields should be designed to withstand ground velocities of 4 m/s (Figure 3, right-hand scale). Moreover, this stope support should be able to accommodate displacements of at least 400 mm (Figure 2). These conclusions are not very original to this report inasmuch as Wagner (1984) made much the same recommendations. The observations reviewed here, however, permit an improved understanding of the factors motivating these recommendations.

ACKNOWLEDGMENTS

This work was partly sponsored by the Air Force Technical Applications Center and the Air Force Geophysical Laboratory, Earth Sciences Division. I thank A. Rossouw, P. de Jong, and S. M. Spottiswoode of the Chamber of Mines of South Africa, R. W. E. Green of the University of the Witwatersrand, and A. van Zyl Brink and P. Mountfort of Western Deep Levels Gold Mine for considerable field support as well as mine seismic network data. J. Bicknell, E. Sembera and R. Grose played critical roles in the data acquisition phase of this study. A. Frankel and D. Boore provided insightful reviews of this manuscript. I thank C. Sullivan for editorial assistance, R. Eis and E. Dingel for excellent drafting, and R. P. Young for encouraging me to submit this report.

REFERENCES

Aki, K. 1966. Generation and propagation of G waves from the Niigata earthquake of June 16, 1964, 2, Estimation of earthquake moment, released energy, and stress–strain drop from the G wave spectrum. *Bull. Earthquake Res. Inst.* Tokyo Univ. 44: 73–88.

Aki, K. and P. Richards 1980. *Quantitative Seismology: Theory and Methods.* Freeman, Cooper, San Francisco.

Borcherdt, R. D., J. B. Fletcher, E. G. Jensen, G. L. Maxwell, J. R. van Schaack, R. E. Warrick, E. Cranswick, M. J. S. Johnston and R. McClearn 1985. A general earthquake-observation system (GEOS). *Bull. Seismol. Soc. Am.* 75: 1783–1825.

Brummer, R. K. and A. J. Rorke 1990. Case studies on large rockbursts in South African gold mines. *Rockbursts and Seismicity in Mines*, Editor C. Fairhurst, Balkema, Rotterdam: 323–329.

Brune, J. N. 1970. Tectonic stress and the spectra of seismic shear waves from earthquakes. *J. Geophys. Res.* 75: 4997–5009. (Correction 1971. *J. Geophys. Res.* 76: 5002.)

Churcher, J. M. 1990. The effect of propagation path on the measurement of seismic parameters. *Rockbursts and Seismicity in Mines*, Editor C. Fairhurst, Balkema, Rotterdam: 205–209.

Cook, N. G. W. 1963. The seismic location of rockbursts. *Proc. Fifth Rock Mechanics Symposium*, Pergamon Press, Oxford: 493–516.

Dieterich, J. H. 1979. Modelling of rock friction, 1. Experimental results and constitutive equations. *J. Geophys. Res.* 84: 2161–2168.

Feignier, B. and R. P. Young 1992. Moment tensor inversion of induced microseismic events: Evidence of non-shear failures in the $-4 < M < -2$ moment magnitude range. *Geophys. Res. Letters* 19: 1503–1506.

Fernandez, L. M. and P. K. Van der Heever 1984. Ground movement and damage accompanying a large seismic event in the Klerksdorp district. *Rockbursts and Seismicity in Mines*, Editors N. C. Gay and E. H. Wainwright, S.A.I.M.M., Johannesburg: 193–198.

Gay, N. C. and W. D. Ortlepp 1979. Anatomy of a mining induced fault zone. *Geol. Soc. Am. Bull.* 90: 47–58.

Gibowicz, S. J. 1990. The mechanism of seismic events induced by mining. *Rockbursts and Seismicity in Mines*, Editor C. Fairhurst, Balkema, Rotterdam: 3–27.

Hanks, T. C. and D. A. Johnson 1976. Geophysical assessment of peak accelerations. *Bull. Seismol. Soc. Am.* 66: 959–968.

Hanks, T. C. and H. Kanamori 1979. A moment magnitude scale, *J. Geophys. Res.* 84: 2348–2350.

Joyner, W. B. and D. M. Boore 1981. Peak horizontal acceleration and velocity from strong-motion records, including records from the 1979 Imperial Valley, California earthquake. *Bull. Seismol. Soc. Am.* 71: 2011–2038.

Lachenbruch, A. H. and A. McGarr 1990. Stress and heat flow. *U.S. Geological Survey Professional Paper* 1515: The San Andreas Fault System, Editor R. Wallace: 261–277.

Lockner, D. A. and P. G. Okubo 1983. Measurements of frictional heating in granite. *J. Geophys. Res.* 88: 4313–4320.

McGarr, A. 1971. Violent deformation of rock near deep-level tabular excavations—seismic events. *Bull. Seismol. Soc. Am.* 61: 1453–1466.

McGarr, A. 1976. Seismic moments and volume changes, *J. Geophys. Res.* 81: 1487–1494.

McGarr, A. 1984. Scaling of ground motion parameters, state of stress and focal depth. *J. Geophys. Res.* 89: 6969–6979.

McGarr, A. 1991. Observations constraining near-source ground motion estimated from locally recorded seismograms. *J. Geophys. Res.* 96: 16,495–16,508.

McGarr, A. 1992. Am implosive component in the seismic moment tensor of a mining-induced tremor. *Geophys. Res. Letters* 19: 1579–1582.

McGarr, A. 1993. Moment tensors of ten Witwatersrand mine tremors. *Pure and Applied Geophysics*, in press.

McGarr, A. and N. C. Gay 1978. State of stress in the earth's crust. *Annu. Rev. Earth Planet. Sci.* 6: 405–436.

McGarr, A. and J. Bicknell 1990. Estimation of the near-fault ground motion of mining-induced tremors from locally recorded seismograms in South Africa, *Rockbursts and Seismicity in Mines*, Editor C. Fairhurst, Balkema, Rotterdam: 245–248.

McGarr, A., S. M. Spottiswoode and N. C. Gay 1975. Relationship of mine tremors to induced stresses and to rock properties in the focal region. *Bull. Seismol. Soc. Am.* 65: 981–993.

McGarr, A., S. M. Spottiswoode, N. C. Gay and W. D. Ortlepp 1979. Observations relevant to seismic driving stress, stress drop, and efficiency. *J. Geophys. Res.* 84: 2251–2261.

McGarr, A., R. W. E. Green and S. M. Spottiswoode 1981. Strong ground motion of mine tremors: Some implications for near-source ground motion parameters. *Bull. Seismol. Soc. Am.* 71: 295–319.

McGarr, A., J. Bicknell, E. Sembera, and R. W. E. Green 1989. Analysis of exceptionally large tremors in two gold mining districts of South Africa. *Pure Appl. Geophys.* 129: 295–307.

Ortlepp, W. D. 1978. The mechanism of a rockburst. *Proc. 19th U.S. Rock Mechanics Symposium*, University of Nevada, Reno: 476–483.

Ortlepp, W. D. 1984. Rockbursts in South African gold mines: a phenomenological view. *Rockbursts and Seismicity in Mines*, Editors N. C. Gay and E. H. Wainwright, S.A.I.M.M., Johannesburg: 165–178.

Richter, C. F. 1958. *Elementary Seismology*, W. H. Freeman, San Francisco.

Spottiswoode, S. M. 1980. Source mechanism studies on Witwatersrand seismic events. *Ph.D. Thesis*, University of the Witwatersrand, Johannesburg.

Spottiswoode, S. M. and A. McGarr 1975. Source parameters of tremors in a deep-level gold mine. *Bull. Seismol. Soc. Am.* 65: 93–112.

Spottiswoode, S. M. and J. M. Churcher 1988. The effect of backfill on the transmission of seismic energy. Backfill in South African Mines, *S.A.I.M.M.*, Johannesburg: 203–217.

Stump, B. W. and L. R. Johnson 1977. The determination of source properties by the linear inversion of seismograms. *Bull. Seismol. Soc. Am.* 67: 1489–1502.

Trifu, C. I., T. I. Urbancic, and R. P. Young 1993. Estimates of near-source ground motion parameters and source complexity supporting inhomogeneous faulting. Submitted to *Bull. Seismol. Soc. Am.*

Tyser, J. A. and H. Wagner 1977. A review of six years of operations with the extended use of rapid-yielding hydraulic props at the East Rand Proprietory Mines, Limited, and experience gained throughout the industry. Papers and Discussions, 1976–1977, Association of Mine Managers of South Africa, Johannesburg: 321–341.

Van Antwerpen, H. E. F. and M. G. Spengler 1984. The effect of mining-related seismicity on excavations at East Rand Proprietory Mines, Limited. *Rockbursts and Seismicity in Mines*, Editors N. C. Gay and E. H. Wainwright, S.A.I.M.M., Johannesburg: 235–243.

Van Aswegen, G. 1990. Fault stability in SA gold mines. Paper presented at the International Conference on the Mechanics of Jointed and Faulted Rock, Tech. Univ. of Vienna, Vienna.

Wagner, H. 1984. Support requirements for rockburst conditions. *Rockbursts and Seismicity in Mines*, Editors N. C. Gay and E. H. Wainwright, S.A.I.M.M., Johannesburg: 209–218.

Rockbursts and Seismicity in Mines, Young (ed.) © 1993 Balkema, Rotterdam, ISBN 90 5410 320 5

Keynote address: Support of tunnels in burst-prone ground – Toward a rational design methodology

P.K. Kaiser*

Geomechanics Research Centre, Laurentian University, Sudbury, Ont., Canada

ABSTRACT: A rationale for designing support systems to resist rockburst damage is presented. It involves a proper definition of microseismic activity centre, the identification of likely modes of failure at the target, the determination of trigger levels for rock fracturing and rock ejection, and the determination of support survival limits to either maintain a continuous rockmass by reinforcement or to retain and hold broken or ejected rock in place.

Starting from a given, well defined microseismic activity centre in terms of magnitude and source location, this report deals primarily with damage prediction and support selection to contain rockburst damage. The practical implications emerging are presented at the end of each section but three main conclusions stand out:
(1) The damage assessment procedure developed at the Geomechanics Research Centre provides an effective means for data collection. The damage records can be used in combination with source location data to identify the seismic event actually causing the damage;
(2) Damage trigger levels can be predicted from empirical relationships assuming geometric spreading for targets that are in the far-field. For excavations in close proximity to a seismic event, it is mandatory to design for maximum ground motion (2 to 4 m/s); and
(3) For the less critical but more common situation of supporting rock in the far-field, the procedure developed in this report should be applicable and the design charts developed based on energy considerations and displacement criteria are found to be consistent with qualitative observations from Canadian mines.

1. INTRODUCTION

It is the intent of this address to focus on several specific aspects of support in burst-prone ground. The data presented here and the related discussions are of particular relevance to Canadian conditions and cannot be generalized without careful adaption. This report reflects our current research goals and largely presents work in progress. Therefore, it contains concepts that, while not fully verified, are strongly supported, at least in a qualitative manner, by the data presented. The author has attempted to clearly differentiate between factual and speculative information.

The problem of containing rockburst damage is highly complex and only an integrated approach including risk assessment, strategic and tactical measures, in situ monitoring and proper ground control will provide acceptable solutions. Despite rapid progress in microseismic monitoring, the complexity of the issues still forces us to adopt empirical approaches (based on rapidly improving data), to revert to conceptual and analytical models in an effort to identify new hypotheses, and to follow an observational design approach.

World experience with rockbursts and support of bursting ground is substantial and much progress has been made in understanding dynamic failure processes and in developing burst-resistant supports. Economic, social and technical considerations, however, render solutions that are appropriate for one mining camp but unacceptable for others. In an effort to learn from past experiences, an extensive study of current practices was completed by McCreath and Kaiser (1992).

1.1 Current support practice in burst-prone ground

A compilation of current methods of support of burst-prone ground in hard rock mines was assembled by McCreath and Kaiser (1992) as part of the Canadian Rockburst Research Project.

Based on South African experiences, mine openings can be designed to survive fairly large rockburst events, using cable lacing support systems. Such systems are labour intensive to install and therefore extremely expensive in a Canadian context. Consequently, there is a strong incentive to develop a range of intermediate methods of support which will be effective in limiting and controlling the consequences of medium-level rockburst activity, but

with contributions by R.K. Brummer, P.Jesenak, D.R.McCreath, D.D.Tannant & X.Yi

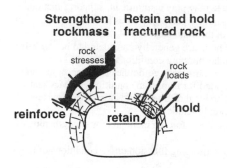

Strengthen rockmass | Retain and hold fractured rock

rock stresses

rock loads

reinforce   retain   hold

Figure 1.1 Primary functions of support in burst-prone ground

at a cost substantially less than a full cable lacing system.

A common thread which emerged is that the Canadian mining industry does not yet have a well developed, consistent methodology for the selection of support in burst-prone ground.

A support system in burst-prone ground has two main functions as indicated by Figure 1.1.

The purpose of strengthening a jointed or fractured rockmass by reinforcement is to form a rock arch capable of carrying the induced stresses, i.e., "to help the rockmass support itself". In addition, it is necessary to retain any broken rock and to hold this material in place by tying it back to some deeper-lying stable ground. As imposed stress levels and deformations increase the "retain/hold" function becomes more critical, and the "reinforce/strengthen" function diminishes.

In Canadian mines, the approach and thinking is generally based on the "reinforce/strengthen" model. Under extreme conditions, i.e., where a serious rockburst risk exists, cable lacing may ultimately be installed.

South African gold mining practice differs significantly from Canadian metal mining practice in that the selection of support recognizes that most mine openings in South Africa are surrounded by a deep envelope of fractured rock. Consequently, the supports must accommodate significant deformation of the broken ground. This leads to the interesting fact that support elements and systems which are suitable under these static conditions will also have inherent characteristics, such as ductility and the ability to retain broken rock which are useful and important under rockburst conditions. If the support systems installed in areas of burst-prone ground are simply based on adding more of the same type of support used in static conditions, which is a common approach in both South Africa and in Canada, then the South African systems are likely to have a greater burst resistance than the Canadian systems.

Broadly, there are two types of design issue that arise in burst-prone ground. The first relates to the risk of seismically induced falls of ground and the second deals with the general response of the overall rockmass immediately surrounding the excavation as well as the means by which support systems can be effective in limiting the consequences of rockburst damage within this zone. Some key observations have emerged from this study:

- Full areal integrity of the retaining and holding elements must be maintained.
- Well designed connections between the retaining and the holding elements are essential.
- Mechanical (end-anchored) bolts generally lose their effectiveness as reinforcing elements but continue to play an important role as part of the retaining/holding system.
- Grouted deformed re-bars often fail immediately beneath the plate but appear to play a continuing, useful strengthening or reinforcing role.
- Smooth-shanked bars, recommended in South Africa, provide ductility to the "holding" elements.
- Chain-link mesh is preferred to welded wire screen as a retaining element under bursting conditions.
- Whereas experience concerning the contribution of reinforced shotcrete in burst-resistant support systems is still limited, there is increasing evidence from Chile, and to a lesser degree from Canada, that shotcrete may play a significant support function.
- Split Set® bolts apparently perform well as short-term support in providing ductility in the axial direction of the bolt.
- Standard Swellex® bolts are generally considered to be too stiff to provide ductility under bursting conditions.

Based on this study, preliminary guidelines for support in burst-prone ground are provided for Canadian conditions by McCreath and Kaiser (1992). The ideal characteristics for support systems placed in bursting ground include:

- High initial stiffness of reinforcing elements for strengthening of the rockmass;
- Maintenance of the reinforcing function under conditions of large deformations;
- Enhancement of support system ductility;
- Maintenance of the integrity of full areal coverage;
- Strong connections between retaining and holding elements;
- Efficient integration of elements comprising low level support systems into higher level support;
- Presence of multiple lines of defence within the support system; and
- A wide range of applicability for any one level of support system, to avoid the need for either accurate prediction of the rockburst potential or for detailed determination of the rockmass characteristics and response to dynamic loading.

The study concluded that, under rockburst conditions, it becomes increasingly important for the support system to provide and maintain complete areal coverage of the retaining/holding function, in order to avoid violent unravelling of the rockmass. Special attention must be paid to the connections between the retaining elements and the holding elements and to the introduction of sufficient ductility to allow the combined retaining-holding system to yield. In Canadian hardrock mines, the use of shotcrete seems promising as an integral part of a spectrum of rockburst resistant support systems. However, there is currently very little experience with the performance of engineered shotcrete systems under bursting conditions.

1.2 State-of-the-art in support design for burst-prone ground

1.2.1 Current design rationales

St. John and Zahrah (1987) presented a comprehensive study on the aseismic design of subsurface excavations and underground structures subject to seismic loading from natural and artificial sources. They reviewed the seismic environment and how structures are designed against earthquake hazards. Fundamentally, a design consists of estimating a design magnitude based on an empirical, site specific attenuation relationship defining the intensity of the ground motion anticipated at the site some distance from the seismic source. Of the parameters describing the intensity of ground motion, the peak particle acceleration ppa and peak particle velocity ppv are the most widely used measures. Seismic regionalization maps, providing effective ppa and effective ppv, are recommended for design by the Applied Technology Council (St. John and Zahrah, 1987).

Damage was attributed to three factors: fault slip, ground failure, and shaking. The latter reflects the recognition that repeated and lasting shaking with strong energy content may cause more damage than a single high ppa or ppv pulse. Because most of the data needed for a detailed analysis of damage is often not obtainable, the need for empirical damage models is considered to be a first and crucial step in any seismic design. Despite the availability of relatively sophisticated methods for modelling the dynamic response of underground openings, St. John and Zahrah (1987) concluded that any design should start with simple empirical methods followed by more rigorous analyses if justified. A similar approach is currently being followed at the GRC for the design of underground excavations in burst-prone ground.

From an empirical data base, Owen and Scholl (1981) established a "no damage limit" at < 0.2 g or < 200 mm/s, and a "major damage limit" at > 0.5 g or > 900 mm/s. They pointed out that less confined conditions near portals are more susceptible to damage. Furthermore, experience from mining (McGarr, 1983) indicates that ppv limits correlate better with damage than ppa limits.

St. John and Zahrah (1987) also pointed out that very high ppa or ppv pulses from tests using high explosives inflicted much less damage than earthquake shaking because the dynamic loading essentially comprised a single (compressive) pulse lasting only some tens to hundreds of milliseconds. Dowding et al. (1983) suggest that the number of stress cycles is critical for damage accumulation.

In summary, the design logic follows four steps:

- the seismic activity centre is defined (location and character);
- an empirical relationship to assess energy attenuation between source and target is established;
- damage mechanisms and limiting values for damage related parameters are determined; and
- the structure is designed to resist the dynamic loads.

This keynote paper follows the same pattern, and this reflects the approach adopted by the GRC to arrive at a rational design methodology.

Ortlepp (1992) points out that in mining, the conventional engineering design approach cannot be followed because of poorly defined load conditions and uncertainties with respect to design parameters. He concludes that a rational design procedure must include four steps:

- determination of most probable mode(s) of failure;
- establishment of a representative model for these failure mode(s) to determine how failure can be prevented;
- determination of material strength for the model(s); and
- selection of a safety factor, often empirically determined, to ensure an acceptably low probability of failure.

Recognition of probable failure modes as the basis for a rational engineering approach to support design for the containment of rockbursts is extremely important. First, it is consistent with conventional geotechnical design methodologies, whereby the stability of a structure and required remedial measures are determined for an assumed mode of failure (e.g., slip circle for slope failure in soil or wedge for rock slopes). Second, it requires that design methodologies must be grouped according to damage mechanisms, i.e., rockburst mechanisms, and limits of applicability must be well defined to ensure that the most appropriate model is chosen for design.

Ortlepp (1992) provides groupings of five source mechanisms (strain-bursting, buckling, pillar or face crushing, shear-rupture, and fault-slip) and four related damage or failure mechanisms (self-explosion of rock for the first two source mechanisms, ejection by seismic wave, displacement by inertia, and seismically triggered, gravity driven falls of ground).

From the perspective of support design, these five mechanisms fit into three categories:

Mode I - self-actuated ejection of fractured rock (strainburst or buckling with seismic source inside the failure mechanism)

Mode II - ejection of part of a fractured, broken or jointed rockmass (driven by inertia or stress waves)

Mode III - displacement of broken rock with gravity as dominant driving force component (seismically triggered falls of ground or enhanced gravity condition)

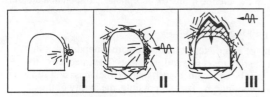

Figure 1.2 Three categories of failure at the target (Modes I to III)

Damage assessment and rational support design procedures must be tailored to deal with these three modes of failure. Unless otherwise indicated, failure mechanisms throughout this paper refer to the target and not the seismic source. In Mode I, the source is at the target.

Ortlepp (1992) subscribes to the same support functions, holding and retaining, as introduced by McCreath and Kaiser (1992) and points out that cladding elements must be designed to retain broken rock, and that holding elements (tendons), acting in tension or shear, provide the holding function. He stresses that a holding element's ability to yield is the essential characteristic of a support to contain violent rockbursts.

In South Africa, stopes are subjected to sufficiently high stresses such that the rock is fractured well ahead of the face and around tunnels (Jager, 1992). Consequently, conditions requiring support often fall into Mode II and normally into Mode III. Works by Wagner (1982/84) and Roberts & Brummer (1988) on support requirements for rockburst conditions, Jager (1992) on yielding hydraulic props and conebolts, and Wilson and Emere (1975) on crib and mat-packs, among others, are primarily concerned with these modes of failure. Because their goal is to hold broken rock, these investigators have adopted design models based on momentum transfer or energy balance considerations to determine the capacity and yieldability of supports to resist the movement of ejected blocks. Hedley (1992) adopted these rather successful models to provide a relative rating of one-tier to three tier support systems commonly used in Canada for support in hard rock mines and in burst-prone ground.

In Canadian mines, the stresses are still fairly moderate and the severity of bursting is fortunately not as extreme as in South Africa. Consequently, there is a need to develop a support design rationale for intermediate levels of damage potential and severity (≤ RDL 3; Kaiser et al., 1992) as well as for major rockbursts (RDL 4 and 5).

As part of the second five-year phase of the Canadian Rockburst Research Program (CRRP), the Geomechanics Research Centre has been contracted to develop a design rationale for support in burst-prone ground with particular reference to conditions representing an intermediate level of risk in terms of frequency of occurrence and magnitude (M_{Nuttli} < 4) or rockburst severity. It is the goal of this study to develop design charts relating source characteristics to support requirements, including both the retaining and holding function of the support. This keynote address is presented at the mid point of this project and hence represents in many ways a report on the current status of our work.

2. CONCEPTUAL DESIGN CONSIDERATIONS

2.1 Engineering design approach

The conventional design approach as outlined above essentially consists of (1) the identification of potential failure modes and (2) a comparison of the available resistances or capacity with the driving forces or demand (including dynamic components). By calculating factors of safety or the probability of failure, it is then determined how much artificial support is required. Unfortunately, there is, at present, insufficient knowledge to quantify rockburst failure mechanisms, to calculate the forces induced by failure and to determine, with sufficient accuracy, the capacity required to resist these forces. Hence, the conventional engineering design rationale cannot be directly adopted and empirical approaches must be used. The fundamental components of engineering design must, however, be incorporated implicitly into such empirical designs.

We must be able to identify areas where damage to an underground excavation is anticipated, including extent, severity and violence to enable selection of the most appropriate support system to prevent or at least minimize this damage.

In short, a proper design consists of the following elements:

- identify **what kind of failure** mechanism is anticipated. Ortlepp (1992) identified five dominant rockburst types.
- eliminate potential failure processes by removing the cause (seismic events) or by preventing the identified failure mechanisms (by rock reinforcement).
- identify **where such failures are anticipated** and whether they could be triggered. Most investigators start from the premise that a failure does occur and needs to be resisted. More work is needed to identify areas prone to violent failure and to determine the anticipated violence of the failure. This implies that we need to obtain a better understanding of the source characteristics in the near field.
- select support system to **survive the impact** (survival = retention of functionality even though reconditioning may be required). This is mostly done based on energy dissipation considerations.
- identify the "maximum practical support limit" (MPSL) beyond which elimination of the cause for damage constitutes the only viable solution.

Consequently, the goal of ongoing research is to develop design charts relating, for given failure mechanisms, the distance from the source and event intensity (magnitude) to a "support survival limit". Such charts, as illustrated schematically by Figure 2.1, will help the ground control engineer to determine whether extra support is needed and to select the most appropriate support system once a microseismic activity centre has been identified and characterized.

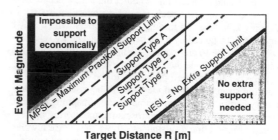

Figure 2.1 Conceptual design chart for support selection (survival limits)

If a burst-resistant support is required, two alternate methods of hazard reduction should be considered:

- prevent seismic events or reduce the magnitude of energy release, e.g., by better sequencing; or
- minimize the extent and violence of rockburst damage, e.g., by destress blasting.

Next, it must be assessed whether the integrity of the fractured rock can be maintained (by reinforcing and strengthening) or the post-peak failure characteristics can be changed (by adding confinement to the rockmass). If reinforcement efforts are inadequate and rock ejection cannot be prevented, then retaining and holding elements are needed to prevent and control rock ejection. In this case, the survival limit must be determined based on design loads defined for an appropriate failure mechanism (depth of failure, ejection velocity or kinematic energy content) or allowable deformations for the support components (yield capacity).

2.2 Factors to consider for support design

An aseismic design process includes a proper understanding of the source, the wave transmission, and the response of the target (the underground opening and its support). This is schematically illustrated by Figure 2.2. Even though it is the goal of this keynote to deal with support, it is necessary to consider the other aspects of design: the source and the transmission. Obviously, it is not possible to elaborate on all factors that affect the design and the following sections are intended to list issues that must be addressed and assumptions that are made to simplify the discussion on support design.

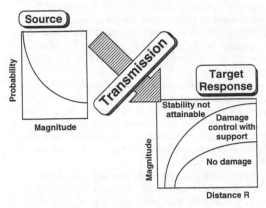

Figure 2.2 Aseismic design sequence

2.2.1 Seismic activity centre

A seismic activity centre must be well defined in terms of probability of occurrence or event frequency, expected magnitude, and spatial distribution. This includes the source characteristics described by the seismic moment tensor.

Despite enormous advances in seismic monitoring, it will be shown that the event causing the damage cannot yet be easily identified. For the establishment of empirical design guidelines, it is, however, essential that the damage-causing event is properly identified. It is crucial to determine whether damage was caused by the main event, or by an aftershock or a sub-source. The most critical aftershock, in terms of mining, is a strainburst which may be a secondary event, triggered by a far-field seismic energy release, but which is very close to an opening. While these aftershocks may be of much lower magnitude than the main event, they may cause serious damage because of their proximity to the opening.

From a design strategy point of view, it will be assumed that a certain activity centre, with given characteristics (magnitude, event type, etc.) can be well defined by microseismic monitoring at the mine. Instead of predicting a seismic event, we design for a well defined "design seismic event" (e.g., an event of given magnitude with a given reoccurrence rate). Consequently, a mine must first develop a strategy to define a "seismic design event" in terms of its magnitude, occurrence frequency, and spatial distribution. The event magnitude and location are subsequently used as two fundamental design parameters.

2.2.2 Energy transmission

Whereas much effort has been expended over the last few years to better understand microseismic source mechanisms, little is still known about wave attenuation in mining environments, where open or backfilled stopes, fractured and broken rockmasses, and major geological structures influence wave propagation. The data presented by Jesenak et al. (1993) suggests that shielding of waves between source and target is a dominant factor in controlling damage. A better understanding of transmission processes is crucial for progress in hazard assessment.

2.2.3 Damage mechanisms

The effect of a seismic event on the target (an underground excavation) obviously depends foremost on what arrives at the target rather than what occurs at the source. Unfortunately, most dynamic measurements are only collected at locations that are remote from the damage locations. Hence, we are faced with the challenging task of predicting what happens at the target from measurements remote from it.

The assessment of damage-potential at a given location by a well defined activity centre is therefore primarily a matter of extrapolation from the source or a remote monitoring locations to a potential target. It is for this reason that the study by Jesenak et al. (1993) focussed on the rate of *ppv* decay to establish whether geometric spreading is a valid assumption for damage assessment. Whether extrapolation from far-field observations to the near-field is acceptable and where the limits for acceptable extrapolation are must be addressed.

2.2.4 Support function and response

Before damage is inflicted to an opening, the support must strengthen the rockmass by reinforcement. The integrity of the rockmass must be maintained such that the rockmass and support respond as a continuum. If at all possible, the ability of this reinforced system to absorb seismic energy must be enhanced.

While a continuum is maintained, stability can be assessed by a quasi-static approach of determining the threshold for fracturing the reinforced rockmass.

Once rock is broken and damage has occurred, the support has to retain the broken rock and hold the support system with the retained broken rock in place. In this role, the support must be able to dissipate the kinetic energy of the ejected rock while maintaining its ability to withstand the static, gravitational forces. The support must be able to reestablish a static equilibrium after an acceptable amount of displacement.

It is intuitively evident, that a support's ability to resist dynamic forces and to reestablish equilibrium must depend on the effectiveness of the support before the event. This is best described by the static factor of safety, a measure of how far an opening exists from failure. The safety margin to initiate or trigger a failure can be assessed by comparing support capacity with the demand placed on the support. After a failure has been triggered and the equilibrium disrupted, energy considerations must be invoked to determine whether an new equilibrium can be achieved. It will also depend on the static factor of safety. This sequence of events is primarily applicable for seismically induced falls and will be discussed in more detail in the design section.

Various support types have been developed to provide the best characteristics for the individual functions described above. For support in burst-prone ground, ductility is clearly an essential characteristic and this has led to the development of ductile retaining systems, such as lacing, and ductile holding elements, yielding props and cone bolts (Jager, 1992), and friction anchored cables (Tannant and Kaiser, 1993). Ideally, a holding element should have a large ultimate yield limit and a high ultimate strength. The high ultimate strength is needed to maintain the capacity of a support after a new equilibrium as been established.

2.2.5 Risk-Cost-Benefit

Civil engineering practice makes extensive use of reliability based design (Harr, 1987). In most cases, the methodology is to use these techniques to determine the variability in the output from a calculation based on the variability in the input parameters. Rockburst damage and seismic data is inherently highly variable and invites analysis by means of reliability based design methodology.

Although most burst-prone mines make use of microseismic monitoring systems, no uniform, industry-wide method for assessing rockburst or seismic hazard has ever been adopted by them, nor is risk assessment used in making support design decisions. It seems logical, therefore, to integrate some measure of rockburst hazard assessment into a decision-making procedure for selecting rockburst-resistant tunnel and stope support.

Based on this logic, the GRC plans to develop a Rockburst Risk Index which makes use of the rockburst, accident and damage report databases to develop a statistical model for the occurrence of rockbursts, based on the double-truncated Gutenberg and Richter frequency-magnitude relationship for earthquakes. This relationship will be extended to relate the risk of occurrence of a rockburst to one or more mining parameters (e.g., tonnage mined, change in stress level in remaining rock and ore, or stored strain energy in the surrounding rock, and the past seismic history of the mine). Detailed primary support and rehabilitation costs will be assembled for use in the decision process.

In this manner, a support design procedure will emerge which balances the extra cost for upgraded support against the benefits which will accrue due to reduced risk of damage, loss of production, and safety issues. This method is expected to represent an improvement on the existing, often ad-hoc, methods of decision-making regarding rockburst-resistant support selection.

2.3 Support design rationale

The preceding review demonstrates that a proper design rationale for support to resist rockburst damage must consist of four steps:

* Define microseismic activity centre; define dynamic loading environment;
* Identify most likely mode of failure at target area, Mode I to III;
* Determine whether rockmass will be disrupted or whether a continuum can be maintained; fresh fracturing or trigger of block ejection; and
* Design or select support to either: (a) maintain continuous rockmass by reinforcement or (b) retain and hold broken or ejected rock; maintain equilibrium or achieve new equilibrium by dissipation of kinetic energy.

More specifically, these four steps involve:

Step 1 -- Load definition: The microseismic activity centre needs to be defined in terms of probability of event occurrence, intensity, failure type, and spatial distribution. Whereas more and more detailed knowledge about the source mechanisms is obtained by careful seismic monitoring, event magnitude M and location R are still the only two parameters that can be readily obtained at Canadian mines. Hence, it is assumed that the mine staff can determine, based on past experience and monitoring, the location and magnitude of a "design event", i.e., the largest seismic event that needs to be survived.

With M and R given, the damage-potential, the expected failure mode, and the required support must be determined.

Step 2 -- Failure mode: Ground conditions near underground openings can basically be divided into six categories (Hoek, 1992) using the degree of jointing and stress level as the prime parameters distinguishing typical ground behaviour. The matrix presented in Figure 2.3, adapted for rockburst conditions from Hoek's matrix for static conditions, provides a means for identifying the anticipated failure mode.

		More Jointing →	
	Massive	**Moderately jointed**	**Heavily jointed**
Low stress / No fractures	*Massive rock with few joints* • no burst potential	*Discontinuous jointing* • no burst potential	**Mode III** *Many joints with low stress favouring key block failures* • potential for falls of ground
High stress / Fractured rock	**Mode I** *Massive but fractured rock near opening with few joints* • strainburst potential	**Mode I or II** *Discontinuous jointing but fractured rock near opening* • buckling and shear burst potential	**Mode III** *Many joints and fractures with potential for stress driven block failures* • potential for falls of ground

(left axis label: More Stress, More Fracturing ↑)

Low stress is defined as $\sigma_1 < 0.3\ \sigma_c$(Lab) and the failure modes are described in Section 1.2.1.

Figure 2.3 Matrix for failure mode identification

Step 3 -- Trigger assessment: To determine whether damage to an opening will be inflicted by a seismic event, quasi-static stress analyses can be conducted to determine the potential for fracturing of massive rock (Mode I and II). If a fall of ground is anticipated, the trigger level for such failures must be determined (Mode III) and if the rockmass is damaged and the continuum is disrupted, the vigor of the event must be established. In particular, the ejection velocity must be predicted for use in energy-based support design procedures. In this case, near- and far-field conditions and seismic wave or self-induced (strainburst)

rock ejection modes must be considered.

Step 4 -- Support design:

(a) if a continuous rockmass is maintained around the opening and a failure is not triggered, conventional, static support design methodologies apply and energy based design methods should not be applied. Measures that strengthen the rockmass, enhance energy dissipation, and promote more ductile post-peak failure behaviour of the reinforced rockmass, are desired. Well reinforced, destress blasted rock falls into this category.

(b) once a failure is triggered, the rockmass is disrupted and the broken rock must be retained and held in place. The kinetic energy contained in the ejected rock must be dissipated and a new, stable static equilibrium must be achieved.

This design rationale is developed further in the following sections on damage assessment and support design. In particular, the following deliberations focus on:

* What triggers failures and how far from the source can damage occur ?
* How can the source causing the damage be properly identified ?
* How violent are failures expected to become ?
* Whether special support is needed and what are its survival limits ?

3. DAMAGE-POTENTIAL ASSESSMENT

3.1 Empirical approach and data base

Early in our research on support for burst-prone ground, it was realized that there is essentially no reliable database on rock and support damage inflicted by seismic activity. Hence, it was necessary to develop a rational procedure for unbiased damage data collection. The procedure is described by Kaiser et al. (1992) and an update with more recent data is presented by Jesenak et al. (1993). Based on data from Creighton mine, the following relationship was developed:

$$ppv = C\ SD^{-1} \quad \text{with} \quad SD = R/10^{0.57M} \tag{1}$$

where SD = scaled distance; C = 460 for 50% and C = 670 for 95% confidence; ppv in mm/s with SD in m.

These relationships with the parameters given is only applicable to Creighton mine and, as will be demonstrated below, for far-field conditions only.

Empirically, damage was related to ppv by:

$$ppv = K\ 2^{(RDL-1)} \tag{2}$$

where K = 50 mm/s for Creighton mine.

By equating Equations (1) and (2), the damage limit presented by Jesenak et al. (1993) as a function of source magnitude was established:

$$R = \frac{C}{K} 10^{0.57M} 2^{(1-RDL)} \tag{3}$$

Again, since the database exclusively contains data from Creighton mine, Equation (3) should not be applied to other mines without prior calibration. In particular, for support conditions that differ from Creighton support standards, adjustments must be made. Furthermore, as demonstrated by Jesenak et al. (1993), gravity dominated failures (Mode III) with marginal static safety margins need to be treated separately (see Section 4).

For comparisons with data from other parts of the world, Nuttli magnitudes had to be related to Richter magnitudes. For this paper, as well as for Jesenak et al. (1993), seismic energy equations presented by Hedley (1992) were equated, leading to:

$$M_{Richter} = 0.87\ M_{Nuttli} - 0.37 \tag{4}$$

(Note: For high magnitudes this leads to significantly higher Richter magnitudes than if the seismic moment equations presented by Hasegawa (1983) and Spottiswoode and McGarr (1975) were equated).

When assessing rockburst damage, it is necessary to differentiate between near- and far-field behaviour. Near the source, strain concentrations at the fracture front or near fault asperities may lead to very high ground motion parameters when compared to those predicted by far-field equations. This is primarily due to the fact that the assumption of energy radiation from a point source is violated within the proximity of a finite source.

3.2 Damage near faults or close to a source

Because mine excavations are often close to or even inside the source region and ground motions are largest in this region, near-field ground motions are critical for such situations.

McGarr and Bicknell (1988) have studied seismic data from underground networks in South Africa and found that peak particle velocities near faults are typically on the order of 2 m/s (maximum 4 m/s). These velocities must be anticipated within Zone A (Fig. 3.1). By evaluating the far- and near-field contributions for conditions typical for Creighton (7000' depth), it can be shown that the near- and far-field velocities, v_{nf} and v_{ff} respectively, are equal for a scaled distance of about 1 to 2. This region is also shown in Figure 3.1. It separates the remaining area into two zones, B and C. Since near- and far-field terms have different time histories and are, therefore, not additive, it is reasonable to assume that far-field equations are applicable where $v_{nf} < v_{ff}$, i.e., in Zone C. In Zones A and B, the actual peak velocities are expected to be higher than predicted by far-field equations.

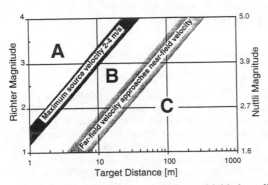

Figure 3.1 Zoning of design chart for unshielded conditions at Creighton mine: (A) region with maximum ground motion parameters; (B) region with elevated ground motion due to near source factors; (C) region with far-field ground motion characteristics

Figure 3.2 presents data from minor and severe rock damage observations (Jesenak et al. , 1993) for comparison with the approximate limit for equal far- and near-field velocity; $v_{nf} = v_{ff}$ (Fig. 3.1). Most data from our database fall to the right, in Zone C, where the far-field theory should be applicable.

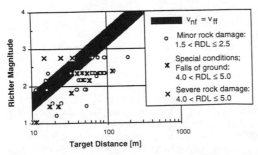

Figure 3.2   Comparison of damage data with far-field limit

Practical implications

In Zone C (Fig. 3.1), the empirical relationships for damage assessment (Eqns 1 and 2) and support design, presented later, should be applicable. In Zone B, more severe damage must be anticipated and extra support will be required. In Zone A, openings must be designed for maximum peak particle velocities. Extremely severe damage must be expected for openings that fall within Zones A and B. In these zones, burst-resistant supports will be loaded to capacity and will not be able to withstand the dynamic impact unless extremely deformable.

3.3 Identification of source causing damage

At Canadian mines, source locations and source magnitudes are determined independently. Because of clock differences and because several seismic events are often recorded within a very short time period, it is necessary to verify whether the event actually causing the damage has been properly identified. For the data presented by Jesenak et al. (1993), it was assumed that the events and damage were properly correlated. The data points (e.g., Fig. 3.2) are plotted for magnitudes obtained from Ottawa and locations provided by the mine.

Whether this procedure, commonly adopted for the development of empirical damage criteria, is justified and the 'correct' source causing damage was identified is explored next.

Typically, a large seismic event results in seismically induced stress redistribution and "aftershocks". The term aftershock is used here in the mining context and refers to any event occurring after the main event. These aftershocks typically have smaller magnitudes but it is possible that they create the damage by either repeated shaking or by being critically located. Therefore, before damage can be properly related to a source, a means has to be found to identify the actual damage causing event.

Assuming the preliminary relationships between *ppv* and damage levels (Eqn. 2) to be valid, limits of damage were established and compared with observed damage levels. This work has been extended and an update is provided by Jesenak et al. (1993). It is still assumed that all damage is caused by the event identified by the mine.

By reversing the procedure used by Kaiser et al. (1992) to relate damage levels to magnitude and assuming that the established empirical relationships (Eqns 1 to 2) are valid, a magnitude M^*, theoretically required to cause the damage at the observed locations, can be back-calculated for each recorded event location:

$$M^* = \frac{1}{0.57} \log \left(R \frac{K}{C} 2^{(RDL-1)} \right)$$

(5)

where K and C are constants (Jesenak et al., 1993), R is the distance from the event to the damage location, and *RDL* is the rock damage level determined by the procedure established by Kaiser et al. (1992).

If M^* is equal to the observed magnitude, the identified event fits the damage; if M^* is much larger, the event could not have caused the damage; and if M^* is smaller, the event may have been shielded or had to be of magnitude M^* to cause the damage.

Figure 3.3 presents M^*-values for rockburst damage associated with a recorded event of magnitude $M_{Richter} = 1.9$. These M^*-values are average values for the seven damage locations recorded for this rockburst.

Only if the actually observed magnitude is greater than M^*, could the event have caused the damage. Otherwise, the event is too far from the damage location. According to Figure 3.3, the seven

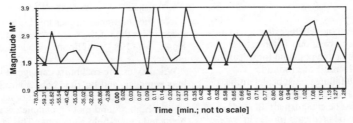

Figure 3.3 Back-calculated magnitudes M^* based on seven observed damage locations for rockburst #1060 at Creighton mine (November 23, 1990; K = 50, C = 460, rock damage levels between 2 and 3)

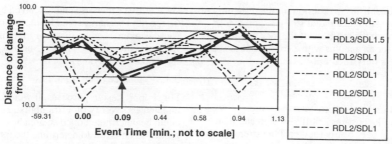

Figure 3.4 Distances from event to observed damage location for rockburst #1060 at Creighton mine (legend gives recorded rock and support damage levels)

events marked with a triangle could have caused the damage within the given time window (time 0.00 marks the event given by the mine). However, since it is not likely that timing errors exceed more than a fraction of a minute, events (0.00) and (0.09) are the most likely sources for damage.

In Figure 3.4, the recorded distances from these seven events to the seven recorded damage locations are plotted.

The event at (0.00), with less damage closer to the event, is not consistent. Of the events with major damage close to the source, event (0.09) is most consistent. Event (0.09) rather than (0.00) must therefore be the damage causing event. The implication of this error is that the event-to-damage distance reduces on average from 37.4 to 28.8 m (individual data points shift between -26.5 and +25.5 m). More important, the average *ppv* calculated for locations with RDL 3 increases from an unrealistically low value of 122 mm/s to 284 mm/s and decreases for RDL 2 from an unrealistically high *ppv* of 240 to 175 mm/s.

Examples with higher damage levels were also analyzed and confirm these findings. Unfortunately, at the time of printing, this procedure has not yet been applied to the data presented by Jesenak et al. (1993) or the data shown in Figure 3.2.

Practical implications

The empirically established relationship between damage levels and *ppv* has been confirmed for the range RDL 2 to 5 and a means to identify the source actually causing damage has been developed. This procedure will be adopted to identify erroneous data in the damage database.

3.4 Damage predictions in far-field

As outlined in Section 2.3 on design rationale, it must be determined whether the rockmass will be disrupted or a continuum can be maintained by proper rock reinforcement. It must be established whether a failure is being triggered and how violent the failure process is.

3.4.1 Failure triggered by exceeding rock strength
        (quasi static stress approach)

As a low frequency p-wave propagates through an elastic medium, it creates a compressive stress wave of $\Delta\sigma = \rho\, c_p\, v_p$ in the direction of wave propagation and of $\Delta\sigma\,(v/(1-v))$ perpendicular to it (where ρ = rock density; c_p = p-wave propagation velocity; and v_p = particle velocity). This stress increment is superposed on the virgin stress field and is magnified near an opening (Dowding et al. 1983). Figure 3.5 illustrates, for the empirical relationship established for Creighton mine (C = 460; Eqn. 1), how far from a source a specified dynamic field stress change must be anticipated. Whether this stress change, when magnified at the wall of an opening, causes any damage depends on the static stress level (or strength factor). For a circular opening, the dynamic tangential stress change at the wall would typically range from 0 to 2.7 times the values shown in Figure 3.5.

Practical implications

For a highly stressed, brittle rockmass close to failure, stress changes of a few MPa may trigger a violent failure process (strain burst). Hence, relatively small seismic events (M = 1 to 2) could trigger a burst at 50 to 100 m from the source. This type of

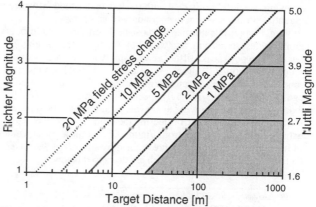

Figure 3.5 Dynamic rockmass field stress limits (C = 460; SD = R/10$^{(0.57M)}$; c_p = 6000 m/s; ρ = 2700 kg/m^3)

rockmass damage due to dynamic stresses must be anticipated at fairly remote locations.

For Creighton mine, a comparison of Figure 3.2 with collected damage data (Jesenak et al., 1993) shows that no damage was observed at locations to the right of the 1 MPa field stress change limit. Hence, no extra support seems to be required where the dynamic field stress change due to a passing p-wave is less than 1 MPa (shaded area).

3.4.2 Stable versus unstable failure process (strain burst potential)

Once it has been identified that the strength of the rockmass near an opening will be exceeded by static or dynamic loading, one needs to determine whether failure could occur in an unstable, violent manner. If there is a high potential for strain bursting, it is also desirable to determine the depth of a potential fracture annulus and expected block ejection velocities. Whereas it is not possible to predict these factors accurately, we can examine those parameters promoting unstable failure processes using the example of a hydrostatically stressed circular opening in elastic rock.

From the violent response of rock samples in soft testing machines, it has been learned that the failure mode, in terms of violence, depends on the unloading stiffness of the machine k and the slope of the post-peak strength curve of the rock sample λ, i.e., the unloading stiffness of the rock. Only stiff loading systems can track the post-peak failure in a controlled manner, i.e., $|\lambda| < |k|$ is a condition for stability.

The same concept has been applied to assess the stability of pillars, by replacing the testing machine by the mine and the sample by the pillar (e.g., Hedley, 1992). In a similar manner, the violent energy release during fault slip processes can be explained. This concept for stability assessment by considering the relative stiffness of the loading system and the failing rock can also be applied to single openings (Jaeger and Cook, 1969). By comparing the stiffness of a fractured rock annulus (corresponding to the sample stiffness) with the radial loading stiffness of the surrounding rock, conditions for unstable failure can be identified and it can be determined whether a strainburst is anticipated (see Appendices A and B for comments on alternate means for strainburst potential assessment).

Jaeger and Cook (1969) established a criterion for the instability of a failed rock annulus by comparing the radial stiffness of the surrounding elastic rock k with the stiffness of the rock annulus. They found that instability for an annulus with a constant negative post-peak stiffness λ occurs only if $\sigma_c / p < 2$ and:

$$\left| \frac{2G}{\lambda} \right| < 1 \tag{6}$$

where G = shear modulus and p = hydrostatic pressure.

Instability is anticipated if the post-peak stiffness is greater than the pre-peak modulus, i.e., in very brittle rock or if the strength does not increase with confinement.

The greatest uncertainty in this stability assessment process is the post-peak stiffness of the failed rockmass. From laboratory tests, it is known that λ is affected by confinement which can be derived from a confining pressure or from boundary shear stresses. In the following, we will briefly deal with the less understood influence of tangential shear stresses at the boundary of an opening.

Figure 3.6 presents normalized post-peak stiffness data from two laboratory test series with and without shear under the loading platens. The post-peak stiffness rises rapidly as the shear restraint is reduced (the ratio $2G/\lambda$ drops), i.e., the material becomes more prone to bursting for a given loading system. The extrapolation to negative shear stresses is purely speculative even though, there is laboratory evidence showing that brittle failure is promoted by soft platens (outward shear). Considering the criteria given above for unstable failure, it follows from Figure 3.6 that these rocks would fail in an unstable manner if unrestrained by shear (< 20%). The practical implications of this graph are discussed below.

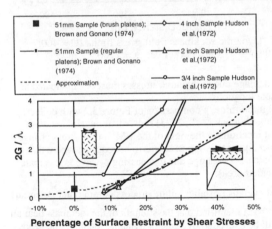

Figure 3.6 Normalized post-peak stiffness for rocks tested by Hudson et al. (1972) and Brown and Gonano (1974)

Practical implications

The significant influence of shear restraint on the post-peak behaviour of fractured rock is of practical significance. Frictional bolts, for example, provide restraining shear stresses to the broken rock in a radial direction and thus increase the rock's ductility. As will be discussed in a later section, a similar effect is expected from shotcrete if applied to fractured rock.

For openings in a non-uniform stress field and openings with irregular shapes, shear stresses exist near the surface of the opening. This leads to a non-uniform distribution of strainburst proneness around the opening.

If a weakness plane or a fault intersects an opening such that there is a tendency for shear along the fault, the rock on one side of the fault will be restrained by shear (thus more stable) whereas shear stresses on the other side promote brittle failure (thus less stable). There is ample field evidence demonstrating that burst damage drastically differs on opposite sides of shear zones.

When entering a stiff burst-prone dyke, the softer rock will deform and impose an inward shear stress on the dyke, thereby rendering the dyke rock more brittle and prone to bursting. When the drift emerges from the dyke, the shear stresses are initially smaller because of the face effect and the dyke is less prone to failure. However, after the drift is advanced for a few rounds, the shear stresses increase and a delayed burst may occur in the dyke. The second scenario could be more dangerous as the burst will cause damage at some distance from the face. This example also illustrates that mining induced shear at dyke/rock interfaces enhances the tendency of dykes to burst.

3.4.3 Violence of strain bursts

For support design purposes, it would be highly desirable, if the anticipated severity of a strainburst, i.e., the ejection velocities, could be determined once an area has been identified as strainburst-prone. In an effort to determine the most relevant factors controlling the severity of strainbursts, some extreme simplifications are made to estimate the released energy.

By assuming that all released energy, due to the enlargement of an opening from a radius a_1 to a_2, is converted into kinetic energy, an upper limit for the anticipated ejection velocities can be determined. Jaeger and Cook (1969) present energy release equations for driving a circular tunnel in a uniform stress field. The change in strain energy per unit length of annulus due to the widening is:

$$\Delta W = \frac{\pi p^2}{2G} (a_2^2 - a_1^2) \tag{7}$$

If this energy is fully converted into kinetic energy, the maximum possible ejection velocity is:

$$v = \sqrt{\frac{2\,\Delta W}{m}} \tag{8}$$

where the mass m is the mass of the ejected part of the fractured rock annulus.

If only part of the annulus is ejected, the energy release may be derived from either part or the entire annulus interface. For a preliminary assessment, it is assumed that the entire energy is released even if only part of the annulus is ejected. In this manner, ejection velocities are obviously overestimated for partial ejections. Figure 3.7 presents the theoretical ejection velocities for a deep drift in a hydrostatic stress field. The ejection velocity depends on the post-peak stiffness $\lambda = \alpha \lambda_{crit}$, shown as a multiple of the critical post-peak stiffness ($\lambda_{critical}$ derived from Eqn. A.2), and the percentage of ejected rock annulus. For full ring ejection (100%) velocities up to 4 m/s are predicted for very brittle rock and much lower values are expected only if the post-peak stiffness of the broken rock annulus is close to the critical stiffness. While the partial ejection velocities, shown in Figure 3.7, must be viewed as extreme, upper level approximations, it can be seen that very high velocities must be anticipated during strainbursts.

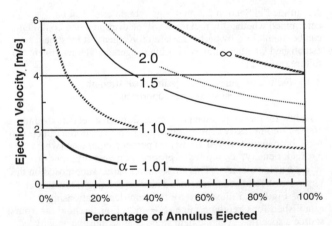

Figure 3.7 Theoretical ejection velocities for strainbursting fractured rock annulus ($\lambda = \alpha \lambda_{critical}$ from Equation A.2; E = 70 GPa; ν = 0.25; σ_o = 50 MPa)

Practical implications

Ejection velocities in excess of 2 to 3 m/s must be expected during strainbursts, even for post-peak stiffnesses near critical. These velocities may well exceed those observed during fault slip events and, thus, support design to retain strainburst damage presents a special challenge.

In very brittle rocks and/or for situations where only part of the drift wall is ejected, very high ejection velocities result from strainbursts. Because velocities in excess of 2 to 3 m/s are extremely difficult to resist with artificial support measures, strainburst must be prevented by other means such as destress blasting to soften the wall. If they cannot be prevented, the goal of the support must be to render the fractured rock annulus more ductile, meaning close pattern reinforcement with frictional or yielding support elements. The maximum practical support limit (MPSL) will, however, be rapidly reached.

3.4.4 Triggering of a failure by exceeding the support capacity (acceleration approach)

Before damage in supported rock can occur, a failure must be triggered by overcoming the reserves in the support system (step 3 of design rationale).

In this context, the term "support" is understood to mean any force available to resist failure initiation. For a block without frictional or cohesive retaining forces (e.g., an unconstrained wedge in the back), the only "support" forces result from the artificial support elements. In more complex situations, support may also be derived from rock bridges, side wall friction, and other restraining forces. Ideally, all of these forces must be considered when determining the static support capacity or the static factor of safety FS_S.

For the following, we will consider the simplest case of an artificially supported rock wedge, without side friction, in the back of a drift. The derivations of the trigger limits are given in Appendix C. From Equation C.2., it follows that the trigger limit depends on the frequency at the maximum ppa. Since no reliable ppa measurements are available, it must be inferred from ppv measurements.

Using somewhat arbitrary but meaningful values of n = 1.5, a static factor of safety of 1.20, and a dominant trigger frequency of 10 Hz in Equation C.2, the trigger limit shown in Figure 3.8 (labeled GRC -- 10 Hz) was determined.

McGarr et al. (1981), by fitting data from mining induced tremors and earthquakes (range:50 m to 1.6 km), derived a relationship for the dynamic shear-stress difference as a function of event magnitude. Using the accelerations from this relationship, a trigger limit that should be applicable for South African conditions was calculated and is plotted in Figure 3.8 (labeled McGarr (1981)

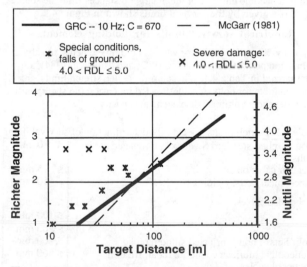

Figure 3.8 Trigger limits for gravity driven failures of the back, determined by Equation C.2 (shown for FS_S = 1.20; n = 1.5)

The location data from all cases with severe damage by seismically induced falls of ground (RDL 4 and 5; recorded in GRC's database; Jesenek et al. 1993) fall to the left of both trigger limits, implying that these falls should have been triggered, on average, by low frequency waves of about 10 Hz or less if FS_S = 1.20 and n =1.5. This frequency is sufficiently low to immediately accelerate the entire volume of rock involved in a fall.

Despite the severely limiting assumptions made for the derivation

of the trigger limits, the limit shown in Figure 3.8 appears to be realistic and suggest that low frequency waves can easily trigger falls of ground at large distances from the source unless the static factor of safety is sufficiently high.

Practical implications

Until strong motion measurements provide a means to verify Equation C.2, the limit shown by the full line in Figure 3.8 provides some guidance to determine whether a fall may be triggered. Instabilities triggered by seismically enhanced gravity are not expected at locations to the right of this trigger limit. If a target is shielded by broken rock, back fill, or mined out areas, the limit determined by Equation C.2 should fall on the conservative side.

Once a failure has been initiated or triggered, a block will move away from the wall or back and, if well supported, decelerate to reach a new equilibrium position. Whether a new equilibrium can be obtained, is analyzed in Section 4.

4. SUPPORT DESIGN AND PERFORMANCE

4.1. Introduction to Support Design

The previous section dealt with the issues of rockburst damage by fracturing and by triggering the ejection of blocks of rock. From a support design point of view, it is then a question of whether this damage can be prevented, either by strengthening the rockmass to a point where it can no longer be fractured, or by eliminating block ejections. In the former case, the rockmass is not disrupted and continues to behave as a "continuum". The limits for instability and the propensity for unstable failure can be assessed in the manner described in the previous section.

Once it has been established that damage will be inflicted upon the rockmass, it must be assumed that the reinforcement function of a support was inadequate and that the support system now must retain and hold the unstable material in place. The support must now dissipate the energy trapped in the ejected rock.

Unfortunately, it is still impossible to accurately predict ejection velocities at potential damage locations, primarily because of a lack of measured ejection velocities near damaged excavations. To circumvent this deficiency, a displacement-based design approach is adopted. First, "allowable" ejection velocities are determined as a function of allowable deformations and then, making assumptions about the relationship between *ppv* and ejection velocities, survival limits are established for various support types.

For a proper support system, two aspects must be considered: (a) survival of the retaining elements (mesh, shotcrete, lacing), and (b) survival of the holding elements (bolts, cables, etc.). In the following sections, holding elements are considered first in Sections 4.2 to 4.4 and the special role of shotcrete as a retaining element is discussed in Section 4.5.

4.2 Allowable block ejection velocities (seismically induced falls; Mode III)

Several assumptions are made to determine the allowable ejection velocity for holding elements:
- only the conditions in the back of a drift, with full gravitational effects, are analyzed;
- the damaged rock is completely separated from the surrounding rockmass;
- the function of the support is to hold the broken rock in place;
- the support elements respond in a perfectly plastic manner; and
- the retaining components of the support system do not fail under the impact.

The work W_b done by a block of mass m moving a distance h, vertically down, after ejection at a velocity v is:

$$W_b = m\,g\,h + 0.5\,m\,v^2 \qquad (9)$$

The work done by a perfectly plastic, yielding support W_{sup} resisting the block motion is:

$$W_{sup} = n\,FS_S\,m\,g\,h \qquad (10)$$

where n = dynamic strength factor, FS_S = static factor of safety, m = mass of broken rock, and g = gravitational acceleration.

After the rock has moved a distance h_{ult}, a new equilibrium can be reached if $W_b = W_{sup}$. Hence, the allowable ejection velocity v_{all}

as a function of the ultimate displacement h_{ult} is:

$$v_{all} = \sqrt{2gh_{ult}(n\,FS_s - 1)} \qquad (11)$$

If the yield strength of the support is independent of the loading rate, n = 1. Of course, if n = 1 and FS_s = 1, no extra forces can be withstood by the support and v_{all} = 0. Equation 11 is illustrated by Figure 4.1 for four static factors of safety and an n-value of 1.15. A dynamic strength increase of 15% is commonly encountered. For example, Jager (1992) presented rapid pull test data from conebolts showing a loading rate dependent, temporary yield strength increase of about 15%.

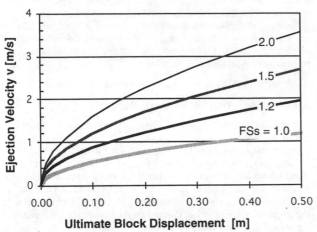

Figure 4.1 Allowable ejection velocities as a function of allowable ultimate rock displacement (for n = 1.15 and variable static factor of safety FS_s)

These curves demonstrate that allowable ejection velocities are very small if the ultimate displacement is small. Unless a support can yield for several centimeters, is designed for a high static factor of safety, or exhibits a high dynamic strength (n >> 1), the allowable ejection velocities are very low (e.g., ≤ 1 m/s).

Wagner (1982/84) and Jager (1992) suggest, as desired support characteristics for heavy bursting conditions, a yield limit of 0.3 to 0.5 m at an ejection velocity of 2 to 3 m/s. Figure 4.1 illustrates that a support with a 15% rate dependent dynamic strength margin, n, would have to be designed for a static factor of safety between 1.2 and 2 to satisfy this criteria.

Practical implications

For stiff and brittle support systems the allowable ejection velocity is very low unless the static factor of safety is high. For example, if a failure is triggered, fully grouted, rough rebars will tend to break when exposed to relatively small events causing low ejection velocities. For marginally stable situations, with low FS_s, the ultimate displacement is extremely sensitive to small differences in ejection velocity. Thus, unless a support is very ductile, means must be found to keep the anticipated ejection velocities low, e.g., on the order of ≤ 400 mm/s for h_{ult} < 20 mm.

4.3 Empirical displacement-based survival limits (seismically induced falls)

After a failure has been triggered and the rockmass disrupted, rock will be ejected. Support will only survive if the ejection velocity is less than the allowable velocity. Whereas it is desirable to directly relate the support properties to ground motion parameters, unfortunately, ejection velocities cannot be predicted accurately. It is, therefore, necessary to invoke some relationship between ejection velocity and source parameters (ppv, M and R).

Conventional theoretical considerations, as well as field observations suggest that the ejection velocity is related to ppv, i.e., the ejection velocity v_e = ppv/m', where the factor m' depends primarily on the seismic wavelength or frequency and the block size. Yi (1993) and Yi and Kaiser (1993) conclude, from analytical considerations, that for relatively large unstable volumes of rock, triggered by low frequency waves, m' should be ≥ 1 for most practical situations(m' = 1 is a conservative assumption).

Adopting the empirical relationship for ppv of Equation 1 and setting ppv equal to (m' v_e) provides empirical support survival limits as a function of the support's ultimate displacement capacity h_{ult}:

$$R = \frac{C}{m'\sqrt{2gh_{ult}(n\,FS_s - 1)}} 10^{0.57M} \qquad (12)$$

where M = Richter magnitude, C in mm/s, and all else in m and s.
This functional relationship is illustrated by Figure 4.2.

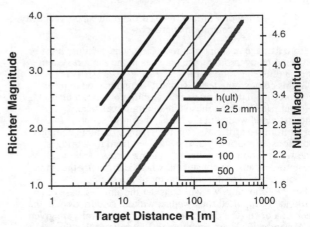

Figure 4.2 Survival limits for rock support as a function of yieldability h_{ult} (C = 460, FS_s = 1.2, n = 1.15 and m' = 1.1)

Practical implications

Figure 4.2 illustrates that a support that can yield 100-times more will survive at one tenth of the distance from the source. The survival limits presented in Figure 4.2 obviously depend on the design parameters C, FS_s, m, and n. Changing any of these parameters simply leads to a lateral shift of the survival limits.

The location of the survival limit is very sensitive to low static factor of safety, e.g., FS_s ≤ 1.3, and to the ppv-constant C in the range of 0 to 1000. This implies that the limits shown in Figure 4.2 are conservative for well supported conditions. From a practical standpoint, it is most important to realize that marginally supported rock will become unstable at much larger distances from the source than those given by the survival limits shown in Figure 4.2.

4.4 Empirical survival limits for holding elements

Typical yield or slip limits for various holding elements, mostly from static pullout tests (Hedley, 1992; and Stillborg, 1984), are summarized in Table 4.1. These limits, called stretch limits, are representative for bolts or cables loaded in tension (no shear) with a rate-independent ultimate displacement capacity.

Table 4.1 Stretch limits selected for determination of survival limits

Rough rebar, standard Swellex®, birdcaged cable bolts		< 10 mm
Regular cable bolts		≤ 25 mm
Mechanical bolts, yielding Swellex®		≤ 50 mm
Debonded cables:	1 m	≤ 50 mm
	2 m	≤ 100 mm
	4 m	≤ 200 mm
Split Sets®, non-lubricated smooth rebar		≤ 200 mm
Conebolt®, lubricated smooth rebar, friction anchored cables		≤ 500 mm

4.4.1 Survival limit chart

Combining these stretch limits with the survival limits determined in terms of ultimate displacement h_{ult} for the back of underground openings, the preliminary design chart shown in Figure 4.3 for holding elements was established. Current research work is aimed at the verification and fine-tuning of this chart. Similar charts will eventually be developed for retaining elements. Field trials are currently underway to establish the survival limits for reinforced

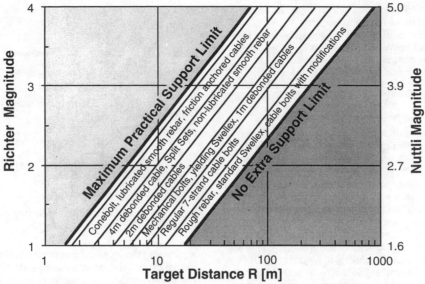

Figure 4.3 Recommended survival limits for holding elements at Creighton mine (C = 670 for 95% confidence; m = 1 or v_e = ppv; n = 1; FS_S = 1.2; the survival limits must be shifted if design parameters differ)

shotcrete. The author would appreciate receiving comments and information that could assist in verifying the boundaries presented in Figure 4.3.

Practical implications

Figure 4.3 suggests that no extra support is needed to the right of NESL and that it is practically impossible to hold rock in place for locations closer to a source than described by the MPSL. Between these two extremes, the survival limits for six groupings of holding elements are provided. Survival in the context of Figure 4.3, implies that significant or major damage to the opening occurs but the functionality of the support is maintained. Minor damage will occur at more remote locations.

If design parameters differ from those used, the survival limits must be modified. Figure 4.3 is strictly applicable only in the far-field as defined by Figure 3.1. On the other hand, very close to fault slip sources, the maximum *ppv* should be between 2 and 4 m/s. Hence, burst-resistant support in the near field should be designed for these limiting values.

4.4.2 Application to Sudbury Neutrino Observatory (SNO)

The SNO cavern with a diameter of 20 m and height of 30 m is currently under construction at the 6800' level of Creighton mine (Oliver, 1992). This cavern was designed to withstand a seismic event of M_{Nuttli} = 4.5 at a distance of 305 m. According to Figure 3.1 the cavern is in the far-field and Figure 4.2 suggests that some minor damage might be anticipated at this location. The cavern is positioned just inside the trigger limit (Fig. 3.8). An event of M_N = 4.5 should, however, only create a field stress change of about 2 to 3 MPa (Fig. 3.5) leading to a stress change at the cavern wall of 5 to 8 MPa due to stress concentration. Assuming that the cavern is designed for a static factor of safety of 2.0, the cavern falls just inside the NESL (Fig. 4.4) and could be supported with regular cable bolts at a slightly above standard design pattern. Because of the extreme sensitivity of the SNO facilities, a special debonded cable support was designed (Oliver, 1992; pers. com.). According to Figure 4.4 this support with debonded cables should survive an event of M_N = 4.5 at about 50 m from the cavern.

This example demonstrates how the charts presented in this paper can be applied to assess the support needed in burst-prone ground. For applications in conditions that differ from those at Creighton mine, revised charts must, however, be developed.

4.5 Shotcrete in burst-prone ground
-- A qualitative assessment --

As was pointed out in the introduction, a support system must contain rockburst damage by retaining the broken rock and holding

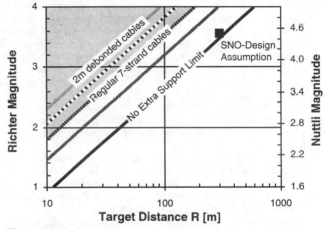

Figure 4.4 Sudbury Neutrino Observatory design assumption M_{Nuttli} = 4.5 at 305 m (C = 670; FS_S = 2; n = 1; m' = 1; shaded area indicates near-field domain)

it in place. The previous sections dealt with the causes of damage, damage localization, and support design to hold the rock. For these purposes, it was assumed that the retaining system was fully functional.

In this last section, an attempt is made to shed some light on the potential role of shotcrete as a component in a support system to contain intermediate level rockburst damage.

The mode of behaviour of shotcrete and its effect on stabilizing the ground when applied in relatively thin layers (≤ 0.1 m) to irregular rock surfaces is not well understood even under static conditions. Therefore, to claim full understanding of shotcrete behaviour under dynamic loading would be overstepping current knowledge boundaries by a wide margin. Hence, the following reflects our current interpretation of how shotcrete might interact with the rockmass and when it should be beneficial as a support component to contain rockburst damage. Much research will be required over the coming years to confirm the following interpretation of shotcrete performance.

Various support mechanisms have been proposed to explain shotcrete action:
- shotcrete acts as a support ring providing internal support pressure (convergence - confinement concept; e.g., Hoek, 1980);
- it provides punch resistance against blocks of rock (shear strength of shotcrete; e.g., Fernandez (1976) and Holmgren (1983));
- it acts as a supermesh (McCreath and Kaiser, 1992); and
- it restrains rockmass dilation by tangential or lateral shear resistance.

This latter mode of shotcrete/rock interaction has so far been ignored but is considered in the following section.

In practice, a wide spectrum of ground conditions between two extremes (unfractured and heavily fractured rock) is encountered. The two ends of this spectrum are discussed separately. Furthermore, two support functions (retention of loose and alteration of wall-rock properties) can be distinguished and are considered.

4.5.1 Shotcrete on unfractured rock

Intact or unfractured rock must fail by fracturing of the rock near an opening (strain bursting). During this process, the rock is broken into small rock blocks, i.e., the rock self-explodes causing high block ejection velocities. The sudden failure is associated with relatively little pre-event dilation before the peak strength of the rock is reached. The shotcrete skin will deform with the rock. When the rock fails, the shotcrete will be ejected with the rock unless it acts as a retaining element (as a supermesh).

Consequently, shotcrete sprayed on unfractured rock is not expected to drastically alter the stability of the opening; e.g., a wall that is prone to fail in an unstable manner will remain unstable even if covered with shotcrete. Shotcrete, in this form, will not be an effective support component unless a stable rockmass fracturing process can take place behind the shotcrete. It can, however, act as a retaining component if properly integrated into the support system.

4.5.2 Shotcrete as a supermesh

Maintaining the integrity of a support system to retain small and large pieces of broken rock was clearly identified by McCreath and Kaiser (1992) as one of the primary functions of a support system in containing rockburst damage. It was concluded that well reinforced shotcrete, held in place by appropriate holding elements, acts as a "supermesh", providing a superior retaining capability. This supermesh serves several purposes and is effective for the following reasons:

Shotcrete closes mesh to retain even the smallest pieces of rock: This is desired because even small pieces can cause serious injuries and loss of small key blocks may start an undesirable loosening process.

Shotcrete protects the mesh: The steel wires are protected from localized impact forces and from corrosion. The former prevents breakage of individual wires which could allow bigger blocks to fall through or start an unzippering process. The latter extends the mesh life while maintaining its full tensile strength.

Shotcrete strengthens mesh at overlaps: The opening up of mesh overlaps is prevented because the mesh would have to shear through shotcrete.

Shotcrete distributes impact loads: Because of the bending stiffness of reinforced shotcrete, impact loads are more evenly distributed to a larger number of holding elements (rock bolts) and, more importantly, the bolts are more directly loaded in tension rather than in shear (lateral pull reduced). Consequently, the functionality of the holding elements is optimized.

Shotcrete acts as a "super plate": If the shotcrete and mesh is well connected to the rock bolts (ideally by bolting through reinforced shotcrete), the full load can be transferred to the bolt and the commonly observed failure mode of mesh tearing over the plate is prevented.

Such a supermesh constitutes an excellent retaining system and, if kept in contact with the rockmass, it also provides a shear or interlock interface which is essential for the optimal performance of shotcrete in fractured ground.

4.5.3 Shotcrete on fractured rock

When the rockmass near an opening is intentionally fractured by destress blasting, or when a fracture zone is created by high stress concentrations near an opening, this broken rockmass will normally dilate as it is deformed under static or dynamic loads. During this dilation process, the rock moves radially into the opening and, more importantly, individual rock blocks move laterally relative to each other in the rearrangement process (see Fig. 4.5), causing relative tangential motion around the opening periphery. If there is a shotcrete layer placed at the rock surface, the relative lateral shear displacement will be resisted by a shear force at the shotcrete to rock interface. This interface shear will resist the rock movement by

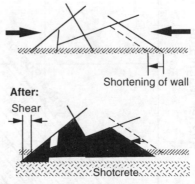

Figure 4.5 Schematic diagram showing relative movement of rock and shotcrete

suppressing the ability of the rock particles to freely dilate, even though the radial support pressure at the wall is minimal.

Recognizing that rock fragments must move laterally, relative to each other and relative to the shotcrete, during the dilation process, is most relevant for a proper understanding of how and when shotcrete can be used effectively to contain rockburst damage. The induced shear can be resisted by either compressive or tensile forces in the shotcrete. Hence, the tensile strength capacity of shotcrete at large deformations is an important characteristic.

Shotcrete strengthens rockmass near wall

From laboratory tests on rock in uniaxial compression, we know that friction under the loading platens increases the strength of the sample (Hudson et al. 1984; Brown and Gonano 1974). Applying shear stresses to the boundary of a rock element has the same effect as increasing the confining pressure. These shear stresses will suppress or reduce dilation leading to a strength increase. This is illustrated by Fig. 4.6, summarizing data from the above mentioned tests.

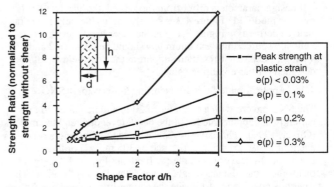

Figure 4.6 Effect of interface shear on compressive strength (data from Hudson et al. 1984; d = diameter and h = height of sample)

As the shape factor increases, the sample becomes more "squat", and the effect of shear or the amount of shear per unit sample volume increases. Figure 4.6 illustrates that the peak strength (i.e., at low plastic strains) approximately doubles. The effect is more drastic on the post-peak strength at large plastic strains; it increases by more than one order of magnitude at 0.3% non-elastic strain).

Consequently, if the shotcrete is capable of providing an interface shear restraint, the fractured rockmass immediately behind the shotcrete is strengthened significantly (within the zone of influence of the shear forces).

Shotcrete makes fractured rock near wall more ductile

As shown by Figure 4.6 the strengthening effect of shear restraint is greater at large deformations. This implies that interface shear alters the post-peak slope of the rock's stress-strain characteristics, and the rock becomes more ductile as the effect of restraining interface shear is increased. This is demonstrated by the post-peak

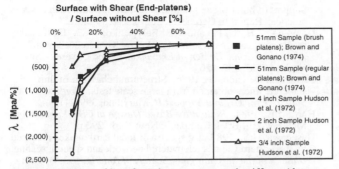

Figure 4.7 Effect of interface shear on post-peak stiffness (data from Hudson et al., 1984 and Brown and Gonano, 1974)

sample stiffness λ plotted in Figure 4.7 (same data as used for Fig. 4.6).

As discussed in Section 3.4.2, unstable failures are anticipated when the loading system's stiffness | k | is less than the rock's post-peak stiffness | λ |. Since interface shear drastically reduces λ, it follows that shotcrete will positively affect the stability of an opening by increasing the ductility of the wall rock, i.e., by reducing the potential for the wall rock to fail in an unstable manner.

These inferences from laboratory tests, have not yet been proven by field testing, but qualitative field observations from Canadian mines generally support this interpretation. Research is currently underway to investigate this effect.

4.5.4 Practical implications for use of shotcrete in burst-prone ground

Shotcrete if applied to contain rockburst damage has two primary functions:

1 - mesh reinforced shotcrete acts as a supermesh to optimize the retaining function of a support system.

The benefit of shotcrete in this case depends on the proper execution of design details. The support is only as strong as the weakest link in the system. This is typically the connection of shotcrete to the holding elements (bolts). Ideally, if a double-pass support system can be justified, bolts should be installed through mesh-reinforced shotcrete. If a single-pass support system is desired, special attention must be paid to the connection of shotcrete to the bolts. Figure 4.8 presents one possible solution with double-plated bolts. The mesh connected to the second plate must be relatively small and stiff to minimize vibration while shotcreting through it.

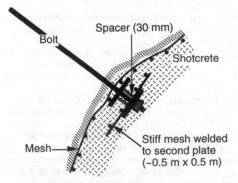

Figure 4.8 Bolt connection detail for single-pass shotcrete support system

Rock bolts used in connection with shotcrete must be partially debonded or able to yield, such that strain concentrations at the shotcrete rock interface are prevented. Hence, frictional bolts, smooth bars, or special yielding bolt types are recommended.

2 - mesh reinforced shotcrete acts most effectively when resisting lateral block movements during dilation of the fractured rock.
In this case, it is important to maintain the tensile strength of the shotcrete during large displacements (needs wire mesh reinforcement) and to maintain interlock and contact between the shotcrete and the rock (friction and interlock). Adhesion of

shotcrete to rock is highly beneficial but not crucial. Shotcrete remains effective as long as it is kept in good physical contact with the fractured rock.

After a support system with shotcrete has been deformed under static or dynamic loads, shear at the interface between shotcrete and rock will strengthen the broken rock as shown by the shaded area in Figure 4.9. These zones will also respond in a more ductile manner when deformed beyond peak strength. It is important to realize that these desirable effects are only achieved after sufficient deformations have mobilized shear. Hence, shotcrete will likely be cracked and fractured when the shear resistance required to strengthen the rockmass and to render it more ductile has actually been mobilized. This observation suggests that the cracking of shotcrete, which may destroy its ability to act as a conventional structural support ring, does not necessarily mean that the shotcrete is ineffective.

If the walls are to be protected against rockburst damage, it follows that reinforced shotcrete should be applied to the entire wall, including the lower corners. Otherwise, an unravelling process will start from the floor behind the shotcrete, leading to the formation of voids and loss of intimate contact between the shotcrete and the rock.

When the shotcrete is applied to flat surfaces, such as pillar walls, shear at the shotcrete/rock interface will only occur when the rock wall dilates as the pillar yields. Thus, shotcrete will generate shear forces at the interface as shown in Figure 4.10. On the two faces without shotcrete, the rockmass strength will deteriorate due to post-peak strength loss and the rock may fail in an unstable manner. In contrast, the shear stresses mobilized by interface shear at the other two faces will strengthen the fractured rock wall in the shaded areas and render it more ductile.

Whereas the shape of the zone influenced by interface shear is unknown, the depth of this influence zone will generally depend on the area over which shear can develop. This is illustrated in Figure 4.10 where it is assumed that the shotcrete on the upper face is heavily reinforced such that no extensive cracks can develop in the shotcrete layer. Consequently, a deep strengthening effect is

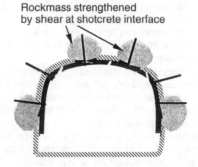

Figure 4.9 Tunnel with highly deformed, broken shotcrete layer. Rock in shaded areas is strengthened by interface shear forces and its ductility is enhanced.

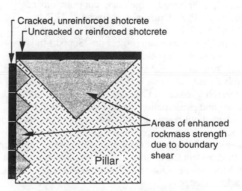

Figure 4.10 Plan view of a pillar with reinforced shotcrete at upper face and cracked, unreinforced shotcrete on left face (it is implied that the shotcrete is held in place by an appropriate bolting pattern; shear forces acting on rock due to shotcrete are shown for top face only)

achieved. On the left pillar face, the shotcrete is assumed to have failed in tension (due to lack of reinforcement). As long as the individual slabs are kept in place, a shallow strengthening effect is still achieved. Under large deformations, the slab of rock and shotcrete will fail unless anchored deep into the pillar.

5. CONCLUSIONS

A rationale for designing support systems to resist rockburst damage is presented. It is, to some extent, based on work by previous researchers (e.g., Wagner, 1982/84) on the same topic and involves four steps:

1 • Definition of microseismic activity centre;
2 • Identification of the likely mode(s) of failure at the target;
3 • Determination of trigger levels for rock fracturing and rock ejection; and
4 • Determination of support survival limits to either: (a) maintain a continuous rockmass by reinforcement or (b) to retain and hold broken or ejected rock.

Starting from a well defined microseismic activity centre, with given magnitude and source location, this report deals primarily with damage prediction and support selection to contain rockburst damage. It is presented at the mid-point of a five year research project on "rock support in burst-prone ground". Consequently, some of the work is of a speculative nature and awaits verification during further research. The practical implications emerging from this report are presented at the end of each section under the heading "Practical Implications" and are not repeated here. However, three main conclusions stand out:

1) The damage assessment procedure, developed at the GRC to collect detailed rockburst damage records, provides an effective means for data collection. The damage records can be used in combination with source location data to identify the seismic event causing the rockburst damage.

2) Damage trigger levels can be predicted from empirical relationships assuming geometric spreading (1/R decay) for targets that are in the far-field. To improve the prediction capability in the near-field, strong motion data will be required. However, for current levels of activity ($M_{Richter} < 3$) near-field conditions are limited to less than about 30 to 40 m from the source. For excavations in such close proximity to a seismic event, it is currently mandatory, and most likely will always be necessary, to design for maximum ground motion (2 to 4 m/s).

3) For the less critical but more common situation of supporting rock in the far-field, the procedure developed in this report looks promising and the design charts developed based on energy considerations and displacement criteria seem to be consistent with qualitative observations from Canadian mines.

Because of the developmental status of our research, it is anticipated that some of the work presented here will require modification over the next few years. The reader is encouraged to send comments and supporting or contradicting evidence to the author.

ACKNOWLEDGEMENTS

The author wishes to acknowledge the many contributions of GRC's staff, foremost R.K. Brummer, P. Jesenak, D.R. McCreath, G. McDowell, D. Tannant, and X. Yi, who are all heavily involved in GRC's research project on "Support in Burst-Prone Ground". The research presented in this paper was made possible by funds from the Canadian mining industry through MRD (Mining Research Directorate) and NSERC (Natural Sciences and Engineering Research Council of Canada). Industrial collaboration during field studies and participation in data interpretation has significantly enhanced our research work and ongoing discussions with industrial partners have assisted us greatly. In particular, valuable contributions through technical discussions with members of the Project Management Committee and review comments from S. Maloney, D. Morrison, P. Oliver, C. Langille and many others are thankfully acknowledged.

6. REFERENCES

Brown, E.T. and L.P. Gonano, 1974. Improved compression test technique for soft rock. ASCE, *Journal of Geotechnical Engineering Division*, 100: 196-199.

Brummer, R.K, 1991. Canadian Rockburst Research Project: Support of burst-prone ground; A review of South African practice. Report to Geomechanics Research Centre.

Dowding, C.H., C. Ho and T.B. Belytschko, 1983. Earthquake response of caverns in jointed rock: effects of frequency and jointing. *Seismic Design of Embankment and Caverns*, New York, ASCE, 142-156.

Fernandez-Delgado, G., 1976. Structural behaviour of thin shotcrete liners obtained from large scale tests. *Conf. on Shotcrete for Ground Support II*, Maryland, 399-441.

Harr, M.E., 1987. *Reliability-Based Design in Civil Engineering*. McGraw-Hill Book Company, New York, 290 p.

Hasegawa, H.S., 1983. Lg spectra of local earthquakes recorded by the Eastern Canada telemetered network and spectral scaling. *Bulletin of the Seismological Society of America*, 73(4): 1041-1061.

Hedley, D.G.F., 1992. *Rockburst Handbook for Ontario Hardrock Mines*. CANMET Special Report SP92-1E, 305 p.

Hoek, E., 1980. *Underground Excavations in Rock*. Institution of Mining and Metallurgy, London, England, 527 p.

Hoek, E., 1992. Support design for hard rock mining -- A progress report. *International Symposium on Rock Support, Sudbury*, 3-15.

Holmgren, B.J., 1983. Tunnel linings of steel fibre reinforced shotcrete. *5th Congress of th ISRM*, Balkema, Rotterdam, D311-314.

Hudson, J.A., E.T. Brown and C. Fairhurst, 1972. Shape of the complete stress-strain curve of rock. 13th U.S. Rock Mechanics Symposium, ASCE, New York, 773-795.

Jaeger, J.C., N.G.W. Cook, 1969. *Fundamentals of Rock Mechanics*. Methuen &Co Ltd., London, 513 p.

Jager, A.J., 1992. Two new support units for the control of rockburst damage. *International Symposium on Rock Support, Sudbury*, 621-631.

Jesenak, P., P.K. Kaiser and R.K. Brummer, 1993. Assessment of rockburst damage-potential. *3rd International Symposium on Rockbursts and Seismicity in Mines*, Kingston, August, 6 p.

Kaiser, P.K., D. Tannant , D: McCreath and P. Jesenak, 1992. Rockburst damage assessment procedure. *International Symposium on Rock Support*, Sudbury, 639-647.

McCreath, D.R. and P.K. Kaiser, 1992. Evaluation of current support practices in burst-prone ground and preliminary guidelines for Canadian hardrock mines. *International Symposium on Rock Support*, Sudbury, 611-619.

McGarr, A. and J. Bicknell, 1988. Estimation of the near-fault ground motion of mining-induced tremors from locally recorded seismograms in South Africa. *2nd International Symposium on Rockbursts and Seismicity in Mines*, Minneapolis, Minnesota, 379-388.

McGarr A., R.W.E. Green and S.M. Spottiswoode, 1981. Strong ground motion of mine tremors: some implications for near-source ground motion parameters. *Bulletin of Seismological Society of America*, February, 71(1): 295-319.

McGarr, A., 1981. Analysis of peak ground motion in terms of a model of inhomogeneous faulting. *Journal of Geophysical Research*, May, 86(B5): 3901-3912.

McGarr, A., 1984. Some applications of seismic source mechanism studies to assessing underground hazard. *1st International Congress on Rockburst and Seismicity in Mines*, Johannesburg, 199-208.

McGarr, A.,1983. Estimating ground motions for small nearby earthquakes. *Symposium on "Seismic Design of Embankments and Caverns"*, ASCE, Philadelphia, Pennsylvania, 113-127.

Oliver, P.H., 1992. The Sudbury Neutrino Observatory (SNO) Project. *16th Canadian Rock Mechanics Symposium*, Sudbury, 593-609.

Ortlepp, W.D., 1992. The design of support for the containment of rockburst damage in tunnels - an engineering approach. *International Symposium on Rock Support*, Sudbury, 593-609.

Owen, G.N. and R.E. Scholl, 1981. Earthquake engineering of large underground structures. LAB-7821. San Francisco: URS/John A. Blume.

Roberts, M. K. C. and R. K. Brummer, 1988. Support requirements in rockburst conditions. *Journal of South African Institute of Mining and Metallurgy*, March, 88(3): 97-104.

Spottiswoode, S.M. and A. McGarr, 1975. Source parameters of tremors in a deep-level gold mine. Bulletin of the Seismological Society of America, 65: 93-112.

St. John, C.M. and T.F. Zahrah, 1987. Aseismic design of underground structures. *Tunnelling and Underground Space Technology*, 2(2): 165-197.

Stillborg, B., 1984. *Experimental Investigation of Steel Cables for Rock Reinforcement in Hard Rock.* Doctoral thesis, 1984: 33 D, Lulea University, 127 p.

Sulem, L, 1992. Application of bifurcation theory in rock mechanics. *4th Conference on Rock Mechanics in Italy,* Torino, Ch.9, 1-11.

Tannant, D.D. and P.K. Kaiser, 1993. Friction bolt anchored wire rope for support in rockburst-prone ground. Submitted to *CIM Bulletin,* January 1993.

Vardoulakis, I., 1984. Rock bursting as an instability problem. *International Journal for Rock Mechanics and Mining Sciences,* 21: 137-144.

Wagner, H., 1982/84. Support requirements for rockburst conditions. 1st International Congress on Rockbursts and Seismicity in Mines, Johannesburg, SAIMM, 1982, Gay and Wainwright (eds) 1984, 209-218.

Wilson, J.W. and G.T.G. Emere, 1975. The laboratory testing of stope support. Association of Mine Managers of South Africa, incl. discussions, 55-105.

Yi, X., 1993. *Dynamic Response and Design of Support Elements in Rockburst Conditions.* Ph.D. Thesis, Department of Mining Engineering, Queen's University, 272 p.

Yi, X. and P.K. Kaiser, 1993. Mechanisms of rockmass failure and prevention strategies in rockburst conditions. *3rd Int. Symp. on Rockbursts and Seismicity in Mines,* Kingston, 5 p.

APPENDIX A

Strainburst Potential Assessment

Jaeger and Cook (1969) established a criterion for the instability of a failed rock annulus by comparing the radial stiffness of the surrounding elastic rock k with the stiffness of the rock annulus. They found that instability for an annulus with a constant negative post-peak stiffnes λ occurs if:

$$\frac{t}{a} > \left| \frac{2G}{\lambda} \right| \qquad (A.1)$$

where $2G = \dfrac{E}{(1+v)}$, a = tunnel radius, t = width of annulus.

Consequently, if only a small annulus fails (t --> 0), it is "soft" relative to the surrounding rock and will fail in a stable manner unless λ is positive as for a Class II material. Even if the above condition for instability is satisfied, spontaneous instability will only arise if the cohesion of the failed annulus can be completely destroyed by sufficiently large strains. Using a Coulomb failure criterion ($\sigma_1 = \sigma_c + q\sigma_3$), where $q = \tan^2(45+\phi/2)$, Jaeger and Cook (1969) established the following instability criterion.

$$\frac{\sigma_c}{p} \left[2(1-v) - \left(\frac{2 - \frac{\sigma_c}{p}}{q+1} \right) \right]^{-1} \geq \frac{\left(\left| \frac{2G}{\lambda} \right| + 1 \right)^2}{\left| \frac{2G}{\lambda} \right| + \frac{1}{(1+v)}} \qquad (A.2)$$

For a circular tunnel in a hydrostatic stress field, failure occurs only if $\sigma_c/p < 2$, thus Equation (B.2) can only be satisfied and instability occur if:

$$\left| \frac{2G}{\lambda} \right| < 1 \qquad (A.3)$$

or q is very small.

APPENDIX B

Comments on strainburst potential

The problem of stability assessment must obviously be addressed from a system perspective as both the loading system and the material properties must be considered. Over the last decade, several attempts have been made to apply the bifurcation theory to rock mechanics. Vardoulakis (1984) has investigated rock bursting as a stability phenomenon and found that buckling may proceed shear failure depending on the rock's tensile to compressive strength ratio as well as the curvature of the stress-strain curve. In a more recent study, Sulem (1992) has applied the bifurcation theory to underground openings and demonstrated that the mode of failure depends on the plastic strain, i.e., it is stress path dependent, and on geometric factors such as slenderness ratio, i.e., the boundary conditions. He showed that the critical plastic strain for slender rock samples (h/d > 1) is about half that of very short samples (h/d < 0.3). Whereas the bifurcation theory does not yet provide a practical framework to determine the strainburst potential, it definitely provides some promising insight. Most importantly, it demonstrates that geometric effects must be considered when assessing the stability of an opening in brittle rock.

APPENDIX C

Derivation of Trigger Limits

The static factor of safety can be defined as $FS_s = C_s/D_s$, where C_s = static capacity or support strength, and D_s = static demand or weight of wedge mg. The dynamic factor of safety can be defined similarly as $FS_d = C_d/D_d$, where C_d, the dynamic capacity of the support during rapid loading, is equal to n-times the static capacity C_s of the support ($C_d = n\, C_s$), and D_d = dynamic demand or weight of wedge $m\,(g+a)$ with a = acceleration due to dynamic loading.

By vertical force equilibrium, the acceleration a_t at which a block will separate from the surrounding rockmass, i.e., a failure is triggered or initiated, can be determined as:

$$a_t = g\,(n\,FS_s - 1) \qquad (C.1)$$

For sinusoidal wave forms, ppa = $(2\pi f)$ ppv = $(2\pi f)$ C SD^{-1}. This equation represents an extreme simplification, ignoring the influence of frequency dependent attenuation effects, but it is used as an approximation for lack of *ppa*- measurements.

By setting ppa = a_t, the following trigger limit can be established:

$$R = \frac{2\pi f C}{1000 g\ (n\,FS_s - 1)} 10^{0.57M} \qquad \text{(C in mm/s for R in m)} \qquad (C.2)$$

This equation suggests that high frequency events cause damage at a large distance. This is, of course, incorrect because high frequencies attenuate more rapidly and the high frequency component of a wave contains much less energy. Equation (C.2) should, therefore, only be applied for low frequencies.

27

Rockbursts and Seismicity in Mines, Young (ed.) © 1993 Balkema, Rotterdam, ISBN 90 5410 320 5

Preconditioning – A technique for controlling rockbursts

D.J.Adams & N.C.Gay
Chamber of Mines Research Organisation of South Africa, Johannesburg, South Africa

M.Cross
Blyvooruitzicht Gold Mining Company, South Africa

ABSTRACT: Preconditioning blasting is a potential proactive method of minimising the occurrence of rockbursts and the damage associated with face bursting. Extensive underground trials have resulted in a method of preconditioning which may be integrated into a normal mining layout and cycle. The method involves drilling preconditioning blast holes parallel to the dip of the reef, 3-5 m ahead and parallel to the mining face. The effects of preconditioning have been assessed by means of a seismic network covering each preconditioning site, by instrumentation for measuring closure and ride in these stopes and by means of ground penetrating radar which quantifies the amount of fracturing in the rock ahead of a working face. The benefits to production and safety of preconditioning are shown.

1 INTRODUCTION

Preconditioning as presently practised by COMRO's rockburst control project is a technique which utilises explosives to minimise high stresses developing ahead of mining faces which may result in rockbursts. This is achieved by drilling holes into the rock ahead of the stope face and detonating explosives in rock which is confined enough to prevent an advance of the stope during the blast but unconfined enough so that existing fractures can be opened up by the blast gases or even new fractures may be formed.

The cause of face bursting may be due to the build-up of stresses ahead of the face. Such stresses may occur as a result of locking up of existing fractures and a lack of the development of new fractures ahead of the face to maintain a fractured zone proportional to the stoping width and the mining span when the ground becomes loaded by further mining. When the stresses reach high enough levels, the rock fails catastrophically, causing severe damage to the face area where the workforce is concentrated. By subjecting the rockmass to the explosive shock energy and gas pressure, any areas of high stress concentration ahead of the mining face can be reduced.

2 HISTORY OF PRECONDITIONING

Preconditioning was first used at the East Rand Propriety Mines during the 1950's to address the problem of rockbursting in the longwall stopes. The technique was referred to as destressing (Roux et al., 1957). At the ERPM sites three metre holes were drilled perpendicular to the east and west advancing breast panels between the normal production blast holes. Very positive results were reported regarding safety and production by Roux et al. (1957). Production rose by approximately 5 % during preconditioning compared to mining similar areas without preconditioning. Roux et al. (1957) demonstrated that in mining approximately 6000 centares the number of rockbursts and severe rockbursts in particular decreased. They also show that the casualty rate dropped from 44 with 6 fatal accidents to 1 casualty. Problems with drilling eventually stopped the further use of the destressing at ERPM.

Interest in preconditioning the stope face was resuscitated in the late 1980s when preconditioning trials were started by COMRO at West Driefontein Gold Mine where preconditioning holes were drilled parallel to breast panels. At this site the technique of drilling preconditioning holes parallel to breast panels was proved and significant increases in production rates were recorded, which will be discussed later.

3 CURRENT PRECONDITIONING TRIALS

More recently three preconditioning sites have been available at Blyvooruitzicht Gold Mine to evaluate this rockburst control technique. Blyvooruitzicht is the oldest and one of the most successful mines in the Carletonville gold mining area of South Africa. Most of the Carbon Leader Reef has been extracted on this mine leaving only stabilising pillars, pillars which lie adjacent to faults and dykes, and reef which in the past had proved difficult to extract due to bad ground conditions, high rates of seismicity and rockburst incidents or a combination of all three. As a result of the mining being almost exclusively in pillars and because of a wealth of experience with up dip mining on the mine, the preconditioning projects sites at Blyvooruitzicht Gold Mine initially involved drilling holes perpendicular to the up-dip faces. At three sites 10 metre preconditioning holes were fanned into the face from the gullies which were oriented on dip and followed the advancing up-dip face. The extent of the project undertaken at Blyvooruitzicht Gold Mine with COMRO at the two most recent sites is indicated in Table 1 below. Most of the 57 preconditioning blasts have been holes which have been fanned into the up-dip faces. Although success was achieved with this method of preconditioning (Rorke et al., 1990) in extracting a highly stressed remnant adjacent to a faulted dyke at Blyvooruitzicht, the localised influence of such blasts relative to an entire panel was recognised by Adams and Stewart (1992). They also suggested that preconditioning and therefore softening of a portion of the face may lead to other areas of the face becoming more highly stressed and therefore increase the risk of crush type events close to the face. Drilling face parallel holes means that an entire face is affected by the preconditioning blast and no high stress regions remain close to the face. An attempt was therefore made to drill holes parallel to the up-dip faces but, drilling difficulties and ground control problems around the collar of the blast holes mitigated against this method of preconditioning. However, more recently holes at a strike stabilising pillar site have been drilled parallel to dip, and to the breast panels by means of which the stabilising pillar is now being extracted.

During the preconditioning experiment at Blyvooruitzicht Gold Mine nearly 1,5 km of 76 mm and 89 mm diameter preconditioning blast holes have been drilled and charged with approximately 6 tons of explosive as shown in Table 1.

4 ORIENTATION OF PRECONDITIONING HOLES

It is important to integrate preconditioning into deep level mining cycles in South African gold mines and Figure 1 summarises in schematic form the different orientations of preconditioning holes relative to the face which have been mined using this technique. The

Table 1. Number of preconditioning blasts, total length of holes drilled and total quantity of explosives used at the two preconditioning sites B17 and B30 at Blyvooruitzicht Gold Mine up to 30/11/92.

	Number of preconditioning blasts	Metres drilled	Explosives used (kg)
17 Level	25	574	2600
30 Level	32	879	3600
Totals	57	1453	6200

holes marked A and B in Figure 1 represent the layout of preconditioning holes parallel to breast and up-dip panels respectively. The entire preconditioning hole can be drilled in fractured ground in both cases. Holes marked C and D represent the layout of preconditioning holes drilled into up-dip and breast panels respectively. The first portion of these holes are drilled in fractured ground but the last section penetrates highly stressed unfractured ground. The holes designated at C are similar to those used at ERPM, and because of their 3 m length are totally in fractured ground

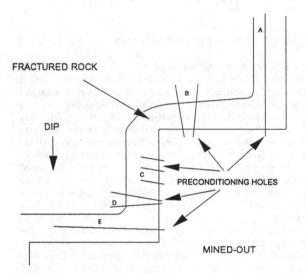

Figure 1. Schematic orientations of preconditioning holes relative to the up-dip or breast panels.

Table 2 shows the drilling rates which were achieved with various drilling orientations. It is clear from the data that the drilling rates were quickest when the preconditioning blast holes were drilled parallel to the dip of the reef 5-6 metres ahead of an advancing breast face. Only 20 % of the holes drilled in this way experienced any blockages. These holes are drilled in fractured rock which has experienced some stress relief and therefore apart from loose chips of rock dislodging into the hole and blocking the hole, there are no other reasons to cause jamming of the drill strings in the hole.

Holes drilled parallel to dip, perpendicular to an up-dip face showed the next best drilling times. The reason for the relatively good drilling rates is mainly due to the fact that these holes were generally less than 10 metres in length. The first 5-6 m of each hole was drilled in fractured ground similar to that in which holes parallel to the breast panels were drilled. The drilling rates for this section of each hole was similar to drilling rates for holes parallel to breast faces. However, beyond 6 m, the perpendicular holes penetrated the unfractured rock ahead of the stope. Drilling penetration rates decreased as the high stresses in the rock clamped the drill bits. In 50% of the holes drilled in this orientation blockages of the holes occurred. Most of these blockages were in the portion of the hole deeper than 6 metres.

5 EXPLOSIVES

The experience from a complementary project in which preconditioning is being carried-out in confined rock ahead of a

development end has assisted in establishing that Anfo type of explosives are more suitable for preconditioning because of their high gas content and the fact that they maintain high accelerations for greater distances from the blast than do emulsion type of explosives. Emulsion type explosives do have higher accelerations than Anfo type explosives for approximately 0,5 m from the hole. The velocities of detonation for Anfo type explosives used at the sites are approximately 4700 m/sec compared with velocities of detonation of 6000 m/sec for emulsion type of explosives. Use of the Anfo type explosives result in fracturing of the rock rather than crushing which is associated more with higher shock explosives such as emulsions.

6 STEMMING

Crucial to the effectiveness of a preconditioning blast is the stemming used in the blast hole. Too little stemming or incorrect stemming is displaced from the hole at the time of the detonation of the explosive. At least 5 m of a mixture of clay and gravel stemming has been found to be necessary to impart as much explosive energy into the rock mass as possible. The gravel ensures that the stemming is not removed by locking itself into the hole and the clay stops the escape of gases out of the hole by providing an impermeable barrier. The experience at a number of the experimental sites has been that if too little stemming is used, the explosive simply displaces the stemming from the hole. If the rock in the vicinity of such a hole is in any way fractured, it is likely that the rock around the stemmed section of hole will be ejected as well, leaving a cavity at the collar of the hole. When this happens, most of the energy of the preconditioning blast is lost out of the collar of the blast hole and little is done to precondition the rock ahead of the face.

7 QUANTIFYING THE EFFECTS OF PRECONDITIONING

The effects of preconditioning blasting on the rock mass and a better understanding of the mechanisms operating in the rock mass which lead to rockbursts in the working face area may only be understood by quantifying the behaviour of the rock. A better understanding of the mechanisms will ultimately improve the design of preconditioning blasts. Therefore, much effort has gone into installing instrumentation in the stope and in the rock mass surrounding the preconditioning site, which will record data before and after preconditioning blasts and seismic events or rockbursts. The most effective means of gaining an understanding of the rock mass as a whole is by using seismic data and therefore, the experimental preconditioning sites have all been covered by dedicated mini seismic systems. In all cases the COMRO developed Portable Seismic System (PSS) has been used for this purpose. Other instrumentation used in the stope for collecting data include closure-ride stations, which consist of three pegs installed in the hangingwall and one peg in the footwall at each station, and from which it is possible to calculate the normal closures and the relative movements of the footwall and hangingwall in the dip and strike directions in the stopes. Closure is also measured continuously by means of a mechanical clockwork closuremeter, where changes in closure are inscribed on a rotating drum, and by means of a potentiometer which is linked into the PSS. Fractures are logged in the stope by means of stereo photography and with the aid of ground penetrating radar (GPR). An attempt has been made to measure changes in strain and stress in the rock before and after preconditioning blasts but the very high stresses in the rock have made this operation very difficult.

Table 2. Drilling times for 76 mm and 89 mm diameter preconditioning holes oriented on dip and strike parallel and perpendicular to the mining face.

	Holes parallel to the face		Holes perpendicular to the face	
	Dip	Strike	Dip	Strike
Metres drilled	118	76	1190	40
Time (hours)	48	119	623	79
Metres/hour	2,4	0,6	1,9	0,5
Holes blocked(%)	20	75	50	50

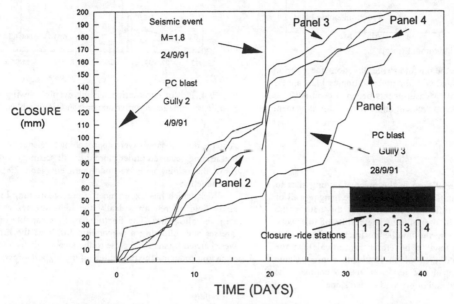

Figure 2. Closures measured at four closure-ride stations in an up dip mining situation at which preconditioning blasting was used. The localised effect of two preconditioning blasts and the regional effect of a seismic event are shown.

In order to assess the data from the preconditioning sites a control site has been established in a stope with similar conditions to those in the stopes where preconditioning is occurring. Similar instrumentation to that at the preconditioning sites has been installed and data are collected regularly.

7.1 Closures

A number of different means of measuring closure have shown that at the time of a preconditioning blast additional closure to the normal closure rate takes place. However, the influence of preconditioning on closure is localised to the area immediately surrounding the position of the blast. Figure 2 is a plot of the closures measured in four up-dip panels where preconditioning has been used. The closures were measured within 5 m of the face and the preconditioning holes were drilled as fans from the gullies into the up-dip face. The preconditioning blast on 4/9/91 was drilled out of Gully 2 and the closures recorded were most at closure-ride Station 1 which happened to be closest to the preconditioning site. Closure-ride Station 2 recorded some closure but, Stations 3 and 4, further from the preconditioning blast registered no perceivable change in their closure rate. The second preconditioning blast, indicated in Figure 2, shows that only closure-ride Station 2, again the closest station to the blast, recorded an increased closure rate directly as a result of the preconditioning blast. Panels 3 and 4 continue at the same rate of closure and Panel 1 only shows a sharp change in closure 2 days after the preconditioning blast. The influence of a magnitude 1,8 event on the closures in all four panels is evident on the graphs. Although closures of up to 50 mm were recorded at the time of the seismic event, the closure rates at all four stations immediately before and after remained almost constant. In the most recent preconditioning blasts where the holes have been positioned parallel to the face, in the region 3-5 m from the face, up to 100 mm of closure has been measured as a direct result of the blast.

7.2 Ride

Data from the closure-ride stations show that often the ride in the dip and strike directions changes by large amounts at the time of a preconditioning blast but, it has not been possible to predict or determine any pattern of behaviour to date. Figure 3 shows the rides plotted in the dip and strike directions over a period of 35 days. The movement recorded immediately after the preconditioning blast is far greater than any of the other daily movements.

7.3 Seismicity

The seismic networks associated with the preconditioning sites have been invaluable in helping to assess the effect of the preconditioning blasts. It is usual and expected that seismic events occur at the time of a preconditioning blast and the seismic systems are sensitive enough to monitor such events.

Following the seismic event in the magnitude range 0,5 to 1,0 at the instant of detonation of the preconditioning blast, are usually a number of smaller events. Table 3 shows the number of preconditioning blasts which detonated seismic events in a particular magnitude range, the total number of seismic events which followed immediately after and were associated with the blast, and the total energy involved.

Figure 4 shows the cumulative number of seismic events during a time period in which two preconditioning blasts were detonated. Immediately following a preconditioning blast the number of events increases markedly but the rate quickly returns to that before the blast, indicated by the slope of the curve.

31

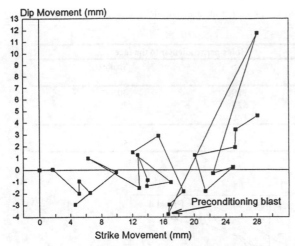

Figure 3. Plot of the rides in the dip and strike directions for a period of 35 days. The rides associated with the preconditioning blast are far greater than on any other day. Measurements began at position 0 , 0 and end after 28 mm of strike movement and 5 mm of dip movement had been recorded.

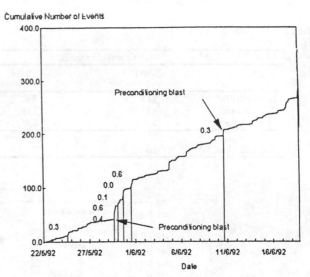

Figure 4. Plot of cumulative seismic events showing the increase in seismicity immediately after one preconditioning blast.

7.4 *Fracture mapping*

Stereo photography has been utilised in the preconditioning sites to record the number and orientation of fractures in the hangingwall of the stope. Because of the difficulty of access in the panels, the photography is carried out in the gullies. The results of this work shows that with preconditioning the number of fractures in the rock mass do increase but they tend to be closer to vertical than the fractures in unpreconditioned sites which results in a more stable hangingwall beam. The control site has been use to compare the fracturing in a preconditioned and unpreconditioned stope.

7.5 *Ground Penetrating Radar (GPR)*

Ground penetrating radar has verified that preconditioning blasts do result in more fracturing of the rock mass even though the blasts are semi-confined. Radar scans have been carried-out before and after preconditioning blasts in order to quantify the changes in fracturing. Figures 5 and 6 show a radar scan before and after a preconditioblast. The preconditioning holes were about 8 m long and fanned into an up-dip panel. The radar was able to penetrate to a depth of approximately 10 m. Before the preconditioning blast a number of fractures are visible. After the blast these fractures are again identifiable but, in many cases have extended in length. In addition a number of new fractures are visible in Figure 6 particularly close to the positions of the preconditioning blast holes.

7.6 *Production rates*

To evaluate the effect of preconditioning on production, production rates in terms of centares broken during a 15 month period of conventional mining were compared with a consecutive 13 month period of preconditioning mining using long holes drilled on dip parallel to the face. During both periods the call and complement of workers remained the same, although the mining during preconditioning took place further from the centre line of the stope and therefore, material handling and scraping was more difficult.

Also the mining panels were approaching a seismically active dyke so that mining occurred under more difficult seismic conditions. Table 4 shows that mining with preconditioning enabled a 29 percent increase in production.

Rock which has experienced a preconditioning blast has had its stress field altered and ideally the high stresses reduced, but not removed altogether. The fractured rock is capable of carrying high stresses and following a preconditioning blast the high loads which were carried by the rock close to the face are distributed over a wider zone thus reducing the magnitude of the highest stress.

7.7 *Safety*

Preconditioning is seen as a potentially viable tool for addressing the rockburst problem in stopes. To date each site at which preconditioning trials have been undertaken has been associated with a high risk of seismic activity and rockbursting. However, where comparisons have been made between similar mining situations there has been an improvement in safety. Since the reintroduction of preconditioning in the late 1980s more than 6000 centares have been mined with preconditioning and it has been shown that the technique can be carried-out safely.

8 DISCUSSION

The preconditioning technique is sensitive to the position of the blast hole relative to a free surface. If a preconditioning blast hole is drilled parallel to dip in front of a breast face, and is positioned in the zone 3-5 m ahead of the face with sufficient stemming, then there is every possibility of the blast being successful.

At the time of a preconditioning blast, existing fractures will be extended, new fractures will be created, seismic energy will be released and closure will occur in the stoped out area. The preconditioning technique seeks to address the rockburst problem associated with the stope face area where there is a concentration of personnel. Closure data collected at the time of seismic activity indicates that sudden closure does occur at the time of these events. Similar localised closures accompanied by seismicity occur during

Table 3. Data related to seismic events associated with preconditioning blasts. The total energy released during this seismicity and the average quantity of explosives used in the blasts are shown.

Magnitude range of seismic events	No. of seismic events	No. of seismic events following	Total energy (MJ)	Average Explosive used (kg)
< 0,5	8	86	1,26	78
0,5-1,0	15	148	16,6	138
>1,0	5	2	14,3	170

Table 4. Figures for the periods without preconditioning and with preconditioning showing the total centares (m^2) broken and the average monthly centares broken for the two periods.

	Time (months)	Total Centares Broken (m^2)	Centares broken per month (m^2)
Without Preconditioning	15	5946	396
With Preconditioning	13	6660	512

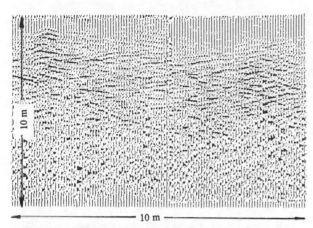

Figure 5. ground penetrating radar scan into the face before the preconditioning blast.

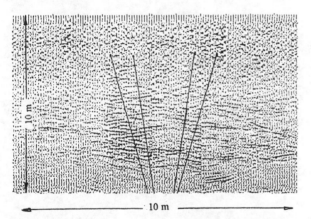

Figure 6. Ground penetrating radar scan into the face after a preconditioning blast.

preconditioning blasts, which indicates that the same mechanism responsible for seismicity and related rockbursts is possibly being generated in a controlled way by the preconditioning blast. Fracture mapping and ground penetrating radar indicate an increase in fracturing following preconditioning blasts, and production and safety figures suggest that there are improvements due to this rockburst control technique.

9 CONCLUSIONS

The following conclusions arise from the work which has been carried-out at preconditioning sites:

1. The most favourable mining layout for preconditioning is breast panels where preconditioning holes can be drilled parallel to dip.

2. Holes should be positioned 3-5 m ahead of the face if drilled parallel to the breast panel face.

3. New fractures are created and existing fractures are extended by preconditioning blasting creating a zone of crushed rock ahead of the face.

4. Stemming should be > 5 m long and consist of gravel and clay.

5. Closures are induced locally in the stope at the time of preconditioning blasts.

6. Production rates can be increased by preconditioning.

7. It has been shown that preconditioning can be carried-out safely.

REFERENCES

Adams, D.J. & Stewart, R.D. 1992. Analysis of a rockburst caused by a 2,4 magnitude seismic event on 30 level, Blyvooruitzicht Gold Mine, 11 October 1991. COMRO Reference Report No. 9/92.

Rorke, A.J., M Cross, H.E.F. van Antwerpen & K. Noble 1990. The mining of a small up-dip remnant with the aid of preconditioning blasts. *International Deep Mining Conference: Technical Challenges in Deep Level Mining*. Johannesburg, SAIMM.

Roux, A.J.A., Leeman, E.R., & Denkhaus, H.G., 1957. De-stressing: A means of ameliorating rockburst conditions. *Journal of the South African Institute of Mining and Metallurgy* Vol 58: 101-119.

Rockbursts and Seismicity in Mines, Young (ed.) © 1993 Balkema, Rotterdam, ISBN 90 5410 320 5

Modelling of a destress blast and subsequent seismicity and stress changes

F. M. Boler & P. L. Swanson
US Bureau of Mines, Denver, Colo., USA

ABSTRACT: Methods of destressing and/or preconditioning rock masses to reduce rockburst hazards are in regular use in some rockburst-prone mines in the Coeur d'Alene district, Idaho (USA). Quantitative measures of the effectiveness of these techniques are lacking. In this paper, we explore the use of several measurement and analysis tools for evaluating destress treatment effectiveness in a case study of an attempt to destress a sill pillar by blasting. A digital seismic array and an array of borehole pressure cells were installed in the vicinity of a stope prior to a blast. Several damaging seismic events occurred in the days following the destress blast. Seismic data analyses are used to constrain rupture plane locations and orientations. Boundary element and dislocation modeling of the destress and seismic events are used to elucidate the interaction of stress changes due to destressing and the subsequent seismicity.

1 INTRODUCTION

Destressing or preconditioning methods are often used in an attempt to reduce the rockburst hazard near the working face (Blake, 1972; Board and Fairhurst, 1983; Rorke and Brummer, 1990). Destressing is expected to (1) trigger a seismic event large enough to release locally stored strain energy when no miners are present, or (2) soften the face pillar and/or to reduce its load by transferring the stress elsewhere. It would be desirable to be able to tailor the design of a particular destress to the site characteristics to maximize its effectiveness, and to have a quantitative means to evaluate the effectiveness of a destress blast before work is resumed in the stope.

The purpose of this paper is to show how quantitative evaluation of the effectiveness of a destress is possible. Through seismic and stress observations, and dislocation and boundary-element modeling, the degree of stress interaction among the destress and the damaging seismic events following a destress is examined. Digitally recorded accelerometer waveforms are used to constrain focal mechanisms for damaging seismic events and several associated microseismic events. These mechanisms, along with the event locations, are used to define fault planes on which dislocation slip, constrained by stress change measurements, is determined.

Rockbursts are often triggered in the Coeur d'Alene mining district of northern Idaho when min-ing into remnant (sill) pillars produced by the commonly used overhand cut-and-fill mining method (Blake, 1972). Pillar destressing has been undertaken frequently at the Galena lead-zinc-silver mine, Wallace, Idaho, the site of the present study. Figure 1 shows a location map for the Galena Mine.

On February 2, 1990, the 21-m high pillar in the 46-99 stope of the Galena Mine was destress blasted. Eight 10-m boreholes and three 4-m boreholes were filled with a total of 125 kg of explosives for the blast. The largest seismic event occurring within seconds of the destress blast was only Richter magnitude (M) 0.4. Based on the mine personnel's experience, a significantly larger event (M > 1.5) is expected for effective destressing. While no such events happened immediately after the destress blast, there was a general increase in seismicity during the next 18 days. Figure 2 shows the events of M > 0 in the 46-99 stope from the period January 1989 through October 1990. The concentration of seismic events from February 4 - 19, following the destress of February 2, 1990, is noteworthy. Two events on February 7 caused damage. One occurred at 03h45m00s (034500) PST (M 2.9); the second occurred at 122020 PST (M 0.9). Seismological and pressure change data collected in association with these two events form the basis for the analyses presented in

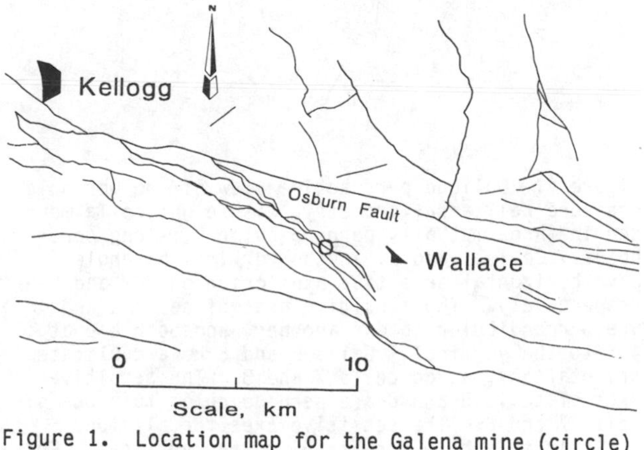

Figure 1. Location map for the Galena mine (circle) in the faulted Coeur d'Alene mining district.

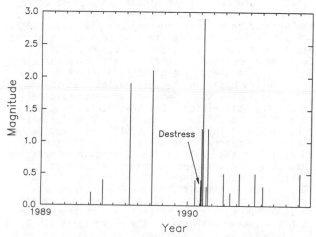

Figure 2. Seismic activity in the vicinity of the 46-99 stope during the period from January 1989 through October, 1990.

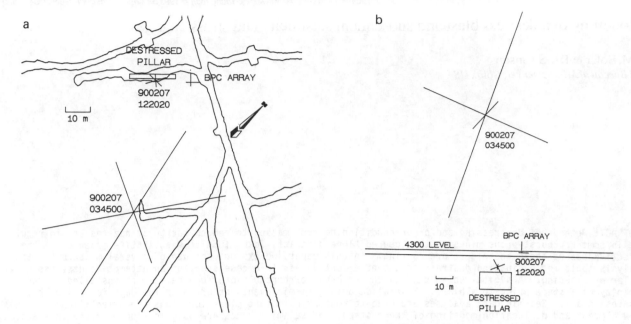

Figure 3. (a) Plan view and (b) vertical section looking N45W of mine openings, destress volume, BPC array, and the locations of two damaging events of February 7. Axes show 95% confidence ellipsoid. The large uncertainty for the 034500 event reflects its location outside of most of the array.

this paper. Figure 3 shows the locations of these two events and the destress volume.

The paper is organized as follows: instrumentation for seismic and rock pressure change monitoring is described; mechanisms for the two damaging events on February 7 are estimated; pressure change data for the destress and the damaging February 7 events are presented; and boundary element and dislocation models are used to describe the stress interaction among the destress and damaging events.

2 INSTRUMENTATION

2.1 Stress monitoring instrumentation

The Bureau's borehole pressure cell (BPC), described in detail in Haramy and Kneisley (1991), was used to monitor stress changes caused by destressing and seismic events. The BPC consists of a stainless steel oblong pancake-shaped bladder encased in a cylinder of grout, molded to be just under the size of the borehole (60 mm). Cell emplacement is accomplished by inserting the package in the borehole and aligning, and pumping the bladder to a preselected setting pressure, typically selected to approximate the overburden pressure (Haramy and Kneisley, 1991).

Figure 4 shows the position and orientation of the eight BPC's installed at the end of July 1989. Cells were emplaced in three mutually perpendicular 6-m long boreholes. The geometry of the BPC array was selected based on the observation that the majority of faults trend N45W and are near vertical. Laboratory observations show that faults tend to form at 30° to 45° to the maximum principal stress direction. If the fault orientation is representative of the present stress field, then BPC's oriented at 45° to the fault will be sensitive to pressure changes in the principal stress orientations.

The overburden pressure in the vicinity of the 46-99 stope is estimated at 34-37 MPa (from the overburden thickness of 1460 m and the rock density of 2400-2600 kg/m^3). A nominal setting pressure of 35 MPa was used with six of the eight cells (1, 2, 3, 4, 6, and 7). In addition, two cells (5 and 8 in Figure 4) in redundant orientations to two other cells (6 and 7) were emplaced with nominal setting pressure of 28 MPa as a check on the effect of the

absolute setting pressure on cell output.

Prior to February 1990, dial pressure gauge readings were manually recorded. Digital logging of data began in February 1990, with readings logged for each cell every 10 minutes. Cell pressures were converted to voltages using Bourns 35-MPa pressure transducers. (Use of brand names does not imply endorsement by the U.S. Bureau of Mines.) (Pressure gauges with a larger maximum pressure than 35 MPa were not available. However, between installation

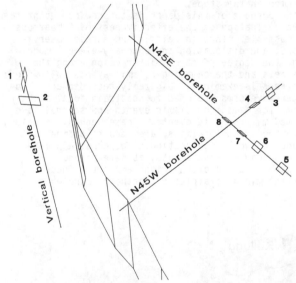

Figure 4. Oblique perspective view of the borehole pressure cell array geometry, with eight cells mounted in three mutually perpendicular 6-m-long boreholes. Cells 1 and 2, in the vertical borehole, have horizontal sensitive axes oriented E-W and N-S respectively. The sensitive axes of cells 2 and 4 are perpendicular to one another, and both are at 45° to the vertical. Cells 5 and 6 have duplicate orientations, as do cells 7 and 8. The sensitive axes of cells 5 and 6 are perpendicular to those of cells 7 and 8. The sensitive axes for all four cells in this hole are at 45° to the vertical. For scale, boreholes are 4.6 m long.

36

in July, 1989 and commencement of digital recording, absolute pressure on all 8 cells had fallen to 32 MPa or less.) Gauge outputs were recorded by a 32-channel digital voltmeter module, installed in a Computer Automated Measurement and Control (CAMAC) data acquisition system (Boler and Swanson, 1990). The pressure transducer installed at cell 2 did not return reliable readings due to electrical problems, and will not be discussed further.

2.2 Seismic instrumentation

The 46-99 stope is one of eight stopes that are each monitored with a stope-level "routine monitoring" accelerometer array with a dimension of approximately 150 m horizontally and 200 m vertically. The 46-99 array consists of 16 rib-mounted Wilcoxon 793 M40 accelerometers with 40 V/g sensitivity over 3 to 6000 Hz. Arrival times of events are picked in hardware when a floating threshold is exceeded. These relative times are transmitted to a computer which calculates and displays locations in near real-time for examination by mine personnel.

For digital recording, the routine monitoring stope-level array was augmented by two additional rib-mounted accelerometers and four borehole-mounted three-component accelerometer packages. The analog signals from these accelerometers were A/D converted at 50 kHz using the same CAMAC-based data acquisition system as used for BPC output recording. The digitizer resolution was 12 bits. The system allows for on-scale recording of events in the approximate magnitude range M -5 to -2.

The arrival times from the routine monitoring system were used to obtain locations in a few cases in which the digital system did not record a particular event of interest.

3 SEISMIC EVENT MECHANISMS

Although it is well-known that the velocity structure in the vicinity of the mine is strongly heterogeneous (and possibly weakly anisotropic) (Estey et al., 1990) an accurate velocity model is not yet available. In lieu of such a model, a constant velocity of 5.02 km/s (Estey et al., 1990) was used for both the location of events and for focal mechanism determinations. The inaccuracy of the velocity model means that individual computed take-off angles and azimuths are inaccurate by an unknown amount corresponding to the difference between the constant velocity model used and the true velocity structure.

Focal mechanisms were estimated for the damaging events at 034500 (M 2.9) and 122020 (M 0.9). It is only possible to estimate event mechanisms for those events which were recorded by the digital system, since the routine monitoring system does not retain first arrival polarity information. Unfortunately, the 034500 event was not recorded by the digital system. To get an idea of a possible fault plane solution for the 034500 event, a foreshock at 030326 (M -0.8), for which digital data were available, was used. Within the location uncertainty, this foreshock was located in the same place as the 034500 event.

The mechanism for the foreshock to the 034500 event, and by inference the main shock mechanism as well, shown in Figure 5a, is poorly constrained. There are many possible solutions consistent with the limited data. The steeply dipping N55W nodal plane is consistent with the plane locus of large seismic events occurring throughout the mine identified by Swanson and Sines (1991). The auxiliary nodal plane (N20E) was obtained through a fit of BPC data to a dislocation model of the event (described below in the section on dislocation modeling).

The 122020 event mechanism is inferred from a composite of six events, including the 122020 event. This composite, shown in Figure 5b, indicates a well-constrained mechanism. The northwest-striking nodal plane is consistent with the N45W predominant strike of mapped faults in the stope area. Given the uncertainty in the velocity structure, and the limitations of the constant velocity model used, it is encouraging to see that this composite can define an acceptable solution.

4 BOREHOLE PRESSURE CELL OBSERVATIONS

Figure 6a shows the time history of BPC measurements from the beginning of February through February 8. There is a certain amount of "cultural" noise which occurs when shift crews are working (probably electric motor noise). This is evident in comparison with times like the weekend from February 3 to 4, where the signals are significantly quieter. This noise limits the resolution of offsets due to specific seismic events.

Figure 6b shows a 4-hour time period around the destress blast on February 2. Offsets larger than the 100 kPa noise level are difficult to confirm. Figures 6c and d show the BPC measurements for the two events occurring on February 7. Changes coinciding in time with the observed events, within the 10-minute resolution of the measurements, are observed for both of the damaging event sequences (034500 and 122020). All of the pressure changes observed are positive for the 034500 event. For the 122020 event, a decrease in pressure was observed for cells 5 and 6 (which have identical orientations). The magnitude of pressure changes for both events on all the cells is a few hundred kPa maximum. Although the 034500 event had a much larger magnitude, the 122020 event was much closer to the BPC array; hence the resultant observed pressure offsets are of the same magnitude for both events. From a comparison of Figures 6b, 6c, and 6d, the seismic events caused a far larger coincident change in stress than did the destress blast itself.

5 BOUNDARY ELEMENT MODELS OF THE BLAST DESTRESSING

Two-dimensional boundary element models were used to estimate the change in slip potential on vertical faults perpendicular to the stope caused by the destress blast. Applied far-field stresses of 50.5 MPa perpendicular to and 38 MPa parallel to the long axis of the pillar were obtained using formulas reported by Board et al. (1981). Figure 7 shows contours of change in left-lateral the slip potential in the vicinity of the hypocenters of the 034500 and 122020 events locations caused by a 50% reduction in pillar modulus. Based on a Coulomb

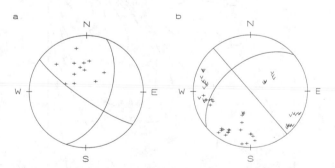

Figure 5. Lower hemisphere projection fault plane solutions. + indicates compression; v indicates dilatation. (a) A possible fault plane solution for the 030326 foreshock to the 034500 event. Readings with angle of incidence > 10° are not shown. (b) Composite fault plane solution for the 122020 event, consisting of readings for that event and 6 other co-located events. Readings with angle of incidence at accelerometer > 10° are not shown.

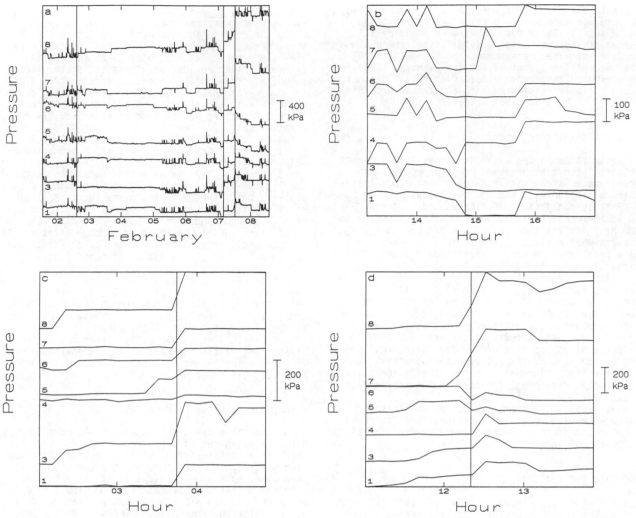

Figure 6. Digitally logged BPC data. Curves are labeled with cell numbers. The times corresponding to the destress on February 2 and the damaging events at 034500 and 122020 on February 7 are shown as vertical lines. (a) For February 1-9, 1990. The period of less noisy data on February 3 and 4 is the weekend, when minimal crews were working. (b) For the 4-hour period surrounding the February 2 destress. (c) For the 3-hour period surrounding the 034500 event. (d) For the 3-hour period surrounding the 122020 event.

shear stress failure criterion, the slip potential change, or change in the Coulomb failure function (CFF), is defined as (Stein & Lisowski, 1983; Oppenheimer et al., 1988)

$$|\Delta\tau| + \mu \, \Delta\sigma_n$$

where $\Delta\tau$ is the change in shear stress, μ is the coefficient of friction, taken as 0.6, and $\Delta\sigma_n$ is the change in normal stress (positive for a decrease in compressive normal stress). (Dip-slip potential change could not be computed with this 2-d model.) To obtain these contours, stresses were applied to a 20-m long boundary element representing the destressed pillar. The stresses applied to the boundary element were reduced relative to the uniform far-field applied stress field to simulate a 50% reduction in the pillar modulus. At the site of the 034500 hypocenter, the increase in left-lateral slip potential is 40 kPa. The increase in slip potential could be greater than 0.5 MPa if the hypocenter location uncertainty (Figure 3) is considered. (It may be significant that the change in slip potential is such to decrease the likelihood of both left-lateral and right-lateral slip at the 122020 hypocenter. The influence of the 034500 event on the 122020 slip plane is discussed below.)

The results are sensitive to the angle that the potential fault plane makes with the principal

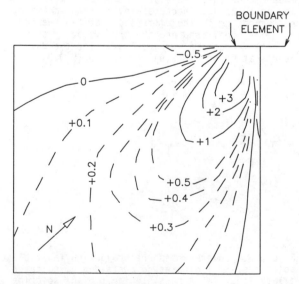

Figure 7. Plan view of change in slip potential for the vicinity of the damaging events resulting from the blast destressing as computed using 2-d boundary-element modeling. The boundary element corresponds to destressed pillar in Figure 3a. The rectangle is 100 by 100 m. Contours are MPa.

38

stress directions, and the angle between the principal stress direction and the pillar strike. For example, shifting the fault plane orientation 10° west of the maximum principal stress direction decreases rather than increases the left-lateral slip potential at the 034500 hypocenter. This is not consistent with the slip inferred from Figure 5a. However, the uncertainties in principal stress and fault plane orientations are large enough that the best interpretation may be that the faults in the vicinity are oriented such that shear stresses are in general low. When perturbing events such as the destress occur, sufficient increase in shear stress combined with decrease in normal stress occurs to induce slip on favorably oriented faults. This fault slip ultimately serves to facilitate stope closure.

6 DISLOCATION MODELS OF FAULT OFFSET

Using 3-dimensional dislocation models of seismic sources, static changes in elastic field values (after minus before) caused by fault slip can be computed (e.g., Chinnery, 1963). These models are based on analytic expressions for the elastic fields developed by a rectangular dislocation (Mansinha and Smyllie, 1971). Computations in this paper were done with a computer program described in Erickson (1986). Such models were used to find the amount of fault slip necessary to fit the observed pressure offsets at the BPC array. The available seismic measurements (event location, fault plane solution) were used to constrain the location, size, orientation, and sense of slip for the dislocation models. Dislocation models were also used to evaluate the effect of the 034500 event on the fault plane site of the 122020 event.

6.1 Fits to BPC data

The focal mechanisms determined above were used to define fault planes for each of the 034500 and 122020 events. Magnitudes for these two events were obtained using the coda duration (of seismic signal after first arrival) fit from Swanson and Sines (1991). Slip areas were estimated using magnitude-area relations (Utsu and Seki, 1955). Fault plane solution information was used to constrain the possible sense of slip for the 034500 event, but not the relative amounts of strike slip and dip slip, since these were not constrained by the available polarity data. For the 122020 event, the fault plane solution was used to constrain the relative amounts and sense of strike slip and dip slip. The magnitude of slip was then determined by the best fit to the measured BPC offsets. Figures 8a and 8b show the observed BPC offsets for the 034500 and 122020 events along with the best fit pressure offsets (least squares) expected using dislocation slip models. For the redundantly oriented BPC's, only cells 6 and 8 were used in the fits since their total pressures were closest to the average total pressure for the other cells of the array. The 034500 event was modeled as a vertical rectangular dislocation 297 m on a side, centered at the located event. The model for this event allowed the relative amounts of strike slip and dip slip to be obtained from the fit. The best fit at 95% confidence has 1.09 (± 1.18) mm left-lateral strike slip and 1.6 (± 0.99) mm NE-downward dip slip. The auxiliary nodal plane shown in Figure 5a corresponds to this slip vector. The large uncertainty in fit can be attributed to measurement error and to inadequacy of the homogeneous elastic media of the dislocation model for describing the actual response of the fractured rock mass.

The 122020 event (Figure 8b) was modeled as a vertical rectangular dislocation, 28 m on a side, centered at the location for that event. The dislocation is oriented N40W, consistent with the vertical nodal plane of the mechanism. (The other nodal plane would require right-lateral strike slip motion only. Such motion results in BPC offsets in the opposite sense from those observed for all but cells 5 and 6.) The orientation and relative amounts of NE-downward dip slip and left-lateral strike slip were determined by the composite focal mechanism shown in Figure 5b. The total amount of slip which best fit the offsets at 95% confidence was 0.45 (± 0.55) mm.

There are several possible reasons for the large uncertainties in the fit for the 122020 event: (1)

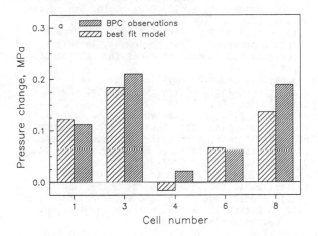

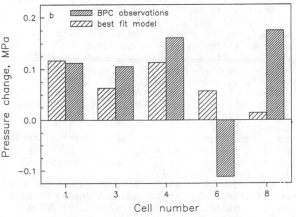

Figure 8. Results of dislocation modeling of (a) the 034500 event and (b) the 122020 events to fit observed BPC pressure changes.

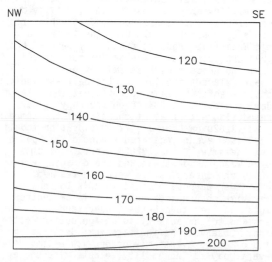

Figure 9. Change in slip potential on the 122020 fault plane from dislocation modelling of the 034500 event. The rectangle is 28 by 28 m. Contours are kPa.

There may be measurement error in the BPC offsets.
(2) The homogeneous elastic model may not adequately
represent the complex stope geometry in the vicinity
of this event. (3) There may have been a localized
inelastic response of the rock mass at the BPC array
due to its close proximity (20 m) to the 122020
event. (4) The proximity of the cells to the event
hypocenter makes them sensitive to the details of
the rupture geometry. Similarly, slight changes in
the model geometry have a large impact on the computed stress changes. (5) Deviation from the assumed planar slip surface may play a role. For
example, the orientation of the fault may have
shifted away from the fault plane solution after
slip began.

6.2 Event interaction

The dislocation parameters which fit the BPC data
for the 034500 event were used to calculate the
induced changes in stress field components in the
vicinity of the 122020 event. The results are used
to determine if the stress changes produced by the
034500 event inhibit or enhance the prospects for
slip of the type observed for the 122020 event.
This dislocation results in a decrease in normal
stress of somewhat less than 0.2 MPa across the
122020 fault, and an increase in shear stress favoring SW-downward dip slip of 0.4 MPa (maximum) with a
decrease in shear stress favoring left-lateral
strike slip of 0.3 MPa (maximum). Figure 9 shows
the change in slip potential in the direction of
observed slip on the 122020 fault induced by the
dislocation described above on the 034500 fault.
The slip potential change ranges from 120 to 200 kPa
across the 122020 slip plane. The change in slip
potential at the few hundred kPa level is similar in
magnitude to changes observed to enhance aftershock
activity for certain California earthquakes (Stein
and Lisowski, 1983; Oppenheimer et al., 1988; Simpson et al., 1988).

7 CONCLUSIONS

Seismic observations and measurements of changes in
ground pressure were made in conjunction with a
destress of the 46-99 stope of the Galena Mine.
Similar observations were made during a period of
elevated seismic activity during the 18 days following the destress.
Figure 6b showed that pressure changes due to the
destress as monitored at the BPC array were no more
than the noise level, which was at about 100 kPa at
that time. Given the intense seismic activity during
the next 18 days, this absence of observable pressure change is surprising. Over the 18 days of
seismic activity, pressure changes of several hundred kPa are observed on the array; most of this
change is coincident with seismic events.
Boundary element modeling of the destress indicated possible enhanced slip potential at the site of
the damaging event at 034500 on February 7, but
decreased slip potential at the site of the damaging
event at 122020 on February 7. Dislocation modeling
of the 034500 event indicated enhanced slip potential at the site of the 122020 event. The time
delay between the destress and the damaging events
is cause for concern, since work had resumed.
The broader purpose of this work is to show how
quantitative observation and analysis of destressing
attempts can lead to a better understanding of the
process and to the ability to design better destress
methods. Observations indicate that destressing-
related damaging events may occur with a time delay
of several days. The ability to determine the effectiveness of the destress and the safety of workers returning to the destressed area depends on
additional monitoring of the rock mass, whether
through ground pressure or microseismic monitoring.

8 ACKNOWLEDGMENTS

The authors acknowledge ASARCO Inc., for their cooperation with the Bureau in providing information,
access, and assistance for the data-gathering phase
of this research. C. D. Sines provided invaluable
on-site expertise and assistance. R. W. Simpson of
the U.S. Geological Survey provided the 3-D dislocation code.

REFERENCES

Blake, W. 1972. Destressing test at the Galena
Mine, Wallace, Idaho. Soc. Min. Eng. AIME Trans.
252: 294-299.
Board, M., E. Hardin, N. Barton, and M. Voegele
1981. Instrumentation and evaluation of the 2400-
ft level of the silver shaft, Mullan, Idaho.
Terra Tek Inc. Tech. Rept. 81-62, Salt Lake City,
UT.
Board M., and C. Fairhurst 1983. Rockburst control
through destressing - a case example. In Rockbursts: Prediction and Control, the Institution
of Mining and Metallurgy, p. 91-102. London.
Boler, F.M., and P.L. Swanson 1990. Computer
Automated Measurement and Control based workstation for microseismic and acoustic emission research. U.S. Bureau of Mines Information Circular
9262.
Chinnery, M.A. 1963. The stress changes that
accompany strike slip faulting. Bull. Seismol.
Soc. Am. 53: 921-932.
Erickson, L.L. 1986. A three-dimensional
dislocation program with application to faulting
in the earth. Stanford Univ. M.S. Thesis.
Estey, L.H., P.L. Swanson, F.M. Boler, and S.
Billington 1990. Microseismic source locations: a
test of faith. Proceedings of the 31st Symposium
on Rock Mechanics, p. 939-946. Colo. Sch. Mines,
Golden, CO.
Haramy, K.Y., and R.O. Kneisley 1991. Hydraulic
borehole pressure cells: equipment, technique, and
theories. U.S. Bureau of Mines Information Circular 9294.
Mansinha, L., and D.E. Smyllie 1971. The
displacement field of inclined faults. Bull.
Seismol. Soc. Am. 61: 1433-1440.
Oppenheimer, D.H., P.A. Reasenberg, and R.W. Simpson
1988. Fault plane solutions for the 1984 Morgan
Hill, California, earthquake sequence: evidence
for the state of stress on the Calaveras Fault.
J. Geophys. Res. 93: 9007-9026.
Rorke, A.J., and R.K. Brummer 1990. Use of
explosives in rockburst control. In C. Fairhurst
(ed.), Proceedings of the 2nd International Symposium on Rockbursts and Seismicity in Mines, p.
377-386. Rotterdam, Balkema.
Simpson, R.W., S.S. Schulz, L.D. Dietz, and R.O.
Burford 1988. The response of creeping parts of
the San Andreas Fault to earthquake on nearby
faults. PAGEOPH 126: 665-685.
Stein, R.S. and M. Lisowski 1983. The 1979
Homestead Valley earthquake sequence, California:
control of aftershocks and postseismic deformation. J. Geophys. Res. 88: 6477-6490.
Swanson, P.L., and C.D. Sines 1990. Repetitive
seismicity (M ~ 2-3) and rock bursting along a
plane parallel to known faulting in the Coeur
d'Alene district, ID. EOS Trans. Amer. Geophys.
Union 71: 1453.
Swanson, P.L., and C.D. Sines 1991. Characteristics
of mining induced seismicity and rock bursting in
a deep hard-rock mine. U.S. Bureau of Mines Report of Investigation 9393.
Utsu, T., and A. Seki 1955. A relation between the
area of after-shock region and the energy of main
shock. J. Seismol. Soc. Japan (Zisin) 7: 233-240
(in Japanese).

Rockbursts and Seismicity in Mines, Young (ed.)© 1993 Balkema, Rotterdam, ISBN 90 5410 320 5

Ground velocity relationships based on a large sample of underground measurements in two South African mining regions

A.G. Butler & G. van Aswegen
ISS International Limited, Welkom, South Africa

ABSTRACT: Measurements of peak ground velocity have been recorded at two different South African gold mining regions (Orange Free State: Welkom and Western Deep Levels) currently monitoring seismicity with the Integrated Seismic System (ISS). Relationships between peak ground velocity (v), distance (R) and each of the parameters; local magnitude (M), radiated seismic energy (E), seismic moment (M_o), and apparent stress (σ_a) for the relative near-field and far-field are presented. The far-field relations greatly overestimate the ground velocities measured in the relative near-field. The resulting relationships are used for estimating magnitudes of seismic events using the "quick location" facility in the ISS, and also form an integral part of seismic hazard evaluation using the Volume of Ground Motion concept (VGM) (Mendecki 1985, Kijko and Funk 1992). The relationships are also useful for estimating the peak ground velocity at sites of falls of ground occurring some distance from a seismic event. Differences between the relationships obtained in the two mining regions are also discussed.

1 INTRODUCTION

Relations between peak ground velocity (v), distance (R), and some measure of magnitude are of interest to rock mechanics engineers designing mine layouts and support in rockburst prone mines. The majority of published papers examining such relationships have concentrated on large magnitude seismic events, with measurements made at large distances from the seismic source [(Atkinson 1992, Boore and Atkinson 1987, Boore and Joyner 1982, Burger et. al 1987, Campbell 1981, Espinosa 1979, Hanks and Johnston 1992, Hanks and McGuire 1981, Hasegawa et. al 1981, Hermann and Goetz 1981, Joyner and Boore 1984, Sabetta and Pagliese 1987, Street 1982)]. Very few publications deal with peak ground velocity measured relatively close to the seismic source (<10 kilometres), largely due to lack of measurements. McGarr et. al (1981) provided the first relationship for mine tremors based on a limited number of measurements (12 events) made close to the seismic sources and this relationship has been used by Wagner (1984) for estimating support requirements for rockburst conditions. Gibbon et. al (1987) measured ground velocities in stopes exposed to rockburst conditions and compared these with a relation from Spottiswoode (1984). Hedley (1990) provided a different form of relationship based on a velocity relationship used in the blast monitoring environment for 25 rockbursts in several Canadian mines.

A summary of these published relations for the mining environment is as follows;

1. (McGarr et. al 1981): $LOG(Rv)=0.57M+1.95$

2. (Spottiswoode 1984): $LOG(Rv)=0.5M+2.81$

3. (Hedley 1990): $v = 4000\left(-\dfrac{R}{10^{M/3}}\right)^{-1.6}$

where R is in metres and v is in mm/s.

This paper examines relations between v, R, and several source parameters from two gold mining regions in South Africa. Differences between the regions and between far-field and relatively near-field observations are explained and applications discussed.

2 METHOD OF ANALYSES

The Integrated Seismic System (ISS) has been operating in the Orange Free State: Welkom Region (ISS-OFS) in more or less its current form since October 1988 and at Western Deep Levels (ISS-WDL) since October 1991. Details of each seismic system and the data used in this analyses are provided in Table 1. The ISS system response is governed at low frequencies by the geophones which are damped at 70% of critical for the flattest response. At high frequencies the anti-aliasing filters dominate. Fifth order Bessel filters are used for minimum distortion of the signal in the time domain, with cutoff frequency 20% of the sampling frequency. Mendecki (this volume) and Mendecki et. al (1990) provide more details about the ISS. These seismic systems provide an abundance of data used by the host mines for the monitoring of rock mass behaviour (e.g. van Aswegen and Butler this volume, Dennison and van Aswegen this volume, Mendecki this volume).

Several seismic parameters were examined to determine which would provide the best correlation with peak ground motion and distance. The seismic moment (M_o) and radiated seismic energy (E) as derived from seismograms and the calculated apparent stress (σ_a) for each individual measurement recorded were used. The local magnitude (M) as determined for each individual seismic station was also used. The local magnitude is determined by the relation;

$$M=A\{LOG(E)\}+B\{LOG(M_o)\}+C$$

based on correlation with published magnitude data from the Geological Survey of South Africa (Butler 1992). The seismic networks consist of three component geophone stations underground (majority) and at surface. Only underground measurements for seismic events containing a minimum of five stations in their solution were used in the analyses. This reduces both the surface effects and the effect of inaccurately located events on the results. Measurements between 100m and 10000m from the source were used initially to reduce near-field effects (radiation pattern), and to reduce large distance attenuation and raypath effects. The lower limit was then increased to 1000m for reasons explained below. The upper limit is also a practical constraint as falls of ground associated with seismic events rarely occur at distances greater than 10 kilometres.

Originally, the form of peak ground velocity-distance-magnitude relationship as developed by McGarr et. al (1981) for the mining environment was used in the analyses. However, it was found that the following form provides a better fit for all seismic parameters;

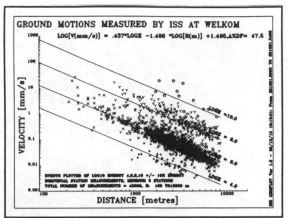

(a) OFS: LOG Energy

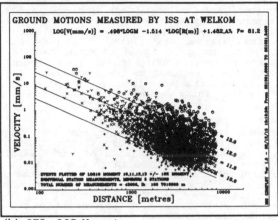

(b) OFS: LOG Moment

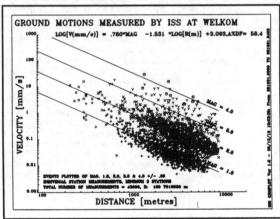

(c) OFS: Magnitude

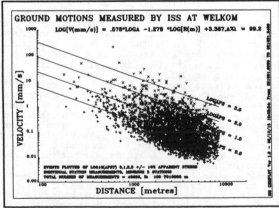

(d) OFS: LOG Apparent Stress

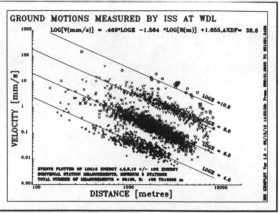

(e) WDL: LOG Energy

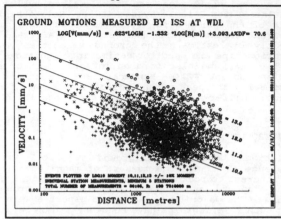

(f) WDL: LOG Moment

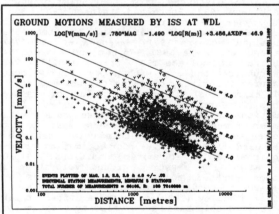

(g) WDL: Magnitude

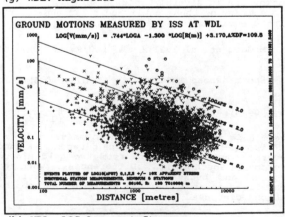

(h) WDL: LOG Apparent Stress

Table 1. Seismic Network and data details

	Number current stations	Area covered	Average density	Number seismic events	Sampling rate (Hz)	Time period
OFS	47	400km²	0.12 sta/km²	8386	500-2000	88/10/01-92/10/31
WDL	25	24km²	1.04 sta/km²	10340	2000	92/01/01-92/10/31

$$LOG(v) = A(\text{event parameter}) + B\{LOG(R)\} + C$$

Street (1982) used a similar form when examining ground velocities due to a large seismic event in the eastern United States.

A first norm fitting method was used to minimize errors in velocity. A meaningful indicator for describing the "goodness of fit" called the percentage difference (%DF) was also developed. The %DF value is the percentage difference between the measured peak ground velocity and the predicted peak ground velocity using the best fitting relationship. The A%DF is the average %DF for a given data set, with equal weighting to each individual measurement in the database.

3 RESULTS & DISCUSSION

Distance 100-10000m

Ground velocity relationships for OFS:WELKOM and WDL are provided in Figure 1 for all the seismic parameters; LOG(E), LOG(M_o), σ_a and M, for both regions. LOG(E) provides the best correlation for both regions, with an A%DF less than 50%. The relatively poor correlation with M_o was also found by McGarr (1986) for mine tremors. The good correlation of LOG(E) with v compared to that of LOG(M_o) is readily explained by the definition of the two source parameters. Radiated seismic energy is directly dependent on velocity of ground motion while M_o is dependent on the low frequency content of ground motions. The generally higher velocities experienced for events of equal M_o at WDL compared to the OFS shows that generally more energy is released per unit of seismic deformation at WDL. This, of course is also clear from comparisons of apparent stress (Mendecki this volume), and confirms the higher state of stress in the rock mass in the deeper mining environment. Since σ_a and M are functions of both E and M_o, it can be expected that the correlations should not be as good as LOG(E) alone.

Figure 2a shows the distribution of individual measurements with distance for the two mining regions in the study. Only measurements made with distances from 100 - 10000m were used. The distribution of magnitudes estimated from each measurement is provided in Figure 2b. Although limitations exist with respect to the calculation and usefulness of magnitudes and corner frequencies, these parameters are still widely used in describing seismic event characteristics. Figure 2c uses these parameters to show the difference in frequency content for a given size of seismic event between the two regions. Figure 2d shows the A%DF variation with distance, and

Figure 2e gives the measurement distribution of A%DF.

The best correlated relations for each region are as follows;

$$OFS: LOG(v) = 0.437\{LOG(E)\} - 1.498\{LOG(R)\} + 1.495, A\%DF = 47.5\%$$

$$WDL: LOG(v) = 0.459\{LOG(E)\} - 1.564\{LOG(R)\} + 1.655, A\%DF = 38.6\%$$

The majority of measurements of peak ground velocity occur within 5000m from the source. The high average density of stations at WDL coupled with the high sampling rate results in the majority of events for the study coming from the magnitude 0-1 range. The lower average station density in the OFS region allows resolution down to magnitude 1.0 over the entire network area. It should be noted that important areas in each region have a higher station density than the average.

The average corner frequency vs. magnitude plot (Figure 2c) shows significant differences between the two regions. These differences are due largely to geological variations and stress regime. The geology of the Witwatersrand succession in the OFS gold field is described by (Minter et. al 1986). The quartzites are disrupted by a large number of faults and dykes enforcing the scattered mining method. Total displacement along some faults exceeds 1000m. Geology at WDL is described by Engelbrecht et. al (1986). The tabular ore bodies are significantly less disturbed than in the OFS, allowing longwall mining methods for both the Ventersdorp Contact Reef (1900-2600m below surface) and the Carbon Leader Reef (2600-3600m below surface). These and other differences betweeen the two mining environments are summarized in Table 2.

1000m-10000m (OFS), 1000M-7000M (WDL)

The A%DF vs. Distance plot (Figure 2d) shows that the attenuation relationship for the OFS region is very good from 1000m to 10000m. The WDL relation breaks down at approximately 7000m with predicted ground velocities being higher than those actually experienced at these distances. The higher average corner frequencies can partly explain this, as higher frequencies are attenuated more quickly with distance (i.e. radiated energy of a seismic event of higher dominant frequency will be attenuated more than that of an event of lower dominant frequency). The major reason however is due to the attenuation factor, Q, used at WDL. A Q value of 300 is used for the quartzites in the immediate vicinity of the mining horizons. For seismic events within the

Figure 1. Velocity-distance-seismic parameter relations for OFS and WDL regions. All relationships are based on individual ground velocity measurements made with 3 component geophones. Only seismic events with a minimum of 5 underground stations used in their solution were used in the analyses. Data plotted for energy, moment, and apparent stress, are only those measurements within ±10% of the value indicated on the lines (for presentation purposes). Measurements plotted for the magnitude relations are within ±.03 of the magnitudes shown.

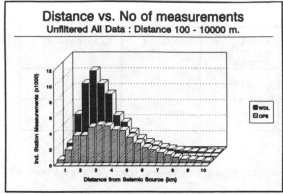

(a) Distribution of number of measurements with distance

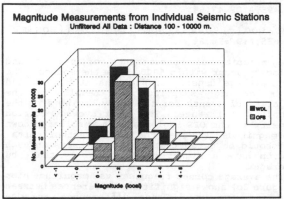

(b) Distribution of number of measurements per magnitude range

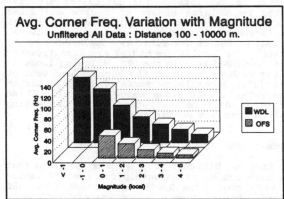

(c) Average corner frequency variation per magnitude range

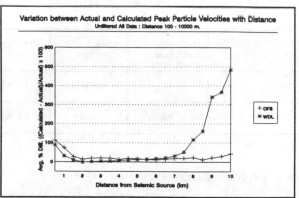

(d) Average percentage difference variation with distance

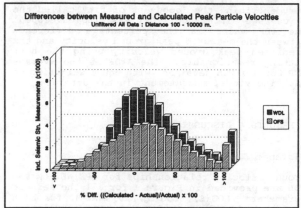

(e) Distribution of average percentage differences

Figure 2. Details of dataset used for OFS and WDL regions. All measurements have source-geophone distance from 100-10000m measured with three-component geophones installed underground. All magnitudes are local regional magnitudes as determined by Butler (1992). Limitations of the use of magnitude and corner frequency as seismic parameters are known but they still prove useful for demonstration purposes as they remain widely used for describing seismic event characteristics. Percentage difference is an indicator of the "goodness of fit".

Table 2. Geology and mining details at OFS and WDL.

OFS	WDL
Intermediate depth (1500-2500m)	Great depth (1600-3600m) Current mining at deeper levels
High frequency of faults & dykes; significant displacement	High frequency of faults & dykes; minimal displacements
Faults at mining horizon characterized by soft filling (pyrophyillite)	Faults at mining horizon characterized by hard filling (vein quartz, pseudotachyllite)

mining area, generally it will be this rock mass the seismic wave travels through. At long distances, however, it is most likely that the waves actually reaching the seismic stations travel mostly through the faster, more solid overlying lava. Only in recent months have we been adjusting the Q value to 500 for this

situation. The low Q value used in the past results in an overestimation in the predicted ground velocities. The seismic events in the database are being reprocessed using the new Q value where required. This has little effect on events within WDL, but does affect the source parameter calculations for events located beyond the WDL boundaries. These seismic events in the database will be reprocessed using the new Q value where necessary. A procedure for automatic Q variation is currently being developed.

Both regions show a poor fit with distances less than 1000m. Generally, the relation predicts a higher value of ground velocity than is experienced, indicating a flattening of the ground velocity relation at smaller distances. To eliminate this effect and the long distance effects at WDL, the analyses were repeated for distances of 1000-10000m (OFS) and 1000-7000m (WDL).

Figure 3a shows the average percentage difference variation with distance for the new data set and Figure 3b shows the distribution of average percentage differences. The following relations for the new data set are shown in Figure 4;

OFS:LOG(v)≈0.439{LOG(E)}-1.497{(LOG(R)}
+1.483, A%DF=44.3%

WDL:LOG(v)=0.480{LOG(E)}-1.617{(LOG(R)}
+1.779, A%DF=32.7%

As expected, only the WDL relationship changed significantly by not including the near-field and very far-field data. The A%DF for both regions show that the predicted ground velocity using the above relations agree with actual measurements underground, to a very large degree.

1000m-10000m(OFS), 1000m-7000m (WDL)
Filtered Final Analysis

The ground velocity relationships can be used as a form of seismic station monitor. By examining the individual measurements for which the %DF was high (i.e. >200%), it was possible to determine if one or several of the seismic stations at each region were responsible for the poor fitting data. The results of the analysis are shown in Figure 5. For each region, specific seismic stations are responsible for ill-fitting data points. It was found that these particular stations were not perfectly calibrated, generally resulting in an underestimation of actual ground velocities. Although on a percentage basis, the number of measurements with a %DF greater than 200 was quite small, these data points are introducing some error into the final relationships.

Using the data set described immediately above and the resulting ground velocity relations, all individual measurements with a %DF greater than 200% were eliminated from the data set. Figure 6a gives the final average percentage deviation with distance, while Figure 6b gives the final variation of percentage differences. Using this 'filtered' data set, new final ground velocity relations were developed. The following relations are shown in Figure 7;

OFS:LOG(v)=0.453*{LOG(E)}-1.502*{LOG(R)}+1.420,
A%DF=37.2%

WDL:LOG(v)=0.481*{LOG(E)}-1.608*{LOG(R)}+1.747,
A%DF=38.6%

Seismologically the differences in the A, B and C parameters between the OFS and WDL regions, can all be related to the difference in frequency content of the velocities. At higher frequency a higher velocity amplitude is required to yield the same energy, explaining the higher value of C in the WDL case. The numerically lower value of B for WDL is directly attributable to the fact

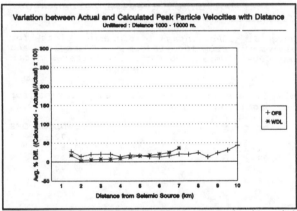

(a) Average percentage difference variation with distance

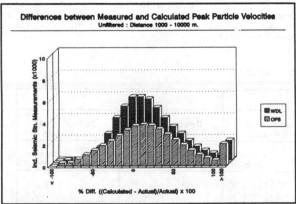

(b) Distribution of average percentage differences

Figure 3. Percentage difference plots for measurements from 1000-10000m at OFS region, and from 1000-7000m at WDL.

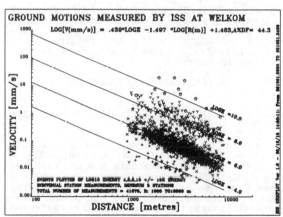

(a) Velocity-distance-LOG(E); OFS 1000-10000m

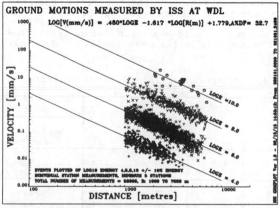

(b) Velocity-distance-LOG(E); WDL 1000-7000m

Figure 4. Velocity-distance-LOG(E) relationships with near-field (100-1000m) data excluded for both regions and extreme far-field (>7000m) excluded for WDL.

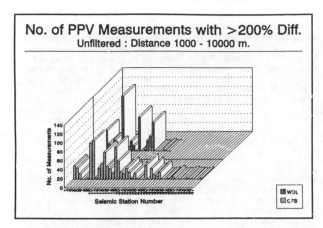

Figure 5. *Number of measurements with %DF above 200 for individual seismic stations*

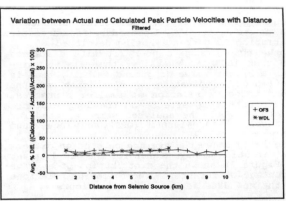

(a) *Average percentage difference variation with distance*

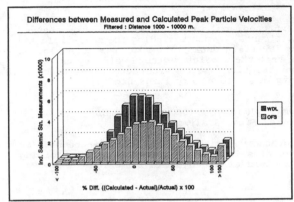

(b) *Distribution of average percentage differences*

Figure 6. *Percentage difference plots for measurements from 1000-10000m at OFS region, and from 1000-7000m at WDL for Final Filtered data.*

that attenuation increases with frequency. The higher A value for WDL indicates that the difference in dominant frequency between events there and events in the OFS region is more significant for larger events than smaller events. The ratio of cf WDL/cf OFS for the different magnitude ranges shown in Figure 2c are given in Table 3.

An example are the tremor at WDL on 7 March 1992 (M=4.7) with a corner frequency of approximately 21Hz and that in the OFS region of 6 June 1992 (M=4.7) with a corner frequency of approximately 10Hz.

50-1000m: Near Field Data

The final far-field ground velocity relations significantly overestimate ground velocities in the near field (Figure 2d). Treating the near-field data (50-1000m) on its own, results in the following near-field ground velocity relationships shown in Figure 8;

OFS:LOG(v)=0.409{LOG(E)}-1.447{LOG(R)} +1.535,
A%DF=109.5%

WDL:LOG(V)=0.472{LOG(E)}-1.413*{LOG(R)} +1.143,
A%DF=52.7%

The A%DF for each relation is significantly higher than the far-field data, indicating a poorer quality of fit. The data was not filtered because these effects are believed to be real and not solely due to miscalibrated stations. The large A%DF values are to be expected in the near-field, with source effects such as radiation pattern and rupture propagation greatly influencing the experienced ground velocities. The near-field relations do however, show that there should be a flattening of the far-field relations if they are to be extrapolated to the near-field. Table 4 shows the differences between the far-field and near-field equations when predicting ground velocities in the near-field. It is clear

Table 3. *Average corner frequency ratios for given magnitude ranges.*

Magnitude	CF WDL/CF OFS
0-1	1.72
1-2	1.83
2-3	2.19
3-4	2.74
4-5	2.69

that significant overestimations of ground velocities can be expected if far-field relations are extrapolated to the near-field. This observation has serious implications on the design of support in under ground excavations and on the estimation of the volume of rock mass around a seismic source that experiences very strong ground velocity.

Table 4. *Comparison of far-field vs near-field predicted results*

			OFS		WDL	
			Near-field relation	Far-field relation	Near-field relation	Far-field relation
LOG(E) = 4 MAG≈ -0.5	DIST =	10m 100m	53mm/s 2mm/s	54mm/s 2mm/s	42mm/s 2mm/s	116mm/s 3mm/s
LOG(E) = 6 MAG≈ 1.0	DIST =	10m 100m	348mm/s 12mm/s	433mm/s 14mm/s	365mm/s 14mm/s	1.1m/s 26mm/s
LOG(E) = 8 MAG≈ 2.0	DIST =	10m 100m	2.3m/s 82mm/s	3.5m/s 110mm/s	3.2m/s 124mm/s	9.7m/s 239mm/s
LOG(E) =10 MAG≈ 3.2	DIST =	10m 100m	15.1m/s 538mm/s	28.1m/s 883mm/s	28.1m/s 1.1m/s	88.9m/s 2.2m/s

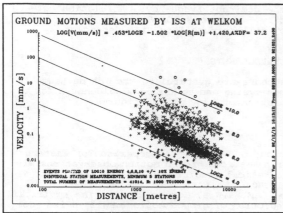

(a) *Velocity-distance-LOG(E); OFS 1000-10000m; final filtered data*

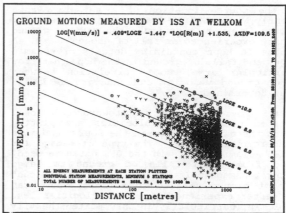

(a) *Velocity distance-LOG(E); OFS 50-1000m*

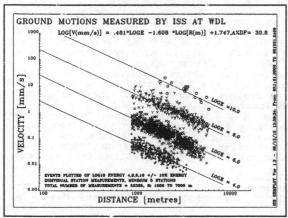

(b) *Velocity-distance-LOG(E); WDL 1000-7000m; final filtered data*

Figure 7. Velocity-distance-LOG(E) relationships with near-field (100-1000m) data excluded for both regions, extreme far-field (>7000m) excluded for WDL and data with a %DF >200 excluded.

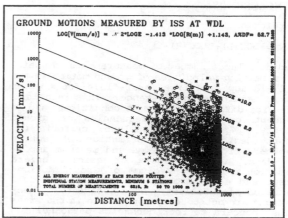

(b) *Velocity-distance-LOG(E) WDL 50-1000m*

Figure 8. Velocity-distance-LOG(E) relationships for OFS and WDL regions using near-field (50-1000m) measurements only. All data points used to obtain the fit are plotted.

Because measurements are rarely made in the immediate vicinity of a seismic source, the velocity at which a mine opening closes is estimated from extrapolations of ground velocity relations such as those presented above. Although relatively near-field relations for ground velocity have been presented, the data from which they are derived is still not close enough to the area of maximum ground velocity within the seismic source regions to predict this maximum value. Although these relations are therefore not totally applicable to the design of support for "face bursting" situations which are largely in the immediate near-field, they can be used for support design in the relative near-field (e.g. excavations near major geological structures with the potential for yielding major seismic events).

Quick location and magnitude estimation

One of the major requirements of the ISS system is to provide fast and accurate seismic event locations and estimates of magnitude and energy index (van Aswegen and Butler, this volume) to rock mechanics engineers and Management. The time for an event to be automatically located varies depending on the number of seismic stations associated, but is generally between 3 to 15 minutes. Automatic P and S picking is accomplished usually within one minute. The location and size of a seismic event can, however, be determined much more quickly and need not require full waveform transfer. Trigger times recorded at each individual station are sent immediately to surface along with the value of maximum velocity experienced. A "quick location" is

provided assuming the trigger time as the arrival of the P wave. The distance from the seismic source to each station is calculated and knowing the maximum ground velocity experienced at each station, a magnitude can be estimated using the magnitude-distance-velocity relation applicable to the mining region of interest. The entire "quick location" process provides an estimated location and magnitude generally within 15 seconds after the seismic event is detected.

Volume of ground velocity concept (VGM)

The "Volume of Ground Motion" (VGM) concept (Mendecki, 1985, Kijko and Funk 1992) provides an estimate of seismic hazard in a particular area in the form of a single number combining the effects of several parameters. The VGM concept integrates the probabilities of occurrence and the rate of events, and through the relations described above, the total rock mass volume affected by peak ground velocities greater than some critical potentially damaging value over a given time period. The main application of the VGM concept is to provide a routine estimate of the variation in hazard levels in both space and time. Within a particular mining region, different areas can be evaluated in terms of potential hazard for a given time period, sometimes providing explanations for seismicity related production or rock mechanics problems in a given area.

Ground velocities experienced at fall of ground sites

Falls of ground often occur some distance away

from the source location of a seismic event. Currently, the LOG(E)-distance-ground velocity relation is used to estimate the maximum ground velocity experienced at fall of ground sites. A database is kept containing details of seismic damage, and falls of ground in the hope that, for a particular area, some correlation between maximum ground velocity and underground damage can be determined.

Figure 9 provides a first step towards this goal. Estimated peak ground velocities are plotted against the distance between the source and site of underground damage. The type of damage for the 142 seismic events in the OFS database used, ranged from small falls of ground to large rockburst damage. It was found that significant falls of ground related to seismic events can occur after experiencing ground velocities less than 1mm/s, particularly in unsupported, heavily fractured rock masses.

4 CONCLUSIONS

The relation;

$$LOG(v) = ALOG(E) + BLOG(R) + C$$

based on a large number of measurements is useful for estimating velocities of ground motion in the far-field and "relative" near-field for a given seismic magnitude and distance. This relation varies for different rock mass conditions and is dependent on dominant frequencies of ground velocity. Applications include; criteria for support design, quick estimation of source parameters from trigger-time and maximum amplitude data, a new measure of seismic hazard estimation using the "Volume of Ground Motion" concept and estimation of velocities of ground motions causing fall of ground accidents. These relations are not totally applicable in the areas of most concern to rock mechanics personnel and support design people, that is, near the working faces close to the sources of the majority of seismic events. To develop an adequate support methodology and support components to protect workers in these areas, near field estimates of ground velocity are required. It has been shown that far-field extrapolations to the near-field results in significant over-estimations of ground velocities and relative near-field equations have been presented. As the density of the ISS networks in the three mining regions is increased (75 seismic stations at WDL within two years, decentralization to provide increased density of coverage for specific regions in the OFS) more information will become available from the near field, allowing more study into this important issue. Future research should include instrumentation of stopes at rockburst prone sites to include immediate near-field measurements. The velocity-distance relation should also incorpor-

ate the dominant frequency. A relation between velocity, distance, LOG(E), and dominant frequency will be much less site dependent.

ACKNOWLEDGEMENTS

The authors are grateful to Mrs Ida de Lange for tedious typing and to Mr Doons Geldenhuys for the preparation of many of the figures.

REFERENCES

Atkinson, G.M. 1990. *A comparison of Eastern North American ground motion observations with theoretical predictions.* Seism. Res. Let. 61: 171-180.

Boore, D.M. & W.B. Joyner, 1982. *The empirical prediction of ground motion.* Bull. Seism. Soc. Am. 72:S43-S60.

Boore, D.M. & G.M. Atkinson 1987. *Stochastic prediction of ground motion and spectral response parameters at hard-rock sites in Eastern North America,* Bull. Seism. Soc. Am. 77:440-467.

Burger, R.W., P.G. Somerville, J.S. Barker, R.B. Herrmann & D.V. Helmberger 1987. *The effect of crustal structure on strong ground motion attenuation relations in Eastern North America,* Bull. Seism. Soc. Am, 77: 420-439.

Butler, A.G. 1992. *Magnitude evaluation for mining regions using the ISS. Internal Report,* ISS International, Welkom S.A.

Dennison, P.J.G. & G. van Aswegen, 1992. *Stress modelling and seismicity on the Tanton fault: a case study in a South African gold mine.* Submitted to 3rd Symposium on Rockburst and Seismicity in Mines, Kingston, Ontario, Aug. 1993.

Engelbrecht, C.J. & G.W.S. Baumbach, J.L. Matthysen & P. Fletcher, 1986. *The West Wits line.* In: Anhaeusser, C.R. and Maske, S. (eds), Mineral Deposits of Southern Africa, Geol. Soc. S. Afr., Vol.I: 599-648.

Espinosa, A.F., 1979. *Horizontal particle velocity and its relation to magnitude in the Western United States.* Bull. Seism. soc. Am. 69: 2037-2061.

Gibbon, G.J., A. de Kock & J. Mokebe, 1987. *Monitoring of peak ground motion rockbursts.* IEEE Transactions on Industry Application. IA23(6): 1094-1098.

Hanks, T.C. & R.C. McGuire 1981. *The character of high-frequency strong ground velocity.* Bull. Seism. Soc. Am. 71: 2071-2095.

Hanks, T.C. & A.C. Johnston 1992. *Common features of the excitation and propagation of strong grounds motion for North American earthquakes.* Bull. Seism. Soc. Am. 82:1-22.

Hasegawa, H.S., P.W. Basham & M.J. Berry 1981. *Attenuation relations for strong seismikc ground motion in Canada.* Bull. Seism. Am. 71: 1943-1962.

Hedley, D.G.F. 1990. *Peak particle velocity for rockbursts in some Ontario mines.* In C. Fairhurst, (ed): *Rockbursts and seismicity in mines.* Balkema, Rotterdam: 345-348.

Hermann, M.R.B. & M.J. Goetz 1981. *A numerical study of peak ground velocity scaling.* Bull. Seism. soc. Am. 71: 1963-1979.

Joyner, W.B. & D.M. Boore 1981. *Peak horizontal acceleration and motion from strong-motion records including records from the 1979 Imperial Valley, California, earthquake.* Bull. Seism. Soc. Am. 71: 2011-2038.

Kijko, A. & C. Funk 1992. *Seismic hazard assessment in mines.* Submitted for publication. Pageoph.

McGarr, A/. R.W. Green & S.M. Spottiswoode, 1981. *Strong ground motion of mine tremors: some implications for near source ground motion parameters,* Bull. Seism. soc. Am. 71: 295-319.

McGarr, A. 1986. *Some observations indicating complications in the nature of earthquake scaling.* Earthquake Source Mechanics. Geophys. Monograph 37 (Maurice Ewing 6) S. Das, J. Boatwright, C. Scholtz (ed): 217-225.

Mendecki, A.J., G. van Aswegen, J.N.R. Brown & P. Hewlett 1990. *The Welkom Seismological Network.* Balkema, Rotterdam: 237-243.

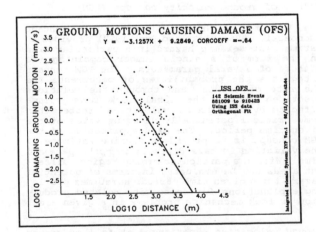

Figure 9. Estimated peak ground velocity versus distance between seismic source and site of underground damage.

Mendecki, A.J. 1985. *An attempt to estimate seismic hazard on the AAC Gold Mines.* AAC Internal Report SP8/85.

Minter, W.E.L., W.C.N. Hill, R.J. Kidger, C.S. Kingsley & P.A. Snowden, 1986. *The Welkom goldfield.* In: Anhaeusser, C.R. & Maske, S. (eds). *Mineral Deposits of Southern Africa,* Geol. Soc. S. Afr. Vol.I: 497-540.

Sabetta, F. & A. Pugliese 1987. *Attenuation of peak horizontal acceleration and velocity from Italian strong-motion records.* Bull. Seism. Soc. Am. 77: 1491-1513.

Spottiswoode, S.M. 1984. *Underground Seismic Networks and Safety. In Monitoring for Safety in Geotechnical Engineering.* SANGORM, Johannesburg.

Street, R. 1982. *Ground motion values obtained for 27 July 1980 sharpsburg, Kentucky, earthquake.* Bull. Seism. soc. Am. 72: 1295-1307.

van Aswegen, G. & A.G. Butler 1992. *Application of quantitative Seismology in mines.* Submitted to 3rd Symposium on Rockburst and Seismicity in Mines, Kingston, Ontario, Aug. 1993.

Wagner, H. 1984. *Support requirements for rockburst conditions.* Proc. 1st Int. Congr. *Rockburst and Seismicity in Mines.* S. African Inst. Min. Met., Johannesburg. Gay, N.C. & E.H. Wainwright (eds): 209-218.

Rockbursts and Seismicity in Mines, Young (ed.) © 1993 Balkema, Rotterdam, ISBN 90 5410 320 5

Extraction of a highly stressed pillar at Mount Isa

E. Dailey
Mount Isa Mines Ltd, Qld, Australia

Abstract: K705, a regional pillar in the lower northern open stoping block of 7 orebody, was being maintained to ventilate footwall mining operations through to an exhaust shaft located beyond 5 orebody. Subsequent development of 5 orebody eliminated the requirement to retain K705. Numerical modelling indicates that high stress levels would be experienced within K705 especially during the later stages of extraction.

Initially cut and fill mining was employed to recover the lower portion of K705 but stress related problems in the "leading" footwall orebody forced a switch to open stoping. Knowing high stress levels would be experienced, during open stoping, an extraction philosophy was devised which minimised K705 development. Extraction of K705 was safely completed in two stages which resulted in considerable seismic activity. Rock noise generated by K705 was monitored with accelerometers. As predicted, by numerical modelling, significant stress redistribution to 7 orebody abutments resulted from K705 extraction.

1 INTRODUCTION

Mount Isa Mines Limited produces over 4.5 million tonnes of silver-lead-zinc ore from both Mount Isa and Hilton mines. This ore is produced via a combination of open stoping and bench stoping methods from stratiform orebodies.

Throughout the past 25 years rock mechanics studies have played a major role in the efficient recovery of this ore. To assist in this process Mount Isa Mines employs a number of rock mechanics engineers who provide specialist input to the operations. Numerical modelling which is routinely employed to assess mining sequences as proposed in production schedules is one service they provide.

This paper details rock mechanics aspects of the recovery of K705 pillar at Mount Isa which was predicted by numerical modelling to be highly stressed. This prediction, combined with a knowledge of ground conditions, allowed production personnel to formulate an extraction strategy to extract this highly stressed pillar.

2 GEOLOGICAL SETTING

The Urquhart Shale Formation contains all the known economic mineralisation at Mount Isa and Hilton. Bedding in the shale strikes North-South and generally dips at 65° to the West at Mount Isa, Figure 1. The silver-lead-zinc (lead) orebodies are disposed in a roughly en-echelon pattern both along strike and down dip (Bartrop and Sims 1990).

The lead orebodies are naturally divided into two groups:

1. The Blackstar orebodies which are wide and lie to the hangingwall (west) of the sequence. These contain 1 through to 5 orebody which tend to be high in pyrite with fair to good ground conditions.
2. The Racecourse orebodies which are narrower and lie to the footwall (east) include 5/60 through to 14/10-30 orebody with fair to poor ground conditions.

3 MINING METHODS

Generally the width of the orebodies determines the mining method to be employed at Mount Isa. Where orebodies exceed 10-12m in width, and ground conditions allow, sub-level open stoping is employed. Previously cut and fill but now a variation of it termed benching is used to extract the narrower ore.

Over the past 25 years during which multiple orebody mining has been carried out in the lead orebodies, rock mechanics studies have played an important role in understanding the behaviour of the rock mass (Lee & Bridges 1980). It was noted that mining operations in one orebody have a major influence on ground conditions in adjacent orebodies, both to the footwall and hangingwall. This inter-relationship between orebodies combined with a knowledge of the in-situ stress field has allowed sequencing of mining operations such that extraction generally

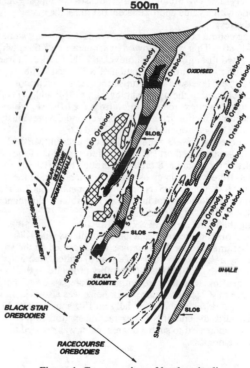

Figure 1. Cross section of lead orebodies

occurs in a low stress environment. A typical example is 7 Orebody, which has been mined extensively in the past, either as a member of the cut and fill sequence (MICAF) or as an open stoping source in tandem with 8 Orebody. Figure 2 is a long section of 7 Orebody showing areas extracted by the different mining methods. Generally extraction of this orebody has been carried out in a low stress environment but the converse was true for the northern lower open stoping block.

4 OPEN STOPING IN LOWER 7 OREBODY BLOCK

This area lies below 16B sublevel and is located between 6950N and 7135N, Figure 2. The extraction horizon for the open stopes was either 18B sublevel or 19C sublevel with the latter being located approximately 920m below surface. At this depth the virgin major principal stress has been determined to be around 35MPa. As part of the overall rock mechanics program absolute stress measurements have been carried out over the past 15 years

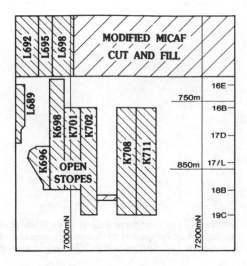

Figure 2. Longsection of 7 Orebody

to determine the virgin stress field in the lead orebodies. The results indicated that σ_1, the major principal stress, acts normal to bedding, with σ_2 and σ_3 acting down dip and along strike, respectively. The two minor principal stresses are almost equal in magnitude with a value approximately 0.7 times the major principal stress.

Mining commenced with extraction of K702 primary stope in the mid-1980's. This was followed in sequence by K698, which was extracted through to 15B sublevel, K701, K708, K711 and finally K705 pillar. During extraction of K701 and K708 stopes increased stress levels, in the form of rock noise and minor spalling, occurred in the upper sublevels, i.e. 16B and 17D. This was the result of stress redistribution from the MICAF, which commenced production from 15B sublevel, and the previously extracted 7 Orebody open stopes K702 and K698.

During the mid-1980's a mechanised cut and fill (MECAF) horizon was established on 19C sublevel to extract the footwall Racecourse orebodies i.e. 8 orebody through to 14/10-30 orebodies. Here two main crosscuts were established in the MECAF at 6751N and 7051N. On the major sublevels 7051 was the main ventilation crosscut linking all footwall development to H70 exhaust shaft, Figure 3.

During the initial stages of K711 extraction, the fifth stope in the lower 7 orebody sequence, stability problems with K70 orepass became evident. K70 orepass was accessed on the various sublevels by a short LHW(long hole winze) drilled from a tipple access just off the 7051 crosscut. Stability problems in the form of sidewall spalling, was detected within K70 orepass even in the early stages of K711 extraction. Rock noise was also evident in 7051 crosscut development especially on 18B sublevel and 17 Level. Towards the end of K711 extraction spalling in K70 orepass had increased to such a degree that it was decided to abandon the orepass.

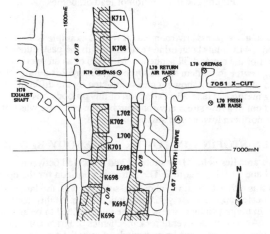

Figure 3. Plan of 17 level

5 ROCK MECHANICS OF THE LEAD OREBODIES

As part of the commitment to have rock mechanics input to routine mining operations, production schedules are evaluated to assess the effect of sequential mining. As mentioned previously mining operations in one orebody influences the ground conditions in adjacent orebodies. Hence any rock mechanics assessment of production schedules has to include a prediction of expected ground conditions. To assist in this assessment process extensive use is made of empirical and numerical models. Rock mechanics instrumentation has played a major role in the development of empirical models to assess ground behaviour. Instrumentation results have also allowed validation of numerical modelling codes which are routinely used to assess production schedules. To simulate multiple orebody extraction a numerical modelling code named NFOLD is employed (Bywater, Cowling & Black 1983). This code has been used routinely over the past 10 years at Mount Isa and elsewhere to evaluate extraction of single and multiple orebodies (Hammett & McKervey 1983).

NFOLD is a pseudo three dimensional program which is based on the displacement discontinuity method (Sinha 1979). Pre and post processing programs have been developed; the former to allow any changes in geometry, as defined in mining schedules, to be easily input and the latter to visually assess changes in stress and/or displacement. To allow the input and output of NFOLD to be easily understood pre and post processing programs produce graphical plots which conform to long sections of the orebodies.

From back analysis of documented ground problems it was noted that once the major principal stress exceeds 80MPa or when sudden stress increases of 10 to 15MPa are experienced, at any absolute stress level, the rock mass will exhibit signs of failure. NFOLD is ideally suited to simulate changes in normal stress, i.e. the stress which acts perpendicular to bedding. This allows simulation of pillar loading/unloading, as multiple orebody extraction proceeds. In this role NFOLD is routinely employed to highlight potential problem areas with proposed mining sequences.

During the mid-1980's, prior to the commencement of open stoping in the northern lower open stoping block, the mining sequence was evaluated using NFOLD (Bywater & Cowling 1985). In this analysis open stoping in 5, 7 and 8 orebodies was modelled. The results indicated that isolating K705 would cause its normal stress, especially above 18B sublevel, to rise significantly. Initial open stoping in 5 orebody resulted in only a minor increase in normal stresses.

In 1986 a block design was issued which incorporated these NFOLD results. By pushing ahead with the northern 5 orebody open stopes, alternative ventilation circuits would need to be established making K705 redundant as a regional pillar. K705 had dimensions of 140m high, 10m wide and 30m along strike which contained over 200,000 tonnes of lead ore. Further planning work in 1987 examined two options to extract K705:

Option 1. Using cut and fill by accessing from the on-going MECAF operations to the footwall.

Option 2. Continue MECAF to 15B sublevel and then recover K705 by stoping from sublevel to sublevel.

Following Option 2. the relaxed ground within K705, after 15 years of MECAF, could result in operational problems when open stoping. Hence it was decided to pursue Option 1.

6 CUT AND FILL MINING IN K705

Following completion of Lift 4 MECAF, during 1987, 7051 crosscut was extended by "ramping up" to K705 pillar, Figure 4.

When the hangingwall of 7 orebody was reached K705 was silled out. This created an opening 26m long by 10m wide and extensive cable support was installed. When Lift 5 MECAF was finished K705 was again accessed from 7051 crosscut. At this time K711 was nearing completion. There were no stress related ground problems within the silled-out portion of K705 due to the shielding effect of previous MECAF lifts (Lee & Bridges 1980).

During the break-off for Lift 6 MECAF stability problems were experienced in 14/10-30 orebody especially around 7051 crosscut. With 14/10-30 being the "leading" orebody higher stresses were expected but these combined with adverse structure around 7051 crosscut, and stress redistribution from K711, resulted in severe stability problems. At this time it was decided to review extraction of K705 using cut and fill. It was proposed that by extracting K705, as an open stope, 7051 crosscut would be "shielded" and hence no further stress related problems should be encountered. There would be two stages of extraction:

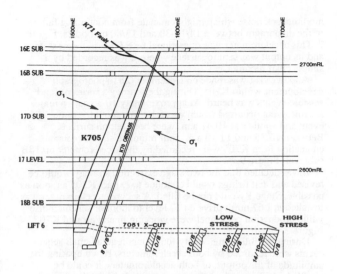

Figure 4. Cross section through K705 showing MECAF

Stage 1. Lift 6 MECAF to 18C sublevel.
Stage 2. 18C sublevel to 16B sublevel.

As part of the normal review process NFOLD was used to model the critical extraction stages of K705. It was concluded that sequential extraction of K705, at the completion of Stage 1, and subsequent extraction of the cut-off up to 16B sublevel, would result in very high stress levels. Extraction of this pillar would cause major stress redistribution to the 7 orebody abutments. Beneficial shielding of the footwall orebodies around 7051 crosscut would result from K705 extraction. Figures 5 & 6 show the normal stresses at the completion of Stages 1 and 2 in 7 and 14/10-30 orebodies. These figures are long sections showing contours of normal stress which range from a wide polka dot pattern (low stress <20MPa) to a close polka dot pattern (very high stress >90MPa). Usually colour plots are produced for in-house presentations.

Reviewing NFOLD results with production personnel it was agreed to proceed with open stope extraction. Due to the highly stressed nature of K705 the stope design called for development

within the pillar to be minimised. This meant that the 4.0m wide 7051 crosscut was to be used as the stope "ballroom" and it could not be stripped to the standard 4.6m width. Above 18B sublevel the stope was designed with drill drives mined in 6 orebody with drillholes fanned towards the footwall. On 18B sublevel a footwall drill drive had to be developed since there was no hangingwall access, Figure 7. The cut-off raise was to be a single raisebored hole down the hangingwall from 16B sublevel to Lift 6 MECAF. Main rings were designed with an average burden of 3.4m. With 7051 crosscut not being central in the pillar only two main rings were designed south of the cut-off, Figure 8. Additional ground support would be installed in all development to contain the expected increase in stress levels. If stress related ground problems were to occur mining operations would be suspended until conditions improved.

7 PREPARATIONS FOR OPEN STOPING K705

Cut-off drilling commenced early in 1989 but problems arose while raiseboring the final section of the raise between 17D and 16B sublevels. It was decided to abandon raiseboring and to drill a LHW to complete the cut-off raise. A meeting was held to finalise the extraction strategy and the main points are summarised here:

1. Cut-off drilling to be completed on all sublevels before firing commences.
2. Stage 1 to be filled before Stage 2 firing commences.
3. Main rings for Stage 2 not to be drilled before Stage 1 firing completed.

It was decided to cablebolt intersections of 7051 crosscut with 8 orebody development as well as drill drives in 6 orebody since they would suffer large stress re-adjustments during K705 extraction. Ten metres above 16B sublevel K71 fault, a shallow ESE dipping structure, intersected 7 orebody. It was postulated that movement of K71 fault would cause crown instability during K705 extraction. Cablebolts were installed down through the floor of 7051 crosscut on 16E sublevel to intersect K71 fault.

8 OPEN STOPING WITHIN K705

8.1 Extraction of Stage 1

This commenced, from Lift 6 MECAF to 18C sublevel, with cut-off firings at the end of October 1989. Main ring firings for Stage 1 were completed on the 21st of November, Figure 8.

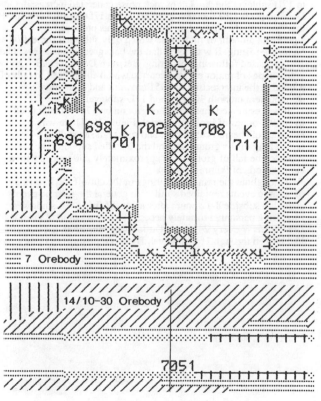

Figure 5. Normal stress at completion of Stage 1

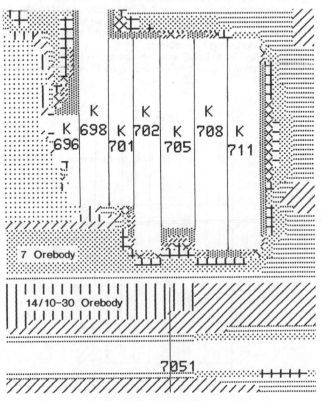

Figure 6. Normal stress at completion of Stage 2

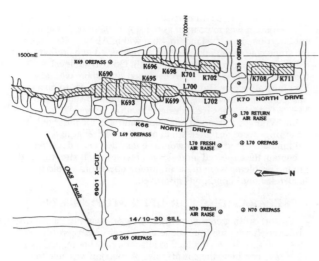

Figure 7. Plan of 18B sublevel

Minor rock noise was encountered in the upper sublevels of K705 but overall little evidence of major stress re-adjustment was observed.

On December 4th at 7.10 p.m. a seismic event occurred which was felt both underground and on surface. A ground support crew working in 14/10-30 orebody, at approximately 6920N, in the MECAF reported that their rockbolting rig had been shaken by movement of the O68 fault. Other crews working in the vicinity of the O68 fault, a major fault structure intersecting the MECAF area, also reported hearing the event. Rock noise within K705 was also reported at this time.

The following day a visit was made to the MECAF. O68 fault was just to the south of the rockbolting rig and an inspection of this fault revealed that no displacement had occurred. With the diaphragm pillar between 13/80 and 14/10-30 orebodies being at the critical 1:1, height to width, ratio it was concluded that bedding plane slippage within the pillar had now occurred (Lee & Bridges 1980).

K705 was also visited to investigate reports of rock noise. On 18B sublevel no rock noise was evident and both the 7051 crosscut access and drill drive were in good condition. Above on 17 level it was observed that the sidewalls of some cut-off drillholes were beginning to spall. Further indications of high stress were minor spalling within the raisebore hole as well as from the bottom corners of the ballroom. Initial signs of spalling were also evident on 17D sublevel.

It was decided to install an accelerometer on 18B sublevel, at the intersection of 6901 crosscut and 14/10-30 orebody, to

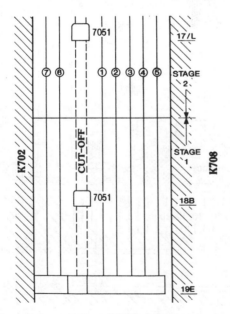

Figure 8. Long section of K705 showing main rings

monitor rock noise which might emanate from continuing failure of the diaphragm between 14/10-30 and 13/80 orebodies, Figure 7. This accelerometer was operational by 8.00 a.m. the next day and its output was sent to surface where it was recorded by a simple chart recorder.

Another visit was made on December 6th to photograph development within K705. During the journey a series of loud seismic events was heard. At approximately 10.35 a.m. a major seismic event occurred which was felt throughout the mine. This event and another at 10.41 a.m. were both felt on surface. Between 10.35 and 11.10 a.m. four major seismic events emanating from K705 were recorded by the accelerometer on 18B sublevel, Figure 9.

It was decided that the two stage extraction strategy would be revised and that firings would continue to extract K705 as soon as possible; Stage 1 would not be filled. A second accelerometer was installed in L67 north drive on 17 level to monitor seismic noise while production personnel were working in and around K705, Figure 3.

During the following twenty four hours regular small seismic events were recorded by both accelerometers. By comparing the amplitude of the output, of both accelerometers, it could be established when major events were recorded i.e. full scale deflection. On December 7th a visit was made to K705 and it was noted that significant deterioration had occurred in the ballroom on 17 level. There was no apparent change in and around K705 on 17D sublevel. During cable bolt drilling, on December 9th, personnel reported hearing a series of loud rock noises on 16B sublevel which were audible above the sound of drilling. It was also observed that the back was spalling while collaring holes for cablebolts. Checking the accelerometer output it was confirmed that a series of seismic events had occurred. By December 15th cablebolt installation was complete on 16B sublevel and all sublevels had been ring drilled. Throughout this period numerous small seismic events were reported but drilling noise masked most of the recorded events.

8.2 Extraction of Stage 2

Restricted firing of the cut-off for Stage 2 commenced on December 16th. Sporadic seismic events were recorded as the cut-off was opened up to 17 level. The final firing of 17 level cut-off, combined with the first firing in the cut-off to 17D sublevel, produced considerable rock noise which took over four hours to subside. It was noted that a large wedge had been displaced from main ring #6 below 17 level. As the cut-off firings proceeded up to 17D sublevel the amount of rock noise generated increased. On December 23rd the remaining collars in the 17D cut-off were fired together with the first cut in 16B LHW. This produced a significant amount of rock noise which only subsided some 12-18 hours after firing. It was noted that the background noise level had also increased following this firing. Between December 23rd and 26th a series of major events occurred which culminated in falls of ground at the intersection of 7051 crosscut and 8 orebody development on both 17 level and 17D sublevel.

It was estimated that the fall of ground on 17 level contained over 500 tonnes of broken material, Figure 10. There was evidence of sheared cablebolts, and rockbolts, as well as unravelling of the ground around the installed support. On 17D sublevel the fall of ground was approximately 200 tonnes with a similar mode of failure.

Throughout the remaining firings of the cut-off the frequency of seismic events remained similar. In general the seismic events started to subside 8-12 hours after a cut-off firing. This allowed ring firing personnel to safely access 7051 crosscut to load and fire the LHW every 24 hours. On January 5th, 1990, the cut-off had broken through 16B sublevel. By this time most of main rings

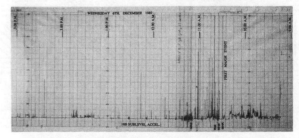

Figure 9. Accelerometer trace December 6th, 1989

#1 and #6 had disappeared and only 70% of ring #7 remained to the south of the cut-off.

On January 9th K705 was mass fired at 4 p.m. and a significant amount of rock noise was generated which culminated in a major seismic event approximately 4 hours later that was felt 5km from the mine, Figure 11. Within 12-18 hours seismic noise had reduced to background levels and by the following day only muck drawdown was triggering the 17 level accelerometer.

On January 10th a visit was made to investigate reports of damage caused by K705 mass blast. On 17 level there was little change in 7051 crosscut. An extensive fracture appeared in the footwall access to K695 in 8 orebody. L67 north drive suffered hangingwall relaxation for 30m south of 7051 crosscut. On 17D sublevel the crown pillar above K696, in 7 orebody, exhibited signs of floor heave as well as damage to an adjacent brick bulkhead. 7051 crosscut showed further deterioration in the vicinity of the previous fall of ground. In 7051 crosscut on 16B sublevel an open fracture 5cm wide appeared on the footwall of K705. The crown pillar area above K696 also indicated signs of floor heave on 16B sublevel. K71 fault had failed above the crown of K705. On 16E sublevel it was found that K71 fault had been displaced up to 5mm in 7051 crosscut. Finally 15 level haulage experienced buckling of railway lines in the vicinity of K696 crown pillar.

From an assessment of damage, following the K705 mass blast, it was concluded that the stress carried by K705 had been shed to the abutments. The area above K696 suffered extensive damage as did the crown pillar above K705. Comparing the outcome of the mass blast with the previously completed NFOLD modelling confirmed the stress transfer to these areas.

Mucking of K705 continued over the following two months with no stress related ground problems reported around 7051 crosscut. On completion of mucking K705 was backfilled, with aggregate sized material, directly from surface.

9 CONCLUSIONS

K705 pillar was predicted, using numerical modelling, to be a highly stressed pillar. This prediction assisted in the determination of an extraction sequence which ultimately led to the safe recovery of K705. The deterioration in ground conditions within K705 was expected but not at the completion of Stage 1 extraction. At this time K705 pillar must have been on the edge of the stable/unstable region. The extensive ground support installed throughout K705 performed well but failures at the intersection of 7051 crosscut with 8 orebody, on both 17 level and 17D sublevel, were a

concern. In hindsight a combination of shotcrete and heavy cable support could probably have prevented these falls of ground. The use of accelerometers to monitor rock noise was an integral part of the monitoring program which provided immediate feedback on ground conditions, existing within K705, to operating personnel. The two accelerometers performed well throughout but there was no indication of where or how large the seismic events were. A small microseismic system, with sensors located around K705 on the various sublevels, would have allowed source location/magnitude calculations to be carried out.

ACKNOWLEDGMENTS

The author would like to thank the management of Mount Isa Mines Ltd. for permission to publish this paper.

REFERENCES

Bartrop, S.B. & Sims, D.A. 1990. Mining geology for silver-lead-zinc orebodies at Mount Isa. Aust. Inst. Min. & Met. Mine Geologists Conference, Mount Isa: 129-135

Bywater, S. & Cowling, R. 1985. Application of stress analysis techniques to planning and scheduling of Pb/Zn/Ag orebodies at Mount Isa Limited. A.M.I.R.A. Annual General Meeting, Brisbane.

Bywater, S., Cowling, R. & Black, B.N. 1983. Stress measurement and analysis for mine planning. Proc. 5th Intnl. Cong. on Rock. Mech., I.S.R.M.: D29-D37

Hammett, R.D. & McKervy, G.W. 1983. Application of three-dimensional stress analysis to optimising pillar stability and mining layout at Copperhill, Tennessee. SME-AIME Annual Meeting.

Lee, M.F. & Bridges, M.C. 1980. Rock mechanics of crown pillars between cut and fill stopes at Mount Isa mine. Proc. Appl. of Rock. Mech. to Cut & Fill Mining, I.S.R.M.: 316-329

Sinha, K.P. 1979. Displacement discontinuity technique for analysing stresses and displacements due to mining in seam deposits. Ph.D. Thesis, Univ. of Minnesota. (unpublished)

Figure 10. Fall of ground 7051 crosscut 17 level

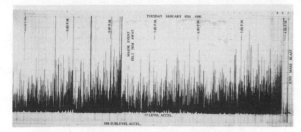

Figure 11. Accelerometer trace of K705 mass blast

Rockbursts and Seismicity in Mines, Young (ed.) © 1993 Balkema, Rotterdam, ISBN 90 5410 320 5

Mining in the vicinity of geological structures – An analysis of mining induced seismicity and associated rockbursts in two South African mines

N.C.Gay
Chamber of Mines Research Organisation of South Africa, Johannesburg, South Africa

ABSTRACT: This paper is concerned with seismic events and associated rockbursts which occurred during a period of 15 months on two mines, one in the Carletonville district and the other in the Klerksdorp district. The frequency of seismic events in the Carletonville mine is 7-8 times that in the Klerksdorp mine, although proportionately more large events (Magnitude > 3) occur in the Klerksdorp mine. In both data sets, the average frequency of rockburst occurrence is 1 rockburst for every 7-8 seismic events with M > 0. Criteria, such as production days lost and length of stope face closed, are used to evaluate the damage caused by rockbursts. The greatest damage results from seismic events of Magnitude 2-3. The nature of the damage is such that it could be readily prevented by utilizing better support systems.

1   INTRODUCTION

Seismicity has been associated with mining activities in South African Witwatersrand gold mines since the early 1900's. Initially this seismicity could not be associated with any particular mining activity except for a tenuous correlation with failure of support pillars. However, as mining went deeper, it became clear that dykes were particularly prone to rockbursting. (Krause, 1933).

The first detailed attempt to correlate rockburst occurrence and geology quantitatively was made by Hill (1954) on the East Rand Proprietary Mine (ERPM) between 1948-1953. During this 5 year period, 641 rockbursts occurred on the mine, of which, 471 were located in the Central Section with 122 being in stopes where the whole, or a portion, of the face was in dyke. This indicates a frequency of rockbursting 2,5 times greater than in stopes free of dyke.

A similar exercise carried out in the late 1950's confirmed this correlation (Cook et al 1966). However, Jeffery (1975), who also worked at ERPM, was unable to establish a statistically significant increase in rockburst occurrence when mining through dykes or faults, except when mining took place within 20 m of one particular 40 m wide dip dyke. Seventeen per cent of the rockbursts which occurred during a period of 7½ years were located close to this dyke. Dykes sub-parallel to the strike of the bedding were mined out virtually without giving rise to rockbursts.

Cook (1962) and Joughin (1966) also pioneered the use of close-in seismic networks to monitor seismic events and associated rockbursts. This work showed that most seismic events did not cause rockbursts, although the likelihood of a rockburst occurring increased with seismic event magnitude.

It is the objective of this paper to evaluate the association between rockbursts, i.e. seismic events which cause damage to mine excavations, and seismicity accompanying the two main mining methods utilized in the deep South African mines, namely scattered mining and longwall mining. Attention is also given to the influence of geological structures on rockbursting, particularly dykes, faults and joints, and quantification of the resultant damage.

2   SEISMICITY AND ASSOCIATED ROCKBURSTS

In this section the seismic and rockburst history over a period of 15 months is analysed for two mines with approximately equivalent production rates, one in the Klerksdorp district, where scattered mining predominates, and the other in the Carletonville district where longwall mining is practised. The depth at which mining took place was similar in both mines, i.e. 1900 - 2400 m.

The data used for the analysis was obtained from reports of rockburst damage at each mine as well as information provided by the regional or mine-wide seismic networks. Table 1 summarizes these data and indicates the frequency of rockbursting for the various ranges of seismic event magnitude.

The data in Table 1 show that about 7-8 seismic events occur in the relatively high ERR longwall mine for every one event in the low ERR scattered mine although the proportion of large ($M_L > 3$) events in the scattered mining situation is approximately 5 times that in the longwall situation. This is due to the presence of large faults, which necessitate scattered mining in the Klerksdorp district, but which also are associated with the largest seismic events.

57

Table 1   A comparison between seismicity and rockbursting in a scattered and a longwall mine.

	CARLETONVILLE MINE (Longwall)			KLERKSDORP MINE (Scattered)		
Magnitude range	No. of events	No. of rockbursts	%	No. of events	No. of rockbursts	%
0-0,9	569	2	0,3	65	2	3,1
1-1,9	746	83	11,1	99	7	7,1
2-2,9	286	95	33,2	42	10	23,8
3-3,9	21	16	76,2	15	11	73,0
> 4	0	0	-	1	1	100,0
Total	1 622	196	12,1	222	31	14,0

Comparing the occurrence of rockbursts in each sample, it appears that, in both mining configurations, one rockburst occurs for every 7-8 seismic events of magnitude > 0. Moreover, the percentage of events which result in rockbursts in a given magnitude range is approximately the same. However, there are significant differences, in the longwall mining sample, 42%, 40% and 8% of all the rockbursts result from events of magnitude 1-1,9; 2-2,9 and 3-3,9 respectively, compared to 23%, 32% and 35% for the scattered mining sample. This confirms that in the Klerksdorp area a disproportionately large number of very large events and corresponding rockbursts occur. In both configurations about 75 per cent of events of magnitudes greater than 3 result in rockbursts.

The frequency of occurrence of seismic events and rockbursts in the two mines during the 15 month sampling period is plotted in Figure 1 against magnitude. In the longwall mine, 90 per cent of the rockbursts result from moderate sized events (M_L = 1 - 2,9) while in the Klerksdorp scattered mining sample, most rockbursts (70 per cent) are associated with larger seismic events (M_L > 2).

3   THE INFLUENCE OF GEOLOGICAL STRUCTURES ON ROCKBURST FREQUENCY AND DAMAGE

To assess the influence of geological structures on rockburst frequency and damage each rockburst was identified from the rockburst damage reports as being associated solely with either mining activity or with mining close to a fault or a dyke. Damage was estimated using criteria such as length of stope face or tunnel closed and production days lost.

3.1   Longwall Mining

In the longwall mining sample, an estimate of production days lost, based on an early assessment of damage was used by mine personnel to quantify damage. On this basis the total number of days lost for the sample of 196 rockbursts was 1 626. Figure 2 shows the days lost with the type of rockburst (mining, dyke, fault, etc. as reported by the mine overseer) and the magnitude of the causative seismic event.

About 90 per cent of mining took place in stopes free of easily observed geological structures and about 70 per cent of rockbursts occurred in these areas with an average damage rate of nine days production loss per rockburst. Dyke associated rockbursts were next most common, while fault rockbursts were fewer and caused least damage. Most rockburst damage resulted from seismic events of magnitude 2-3.

The data in Figure 2 were not normalized with respect to the amount of mining in the area where the rockbursts occurred, since this information could not be readily obtained. However, an indication of the importance of such a correction can be obtained from the rockburst records for the whole mine. Figure 3 shows the cumulative percentage increase in number of rockbursts, damage and area mined, plotted against distance from a geological

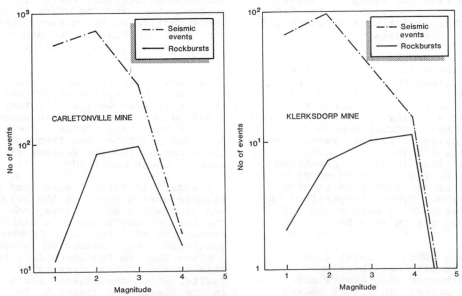

Figure 1   Frequency - magnitude curves of seismic events and rockbursts in two Klerksdorp and Carletonville mines.

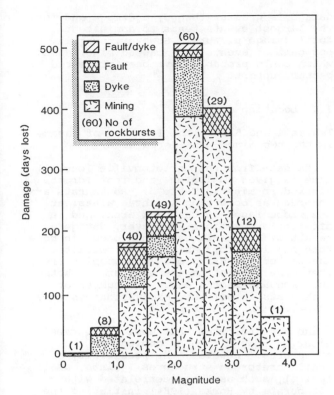

Figure 2   Correlation between rockburst damage and magnitude of seismic events for various types of seismic events or rockbursts - Carletonville mine.

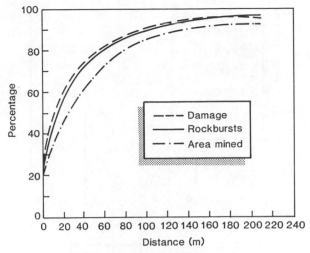

Figure 3   Cumulative frequency curves of the variation in rockburst occurrence, damage, and area mined with distance away from a geological structure. Carltonville Mine (After Spottiswoode, 1986).

structure. There is a good correlation between rockbursts and damage and more than half (57 per cent) of the rockbursts occur within 20 m of a geological structure although only 45 per cent of mining takes place in this area.

3.2 Scattered Mining

The rockburst frequency in the scattered mine was normalized with respect to area mined, since the total area mined during the sample period was known. Table 2 summarizes the results obtained from this exercise.

The table clearly shows that rockbursts occur far more frequently close to geological structures than in areas distant from such structures.

Table 2   Frequency of occurrence of rockbursts in geologically disturbed and undisturbed ground

Associated Structure	Number of Rockbursts	Frequency (rockbursts/1 000 m²)
Fault	14	0,05
Fault/Dyke	7	0,03
Dyke	5	0,02
All structures	26	0,04
Mining (no structure)	5	0,01

The criterion used for assessing damage due to rockbursts in this scattered mining area is based on a method devised by Hepworth (1984). This criterion considers not only days lost, but also length of face or tunnel affected, amount of closure, and performance of support units. The damage rating varies from 0-8 for relatively slight damage such as that due to seismically induced falls of ground, and from 8-16 for more severe, dynamic damage.

Table 3 and Figure 4 summarize the results of the analysis of damage due to rockbursts. The data include not only the basic set of data of seismic events and

Table 3   Klerksdorp mine rockburst - damage correlation for seismic events of various magnitude and different structures (N - number of rockbursts, D.R. - damage rating).

Magnitude	Fault		F/D		Dyke		All Structures		Mining		Total Average	
	N	D.R	N	D.R	N	D.R	N	D.R	N	D.R	N	D.R
0 - 0,5	-	-	-	-	2	10	2	10	-	-	2	5,0
0,6 - 1,0	2	12	-	-	1	8	3	20	2	6	5	5,2
1,1 - 1,0	1	4	1	6	3	16	5	26	4	12	9	4,2
1,6 - 2,0	3	28	3	20	1	10	7	58	4	20	11	7,1
2,1 - 2,5	3	22	1	8	1	10	5	40	5	44	10	8,4
2,6 - 3,0	2	18	3	26	2	18	7	62	2	18	9	8,9
3,1 - 3,5	7	36	-	-	-	-	7	36	7	54	14	6,4
3,6 - 4,0	4	42	-	-	-	-	4	42	1	10	5	10,4
> 4,0	-	-	-	-	-	-	-	-	1	12	1	12
Total Ave D.R	22	7,4	8	7,5	10	7,2	40	7,4	26	6,8	66	7,1

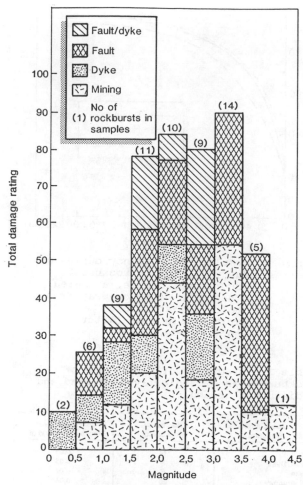

Figure 4 Correlation between rockburst damage and magnitude of seismic events for various types of seismic events or rockbursts - Klerksdorp mine.

damage is greater than dyke-damage and for the largest events tends to dominate the total damage picture. However, overall, 62 per cent of events only caused minor damage which could probably have been prevented by better support.

3.3 Location of damage

Tables 4 and 5 list the location of damage in the two mines considered here.

The data from the Carletonville longwall mine is given in Table 4 and that for the Klerksdorp mine in Table 5. Rockbursts are recorded as occurring in three areas: at the stope face, in the back area, and in off-reef excavations. However, damage as a result of a seismic event can occur in more than one locality, and this is reflected in the table; for example in the stoping area, where the statistics are similar for both the face and back areas. The number of rockbursts in off-reef excavations is very much less than in stopes.

Table 4 indicates that most rockbursts occur in stopes free of geological structures. However, the possibility of small unrecorded structures (joints, small faults), such as those correlated with rockbursts by Rorke (1984) initiating the rockbursts, cannot be discounted.

Table 5 summarizes the data obtained from the Klerksdorp scattered mining district. Most rockbursts occur in the stope area. However, if all structures are considered, the likelihood of a rockburst occurring in an area close to a geological structure is greater than that in an area far from a geological structure. This conclusion is reinforced when it is considered that most mining takes place in areas free of geology.

4 DISCUSSION

Most seismic events occurring in the mines considered here do not result in rockbursts. However, the proportion of events which cause rockbursts, increases sharply with event magnitude. This appears to be true in all seismogenic mining situations, although the number of seismic events, and rockbursts, is much greater in deep longwall mining operations than in scattered mining operations at equivalent

associated rockbursts, but also some additional rockbursts which occurred outside the initial 15 month period.

Most rockbursts are associated with geological structures. The average damage rating appears to increase with the size of the seismic event and damage due to rockbursts associated with geological structures, is slightly greater than that due to mining rockbursts. Fault-related

Table 4 Rockbursts in the Carletonville longwall mining sample, with respect to location, geological structure, and event magnitude.

| Location | Magnitude | Number and type of rockburst | | | | |
		Mining	Fault	Dyke	Fault/Dyke	All structures
Face area	0 - 0,9	-	-	3	-	3
	1 - 1,9	46	16	9	2	27
	2 - 2,9	65	8	10	3	21
	3 - 3,9	13	2	2	-	4
Back area	0 - 0,9	-	-	3	-	3
	1 - 1,9	43	13	8	2	23
	2 - 2,9	60	8	12	3	23
	3 - 3,9	11	2	2	-	4
Off-reef	0 - 0,9	-	-	1	-	1
	1 - 1,9	7	3	4	1	8
	2 - 2,9	15	1	3	1	4
	3 - 3,9	1	-	1	-	1

Location	Magnitude	Number and type of events				
		Mining	Fault	Dyke	Fault/Dyke	All structures
Stopes	0 - 0,9	1	1	2	-	3
	1 - 1,9	5	2	4	1	7
	2 - 2,9	11	3	2	3	8
	3 - 3,9	8	8	4	3	15
	> 4	2	2	-	-	2
Gullies	0 - 0,9	-	-	-	-	-
	1 - 1,9	2	-	1	1	2
	2 - 2,9	2	2	-	-	2
	3 - 3,9	5	2	3	-	5
	> 4	1	2	-	2	4
Haulage X-cuts	0 - 0,9	1	1	2	-	3
	1 - 1,9	1	-	2	-	2
	2 - 2,9	3	4	3	-	7
	3 - 3,9	6	6	2	1	9
	< 4	3	3	-	1	4

depths. This is presumably due to the higher ERR levels associated with the larger spans in typical longwall mining situations, and also reflects the increasing frequency of stabilizing pillar foundation failures.

With respect to geological structures and the occurrence of seismic events and rockbursts, the data given here agree with previous results reported by Gay et al (1984) for the Klerksdorp district and by Lenhardt (1988, 1992) for the Carletonville area. Dykes are identified as the dominant seismic structures in both districts with faults also being significant. Small faults or geological joints across which small displacements have occurred have also been identified as foci for events causing rockbursts by Rorke (1984).

Faults and dykes are both prone to rockbursting when mining is carried out in their close proximity. In the Carletonville district the critical distance from a major structure is 20 m (Fig. 2). Jeffrey (1975) reported a similar distance for rockbursting close to dykes at ERPM and noted that the frequency of fault-associated rockbursts was half that of dykes.

Quantitative information of this nature, together with information on the likelihood of large seismic events is required for the successful implementation of rockburst control strategies in geologically disturbed ground, particularly with respect to mine layout design, extraction sequences and the implementation of rockburst resistance support systems for the stope face area. However, a major question which still has to be answered is why do some seismic events cause rockbursts but most do not?

The answer requires reliable information on the state of stress and rock deformation properties. In-situ stress measurements have shown that in the Klerksdorp area abnormally high horizontal stresses can exist (Gay and van der Heever, 1982), suggesting that anomalous tectonic strain energy may be stored in the rocks adjacent to faults. Evidence for anomalously high stresses has also been found in a dyke (Gay, 1979). The high horizontal stresses

should assist in clamping fault planes. However, the variability in virgin stress levels may also be high in geologically disturbed ground and localized areas of low clamping stresses could exist in places. For example, stress measurements in the vicinity of a large fault in the Orange Free State gold mining district indicated a state of unstable equilibrium close to the fault.

Because faults extend over very large distances, movement on the fault plane can result in the release of strain energy over large areas, causing damage wherever the fault intersects an excavation. Similarly, dykes also traverse large distances and because of their different elastic properties from quartzites and their mode of emplacement, may also be highly stressed and, when disturbed by mining, can release strain energy violently over large areas of the dyke.

Methods of identifying inherently unstable geological and mining situations need to be developed, as do mining procedures to reduce the likelihood of sudden failure occurring. These methods should include routine in situ stress measurements, monitoring and analysis of seismic activity, back-analysis of previous rockburst occurrences, utilizing all available seismic information, and, the use of suitable modelling procedures such as Excess Shear Stress (Ryder, 1988) and Volume Excess Shear Stress (Spottiswoode, 1990). However, both these numerical models still require calibration to ensure their realistic application.

5   CONCLUSIONS

i)   Most seismic events in both longwall mining and scattered mining situations do not cause rockbursts. However, the frequency of rockbursts increases with event magnitude.

ii)   The frequency of rockbursts increases when mining takes place close to geological structures.

iii)   Damage to mining excavations during a rockburst in general increases with the magnitude of the associated seismic event.

iv) Quantification of damage using suitable criteria (e.g. Hepworth, 1984) shows that 100 - 1000 fold increases in seismic energy in large events only lead to increases of +/- two units of Hepworth's damage index. i.e. a large event is more likely to cause damage over a larger area, but not more intensive damage. This indicates that a lot of rockburst damage can be reduced by installing better energy absorbing support systems, which also provide good area coverage.

v) Because of the general unpredictability of rockburst-prone situations, methods for identifying inherently unstable geological and mining situations need to be established. These methods will include studies of previous rockbursts, stress and seismic analyses, numerical analyses of the current mining situations, monitoring of seismic activity, and information on the prevailing in situ stress state.

6   ACKNOWLEDGEMENT

The work described in this paper forms part of the rockburst research programe carried out by COMRO on behalf of the South African Mining Industry. John Ryder, Steve Spottiswoode and Tony Jager are thanked for constructive reviews.

7   REFERENCES

Cook, N.G.W. 1962. A study of failure in rock surrounding underground excavations. PhD. Thesis, University of the Witwatersrand Johannesburg.

Cook, N.G.W., Hoek, E., Pretorius, J.P.G., Ortlepp, W.D., and Salamon, M.D.G. 1966. Rock mechanics applied to the study of rockbursts. J. S. Afr. Inst. Min. Metall., 60: 435-538.

GAY, N.C. 1979. The state of stress in a large dyke on ERPM, Boksburg South Africa. Ist. J. Rock Mech. Mn. Sc. and Geomech. Abstr., 16: 179-189.

GAY, N.C., and VAN DER HEEVER, P.K. 1982. In-situ stresses in the Klerksdorp gold mining district, South Africa. Proc. 23rd U.S. Symp. on Rock Mechanics. A.I.M.E, 176-182.

HEPWORTH, N. 1984. Personal communication.

HILL, F.G. 1954. An investigation into the problem of rockbursts. An operational research project. Part I: J. Chem. Metall. Mining Soc. S. Africa, Oct.: 63-72.

JEFFERY, D.G. 1975. Structural discontinuities in the Witwatersrand group on the ERPM Mine: their geology, geochemistry and rock mechanics behaviour. M.Sc Dissertation, University of the Witwatersrand.

JOUGHIN, N.C. 1966. The measurement and analysis of earth motion resulting from underground rock failure. Ph.d Thesis. University of the Witwatersrand.

KRAUSE, H.L. 1933. Contribution to discussion on the paper by B.T. Alston. IN: Some aspect of deep level mining on the Witwatersrand gold mines with special reference to rockbursts. Assoc. Mine Managers Tvl: 79-81.

LENHARDT, W.A. 1988. Some observations on the influence of geology on mining induced seismicity at Western Deep Levels Limited. Rock Mechanics in Africa, SANGORM, S. Africa: 45-48.

LENHARDT, W.A. 1992. Seismicity associated with deep level mining at Western Deep Levels Limited. J. S.Afr. Inst. Min. Metall. 92: 113-120.

RORKE, A.J. 1984. A seismically orientated study of mining induced fracturing around a deep level gold mine stope. M.Sc Dissertation, Rand Afrikaans University.

RYDER, J.A. 1988. Excess shear stress in the assessment of geologically hazardous situations. J. S Afr. Inst. Min. Metall. 88: 27-39.

SPOTTISWOODE, S.M. 1986. Personal communication.

SPOTTISWOODE, S.M. 1990. Volume excess shear stress and cumulative seismic moments. Rockbursts and seismicity in mines. Fairhurst (Ed). Balkema, Rotterdam: 39-43.

Rockbursts and Seismicity in Mines, Young (ed.) © 1993 Balkema, Rotterdam, ISBN 90 5410 320 5

A practical engineering approach to the evaluation of rockburst potential

Denis E. Gill, Michel Aubertin & Richard Simon
École Polytechnique, Montréal, Que., Canada

ABSTRACT: This paper presents a methodology to evaluate the rockburst potential of underground excavations. It is based upon routine mining engineering, and makes use of rock mechanics tools such as zoning, geomechanical classifications, intact rock properties and numerical stress analysis. It mainly consists in performing stability analysis for the rock structure (or/and of any nearby major geological discontinuity) under consideration; if the rock structure (or the nearby discontinuity) is found to be unstable, then the methodology gives means of determining if this instability will be gradual or sudden. In order to show how to use the methodology, a back analysis of a specific rockburst case is presented and discussed.

1 INTRODUCTION

For many decades, rockbursts have been one of the major problems for the mining community all around the world. Rockbursting is usually associated with deep underground openings in hard rock, but it can also occur in open pits or in soft rock mines, tunnels, and other underground excavations (Brown, 1984). In Canada, the problem is encountered in many hard rock mines, especially those going deeper than about 500 to 1000 meters. The problem has received much attention, particularly in the Province of Ontario where it is more acute because of the size and depth of the mine openings. For large mining operations, various approaches have been used to deal with rockbursts, such as seismic monitoring techniques, which have retained the attention of much of the scientific and engineering communities in recent years (Hedley, 1991, 1992).

But rockbursting can also be a problem for small mines, such as those found in the Abitibi region, in the northwest part of the Québec Province (Mottahed, 1992). Because of the relatively small size of these mines and their limited financial capabilities, the rockburst "know-how" developed elsewhere is not always applicable. Furthermore, because rockbursts are less frequent, appropriate care has not always been paid to the problem before a first event has occurred.

Convinced that proper engineering design of rock structures should include the evaluation of its rockburst potential, Gill and Aubertin (1988) - see also Aubertin et al. (1992) - have extended a methodology presented in Gill (1982) lecture notes, which allows to achieve this objective. The purpose of this paper is to present, with some details, this methodology, which can be easily integrated to routine mining engineering. So as to show how to use the methodology, a back analysis of a specific rockburst case is also presented and briefly discussed.

2 CLASSIFICATION OF ROCKBURSTS

A rockburst is a sudden rock failure characterized by the breaking up and expulsion of rock from its surroundings, accompanied by a violent release of energy (Blake, 1972). According to Salamon (1983), there can be three sources for this release of energy:

i) The strain energy stored in the surrounding rock mass;
ii) The change in the potential energy of the rock mass;
iii) The slippage along rock wall contact.

Rockbursts induced by mining are associated to unstable equilibrium states which may involve (Brown, 1984):

i) Slip on pre-existing discontinuities;
ii) Fracturing of the rock mass.

This leads to the definition of two broad classes of rockbursts:

i) Type I rockbursts, resulting from fault-slip events;
ii) Type II rockbursts, resulting from the failure of the rock mass itself, these including the so-called strain bursts and pillar bursts.

These two classes are considered in the proposed methodology, presented later.

3 ROCKBURST MECHANISM

Creating a new underground opening or modifying an existing one causes stress changes in the surrounding rock mass. These stress changes can induce failure of the rock mass about the opening, or trigger a fault-slip on a pre-existing discontinuity.

In the late fifties and early sixties, various authors have proposed a simple analogy between the violent failure of a rock sample in a soft-testing machine and the rock dynamic fracture during a rockburst (e.g., Petukhov, 1957; Cook, 1965; Diest, 1965). With this analogy, the rock specimen under a uniaxial compression state of stress acts as the fractured rock and the loading system, as the surrounding rock mass. Figure 1 presents the analogy which holds for type II rockbursts; here K_r is the pre-peak stiffness of the rock, K_r' is its post-peak stiffness and K_{ls} is the loading system stiffness. If the value of K_{ls} (Figure 1a) is less than the value of K_r' (in absolute values), the amount of stored strain energy in the loading system is larger than the amount of energy that the rock can dissipate in its post-peak

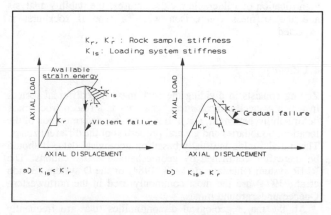

Figure 1. Influence of the relative stiffness of the loading system and the rock sample for an uniaxial compressive test. a) Violent failure. b) Gradual failure. (after Cook, 1965).

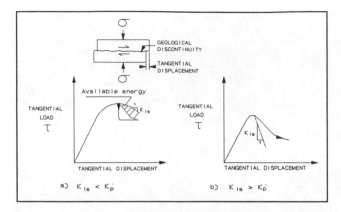

Figure 2. Influence of the relative stiffness of the loading system and the discontinuity for a direct shear strength test. a) Violent failure. b) Gradual failure. (after Salamon, 1974).

phase. When such a condition is reached, the equilibrium state becomes unstable and the failure of the rock sample is violent. Otherwise, when the absolute value of K_{ls} is higher than the absolute value of K'_r, the equilibrium state is stable and the failure is gradual (Figure 1b). This analogy has been integrated in the local mine stiffness coefficient approach for predicting stability of mine pillars (e.g., Starfield and Fairhurst, 1968; Starfield and Wawersik, 1968; Salamon, 1970).

A similar analogy to that of Figure 1 has also been proposed by Salamon (1974) to explain the mechanism for type I rockbursts. Figure 2 shows two tangential load-tangential displacement curves for direct shear strength tests on a discontinuity, with a low normal stress. Here K_p is the shear plane pre-peak stiffness, K'_p, is its post-peak stiffness, and K_{ls} is the loading system stiffness. Again, if the absolute value of K'_p is larger than the absolute value of K_{ls}, failure produces a sudden and violent release of energy (Figure 2a). Otherwise, the equilibrium state is stable and failure is gradual (Figure 2b). Obviously, the application of this model to in situ conditions necessitates the knowledge of the loading system (rock mass) stiffness. Moreover, the complex normal stress-shear displacement-normal displacement-shear strength behaviour of the discontinuity (e.g. Goodman, 1980; Fortin et al., 1990) needs also to be known. Unfortunately, this aspect of the problem is still partly unsolved.

4 PROPOSED METHODOLOGY

The methodology presented here was developed to evaluate the rockburst potential of underground excavations, starting from routine mining and ground control engineering. As shown in the diagram of Figure 3, it includes up to four steps: the zoning, the identification of vulnerable rock structures, the stability analysis and the stiffness comparison when a type II rockburst is expected.

4.1 Zoning

Zoning consists in dividing the rock mass into different sectors in which a specific mechanical behaviour is foreseen (rock mass deformability, strength, etc.). It includes the determination of the location, boundaries and general properties of the different zones. The initial zoning is usually based on geological data. It should be thereafter confirmed by geomechanical classifications. The RMR system (Bieniawski, 1973, 1984) or the Q system (Barton et al., 1974) are the most commonly used in the northwestern Québec underground mines.

All the major geological discontinuities (that are frequently delineating the different zones) must also be identified and located at this stage.

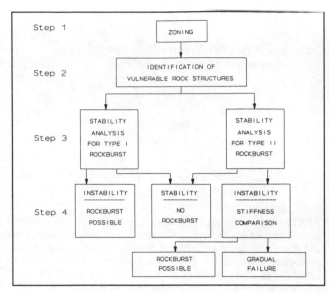

Figure 3. Diagram showing the proposed rockbursts potential evaluation methodology (after Gill and Aubertin, 1988)

4.2 Identification of vulnerable rock structures

The methodology considers three broad categories of potentially vulnerable rock structures, namely:

i) An excavation that approaches a major geological discontinuity, as shown in Figure 4a. Here, the stress changes induced by the excavation can increase the shear stress and/or reduce the normal stress along the discontinuity. Any of these can provoke a shear-slip, which can result into a type I rockburst;

ii) An excavation that goes through a major geological discontinuity or through a zone boundary; this case is illustrated in Figure 4b. If part of the rock mass located close to both the zone and the excavation boundaries in one of the zones, is brought to its failure state, a sudden and violent failure is possible, depending on the deformational properties of the rock mass of the other zone. This can then lead to type II rockbursts;

iii) An excavation that follows a major geological discontinuity or a zone boundary. A typical example of this case is a mine pillar, as shown in Figure 4c. If the pillar (zone C) fails, its failure can be violent if the deformational properties of zones A and/or B satisfies certain requirements. This category includes any isolated structure which may present some differences in mechanical properties due to local heterogeneities in the rock mass; it also includes isolated structures that may show pronounced geometrical irregularities.

4.3 Stability analysis

In routine mining engineering, stress analyses are usually performed using an elastic constitutive model for the rock mass; this has been proven to be an adequate approach for rockburst situation (e.g., Ortlepp, 1983). Rock properties are generally obtained through standard laboratory tests conducted on specimens prepared from appropriate rock samples. Rock mass properties are extrapolated from rock properties by using various relationships that take into account the mechanical effects of geological discontinuities; relationships based on geomechanical classification ratings are often used for that purpose (Hoek and Brown, 1980; Bieniawsky, 1984). The knowledge of the pre-mining state of stress results from in situ measurements or from empirical relationships such as those proposed by Herget (1987) for the Canadian Shield underground mines.

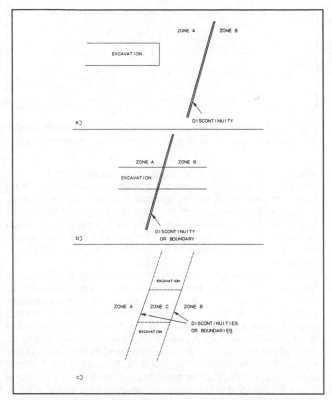

Figure 4. Vulnerable rock structures: a) Excavation that comes near a major geological discontinuity: potential for a type I rockburst; b) Excavation that goes through a major geological discontinuity or zone boundary: potential for a type II rockburst; c) Excavation that follows major geological discontinuities or zone boundaries: potential for a type II rockburst (after Gill and Aubertin, 1988).

4.3.1 Type I rockburst

This type of rockburst has been defined as a sudden slip on a pre-existing discontinuity. Unfortunately, very little information on post-peak behavior of geological discontinuities is available. In addition, there seems to be no recognized method to evaluate the equivalent local stiffness of the rock mass on both sides of the discontinuity. This is why the proposed methodology recommends (at this stage of its development) that if the stress conditions are such that a slip on the discontinuity is possible, then it should be considered that the equilibrium state is unstable and that there is a potential for rockbursting. However, it should also be considered that if the normal stress is approaching the uniaxial compressive strength of either one of the rock masses bordering the discontinuity, the failure can be gradual.

To model the peak shear strength of the discontinuity, the authors suggest using Barton's equation (Barton and Choobey, 1977; Bandis et al., 1981), for its relative simplicity and broad applicability.

4.3.2 Type II rockburst

This type of rockburst has been defined as the brittle failure of a certain volume of rock. Unlike the situation described in section 4.3.1, it is possible, here, to be more specific about whether the failure is violent or gradual.

If, while performing the stability analysis, expressing the rock mass strength through its uniaxial compressive strength is not adequate, the authors suggest using the well known Hoek and Brown (1980) failure criterion.

With mine pillars, both size and shape effects should be considered; these affect the peak strength as well as the pre-peak and the post-peak parts of the stress-strain relationship. Such

effects can be introduced into the stress-strain relationship through geomechanical classification ratings (e.g., Särkkä, 1984), empirical formulas (e.g., Bieniawski, 1975; Barron and Yang, 1992) or confined core concept (e.g., Wilson, 1972).

4.4 Stiffness comparison

At this point, this final step only applies to type II rockbursts. Two different situations are dealt with here: (i) mine pillars; (ii) other rock structures.

4.4.1 Mine pillars

Let us consider a mine pillar which is axially loaded, as it is postulated with the tributary area theory (Brady and Brown, 1985) or with the pillar loading theory proposed by Coates (1965). It can be shown that the pre-peak stiffness coefficient, K_{pr}, for a "long" pillar (plane strain conditions) is, considering a unit thickness and idealizing the stress-strain relationship:

$$K_{pr} = \frac{EB}{H(1-\nu^2)} \qquad (1)$$

where E is the pre-peak rock mass elastic modulus, ν is the Poisson ratio, B is the pillar width, and H is the pillar height. The post-peak stiffness coefficient, K'_{pr}, is obtained by substituting E', the post-peak rock mass elastic modulus, into equation (1). It should be noted that stiffness coefficients are expressed, in the present paper, as a force per unit length (e.g., pounds per inch or meganewtons per metre) as it is the case in most publications.

If the post-peak modulus of the rock mass involved is unknown, empirical relationships that have been proposed for rocks can be used, such as the relationship proposed by Brady et Brown (1981) for instance.

On the other hand, the stiffness coefficient of the country rock mass, K_e, can be determined in a number of ways. The analytical models proposed by Starfield and Wawersik (1968) and by Salamon (1970) could be used as such or implemented into a variety of numerical stress analysis methods. The authors of the present paper rather favor a more simple approach which is more easy to incorporate into routine engineering. It consists in performing numerical stress analyses following the process described by Hoek and Brown (1980) for obtaining ground characteristic lines with the convergence-confinement method as applied to pillar design. To illustrate this approach, let us consider the schematic single symmetrical pillar model shown in Figure 5a. A uniformly distributed stress σ_p is applied at the pillar location over a strip of width B and the relative displacement Δ of points A and A' along the pillar axis (or the relative average displacement along the pillar-country rock interfaces) is computed using any two-dimensional numerical code. This analysis is repeated for different values of σ_p and the results are plotted on a $B\sigma_p$ versus Δ diagram (Figure 5b). It can be shown that the slope of the line so obtained is the local mine stiffness coefficient K_e for the pillar under investigation, as defined by Starfield and Fairhurst (1968) for instance.

For pillars with finite cross-sectional dimensions, plane stress conditions have to be assumed and it can be demonstrated, for an idealized stress-strain relationship, that their pre-peak stiffness coefficient, K_{pr}, is:

$$K_{pr} = \frac{EA}{H} \qquad (2)$$

In this equation, E is the pre-peak rock mass modulus and A, the cross-sectional area of the pillar. Again, the post-peak stiffness coefficient, K'_{pr}, can be obtained by substituting E', the post-peak rock mass elastic modulus, into equation (2).

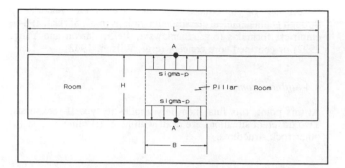

Figure 5a. Model used to evaluate the equivalent local stiffness of the rock mass surrounding a pillar (after Hoek and Brown, 1980).

The determination of local mine stiffness coefficients (K_e) can be done as suggested above (Figure 5a), using in these cases a three-dimensional stress analysis code. For a given pillar, the results are plotted on a $A\sigma_p$ vs Δ diagram; the slope of the line so obtained is the local mine stiffness coefficient for that pillar. Starfield and Wawersik (1968) model has been recently implemented into a three dimensional boundary elements code by Zipf (1992); this code should be available shortly and it is hoped that it could be easily integrated into routine mining engineering.

It is recalled that if the pillar should fail, and if K'_{pr} is larger than K_e (in absolute values), then the failure would be sudden and violent, leading to a type II rockburst. This comparison is usually done on what has been called force-convergence diagram by Starfield and Fairhurst (1968). The diagrams to which it has been referred above for finding the local mine stiffness coefficients are such diagrams and they can be used for that purpose as illustrated in Figure 5b. Curve (a) in this figure is an idealized reaction curve for a pillar which should fail and potentially burst while in the case of the pillar with reaction curve (b), the failure should be gradual; curve (c) stands for a pillar that should not fail.

4.4.2 *Other rock structures*

An approach similar to that describes for mine pillars is suggested for other rock structures. However, in such cases, it is more convenient to deal with a stress-strain diagram rather than with the conventional force-convergence diagram. Consequently, the comparison is involving moduli instead of stiffness coefficients.

It has been recognized, when the rock mass is assumed to be linear elastic and homogeneous, that the stresses known to trigger the failure of any unsupported underground excavation are those at boundary of the opening. In routine mining engineering work, failure criteria generally used, when performing stability analyses, involves only the two extreme principal stresses, one of

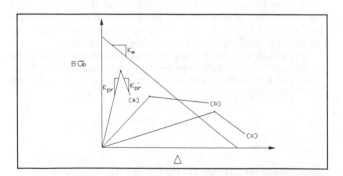

Figure 5b. Force - convergence diagram usually used to state on the nature of pillar failures; (a): violent failure; (b) gradual failure; (c): no failure. (after Starfield and Fairhurst, 1968).

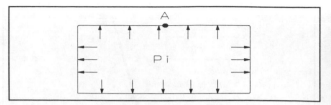

Figure 6. Model used to estimate the equivalent local stiffness for the stable surrounding rock mass (after Gill and Aubertin, 1988).

which is often zero (the stress component normal to the boundary). Moreover, the stress component normal to the cross section of openings is usually ignored. With these simplifications in mind, one is then justified to define the pre-peak modulus (E) and the post-peak modules (E') of the rock mass portion which is brought to its failure state from the uniaxial rock mass stress-strain diagram. No shape effect on the post-peak modulus is considered.

The local excavation modulus for the stable surrounding rock mass can be found by performing numerical stress analyses, as illustrated on the model shown in figure 6, where the dimensions of the model are the same as those of the excavation. A normal internal pressure P_i is applied to the wall and the tangential deformation (ϵ_t) at point A (where the instability is foreseen to be initiated) is computed. Analyses are performed for different values of P_i; the slope of the graph of P_i versus ϵ_t is the local excavation modulus (E_e) for the stable surrounding rock mass and it can be shown that the latter can be used in a way similar to that of the mine local stiffness coefficient. The local excavation modulus is expressed in stress units (e.g., pounds per square inches or gigapascals).

As usual, if it is found that failure can occur and that the value of E' is larger than E_e (in absolute values), then a type II rockburst can occur.

5 APPLICATION OF THE PROPOSED METHODOLOGY

To shown how to use the proposed methodology and, in a way, to go one step further in validating the approach, a back-analysis of one rockburst, which has occurred in a northwestern Québec mine is briefly reported here; more details on this case study can be found in Simon (1992). It should be mentioned that the methodology has also been applied to two cases where the rockburst potential was found to be negligible; the validity of these conclusions were confirmed by the fact that the studied openings have not shown any sign of rockbursting problems even after many years (Gill and Aubertin, 1988). The rockburst back-analyzed here occurred on February 23, 1990, in a shaft station.

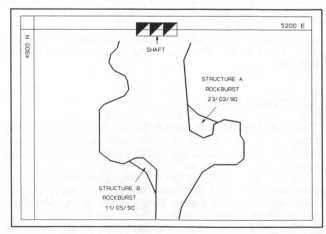

Figure 7. Shaft station (after Simon, 1992).

- Zoning

The shaft station was excavated, for the most part, in a massive basaltic-tuff rock mass containing several small carbonatite inclusions. The area where the bursting had occurred has been concreted shortly after, so direct observation of local conditions has not been possible at the time this study was done. Because there was no geomechanical rating available for the station, data from a survey conducted in a nearby ramp segment were considered instead. This segment of the ramp is located in a rock mass portion showing a geology similar to that of the station. The Q geomechanical classification system (Barton et al., 1974) was employed, giving a mean rating of 73.

Uniaxial compressive and brazilian tests were performed on intact rock samples in order to determine the values of Co (uniaxial compression strength), E (elastic modulus), ν (Poisson ratio) and To (tensile strength). Results of this testing program are shown in Table 1.

Table 1. Some mechanical properties of the tuff.

Property	Mean values
Co	229 MPa
E	97,0 GPa
ν	0,30
To	18,9 MPa

- Identification of vulnerable rock structures

A plan of the shaft station is shown in Figure 7. The rock structures involved fall into the third category (see section 4.2 - opening with an irregular geometry).

- Stability analysis

Local narrowing of an opening as it is shown on the plan in Figure 7 (structures A and B) has been known to lead to type II rockbursts (Coates and Dickhout, 1970). In the stability analysis conducted, it was assumed is that these structures act as "pillars" submitted to high stress concentration. A two-dimensional finite elements model was constructed to evaluate the state of stress in these structures. To simulate in 2-D the high stresses in these 3 D "pillars", the choosen strategy was to raise their elastic modulus by a value equal to the ratio of their area in the model on their real area.

The m and S parameters of the Hoek and Brown failure criterion were determined from the Q rating; it yielded an m value of 11, and an S value of 0,071. The elastic modulus used for the rock mass was determined by the empirical relationship of Nicholson and Bieniawski (1990) which makes use of the RMR rating. The Poisson ratio of the rock mass was considered identical to that of the intact rock samples. Results of the numerical stress analysis indicated that structure A was unstable.

- Stiffness comparison

Since failure was obtained, a comparison between the post-peak modulus of the failed rock and the local modulus of the surrounding stable rock mass had to be made.

For the local modulus of the surrounding stable rock mass, numerical stress analyses were performed on a model similar to the one shown in figure 6 and adapted to the geometry at hand; the results have been plot on a graph of P_i vs ϵ_t. This gives a local modulus of the stable surrounding rock mass of 124 GPa.

The post-peak modulus of the tuff being not available, the relationships presented by Brady and Brown (1981) for various rocks were used, and lead to the values presented in Table 2.

It can be seen that the range of values for E' extents beyond that calculated for K_s (in absolute values), so this leads to the conclusion that there was a rockburst potential in this case.

Table 2. Post-peak modulus E' (calculated from the graphical relationships proposed by Brady and Brown, 1981)

	E'
Minimum	76 GPa
Mean	81 GPa
Maximum	210 GPa

6 DISCUSSION

When elaborating the proposed methodology, the authors have kept in mind the following premises: proper routine engineering design of rock structures should include the evaluation of its rockburst potential. The authors are of the opinion that much more back-analyses are obviously needed before stating that the methodology described in this paper should be considered validated. This proposed methodology, which satisfies the above mentioned premises, should be regarded, for the time being, as a first proposal which will be revised if ever required.

The authors have not discussed what should be done whenever it is concluded that there is a rockburst potential as this is considered to be beyond the scope of this paper. Literature dealing with this subject is abundant.

It should be nevertheless recognized that there are rockburst situations which cannot be foreseen by applying the methodology, especially those leading to rockbursts involving a few cubic metres of broken rock. Geological features can be ignored because of the scale at which the surveys leading to zoning are conducted and such features are sometimes responsible for rockbursting at this small scale.

From a practical point of view, most of the pillars encountered in northwestern Québec metal mines are long compared to their other cross sectional dimension (sill and transverse pillars); the method suggested for determining their local mine stiffness coefficients using simple two-dimensional stress analysis codes has many advantages compared to the use of Starfield and Wawersik (1968) or Salamon (1970) models.

The approach suggested for stiffnesses comparison when dealing with rock structures other than mine pillars need to be studied with more details.

According to our experience, lack of specific geomechanic data pertaining to rockburst in mines is very often the major problem encountered while performing back-analyses. The case reported in this paper is no exception. In fact, most of the data used to perform this back-analysis were obtained from other areas of the mine where geology was considered comparable. This led to several imprecisions in the evaluation of the rock mass mechanical properties. However, sensitivity analyses were performed to cover most of the envisioned properties spectrum, and failure was obtained even for large variations in the various property values (Simon, 1992).

The evaluation of the post-peak modulus for the unstable zone was also a problem, since the post-peak modulus of the tuff was not determined in the laboratory testing program. The use of empirical relationships obviously affect the results precision. The post-peak modulus of the unstable rock, based on the Brady and Brown (1981) empirical relationship, was up to 170% higher than the local modulus of the stable surrounding rock. Furthermore, according to some work done by Simon (1992) along this line, this percentage is rather two or three times larger; even the minimum value of E' for the unstable zone proposed by Simon (1992) is larger that the value calculated for K_s.

The back-analysis reported in the paper has demonstrated that the methodology would have been able to identify a rockburst potential for the structure analyzed, using only routine data and tools that are available to mining engineers.

7 CONCLUSIONS

The proposed methodology is based upon routine mining engineering and allows the engineers to evaluate the rockburst potential before the rock structure is created. The goal of this methodology is to assess whether there is a risk or not of rockbursting.

The back-analysis presented here to show how to use the methodology, has demonstrated that the methodology can be used to evaluate a priori if the failure of a rock structure will be violent or gradual.

This methodology, which could be used for any kind of underground excavations, gives the engineers a new means, with logical steps, for the assessment of rockburst potential.

Even though the methodology is not in its final stage of development, it still can be a very useful tool for small mines, where other techniques of evaluation are not readily accessible. Obviously, assessments made with the proposed methodology should be regarded, for the time being, as preliminary.

ACKNOWLEDGEMENTS

The authors gratefully acknowledge the financial contribution of the IRSST (Institut de Recherche en Santé et Sécurité au Travail du Québec).

REFERENCES

Aubertin, M., D.E. Gill, R. Simon 1992. Évaluation du Potentiel de Coups de Terrain dans les Mines; Phase II - Validation de la Méthodologie Proposée. Rapport Final présenté à l'Institut de Recherche en Santé et Sécurité au Travail (IRSST), Septembre 1992.

Bandis, S.C., A.C. Lumsden, N.R. Barton 1981. Experimental Studies of Scale Effects on the Shear Behaviour of Rock Joints. Int. J. Rock Mech. Min. Sci., 18, 1, 1-21.

Barron, K., T. Yang 1992. Influence of Specimen Size and Shape on Strength of Coal. Proc. Workshop on Coal Pillar Mechanics and Design, Santa Fe, USBM IC 9315, 5-24.

Barton, N., V. Choobey 1977. The Shear Strength of Rock Joints in Theory and Practice. Rock Mech., 10, 1, 1-54.

Barton, N., R. Lien, J. Lunde 1974. Engineering Classification of Rock Masses for the Design of Tunnel Support. Rock Mech., 6, 4, 189-239.

Bieniawski, Z.T. 1973. Engineering Classification of Jointed Rock Masses. The Civil Engineer in S. Afr., 15, 335-344.

Bieniawski, Z.T. 1975. The Significance of In Situ Test on Large Rock Specimens. Int. J. Rock Mech. Min. Sci., 12, 101-113.

Bieniawski, Z.T. 1984. Rock Mechanics Design in Mining and Tunnelling. A.A. Balkema.

Blake, W. 1972. Rockburst Mechanics. Quarterly of the Colorado School of Mines, 67, 1, 1-64

Brady, B.H.G., E.T. Brown 1981. Energy Changes and Stability in Underground Mining: Design Application of Boundary Elements Methods. Trans. Int. Min. Metall., 90, A61-A68.

Brady, B.H.G., E.T. Brown 1985. Rock Mechanics for Underground Mining, George Allen and Unwin.

Brown, E.T. 1984. Rockbursts: Prediction and Control. Tunnels and Tunnelling, April, 17-19.

Coates, D.F., M. Dickhout 1970. Elements of Planning in Deep Mines. Can. Min. J., 91,9,74-78.

Coates, D.F. 1965. Pillar Loading Part I: Literature Survey and New Hypothesis. R168, Mines Branch (CANMET) Research Report, Queen's Printer, Canada

Cook, N.G.W. 1965. A Note on Rockbursts Considered as a Problem of Stability. J. S. Afr. Inst. Min. Metall., 65, 437-446.

Diest, F.H. 1965. A Non-linear Continuum Approach to the Problem of Fracture Zones and Rockbursts. J. S. Afr. Inst. Min. Metall., 65, 502-522.

Fortin, M., D.E. Gill, B. Ladanyi, M. Aubertin, G. Archambault 1990. Simulating the Effect of a Variable Normal Stiffness on Shear Behavior of Discontinuities. Mechanics of Jointed and Faulted Rock, Rossmanith (ed.), Balkema, 381-388.

Gill, D.E. 1982. Mécanique des Roches (6.502). Notes de cours (non publiées), Département de génie minéral, École Polytechnique de Montréal.

Gill, D.E., M. Aubertin 1988. Évaluation du Potentiel de Coups de Terrain dans les Mines d'Abitibi. Rapport de recherche de l'URSTM présenté à l'Institut de Recherche en Santé et Sécurité du Travail (IRSST).

Goodman, R.E. 1980. Introduction to Rock Mechanics. John Wiley & Son.

Hedley, D.G.F. 1991. A Five-Year Review of the Canada-Ontario Industry Rockburst Project. CANMET Special Report SP90-4E.

Hedley, D.G.F. 1992. Rockburst Handbook for Ontario Hardrock Mines. CANMET Special Report SP92-1E.

Herget, G. 1987. Stress Assumptions for Underground Excavations in the Canadian Shield. Int. J. Rock Mech. Min. Sci., Vol. 24, 95-97.

Hoek, E., E.T. Brown 1980. Underground Excavations in Rock. Inst. Min and Metall, London.

Mottahed, P. 1992. Atelier sur les Coups de Toît. Organisé par CANMET et l'AMQ, Val-d'Or.

Nicholson, G.A., Z.T. Bieniawski 1990. A Nonlinear Deformation Modulus Based on Rock Mass Classification. Int. J. Min. Geol. Engng., 8, 3, 181-202.

Ortlepp, W.D. 1983. The Mechanics and Control of Rockbursts. Rock Mechanics in Mining Practice. Budavari (ed.), SAIMM, Johannesburg, 257-282.

Petukhov, I.M. 1957. Rockbursts in Kizel Coalfield Mines (in russian). Perm. Publ., Perm. (cited in Pethukov and Linkov, 1979).

Petukhov, I.M., A.M. Linkov 1979. The Theory of Post-Failure Deformations and the Problem of Stability in Rock Mechanics. Int. J. Rock Mech. Min. Sci. & Geomech., 16, 57-76.

Salamon, M.D.G. 1970. Stability, Instability and Design of Pillar Workings. Int. J. Rock Mech. Min. Sci., Vol. 7, 613-631.

Salamon, M.D.G. 1974. Rock Mechanics of Underground Excavations. Advances in Rock Mechanics, Proc. 3rd Cong. Int. Soc. Rock. Mech., 1-B, 951-1099.

Salamon, M.D.G. 1983. Rockburst Hazard and the Fight for its Alleviation in South African Gold Mines. Proc. Rockbursts Prediction and Control, IMM, London, 11-36.

Särkkä, P.S. 1984. The Interactive Dimensionning of Pillars in Finnish Mines. Proc. 2nd Int. Conf. Stability in Underground Mining, Lexington, 71-84.

Simon, R. 1992. Validation d'une Méthodologie d'Évaluation du Potentiel de Coups de Terrain dans les Mines. Mémoire M.Sc.A., Département de génie minéral, École Polytechnique de Montréal.

Starfield, A.M., C. Fairhurst 1968. How High-Speed Computers Advance Design of Practical Mine Pillar Systems. Engng. Min. J., 169, 78-84.

Starfield, A.M., W.R. Wawersik 1968. Pillars as Structural Components in Room-and-Pillar Mine Design. Proc. 10th US Symp. Rock Mech., Austin, 793-809.

Wilson, A.H. 1972. A Hypothesis Concerning Pillar Stability. Min. Eng., Vol. 131, 407-419.

Zipf, R.K. Jr 1992. Analysis of Stable and Unstable Pillar Failure Using Local Mine Stiffness Method. Proc. Workshop on Coal Pillar Mechanics and Design, Santa Fe, USBM IC 9315, 128-143.

Rockbursts and Seismicity in Mines, Young (ed.) © 1993 Balkema, Rotterdam, ISBN 90 5410 320 5

Excavated volume and long-term seismic hazard evaluation in mines

Ewa Glowacka
CICESE, Ensenada, Mexico (On leave from: IGF, Polish Academy of Sciences, Warsaw, Poland)

ABSTRACT: The probabilistic dependence of seismic energy on the excavated volume is applied to the continuous evaluation of seismic hazard in mines. It is assumed that the distribution of the sum of seismic energy is lognormal and that the energy accumulation and relaxation in time are described by a Kelvin-Voigt model. The algorithm was tested in four different mining areas (in Wujek and Gottwald mines in Upper Silesia, Poland, and in Doubrava, (Ostrava), and Gottwald, (Kladno, Czechoslovakia) excavated with caving during about 5 years each. The correlation coefficient between the maximum excess of accumulated energy ΔE_{max} and seismic energy E released by strong tremor equals 0.79, and $E/\Delta E_{max}$ =0.96. For two areas (in Wujek and Gottwald, Poland) where neither edges nor old workings occur in the studied area, the relation between logarithm of energy released by strong shocks and time of energy accumulation t_E, has the form $\log E = 5.5 + 0.07\ t_E$, (t_E in months), with a correlation coefficient equal to 0.96, while the relation between E and the measure L of the area of the energy accumulation has the form: $\log E = -1.29 + 3.1 \log L$, (L in meters), with correlation coefficient equal to 0.96.

1 INTRODUCTION

The qualitative dependence between seismic activity and extracted volume has been known for a long time (e.g. Sibek 1963, Hodgson and Cook 1971, Bober and Kazimierczyk 1979). However there are many geological and mining conditions which influence the local state of stresses (e.g. McGarr et al. 1989, Wierzchowska 1989,) and the current seismicity of a mine; for a long range it was shown (McGarr 1976, McGarr and Wielbos 1977) that for a deep mine the cumulative seismic moment is quantitatively related to the increase in volume of elastic convergence of the mined-out area, and practically to the volume of the extracted rock. Energy Release Rate (ERR), the spatial rate of energy released, closely allied to volumetric closure, is currently the most widely used parameter in designing mine layouts to reduce seismicity in South Africa (Spottiswoode 1990).

From recent studies follows that the strongest tremors are not always associated with the extraction on a particular stope face but they have more regional origin, probably associated with movement on major geological discontinuities (Kijko et al. 1979, McGarr et al. 1989, Stankiewicz 1989, Gibowicz 1990). In the magnitude-frequency distribution they give a bimodal character to the curve. This paper will discuss one of these cases below.

Deformation of the earth crust, measured directly and indirectly as changes of seismic regime and changes of different geophysical field, are used for earthquake prediction. In mining practice deformation and convergence measurements are rarely done, so extracted volume can be a very important parameter for long-term prediction in mines.

An attempt to apply the probabilistic dependence of seismic energy on the excavated volume to continuous seismic hazard evaluations in mines is being done in Poland since 1985. In 1985, Kijko, using the earlier solution of McGarr (1976) introduced the deterministic dependence of seismic energy on extracted volume. In 1987 (Glowacka et al. 1987) this dependence was interpreted and applied as a probabilistic relation for the Wujek, (Poland) coal mine, and later for another two mines in Poland and Czechoslovakia (Glowacka et al. 1988 a,b, Glowacka and Kijko 1989a). The results of continuous seismic hazard evaluation done with a normal energy distribution show that strong tremors were preceded by a several months long period of energy accumulation and increasing probability. For some tremors the energy accumulation was observed in all working longwalls at the same time. In 1989 (Glowacka and Kijko 1989b, Glowacka et al. 1990) the results of using a bimodal distribution were taken into account for the relationship between seismic energy and extracted volume. In following years, the lognormal distribution of energy and energy release time was used for continuous long-term seismic hazard evaluation, and the examples of how to calculate the synthetic probability were published (Glowacka and Pilecki 1991, Glowacka and Lasocki 1992, Glowacka 1992b).

The purpose of this work is to present a method of continuous evaluation of seismic hazard, and to show the results of reinterpretation obtained for different mining regions. The energy accumulation and relaxation with time are described by a Kelvin-Voigt model, and it is assumed that the distribution of the sum of seismic energy is lognormal.

All results presented in this study were obtained retrospectively for different underground coal mines in Poland and Czechoslovakia.

2 PROBABILISTIC DEPENDENCE BETWEEN SEISMICITY AND EXTRACTED VOLUME

The method is based on a quantitative relation between seismic moment and extracted deposit volume (McGarr 1976), and a statistical relation between released seismic energy, and extracted volume (Kijko 1985):

$$\Sigma E = C\ V^B, \qquad (1)$$

where

$$B = (\beta - b) / (d-b), \qquad (2)$$

and b, β, d are parameters from the Gutenberg-Richter relation, and from relations between magnitude and energy, and between seismic moment and energy, respectively. More details about (1) and its applications can be found in (Glowacka and Kijko 1989a, Glowacka 1992a). The origin time, location, and value of seismic energy (calculated from duration or from maximum amplitude), together with the value of extracted volume, are used for the continuous, time-dependent seismic hazard evaluation.

Let t be elapsed time since the extraction began, and Δt the time interval under consideration, a month or a day, say, for which the seismic hazard is evaluated. Assume that for time $t - \Delta t$ (i.e. for the past) all information about seismicity and excavation is known.

The most probable value of the sum of seismic energy $< \Sigma E(t - \Delta t, t >$ resulting from the excavation of the planned volume $\Delta V = V(t) - V(t - \Delta t)$, that will be released in unit time Δt, is expressed by the equation:

$$< \Sigma E(t - \Delta t,\ t) > = \{\Sigma E_T(t - \Delta t, t)e(\Delta t, b) + \Delta E(t - \Delta t)\}\ e(\Delta t, bt), \quad (3)$$

where

$$\Sigma E_T(t - \Delta t,t) = C[V^B(t) - V^B(t - \Delta t)], \qquad (4)$$

and parameters C and B are calculated on the basis of formula (1). Parameters C and B of equations (1)-(3) depend on the mining and geological situation of the rockmass and, since this situation is changing during the course of extraction, they are functions of time, i.e. C=C(t), B=B(t).

The expression $e(\Delta, b_t) = 1-\exp(\Delta t\, b_t)$ characterizes energy loading and relaxation with time; happens to be equivalent to the assumption that the medium is described by a Kelvin-Voigt model. The value of parameter b_t was calculated for the longwall J-5, Marcel mine, Poland, (Glowacka and Pilecki 1991) for a period when works on that longwall were already finished, and seismic tremors caused by release of seismic energy accumulated in the rockmass were still being recorded. It was calculated that b_t, which characterizes the energy release time, is equal to 0.2/day; which agrees with laboratory results (Scheidegger 1982). The parameter $b_t = 2/\tau = 2\mu/\eta$, where τ is strain release time in a Kelvin-Voigt model, and μ is rigidity and η is viscosity of the medium. Approximation of the rockmass by a Kelvin-Voigt model is a great simplification, but there is not enough data to warranty use of more complicated models.

The term $\Sigma E_T(t-\Delta t, t) \cdot e(\Delta t, b_t)$ of (3) describes energy loading in the rockmass as a result of extraction of ΔV , with loading function $e(\Delta t, b_t)$. The value $\Delta E(t-\Delta t)$ is the excess of seismic energy accumulated in the rockmass at time t-Δt, and is evaluated at every stage of the calculations as the difference between the expected, $<\Sigma E(t-\Delta t)>$ and observed, $\Sigma E(t-\Delta t)_o$ sums of released seismic energy:

$$\Delta E(t-\Delta t) = <\Sigma E(t-\Delta t)> - \Sigma E(t-\Delta t)_o; \qquad (5)$$

it is based on the assumption, that the seismic energy is not be dissipated in the rockmass. The above mentioned assumptions are true if the value of the seismic efficiency μ_s is constant. From literature results, μ_s varies between 0.1% and 0.5% (McGarr 1976, Gibowicz et al. 1979), so for our purposes, we can assume that changes of μ_s are slow with time, which would be reflected in C=C(t). If $\Delta E(t-\Delta t) < 0$, then it is assumed, that $\Delta E(t-\Delta t)=0$.

Finally, the term $\{\Sigma E_T(t-\Delta t, t)e(\Delta t, b_t)+\Delta E(t-\Delta t)\}$ in equation (3) is the total seismic energy available for release with time release function $e(\Delta t, b_t)$, in time interval Δt.

The most probable sum of seismic energy is calculated twice for every stage of seismic hazard evaluations; once from equation (3), and a second time assuming $\Delta E=0$ in (3) (i.e. no excess energy accumulation), expressed by:

$$<\Sigma E(t-\Delta t, t)>' = \{\Sigma E_T(t-\Delta t, t)e(\Delta t, b_t)\}\, (e(\Delta t, b_t)), \qquad (6)$$

so the difference between values (3) and (6) can be easily interpreted as the excess energy accumulation (multiplied by $e(\Delta t, b_t)$). In figure 1 the separation between curves 4 and 4a is a measure of excess energy accumulation.

The value of excavated area S was used, instead of excavated volume V , for the cases in which the height of workings is constant. At every stage of calculations, C, B and ΔE are evaluated from previous history of the dependence between seismic energy and the extracted deposit volume.

Equation (3) gives the most probable value of energy which is released during time interval Δt . It was assumed that the distribution of the released seismic energy, about the most probable energy described by formula (3), is lognormal (Glowacka 1992a):

$$f_E(E) = \left[\frac{1}{E\sigma_x(2\pi)^{\frac{1}{2}}}\right]\exp\left\{-\frac{1}{2}[(\ln E - m_x)/\sigma_x]^2\right\} \qquad (7)$$

where σ_x is the standard deviation of $\ln(E)$ and m_x is the most probable value of $\ln(E)$ (Benjamin and Cornell, 1970).

Let $P(E_j)$ denote the probability that the expected sum of seismic energy exceed the predetermined threshold E_j , i.e. $P(E_j)=P(\Sigma E \leq E_j)$ then

$$P(E)_j = 1 - \int_0^{E_j} f_E(E)\, dE. \qquad (8)$$

Three measures of seismic hazard are evaluated simultaneously, for every time interval Δt , using information from the past: the most probable sum of seismic energy - $<E(\Delta t)>$, which can be

released during time interval Δt, expressed by (3), the excess of accumulated energy ΔE- (5), and the probability $P(Ej)$ that the amount of seismic energy released per unit time Δt exceed the threshold Ej - (8).

3 RESULTS

The algorithm was tested in four different mining areas excavated with caving during about 5 years each. The studied areas are situated as following: 1- five longwalls in seam 501 in Gottwald[1] mine (Poland, Upper Silesia), where always two longwalls were extracted simultaneously, keeping the outstripping character of their fronts; 2- five longwalls in seam 501 in Wujek (Poland, Upper Silesia), two or three longwalls worked simultaneously, keeping outstripping character of their fronts; 3- field Robert in Gottwald (Kladno, Czechoslovakia); 4- longwall 4 3765 in Doubrava (Ostrava, Czechoslovakia).

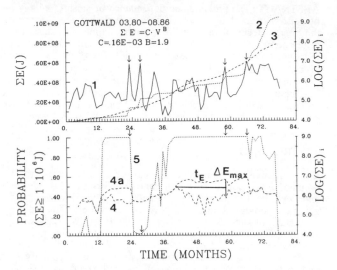

Figure 1. Seismic energy and seismic hazard versus time (in months) for the chosen region (five longwalls/seam 501) of the Gottwald Coal Mine, for the period 1.03.1980-31.08.1986. Arrows mark the strong tremors, with energy greater than 5×10^6J. Curves: 1-- monthly sums of seismic energy, referred to the right vertical axis; 2-- cumulative seismic energy, referred to the left vertical axis; 3-- energy calculated from relation (1); 4-- predicted seismic energy calculated from relation (6), 4a-- predicted seismic energy (3); both 4 and 4a curves referred to the right vertical axis; 5-- probability of exceeding the energy $1\ 10^6$J (5). Modified from Glowacka, Korotynski & Syrek, (1988).

As an example, figure 1 illustrates results obtained for the Gottwald (Poland) mine. Only tremors with energy smaller then 10^7J were taken into account. According to Stankiewicz (1989), for Gottwald mine, energy 10^7J is a boundary between events of regional origin, not associated with extraction in the particular workings and events which result from the present mining activity. During the studied period, two tremors with energy larger than 10^7J occurred in Gottwald mine: $9.5\ 10^7$J in February 1981 and $1\ 10^9$J in October 1985. For the chosen longwall in Doubrava mine (Ostrava, Czechoslovakia) the problem of bimodal distribution was studied before (Glowacka et al. 1990, Glowacka 1992a) and, since there was not clear where the boundary of bimodality is, in this paper all events are assumed to belong to the class related to extraction. In Wujek mine energy of studied events was below the boundary of bimodality (Stankiewicz 1989); and tremors from Kladno, analyzed in this study, did not indicate a bimodal distribution of energy, so all seismic events, which occurred in this mine were taken into account.

For this paper we define the strong shock, as a tremor assumed to be associated with extraction in the studied area, but the energy of which results from extraction in a time period longer than Δt. Thus the strong shock is that which the energy is larger than the mean expected energy (6), expressed with half an order of magnitude accuracy. The definition of strong tremor for each particular mine is presented in table I.

Figure 1 shows observed and predicted values of released seismic energy and probability, for monthly intervals. Curve 1 illustrates the monthly energy sum (in joules). Curve 2 represents the total energy sum, and curve 3 the theoretical relation (1) plotted for the last stage of calculations (that is, for t-Δt). Curves 4, 4a and 5 illustrate the evaluation of seismic hazard: the most probable value of the energy sum expressed by relation (6), when ΔE=0 (curve 4); the expected value of energy (3), taking energy accumulation into account (curve 4a); and the probability (8) that energy 1 10^6J is exceeded (curve 5).

In fig.1. the maximum excess of accumulated energy ΔE_{max} and duration of energy accumulation t_E for one event are indicated. Parameter ΔE_{max} is defined as the maximum value of ΔE during time t_E. Since the curves 4 and 4a in figure 1 are plotted on a logarithmic scale, the value of ΔE_{max} is always equivalent to the local maximum of curve 4a, not to the largest separation between the 4 and 4a curves.

Conspicuous probability $P(E_j)$ increase (curve 5) and energy accumulation (where curves 4 and 4a separate) precede three out of the four strongest events (indicated by arrows). Similar results were obtained for three other mines.

For all analyzed areas, 14 out of 18 strong events were preceded by conspicuous energy accumulation, 4 events were not preceded by energy accumulation, and there were 5 false alarms (Table I). Thus, the efficiency of ΔE as a precursor is equal to 0.78. The correlation coefficient between ΔE_{max} and energy E released in strong tremor equals 0.79, and $E/\Delta E_{max}$=0.96 (fig.2). Therefore, we conclude that the value ΔE_{max} is a good predictor for the magnitude of a forecoming tremor. However, energy accumulation can go on for several months (the time of energy accumulation covers about half of the analyzed period), so the probability of having a strong tremor with energy E=~ΔE is only about 0.06/month, for months when accumulation occur. Whereas the mean probability of having a strong tremor, during any month, is 0.04, which gives a probability gain equal to 1.5.

TABLE I
PREDICTION ON THE BASIS OF ENERGY ACCUMULATION

Mine	E(J)	Number of tremors	Success	Failure	False alarm
1	$10^7 \leq E$	2	2	0	0
2	$5 \cdot 10^6 \leq E \leq 10^7$	4	3	1	1
4	$10^5 \leq E$	6	4	2	2
3	$10^7 \leq E$	6	5	1	2
		18	14	4	5
2a	$10^7 < E$	2	1	1	

1. Wujek, Poland, 2. Gottwald, Poland, $E<10^7$ J, 2a.Gottwald, all tremors, 3. Doubrava, Ostrava, (CS), 4. Gottwald, Kladno, (CS)

If tremors with energy stronger than 10^7J are taken into account for the Gottwald mine, one tremor with E=10^9J is preceded by ΔE five times smaller than the tremor energy (Fig.2). We define the parameter 1 as

$$1 = E/\Delta E \qquad (9)$$

and 1=5 for this particular case. This results probably from the fact that the real area of energy accumulation for this particular event covers an area about five times larger than the analyzed one. The existence of an area five times larger than the studied area is theoretical only because the workings in this mine are situated in different seams, and no mining area, five times larger than the studied area exists in Gottwald mine. On the other hand, the possibility of interaction between mine-induced and residual tectonic stresses for very strong events cannot be excluded (Gibowicz 1990).

The last line of table I includes the results for Gottwald mine when two events with energy greater than 10^7J are taken into account. It seems, however, that the limited analyzed area does

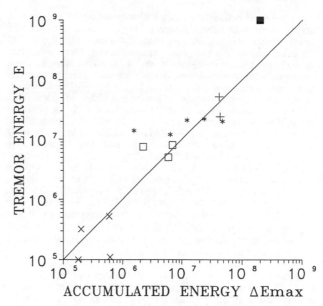

Figure 2. Tremor energy E versus accumulated energy ΔE_{max} Mines: Wujek (PL) + , Gottwald (PL)☐ , Gottwald (Kladno CS) X, Doubrava (Ostrava CS) *. Tremor with energy 1 10^9J (Gottwald mine ■) was not taken into account.

not allow to evaluate the probability of occurrence of very strong mining tremors. To evaluate the probability of regional tremors larger analyzed areas and greater number of strong tremors are required. Also, corrections to equations (1),(3),(4) and (6), to take into consideration a bimodal character of the energy distribution, are necessary. An attempt at such evaluations were done in (Glowacka and Kijko, 1989b), and (Glowacka et al. 1990).

For two regions (Wujek and Gottwald, Poland) where no edges nor old workings occur in the analyzed area, the relation between the logarithm of energy released in strong shocks and time of energy accumulation t_E, has the form: logE = 5.5+0.07 t_E (t_E in months), with a correlation coefficient equal to 0.96 (fig 3). The relation between the magnitude M and $\log t_E$ has the form: $\log t_E$=0.56+0.75 M, (t_E in days) with correlation coefficient of 0.89. The factor 0.75, mentioned above, agrees with results obtained by Niazi (1985) for precursors of earthquakes with high correlation coefficient, but the time t_E for any particular magnitude is considerably larger then the precursory lead time.

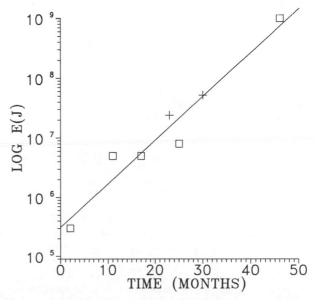

Figure 3. Logarithm of energy versus time of energy accumulation t_E for strong tremors in Wujek +, and Gottwald ☐ mines (PL).

For the same two regions, the relation between the energy E of strong tremors and the measure L of the area of the energy accumulation was studied. Measure L (in meters) is the sum of length of longwalls which were extracted during time t_E preceding every strong seismic event. If the 4 strongest tremors from Gottwald mine, and two from Wujek mine are taken into account the relation has the form: $\log E = -1.29 + 3.1 \log L$ (fig. 4, solid line), with correlation coefficient equal to 0.96. This means that the energy is roughly proportional to L^3, a common relation in seismology. (e.g. Kanamori and Anderson 1975). Using the relation $\log E = 1.8 + 1.9 M$ (Dubinski and Wierzchowska 1973) the dependence between magnitude and L takes the form $M = -1.63 + 1.63 \log L$. Including the strongest microearthquake with $E = 1\ 10^9 J$, from Gottwald mine, with an area of accumulation five times bigger then area under study (9), the relation between E and L has a form $\log E = 0.41 + 2.5 \log L$ (fig. 4, dashed line), which result in $M = -0.73 + 1.31 \log L$.

The above mentioned relations between tremor energy E and energy accumulation ΔE_{max}, time t_E and area of energy accumulation L^3, suggest that the energy accumulation evaluated by the method has a physical meaning.

An attempt to use equations (3),(6) and (8) for seismic hazard evaluation for every longwall separately gives relatively good results for Wujek mines. This means, that the mean predicted energy value agrees with the mean observed sum of energy released, for the case when only tremors resulting from extraction in this particular workings during time period Δt, occur. In case of occurrence of strong events, the probability increase and energy accumulation preceding some of the strong tremors were observed for every longwall separately. For Gottwald mine the results were not satisfactory. This lead to the conclusion that for the presented method, all workings, if situated near each other extracted at the same time, should be taken into account.

The results from Doubrava mine, (Glowacka et al. 1990), show, that the presence of edges influences the prediction possibility very strong, as was expected. The only possibility to resolve this problem, is to incorporate, in the present algorithm, results of other geophysical measurements or evaluations, presented in probabilistic form.

Table II presents values for parameter B calculated from (1) and for parameter b of the Gutenberg-Richter relation, calculated both from (2) and from catalogs, using the maximum likelihood method (Lasocki 1989). Values $d = 1.5$ (Hanks, Kanamori), and $\beta = 1.9$ (Gibowicz 1963, Dubinski and Wierzchowska 1979) were used for equation (2). For the Wujek, Gottwald (Poland) and Doubrava (Czechoslovakia) mine, all located in Silesia, the B parameter values are very similar, and are in the range between 1.56-2.3. For Gottwald (Kladno, Czechoslovakia) mine the B value is

smaller, which can result from different mining techniques; the coal is extracted here using a pillar-chamber method, while in the other three mines the deposit is extracted in longwalls. Values of the b parameter for the Gottwald mine, where the catalog has the largest magnitude range, are almost identical for both methods used. This probably results from the assumption made for evaluating (1) and (2) by (Kijko 1985), that the difference between E_{max} and E_{min} has to be large, that is satisfied for the Gottwald mine. Time variations of the B parameter need additional research.

TABLE II

Mine	B			b		Catalog extension	
	Time var.	Mean value	(2)	m.l.m	min E/M	max E/M	
1	1.56-2.3	1.56	0.83	1.47	$10^5 J/1.68$	$5\ 10^7 J/3.1$	
2	1.61-2.2	1.87	1.06	1.17	$10^9 J/0.7$	$8\ 10^6 J/2.5$	
2a	1.61-2.3	2.3	1.19	1.17	$10^9 J/0.7$	$1\ 10^9 J/3.5$	
3	1.7-2.2	2.01			$10^2 J$	$1\ 10^7 J$	
4	0.75-0.95	0.83			$10^2 J$	$5\ 10^5 J$	

1. Wujek, Poland, 2. Gottwald, Poland, $E > 10^7 J$ 2a. Gottwald, all tremors, 3. Doubrava, Ostrava, (CS) 4. Gottwald, Kladno, (CS), E--seismic energy, M--local magnitude, m.l.m--max. likelihood method.

The on-line, time dependent method presented in this study is en example of long-term seismic hazard evaluation. It gives the best results for areas separated from other workings, in not complex mining conditions. For short longwalls or longwalls excavated in complex conditions the method with constant parameters has been proposed (Glowacka & Pilecki 1991, Glowacka & Lasocki 1991), and good results were achieved.

In order to take into account more complex mining conditions (which prevail in mining), and to evaluate short-term seismic hazard (which is quite important) additional information has to be used. An example of hazard evaluation based on probabilistic synthesis is shown in fig.5. The *a posteriori* probability (p) is a combination of the *a priori* probability (p_o), the probability based on extracted volume (p_v -normalized $P(E_j)$ to one hour), and the probability evaluated using seismoacoustic activity (p_a). Probability maxima before the strongest events ($E \geq 10^6\ J$) can be seen. The probability gain before strong events is equal to 5, which resulted from the occurrence of strong tremors in periods when p_v had maxima. Examples of how to calculate synthetic probability are presented in (Glowacka and Pilecki 1991, Glowacka and Lasocki 1992, Glowacka 1992b).

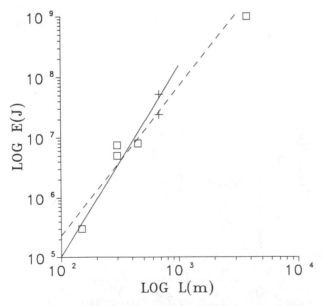

Figure 4. Logarithm of E versus logarithm of L for strong tremors in Wujek +, and Gottwald □ mines. (PL). Solid line: without $1\ 10^9 J$ taken into account, dashed line: with $1\ 10^9 J$ taken into account.

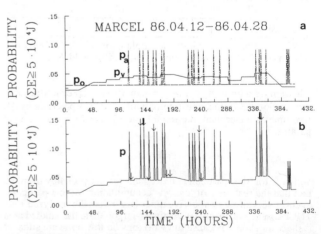

Figure 5. Results of seismic hazard evaluation for J5/707 longwall in the Marcel Coal Mine (PL). Modified from Glowacka & Pilecki (1991). a) p_o - *a priori* probability, p_a - probability evaluated using seismoacoustic activity, p_v - probability based on extracted deposit volume (3), normalized to 1 hour, b) p - '*a posteriori*' probability. Arrows mark the tremors with energy not less than $5.10^4 J$. The bold arrows mark tremors with energy not less than $10^6 J$.

4 CONCLUSIONS

The presented method of continuous, time-dependent, on-line seismic hazard evaluation, based on the dependence of seismic energy on the excavated deposit volume, gives results expressed in probabilistic form. This allows easy incorporation of outcomes from other methods of probabilistic seismic hazard evaluations.

For the studied areas probability increase and accumulated energy excess preceded most of strong tremors. A good correlation between the excess of accumulated energy and seismic energy released in strong tremors can be noticed. However, the excess of accumulated energy seems to be a good predictor of the energy of the forecoming strong shock; the long-term character of the presented method causes the probability gain of this parameter to be 1.5 only.

For regions where neither edges nor old workings occur, good correlation between the logarithm of seismic energy and time of energy accumulation can be noticed. The relation between logarithm of seismic energy and logarithm of L shows that the energy is roughly proportional to the volume of energy accumulation.

The above mentioned relations suggest that excess of energy accumulation, evaluated by the method has a physical meaning.

The parameter B values are similar for three areas situated in Upper Silesia.

The results obtained for the strongest tremor, which has regional origin, show that the occurrence of this tremor can be probably explained if a larger mining area, and a longer mining period are used for evaluations.

The method gives the best results for areas separated from other workings, therefore, if more than one longwall is extracted at the same time, this has to be taken into consideration during computations. Only events which result from deformation caused by the extraction of studied area are to be considered.

For taking into account more complex mining conditions, and for evaluating short-term seismic hazard, additional information should be incorporated, using methods of probability synthesis.

ACKNOWLEDGEMENTS

Sincere thanks to Prof. Andrzej Kijko (IGF Pol. Ac. Sc., Poland) for his helpful guidance. I particularly thank Prof. F. A. Nava from CICESE (Mexico) for reviewing the manuscript and for his helpful comments. I am also grateful to Profs. S.J. Gibowicz (IGF Pol.Ac. Sc., Poland) and M. Grad (UW, Poland) for their valuable remarks, which allow me to obtain some of the results.

REFERENCES

Benjamin, J.R., Cornell, C.A., 1970, *Probability, Statistics, and Decisions for Civil Engineers.* McGraw-Hill, New York

Bober A., Kazimierczyk M., 1979. Seismic activity and extraction with caving in Lubin mine. *Publ. Inst. Geoph. Pol. Ac. Sc* M-2 (123) p. 271-288

Dubinski J., Wierzchowska Z., 1973 Methods of calculation of seismic energy for mining tremors (in Polish) *Prace GIG Komunikat* GIG 591

Gibowicz S. J., 1963, Magnitude and energy of subterrane shocks in Upper Silesia., *Studia Geophys. et Geol.*, 7, p.1 - 19

Gibowicz, S.J.: 1990, Seismicity induced by mining. *Advances in Geophysics,* vol.32, p.1-74

Gibowicz, S.J., Bober A., Cichowicz A., Droste Z., Dychtowicz Z., Hordejuk J., Kazimierczyk M., Kijko A., 1979, Source study of the Lubin, Poland, tremor of 24 March 1977., *Acta Geophys. Pol.*, 27, p 3-38

Glowacka E.,1991, Prediction of seismic activity of rockmass in view of course of exploatation. Ph.D. Thes. (in Polish). Inst. Geophys. Pol. Ac. Sc.

Glowacka, E.: 1992 a, Probabilistic seismic hazard evaluation in underground mines. *submitted in Natural Hazards*

Glowacka, E., 1992 b, Application of the extracted volume as a measure of deformation for the seismic hazard evaluation in mines. *Tectonophysics* 202, p.285-290.

Glowacka, E., Kijko, A.: 1989a Continuous evaluation of seismic hazard induced by the deposit extraction in selected coal mines in Poland, *Pure and Applied Goephys.* Vol.129 no 3/4, 523-533.

Glowacka E., Kijko A., 1989 b, Continuous Evaluation of Seismic Hazard Induced by the deposit Extraction in Selected Mines in Poland, *Proc. XXI conference of ESC*, Sofia 23-27 1988

Glowacka, E., Lasocki, S.: 1991, Probabilistic synthesis of the seismic hazard avaluation in mines. *Acta Montana* (in press)

Glowacka, E., Pilecki, Z.: 1991, Seismo-acoustic anomalies and evaluation of seismic hazard at the "Marcel" Coal Mine. *Acta Geophysica Polonica* Vol. XXXIX, Nr 1, 47-59

Glowacka E., Korytynski A., Syrek A., 1988, Continuous assessment of seismic hazard resulting from mining in the Gottwald colliery, Upper Silesia. (in Polish with English abstract) *Publ. Inst. Geophys. Pol. Acad. Sc.*, M-10 (213)p.321-337

Glowacka, E., Rudajew, V., Bucha, V.: 1988, An attempt of continuous evaluation of seismic hazard induced by deposit extraction for the Robert Field in Gottwald Mine in Kladno, Czechoslovakia, *Publ. Inst. Geophys. Pol. Acad. Sc.*, M-10 (213)p.311-319

Glowacka, E., Stankiewicz, T., Holub, K.: 1990, Seismic Hazard Estimate Based on the Extracted Deposit Volume and Bimodal Character of Seismic Activity. *Gerlands.Beitr. Goephysic.* spec. issue "Induced Seismicity" 99, pp 35-43

Glowacka, E., Syrek, B., Kijko, A.: 1987, Dynamic evaluation of seismic hazard in selected area of the "Wujek" coal mine.(in Pol.) *Acta Montana* 75: 5-20

Hanks T. C., Kanamori H., 1979, A moment magnitude scale *JGR.* vol. 84,no B5

Hodgson K., Cook N. G. W., 1971, The mechanism, energy content and radiation efficiency of seismic waves generated by rockbursts in deep level mining. D. A. Howells et. al.-editors. *Dynamic waves in civil eng.* Willey Intersciences N. Y. p.121 - 135

Kanamori, H., Anderson D.L. 1975, Theoretical Basis of some empirical relations in seismology. *Bull. Seismol. Soc. Am.* vol 65, No5, pp 1073-1095.

Kijko, A.: 1985, Theoretical model for a relationship between mining seismicity and excavation area, *Acta Geophys.* Pol., vol 33, no 3 231-242

Lasocki, S.: 1989, Some estimates of rockburst danger in underground coal mines based on the energy of microseismic events, in: Hardy H.R.Jr. (ed), *Fourth Conference on Acoustic Emission/Microseismic Activity in Geologic Structures and Materials, Pennsylvania State University 22-24.10.1985, Proceedings*, Trans Tech Publs, Clausthal, 617-633

McGarr, A.: 1976, Seismic moments and volume changes, *J. Geoph. Res.* 81, 1487-1494

McGarr A., Bicknell J., Sembera E., Green R.W.E., 1989 Analysis of exeptionaly large tremors in two gold mining district of South Africa. *Pure and Applied Geophys.* vol 129 No 3/4, p. 295 - 308

McGarr A., Wielbos G. A., 1977, Influence of mine geometry and closure volume on seismicity in a deep - level mine. *Int. J. Rock Mech. Min. Sci. Geom.* p.139 - 145

Niazi M., 1985, Regression analysis of reported earthquake precursors. 1. Presentation of Data. *Pure and Applied Geophys.* vol. 122, p. 966-981

Scheidegger A. E.,1982, *Principles of Geodynamics.* Springer-Verlag. Berlin Heidelberg New York

Sibek V., 1963, On securing the safety of operating ore mines exposed to rockburst hazard. *Prec. 3rd Int. Min. Congr. Saltzburg - Safety in Mining*

Spottiswoode S.M., 1990, Volume excess shear stress and cumulative seismic moments. *Rockbursts and Seismicity in Mines.* 39-43. Rotterdam: Balkema

Stankiewicz T., 1989, Stochastic model of seismic activity and its application to seismic hazard estimates in mines. (in Pol.) Ph.D. Thesis, Inst. Geophys. Pol. Ac. Sci., Warsaw.

Wierzchowska Z., 1989, Aplication of seismic methods for the rockburst hazard control in mines. (in Polish) *Bezpieczenstwo Pracy w Gornictwie*, nr 2.

---

1 At present the Gottwald mine (Poland, Upper Silesia) has a name Kleofas, and in references can appear with both names.

The effect of backfill on ground motion in a stope

D.A. Hemp & O.D. Goldbach
Chamber of Mines Research Organisation of South Africa, Johannesburg, South Africa

ABSTRACT: Seismic waveforms from mine tremors with magnitudes between -1,9 and 3,0 were recorded in the gullies of backfilled and unfilled tabular stopes, as well as inside backfill bags, at two sites. In this study the waveforms have been compared to determine the effect of backfill on ground motion inside a stope. In addition, off-reef ground motion was recorded for comparison with that recorded on-reef. Analysis of the events shows that:-
i) Backfill causes a reduction in peak ground velocities and accelerations;
ii) The hangingwall beam in a backfilled stope resonates at higher frequencies than in an unfilled stope;
iii) Backfill causes a reduction in vibration times;
iv) Backfill causes a reduction in low frequency surface wave development.

1 INTRODUCTION

Considerable research has been carried out into the in-situ behaviour of backfill. In general it has been observed that backfill causes a reduction in rockfall accidents and rockburst damage. However, our understanding of the ground motion in a stope during a rockburst is limited. This research programme was set up in order to increase this understanding and to try and explain the observed reduction in damage in backfilled stopes.

The work was carried out in two stages. Stage I involved the analysis of waveforms recorded for small magnitude (M_L = -1,9 to 0,5) seismic events. Stage II involved the analysis of seismograms for events with larger magnitudes (M_L = 0,5 to 3,0).

Early work in this area (Spottiswoode and Churcher, 1988) predicted that backfill would reduce the length of the hangingwall beam and therefore:-
i) increase the resonant frequency of the hangingwall
ii) reduce absolute vibration levels
iii) reduce the differential movement between the roof and the floor and thereby reduce the cyclic strain on the backfill.

Adams et al. (1990) analysed a number of seismic events with magnitude less than 0,2 recorded in a backfilled stope and found that in general:-
i) ground motion is damped in the presence of backfill
ii) backfilled stopes experience higher dominant frequencies than unfilled stopes
iii) backfill restricts the zone of small compressive and tensile stresses above the gully.

In this study events of larger magnitude are recorded and a site has been selected where the same event can be recorded in equivalent filled and unfilled areas for comparison.

2 SITE DESCRIPTION

Two sites were chosen for recording the ground motion during mine seismic events in backfilled and unfilled stopes. At both these sites transducers were installed in the stopes which were high ERR (Energy Release Rate) areas with intense fracturing in both the hangingwall and footwall.

Two COMRO designed Portable Seismic Systems (PSS) (Pattrick et al. 1990) were used for monitoring the ground motion at both these sites.

Correlation between event locations and magnitudes recorded by the PSS and events recorded by mine wide regional networks was good but for the sake of consistency all magnitudes and locations used in the analysis were calculated from the data recorded by the PSS. Locations were determined from P- and S- wave arrival times and back azimuths from P-waves recorded at sites located in

relatively unfractured ground below the reef. Energy magnitudes were calculated from the signals recorded at these off-reef sites (Adams et al. 1990).

At both sites signals were sampled at 10000 samples per second with anti-alias filters set to 2500 Hz. The off-reef signals were amplified by a gain 5 times larger than that used in the stope in order to normalise for amplification of signals in the stope.

Transducers were mounted in aluminum boats and attached directly to the skin of the footwall or hangingwall. Bowden cable was used to protect co-axial accelerometer cable.

Vibrometer CE508 accelerometers which have a flat response between 6 Hz to 10 kHz were used at both sites.

2.1 Mine A

The first site chosen for recording data for this research programme was located in the NW corner of a shaft pillar. This site was at a depth of approximately 2250 m. Classified tailings were being placed in all panels and approximately 70 per cent filling of mined out areas was achieved. The backfill was placed approximately 7 m behind the face in conjunction with one row of rapid yielding hydraulic props at the face and timber packs on both sides of the gullies. The average dip of the Carbon Leader reef in this area was 23°. The stoping width was approximately 1 m. Mining proceeded in an easterly direction towards a NE-SW trending seismically active dyke with a throw of 45 m.

Data from this site was collected in two stages. During the first stage small events (up to magnitude 0,5) were recorded on-reef using geophones and in the second stage larger events (up to magnitude 3,0) were recorded using accelerometers. For both stages of the experiment a triaxial set of geophones, installed off-reef in relatively unfractured ground about 60 m below reef, was used to monitor off-reef ground motion.

Different transducer configurations were used for the two stages of data collection. During the first stage four triaxial sets of geophones were installed in the stope, two sets were installed in a dip gully, one on the hangingwall and one on the footwall; a third set was installed 10 m inside a backfill bag and the last set on the hangingwall of a strike gully close to the face.

During the second stage four triaxial sets of accelerometers were installed in the stope. Two were placed 5 m inside a 30 m long backfill bag - one set on the footwall and the other on the hangingwall. The third set of accelerometers was placed on the hangingwall in a dip gully adjacent to backfill, while the last set was placed on the hangingwall of a strike gully and moved forward as mining progressed (Figure 1a). This was done in order to examine the ground motion at various points in the backfilled stope relative to the off-reef ground motion.

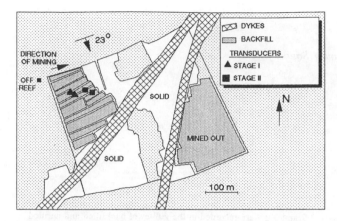

Figure 1a. Mine layout at mine A showing transducer positions for Stages I and II.

Of the approximately 200 events recorded by the system 64 were selected for analysis. 24 of these had magnitudes less than 0,5 and 40 had magnitudes between 0,5 and 3,0. The event magnitudes were calculated from the off-reef signals and the events located between 24 and 461 m from the on reef transducers. During the recording period, hydraulic stress meters installed inside a backfill bag revealed that the vertical stress inside the backfill had increased to from 6 MPa to 26 MPa with approximately 21 m of face advance.

2.2 Mine B

At the second site the Ventersdorp Contact Reef (VCR) was mined at a depth of approximately 2400 m. In the area under consideration a number of panels were mined with backfill and a number were mined using conventional support. The area was mined with a double cut and filled with cemented classified tailings. The total stoping width was 3,5 m.. Timber packs were used on both sides of the gullies .

The average dip of the VCR in this area was 23°. Mining proceeded in an easterly direction towards a NE-SW trending seismically active dyke with a throw of 40 m.

Five triaxial sets of accelerometers were installed. The first triaxial set was installed approximately 45 m below reef and was therefore ideal for recording off-reef behavior in unfractured rock.

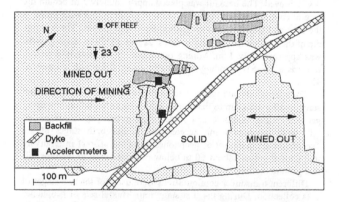

Figure 1b. Mine layout at mine B including transducer positions.

The remaining four triaxial sets of accelerometers were installed in the stope. Two of them were installed in a strike gully adjacent to a filled area and two in a strike gully adjacent to an unfilled area. In both areas one set was placed on the footwall and one on the hangingwall. Where possible the sets were moved forward as mining progressed. This layout of transducers allowed the study of the relative response of the hangingwall and footwall close to the face of filled and unfilled panels in the same stope subjected to the same event (Figure 1b).

Out of a total of approximately 200 seismic events, 32 events were selected for analysis. These had magnitudes ranging from 0,5 to 2.5 and they located between 56 and 414 m away from the on-reef transducers. Event locations and magnitudes were calculated from the off-reef signals in order to avoid over estimation of the magnitudes because of in-stope amplification.

The setup at mine B therefore provided data for stage II of the analysis.

3 RESULTS

Of the 24 events recorded during Stage I of the experiment approximately half located in the hangingwall and half in the footwall. The majority of the events recorded during Stage II located in the hangingwall.

The waveforms of all events have been analysed according to :-
1) peak ground velocities and accelerations
2) durations
3) frequency content

In addition instantaneous amplitude traces have been calculated for the larger events in an attempt to determine the energy distributions. The effect of backfill on seismic attenuation values has also been assessed.

3.1 Peak ground velocities and accelerations

Peak accelerations and velocities recorded during this study at both sites are given in Table 1. To normalize for variations in hypocentral distance, the peak velocities (v in mm/s) and accelerations (a in m/s^2) were multiplied by the hypocentral distance (R in m).

Table 1. Peak velocity and acceleration values recorded during the study.

	Mine B		Mine A	
	Off-reef	On-reef	Off-reef	On-reef
Peak velocity (mm/s)	18	453	40	368
Rv (m^2/s)	3,9	45,0	11,3	56,7
Peak acceleration (m/s^2)	8	203	-	277
Ra (m^2/s^2)	2,6	22,2	-	42,7

From these values it is evident that the ground motion recorded on reef is considerably higher than that recorded off-reef. This is a result of seismic amplification caused by the intense fracturing which typically occurs around deep level stopes. The median value of amplification of in-stope to off-reef ground velocities (corrected for distance) at mine A and at the filled area of mine B is 5.3. The value for the unfilled area at mine B is 9.9. Spottiswoode and Churcher (1988) and Adams et al. (1990) obtained on-reef peak ground velocity amplification of 2 to 2,5. The higher values obtained during this study are not considered excessive and may be due to the fact that the off-reef sites in previous work were still close enough to mining to be affected by stope amplification. The lower values recorded at the filled sites than at the unfilled sites indicates that backfill results in a clamping of the fractures around the stope and a subsequent reduction in the amplification of the ground motion.

In the case of random ground motion one would expect the peak velocity to be recorded on a horizontal component for two thirds of the events. This was found to be the case for data recorded at the off-reef sites and at the unfilled site at mine B. On the other hand peak ground velocities, recorded in the stope at mine A and at the filled area on mine B, occurred on the vertical components for the majority of the events, regardless of the events location. Horizontal clamping of fractures by backfill ensures that the stope in a filled area is not free to move in the horizontal direction. The stope, which is essentially two free surfaces is however free to move in the vertical sense, thus causing peak ground motions to be recorded on the vertical components.

Figure 2a shows the log(Rv) values recorded at the filled site plotted against the values recorded at the unfilled site for mine B. It is clear that peak values recorded at the unfilled site are typically twice those recorded at the filled site. Figure 2b shows equivalent data recorded in a gully adjacent to a filled area at mine A plotted against data recorded inside a backfill bag. This data shows a similar trend and data from both mines show that peak ground velocities are highest adjacent to unfilled areas and lowest inside backfill bags.

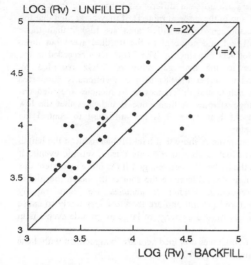

Figure 2a. Log (Rv) values calculated for the filled and unfilled sites at mine B.

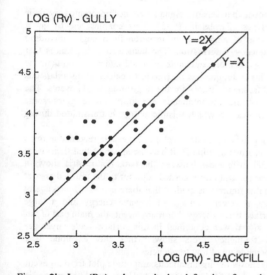

Figure 2b. Log (Rv) values calculated for data from the gully and inside a backfill bag from mine A.

McGarr and Bicknell (1988) interpreted far-field accelerations in order to predict near-field ground velocities. For events with magnitude greater than 2 they proposed values of about 2 m/s adjacent to a fault, with 4 m/s as an upper limit. A number of the events recorded during this study were near-field events, with hypocentral distances less than twice the source radius. A summary of the data recorded for several of these events is given in Table 2.

From this table it is evident that even next to an unfilled area the peak velocities are considerably lower than the values predicted by McGarr and Bicknell. The analysis of waveforms from a magnitude 2,0 which located approximately 35 m from the in-stope transducers illustrates this point. The peak velocity recorded was 125 mm/s 24 m from the event, and 60,9 mm/s 45 m from the event. The values also indicate that the velocities within the source region are not constant. The peak accelerations at these two points were $179 m/s^2$ and $59,6 m/s^2$ respectively. Dynamic shear-stress difference values at these two points were calculated using $\sigma = \rho Ra$ (Hanks and Johnson, 1976). The values obtained were 11,6 MPa and 7,2 MPa. The source radius of 93 m was calculated from Brune's (1970) corner frequency model.

Table 2. Summary data for a number of near-field events

Magnitude (mine)	Source Radius (m)	Minimum distance (m)	Peak velocity (mm/s)	Shear stress difference (MPa)
1,7 (A)	70	63	67	18,3
1,4 (A)	52	56	201	39,6
2,0 (A)	93	24	125	11,6
3,0 (A)	110	144	368	115,2
1,7 (A)	75	96	91	50,3
2,3 (B)	151	164	210	47,4
1,9 (B)	44	64	453	31,6
1,8 (B)	115	115	56	6,4
2,4 (B)	137	250	143	59,9

Peak ground velocities have been related to rockburst damage in underground excavations (Wagner, 1984, Lenhardt, 1988). The higher the peak velocities the more extreme the damage. Lenhardt predicted that a magnitude 3,0 event would result in damage to more than 7 panels. Damage caused by a magnitude 3,0 event that occurred on mine A was limited to only 3 panels. In general the damage was minor and the number of damaged panels is considerably lower than that predicted by Lenhardt. It is most likely that this reduction in damage was a result of the increased stability of the excavation because of the presence of backfill.

Although it has been suggested that damage potential is more accurately assessed using peak ground velocities than accelerations (McGarr et al. 1980) peak acceleration values have been analysed in this study. After the values have been corrected for the distance differential they show a similar trend to that observed with the peak velocities. In other words, peak ground acceleration values are highest adjacent to the unfilled area and the lowest inside the backfill.

3.2 Durations

An event duration was defined as the time taken for a seismic signal to decay to less than 20 per cent of its peak value. Table 3 contains a summary of the average duration values calculated at the two sites.

Table 3. Average duration times recorded at mine A and mine B.

	Mine A	Mine B
Location	Average duration (msecs)	Average duration (msecs)
Off-reef	83	99
Filled-gully	57	87
Inside backfill	46	-
Unfilled	-	104

From this table it is evident that vibration times in backfilled stopes are lower than those recorded adjacent to unfilled areas and lower than those recorded off-reef. The lower average values recorded at mine A are most likely a result of the higher percentage fill in this area. The lowest durations are recorded inside the backfill bag. This is a result of the increased stiffness of the rock surrounding the backfill. The larger stresses provide for efficient transmission of energy in the shortest time, thus making the stope more stable and reducing the potential for damage. In other words, seismic events are dissipated more quickly in filled stopes than in unfilled stopes.

3.3 Frequency content

Fracturing around a stope results in amplification of the signal and a subsequent modification in the seismic signature of the event. This means that spectral analysis no longer provides information related to the nature of the source. For this reason the peak in an in-stope velocity spectrum will be referred to as the dominant frequency of the event and not the corner frequency. Signals recorded at the off-reef sites should yield spectral peaks very close to the corner frequency values.

It has been proposed (Spottiswoode and Churcher, 1988) that the dominant frequency recorded in the stope is controlled by the length of the unsupported hangingwall beam. This means that in a conventionally supported stope the beam would have a length equivalent to the distance between the face and the point at which stope closure has taken place. They calculated a resonating beam length of 30 m in a conventionally supported stope. In a backfilled stope this beam would be considerably shorter having a length equivalent to the fill to face distance. The shorter the beam the higher its resonant frequency will be.

Analysis of the events recorded during stage I of this work showed that the dominant frequencies recorded off-reef varied between 130 Hz and 900 Hz depending on the magnitude of the event. The signals recorded inside the backfill bag yielded similar spectral peaks in the range 100 Hz to 760 Hz. This would imply that because of the high stresses generated inside backfill and the stiff response of backfill, the position inside the bag resembled 'solid rock' behaviour similar to that recorded off-reef. Data recorded in the gullies adjacent to the fill on the other hand were confined to narrow frequency bands, generally between 100 and 350 Hz. The peaks consistently fell into these narrow ranges, regardless of the magnitude of the event. This narrow band frequency response was clearly a site effect.

Dominant frequencies calculated for the larger events are plotted in Figure 3a and 3b as a function of magnitude.

From these figures it is apparent that the trends observed for small magnitude events are consistent for the larger magnitude events.

Dominant frequency values recorded at the off-reef site at mine A are substantially lower than those recorded in the stope. In general the frequencies recorded inside the backfill are lower than those recorded in the gullies. In other words the point inside the backfill bag behaves in a similar way to the solid rock behaviour recorded off-reef. The higher frequencies recorded in the gullies are similar to those recorded for the small events. The characteristic length, L, of a structure that resonates at a frequency f can be estimated from $L \sim V/2f$ where V is the S-wave velocity. The length of beam that

would resonate at these higher frequencies is approximately 10 m. This is roughly the distance behind the face at which placed backfill is taking load and effectively 'closing' the hangingwall and footwall. Results from the analysis of the larger events differ from those obtained for the smaller events in that the magnitudes have some effect on the dominant frequencies of the larger events.

In general, there are no clear differences between the frequencies recorded off-reef and in the stope at mine B. It may be argued that the values recorded adjacent to the filled area are higher than those recorded both off-reef and adjacent to the unfilled areas but these differences are not very convincing. The frequencies recorded in the stope are closer to the average value of 70Hz recorded by Spottiswoode and Churcher (1988) in a conventionally supported stope. The more substantial differences in frequencies recorded on-reef versus in-stope at mine A than mine B indicates that the low percentage of backfill at mine B is not sufficient to control the resonance of the stope.

The results from mine A show that backfill has reduced the length of the stope beam that is able to resonate during seismic events. It would seem that the higher the percentage fill the more consistent this reduction is. This is desirable since the shorter the beam the higher the resonant frequency. Higher frequencies are more rapidly attenuated in fractured ground and are therefore less likely to cause damage. In addition, most the energy of large magnitude events is in low frequencies.

The resonant frequencies obtained here are in agreement with those obtained by Adams et al. (1990).

3.4 Instantaneous amplitudes

It has been shown that seismic signals can be represented by their complex envelopes (Farnbach, 1975). This enables one to get a natural separation of amplitude information from angle (including phase and frequency) information. The benefit of doing this is that often the envelope is more amenable to visual interpretation than the signal itself. This technique has been used to calculate instantaneous amplitude and frequency traces for the larger magnitude events. The general shape of the instantaneous amplitude curve gives some indication of the way in which seismic energy is transmitted during the event.

On comparing the amplitude envelopes for events recorded at mine A with those recorded at mine B, it became apparent that there were fairly significant differences between the two. Figure 4(i) shows a typical envelope for an event recorded adjacent to the unfilled area at mine B. From this figure it is evident that there is a small package of P-wave energy, a larger package of S-wave energy and a large package of surface wave energy. For many events the main part of the energy in the signal was contained in this surface wave envelope. Instantaneous frequency plots showed that this envelope was predominantly made up of low frequencies.

Figure 4(ii) shows the instantaneous amplitude plot from an event recorded inside a backfill bag at mine A. Here very little energy is contained in surface waves; most of the energy is in the S-wave coda. Both traces are from magnitude 0,5 events which located approximately 85 m away.

From this figure it appears that backfill causes a reduction in low

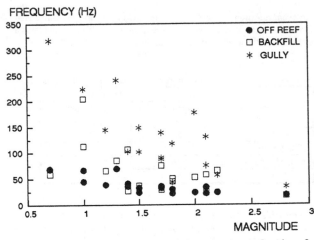

Figure 3a. Dominant frequencies recorded at mine A as a function of event magnitude.

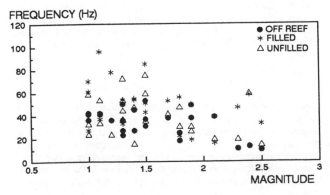

Figure 3b. Dominant frequencies recorded at mine B plotted as a function of event magnitude.

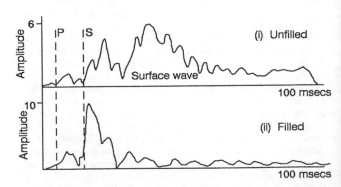

Figure 4. Instantaneous amplitude envelope for a magnitude 0,5 event. Recorded (i) adjacent to an unfilled area on mine B, (ii) inside a backfill bag on mine A.

frequency surface wave development. These results tie in with the concept of a length of unsupported hangingwall beam. In backfilled stopes this is relatively short and surface wave development is inhibited. The length of this beam in unfilled stopes, on the other hand, is relatively long and surface waves are well developed. Backfill also results in a clamping of the fractures surrounding the stope which will in turn lead to a reduction in the extent of the highly fractured low velocity zone around the stope. This reduced vertical extent would explain the low frequencies of the surface waves.

3.5 Attenuation

Intrinsic attenuation can be calculated from body wave spectra of seismic events. If there are no site effects, then the attenuation is given by:

$$A(f) = \frac{A(0).\exp(\kappa.f)}{1+(f/f_0)^2}$$

where $A(0)$ is the spectral plateau and f_0 is the corner frequency. The attenuation factor (κ) is not proportional to distance (D) but can be described by:

$$\kappa = \kappa_o + \frac{\pi D}{vQ}$$

where v = P- or S- wave velocity and $Q = 200$.

It has been proposed (Spottiswoode, 1993) that variations in the value of κ_0 are associated with near-source fracturing. The higher the κ_0 value, the greater the degree of fracturing. κ_0 values were calculated for all signals recorded at the off-reef sites where site effects were minimal. Although these signals were recorded in relatively unfractured ground the majority of the events located in the highly fractured ground surrounding the stope. An analysis of the κ_0 values should therefore give some insight into fracture patterns around the stope.

The average value of κ_0 calculated for events recorded off-reef at mine B is 3.1 msecs, while the average value for off-reef data recorded at mine A is 0.9 msecs. The difference in these values would imply that there is a greater degree of near-source fracturing at mine B than at mine A. Since the majority of the events recorded at both mines located in the stope or close to the face, it can be assumed that the variation is a result of different fracture patterns surrounding the stope. Higher κ_0 values imply a greater degree of fracturing, which at mine B could be explained by the low percentage fill compared to that at mine A. It would appear that backfill causes a reduction in the degree of fracturing by increasing the horizontal stresses and thereby closing up fractures. The more limited extent of fracturing would ensure that the stope was more stable.

4 CONCLUSIONS

In general, the research into ground motion in backfilled stopes has shown that backfill reduces the overall ground motion during seismic events.

In particular:
- backfill causes a reduction in peak ground velocities and accelerations, and an associated reduction in the off-reef to on-reef amplification;
- backfill causes a decrease in the length of the unsupported hangingwall beam, thereby increasing the resonant frequency of the beam. There is less seismic energy available to do damage at these higher frequencies, which are more rapidly attenuated than lower frequencies;
- high backfill stresses cause closure of fractures in the rockmass surrounding the stope. This together with the reduced beam length, provides for efficient transmission of energy. As a result the lowest vibration times are recorded inside backfill and the highest in gullies adjacent to unfilled areas;
- low frequency surface wave development is inhibited in backfilled stopes compared to unfilled stopes;

- analysis of variations in κ_0 values shows that the extent of near source fracturing is reduced in well filled stopes.

From the above points it can be concluded that the hangingwall in backfilled stopes is more stable than in unfilled stopes. Backfilled stopes vibrate at less damaging frequencies and for shorter time periods during seismic events. These factors explain the observed reduction in rockfall accidents and rockburst damage in backfilled stopes.

Acknowledgments

This work forms part of the research programme of the Rock Engineering Division of COMRO. The co-operation of Management and Staff at the mines concerned is gratefully acknowledged. We would like to thank Messrs. A. Ruskovich, D de Beer and M. Gonlag for their assistance during network installation and maintenance.

REFERENCES

Adams, D.J., Hemp, D.A. and Spottiswoode, S.M. 1990. Ground motion in a backfilled stope during seismic events. *Proc. Static and Dynamic Considerations in Rock Engineering*, Brummer (ed.), Balkema, Rotterdam, pp. 13-22.

Drunc, J. N., 1970. Tectonic stress and the spectra of seismic shear waves from earthquakes, *J. Geophys. Res.*, 75, 4997-5009.

Farnbach, J.S., 1975. The complex envelope in seismic signal analysis. *Bulletin of the Seismological Society of America*. Vol. 65, No. 4, pp. 951-962.

Hanks, T.C. and Johnson, D.A., 1976. Geophysical assessment of peak accelerations, *Bull. Seismol. Soc. Am.*, 66, 959-968.

Lenhardt, W.A., 1988. Damage studies at a deep level African gold mine. *Proc. 2nd International Symposium on Rockbursts and Seismicity in Mines*, Minneapolis, Minnesota.

McGarr, A. and Bicknell, J., 1988. Estimation of the near-fault ground motion of mining induced tremors from locally recorded seismograms in South Africa. *Proc. 2nd International Symposium on Rockbursts and Seismicity in Mines*, Minnesota, Minneapolis.

McGarr, A., Green, R.W.E. and Spottiswoode, S.M., 1981. Strong ground motion of mine tremors: some implications for near source ground motion parameters, *Bull. Seismol. Soc. Am.*, 71, 295-319.

Pattrick, K.W., Kelly, A.M. and Spottiswoode, S.M., 1990. A portable seismic system for rockburst applications. *Proc. International deep mining conference*, Johannesburg, SAIMM, pp. 1133-1146.

Spottiswoode, S.M. and Churcher, J.M. 1988. The effect of backfill on the transmission of seismic energy. *Proc. Backfill in South African Mines*, Johannesburg, SAIMM, pp. 203-217.

Spottiswoode, S.M. 1993. Source parameters and seismic attenuation of small seismic events in deep-level mines. *Proc. 3rd International Symposium on Rockburst and Seismicity in Mines*, Kingston, Canada.

Wagner, H. 1984. Support requirements for rockburst conditions. *Proc. 1st International Symposium on Rockbursts and Seismicity in Mines*, Johannesburg: SAIMM.

Rockbursts and Seismicity in Mines, Young (ed.) © 1993 Balkema, Rotterdam, ISBN 90 5410 320 5

Rockburst damage potential assessment – An update

P. Jesenak, P. K. Kaiser & R. K. Brummer
Geomechanics Research Centre, Sudbury, Ont., Canada

ABSTRACT: The paper presents an update on GRC's current research task "Rockburst damage assessment procedure". The previously established relationship between the magnitude of a seismic event, distance and peak particle velocity, applied to the ground motion data from Creighton mine, has been further developed and the term "Target Magnitude" (or Apparent Magnitude) has been introduced to take into account mining related variations in energy loss along the transmission path. Based on the peak particle velocities predicted in this manner, the limits for anticipated level of damage have been estimated. Rockburst damage data collected underground following the previously introduced damage classification procedure are compared to these predicted limits of damage, and the range of applicability of the method (in terms of ground conditions and failure mode) is defined.

1 INTRODUCTION

1.1 Empirical design approach

The conventional engineering design approach essentially consists of (1) the identification of potential failure modes, and (2) a comparison of the available resistances (capacity) to the static and dynamic driving forces (demand). In the keynote address entitled "Support of Tunnels in Burst-Prone Ground -- Toward a Rational Design Methodology", Kaiser (1993) concludes that, for the implementation of this design approach for bursting ground, two fundamental questions need to be addressed if rockbursts cannot be prevented by other means. First, where and what kind of damage is to be anticipated and, second, how violent is the failure and how can rock support systems effectively contain rockburst damage.

As St. John and Zahrah (1987) pointed out, "despite the availability of relatively sophisticated methods for investigating the dynamic reponse of underground structures to seismic loading, design tools still remain relatively simple and uncertainties in the data defining the problem will render apparent improvements offered by detailed analyses illusory rather than real". As a consequence, empirical design and damage assessment criteria will, and should, always constitute a component of design. Despite obvious and severe limitations of such empirical relationships, engineering logic dictates that past experiences are considered during the design process.

Many investigators have developed direct relationships between damage levels and ground motion parameters. However, difficulties resulting from highly subjective assessment criteria and incomplete or questionable peak ground motion data severely limit the applicability of these relationships. This limitation will only be overcome when high quality strong motion data has been collected and properly analyzed. In the meantime, the need for regionally applicable empirical relationships based on less subjective damage assessment procedures and carefully "filtered" ground motion data remains.

Whereas there are extensive databases available in other parts of the world, only limited efforts have been made in the past (Hedley, 1992) to establish a reliable empirical damage model for Canadian hard rock mining conditions. The development of such a model consists of three elements:
- define damage accurately
- locate damage relative to the seismic source that actually caused the damage
- group data based on factors dominating specific response characteristics.

1.2 Damage assessment

As outlined by Kaiser et al. (1992), four factors must be considered during the assessment of the damage-potential to an underground opening: rockwall quality, failure potential or stress level (comparable to RCF introduced by Wiseman, 1979), mine or local rockmass stiffness and support effectiveness.

These parameters are determined as described by Kaiser et al. (1992). This paper presents an up-date on our ongoing effort to develop a rockburst damage database and to interpret the collected data in an empirical manner. The data is still site specific and thus primarily applicable to Creighton mine, Inco Ltd., Sudbury.

Starting from relationships that seem to work elsewhere, i.e., Mercalli Intensity Scale, a method for assessing rockburst damage was developed (Kaiser at al., 1992) and the data was related to the available ground motion data (peak particle velocities; distance and magnitude). As the ground motion data is improving, damage will be compared to increasingly more sophisticated source parameters. Eventually, this data will provide the foundation for a rational support design strategy to contain rockburst damage (Kaiser, 1993).

For support design, it is of primary importance to separate events causing damage from those that are less critical and to establish means to extrapolate to conditions seldom observed, i.e., closer to the source. The challenge is to define how to extrapolate from data collected at rather large scaled distances from the source. Clearly, the frequency dependent effect of wave attenuation and the near-field ($1/R^2$) decay must be considered. In other words, extrapolation based on relationships fitted to entire data sets containing high and low energy events (e.g., Hedley, 1992) as well as far and near-field data is clearly not acceptable. This paper presents our current approach to establish a basis for data extrapolation. As demonstrated by Kaiser (1993) most of the damage locations presented in this paper fall into the far-field region. Far field conditions are encountered at several wavelengths from the source, hence, the extent of the far-field zone depends on the dominant frequency. The established relationships should obviously only be applied in this domains. It must be pointed out that, while we believe the logic is sound, insufficient data is as of yet available to conclusively determine the limitations of the established relationships.

Two very important issues, damage due to aftershocks and conditions near the source (near-field behaviour), are still unresolved and are discussed in more detail by Kaiser (1993). These factors obviously affect the extrapolations toward the source, and strong motion data, when available, must eventually be used to verify or improve the basis for damage prediction.

2 RELATING GROUND MOTION CHARACTERISTICS TO DAMAGE LOCATION

2.1 Data base

Ground motion data from one mine (Inco's Creighton mine) is presented here to avoid misinterpretation due to unknown influences of change in overall conditions between different mines (geology, stress level, mining methods etc.). The data was acquired by CANMET's microseismic monitoring system working with triaxial accelerometers and the magnitudes of the recorded seismic events (determined by the Eastern Canadian Seismic Network) ranged from 1.5 to 2.4 on the Nuttli scale (0.9 to 1.7 on the Richter scale). The magnitude was converted from Nuttli (M_N) to Richter (M_R) scale using a relationship

$$M_R = 0.87M_N - 0.37 \tag{1}$$

which was derived from the radiated seismic energy relationships published by Gutenberg and Richter (1956) and Hedley (1992). The distances between the seismic source and sensors were calculated using the source locations provided by the mine's microseismic system (they are currently being reviewed using the damage-based verification approach presented by Kaiser (1993)).

In order to compare the data from different seismic events, distances between the sensor and the source have been scaled to the magnitude and resulting scaled distances were then related to measured peak particle velocities (ppv). A data base with available ground motion data including the source locations, peak particle velocities, accelerations and displacements, magnitudes, energy numbers and damage descriptions has been created. Rockburst damage data for comparison of predicted and observed damage were collected underground and from the unusual occurrence reports at the mine site, using the procedure introduced by Kaiser et al. (1992).

2.2 Data processing and presentation

For the presentation of the ground motion data, Hedley (1990) proposed the scaled distance relationship

$$SD_N = \frac{R}{10^{0.33M_N}} \tag{2}$$

using the Nuttli magnitude and distance R in meters. He also suggested that the ppv (in mm/s) generated by a seismic event can be determined as a function of scaled distance:

$$ppv = C \times SD^a \tag{3}$$

In order to predict ppv for support design, the goal of the analysis is to define reliable values for both constants C and a in Equation (3), and to establish the range of applicability of these parameters.

A criterion proposed by McGarr et al. (1981)

$$\log Rv = 0.57M_R - 0.05 \tag{4}$$

relates peak particle velocity (v in cm/s), distance (R in meters) and Richter magnitude. Equation (4) leads to a scaled distance with a slightly different exponent (0.57 instead of 0.33)

$$SD_R = \frac{R}{10^{0.57M_R}} \tag{5}$$

The scaled distance defined by Equation (5) was also used for the data analysis presented in this paper. Equation (4) rewritten into the form of the Equation (3) yields constants C = 891 and a = -1.

In order to extrapolate from the available data toward the source, it is most important to establish the slope of scaled distance-ppv relationship. Due to the wide scatter in the data shown in Figure 1, the correlation is so poor that a linear fit to all data is meaningless.

In attempting to define a consistent data set as the basis for regression and extrapolation, it was recognized that the seismic energy arriving at the sensor varies widely and provides a means for grouping ppv - data.

The energy radiated from the seismic source can be determined from the Equation (6) by Gutenberg and Richter (1956)

$$\log E_S = 1.5M_R - 1.2 \tag{6}$$

where E_S is the seismically radiated part of the source energy in MJ and M_R is the magnitude on the Richter scale. The same equation was adopted by McGarr (1984).

The seismic energy recorded by the triaxial sensor (calculated by CANMET from the full wave forms) represents a fraction of the energy radiated from the seismic source and it is referred to as the target energy (energy arriving at the target - measuring device or excavation wall). This target energy fully reflects the magnitude of the seismic source only if there is no energy loss during the wave transmission. Hence, it is highly dependent on the properties of the rockmass between the source and the sensor. Significant geological structures, zones of broken rockmass and intervening mining apparently cause higher rates of attenuation of the seismic wave resulting in lower target energy and peak particle velocity at sites protected in this manner.

Based on a comparison of the theoretical (calculated from Eqn (6)) source energy and the target energy (received at the sensors), the amount of the seismic energy lost during the transmission through the rock medium was estimated. The ground motion data presented on Figure 1 was subdivided into three groups with arbitrary limits for the estimated energy losses set at 30 and 80%. These groups correspond to previously defined types of the ground conditions covering a wide range from minimum to maximum attenuation. A careful study of the mine geometry relative

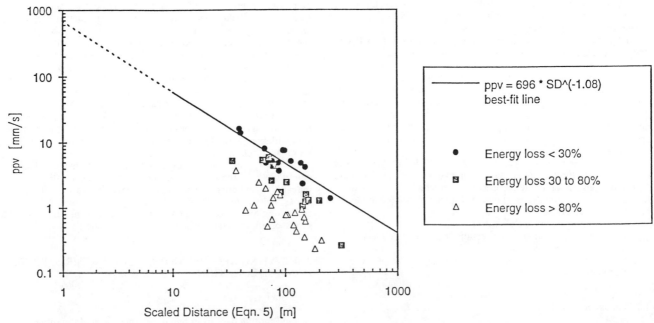

Figure 1 Peak particle velocity versus scaled distance (Eqn. 5) for triaxial data from Creighton mine (data to July 1992) with linear regression line for data with minimum energy loss

to the sensor and event locations demonstrated that high energy losses could primarily be attributed to mined out, backfilled or highly fractured zones (Kaiser et al., 1993).

For the group of data with energy losses estimated to be less than 30% (representing minimum attenuation of the seismic wave) the parameters of the linear regression line for use in Equation (3) are C = 696 and a = -1.08. The slope a, close to -1, agrees well with Equation (4) derived for conditions with essentially solid rock between the source and sensors. Figure 1 shows that the highest energy waves arrive at the target with minimum attenuation. Most importantly, the data demonstrates that radial spreading is an acceptable approximation for conditions where little non-elastic energy loss is encountered. This reflects the most critical condition with respect to support design.

2.3 Effects of differential wave attenuation

The seismic wave spreading from the source is subjected to different degrees of attenuation as it encounters media with varying properties in the transmission path (e.g. major structures, broken rock, backfilled or empty excavations). Therefore the magnitude of the seismic event characterizes the source but does not truly reflect the dynamic conditions at the target (sensor or the affected excavation wall). In order to describe these conditions better, the term "Target Magnitude" (or Apparent Magnitude) is introduced. It is the magnitude as it appears at the target, or the magnitude the target senses (which is normally less than the source magnitude). The target magnitude is the magnitude of an equivalent event, would homogeneous, isotropic, elastic conditions have existed between the source and the sensor. In other words, the target/apparent magnitude is the magnitude of the seismic event normalized relative to the target.

The target magnitude can be determined by Equation (6) with the target energy as input. For every data point in Figure 1, the target magnitude has been determined and used to calculate the target scaled distance SD_t the scaled distance corrected for the energy losses. High energy losses (high attenuation) result in larger SD_t. For the data points with no energy losses (no attenuation) SD and SD_t are equal. In a few instances, the SD_t is smaller because the target energy was actually higher than the estimated source energy. Figure 2 presents the same data as shown in Figure 1, plotted against the target scaled distances.

The scatter in the data is significantly smaller even though three distinct groups can still be identified. The group with minimum energy loss, plotted as black circles, shows almost perfect geometric spreading, resulting in regression line parameters

C = 550 and a = -1.04.

This demonstrates that it is meaningful to assume geometric spreading at Creighton mine for extrapolation within the far-field domain. Furthermore, for damage prediction and support design, it is necessary to consider situations with minimum energy loss. Hence, a forced fit with the slope a = -1 through the low energy loss data is justified and leads to C = 460 (Figure 2). This "forced fit mean" is representative to characterize average dependence (50% confidence) of ppv on the target scaled distance. With the standard deviations calculated on this limited data set, C = 670 at a = -1 was determined for the upper 95% confidence limit on current data plotted as a dashed line on Figure 2. These two limits are used by Kaiser (1993) for support design calculations. Until additional data are available, the relationship is recommended for damage assessment in the far-field domain.

3 PPV AND DAMAGE SCALES

Rockburst damage to the underground openings was classified according to the procedure proposed by Kaiser et al. (1992), where damage to the rock and support is evaluated separately and the initial conditions of the excavation are recorded at the same time. Based on the experience from various damage assessment studies, relating damage level and peak particle velocity, a preliminary empirical relationship was proposed:

$$ppv = K \times 2^{(RDL-1)} \qquad (7)$$

where the constant K = 50 mm/s was found to provide acceptable results for Creighton data and RDL is the rock damage level obtained from the classification system. According to Equation (7), significant damage would be expected between ppv = 400 and 800 mm/s and no damage should be found for ppv below the lower limit of 50 mm/s.

3.1 Comparison with empirical database

Several attempts have been made in the past to develop an empirical relationship between damage levels, usually very subjectively assessed, and ground motion parameters. Experience from mining (McGarr, 1983) indicates that damage to underground openings is best correlated with peak particle velocity.

Analyzing earthquake damage, Owen and Scholl (1981) established a "no damage limit" at ppv lower than 200 mm/s (or 0.2g) and "major damage limit" above 900 mm/s (or 0.5g). They also pointed out that less confined conditions, near portals of road

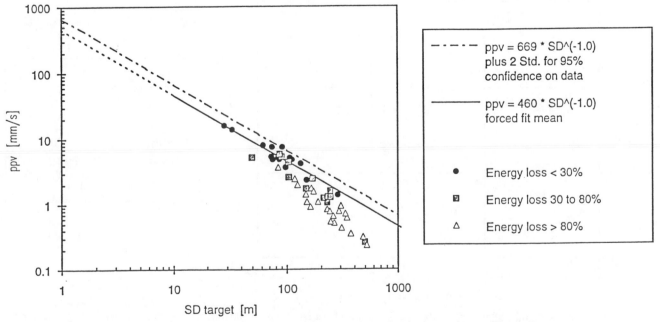

Figure 2 Ground motion data corrected for the energy losses (same data as Figure 1)

tunnels, are more susceptible to damage.

Several empirical design criteria have also been suggested in order to protect excavations from the damaging effects of blasting. Langefors and Kihlstrom (1963) expected rockfalls at ppv of 300 mm/s while ppv of 600 mm/s should cause new cracking in rock. Even higher values were suggested from the results of the Underground Explosion Test Program where cracking of the rock was observed for ppv of 900 mm/s with continuous damage from ppv = 1800 mm/s. These extremely high values of ppv suggested for blasts with one high-peak compressional pulse cause less damage than a seismic event with several cycles and substantially longer vibration durations. Dowding et al. (1983) proposed that the number of stress cycles plays a critical role for the extent of permanent damage to rock. For this reason, correlation of ppv and the blast damage should not be directly applicable for rockburst conditions.

In his recent work Hedley (1992), based on observations from Ontario mines, suggested that no damage should be encountered at ppv below 50 mm/s, falls of loose rock should occur at ppv between 50 and 300 mm/s, falls of ground should be encountered from 300 to 600 mm/s, and severe damage is expected for ppv above 600 mm/s. These limits compare well with the limits defined by Equation (7).

The level of damage to an excavation depends also on the character of the rockmass close to the opening. Three types of wall conditions typically exist:

a)   "intact" or slightly jointed rock which has to be broken prior to ejection (excellent rockmass quality, no destress blasting during excavation),

b)   zone of highly fractured rock around the opening due to stress concentration and/or destress blasting (good to poor rockmass quality), part of the energy is used to create new fractures; and

c)   same as previous case but with well defined weakness planes (structures), creating potentially unstable wedges with a static factor of safety close to unity. These conditions are susceptible to seismically induced falls of ground because they are easily triggered by relatively low accelerations.

Evidence collected from underground observations at Creighton mine suggests that Equation (7) should be applicable to rockburst conditions characterized by conditions (b). Since conditions (c) are more susceptible to failure by seismic triggering, these types of failures are anticipated to occur at larger distances from the source.

4 DAMAGE LEVEL PREDICTION

By setting equations (3) and (7) equal, a relationship between the expected damage level (here represented by ppv limits), Target Magnitude (or Apparent Magnitude) of the event and the distance to the source can be established. C = 460 and 670 provide limits for 50 and 95% respectively. Limits for minor (RDL 2) and severe damage (RDL 5) to the rock, predicted in this manner, are presented in Figures 3 and 4 for comparison with rockburst damage data from Creighton mine. Damage data are plotted at the source magnitude rather than at the (unknown) target magnitude. This increases the scatter. A significant portion of the data would be shifted downward if the target magnitudes were known.

Figure 3 shows excellent agreement between predicted and observed damage. The predicted limits were derived for conditions with minimum losses of energy in transmission path, thus representing the upper limits for the distance R between the source and damage locations. The fact that most data points plot to the left of predicted limits, i.e., closer to the source than expected, suggest that in many cases damage occurred in conditions when a substantial amount of the energy is lost between the source and the target. This suggests that careful mining sequencing (e.g. stope development between seismically active areas and important access drifts) can protect openings from the damaging effects of high ppv.

Most of the data representing severe damage, shown on Figure 4, do not correspond well with predicted limits. Detailed investigations of each case showed that all data points plotting to the right of predicted limits (labeled as special conditions) represent seismically induced falls of ground or combinations of extremely poor ground and ineffective support (conditions (c) above or mode III, (Kaiser, 1993)). Furthermore, the fact that the limits of ppv = 100 mm/s (RDL 2), used to predict minor damage (Fig. 3), would actually bound the data for RDL 5 well, suggest that these severe damages occurred at very low values of ppv. Conditions where falls of ground occur are extremely sensitive to the support's effectiveness and depend to a large degree on the static factor of safety (Kaiser, 1993). These must be treated separately.

5 CONCLUSIONS

This paper presents an update on the current status of GRC's research task on "Rockburst Damage Potential Assessment".

When analyzing the ground motion data, it is important to consider energy losses between source and target which depend on intervening ground conditions. The Target Magnitude (or Apparent Magnitude), introduced to describe the source effectiveness at the target, was found to vary widely for conditions encountered at Creighton mine in dependence on the

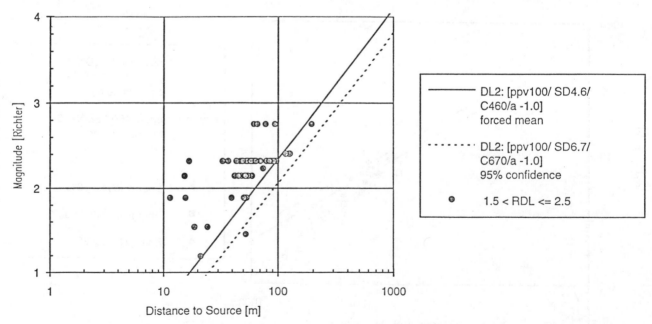

Figure 3 Predicted limits of anticipated damage for damage level DL 2 (ppv = 100 mm/s) compared to minor damage recorded underground

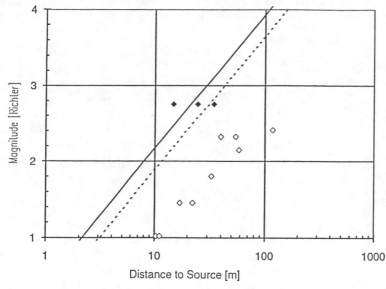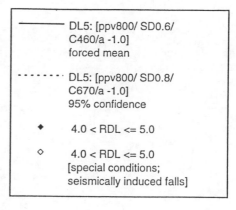

Figure 4 Predicted limits of anticipated damage for damage level DL 5 (ppv = 800 mm/s) compared to severe damage recorded underground

state of mining and geological conditions between the source and the target.

It is concluded that, for the purpose of support design in the far-field domain, the relationship given by Equation (3) is applicable for seismic events large enough to cause damage. For Creighton mine, the constant C = 460 is about 50% of that proposed by McGarr et. al. (1981) for South African conditions.

The relationship between the peak particle velocity and target scaled distance presented by Equation (3) (C = 460, a = -1) is so far only supported by data from one mine and before applied elsewhere, local ground motion data have to be acquired and analyzed in a similar fashion.

The peak particle velocity limits proposed for predicting damage levels (Equation (7)) can only be applied to rockburst damage in statically stable ground conditions described as conditions (b) in section 3.1 with failure mechanism characterized by mode II (Kaiser, 1993). The potential for seismically induced falls of ground representing ground conditions (c) and failure mode III, has to be recognized applying classical geological methods, and trigger limits as well as support survival limits must be determined separately as suggested by Kaiser (1993).

Application of this method for rockburst damage assessment (Kaiser et al., 1992) to Creighton data has confirmed its usefulness and therefore it is recommended for use by ground control engineers at other mines. Since both ground motion and damage data constitute the basis for this damage potential assessment, the reliability of this method will increase as the data base is expanded.

ACKNOWLEDGMENTS

This research is part of the Canadian Rockburst Research Project supported by many mining companies through the Mining Research Directorate (MRD) and was supported by a grant from the Ontario Ministry of Northern Development and Mines. This funding and the extensive cooperation of the staff at the various mines is thankfully acknowledged. In particular, the enthusiastic cooperation of D. O'Donnell, C. Langille, A. Punkkinen and T. Villeneuve during site visits is greatly appreciated. Discussions with various members of GRC's research team, in particular, D. McCreath and D. Tannant were extremely useful and have contributed to the establishment of these research findings.

REFERENCES

Dowding, C.H. and A. Rozen, 1978. Damage to rock tunnels from earthquake shaking. Jrn. geotech. Engng. Div. Am. Soc. Civ. Engrs. 104 (GT2), 175-191

Gutenberg, B. and C.F. Richter, 1956. Seismicity of the Earth and associated phenomena. 2nd. edition, 1965, Hafner Press, New York, 310p.

Hedley, D.G.F., 1992. Rockburst Handbook for Ontario Hardrock Mines, CANMET, 272p.

Hedley, D.G.F., 1990. Peak particle velocity for rockbursts in some Ontario mines. 2nd International Symposium on Rockburst and Seismicity in Mines, Minnesota 1988, Ed. Fairhurst, Balkema, 345-348; discussions 428-429.

Kaiser, P.K., 1993. Support of tunnels in burst-prone ground -- Toward a rational design methodology. 3rd Int. Symposium on Rockburst and Seismicity in Mines, Kingston, 30p.

Kaiser, P.K., P. Jesenak and R.K. Brummer, 1993. Rockburst damage potential assessment. International Symposium on Assessment and Prevention of Failure Phenomena in Rock Engineering, Istanbul, April, Balkema (in press).

Kaiser, P.K., D.D. Tannant, D.R. McCreath and P. Jesenak, 1992. Rockburst damage assessment procedure. International Symposium on Rock Support, Sudbury, Balkema, 639-647.

Langefors, U. and B. Kihlstrom, 1963. The modern technique of rock blasting. John Wiley and Sons, New York; Almquist and Wiksell, Stockholm.

McGarr, A., 1984. Some applications of seismic source mechanism studies to assessing underground hazard. 1st International Symposium on Rockbursts and seismicity in Mines, Johannesburg, 199-208.

McGarr, A., 1983. Estimated ground motion for the small near-by earthquakes. Seismic Design of Embankments and Caverns, New York: ASCE, 113 - 127.

McGarr, A., R.W.E. Green and S.M. Spottiswoode, 1981. Strong ground motion in mine tremors: some implications for near-source ground motion parameters Bulletin of the Seismological Society of America, 71(2): 295-319.

Owen, G.N. and R.F. Scholl, 1981. Earthquake engineering of large underground structures. JAB-7821. San Francisco: URS/John A. Blume.

Roberts, M.K.C. and R.K. Brummer, 1988. Support requirements for rockburst conditions. Journal of South African Institute of Mining and Metallurgy, 88(3), 97-104.

Spottiswoode, S.M., 1984. Underground seismic networks and safety. Symposium on Monitoring for Safety in Geotechnical Engineering, Sangorm, 39-45.

St. John, C.M. and T.F. Zahrah, 1987. Aseismic design of underground structures. Tunneling and Underground Space Technology. 2 (2), 165-197.

Wagner, H., 1984. Support requirements for rockburst conditions. 1st International Symposium on Rockburst and Seismicity in Mines, Johannesburg, 1982, SAIMM, 209-218.

Wiseman, N., 1979. Factors affecting the design and conditions of mine tunnels. Chamber of mines of South Africa. Research Organization, Rept. 45/79, 15p & 14 figs.

Rockbursts and Seismicity in Mines, Young (ed.) © 1993 Balkema, Rotterdam, ISBN 90 5410 320 5

Coal bumps induced by mining tremors

Z. Kleczek & A. Zorychta
Technical University of Mining and Metallurgy, Krakow, Poland

ABSTRACT: The geodynamic phenomena such as coal bumps and mining tremors are very frequent in Polish coal mining. Every year, about 20 coal bumps and about 2500 mining tremors, with their seismic energy greater than 10^5J, are registered. Generally speaking, 60% of Polish coal mines are endangered by the coal bump hazard. Due to the safety and economy, the problem of designing excavations (problem of defining the coal bump and mining tremor criteria) in coal bump conditions is very important. The paper deals with these criteria. Assuming that the mining tremor is caused by the fracture of rock layers above the seam, the sufficient condition for mining tremor, in front of the long wall, will be created. Since tremors result in dynamic loads on the coal seam, the sufficient coal bump criterion is obtained, by assuming that the coal bump is a dynamic loss of stability. This condition defines the possibility of a coal bump occurrence in long wall (long wall with caving or with backfill).

1. INTRODUCTION

Polish coal mining is characterised by a rather unusually high rate of the geodynamic events in the world. Every year there are registered 20 bumps and over 2500 tremors with the energy higher than 10^5 J. This situation exists over couple of decades, inspite of applications of various profilactic methods. This confirms that in order to solve the problem of bumps better understanding of the bump mechanism and formulation of better criteria are nessesery.

Two types of bumps with respect to their mechanism are recognised in Polish mining:
- bumbs induced by a concentration of static stresses; known as the seam bumps,
- bumps induced by the action of dynamic loads created by mining tremors; these are roof bumps.

Since both types of bumps are a result of loss of equilibrium (Salomon 1970; Kleczek, Zorychta, and Cyrul 1986) of the rock mass surrounding the mining exploitation, the criteria are defined on the assumption of a nonstable solution describing the state of stress and strain around mining working. In Polish coal mining 90% of coal comes from the long wall mining. Therefore the presented criteria will be valid only for this type of mining method.

2. ANALYTICAL BUMP CRITERIA

From the longtime experience in Poland it is known that the high energy bumps occur in the cases, when a thick layer of rock with high strength and stiffness is present in the upper strata. Most often, these are thick layers of sandstone. During exploitation, bending of the layer can lead to fracture and the resulting tremor. The condition for the fracture of the sandstone layer whose geometry is shown in Fig.1. is described by the coefficient W (Kleczek and Zorychta 1991; Chudek, Flisiak, Kleczek, and Zorychta 1992):

$$W = \frac{1.325 \left[\dfrac{R_g}{p_z}\right]^2 \dfrac{F}{E_s}}{\dfrac{h}{E_g} + \dfrac{h_1}{E_1}} \qquad /1/$$

where R_g is the bending strength of the layer generating the tremor,
E_s is Young modulus of the layer,
F is the thickness of the layer,
h is the thickness of the coal seam,

p_z is the vertical component of initial state of stress,
E_g is the deformation modulus of the mined strata / for backfill:
E_{gp}; for the roof colapse: E_{gz} /, $E_{gp} > E_{gz}$.

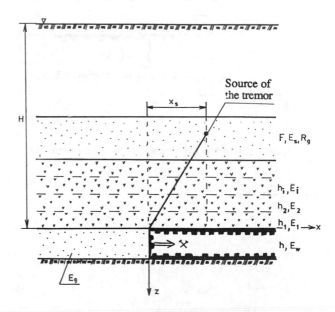

Figure 1. Crossection of the long wall mining.

There are two characteristic cases: when coefficient W is greater then 1 and when coefficient W is smaller then 1.
- for W > 1, the fructure and tremor are impossible to happen,
- for W < 1, as the result of the fructure the tremor will develop with the seismic energy A_s given as:

$$A_s = 10^{-3} R_g^2 F b \sqrt{\frac{F}{E_s} \left[\frac{h}{E_g} + \frac{\sum h_1}{E_{gr}}\right]} \qquad /2/$$

where b is the length of the fracture,
E_{gr} is the replaced deformation modulus:
The replaced deformation modulus is given:

$$E_{gr} = \frac{h + \sum h_1}{\dfrac{h}{E_g} + \sum \dfrac{h_1}{E_1}} \qquad /3/$$

In order to determine the criteria of the bump, which was initiated by tremor, a two dimensional model of the long wall mining was considered (Fig. 2). The following assumptions were made:
- bending of the roof plate with the stiffness EF is caused by shear forces,
- the plate is laying on the non-linearly deforming strata for which the stress-displacement relationship is described by the function (Fig. 3):
- the of mining workings characterised by the external energy $A_z^{(s)}$
- the system is affected by an external load which is a sum of static stresses p_z and dynamic stresses p_d :

$$p = p_z + p_d \qquad /4/$$

The dynamic stress is defined as:

$$p_d = p_i \, \delta(t) + p_s \, H(t) \qquad /5/$$

where $\delta(t)$ is Dirac's delta,
$H(t)$ is Heaviside's function.

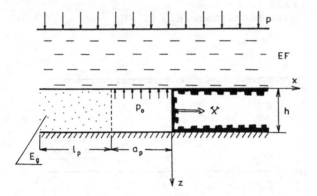

Figure 2. Two dimensional model of the long wall mining.

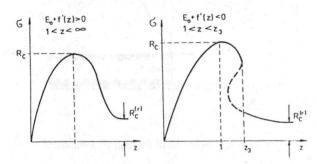

Figure 3. The stress-displacement relationship.

The vertical displacement field in the coal seam is described by:

$$\frac{1}{3} E F z_{cr} \frac{\partial^2 z}{\partial x^2} - \rho_s h z_{cr} \frac{\partial^2 z}{\partial t^2} =$$

$$f(z) - (p_z + p_s) \qquad /6/$$

One can prove that the above equation is quasi-linear hiperbolic (Strauss 1978; Zorychta), which, due to the property of the function $\sigma = f(z)$, is characterised by, so called, instability connected with a rock burst. If the bump is identified as the loss of stability, than it is obvieus, that the investigated criteria will be the condition of the instability of the equation /6/. The criteria are called the bump criteria for the following two cases:
1. The bump is induced by a jump, if /Fig. 4/:

$$\left\{ \begin{array}{l} E_o + \dfrac{df(z)}{dz} < 0 , \qquad 1 < z < z_3 \\[4mm] A_z^{(s)} + A_z^{(d)} > \displaystyle\int_{z_1}^{z_3} \left[f(z) - (p_z + p_s) \right] dz \end{array} \right.$$

$$/7/$$

where $A_z^{(s)}$ is the static energy, which in case of the long wall mining is described by formula /Fig. 4/:

$$A_z^{(s)} = \frac{p_z}{2 E_g \varepsilon_{cr}} \left[\left(1 - \frac{p_o}{p_z} \right) \alpha_p a_p \right.$$

$$\left. + \operatorname{tgh}(\alpha_p L_p) \right]^2 \qquad /8/$$

$A_z^{(d)}$ is the dynamic energy, which value is given by formula (Kidybinski 1988):

$$A_z^{(d)} = 25.7 \frac{\rho}{\varepsilon_{cr}} \left[\frac{\sqrt{\ln A_s}}{r} \right]^{3.1} \qquad /9/$$

where r is the distance between the source (focus) of the tremor and the face of long wall,
L_p is half of the width of the mining working,
E_o is reduced modulus of elasticity of the roof strata (E_r) and floor strata (E_f) when subjected to unloading:

$$E_o = \frac{E_r \, E_f}{E_r + E_f}$$

ρ is the density of the seam,
z_{cr} is the critical displacement given as:

$$z_{cr} = h \, \varepsilon_{cr} ,$$

and α_p and p_s are given in formulae:

$$\alpha_p = \sqrt{\frac{3 E_g}{E F h}}$$

$$p_s = \sqrt{2 R_c A_z^{(d)}}$$

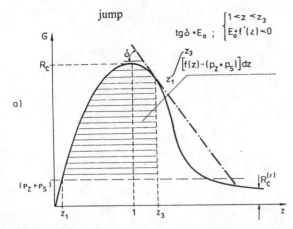

Figure 4. The stress-displacement relationship when the bump is induced by a jump

2. The bump is induced by the loss of load carrying capacity if /Fig.5/:

$$
\begin{cases}
E_o + \dfrac{df(z)}{dz} > 0, \qquad 1 < z < \infty \\[2em]
A_z^{(s)} + A_z^{(d)} > \displaystyle\int_{z_1}^{z_2} \Big[f(z) - (pz + ps) \Big] dz \\[2em]
pz + ps > R_c^{(r)}
\end{cases}
\qquad /10/
$$

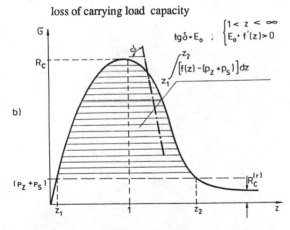

Figure 5. The stress-displacement relationship when the bump is induced by the loss of carrying load capacity.

The defined conditions describe possibility of the bump occurrence in the situation when it is induced by a tremor. It means, it relates to the roof bump. However, if in the formulae /9/ and /10/ the value of $A_z^{(s)}$ is put as equal to zero,

$A_z^{(s)} = 0$,

then, automatically the obtained criteria characterise the possibility of the occurrence of the bump induced by the concentration of the static stresses.

Without going into details one can notice that the bump occurrence is influenced by both the natural causes and the technological parameters as: type of mining (caved roof or with backfill), type of lining, and geometry of the long wall.

3. CONCLUSIONS

The increase of the depth of exploitation increases the danger of occurrance of the bumps. Therefore when chooosing the method of the exploitation it is very neccessery to consider proper prediction method. It means, that the mining engineer should know proper criteria, which could help him in choosing the proper method of mining, and which would minimise the danger of bumping in the given goelogical and mining conditions. The criteria which concern the bumps caused by tremors (roof bumps) and the bumps caused by concentration of the static stresses (seam bumps) have been presented in this paper.

REFERENCES

Salomon M.D.G., 1970. Stability, Instability and Design of Pillar Workings. *Int. J. Rock Mech. Min. Sci.* Vol.7.

Kleczek Z., A. Zorychta, and T. Cyrul, 1986. A Failure Criterion for Coal Seam at Longwall Mining. *The 27-th US Symposium on Rock Mechanics..* Alabama.

Kleczek Z. and A. Zorychta, 1991. Geomechanical Conditions Defining the Mining Tremors Occurence. *Applications of the Geophysical Methods in Mining.* Krakow.

Chudek, M., D. Flisiak, Z. Kleczek and A. Zorychta, 1992. Geomechanical Phenomena Accompanying Underground Mining. The Mining Outlook. *The 15-th World Mining Congress..* Madrid. Spain.

Kidybinski A., 1988. The Basis of Choice of Heading Lining for Regions with Tremors and Crumb Hazard. *Safety in Mining.* No.1.

Strauss W. A., 1978. Nonlinear Invariant Wave Equations. *Lecture Notes of Physics.* Vol. 73. Springer Verlag. New York.

Zorychta A., Criterion of Rock Burst during Exploitation of Coal Seam.

Rockbursts and Seismicity in Mines, Young (ed.) © 1993 Balkema, Rotterdam, ISBN 90 5410 320 5

Effect of artificially induced vibrations on the prevention of coal mine bumps

Ch. Liang, H. Lippmann & J. Najar
Technical University, Munich, Germany

ABSTRACT: Vibrational excitation in a coal seam ahead of a work cavity in deep coal mining is investigated for its value in enhancing coal bursting prevention measures. A mechanical model is proposed, taking into account the interaction of induced steady vibrations of the system seam - confining rock strata with the critical state of stress, predicted by the coal mine bumps theory, [3]. Cyclic loading tests conducted with a model material under confining pressure reveal rate-dependence effects on its strength, as well as low cycle fatigue - induced decay of its burst proneness. Incorporating the data into the vibration model allows for determination of the optimal frequency and location of the vibration source with regards to the longwall cavity.

1 INTRODUCTION

Several papers published in the former Soviet Union in the late 80th indicate that artificial vibrations induced in the coal seam, some meters ahead the longwall or, correspondingly, in the stiffer rock above or below the seam, may enhance the prevention of coal and gas outbursts. However, the evidence collected has been rather vague or, to some part, contradictory.

In the papers [4], [10], [7], [8] and [11] qualitative observations are presented, without relating them to mechanical or material phenomena. In the investigated coal seams, a reduction of the gas content and/or a reduction of the state of stress seems to have been detected as a consequence of the vibration-induced increase in the number and size of internal cracks and fissures, thus leading to increased permeability and drop of the material's cohesion. Laboratory tests aimed at modelling of the effect, [9], have not been conclusive, the more as the natural frequencies of coal specimens used in the tests seem to have had an influence on the results. Therefore, their direct application to the in-situ situation could have not been attempted.

In an unrelated geoseismic set of measurements [5], a substantial reduction of the energy of local quakes has been observed in the vicinity of a large dam in result of the ground vibrations caused by the overflow cascade from the water reservoir, leading to the exclusion of major quakes altogether.

It is the latter observation, which may raise hope that artificially induced vibrations could reduce the danger of initiation of coal mine bumps, particularly of the kind not combined with gas outbursts [1].

In the present investigation the following system, Fig. 1, is being examined. A horizontal infinite coal stratum of thickness $2h_1$ is enclosed between two identical stiffer and stronger rock masses of thickness h_2 each. The seam is intersected by a horizontal cavity, corresponding to a longwall working, a roadway or a gallery, of a width $2b$. The rock is exposed to a constant horizontal primary pressure αq acting at the lateral infinity and to a constant vertical primary pressure q, where $\alpha = const$. Both the coal seam and the enclosing rock are assumed to be homogeneous and isotropic.

At an arbitrary point P_1 in the rock, or P_2 in the seam, a combination of lateral, vertical or horizontal in-depth exciting forces $F_x(t)$, $F_y(t)$ and $F_z(t)$, periodic in time t, induces vibrations in the system, which are to be observed at any other point Q_1 of the seam or Q_2 of the rock. From the charac-

teristics of the vibrational response of the system with given geometrical parameters, in combination with separately determined changes of the material properties under cyclic loading in a multiaxial state of stress, conclusions shall be drawn with respect to possibilities of enhancing conventional coal bump prevention measures.

Correspondingly, three stages of the investigation still in progress are considered:

1. Experiments on the determination of the material properties at cycling loading at confining pressure. Here, a substitute material, known as model-araldite, exhibiting mechanical properties close to those of hard coal, and similar to it regarding burst-proneness, is examined, [12], [2].

2. Theoretical analysis, based on the interaction of the assumed linear elastic vibrations with the measured material properties at cyclic loading, and correlated with the predictions of the coal mine bumps model, [3], in particular regarding the initiation and the expected strength of the event.

3. Experimental laboratory small scale modelling of the system, Fig. 1, with application of force excitations and measurement of the response in the sensitive points predicted by the previous analysis, and investigation of

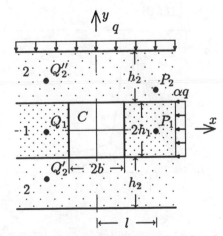

Figure 1: Coal seam (1) intersected by cavity (c) and enclosed in stiff rock layers (2).

their influence on the strength of the laboratory - produced bumps.

In the paper, preliminary results concerning two first stages of the investigation in progress can be presented.

2 MATERIAL PROPERTIES UNDER CYCLIC LOADING

While the rock layers surrounding the coal seam, s. Fig. 1, shall be modelled in the laboratory small scale test by the use of standard materials (steel and ceramics), the coal stratum needs to be fabricated for these tests of model - araldite, as the only material reproducing bursting effects in laboratory. This material has been tested for its mechanical properties in an improved tri-axial cell, [12], under controlled cyclic varying axial pressure $p(t)$, at given constant confining pressures σ. Details of the cylindrical specimens' preparation and changes in the set-up can be found in [2]. The rather stiff oil - protecting casing of the specimen has been replaced here by a flexible foil sheath, allowing for more direct application of the pressure σ.

The tests were conducted in the tri-axial cell, for a range of confining pressures up to 60 MPa, at constant loading stress rates $\dot{p}(t)$, varying between 17.8 and 222 MPa/min, with the axial cyclic load amplitudes p_a corresponding to selected fractions (between 45 and 80 %) of the compressive strength p_m of the material at given confining pressure, determined in separate static tests at the same loading rates.

It has been found, that one may apply to the material's compressive strength p_m the Coulomb-Mohr condition

$$p_m = cot(\frac{\pi}{4} - \frac{\phi}{2})[cot(\frac{\pi}{4} - \frac{\phi}{2})\sigma + 2c] \quad (1)$$

modified by the rate-dependence of the parameters of cohesion c and internal friction ϕ, both of them slightly increasing with the axial loading rate $\dot{p}$, comp. Table 1.

As observed already in [12] and [2], the collapse of the specimens at confining pressures σ below 25 MPa was in static tests characteristically violent and brittle, coresponding to its burst-proneness.

In the dynamic cyclic loading tests, see Fig. 2, the burst proneness vanishes in nearly all tests with some confining pressure σ applied, and even sometimes at $\sigma = 0$, at least for all amplitudes p_a of the cyclic axial pressure p, which were applied in these experiments, i.e. exceeding 45 % of the corresponding compressive strength p_m.

This is an extremely important observation, which can be interpreted on the basis of the development of internal damage at cyclic loading, causing transition from brittle to quasi-ductile response, cf. [6]. It is also indicating that the method of artificial vibrations may be effective in preventing bursting, as far as material properties are concerned. It would be of further importance to check this observation both for lower amplitudes of cyclic loading and for higher loading frequencies, corresponding to the typical technological vibration frequency at $f = 50\ Hz$.

The typical result of the cyclic loading tests with confining pressure σ is shown in Fig. 3, where the number N of the cycles to failure is presented in dependence of σ, axial loading rate $\dot{p}$ and cyclic axial pressure amplitude p_a in %, as related to the compressive strength p_m. They may be well represented by straight lines in a semi - logarithmic scale, corresponding to

$$p = B - Aln(N),$$
$$A = A(\sigma, \dot{p}),\ B = B(\sigma, \dot{p}) \quad (2)$$

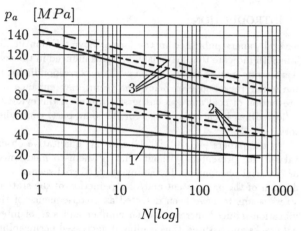

Figure 3: Cycles at failure N in dependence of the pressure amplitude p_a, at various confining pressures: $\sigma = 0\ MPa$ (curve 1), $\sigma = 20\ MPa$ (curve 2), $\sigma = 60\ MPa$ (curve 3), and various loading rates $\dot{p} = 17.8$ (contin.line), 111.1 (dotted line) and 222.2 MPa

These results should also be extended towards lower cyclic load amplitudes and higher frequencies f, better corresponding to mining applications of the vibrational prevention method.

3 THEORETICAL ANALYSIS OF THE LONGITUDINAL VIBRATIONS OF THE SYSTEM

According to the geometrical and mechanical model presented in Fig. 1, consider a source of the artificial vibrations with given circular frequency Ω located in an arbitrary point P within the system and causing its stationary vibrations. The system is represented as a linear visco - elastic system of beams under given vertical and lateral distributed forces q

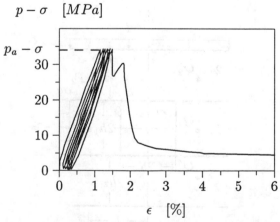

Figure 2: Cyclic loading test with axial pressure amplitude p_a=0.58 p_m, at confining pressure $\sigma = 12\ MPa$ and loading rate $\dot{p} = 17.8\ MPA/min$

Table 1: Cohesion and internal friction angle in dependence of the axial pressure rate

$\dot{p}\ \frac{MPa}{min}$	17.8	111.1	222.2
$c\ MPa$	17.2	20.6	29.6
$\phi\ deg.$	10.0	11.5	12.6

and αq, with corresponding boundary conditions. This beam model can be applied for sufficiently low exciting frequencies Ω obeying the condition

$$\sqrt{\frac{E}{\rho}} \gg \Omega h \qquad (3)$$

where E denotes the elastic stiffnes modulus, ρ the mass density, h the half-height of the beam's cross - section. For practical data (sandstone, coal) the typical values of h of some 10 m or less would correspond to frequencies of 50 Hz or less.

In the following, as a part of the present state of the investigation, only uniaxial vibrations corresponding to lateral loading are considered. Also, the sandwich beam with vanishing width b of the cavity is examined here. The general results for vibrational bending and antiplanar shear, allowing for cavities $b > 0$ shall be presented elsewhere.

3.1 Geometrical and material parameters

In the following, the indices 1 and 2 denote quantities related to the coal seam and to the surrounding rock, while geometrical parameters correspond to those of Fig. 1. Denote further

$$h = h_1 + h_2 \qquad (4)$$

$$E = \frac{h_1 E_1 + h_2 E_2}{h} \qquad (5)$$

$$\eta = \frac{h_1 \eta_1 + h_2 \eta_2}{h} \qquad (6)$$

$$\rho = \frac{h_1 \rho_1 + h_2 \rho_2}{h} \qquad (7)$$

where η_i, $i = 1, 2$, are stretching viscosities of the materials.

3.2 Equations of horizontal motion of the system

Denoting by $u(x, t)$ the horizontal displacements, we obtain for the longitudinal force N the formula

$$N = 2h(E\frac{\partial u}{\partial x} + \eta \frac{\partial^2 u}{\partial x \partial t}) \qquad (8)$$

Dynamic equations of motion in the horizontal direction can be now represented as

$$E\frac{\partial^2 u}{\partial x^2} + \eta \frac{\partial^3 u}{\partial x^2 \partial t} - \rho \frac{\partial^2 u}{\partial t^2} = 0 \qquad (9)$$

The solution is being sought in the form

$$u(x, t) = U(x)e^{i\Omega t} \qquad (10)$$

The ansatz yields the characteristic equation

$$(E + i\eta\Omega)\frac{\partial^2 U}{\partial x^2} + \rho\Omega^2 U = 0 \qquad (11)$$

which can be put in the form

$$(c^2 + iH\Omega)\frac{\partial^2 U}{\partial x^2} + \Omega^2 U = 0 \qquad (12)$$

where $c = \sqrt{E/\rho}$ denotes the wavespeed, while $H = \eta/\rho$ corresponds to kinematic viscosity.

3.3 Solution

To determine the solution denote

$$(\alpha + i\beta)^2 = \frac{\rho\Omega^2}{E + i\eta\Omega} \qquad (13)$$

Here, the parameters α and β can be expressed as

$$\alpha = a\cos\gamma \qquad \beta = a\sin\gamma \qquad (14)$$

while the auxiliary quantities γ and a are defined as

$$a^2 = \frac{\rho\Omega^2}{\sqrt{E^2 + (\eta\Omega)^2}} \qquad (15)$$

$$\cos 2\gamma = \frac{E}{\sqrt{E^2 + (\eta\Omega)^2}} \qquad (16)$$

$$\sin 2\gamma = -\frac{\eta\Omega}{\sqrt{E^2 + (\eta\Omega)^2}} \qquad (17)$$

If the real part of the solution is assumed to remain bounded at infinity, it must consist of two expressions

$$U^- = U^0 e^{\alpha(x-l)}\cos[\beta(x - l)] \qquad (18)$$

for $x < l$, and

$$U^+ = U^0 e^{-\alpha(x-l)}\cos[\beta(x - l)] \qquad (19)$$

for $x > l$, where l denotes the position of the excitation source with regards to the coordinates origin, chosen here at the face of the cavity, and U^0 is a constant, which remains to be determined.

In order to determine the constant we apply the conditions at the point of excitation P, where the jump discontinuity of the normal forces N corresponds to the vibrational force F_x applied. Assuming:

$$F_x - F_x^0 \cos(\Omega t - \phi) \qquad (20)$$

where ϕ denotes the phase shift between the excitation force and the displacement in the coal seam, on gets the following boundary condition at $x = l$:

$$F_x^0 \cos(\Omega t + \phi) = Re[-(N^+ - N^-)] \qquad (21)$$

Here, $N^+ = N(x = l + 0, t)$ and $N^- = N(x = l - 0, t)$ denote the limiting values of the $N(x)$-distributions for $x > l$ and $x < l$ correspondingly, as computed from the equations (8), (10), (18) and (19).

The result is:

$$\phi = 2\gamma \qquad (22)$$

and

$$F_x^0 = 4hU^0\alpha\sqrt{E^2 + (\eta\Omega)^2} \qquad (23)$$

3.4 Discussion

Introducing the dimensionless amplitude amplification factor

$$A = \frac{4hU^0 E}{F_x^0}a \qquad (24)$$

and substituting the definitions (14)-(17) one gets the following expression

$$A = \frac{\cos 2\gamma}{\cos\gamma} \qquad (25)$$

Here, the auxiliary quantity γ is a function of the dimensionless frequency of the problem, defined as

$$\omega = \frac{\eta\Omega}{E} \qquad (26)$$

Equations (16) and (25) render the dependence of the vibration amplification factor A on the frequency ω, Fig. 4.

One observes, in general, a fast decay of the vibrational effect with the increase of the excitaton frequency Ω. Another

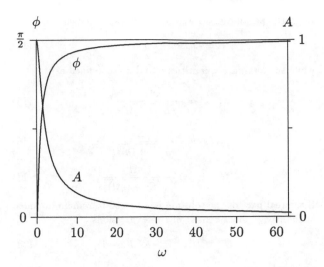

Figure 4: Vibration amplification factor A and the phase shift angle ϕ in dependence on the frequency ω.

question is, however, the determination of the proper location of the source of vibrations with respect to the wall of the cavity, for a given value of Ω.

To this end, introduce another parameter of the solution: the characteristic length of the problem

$$L = \frac{\eta c}{E} \quad (27)$$

By substitution of the parameters (26 - 27) into formula (15), we can represent it in the form

$$a = \frac{\omega}{L(1 + \omega^2)^{\frac{1}{4}}} \quad (28)$$

Consider now the influence of the vibrations at point P, $x = l$, on the displacements at the face of the cavity, $x = 0$. Calculating the longitudinal force N according to the solution (18), (22-23), we can relate its amplitude at $x = 0$ to the amplitude of the excitation force F_x^0, choosing thus the optimal values of l and ω. We may define here the dimensionless vibration influence parameter as

$$v = \frac{max(ReN_{x=0})}{F_x^0} \quad (29)$$

and by substituting (10), (18) und (23) into (8) find, that

$$v = \frac{exp(-\xi)cos(\zeta - \gamma)}{2cos\gamma} \quad (30)$$

where

$$\xi = \alpha l = \frac{l\omega}{L(1 + \omega^2)^{\frac{1}{4}}} cos\gamma \quad (31)$$

$$\zeta = \beta l = \frac{l\omega}{L(1 + \omega^2)^{\frac{1}{4}}} sin\gamma \quad (32)$$

while γ is a function of the dimensionless frequency ω

$$cos2\gamma = \frac{1}{\sqrt{1 + \omega^2}} \quad (33)$$

comp. eq. (16) and (26). A maximum of the function v with respect to ω yields the preferable excitation frequency Ω at given location of the vibrator, while a maximum with regards to l/L renders the preferable position of the vibrator at given excitation frequency.

4 CONCLUSIONS

The method of inducing forced vibrations into the stratum of confining rock layers in order to reduce the danger of coal mine bumps is being investigated theoretically, as well as experimentally. Preliminary results show that a preventive effect can possibly be obtained for sufficiently low exciting frequencies, when the proneness to bursting of the material might diminish due to the cyclic fatigue and brittle - to - ductile transition in a sufficiently large zone next to the cavity. Further investigations are necessary, pending sponsorship applied for. At this point, the Authors wish to acknowledge gratefully the sponsoring of the first part of the work presented here by the Ruhrkohle - Westfalen AG, Dortmund.

REFERENCES

[1] Brauner, G., 1989, *Gebirgsschlaege und ihre Verhuetung*, 2nd ed., Essen, Gluckauf-Verlag.

[2] Kuch, R., Lippmann, H., Zhang, J., 1991, Simulating coal mine bumps with model material, submitted for publication; summary: Zur Modellierung des Gebirgsschlages in Kohlegruben, Sitzungsbericht Bayer. Akad. Wiss, Jg. 1991, 8* − 12*.

[3] Lippmann, H., 1987, Mechanics of bumps in coal mines, *Appl. Mech. Rev.*, 40, 1033-1043.

[4] Mineev, S.P., Repeckij, V.V., 1987, Reduction of the danger of a coal mine burst by vibrational action on the rock, *Ugol Ukrainy*, 10, 39-50, (in Russian).

[5] Mirsoev, K.M., Nigmatullaev, S.Ch., 1990, Effect of mechanical vibrations on the seismic behaviour, *Dokl. A.N. SSSR*, 1, 78-83 (in Russian).

[6] Najar, J., 1987, Continuous damage of brittle solids, *Continuum damage mechanics. Theory and applications*, eds. D. Krajcinovic, J. Lemaitre, Springer Verl., Wien - New York, 233-294.

[7] Poturaev, V.N., Mineev, S.P., 1989, Induction of vibrations in underground rock for improvement of efficiency and safety of coal excavation, *Ugol Ukrainy*, 7, 3-6 (in Russian).

[8] Riasanzev, N.P., 1987, Shock and vibration systems in mining, *FTPRPI*, 5, 51-62 (in Russian).

[9] Rosanzev, E.S., 1980, Investigation of the gas flow in coal seams under the action of induced vibrations, *Gas control in coal mining*, Papers of VNII, Kemerovo, pp. 99-105 (in Russian).

[10] Zorin, A.N., 1985, Vibration method for reduction of the bursting danger in coal mines, *Shacht. Stroit.*, 6, 7-8 (in Russian).

[11] Zorin, A.N., 1987, *Control of underground rock*, Kiev, Nauk. Dumka (in Russian).

[12] Zhang, J., 1990, *Experimente und theoretische Untersuchungen zum Gebirgsschlag*, Dr.-Thesis, Techn. Univ. Munich.

Rockbursts and Seismicity in Mines, Young (ed.) © 1993 Balkema, Rotterdam, ISBN 90 5410 320 5

Quantitative analysis of seismic activity associated with the extraction of a remnant pillar in a moderately deep level gold mine

D.S.Minney
Steyn Gold Mine, Welkom, South Africa

W.A.Naismith
Department of Mining Engineering, University of the Witwatersrand, Johannesburg, South Africa

ABSTRACT: Seismic data obtained during the extraction of a small pillar has been compared with results from numerical modelling of the area in order to identify changes in rock mass condition. A combination of factors occurring over a short time period identified a change which has been interpreted as a move from elastic to non-elastic conditions; It is concluded that further cases must be analysed to confirm links between seismic activity, numerical modelling and rock mass condition.

INTRODUCTION

The President Steyn Gold Mine is situated in the central part of Orange Free State Goldfields of South Africa. The mining method practised is typically the scattered method because of the flexibility it offers to cope with grade fluctuations and geological discontinuities.

The method involves the pre-development of footwall drives and at some distance (typically 150-200m) the breaking away of crosscuts to reef. Once the reef is intersected, on reef raising occurs on the true dip of the ore body.

From these raises stoping occurs. Stoping can be single or double sided; however, regardless of which system is used island or peninsular remnants are created (An Industry Guide to Amelioration of Rockfalls and Rockbursts 1988 Edition).

These remnants became highly stressed and are often the source of much seismic activity.

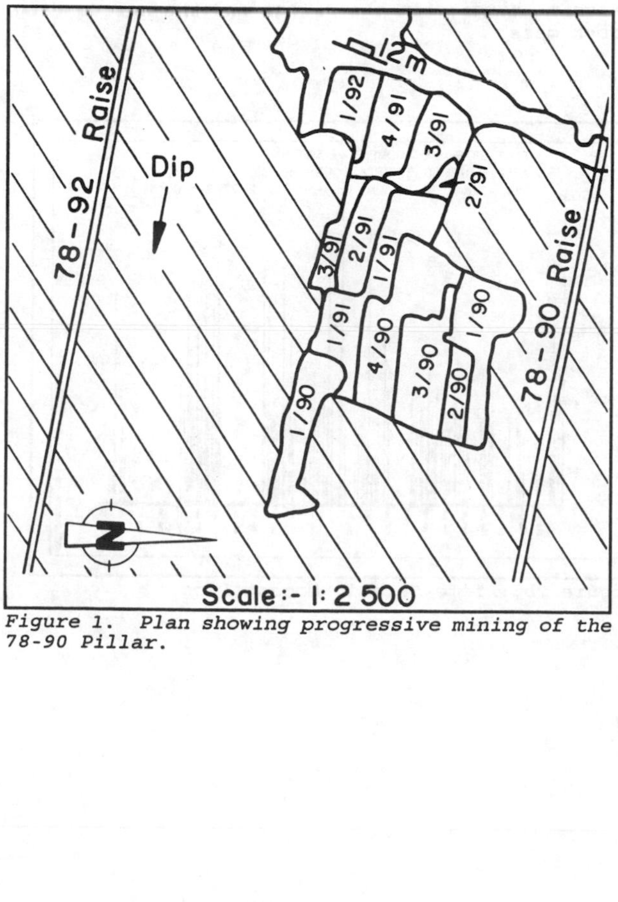

The mining of remnants may be difficult on account of the seismic activity or because of intense fracturing of the rockmass that has occurred. This has implications for the stoping strategy and stope support.

To be able to use seismic data to complement existing design tools in mining, certain indicators need to be readily available; as it is not common to have all mine stopes covered by an accurate and sensitive seismic system at all times.

It would therefore be very useful to have a system that could make use of a "snap shot" methodology to capture specific data that could be analyzed and used to define the current status of the rock mass condition.

Such a methodology, it is anticipated, would entail the temporary monitoring of specific pillars and on the basis of the interpretation of the seismic data and analysis, a quantitative decision can be made for each pillar being mined.

As part of a fault monitoring project, the seismic history of a particular pillar, 78-90 pillar, some 2300m below surface was captured at the President Steyn Gold Mine, as it was progressively mined out.

Some 339 seismic events in the general range from magnitude (ML)-1.0 to 2.0 were recorded from the specific pillar.

By analysing these events it is shown that the pillar and surrounding rockmass changes its physical nature as it is progressively reduced in size and the stress intensities vary.

It is hypothesized that as the pillar is reduced in size, it and the surrounding rockmass stores strain energy to a point when general failure begins and load shedding occurs. This may be likened to the pillar work hardening and then softening.

The Mining

The pillar consisted of an island remnant left abutting against a 12m fault that had its strike parallel to that of the ore body. The throw of the fault was down to the west while the ore body dipped to the east. Immediately above the ore body occurs a pyrophyllite rich shale band the

Figure 1. Plan showing progressive mining of the 78-90 Pillar.

TIME	MINED AREA		APPROXIMATE VOLUME EXTRACTED		VOLUMETRIC CONVERGENCE m³
	m²	%	m³	%	
2nd Qtr 90	829	8.3	984	7.2	279.80
3rd Qtr 90	1193	12.0	1921	14.1	608.70
4th Qtr 90	1583	15.9	2291	16.8	855.40
1st Qtr 91	2138	21.4	2902	21.3	1107.00
2nd Qtr 91	1407	14.1	1699	12.5	1163.40
3rd Qtr 91	1406	14.1	2370	17.4	325.10
4th Qtr 91	593	5.9	830	6.1	516.90
1 Qtr 92	435	4.4	617	4.5	N/A
TOTAL	9584		13614		

khaki shale. The footwall and hangingwall of the ore body was typical Witwatersrand quartzite having an UCS in the range 180 to 240MPa.

The pillar was mined progressively from March 1990 to February 1992.

The Seismic System

The seismic system used for the data collection was the COMRO Portable Seismic System (PSS) (Patrick, Kelly, Spottiswoode 1990). The PSS had been installed as part of a fault monitoring project and therefore had not been sited to specifically monitor the pillar.

The PSS was used for recording the location and magnitude in the first instance and later the data was processed using "Sourceq" (a source parameter routine, personal communication S. Spottiswoode) and for calculating the b-value. The Sourceq output was imported into the Integrated Seismic System (ISS) (A. Mendecki) and subjected to various statistical procedures. The plots for various parameters are shown in Figures 2a to 6.

These parameters were chosen because previous and current work has indicated they offer the best potential for interpretation of physical changes occurring within the rock mass. (Mendecki, Spottiswoode, van Aswegen 1992). Note, gaps in data due to the system being off line for various reasons.

Computer Modelling

The most effective way to model the mining was with Minsim-D, a boundary element computer method developed by COMRO (Napier, Stephansen and Johnstone 1986).

From the modelling it was possible to estimate the volumetric convergence due to mining and also the strain energy stored in the rock mass and released for each time step.

A series of benchmark sheets were placed over the pillar at 5m spacings, with each benchmark at 5m centres. At each benchmark position, the Minsim-D program calculated total and shear stresses and strains. This data was imported into a spreadsheet for the strain energy calculations.

Table 2 is a summary of the stored and released strain energies in the pillar and the surrounding rock mass.

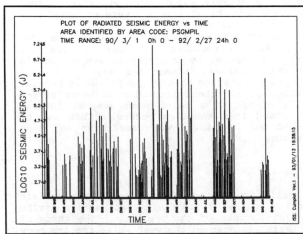

Figure 2a.   Time-radiated Seismic energy Plot

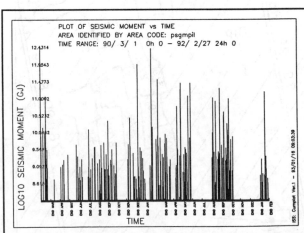

Figure 2b.   Time-Seismic Moment Plot.

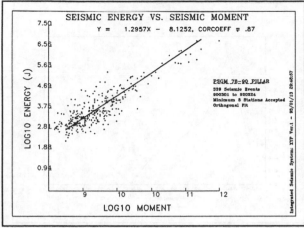

Figure 2c. The relationship between Seismic Energy in Figure 2a and Moment Figure 2b.

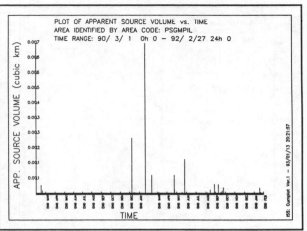

Figure 3c. Time-Apparent Source Volume Plot.

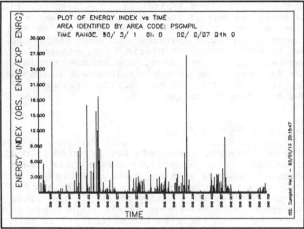

Figure 3a. Time-Energy Index Plot

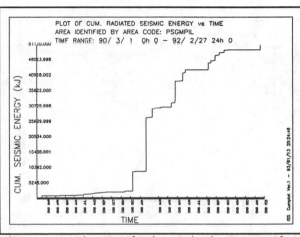

Figure 4a. Time-Cumulative Seismic Energy Plot. Note change in slope of line after December 1990.

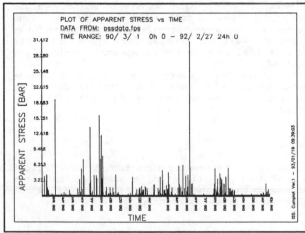

Figure 3b. Time-apparent Stress Plot

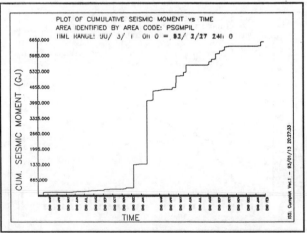

Figure 4b. Time-Cumulative Seismic Moment Plot. Note change in slope of line after December 1990.

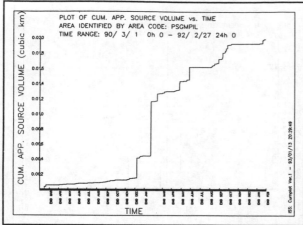

Figure 4c. *Time-Cumulative Apparent Source Volume Plot. Note change in Slope of line after December 1990.*

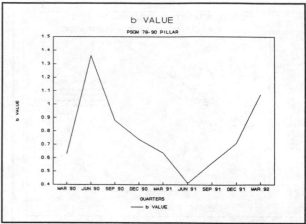

Figure 6. *Time-'b' value plot. Note that the 'b' value was at a minimum at the end of June 1991.*

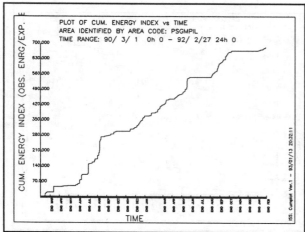

Figure 5a. *Time-Cumulative Energy Index plot. Note change of slope at end of June 1990.*

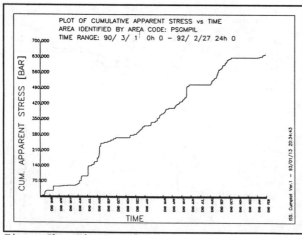

Figure 5b. *Time-Cumulative Apparent Stress plot. Note change of Slope at end of June 1990.*

As vertical symmetry was assumed, the Minsim benchmarks were only placed on one half of the pillar (hangingwall). As there is strain associated with total and shear stresses, the two

components are shown separately. The last column indicates the sum of the two strain components.

The different columns show the difference between the strain energies for each step. It must be remembered that Minsim is an elastic computer program and therefore cannot simulate rock failure. The interpretations are based on elasticity and this is one of the drawbacks of this modelling technique.

Results of Seismic Analysis

The raw data for the events was imported into a spreadsheet and summations made for each chronological time step as shown in Table 3.

From this it can be seen that the pillar and its environs becomes seismically active in the first quarter of 1991 (see Figure 4a etc.) and there is consistency with the significant changes of each parameter.

Evaluation of stress drop and energy index (Figures 3a,b, 4a and 5) would appear to indicate an onset in changed conditions within the pillar from the end of June 1990, some six months ahead of moment, energy and apparent source volume changes, and twelve months ahead of the 'b' value parameter. Investigation into these two parameters will continue as they appear to offer the potential to give the 'snap shot' methodology required.

The cumulative seismic data and the 'b' value appear to offer the most potential in detecting indirectly stress changes in the pillar and its environs.

Summary

To summarise the results thus far, it was decided to rank them by allocating a score of 1 to the value with the highest influence or significance.

The findings are presented in Table 4.

CONCLUSIONS

The analysis carried out to date has indicated the "snap shot" method has potential for defining the onset of a changed rock mass character.

Only with hindsight are the MINSIM results of any material use because the rate of change in stored energy does not give an indication (except

Table 2. Tabulation of stored strain energies for each time step for the pillar and hangingwall portion of the surrounding rockmass.

TIME	NORMAL STRAIN ENERGY (J)	NORMAL STRAIN ENERGY RELEASED (J)	SHEAR STRAIN ENERGY (J)	SHEAR STRAIN ENERGY RELEASED (J)	TOTAL STRAIN ENERGY (J)
2nd Qtr 90	9.78824E17		1.27188E17		1.106012E18
		9.73049E16		-4.46144E15	
3rd Qtr 90	8.81519E17		1.31650E17		1.013169E18
		-5.33761E17		5.64874E15	
4th Qtr 90	1.41528E18		1.26001E17		1.541281E18
		2.56162E17		1.56153E16	
1st Qtr 91	1.15912E18		1.10386E17		1.269506E18
		1.95510E17		1.78709E16	
2nd Qtr 91	9.63608E17		9.25150E16		1.056123E18
		1.17821E17		1.03245E16	
3rd Qtr 91	8.45787E17		8.2190E16		.84578E18
		1.19550E17		7.44643E15	
4th Qtr 91	7.26237E17		7.47441E16		.8009811E18
		1.21921E17		9.72342E15	
1st Qtr 92	6.04316E17		6.50207E16		.6693367E18

Results of Seismic Analysis

The raw data for the events was imported into a spreadsheet and summations made for each time step.

Table 3. Tabulation of seismically released moment, energy and specific volume.

TIME	MOMENT		ENERGY		APPARENT VOLUME	
	JOULES x 10^9	%	JOULES x 10^6	%	Km³	%
2nd Qtr 90	192,58	2,9	1,31	2,6	.0083	4,3
3rd Qtr 90	68,81	1,0	0,87	1,7	.00022	1,1
4th Qtr 90	105,52	1,6	0,53	1,0	.0005	2,6
1st Qtr 91	4183,95	63,2	27,32	53,8	.01101	56,5
2nd Qtr 91	649,63	9,8	9,14	11,6	.00167	10,2
3rd Qtr 91	584,73	8,3	5,87	11,6	.00199	10,2
4th Qtr 91	655,01	9,9	4,22	8,3	.00271	13,9
1st Qtr 92	182,19	2,8	1,51	3,0	.00055	2,8

intuitively) that a peak has been reached.

Further cases must be analyzed to provide more insight into links between stored strain energy as predicted by numerical modelling and seismic activity.

The cumulative seismic plots give a good indication that physical conditions within the pillar are changing. The sudden change of slope indicating an increase in seismic activity during the first quarter of 1991, are significant. The tailing-off of the cumulative plots is less specific, possibly because the pillar maintains a highly stressed core, even though its volume is greatly reduced.

The difference between moment, energy, apparent volume and the "b" value; is that events contributing to the "b" value carry equal weight, and this a possibly why the significant "b" value lags by one quarter that of the other seismic parameters.

ACKNOWLEDGEMENTS

The Authors wish to thank the management of President Steyn Gold Mine for permission to publish this paper. We wish to also acknowledge

Table 4. Ranking of all data collected.

PARAMETER	2Q90	3Q90	4Q90	1Q91	2Q91	3Q91	4Q91	1Q92
"b' Value	8	6	5	3	1	2	4	7
Seismic Moment	5	8	7	1	3	4	2	6
Seismic Energy	6	7	8	1	2	3	4	5
Apparent Volume	5	8	7	1	4	3	2	6
Volume Mined	6	4	3	1	5	2	7	8
Volumetric Convergence	7	4	3	2	1	6	5	N/A
Stored Strain Energy from Minsim	3	5	1	2	4	6	7	8

Steve Spottiswoode for helping with advice on processing the PSS data and to Gerrie van Aswegen for his help and patience with the statistical plots and both for their enthusiasm throughout the project.

REFERENCES

An Industry Guide to Methods of Ameliorating the Hazards of Rockfalls and Rockburst. COMRO 1988.

Patrick, Kelly, Spottiswoode 1990. *A portable Seismic System for Rockburst Applications.* SAIMM, 1990.

Napier, J.A.L., S.J. Stephansen, R.A. Johnstone. *Supplement to MINSIM D users guide,* COMRO Report 19/86, Published Feb. 1986.

A. Mendecki - *Real time quantitative Seismology in Mines* (1993) refer these proceedings.

Rockbursts and Seismicity in Mines, Young (ed.) © 1993 Balkema, Rotterdam, ISBN 90 5410 320 5

High ground displacement velocities associated with rockburst damage

W. D. Ortlepp
Steffen, Robertson and Kirsten Inc., Johannesburg, South Africa

ABSTRACT: The importance of the displacement velocity in determining the intensity of rockburst damage is discussed and recent developments in this area are reviewed. Close study of actual damage in two significant rockbursts provides evidence of the probable occurrence of much higher velocities in the near-source region than have previously been considered.

1 INTRODUCTION

Despite intensive study, the problem of rockbursts remains the single most intractable technical problem wherever it occurs in hardrock mines throughout the mining world. A rockburst is best defined as a seismic event that causes significant damage - Ortlepp (1984).

The need to understand the fundamental energy source has long been the most taxing aspect of the problem. In recent years seismological studies have provided considerable insight into the source mechanism. If the sophisticated digital systems now available were more widely applied and the resulting proper seismological data of significant seismic events analysed closely e.g. van Aswegen (1990), an adequate understanding would soon be developed.

Studies of the mechanism of damage on the other hand, are probably less well advanced - Ortlepp (1992). In the search for some basic parameters which could characterize the damage potential of a seismic event, attention has been focused on peak particle velocity (PPV). It is a ground motion parameter that can be reliably measured by seismic monitoring and it has been used in earthquake engineering and in blasting technology as a predictor of possible damage associated with large mass-blasts in open-stoping operations.

PPV has, to date, proven to be inadequate in two ways. Firstly, most underground seismic measurements have been made with seismometers (velocity transducers) or with recording systems that lack the dynamic range to permit high-fidelity recording very close to source. Secondly there is not an adequate understanding of the scaling laws to permit extrapolation of the amplitude/distance decay laws back to near-source distances.

Ultimately, however, it is the velocity of displacement of the wall-rock forming the surface of the underground opening which has to be known in order to permit the development of an engineering solution to the problem of containing the damage of a rockburst. Probably because of the inherent complexity of the physical processes at the interface, it appears to be impossible to theoretically determine how the PPV becomes translated into displacement velocity of the free surface of the excavation or into expulsion of partially loose rock blocks or fragments.

It is not possible to anticipate where or when a rockburst might occur. Therefore it is not practical to attempt to directly measure the actual velocity of the wall rock of an excavation. Very few attempts to do this have been made anywhere.

Because of these difficulties and the resulting scarcity of useful engineering estimates, it is thought that a direct phenomenological approach may have merit.

It is submitted that valid engineering estimates of damage parameters can be deduced from careful observation of actual incidents. Two such sets of observations are presented in this note as indications that more extreme phenomena are sometimes involved than might hitherto have been believed possible.

2 REVIEW

A convenient and effective way to review the development of ideas concerning the mechanism of damage is via the proceedings of the two international symposia on Rockbursts and Seismicity in Mines which were held in Johannesburg in 1982 and in Minneapolis in 1988. In his preface to the second of these, Dr Paul Young emphasized the need for future work to be concentrated on near-field propagation/attenuation and the relationship between near-field amplitude and damage. It would seem, therefore, that knowledge in this important area was distinctly lacking at that time.

During the previous several years, important pioneering contributions by, notably, McGarr (1984) and Wagner (1984) had pondered the question as to how the high ground velocities necessary to cause intense damage are generated by seismic events that have their origins deep inside the rock mass.

Although he observed that seismic moment M_O was the correct measure of the magnitude of a seismic event, McGarr suggested that the peak velocity, $\bar{v}$, could be related in a useful way to the more empirical measure of event size viz the Richter local magnitude, M_L, and the distance, R, as follows:

$$\log R\, \bar{v} = 3{,}95 + 0{,}57\, M_L$$

In an earlier discussion which assumed homogeneous slip at the source McGarr et al (1981) inferred that the hypo-central distance must be as small as 10 m for an event of M_L 2,4 to produce $\bar{v} = 2$ m/s. McGarr (1984) makes the important observation that "... source processes not accounted for in the Brune 1970 source model play a role in determining $\bar{v}$". He goes on to develop the Rv scaling laws for an inhomogeneous fault source where a strong asperity is surrounded by an annular failed region, the ratio of the size of the asperity to the failed region determining the amplitude of the high frequency radiation. Notwithstanding this reservation the conclusion is made that "... only seismic failure in the immediate vicinity of the workings is likely to generate the peak velocities necessary to cause support failure". The support referred to here is rapid-yield hydraulic props in a deep, extensive tabular stope.

Wagner used the above relationship for $R\bar{v}$ to categorize the spatial distribution of damage observed in eight rockburst events which ranged in size from M_L 1,1 to M_L 5,0. No useful description was given regarding the intensity of the damage. The deduction was made that the peak ground velocity within the zones of extreme stope damage was in excess of 2,5 m/s in all instances.

Among the questions which need to be answered, are:
• what is the essential nature of the "source" which has to be within 10 m or so distance before damage can result?
• what is considered extreme damage?
• how much above 2,5 m/s is a realistic upper limit for the velocity of displacement of the wall rock?

The second symposium on Rockbursts and Seismicity in Mines held in Minneapolis in 1988, provided a further convenient opportunity for marking progress in the development of rockburst damage criteria.

Hedley (1990) showed that peak particle velocities determined from seismograms of 25 rockbursts in Ontario could be characterized by a scaling relationship similar to that of McGarr but with a significantly higher attenuation factor. Thus values of $\bar{v}$ are considerably less than in South Africa, for events of comparable distance and seismic magnitude. The events constituting his data base ranged in magnitude from M_n 0,9 to M_n 3,3. The closest focal distance was 72 m and the highest observed value of $\bar{v}$ was 0,075 m/s.

Blake and Cuvelier (1990) used Hedley's scaling relationship to calculate a value of 0,97 m/s for $\bar{v}$ for a rockburst of M_L 3,0 in the Lucky Friday mine in Idaho at a focal distance of about 25 m. Damage was wide-spread but light and resin-grouted fully-bonded steel bars apparently did not break. A second large event of magnitude M_L 2,7 located somewhat closer to the tunnel, caused serious damage with a calculated $\bar{v}$ of some 1,5 m/s.

Using mainly the S-wave pulse from some 50 rockburst events in South Africa recorded in the far-field by seven GEOS digital event recorders, McGarr and Bicknell (1990) determined source and ground motion parameters by means of spectral analyses and direct measurement of $\bar{v}$. The seismic moments, M_O, of the 50 rockbursts ranged through 9 orders of magnitude. From these data they calculated that peak velocity adjacent to the causative fault was about 2 m/s with an upper bound of 4 m/s.

When discussing the effects of seismic waves on the fractured rock around stopes, Kirsten and Stacey (1990) accept that closure velocities in tabular stopes seldom exceed 3 m/s. They nevertheless advocate the use of a more conservative value for $\bar{v}$ of about 10 m/s.

The previously quoted values for $\bar{v}$ were either inferred from far-field observations or measured inside the rock mass at some distance from the damage site. It appears that very few measurements of strong ground motion at the site of the damage have been made anywhere. Roberts and Brummer (1988) quote four values of peak velocity measured by accelerometers installed 1 m deep in the sidewall of a tunnel that was affected by seismic events ranging from M_L 1,9 to M_L 2,5. In one instance where the distance to the M_L 2,5 event was 70 m, damage resulted at the recording site. The measured value for $\bar{v}$ was 1,1 m/s with a rise time of 0,031 seconds but no description of the resultant damage was given.

In considering the practical implications for the design of tunnel support, Jager et al (1990) refer to estimates of wall rock velocity of 5 to 6 m/s calculated from observations of the distance that identifiable blocks of rock have been projected during a rockburst. However they base their criteria for adequate tunnel support somewhat more conservatively on a velocity of 3 m/s.

More recently, Ortlepp (1992) examined the effect on conventional and yielding tunnel support of explosively-produced ejection velocities of 10 m/s. The visible effects produced were indistinguishable from the damage caused by a severe rockburst which occurred in a nearby, identically-supported tunnel a few weeks later.

3 DISCUSSION

The foregoing review has shown that there is widespread acceptance by researchers that rockburst damage is characterized by large strains, significant displacements of excavation walls and high velocities associated with these displacements or with the ejection of rock fragments.

It is of profound importance for the proper design of support components intended to control or contain these displacements, that the designer should know quantitatively what the typical and upper limit values of the associated velocities are.

Somewhat surprisingly, there has been a reluctance on the part of mine operators and even their geotechnical advisers, to recognize the reality of these dynamic phenomena and, particularly, their vital importance in the design of underground support. Yielding of tunnel support tendons was first said to be necessary (Cook and Ortlepp, 1968) and shown to be effective for containing violent damage (Ortlepp, 1969), nearly 25 years ago. Only now does it appear that the concept is becoming accepted as an indispensable part of the design of deep hard-rock tunnels.

It is the belief of the reviewer that this hiatus is at least partly due to lack of proper communication. The inherent complexity and obscurity of the rockburst phenomenon demands the use of sophisticated observations and computations for its understanding. Consequently mining engineers sometimes have difficulty in drawing the correct practical inferences from published results which are often presented in an esoteric way.

Other engineering disciplines have the distinct advantage of being able to rely on rigorous mathematics, on repeatable experiments in laboratory conditions and on easily *visible* and *demonstrable* effects. Observation of underground dynamic failures, particularly, is made difficult by the typically confused, even chaotic nature of the after-effects, by the extremely restricted, physically uncomfortable conditions for the observer and by the inescapable fact that the surrounding rock space is opaque.

It is possible that simple direct observations, clearly documented photographically, will reduce this communication difficulty and at the same time provide solid proof of the reality of high-velocity phenomena in rockburst damage. It is with this hope that the visual evidence of two important rockbursts is offered in the next section.

It may also stimulate conjecture as to what the upper bounds of displacement velocity might be and what implications this may have for complete understanding of the source processes.

4 OBSERVATION

The Carltonville district in South Africa is noteworthy for the uniformity and planar continuity of its major stratiform ore-body, the Carbon Leader Reef.

The high value and consistency of gold values and the absence of major faults has led to it being mined virtually continuously over a strike length of almost 20 km and for 8 km down-dip at 20° to a depth of over 3500 m. As a consequence, induced seismicity continues at a high level and very severe rockbursts have occurred. The worst of these occurred in 1977 in circumstances which have been described in some detail by Ortlepp (1984). That account mentioned and illustrated the penetration of the floor of the stope by steel support props. It did not explore the full implications of the phenomenon, however. Although brief mention was made, during discussion, of the 'Hilti gun effect' this topic is now thought to warrant closer attention.

The photographs in Figures 1 and 2 illustrate the 'punching' phenomenon, and certain features need to be emphasized.

The steel props which formed the main supporting elements in the stope were of the rapid-yield hydraulic type specially

Figure 1. Enlarged detail of the two 'impacted' props shown in Figure 11 of Ortlepp (1984).

developed to accommodate a rapid closure rate. The resistance to slow closure is designed to be 400 KN with a specified maximum overload of 15% at a closure rate of 1 m/s. The wall thickness of the outer hydraulic cylinder is about 10 mm and the diameter of the inner plunger (which is not visible in any of the photographs) is 110 mm terminating in a footpiece of 130 mm dia. The length of the closed prop is 650 mm and the total available extension of the plunger is 440 mm. In its closed state the prop would represent a very stiff indenter or 'projectile'.

The quartzite in the footwall of the Carbon Leader reef is slightly argillaceous, having a quartz content of 70%. The mechanical properties of the material are : UCS = 200 MPa, E = 70 GPa and ν = 0,2.

Because of the extreme convergence in the stope (between 50% and 100% of working height), it was necessary to cut shallow trenches in the footwall to gain access. The trench in Figure 1 revealed typical face-parallel extension fracturing of the type and intensity normally expected from stoping in high-stress ground. Importantly there was no relative movement or shear displacement along these extension fractures and the footwall surface was smooth and planar right up to where it is was penetrated by the two props. Careful measurement of prop dimension made on the enlarged views of Figure 1, shows that the props penetrated the floor by a minimum of 235 mm for the distant prop and 270 mm for the closer one.

Remarkably and inexplicably, the penetration occurred without obviously fragmenting the rock or causing an indentation 'rim' or bulge around the steel cylinder.

A similar trench several metres away where the convergence was complete, revealed another prop that had penetrated more than 350 mm and again not pulverized the surrounding rock - Figure 2.

The image most easily conjured up to explain these remarkable facts is the one of a pointed, hardened-steel pin driven by a blank cartridge such as is used in a Hilti-type gun in the construction industry.

Realistically, the more challenging questions posed by the above observations are:

• What was the starting velocity at the beginning of the dynamic closure movement?

• How does this velocity relate to the PPV in the rock mass?

• What happens to the rock material that is displaced ahead of the blunt footpiece of the prop 'indenter'?

The most likely estimate for the minimum distance of the M_L 4,0 event was about 20 m (see Figure 9 of Ortlepp, 1984). According to McGarr's R$\bar{v}$ relationship (Figure 2 in Wagner, 1984) the PPV would be close to 10 m/s. An undulation or offset in the dyke contact could conceivably have reduced the source distance to around 10 m which would have produced a value close to 20 m/s for the PPV.

The depth of the trench in Figure 2 and a broader consideration of the overall mechanism (see Figure 19 of Ortlepp, 1984) would appear to rule out the possibility that convergence was due to superficial upward buckling of the floor layer which might leave separational cavities between strata that could accommodate a 'plug' of rock material 'punched-out' by the prop. The second significant event occurred at Durban Roodepoort Deep (DRD) mine at the western end of the central Witwatersrand district. DRD differs from most other large South African gold mines in that the reef dips very steeply over the greater part of the 6 km of strike length. Although fairly extensive stoping has been carried out down to a depth of 2700 m and the quartzites are strong and relatively unfaulted, large rockbursts are rare.

One of these events, M_L 2,1, occurred in August 1977. Its effects and the surrounding geometry and associated stress circumstances were described by Ortlepp (1984). Considerable damage occurred along 90 m of tunnel on 52 level.

Nearly 5½ years later, after relatively small changes in the stope geometry in the vicinity, the same tunnel was severely damaged by an event that registered as M_L 1,8 at the WWSSN station at Pretoria, some 60 km away. As before, severe damage occurred over a 90 m length of tunnel a short distance West of the portion

Figure 2. General appearance, up-dip, of completely-converged stope and footwall rock, and enlarged view of the prop which has penetrated the floor.

that had been re-opened and re-supported after the previous rockburst. Apart from a few large slabs thrust out from the lower corner on the North side, most of the displaced and fragmented rock came from the South side upper quadrant. Relatively easy access was possible along the walk-way between the track and the North side wall, and it was possible to examine and photograph the entire damaged section before any clearing-up was commenced.

Figure 3 a) shows a general view of the scene after the burst while the situation existing before and after, is sketched in 3 b).

The intensity of damage was greatest at the centre and gradually diminished in either direction. Slight scattered falls of rock were evident in other excavations as much as 200 m away.

A strong impression was gained from the way in which the tracks were tilted on end - see Figure 3 b)(ii), that the main shock was directed from the North. The much greater volume of rock that flushed in from the South was largely 'shake-out' of the inadequately supported, previously fractured top corner of the haulage drift. Strangely, almost one-quarter of the periphery of the tunnel from South of the top centre to about 1,5 m above the floor on the North side, remained completely intact.

The most significant indications of high displacement velocity were provided by two pieces of evidence. Towards the top of Figure 3 a) left of centre, a pencil pointer indicates a precariously-balanced piece of concrete with a moulded groove or rebate (detailed in Figure 3 c) which positively proves that the concrete came from the cast drain on the South side of the tunnel. Figure 3 b)(i) shows the original position of the concrete drain and 3 b)(ii) shows the most direct trajectory which would have been required to deposit the concrete fragment where it was found.

A simple calculation shows that an ejection velocity of 8 m/s must have been imparted to the concrete fragment for it to have attained the height indicated.

Several metres towards the West, the lower, northerly quadrant of an insulated chilled-water pipe of the mines cooling system was observed to be encrusted with mud and, in places, penetrated by small fragments of rock - Figure 4 a).

The insulation on the pipe column consisted of about 1,0 mm thickness of a tough PVC-type plastic skin covering the main

insulating layer of about 30 mm thick foamed poly-styrene.

The rock fragment shown in detail in Figure 4 b) was not clean and sharp edged as would have been expected if it had been ejected by a superficial strain-burst from the adjacent solid portion of the tunnel sidewall. Its direction of penetration and dirty appearance suggest that it was a piece of ballast lying by the side of the track.

A few metres away the haulage lighting cable had been chopped through by a somewhat larger fragment of rock impacting it against the insulation. Figure 4 c) shows the cable end re-positioned close to the vertical indentation which is just visible in the PVC-cladding immediately above. The sheared wires of the cable shielding are shown in Figure 4 d).

The smaller fragment of rock 'shrapnel' was removed, weighed and simulated by a carefully-shaped piece of aluminium which closely approximated the relative density of quartzite. Using a compressed air gun, this projectile was fired repeatedly at a target consisting of a portion of the chilled water pipe recovered from underground.

Unfortunately some difficulty was experienced in the laboratory with the velocity measurement so that it was possible to state only that the impact velocity exceeded 50 m/s when the fragment penetrated the lining.

5 CONCLUSIONS

It is intended that the evidence presented in this note will help to create an awareness amongst mine operators and ground control engineers that a rockburst is an extremely complex and obscure phenomenon.

Understanding of this phenomenon is still far from complete but it is believed that it will be facilitated by recognizing the value of separating the total problem into two closely associated but separately identifiable aspects: the *source* mechanism and the *damage* mechanism.

This dichotomy will highlight paradoxes such as, importantly, how the relatively low PPV generated in the rock mass does not reconcile with the high velocity displacements of the wall rock that has so often been observed to break rock support tendons and

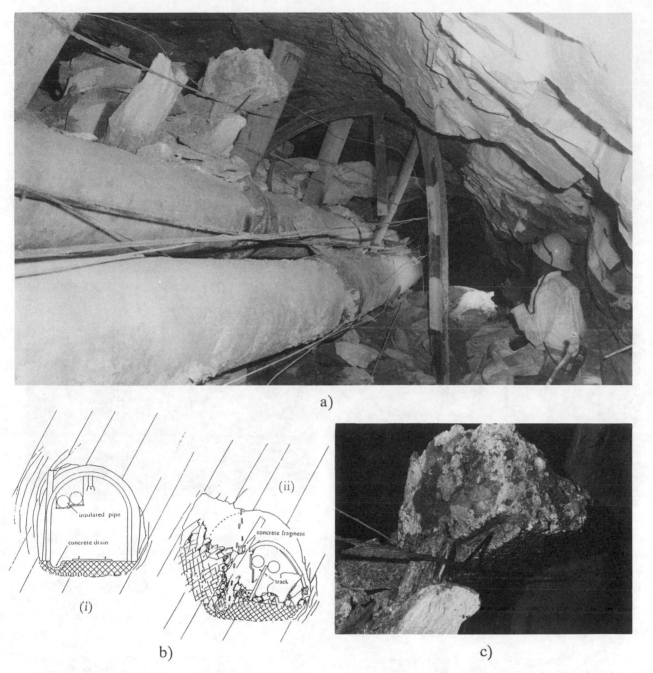

Figure 3. a) General view near the easterly end of the damaged portion of 52 haulage. b) Sketch of haulage section (i) before the event and (ii) after the rockburst. c) Detail of concrete fragment.

produce other phenomena such as described in this paper. There is a strong need to increase efforts to make direct measurements of these parameters.

A second very important benefit that comes from focusing attention on the actual mechanism of damage, is the realization that even severe damage can be controlled simply by ensuring that the steel tendons used in active support of tunnels are designed not to break but to yield. To ensure this, rational design needs to adopt damage criteria based on rock-displacement velocities somewhat higher than have recently been assumed.

The challenge of these paradoxes will, it is hoped, stimulate further conjecture that may draw upon other geotechnical areas or other engineering disciplines to provide more satisfying explanations for the observed phenomena and better support design procedures for the future.

REFERENCES

Blake, W. and Cuvelier, D.J. (1990). Developing reinforcement requirements for rockburst conditions at Hecla's Lucky Friday mine. *Proc. 2nd Int. Symp. on Rockbursts and Seismicity in Mines.* Ed. Charles Fairhurst, Univ. of Minnesota. Balkema pp. 407-409.

Cook, N.G.W. and Ortlepp, W.D. (1968). A yielding rockbolt. *Chamber of Mines of South Africa Research Organization Bulletin No. 14*, August 1968.

Hedley, D.G.F. (1990). Peak particle velocity for rockbursts in some Ontario mines. *Proc. 2nd Int. Symp. on Rockbursts and seismicity in Mines.* Ed. Charles Fairhurst, Univ. of Minnesota. Balkema 1990. pp 345 - 348.

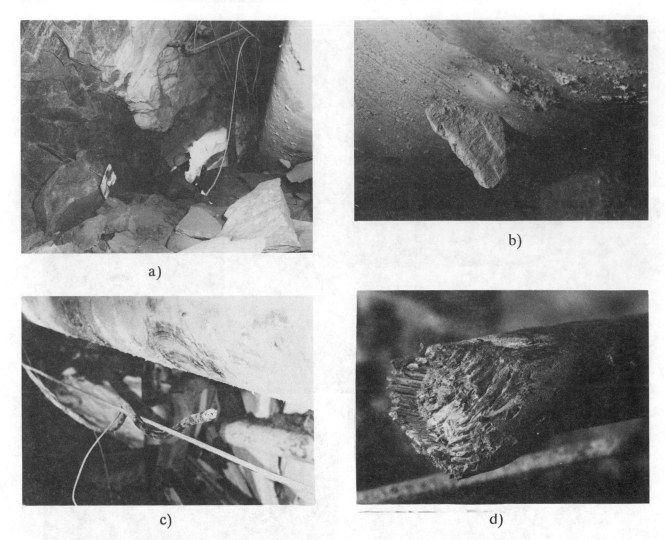

Figure 4. a) General view eastward of worst-affected portion of haulage. b) Detail of embedded rock fragment visible at right in a). c) Close view of surface of insulated pipe and severed cable. d) Detail of sheared wire armouring in severed cable.

Jager, A.J., Wojno, L.Z. and Henderson, N.B. (1990). New developments in the design and support of tunnels under high stress conditions. *Proc. Conf. Technical Challenges in Deep-Level Mining*. SAIMM, Jhb. Sept. 1990 pp. 1155 - 1172.

Kirsten, H.A.D. and Stacey, T.R. (1990). Destabilizing effects of seismic disturbances on fractured rock surrounding tabular stopes. *Proc. 2nd Int. Symp. on Rockbursts and Seismicity in Mines*. Ed. Charles Fairhurst, Univ. of Minnesota. Balkema 1990, pp 345-348.

McGarr, A., Green, R.W.E., and Spottiswoode, S.M. (1981). Strong ground motion of mine tremors : some implications for near-source ground motion parameters. *Bull. Seism. Soc. Am.*, vol 71, 1981, pp 295 - 319.

McGarr, A. (1984). Some applications of seismic source mechanism studies to assessing underground hazard. *Int. Symp. on Rockbursts and Seismicity in Mines*, S. Afr. Inst. Min. Metall. Symposium Series No. 6, Jhb 1984, pp 199 - 208.

McGarr, A. and Bicknell, J. (1990). Estimation of the near-fault ground motion of mining -induced tremors from locally recorded seismograms in South Africa. *Proc. 2nd Int. Symp. on Rockbursts and Seismicity in Mines*. Ed. Charles Fairhurst, Univ. of Minnesota. Balkema 1990. pp. 245 - 248.

Ortlepp, W.D. (1969). An empirical determination of the effectiveness of rockbolt support under impulse loading. *Proc. Int. Symp. Large Permanent Underground Openings*. Oslo, 1970 pp. 197 - 205

Ortlepp, W.D. (1984). Rockbursts in South African gold mines: a phenomenological view. *Int. Symposium on Rockbursts and Seismicity in Mines,* South African Institute of Min. Metall. Symp. Series No. 6, Johannesburg 1984, pp. 165 - 178.

Ortlepp, W.D. (1992). The design of support for the containment of rockburst damage in tunnels - an engineering approach. *Int. Symp. on Rock Support*, Laurentian University, Sudbury, Ontario, Canada p. 606.

Roberts, M.K.C. and Brummer, R.K. (1988). Support requirements in rockburst conditions. *Journal South African Institute Mining Metall.*, vol. 88, no. 3, Mar. 1988, pp. 97 - 104.

Van Aswegan, G. (1990). Fault stability in South African gold mines. *Int. Conf. on Mechanics of Joint and Faulted Rock.* Tech. University of Vienna, Austria, April 1990.

Wagner, H. (1984). Support requirements for rockburst conditions. *Int. Symposium on Rockbursts and Seismicity in Mines*, South African Institute Mining Metallurgy. Symposium Series No. 6, Johannesburg 1984, pp. 199 - 208.

Rockbursts and Seismicity in Mines, Young (ed.) © 1993 Balkema, Rotterdam, ISBN 90 5410 320 5

A quantitative analysis of pillar-associated large seismic events in deep mines

M.U.Ozbay
Department of Mining Engineering, University of the Witwatersrand, South Africa

S.M.Spottiswoode & J.A.Ryder
Chamber of Mines Research Organisation of South Africa, Johannesburg, South Africa

Abstract: Stabilizing pillars used in deep level mining for reducing face seismicity can, in some cases, cause large seismic events in their own right. These events seldom affect the face area, but cases of rockbursts have been reported in footwall haulages as a result of large seismic events (M>2) occurring in the pillar foundations. Slippage along vertically formed rupture planes along pillar edges can lead to increased stope closures, thereby reducing the effectiveness of stabilizing pillars in terms of their ability to reduce Energy Release Rate and face seismicity. The areal extent and amount of slippage along these planes was estimated by numerical modelling and gave good agreement with results from the standard relationship connecting Seismic Moment and Richter Magnitude. Similar procedures can be applied to estimating the potential for pillar-associated seismic events and the effectiveness of stabilizing pillars in reducing Energy Release Rate and face abutment stresses in deep mines.

1 INTRODUCTION

Stabilizing pillars with large width to height ratios (>20) are used in deep gold mines in South Africa mainly for reducing damaging seismic events at the stope face by reducing the release of gravitational potential energy due to stope closure. McGarr and Wiebols (1977) showed that face seismicity was reduced in direct proportion to the reduction in stope closure. Recent experience, however, has shown that the pillars themselves may cause large seismic events. These seismic events rarely affect the face area, but can cause serious damage to footwall haulages running parallel to the pillars (Lenhardt and Hagan 1990). As important, failure in pillar foundations reduces the effectiveness of pillars in resisting elastic closure and in reducing released gravitational potential energy.

Failures in pillar foundations are thought to occur by the formation of sub-vertical rupture planes running parallel to the long axis of pillars, Figure 1. In reality, the rock within some 5 m of the pillar is more severely crushed than is implied in Figure 1, but as an overall mechanism, the concept of a single localized rupture plane is probably a sound one. Although no direct observations of rupture planes daylighting into the stope have been reported, large stope closures extending right up to stabilizing pillars have been observed. Shear planes which resulted from ruptures in intact rock in a pillar foundation have been observed from boreholes drilled perpendicular to the pillar's long axis approximately 20 m in the footwall (Lenhardt and Hagan 1990). Locations of microseismic events also point to the existence of these shear planes (Lenhardt and Hagan 1990).

In this paper, stabilizing pillar foundation failure and associated large seismic events (Richter magnitude > 2) in a section of Blyvooruitzicht Gold Mine are analysed to estimate the extent and magnitude of slippage occurring along rupture planes, using boundary element computer modelling. The results from these studies are further evaluated to determine the relevant seismic moment M_0 and inferred Richter magnitudes compared to those recorded by the seismic networks. Analyses of Excess Shear Stress along potential rupture planes and changes to pillar stresses due to slippage are also carried out.

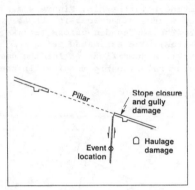

Fig. 1 Stabilizing pillar foundation failure.

2 LARGE SEISMIC EVENTS ASSOCIATED WITH STABILIZING PILLARS

Although the relationships between the occurrence of rupture planes in pillar foundations and large seismic events have been widely reported by various authors (Hagan 1988, Lenhardt and Hagan 1990) the mechanism governing the development of these rupture planes is not well understood. A commonly used criterion in assessing the potential for rupture in pillar foundations is known as "Excess Shear Stress", ESS (Ryder 1988). ESS can simply be defined as the shear stress value acting on an existing or potential rupture plane in excess of the dynamic shear strength of the given plane. The extent and magnitude of ESS on potential shear rupture planes can be estimated using available numerical computer programs such as MINAP (Crouch 1976) and MINSIM (Napier and Stephansen 1987). It is generally accepted that using these models, the extent of actual slippage "overshoots" the positive ESS zone on a given plane (Ryder 1988).

As mining progresses, rupture in pillar foundations is expected to occur once the ESS reaches a certain value on the potential rupture plane. The rupture results in release of seismic energy and the process is repeated with increasing mining, as shown in Figure 2. Figure 3 shows the mining area of Blyvooruitzicht gold mine which is considered in this paper. The progressive seismicity for the section is illustrated in Figure 4. In this figure, the seismicity (M>2) recorded within the area enclosed by the dashed rectangle in

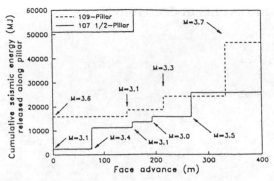

Fig. 2 Seismic energy release as function of face advance for two stabilizing pillars at Western Deep Levels gold mine, after Lenhardt and Hagan (1990).

Figure 3 is plotted against time, after projecting the events onto the section line A-A'. The magnitudes of the events were obtained from the bulletins of the Geological Survey of South Africa, and McGarr (1987) showed that these magnitudes were in good agreement with the Hanks-Kanamori (1979) Moment-Magnitude relationship. Figure 4 also shows the estimated source dimensions for those events having M > 2. The source dimensions (strike rupture length in metres) were estimated using $\log_{10}L = 0,8 + (M/2)$. A justification for the use of this particular relationship is given in the Appendix.

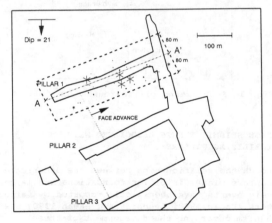

Fig. 3 A schematic view of the stabilizing pillar layout at Blyvooruitzicht showing the locations of seismic events with M > 2 associated with Pillar 1.

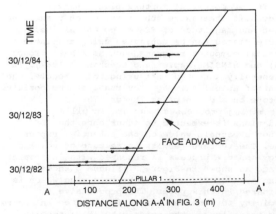

Fig. 4 Source dimensions of the pillar associated seismicity (M > 2) as function of face advance at Blyvooruitzicht gold mine for Pillar 1 in Fig 3.

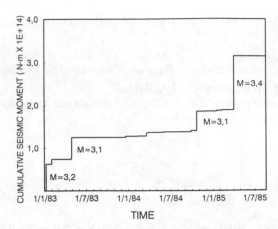

Fig. 5 Cumulative seismic moment as a function of mining progression for Pillar 1 in Figure 3. Events with M>3 are marked.

The cumulative seismic moment for the same area is shown in Figure 5. A similar pattern to that in Figure 2 with regard to "recurrence" of large seismic events can be observed in this figure. The value for the Richter magnitude corresponding to the cumulative seismic moment of $3,2 \times 10^8$ MN-m can be estimated via Hanks-Kanamori relationship to be 3,6.

3 MODELLING

The area of the Blyvooruitzicht Mine selected for analysis is given in Figure 3. The dotted lines indicate approximate initial face positions for January 1983. The stabilizing pillars are 45 m wide with centre-to-centre distance of 300 m. The reef dips 21 degree towards south and is located at depths from 1800 to 2800 m. The stoping width is 1,2 m. For ESS and explicit-slip calculations, a dynamic friction angle of 30° was assumed.

For Average Pillar Stress (APS) and ESS calculations, modelling was carried out using the pseudo three-dimensional boundary element program MINSIM-D. Figure 6 shows the change of APS values along Pillar 1 as a function of the face advance. As expected, the stresses are highest at the "pillar toe", being approximately 600 MPa when the pillar is 100 m long. When the pillar reaches its full length, the APS on the pillar toe is about 660 MPa, which approaches critical stress levels according to the foundation failure criterion proposed by Wagner and Schumann (1971), assuming a uniaxial compressive strength (UCS) for the footwall rock of approximately 200 MPa. The APS value at the middle of the pillar reaches about 450 MPa and despite the fact that this is < 3,5 times the foundation UCS, large seismic events took place in the pillar foundations according to Figures 3 and 5. There are two reasons for this apparent anomaly. Firstly, Wagner and Schumann's criterion was deduced from laboratory tests and in its normal form ignores "scale-effect" weakening, that is, the in situ or "rockmass" strength of the foundations may well be much less than the laboratory UCS value. Secondly, the MINSIM-D modelling ignores the effect of foundation failure events progressively migrating back from the highly stressed toe of the pillar and transferring more weight onto the central areas. When modelling was repeated with an explicit plane located vertically in the footwall, the APS was observed to reduce to 390 MPa, i.e. a 13 per cent reduction in APS compared to elastic modelling.

Figure 7 shows the areal distribution of ESS along a plane parallel to the long axis of the pillar and vertical to the reef, as obtained from MINSIM modelling. At a pillar length of 370 m, the

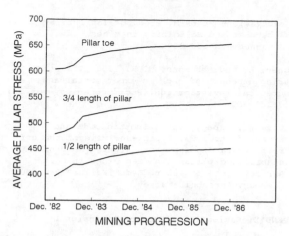

Fig. 6 Average Pillar Stress values calculated at
two different sections of Pillar 1 as function of
face advance (or pillar length)

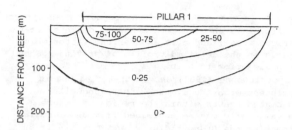

Fig. 7 Areal distribution of ESS (in MPa, shown by
numbers) along the imaginary plane located normal
to reef along the 370 m long stabilizing pillar
(Pillar 1).

ESS distribution is acceptably two-dimensional in
that the minimum positive ESS aligns with the 150 m
depth contour along a large part of the pillar. The
highest values of ESS occur below the "pillar toe"
area, which confirms the increased seismicity
during initial stages of stabilizing pillar
formation (van Antwerpen and Spengler 1984).

A limitation of MINSIM in modelling ESS (and thus
slippage on daylighting slip planes) is that the
stresses are calculated at points normal to grid
points and interpolated, which results in poor
estimates of ESS values very close to reef.
However, since the two-dimensional MINAP program
can satisfactorily calculate near-reef stresses,
test runs were conducted to establish whether
Pillar 1 could be modelled in two dimensions.
Figure 8 shows that the ESS values obtained from
MINSIM and MINAP simulations are almost identical
after the first two elements. The shape of the ESS
lobe in Figure 7 also supports the two-
dimensionality of the layout.

Figure 9 shows the distribution of ride along the
vertical slip plane as obtained from MINAP
simulations. The magnitude of the ride at 5 m from
the reef plane is approximately 0,56 m, which is
close to half of the stoping width, and reduces
with increasing distance from the reef plane.
Similar magnitudes of displacements within 30 m
proximity of the pillars in the footwall were also
reported by Yilmaz et al, (1993). The ride
decreases roughly linearly out to about 200 m from
reef and becomes zero at a depth of approximately
310 m in the footwall. The extent of the ride is
almost three times the extent of the positive ESS
lobe, thanks to the "daylighting" geometry of the
rupture plane, and this represents about twice the
extent of rupture that would be associated with
"Brune-type" isolated seismic events in an intact
rockmass discussed by Ryder (1988).

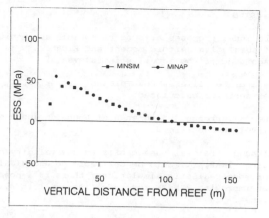

Fig. 8 ESS values determined along a vertical
plane in the foundation of Pillar 1 using MINSIM
and MINAP.

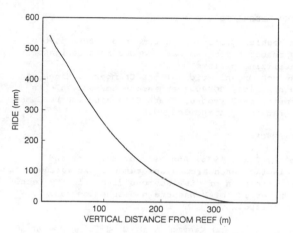

Fig. 9 The ride along the rupture plane below
Pillar 1 as calculated from MINAP modelling.

4 EQUIVALENT SEISMIC MOMENT AND MAGNITUDE

As mentioned previously, a recurring sequence of
pillar-associated seismic events has been observed
to follow the development of the pillar at some
distance back from the face. In the present study,
only two mining face positions, as shown in Figure
3, are considered. The Seismic Moment determined by
means of modelling between the initial and final
stages of Pillar 1 will be compared to the
cumulative Seismic Moment calculated from the
actual seismicity recorded during the same mining
interval.

The Seismic Moment and Richter magnitude of a
single event that would result from the slippage
described in Figure 9 can be calculated as follows:

Volume of ride (area under curve in
Fig. 9) : 68 m^3/m of strike extent.

Approx. strike extent mined : 270 m.

Then, volume of ride V=68x270=18360 m^3.

Seismic moment M_o=GV=29000x18360

$\qquad$ =5,3x10^8 MN-m.

Single event Richter magnitude, according to
Hanks-Kanamori relationship:
$\qquad$ M=(1/1,5)($\log_{10}M_o$-3,1)=3,75.

This compares reasonably well with the measured
equivalent magnitude of 3,6 inferred from Figure 5,
particularly considering the uncertainty of key
friction parameters of the rupture planes, the lack
of true two-dimensionality of Figure 7, and many
other factors.

CONCLUSION

- The model proposed allowed reasonable estimation of cumulative seismic moment and Richter magnitude of the actual seismic events.

- An average pillar stress drop of approximately 13 per cent was calculated.

- The "recurrence" phenomenon of Lenhardt and Hagan (1990) was confirmed.

- Although the test runs confirmed the validity of using MINAP, it would be desirable to use MINSIM for more realistic simulations; there is a need for improving MINSIM for modelling explicit slip planes.

- More detailed studies are needed for better understanding of constitutive behaviour of stabilizing pillars and their effectiveness in controlling ERR.

ACKNOWLEDGMENTS

The results presented in this paper form part of the research program carried out into design of stabilizing pillars by the
Research Organization of the Chamber of Mines of South Africa, COMRO. Permission to publish this material by Blyvooruitzicht Gold Mine and by COMRO is gratefully acknowledged.

REFERENCES

Crouch, S. L. 1976. Analysis of stresses and displacements around underground excavations: An application of displacement discontinuity method. University of Minnesota geomechanics report, Minneapolis.

Hanks, T. C. and Kanamori, H. O. 1979. A moment magnitude scale. J. Geophys. Res. vol.84:2348-2350.

Hagan, T. O. 1988. Pillar foundation failure studies at a deep South African gold mine. Proc. 2nd Int. Symp. on Rockburst and Seismicity in Mines, Minnesota.

Lenhardt, W. A. and Hagan, T. O. 1990. Observations and possible mechanisms of pillar associated seismicity at great depth. Technical Challenges in Deep Level Mining, Johannesburg, SAIMM: 1183-1194.

McGarr, A. 1987. Pers. comm.

McGarr, A. and Wiebols, G. A. 1977. Influence of mine geometry and closure volume on seismicity in a deep-level mine. Int. J. Rock Mech. Min. Sci. and Geomechanics Abstr. 14:139-145.

Napier, J. A. L. and Stephansen S. J. 1987. Analysis of deep level mine design problems using the MINSIM-D boundary element program. APCOM 87. 20th Int. Symp. on the App. Comp. and Math. in the Mineral Ind. vol. 1:3-9

Ryder, J. A. 1988. Excess shear stress in the assessment of geologically hazardous situations. J. S. Afr. Inst. Min. Metall., vol.88, no 1:27-39.

Spottiswoode, S. M. 1984. Source mechanisms of mine tremors at Blyvooruitzicht gold mine. Proc. of 1st Int. Symp. on Rockburst and Seismicity in Mines, Johannesburg:29-37

Van Antwerpen, H. E. F. and Spengler M. G. 1984. Effect of mining Related seismicity on excavations at ERPM. Proc. of 1st Int. Symp. on Rockburst and Seismicity in Mines, Johannesburg:235-243.

Wagner, H. and Schumann, H. R. 1971. The stamp bearing strength of rock:An experimental and theoretical investigation. Rock Mechanics, vol.3/4 no.3:185-207.

Yilmaz, H., Ozbay, M. U., Spottiswoode and Ryder, J. A. 1993. A back-analysis of the off-reef strata displacements within the proximity of stabilizing pillars in deep mines in South Africa. Proc. of Inter. Symp. Failure Phenomenon in Rock, Istanbul, Turkey.

APPENDIX: Estimation of Source Dimension L

The seismic moment M_o from a Brune-type (isolated circular) rupture is given by Ryder (1988) as:

$$M_o = c\tau_e L^3 \qquad\qquad A1$$

where τ_e is the peak ESS on the rupture plane prior to slip; L is the Source dimension (diameter of the ruptured area); and c is a numerical factor which depends on the detailed profile of driving ESS on the plane prior to slip.

Spottiswoode (1984) showed that $\tau_e = 3$ MPa is consistent with the events within longwall associated events within the region of Figure 3. The value of c can be estimated from the formula connecting the Brune radius r_o and τ_e:

$$\tau_e = 7M_o/16r_o^3,$$

substituting $r_o = L/2$ in the above formula,

$$M_o = (2/7)\tau_e L^3,$$

which, through A1, implies c = 0,29.

Taking logarithms of both sides of equation A1, and using the Hanks-Kanamori relationship, the following formula can be arrived at:

$$\log_{10}M_o = \log_{10}c\tau_e + 3\log_{10}L = 1,5M + 3,1,\ \text{or}$$
$$\log_{10}L = 0,5M + (1/3)(3,1 - \log_{10}c\tau_e).$$

For c=0,29 and $\tau_e = 3$ MPa, the above reduces to:

$$\log_{10}L \approx 0,5M + 1,0.$$

For a sequence of foundation failure events, as considered in the present paper, some modifications are necessary; in particular, the value of c for a daylighting event is likely to be at least 4 times higher than for an isolated Brune-type event. On this premise, the equation becomes:

$$\log_{10}L \approx 0,5M + 0,8,$$

which was used to produce Figure 4, and which also generates a realistic source dimension of about 400 m for the observed equivalent cumulative event magnitude of 3,6 inferred from the seismic data.

Rockbursts and Seismicity in Mines, Young (ed.) © 1993 Balkema, Rotterdam, ISBN 90 5410 320 5

Progressive pillar failure and rockbursting at Denison Mine

C.J. Pritchard
CANMET, Energy, Mines and Resources Canada, Elliot Lake, Ont., Canada (Formerly: Denison Mines Ltd, Elliot Lake, Ont., Canada)
D.G.F. Hedley
Elliot Lake, Ont., Canada

ABSTRACT: As a result of increased pillar loading and distribution of seismic activity over a five year period, the 26 Shop area of the Denison Mine in Elliot Lake, Ontario, experienced severe rockbursting ground and an area of spreading pillar collapse.

The dynamic stress waves emanating from these rockbursts pass through large geological and backfilled barriers in the area causing deterioration in weakened pillars. The limits of stress redistribution are defined by the zone of higher areal extraction and reduced pillar strengths. Numerical models of the affected area generally confirm the field observations and the trend toward increased deterioration down dip of large natural abutment pillars. Rockbursting and seismic activity is expected to continue in this area until such time as the stress transfer and pillar collapse zone reaches equilibrium.

INTRODUCTION

A slow chain reaction of pillar failures occurred at Rio Algoms Quirke II Mine in Elliot Lake in the 1980's (Hedley et al. 1983). Over a five year period, over 160 large seismic events up to a magnitude of 3.5 Mn were recorded. An area larger than 70 ha was affected underground. Seismic activity decreased and the affected area is thought to have essentially stabilized after the hangingwall fractured through to surface in 1985 (Hedley 1992).

Denison Mines is directly down dip of the Quirke Mine and has also experienced a slow chain reaction of pillar failures in two separate areas: in the Boundary Pillar area adjacent to the Quirke Mine and around the 26 Shop about 2 km to the east. Seismic activity, and the magnitude of the events and the size of the affected area has been smaller effecting a less extensive area than those at the Quirke Mine. This is partially attributed to the use of cemented backfill to stabilize the pillars in the affected areas at Denison Mine.

This paper describes the sequences of events around the 26 Shop area at Denison Mine.

26 SHOP AREA MINING METHODS

Denison Mines Limited operated in Elliot Lake from 1957 to the spring of 1992. During this time 65 million tonnes had been milled. This large room and pillar operation is located on the North Limb of the Quirke Lake Syncline. In room and pillar mining, ore thickness and dip dictate the limitations of a particular mining strategy. In the 26 Shop area of the Denison Mine the ore body was 12 m thick and consisted of an upper and lower reef. The upper reef was mined first, leaving the lower reef to be mined to maximize the extraction. During the second stage of mining, benching of the footwall and pillar slotting was initiated to complete the extraction. This created post pillars with final width to height ratios being less than 1.0. (This only applies to the immediate 26 Shop area.) The final extractions are in the order of 65 to 75% with factors of safety in the order of 1.3. The pillars created by this mining method did not have the load bearing capacity initially intended and ultimately failed. Non-violent (yielding) and violent pillar failures were initiated approximately 4 to 6 years after mining in the affected area, and progressed to cases of severe rockbursting and pillar failure.

EXTRACTION

Extraction ratios, and average pillar strength and stress estimates and associated factors of safety were obtained from empirical design equations developed for the Elliot Lake Mines (Hedley and Grant, 1972).

The pre-mining perpendicular stress combines components of the gravitational vertical stress and the mainly north-south horizontal stress. The vertical stress increases at a rate of 0.028 MPa/m. Stress measurements have indicated that the north-south horizontal stress is about 1.5 times the vertical stress. The pre-mining perpendicular stress, So, can be expressed by:

$$So = 0.028D \cos^2 a + (8.3 \text{ MPa} + 0.006D) \sin^2 a$$

D = depth (m)   a = dip of orebody 20° ave.
$So = 0.028D + 0.55$, MPa
Pillar stress, Ps, can be expressed by:

$$Ps = So/1\text{-}R \quad \text{or} \quad 0.028D + 0.55/1\text{-}R \quad \text{in MPa}$$

D = depth below surface, m
R = % extraction

From back analysis, average pillar strength, Qu, can be expressed by:

$$Qu = 133 \text{ MPa } W^{0.5}/H^{0.75}$$

where, W = pillar width, m
           H = pillar height, m.

Finally, the pillar safety factors, Sf, are defined as:
Sf = pillar strength (Qu)/pillar stress (Ps).

Empirical design equations (Hedley, 1978) are used by the engineering department in stope layouts for the active mining areas. A minimum safety factor of 1.5 has been adopted for use by the mine, with final pillar aspect ratios of greater than 1.0, in all areas after 1982.

HISTORY OF GROUND CONDITIONS

In two areas of the Denison Mine, the strength of the pillars was exceeded, initiating deterioration, and a progression of violent

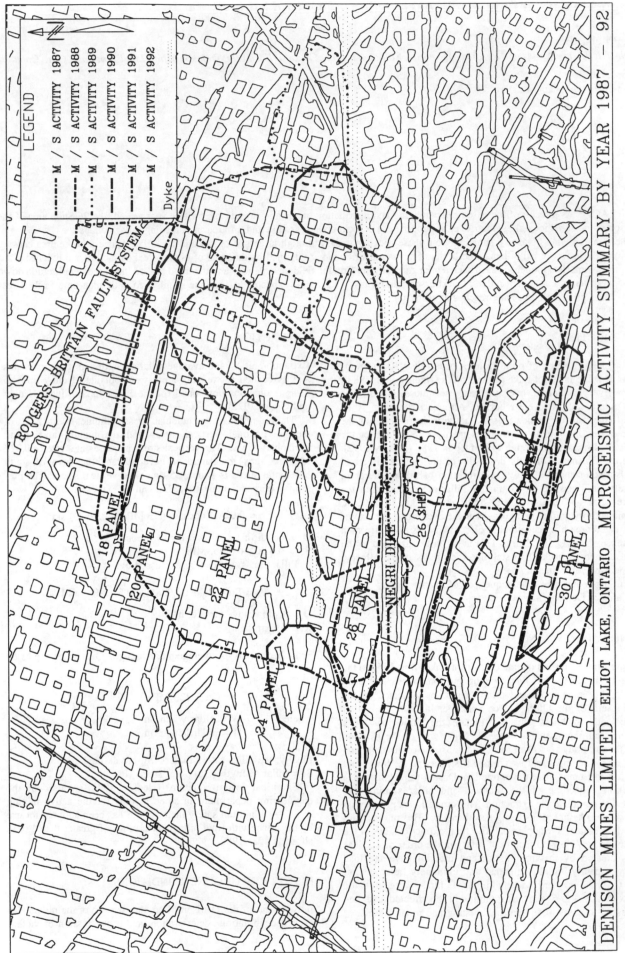

Figure 1: Plan showing microseismic activity by year, and main structural features.

pillar failure. The 26 Shop area and the Boundary Pillar area have a separate origin, with distinct failure mechanisms, with violent pillar bursting being a shared characteristic.

Due to the rapid change in ground conditions backfilling was initiated north of the 26 Shop in 1987 for pillar confinement to stabilize the post yield strength of the failed pillars, and to arrest the spread of the pillar failures that may affect permanent infrastructure in this area of the mine. A plan was devised to confine the failure area with the use of backfilled sections in conjunction with natural abutment pillars. A conveyorway corridor to the west was also utilized, and backfill was placed in these areas to maximizing the peak strength of pillars that had not been affected by spreading pillar collapse. The result was an area containing several large blocks of backfill stabilized pillars.

Due to the backfilling limitations, some sections were left unfilled. This resulted in an unique opportunity to record and document changes in the condition of support pillars in the area. A strong relationship has been developed between the seismic activity in unfilled areas and around the periphery of the filled stopes. After filling, the majority of occurrences were in unconfined pillars. During the placement of fill, a significant increase in seismic activity was documented. The seismic activity appears to be caused by the water reducing the rock strength at the edge of the pillars. However, once the pillars were encased with backfill, seismic activity essentially ceased. In the backfilled area the confined pillars continues to fail, but non-violently and the pillar is stabilized at a realitively high residual strength. Conversely, unconfined pillars tend to fail violently having very little residual strength.

It has been observed that large rockbursts can trigger microseismic activity and pillar deterioration at considerable distance from the source of the main event, see Figure 1. As an example, after a series of large events in the 22 Panel area in November 1987 microseismic activity was initiated in 28 Panel, 400 m away. It is unlikely that this was caused by a static transfer of stress from the failed pillars in 22 Panel, since the 20 m wide Negri Dyke separates the two areas. It was more likely caused by the dynamic stresses emanating from the major rockbursts. When a seismic event occurs, strain waves radiate from the source in a spherical pattern. The peak particle velocity in these waves is the main cause of damage. The peak stress pulse is related to peak particle velocity by:

$$\sigma = VE*(1-v) \, / \, Cs \, (1+v)*(1-2v)$$

where, σ = peak stress pulse
V = peak particle velocity
E = rock elastic modulus
v = Poisson's ratio
Cs = shear wave velocity

These relatively small dynamic stress pulses have no effect on pillars where the static stress is well below pillar strength (i.e., high factor of safety). However when pillar stress is close to pillar strength it appears that spalling could be initiated, as the pillars continue to deteriorate rockburst activity is initiated in these pillars.

STAGES OF PILLAR DETERIORATION:

A pillar deterioration classification was adopted at the mine to categorize unconfined pillars into 6 stages of failure. This classification has been utilized in both of the rockburst areas at Denison Mine.

Figure 2 demonstrates the relationship between the peak strength of the pillar and effective percent extraction at the stages of deterioration. Figures 3 to 8 depict these stages of deterioration.

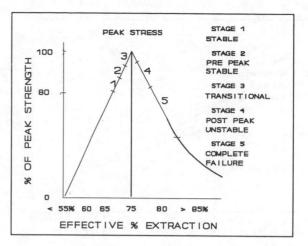

Figure 2. Percent of peak strength versus effective extraction.

As the pillars deteriorate their areal size is reduced and the effective extraction in the surrounding pillars increases.

The first stage of deterioration, Figure 3, shows signs of light to moderate pillar spalling at the pillar and hangingwall contact on the down dip side of the pillar, and at the footwall of the up dip side of the pillar. This is representative of the direction of the maximum principal stress and causes stress concentrations on the side of the pillar at the hangingwall and footwall contacts. This is observed in stages 1 and 2, and is associated with minor microseismic activity.

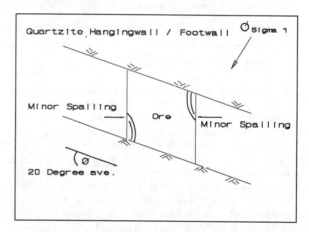

Figure 3. Stage 1 pillar deterioration.

During the second stage, Figure 4, moderate to strong sidewall spalling is located on the up and down dip sides of the pillars. This stage of spalling tends to decrease at the limits of blast damage, normally extending to a depth of 0.3 to 0.5 m. Spalling and local crushing can occur in the ore zone while the massive and competent quartzite in the footwall and hanging wall, is more resistant to increased loads.

In the third stage, Figure 5, strong spalling in conjunction with axial splitting of the pillar is noticed, which progresses inwards to the pillar core. Early signs of pillar shearing along weakened bedding contacts are initiated, this damage occurs typically in the middle section of the pillar. The amount of crushing and buckling of the sidewalls is increasing, and damage to the downdip side of the pillar is generally more pronounced. The pillar has approached or exceeded its peak strength and is entering into the elasto-plastic to plastic range of Figure 2. Signs of stress exploiting structure is evident, and moderate seismic activity, rockbursting and floor heave is well established.

113

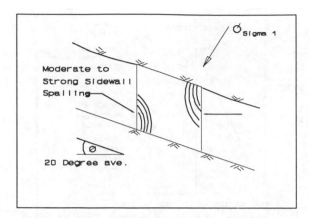

Figure 4. Stage 2 pillar deterioration.

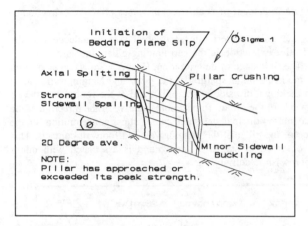

Figure 5. Stage 3 pillar deterioration.

The fourth stage, Figure 6, exhibits signs of extreme spalling and sidewall buckling. Material loss from the pillar in time has significantly reduced its overall size, and shearing of the pillar along prominent bedding slips is evident. Measurable pillar displacements are common, with damage to the pillar being severe in most cases. Behaviour at this point is in the plastic range on the suggested graph. Damage is particulary pronounced when pillars are located on the crest of a broad fold where complex tensional and compressional zones can exist within the same pillars. High stresses have caused strong exploitation of structural features that dissect the pillars, and large structurally controlled blocks can be ejected from the pillar. Locally severe rockbursting is common, and moderate to strong floor heave is evident.

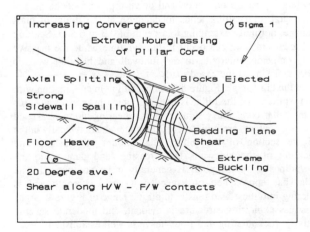

Figure 6. Stage 4 pillar deterioration.

The fifth stage, Figure 7, includes crushing and bursting of the pillar remnants resulting in seismic activity including severe rockbursting. Some pillars exhibit severe signs of shear and displacement. Numerous small rockbursts occur in the affected area while larger pillar bursts were typically in the range of 0.7 Mn to 2.2 Mn. Large pillar bursts upwards of 2.8 Mn have been recorded within the study area, while these large scale bursts are uncommon, they tend to follow periods of intense local activity, with considerable damage done to these sections of the mine. The pillar is in the final stage of its load carrying capability.

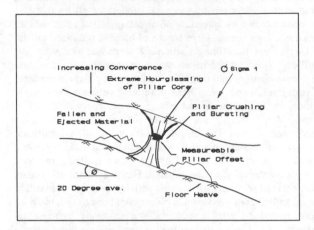

Figure 7. Stage 5 pillar deterioration.

The final stage of failure, Figure 8, appears when the severely damaged pillars have little or no load carrying capabilities. Large roof falls occur when pillar support is diminished and the unsupported hangingwall is free to converge and fail. These large roof falls fail up to prominent structural features and strong bedding slips above the hanging wall. In several locations there is strong doming created between remnant pillar spans where the roof falls have created a classical stable arch reaching 5 to 10 m above the hanging wall. Interestingly, large spans of unsupported hanging wall have remained undamaged and intact in the study area over several years, and microseismic activity has been reduced.

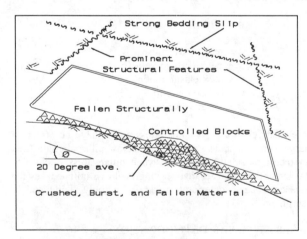

Figure 8. Stage 6 pillar deterioration.

MODELLING

Numerical modelling was contracted to the Itasca Group by Denison Mines Ltd. in December of 1987, to determine numerically the expected benefits of backfill. The purpose was to preform an extensive analysis of the current state of pillars and their performance within the study area. The analysis included two and three dimensional analysis of pillar deformations with and without supporting backfill. Results from 3-D boundary element modelling (BESOL) show that significant increases in pillars residual strength can be obtained from the support provided to a pillar by the enclosing backfill. The increased strength arises from tight confinement, and would not be achieved under standard filling practices unless the fill is placed without a void near the hanging wall interface. This was accomplished at Denison Mine due to the method of hydraulic filling and the nature of the orebody. Results of analysis of the jointed pillar indicate that where bedding slip occurred, very little load bearing capacity can be developed in the pillar.

Two dimensional analysis was accomplished using the FLAC model. The purpose of the large strain elasto-plastic analysis was to determine:

· The stress-strain behaviour of rib pillars using a strain-softened model.
· The effects of hanging wall and footwall slip planes on stress-strain behaviour.
· The effects of backfill on pillar strength.

The FLAC results show that backfill placed prior to the initiation of pillar failure appears to significantly strengthen the post-peak response and have an significant effect on the pillars peak strength. The model also indicated that the yield strength of these pillars may be maintained at strain rates up to 3%.

BESOL analysis indicated that in the study area placement of fill around already yielded pillars would not provide the best return in terms of maintaining pillar strengths. Modelling the effects of a time-dependent reduction in pillar strengths indicated that when minor reductions in peak strength were used in the Boundary Element analysis, the number of yielding pillars increases in the 20-26 panel area and south of the Negri Dyke and 26 Shop (Brady et al. 1988).

MICROSEISMIC ACTIVITY 26 SHOP AREA

The first reports of rockbursting in the 26 Shop area were in August 1984. Upon inspection of the area it became clear that pillar deterioration had taken place. Pillar stability in the area was re-examined and it was discovered that in the immediate area, pillar safety factors were low (Hunt 1984) It was, therefore, not surprising that pillar deterioration had spread over a wide area in a relatively short time. Initially the pillar deterioration was limited to the 22 to 24 panel area immediately north of the 26 Shop and the Negri Dyke. Over the course of the next several years damage spread to the areas bounded by 18 panel to the north and 30 panel to the south of the 26 shop. Microseismic activity steadily increased to a peak of approximately 35 event per day in 1987. The largest event recorded in the area was 2.8 Mn. The deterioration covered approximately 70 ha. A 32-channel Electro-lab microseismic monitoring system was installed under a contract with CANMET Mining Research Laboratory in Elliot Lake to monitor the spread of seismic activity and to assist in the location of spreading pillar deterioration. Under contract with CANMET (Denison Mine, 1988) a project was undertaken to evaluate and demonstrate with a large scale field trial the effects of consolidated tailings backfill in controlling violent pillar failure.

CHRONOLOGY OF MICROSEISMIC ACTIVITY 1984 TO 1992

The locations of microseismic activities recorded from 1987 to 1992 are shown in Figure 1.

1984: First reports of rockbursting in this section of the mine; activity and frequency was random and local.

1985: In conjunction with the start of pillar deterioration microseismic activity was located in a relatively small area, immediately north of the 26 shop area in 26305 roadway. As the frequency and intensity of the activity increased there was a corresponding spread of the damaged area. The spread was up dip and along strike into 24 panel.

1986: Activity progressed from the initial site of the pillar deterioration. Microseismic activity reached into the 22 and 20 panel areas. Activity was still occurring in the 26 and 24 panels. As part of the cooperative research project with CANMET site monitoring was initiated in the fall of 1986. An Electro-lab MP250 was utilized for determining source locations. Since the beginning of 1986, CANMET kept a seismographic record of events for Elliot Lake down to magnitude 0.7 Mn. Any event recorded was checked with the mines to confirm its location.

1987: A spread of the microseismic activity and frequency started in early 1987. The trend in source locations was from the initial site, propagating to the north east into 18 panel. 18 panel is approximately 500 m north of the 26 panel area. Activity and damage was intense, but confined to an area of about 100 m wide. The largest event recorded in the 22 panel area had a magnitude of 2.8 Mn. The damaged area spread from 18 panel in the north into the 26 Shop and the top of 28 panel, its linear extent was approximately 800 m. Reports from the underground workforce were increasing. Numerous large energy events were recorded and felt.

1988: The frequency of microseismic activity decreased slightly from its peak in 1987, but deterioration continued to spread laterally in a westerly direction to cover a much more widespread area with a down dip extent. Damage to the east section of the mine was limited because of the close proximity of a large, stiff, natural abutment pillar, the Rodgers-Brittain Faults system. This abutment effectively limited the transfer of damage in this direction. The progressive westward expansion continued due to the uniform small pillars, general weakened pillar conditions, and absence of large abutment pillars in the area. Dynamic stress transfer and the vibrations from the seismic activity were also contributing to the spread of the damaged area. Some reduction in the amount of seismic activity may be accounted for by the positive effect of backfilling in the area. With the success of the initial fill placement in the area a decision was made to continue backfilling for the purpose of regional stabilization.

1989: A general reduction in the regional activity was noticed and attributed to the large amounts of backfill being placed in the area. There was a noticeable absence of activity in the backfilled panels. Activity was locally strong around the periphery of the backfill during fill placement. Pillar stability around the limits of the activity was changing at a much slower rate than had been previously observed. A trend was emerging that indicated the events were not spreading beyond an area limited to where footwall benching had been done, resulting in slender post pillars. This trend existed into 1992. Instrumentation in the area effectively detected changes in the conditions. Readings from the displacement monitoring instrumentation (Ride instruments) showed changes in the direction and magnitude of relative movement of the hanging wall - footwall. These changes were believed to be in conjunction with the change in the locations of microseismic activity. A change in the trend of the microseismic activity was noticed lining up along what was believed to be a progressive failure front. Activity within the 26 Shop, immediately down dip of a large block of backfilled ground and a natural abutment pillar was limited. This area is believed to be in the shadow of the backfill stabilized area. Consequently, activity was showing signs of increasing in the 28 panel area immediately south of the 26 shop at approximately the same time.

1990: In February a large seismic event, 1.7 Mn, was recorded in the Negri Dyke at the west end of the 26 Shop. This dyke had been acting as an abutment pillar separating the 26 Shop from the seismically active area to the north. Within 3 months of the event there was a significant shift in the Ride instrument reading and seismic activity. All indications were that the previously established trends had changed and a new trend was emerging. Microseismic activity spread to the south of the 26 Shop area into the 28 and 30 panel areas. At this point a redeployment of 8 geophones was initiated to provide coverage of the affected area.

1991: Activity throughout 1991 was predominantly located in the 28 and 30 panel areas. Pillar deterioration had progressed well into 28 panel and was beginning to show signs of stress transfer into 30 panel by the end of 1991. The largest event with extensive damage south of the Negri Dyke in 28 panel occurred in December with a magnitude of 1.9 Mn. Some scattering of activity North of the 26 Shop was evident around the periphery of the failed pillar area. There was large seismically inactive areas within the centre of the pillar failure area. This was believed to be from the complete absence of load bearing pillars.

1992: In general, 1992 was a seismically quiet period. The largest event to occur in the area was in 30 panel with a magnitude of 1.9 Mn. No significant damage was located for this event. All indications suggest that seismic activity may continue to spread down dip into 30 and 32 panel. The possibility of a significant increase in activity existed in this area partly due to the advanced stages of pillar failure, complex mining and pillar geometry, maximum extraction, increasing depth and adverse structural features. Presumably the damage will stop at the Keyes Dyke, approximately 100 m to the south of the 32 panel. The Keyes Dyke is locally 30-50 m wide and is expected to act as a natural abutment pillar and confine the rockburst area due to its width and overall strength characteristics. This type of behaviour was observed in the Rodgers - Brittain Fault area of 18 and 20 panels.

CONCLUSIONS

Experiences gained from changing ground conditions have served as a valuable source of information in the field of stress transfer and pillar deterioration classifications. One of the most significant advances made at the mine has been the use of backfill to control violent pillar failure. Backfill has been used effectively to limit both static and dynamic stress transfer in areas of increased pillar loading by maintaining the pillars peak strength. Underground observations, and laboratory tests conducted (Blight et al., 1983; Swan et al., 1989; Arjang et al., 1991) confirmed that cemented backfill can control violent pillar failure. Field observations in changing ground conditions indicate that backfill is controlling the post-failure behaviour of the pillars resulting in a much higher residual strength compared to unconfined pillars. Seismically active areas have remained quiet after filling. Areas immediately adjacent to these large blocks of filled ground continue to show gradual worsening of conditions, consistent releases of seismic energy, and continual changes to yielding pillars. To date at Denison Mine no event larger than 1.0 Mn has been recorded in an area confined by cemented fill. Events in the nearby unfilled sections have reached magnitudes of 2.8 Mn. Although backfill can stabilize rockburst prone pillars, the possibility still exists that rockbursts can propagate to other areas with weak pillars due to the dynamic stress effects.

ACKNOWLEDGEMENTS

We would like to thank Denison Mines Limited for the free use of the information from the mine and to Mr. Peter Townsend for his critical review and continued encouragement.

REFERENCES

Arjang, B. 1991. Preliminary results of pillar backfill confinement Tests. Division Report MRL 91-063(TR), CANMET, Energy, Mines and Resources Canada.

Blight, G.E., and I.E. Clarke. 1983. Design and properties of stiff fill for lateral support of pillars. Int. Symp. Mining with Backfill, Lulea, June 1983.

Brady, H., M.P. Board, and M.G. Mack. 1988. Analysis of pillar performance in the 20-26 Panel Area. Itasca Consulting Group, Inc., Internal Report to Denison Mines Ltd, March 1988.

Denison Mine. 1988. The use of consolidated fills for controlling violent pillar failure in Ontario mines. Contract Report 23440-6-9010/01-SQ prepared for CANMET.

Hedley, D.G.F. and F. Grant. 1972. Stope and pillar design for the Elliot Lake uranium mines. CIM Trans., LXXV, pp. 121-128.

Hedley, D.G.F. 1992. Rockburst handbook for Ontario hardrock mines. Special Report SP92-1E, CANMET, Energy, Mines and Resources Canada.

Hedley, D.G.F., J.W. Roxburgh, S.N. Muppalaneni. 1983. A case history of rockbursts at Elliot Lake. Division Report MRP/MRL 84-16(OPJ), CANMET, Energy, Mines and Resources Canada.

Hunt. G. August 1984. Personal Notes and Inter-departmental communications.

Swan, G., M. Board. 1989. Fill-induced post-peak pillar stability. Mining with Backfill - Innovations in Mining Backfill Technology, Montreal.

Townsend, P., Personal conversations, Notes, and Inter-departmental communications

116

Rockbursts and Seismicity in Mines, Young (ed.) © 1993 Balkema, Rotterdam, ISBN 90 5410 320 5

Stability of backfilled stopes under dynamic excitation

E. Siebrits, M. W. Hildyard & D. A. Hemp
Chamber of Mines Research Organization, Auckland Park, South Africa

ABSTRACT: Backfill is used extensively in deep-level gold mines on the Witwatersrand in order to reduce rockfall accidents and rockburst damage. This paper shows the effect that backfill has in a tabular excavation (or stope) under dynamic excitation. Two numerical models are used to simulate backfilled and non-backfilled stopes under the influence of dynamic shear sources. These numerical results are compared qualitatively with available seismic data obtained by monitoring backfilled and non-backfilled stopes. The numerical modeling results indicate that backfill does have a beneficial effect in reducing dynamic motions around stopes under seismic influences.

1 INTRODUCTION

Backfill is used in deep-level gold mines as a local and regional support system to alleviate static and dynamic rock pressure problems. As a regional support, backfill is used to reduce stope closures, thereby slowing the build-up of excess shear stress and energy release rate. As a local support, backfill is used to maintain the hangingwall integrity, reduce rockburst damage, and to improve stope face and gully conditions (Gurtunca *et al.* 1989).

In this paper, elastodynamic numerical models are used to determine the relative effects of incoming seismic waves on filled and unfilled stopes. These results are compared qualitatively with results obtained from monitoring programs in filled and unfilled stopes.

2 UNDERGROUND MEASUREMENTS

Instrumentation has been set up on South African gold mines in order to monitor the ground motion in backfilled and non-backfilled stopes (Adams *et al.* 1990, Hemp and Goldbach 1993). Micro-seismic networks were installed underground with seismic transducers (geophones or accelerometers) installed in stopes adjacent to filled and unfilled areas. The first measurements were limited to events with magnitudes less than 0.2 while the most recent measurements have involved events with magnitudes up to 3.0. In these monitoring programs, transducers were attached to the hangingwall or footwall of gullies and were also installed inside backfill bags. Transducers were kept as close to the working face as possible. Source information from seismograms recorded in the highly fractured ground surrounding a stope has not yet been obtained. Amplification of coda due to rock fracturing means that shear waves often cannot be identified.

Transducer measurements are consistent for all magnitude ranges, indicating that backfill has a beneficial effect, and explaining why the hangingwall in backfilled stopes and gullies is more stable during seismic events than the hangingwall in unfilled areas. Peak particle velocities and accelerations measured in backfilled stopes are lower than those measured in unfilled stopes; the shorter effective hangingwall beam in a backfilled stope resonates at higher frequencies, thereby reducing the level of damaging seismic energy; and shorter vibration times occur in a backfilled stope, thereby limiting the duration of possible damage. Analysis of the waveforms for larger events also indicates that backfill causes a reduction in low frequency surface wave development. These benefits are maximized by increasing the percentage of backfilling (Hemp and Goldbach 1993).

Figure 1 shows typical waveforms from backfilled and non-backfilled areas. In the following sections, numerical models will be used to demonstrate and validate the results obtained from seismic data.

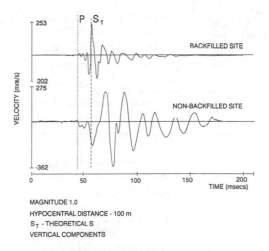

MAGNITUDE 1.0
HYPOCENTRAL DISTANCE - 100 m
S_T - THEORETICAL S
VERTICAL COMPONENTS

Figure 1. Recorded velocity histories from filled and unfilled sites.

3 NUMERICAL MODELS

Numerical modeling allows us to investigate, within the limitations of each model, what effect backfill has in an ideal stope under the influence of dynamic waves. The two numerical models used are an elastodynamic finite difference scheme (WAVE) of Cundall (1990) and an elastodynamic displacement discontinuity scheme (TWO4D) of Siebrits (1992), each capable of modeling stopes, faults, and non-linear internal responses (e.g. seams and various support systems in stope constructs) under transient dynamic excitation. WAVE is capable of modeling any grid-based geometry in two- or three-dimensional space, whereas TWO4D is capable of modeling an arbitrary displacement discontinuity type geometry in two-dimensional (plane stress, plane strain or antiplane strain) space.

Graphics post-processing capabilities allow histories of dynamic displacements, velocities, accelerations, induced stresses, and total stresses to be plotted, as well as snapshots in time of dynamic displacement vectors, velocity vectors, acceleration vectors, and induced and total principal stresses. Stopes are modeled as zero thickness surfaces of displacement discontinuity, and are grid based in WAVE, but can be at any orientation in TWO4D. Similarly, faults, seams, and various support systems are modeled as more complicated variations of stope constructs. Cavity type problems can also be investigated. In WAVE, a cavity is simply a grid with zero elastic constants, and in TWO4D, the cavity boundaries are explicitly defined by displacement discontinuity elements. Both WAVE and TWO4D can model a variety of prescribed transient dynamic displacement and/or traction sources and their interactions with nearby volumetric and/or tabular excavations. The effects of initial static stress conditions can be included in both models. This is used when total stresses are needed, for example, when modeling frictional behavior along faults. Both models use time marching algorithms, thereby ensuring that non-linear material behavior can be included within the stope constructs.

There is almost no data available concerning the dynamic behavior of backfill under laboratory conditions. Backfill is therefore modeled as a linear elastic seam element, containing constant shear and normal stiffnesses.

4 MODEL PROBLEMS

4.1 Displacement Discontinuity Method

The model problem investigated using TWO4D is depicted in Figure 2, which shows the layout of a horizontal stope that is influenced by a nearby vertical shearing source. Both the source and the stope are modeled as displacement discontinuity elements. A parting plane construct, shown above the stope, is also modeled using displacement discontinuity elements.

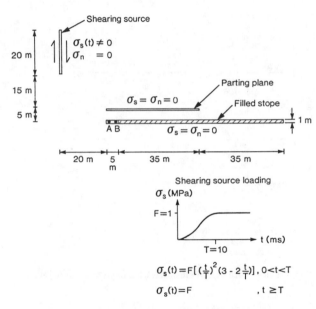

Figure 2. Stope with nearby shearing source and parting plane.

The elastic properties of the quartzitic host rock are assumed to be E = 35 GPa, ν = 0.27, ρ = 2,700 kg/m^3, and the uncemented backfill is assumed to have dynamic shear and normal stiffnesses of K_s = 200 MPa/m and K_n = 400 MPa/m, respectively. The soft Young's modulus is used in order to try to account for the highly fractured nature of the rock mass around a deep-level stope, in an elastic analysis (see Gurtunca and Adams 1991).

Figure 3 shows typical histories of the rate of change of the normal component of displacement discontinuity (or closure rate) at position A (Figure 2) for the cases when the stope is unfilled and filled to within 5 m of the stope face. Notice that the closure rates are lower in the filled case than in the unfilled case.

Figure 4 summarizes the closure rate behavior of the entire stope. This figure is an envelope of the absolute closure rates of all collocation positions along the stope. The filled stope clearly displays smaller closure rates than the unfilled stope. Furthermore, the duration of dynamic activity is less in the filled stope than in the unfilled stope. The time taken for the closure rates to drop to 20 % of their peak values is much less in the filled stope than in the unfilled stope. The backfill reduces the vibration time of the hangingwall "beam." The oscillations evident in Figure 4 are due to the discrete nature of the data; stope closure rates are only sampled at the collocation points of each element.

Figure 5 shows the absolute closure rate envelope if the parting

plane construct is excluded from the model. In this case backfill reduces the peak magnitude of the closure rates, but there is no significant reduction in the vibration times.

Figure 6 is a snapshot in time of the particle velocity vectors plotted on a 150 m by 150 m window which includes the filled stope, parting plane, and shearing source. The largest vectors in Figure 6 are the shear wave effects. The seismic source is of a shearing nature, and the propagated shear wave therefore contains most of the energy. Figures 7 and 8 are also velocity snapshots at the same time step, plotted on a 15 m by 15 m window which includes part of the stope and parting plane. A surface wave effect or eddy is noticeable, traveling between the stope and the parting plane. Figure 7 is the unfilled case and Figure 8 is the filled case. Notice that the eddy travels more slowly in the unfilled case than in the filled case (compare the distance between the eddy centers and the stope face in each case). In the unfilled case the hangingwall skin is a free surface and the eddy travels at the Rayleigh wave velocity, whereas in the filled case, the eddy travels at a velocity closer to the shear wave velocity because the hangingwall skin is no longer a free surface. Furthermore, the peak particle velocity magnitudes are lower in the filled case, indicating that backfill reduces the development of surface wave effects. This agrees with the results of Hemp and Goldbach (1993).

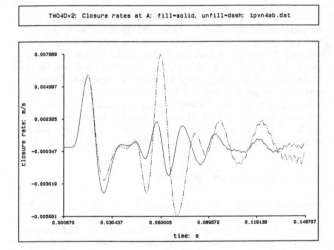

Figure 3. Closure rates at A for filled and unfilled stopes with parting plane.

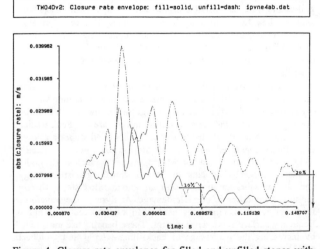

Figure 4. Closure rate envelopes for filled and unfilled stopes with parting plane.

118

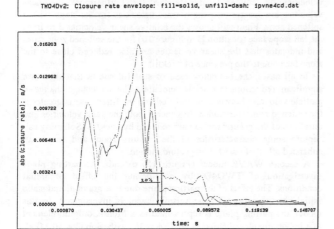

TWO4Dv2: Closure rate envelope: fill=solid, unfill=dash: ipvne4cd.dat

Figure 5. Closure rate envelopes for filled and unfilled stopes without parting plane.

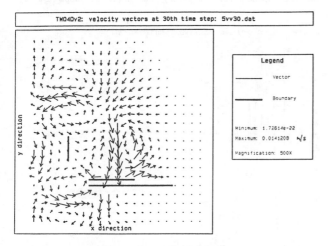

Figure 6. Particle velocity vector snapshot in surrounding rock mass.

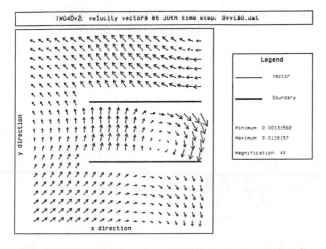

Figure 7. Particle velocity vector snapshot between stope and parting plane for unfilled case.

4.2 Finite Difference Method

A direct comparison between WAVE and TWO4D is presented in Figure 9 to show that the two models produce similar results. Figure 9 is a history of closure rates at position B (Figure 2) in the case of an

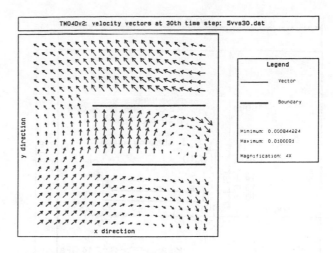

Figure 8. Particle velocity vector snapshot between stope and parting plane for filled case.

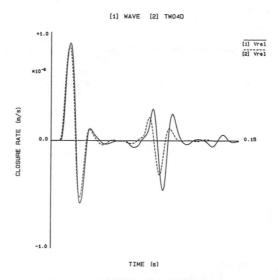

Figure 9. WAVE/TWO4D comparison for unfilled stope without parting plane.

unfilled stope without a parting plane. Slight differences are apparent in the amplitudes of the reflected wave, but the results generally match very well.

WAVE has the capability to allow multiple material types to be modeled. Figure 10 depicts the same geometry and loading as that of Figure 2, but with the additional feature of a local zone around the stope that contains a softer material. This is another way of attempting to capture the highly fractured nature of the rock mass around a stope, within the limitations of linear elastic continuum theory.

The elastic constants used for the rock mass are E = 85 GPa, ν = 0.19, and ρ = 2,700 kg/m^3). The softened zone has elastic constants of E = 20 GPa, ν = 0.4, and ρ = 2,700 kg/m^3). No attenuation is present in this softened zone to account for energy dissipation that would occur naturally in a fractured zone. The stope is modeled as a zero width discontinuity. Backfill is placed up to 5 m from the left stope face.

Three cases of fill are considered: open stope (no fill); fill with stiffnesses K_n = 400 MPa/m, and K_s = 200 MPa/m; and a very stiff fill of K_n = 2 GPa/m, and K_s = 1 GPa/m. Figures 11 and 12 show envelopes of maximum relative normal velocities (closure rates) in the stope for the cases without and with the softened zone, respectively. These results indicate that there are large increases in magnitude and seismic duration due to the presence of the softened zone, and that normal velocities are not decisively reduced by the standard fill. Figure 13 shows an envelope of maximum relative shear velocities (ride rates) in the softened zone case. In this case, shear

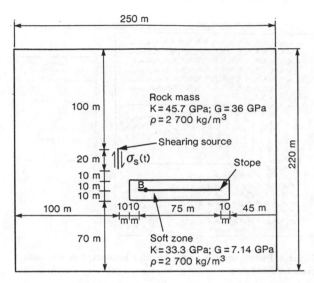

Figure 10. Stope with nearby shearing source and local soft zone surrounding stope.

velocities are reduced by the standard fill. Furthermore, the shear velocities are greater than the normal velocities for the stope with a softened zone. Figure 14 shows the relative shear velocities 3 m from the left stope face (position B in Figure 10) for the softened zone case, and indicates that the shear velocities are also reduced close to the stope face due to the presence of backfill.

In all cases, the bounding case of stiff fill shows that there are significant reductions in particle velocities. The percentage change in particle velocities between the unfilled and the filled case is greater if the softened zone is included. In general, both the peak velocities and durations of the ground motion are reduced by backfill. Softening can capture some characteristics of the fracture zone, and shows an enhanced effect of backfill on ground motion.

A second WAVE model (Figure 15) extends the parting plane investigation of TWO4D, by considering the effect of initial conditions. The effect of the parting plane due to a gravitational static initial condition prior to the seismic event is investigated. In this model, the parting plane is represented as a Mohr-Coulomb element that is initially closed (friction angle $\phi = 30°$, cohesion C = 30 MPa). Static stresses are applied at the finite difference grid boundaries, and the model is equilibrated.

It has been postulated that one of the benefits of backfill is the clamping of fractures in the fracture zone. The WAVE results do, in fact, show opening along the parting plane in the open stope case, but

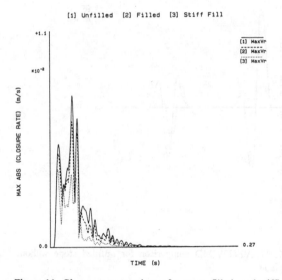

Figure 11. Closure rate envelopes for open, filled, and stiffly filled stope with no softened zone.

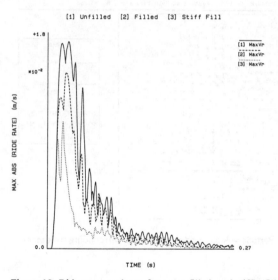

Figure 13. Ride rate envelopes for open, filled, and stiffly filled stope with softened zone.

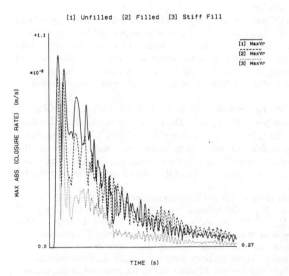

Figure 12. Closure rate envelopes for open, filled, and stiffly filled stope with softened zone.

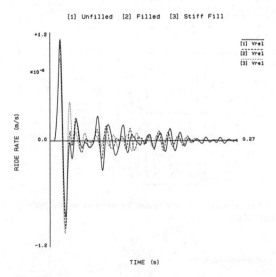

Figure 14. Ride rates at B for open, filled, and stiffly filled stope with softened zone.

clamping throughout in the case of a filled stope. This is demonstrated in Figure 16, which shows the normal traction along the parting plane for the filled and unfilled case, before any dynamic loads are applied. Notice that, in the filled case, compressive (positive) tractions are generated along the entire length of the parting plane, indicating that it is fully clamped. Furthermore, the peak compressive traction is smaller in the filled case because stresses are re-distributed into the backfill. The actual traction magnitudes in Figure 16 are not necessarily accurate, because the boundaries of the WAVE mesh are chosen in order to obtain accurate dynamic results, and not accurate static results (the mesh would be excessive in this case). However, the relative magnitudes of the normal tractions in Figure 16 are comparable with each other.

Once the static solution has been reached, conditions on the fault are relaxed allowing the fault to slip, according to a Mohr-Coulomb failure criterion. Figure 17 shows the normal velocities at position B (Figure 15), indicating a marked reduction for the backfill case.

5 CONCLUSIONS

The numerical modeling supports some of the conclusions drawn by Adams et al. (1990) and Hemp and Goldbach (1993) from the results of underground measurements, viz. backfill reduces the duration of dynamic activity and the peak particle velocities around a stope after a seismic event. It is more difficult, however, to make quantitative

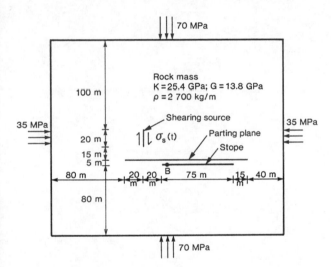

Figure 15. Stope with nearby shearing source and Mohr-Coulomb parting plane construct.

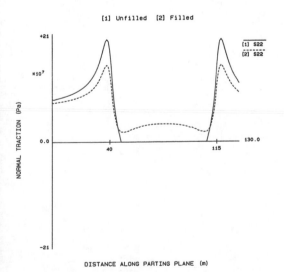

Figure 16. Normal tractions along parting plane for filled and unfilled cases.

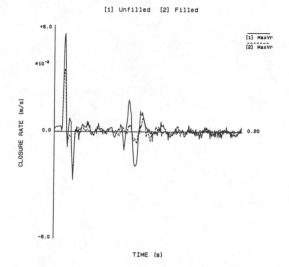

Figure 17. Closure rates at B for open and filled stopes with parting plane.

comparisons between seismic data and these numerical models. This requires two- and three-dimensional case studies and is the subject of current research.

A number of possibilities are being considered for extending the elastodynamic modeling. These include more realistic representations of the fracture zone, and the effect of stope support on these representations, further investigations on the use of a softened zone, the addition of attenuation, the effects of varying the elastic constants and softened zone dimensions, the introduction of multiple fractures surrounding a stope as a comparison with a softened zone approach, and the importance of differences in the initial static stress state, prior to a seismic event.

ACKNOWLEDGMENTS

This work forms part of the research programs into rock mass behavior and regional support of the Rock Engineering Division of the Chamber of Mines of South Africa's Research Organization. Permission to publish is gratefully acknowledged.

REFERENCES

Adams, D.J., Hemp, D.A. and Spottiswoode, S.M. 1990. Ground motion in a backfilled stope during seismic events. Proc. ISRM International Symposium 1990/Swaziland: 13-22. Rotterdam: Balkema.

Cundall, P. A. 1990. personal communication.

Gurtunca, R.G. and Adams, D. J. 1991. Determination of the in situ modulus of the rockmass by the use of backfill measurements. J. S. Afr. Inst. Min. Metall. 91: 81-88.

Gurtunca, R.G., Jager, A.J., Adams, D.J. and Gonlag, M. 1989. The in situ behaviour of backfill materials and the surrounding rockmass in South African gold mines. Proc. 4th Int. Symp. on Mining with Backfill/Montreal: 187-197. Rotterdam: Balkema.

Hemp, D.A. and Goldbach, O.D. 1993. The effect of backfill on ground motion in a stope during seismic events. subm: Proc. 3rd Int. Symp. on Rockbursts and Seismicity in Mines. Rotterdam: Balkema.

Siebrits, E. 1992. Two-dimensional time domain elastodynamic displacement discontinuity method with mining applications. Ph.D. dissertation. University of Minnesota.

Rockbursts and Seismicity in Mines, Young (ed.) © 1993 Balkema, Rotterdam, ISBN 90 5410 320 5

A technique for determining the seismic risk in deep-level mining

R. D. Stewart & S. M. Spottiswoode
Chamber of Mines Research Organisation of South Africa, Johannesburg, South Africa

ABSTRACT: A seismic risk assessment method has been developed for application to rockburst-prone deep-level mining sites which have sufficient seismic coverage. The technique utilizes several seismic hazard parameters (including frequency-magnitude statistics, clustering analyses and stress-drop variations) together, and assesses the seismic risk on a daily basis according to the individual levels of these parameters. Critical levels are determined from investigation of the long-term behaviour of these parameters in relation to recorded potentially-damaging seismic events. The method has been applied to a dataset comprising nearly three years' worth of seismic data, with encouraging results.

1 INTRODUCTION

The South African gold-mining industry experiences difficulties with rockbursts in the underground working areas. These rockbursts are a particular problem in the stopes, where they often occur during the working shift, causing injury to mining personnel and damage to machinery, with consequent loss of production. There is an obvious need to address this problem effectively, either by prevention of the rockbursts themselves, or by adequate warning of their occurrence which enables steps to be taken to prevent the work-force from being present at the time of the rockburst. Certainly, effective support installation is also required, so as to minimise the effects of rockbursts, should they take place despite these measures.

The COMRO Rockburst Control Project is currently engaged in developing a preconditioning technique aimed at prevention of the subset of rockbursts which occur in the immediate vicinity of the stope face. The concept of the preconditioning experiment has been described by Adams et al (1993). The preconditioning experiment is currently being conducted at Blyvooruitzicht Gold Mine, in the Carletonville area, and the experimental sites in use there have also been described by Adams et al.

These sites are being monitored by Portable Seismic System (PSS) networks developed at COMRO (Pattrick et al 1990), as seismic coverage is viewed as being an essential aspect of the determination of the effects and effectiveness of preconditioning at these sites. The seismic data recorded at one of these sites have been used in the development of the seismic risk assessment method which is the subject of this paper, and which is aimed at giving adequate warning of the likely occurrence of a rockburst, so that precautionary measures, including preconditioning, can be taken.

2 THE SEISMIC DATA

The preconditioning site from which the seismic data were recorded is the active stope of a strike-stability pillar which was initially left behind to protect adjacent areas when this part of the mine was originally mined out. Extraction of this planned remnant pillar is presently underway, now that all of the surrounding ground has been removed. The geometry of the pillar, and the positions of the geophone sites of the seismic network which is being used to monitor the seismicity associated with the preconditioning site, are shown in plan in figure 1.

Mining of the pillar initially took place by means of cuts taken out of the eastern edge of the pillar in an up-dip direction, but, since mid-1992, mining has taken place along breast, as shown in figure 1. The geophones of the network are grouted to the ends of boreholes drilled at least 10 metres into the hangingwall or footwall of tunnels above and below the pillar. Figure 2 shows a typical distribution of the locations of seismic events recorded by the seismic network: the preconditioning site appears to be largely isolated from outside seismic influences, as shown by the well-developed tendency towards

clustering of the seismicity about the actively-mined faces. It seems that the seismicity associated with the site is related only to the mining activity at the site and to the response of the pillar to the loading induced by this mining.

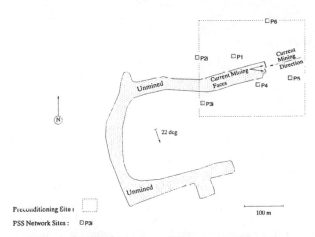

Figure 1. A plan showing the geometry of the remnant pillar and the positions of the geophone sites of the seismic network monitoring the preconditioning site.

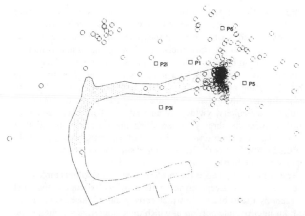

Figure 2. A typical distribution in plan of locations of seismic events recorded over a period of one year by the seismic network monitoring the preconditioning site. Mining during this period was conducted in an up-dip direction.

When the PSS detects a seismic event, it stores the waveforms recorded at the geophone sites in an unique WAVEFORM file for the event, and adds some details of the event to the EVENTS file, which then has a record of the date and time of the event, its location and magnitude, and the peak ground-velocity recorded at each geophone site. The EVENTS file is essentially a catalogue of all events detected by the PSS network. It is the database represented by this EVENTS file which was accessed in the development of the seismic risk assessment method. The seismic events used in this study were all manually-relocated on a daily basis during the routine processing of the seismic data recorded at the preconditioning site.

The seismic network at this site has been in operation from early 1990 to date (end 1992). The frequency-magnitude plot of all events located within the preconditioning site by the seismic network is given in figure 3, in which it can be seen that the minimum magnitude (governed by the lower limit of the sensitivity of the network) of the seismic data is -1.50, while the Gutenberg-Richter b-value (given by the slope of the linear portion of the frequency-magnitude plot) is 1.09, which is typically observed for mining-induced seismicity in the Carletonville area: the seismicity is concentrated ahead of stope faces, and is not significantly influenced by large geological features. There is a small geological fault, oriented sub-parallel to the strike direction, which intersects the top panel of the preconditioning stope. This fault seemed to act as a plane of weakness during the up-dip stage of mining of the pillar, but appears to be aseismic, so that it does not significantly affect the location patterns of the seismicity associated with the preconditioning site.

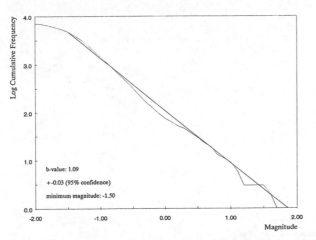

Figure 3. The frequency-magnitude plot for all events located within the preconditioning site by the seismic network. The minimum magnitude is given by the lower limit of the linear portion of the plot. The b-value was obtained via the maximum-likelihood method.

Figure 4 has the diurnal distribution of the seismicity recorded from the preconditioning site by the seismic network: the concentration of seismicity around blasting time on the mine can be clearly seen in the figure. This obvious influence of blasting on mining-induced seismicity should be taken into account in any technique which seeks to quantify the seismic risk of a mining site at any given time. The effect of blasting on the seismicity recorded from the preconditioning site is further illustrated in figure 5, which plots the observed cumulative seismicity, yielded by an analysis of all blast sequences recorded during the three year period, against time after initiation of blasting. The initiation times of these blast sequences were manually identified in the EVENTS file from rapid successions of three or more events at around blasting time on any given day.

The curve in figure 5 is for located events, and exhibits three clear trends: an initial steep linear portion which occupies the first 100 seconds after a blast, and which corresponds to the blast sequence itself; an intermediate portion in which the cumulative seismicity rate follows a trend which is inversely-proportional to time, and which corresponds to the decay of seismicity after the blast (i.e. to the diminishing effect of the blast on the recorded seismicity), which decay is analogous to that observed in aftershock sequences; and a final linear portion, which corresponds to the average background level of seismicity. The return to background levels of seismicity

appears to take place after some 20 minutes (or 1250 seconds) following the blast, which suggests that the immediate effect of the blast on the seismicity, and therefore on the stress distributions ahead of the stope faces, has diminished significantly by this time. Some 41 percent of the seismicity recorded in the 24 hours after blasting was identified on the basis of figure 5 as being directly related to the blasting.

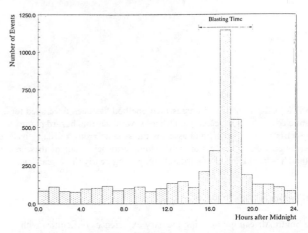

Figure 4. The diurnal distribution of all seismicity recorded from the preconditioning site by the seismic network.

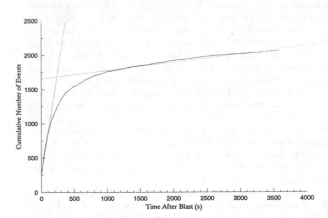

Figure 5. The cumulative seismicity with time after blasting obtained from an analysis of all recorded blast sequences within the preconditioning site. The trends shown in this plot are discussed in the text.

3 OUR APPROACH TO SEISMIC RISK ASSESSMENT

While numerous seismic studies have been carried out in an attempt to arrive at the successful prediction of rockbursts, relatively few have had any significant success. This is due, in part, to the very complicated physical rockburst process, but, also, to the fact that the majority of these studies have concerned themselves with the analysis of short time-periods and single rockbursts: in general, rockbursts do not appear to be accompanied by any observable seismic anomaly (e.g. Xie and Pariseau 1992, Riefenberg 1991). In the spirit of Riefenberg (1991), it is our contention that a clearer understanding of rockburst occurrence can be obtained by statistically investigating many rockbursts over a long time-period. We suggest that, in this way, we are better able to deal with the uncertainties inherent in the process of assigning a level of seismic risk to a given area at a given time.

Riefenberg (1991) characterised rockburst occurrence as related to microseismicity and mining, using the parameter of the number of daily microseismic events. The observed relationship between stress and rockbursts was approximated by using average microseismicity rates (which are observed to be higher when load is applied). Daily microseismicity was found not to be random, but, rather, to be a function of prior seismic activity. Riefenberg also observed spatial clustering of microseismicity in the regions in which active mining was taking place at the time of occurrence of a rockburst, which were interpreted as areas of higher local stress. It was suggested that an analysis of the fractal distributions of microseismicity in time and space might be appropriate.

Xie and Pariseau (1992) suggested that the lack of success in seismic analysis of rockbursts in many studies might be due to incomplete utilisation of initial recorded data. We claim that the uncertainties inherent in the assessment of seismic risk will further be reduced by considering several seismic hazard parameters simultaneously, instead of concentrating on any one parameter in isolation. As stated by Xie and Pariseau, the microseismic noise rate is not a reliable precursor factor when used on its own: the distribution of microseismic event locations is, in itself, a realistic record of the damage evolution process experienced by a rockburst, and should be considered, in combination with the microseismic noise rate, as a seismic hazard indicator.

Indeed, we suggest that any physically-reasonable seismic parameter might be considered in the evaluation of seismic risk, in order to identify parameters that vary in a meaningful way in association with rockburst occurrences. To this end, we have investigated several parameters, and believe that we have identified some that can usefully be incorporated into a seismic risk assessment technique.

During routine application of the technique to recorded seismic data on a daily basis, workers would be withheld from the stope whenever the seismic hazard was determined to be above some established critical level. During this time, preconditioning could be carried out in order to reduce the level of seismic hazard. We shall consider that the technique is a useful one, if a significantly greater proportion of the potentially-damaging seismic events take place during such periods of increased apparent seismic risk than take place outside of those periods. An important consideration when applying the technique would have to be that workers should be withheld from the stope for less than some specified maximum acceptable proportion of time (such as 20 percent of the total time). Thus, we desire that a large proportion of the total number of potentially-damaging seismic events should take place during a small proportion of the total time, which proportion of time is identified by the technique as having increased levels of seismic risk.

4 THE SEISMIC RISK ASSESSMENT METHOD

The seismic risk assessment method was developed by considering the variations in the parameters discussed below, on a daily basis, over the entire dataset of seismic events located within the preconditioning site by the seismic network covering the site, and comparing these observed variations with the recorded occurrences of potentially-damaging seismic events. The parameters used are all assumed to be largely independent of one another, as they reflect variations in different aspects of the physical rockburst process and as they are calculated from different combinations of variables. The calculated parameters for any given day were averaged over the 40 events prior to 06h00 on that day, eliminating periods when the seismic system might have been out of order. This number of events was used in the averaging as it represents about one week's worth of recorded data, and, so, fits within the preconditioning cycle. The averaging is necessary for the accurate determination of the Gutenberg-Richter a- and b-values, and was found to stabilise the variations in the other parameters as well. Median statistics were used in the calculation of all of the parameters, so as to reduce the effects of outliers (seismic events which yield unusually large values of the parameters) on the assessment of seismic risk.

The effect of blasting on the seismic risk assessment was investigated by considering whether or not a blast was to take place on the day for which the risk was being computed, and/or whether or not a blast took place on the previous day. The results of the analysis were considered in these categories, as well. Preconditioning blasts were treated separately, and were excluded from the list of potentially-damaging events against which the risk assessment was compared, as these blasts are generally associated with seismic events of such magnitude that they would fall within the subset of events

considered potentially damaging, but which clearly cannot be predicted on the basis of apparent seismic risk.

4.1 Parameter Pred

The fundamental parameter on which the seismic risk assessment method was constructed is the apparent seismic risk calculated from frequency-magnitude statistics which are well described by the familiar Gutenberg-Richter relationship:

$$\log N = a - bM \qquad (1)$$

in which N is the number of events per unit time of magnitude greater than or equal to M, and a and b are constants. According to McGarr (1984a), in principle, one already has the information necessary to assess the seismic hazard, once a and b are determined for a given region of mining. The apparent seismic risk is calculated from (1) by determining the value of N corresponding to a given critical magnitude M_c. In our study, we have assumed the value for M_c to be zero. Variations in seismicity rate will affect the value of a, and so alter the calculated value of N, thus altering the apparent seismic risk. It is thought that variations in the value of b are largely a response to blasting, but may also reflect changes in the geometry of the fractured zone ahead of the stope faces (e.g. Legge and Spottiswoode 1987). These variations would also have an effect on the apparent seismic risk.

Our parameter, Pred, is calculated from the Gutenberg-Richter a- and b-values (determined for 40 events), and corresponds to the number of events of magnitude greater than or equal to M_c expected for the following day, on the basis of the frequency-magnitude statistics described above.

4.2 Parameters D and Med

Xie and Pariseau (1992) showed that the distribution of microseismic events has a fractal character, and proposed that the fractal dimension of the damage evolution process experienced by a rockburst (described as a fractal cluster of crackings within the rockmass) can be measured directly from the distribution of microseismic event locations. The microseismic events comprising the damage evolution process were observed to follow a fractal clustering geometry, such that, with the approach of a rockburst, greater clustering was observed, so that the fractal dimension was reduced. The occurrence of a minimum fractal dimension of the microseismic event locations was proposed as a reliable precursor to a rockburst.

Our parameter, D, is the reciprocal of the calculated fractal dimension of microseismicity, calculated from 40 events. More distant event-pairs are excluded from this analysis, as they are unlikely to be related to the main clustering mechanism. The reciprocal of the fractal dimension is considered, so that D can be expected to behave in a similar fashion to the other parameters, such as Pred.

Our parameter, Med, is the reciprocal of the calculated median distance between microseismic event pairs, determined from consideration of the distances between all pairs of events within the subset of 40 events. The reciprocal is considered for the same reason as for D. D and Med are effectively independent parameters, as Med describes the clustering distance, which reflects localised failure when significantly reduced, while D describes the dimensionality of the clustering, i.e. whether the events are clustering within a volume or along a plane, etc. Clustering mechanisms which yield lower fractal dimensions (such as clustering along a plane, as opposed to clustering within a volume) are considered the more likely to produce damaging seismic events.

4.3 Parameter Tau

Another parameter which we investigated in terms of its usefulness as a seismic hazard indicator is seismic stress drop. Although McGarr (1984a) stated that stress drops convey little information regarding the state of the rock, we consider stress drop to be a possibly meaningful seismic hazard indicator for the deep-level gold-mining

125

environment for the following reasons. The seismicity associated with deep-level mining (in the absence of significant geological influence) will, on average, locate in the zone of fractured rock, typically five to 10 metres in extent, ahead of the stope face. A seismic event occurring in this zone of previously failed material will, on average, have a low stress drop, as most of the material in this region has low cohesive strength. If, on the other hand, the front of the fractured zone is actively advancing, so that new fractures are forming in previously solid rock, the stress drops of the associated seismicity will, on average, increase due to the higher cohesive strength of the solid material. This solid material just ahead of the fractured zone carries the bulk of the strain energy ahead of the faces, so that this is the region in which a larger, potentially-damaging seismic event (associated with a sudden, large outward movement of the front of the fractured zone into previously unfailed material) is likely to occur.

Our parameter, Tau, is a measure of the seismic stress drop. Tau is calculated from scaling laws, rather than measured directly: routine analysis for stress drops has not been carried out on the majority of the waveforms recorded by the seismic system, and, in any case, according to McGarr (1984b), stress drops inferred from peak ground motion are a truer indicator of stress changes that actually occur in the source region than stress drops inferred from seismic moment and corner frequency. Determination of the latter stress-drop values requires a good spectral shape, as well, which is a somewhat more subjective consideration.

The calculation of Tau proceeds as follows. Seismic energy E_s and seismic moment M_0 are related according to:

$$E_S \propto \Delta\tau M_0 \qquad (2)$$

where $\Delta\tau$ is the seismic stress drop, related to seismic moment and source radius r_0 by:

$$\Delta\tau \propto \frac{M_0}{r_0^3} \qquad (3)$$

Also, from McGarr (1986):

$$Rv \propto r_0 \Delta\tau \qquad (4)$$

where Rv is peak ground-velocity scaled by hypocentral distance. Thus, from (3) and (4):

$$M_0 \propto \frac{(Rv)^3}{(\Delta\tau)^2} \qquad (5)$$

so that, from (2) and (5):

$$\Delta\tau \propto \frac{(Rv)^3}{E_S} \qquad (6)$$

Tau is given by the right-hand side of the proportionality in (6). Thus, Tau is calculated from scaled peak ground-velocities (stored for each geophone channel, for each event, in the EVENTS file database) and energy, which is calculated from the event magnitude (which was determined from the recorded seismic energy, in the first place) via the Gutenberg-Richter relationship (for energy in MJ):

$$\log E_S = 1.5M - 1.2 \qquad (7)$$

The median value of Tau for 40 events is used, to avoid the spurious dominating effects of single large events on the calculated value of Tau.

5 RESULTS

Figure 6 shows a portion of the history of the variations of the four parameters used in the seismic risk assessment method over a period during which a rockburst occurred at the preconditioning site. The values of the parameters are expressed as percentages of the peak values obtained from analysis of the entire dataset recorded from the preconditioning site by the seismic network. Parameters Pred and Tau are calculated in terms of arbitrary units, so that expressing the results as percentages of peak values is the only meaningful way of representing the results. The fractal dimension of seismic event locations (i.e. the reciprocal of parameter D) varied between 0.67 and 1.68, with an average value of 1.12. The median distance between event pairs in each subset of 40 events (i.e. the reciprocal of parameter Med) varied between 12.5 and 142.9 metres, with an average value of 37.0 metres. From figure 6, it is clear that the parameters do all show marked variations from day to day, some of which appear to precede the larger events indicated, which had magnitudes greater than M_c.

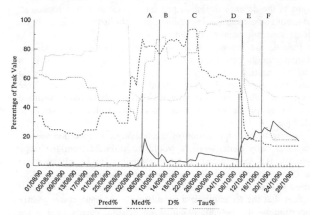

Figure 6. The percentage variations of the four seismic risk parameters over a three-month period. The alphabetically-labelled vertical lines indicate occurrences of potentially-damaging seismic events: the shorter lines are for events which took place on the day of a blast, while the longer lines are for events which took place on non-blast days. Event E was a damaging rockburst.

The values of the parameters shown for any given day in figure 6 were determined by calculations based on the previous 40 events, while the large events indicated in the figure occurred during the 24 hours following the time of determination of the values of the parameters, on the days during which the events took place. Parameter Pred had fairly low values during this period, increasing in value coincidentally with only some of the events, while parameter D had fairly high values, which actually decreased during the time of occurrence of most of the larger events, contrary to its expected behaviour. Parameters Med and Tau seem to have been particularly responsive during this period, both exhibiting markedly increased levels coinciding with the time of occurrence of the bulk of the larger events.

Note that figure 6 does lend support to the approach which we have taken to the development of a useful seismic risk assessment method: analysis over a long time-period was required, in order to determine peak values of the various parameters, and the use of multiple parameters was required -- while the parameters were generally observed to behave in sympathy with one another (as did Med and Tau, in this case), there were occasions on which one of the parameters did not behave as expected (as with D, in this case), and there were also occasions on which one or more of the parameters were only intermittently responsive, and/or responded in a complementary fashion (as did Pred, with events A, E and F, in this case).

5.1 Results from individual parameters

The success of the analysis of the entire dataset recorded from the preconditioning site by the seismic network in terms of the assessment of seismic risk was computed as the ratio of the percentage of the total number of potentially-damaging seismic events successfully predicted by a parameter (in terms of the value of that parameter being above a given critical level) to the percentage of the total number of days for which the value of the parameter was above that critical level (so that mine-workers would have been withheld from the stope for that day). These computations were carried out for the four subsets determined by whether or not blasting was to take place on the day of the analysis and/or whether or not blasting took place on the day before, in order to investigate the effects of these various blasting conditions on the successful application of the seismic risk assessment technique.

Some of the results of these computations are shown graphically in figures 7a to 7d, in which it can be seen that, for relatively large ranges in value of the parameters, a significantly greater percentage of events could have been predicted, compared with the percentage of time for which workers would have been withheld from the stope, for a given critical level of a given parameter. The critical level for each individual parameter was set so as to withhold workers for a maximum of 20 percent of the total time, in the case of an assessment of seismic risk based on that parameter.

In the example shown in figure 7a, if the workers had been withheld from the stope whenever the value of Pred rose above 17 percent of its peak value (indicated by the letter A in figure 7a), 20 percent of the total number of possible working days (indicated by the letter B in figure 7a) would have been lost, but 51 percent of the total number of potentially-damaging seismic events (indicated by the letter C in figure 7a) would have occurred during that time, representing a gain (given by the ratio of the values at C and B in figure 7a) of 2.7 over randomness, which would have a gain of 1. As shown in figures 7b to 7d, similar gains could have been obtained, with various blasting histories, from the other parameters, if the critical values of those parameters had been taken to be those indicated by means of the dotted lines in the examples shown. For nearly every combination of parameter and blasting history, in fact, significant gains could have been made via this seismic risk assessment technique, with less than 20 percent of the total time being lost, which time could have been profitably spent by working in the back areas of the stope and by preconditioning the faces. It should be noted that a preconditioning blast can be set off at the preconditioning site within two days from the time at which the seismic hazard is determined to lie above the critical level. In a situation in which the level of seismic hazard is regarded as unacceptable and in which a normal production blast is scheduled to take place, a preconditioning blast would be set off instead, based on the success of the seismic risk assessment method under conditions in which a blast was scheduled for the day of the analysis.

5.2 Results from combinations of parameters

It was found that the individual parameters tended to identify different, though overlapping, subsets of the total number of potentially-damaging seismic events when applied in isolation to the recorded seismic data. We investigated the performance of the parameters when applied in combination, either in such a way that all of the parameters had to be above their critical levels to identify a coming large event correctly, or in such a way that an event was considered correctly identified if just one of the parameters was above its critical level. The first type of analysis indicates the extent to which the parameters behave in sympathy in accurately identifying periods of increased seismic risk, so that the individual parameters serve to confirm one another. The second type of analysis indicates the extent to which the individual parameters complement one another, so that a larger subset of potentially-damaging events can be identified by applying the parameters in combination. These analyses were again carried out for the four different categories of blasting history discussed previously.

Table 1 has an example of the first type of analysis, in which the values of all parameters must lie above their individual critical levels for an event to be considered correctly identified. The analysis was restricted to just three of the parameters, in this case. The critical levels were adjusted downwards, to correspond to about 75 percent of the total time for each individual parameter, in an attempt to identify more large events when the parameters were combined. The results of the combination are shown in the columns corresponding to ALL in table 1: with 53 percent of the large events (averaged over the four blasting conditions) correctly identified at a gain of 1.6, it is clear that the parameters do behave sympathetically to a large extent, so that each can be used to confirm risk assessments based on other parameters. Note that table 1 does indicate an increase in the

proportion of time (i.e. above the 20 percent level) for which workers would be withheld, so that the critical levels would have to be adjusted to reduce this factor when combinations of parameters are considered in routine analysis.

Table 1. Example of the results of the application of the seismic risk assessment method by means of combining the parameters in such a way that ALL parameters must be above their critical levels for a coming large event to be identified correctly.

Blast?	Yesterday?	No	No	Yes	Yes
	Today?	No	Yes	No	Yes
Pred	% Time[1]	78	74	76	79
	% Events[2]	71	100	78	100
	Gain[3]	0.9	1.4	1.0	1.3
Tau	% Time	71	77	76	77
	% Events	86	80	89	71
	Gain	1.2	1.0	1.2	0.9
Med	% Time	65	71	74	62
	% Events	64	80	100	57
	Gain	1.0	1.1	1.3	0.9
ALL	% Time	38	33	36	29
	% Events	43	60	67	43
	Gain	1.1	1.8	1.8	1.5

[1]% Time: percentage of total time for which value of parameter exceeds critical level.
[2]% Events: percentage of total events occurring during % Time.
[3]Gain: ratio of % Events to % Time.

Table 2 has an example of the second type of analysis, in which an event is considered correctly identified if any one parameter lies above its critical level. The analysis in this case was also restricted to three parameters. The critical levels were adjusted upwards, to correspond to about 15 percent of the total time for each individual parameter, in an attempt to reduce the proportion of time for which workers would have been withheld when the parameters were combined. The results of the combination are shown in the columns corresponding to ANY in table 2: with 48 percent of the large events (averaged over the four blasting conditions) correctly identified at a gain of 1.3, it is clear that the different parameters do also identify different subsets of the large events, so that the risk assessment based on one parameter can be used to supplement that based on another parameter, so as to identify a larger subset of potentially-damaging events. Note that table 2 does indicate an increase in the proportion of time (i.e. above the 20 percent level) for which workers would be withheld, so that, as in the case of the first type of analysis, the critical levels would have to be adjusted to reduce this factor, when combinations of parameters are considered in routine analysis.

Table 2. Example of the results of the application of the seismic risk assessment method by means of combining the parameters in such a way that ANY one parameter must be above its critical level for a coming large event to be identified correctly.

Blast?	Yesterday?	No	No	Yes	Yes
	Today?	No	Yes	No	Yes
Pred	% Time[1]	17	15	18	18
	% Events[2]	0	20	22	29
	Gain[3]	0.0	1.3	1.2	1.6
Tau	% Time	24	12	15	8
	% Events	29	20	44	0
	Gain	1.2	1.6	3.0	0.0
Med	% Time	13	15	16	4
	% Events	7	20	22	14
	Gain	0.5	1.4	1.4	3.5
ANY	% Time	47	35	41	31
	% Events	29	40	78	43
	Gain	0.6	1.1	1.9	1.4

[1]% Time: percentage of total time for which value of parameter exceeds critical level.
[2]% Events: percentage of total events occurring during % Time.
[3]Gain: ratio of % Events to % Time.

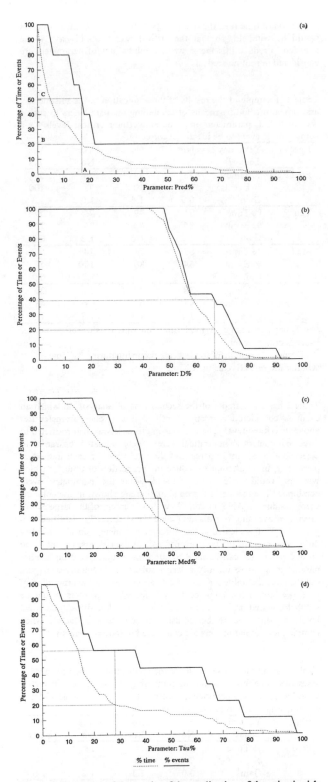

6 CONCLUSIONS

A seismic risk assessment method has been developed for use in deep-level rockburst-prone mines. The method appears to be robust and to offer significant gains in the accurate identification of periods of high seismic risk. The method makes use of physically meaningful seismic hazard parameters, which, while being largely independent of one another, appear to behave in a sympathetic and complementary fashion, allowing for the prediction of a greater percentage of potentially-damaging events, with greater certainty, when these parameters are used in combination. A sufficiently large dataset has been analysed to determine peak values and critical levels of the parameters, so that the method can now be applied routinely to the seismic data recorded from the preconditioning site in order to prove its applicability for use in rockburst-prone mines.

Acknowledgments

This work forms part of the Rockburst Control research programme of the Rock Engineering Division of COMRO. The cooperation of the management and staff of Blyvooruitzicht Gold Mine and their permission to publish the results shown in this paper are gratefully acknowledged.

REFERENCES

Adams, D.J., N.C. Gay & M. Cross 1993. Preconditioning -- a technique for controlling rockbursts. *Proc. 3rd ISRSM.* Rotterdam: Balkema.

Legge, N.B., & S.M. Spottiswoode 1987. Fracturing and microseismicity ahead of a deep gold mine stope in the pre-remnant and remnant stages of mining. *Proc. 6th ICRM:* 1071-1077. Rotterdam: Balkema.

McGarr, A. 1984a. Some applications of seismic source mechanism studies to assessing underground hazard. *Proc. 1st ICRSM:* 199-208. Johannesburg: SAIMM.

McGarr, A. 1984b. Scaling of ground motion parameters, state of stress, and focal depth. *J. Geophys. Res.* 89: 6969-6979.

McGarr, A. 1986. Some observations indicating complications in the nature of earthquake scaling. In *Earthquake Source Mechanics:* 217-225. Washington, D.C.: AGU.

Pattrick, K.W., A.M. Kelly & S.M. Spottiswoode 1990. A portable seismic system for rockburst applications. *Proc. IDMC: Technical challenges in deep level mining.* Johannesburg: SAIMM.

Riefenberg, J. 1991. Statistical evaluation and time series analysis of microseismicity, mining, and rock bursts in a hard-rock mine. *Report of Investigations* 9379. U.S. Bureau of Mines.

Xie, H., & W.G. Pariseau 1992. Fractal character and mechanism of rock bursts. *Proc. 33rd USSRM: Workshop on Induced Seismicity:* 140-156. Pre-workshop volume.

Figure 7. Examples of the results of the application of the seismic risk assessment method. The four graphs compare the percentage of total time for which workers would be withheld from the stope with the percentage of total number of potentially-damaging seismic events which would be detected at a given percentage of peak value of each parameter, over the entire range of each parameter. The dotted lines show the performance for a chosen critical level of each parameter, corresponding to 20 percent of the total time in each case. (a) Parameter Pred, for the case of a blast on the day of the analysis (today), but no blast on the day before (yesterday); (b) D, no blast today or yesterday; (c) Med, no blast today, a blast yesterday; (d) Tau, no blast today, a blast yesterday.

Rockbursts and Seismicity in Mines, Young (ed.) © 1993 Balkema, Rotterdam, ISBN 90 5410 320 5

Ejection velocities measured during a rockburst simulation experiment

D. D. Tannant, G. M. McDowell, R. K. Brummer & P. K. Kaiser
Geomechanics Research Centre, Laurentian University, Sudbury, Ont., Canada

ABSTRACT: The establishment of an energy based support design methodology is one objective of ongoing research into support in burst-prone ground. The rock ejection velocity is an important input parameter for a rational support design methodology for burst-prone ground because kinetic energy is a function of velocity squared. Two blasts were conducted to simulate the ejection of rock by a rockburst. Different techniques were used to measure block ejection velocities during the blasts and these velocities are compared to ejection velocities calculated from peak particle velocities.

Results show that the initial shock wave ejects only a small quantity of rock while the expanding explosive gases propel the bulk of the ejected rock. Ejection velocities calculated from assumptions of gas energy and from the trajectory of the muck pile's centre of mass yield good estimates of the total kinetic energy contained in the moving rock. The ejection velocities obtained from specially designed velocity probes were higher than the overall ejection velocity. Video camera images gave a useful record of the blasts and provided a range of ejection velocities for blocks of different sizes.

1 INTRODUCTION

Ejection of broken rock during a rockburst is one mechanism for damage to underground support. The ejected rock contains kinetic energy that must be absorbed by the ground support if access through the excavation is to be maintained. Therefore, the design of support and rock retaining systems for rockburst prone ground requires an estimate of the velocity at which rock blocks are ejected from the surface of a drift (e.g., Wagner, 1984; Roberts and Brummer, 1988).

If the kinetic energy of the ejected rock can be measured or predicted, then the capacity of the support in terms of strength and yieldability can be determined. The block's kinetic energy is controlled by its mass and velocity. The sizes of ejected rocks depend on the local jointing and fracturing around the excavation. Rockmass classification systems that include an implicit measurement of block size (e.g., Priest and Hudson, 1981) can provide an estimate of the size of the ejected rocks.

Kinetic energy is proportional to the velocity squared and hence, the rock's ejection velocity is more important than its mass for energy based support design. Theoretical estimations of ejection velocities can be made considering the block size and peak particle velocities (Yi and Kaiser, 1993). Unfortunately, there are few measurements of actual ejection velocities during blasts and especially during rockbursts. This paper presents measurements of ejection velocities for two simulated rockbursts. The ejection of rock was accomplished with the use of explosives.

2 BOUSQUET ROCKBURST SIMULATION EXPERIMENT

Field trials to investigate the support resistance of various support types are currently underway at the Bousquet #2 mine in Quebec. During preliminary tests to establish the test procedures, the damage (ejection of rock) caused by a rockburst was simulated by two separate blasts. In each case, a 16 m long, 38 mm diameter borehole was collared at mid-height of the drift wall and drilled horizontally at an angle of 20° to the wall (Fig. 1). The borehole was pneumatically loaded with AMEX II (Table 1). The inclination of the borehole relative to the drift wall was designed to generate a variation in the blast loading intensity along the

drift. The holes were toe primed and the last two metres of the holes were stemmed with gel. The quality of the stemming was poor, especially for the B-wall borehole.

The rockmass at the location of the rockburst simulation experiment consists of a highly foliated sericitic schist with an approximate P-wave velocity of 5.4 km/s. The foliations are nearly vertical and strike perpendicular to the test drift axis. The nature of the rockmass resulted in a predominance of thin rock slabs in the two muck piles. Typical slabs ranged in thickness from a few centimetres to less than about 0.4 m with block volumes generally less than 0.1 m^3.

Table 1 AMEX II properties when pneumatically loaded
(ICI Explosives Canada)

Density	1.00 g/cm^3
Velocity of detonation[*]	3300 m/s
Chemical energy	3.8 kJ/g
* For a 32 mm diameter borehole	

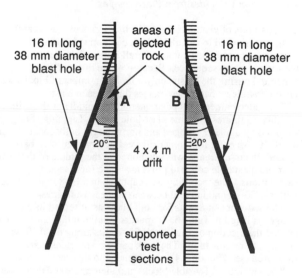

Figure 1 Plan view of the rockburst simulation experiment

The velocities of ejected rock blocks were calculated from video camera images and from ballistic trajectories. During the B-blast, velocities were also measured with 'piston-in tube' velocity probes. Geophones were used to measure particle velocities along the drift wall for both blasts. Figures 1 and 2 show the areas where the bulk of the rock was ejected during each blast. The locations of instrumentation used to measure velocities are shown in Figure 2. A summary of all velocity measurements taken during the two blasts is given in Table 1 and details of these measurements are discussed in the following section.

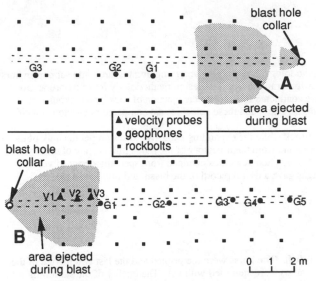

Figure 2 Areas where rock was ejected and locations of velocity measuring instrumentation (Looking at walls A and B)

Table 2 Summary of velocity measurements for the two blasts

	A - Blast	B - Blast
V_e from velocity probes (m/s)	-	10, 21, and 44
V_e from video images (m/s)	4 to 10	4 to 22
V_e from ballistics (m/s)	3 to 7	5 to 10
Total mass ejected (tonnes)	40	25
ppv at drift wall (m/s)	1.1	0.7

3 VELOCITY MEASUREMENTS

3.1 Ejection Velocities from Video Images

The positions of visible rock blocks in a series of video camera images were used to calculate block ejection velocities. The video images provided a good overall impression of the blast and the rock ejection sequence. Because of the dust and gas generated during the blasts, only a limited number of blocks were visible in at least two video images during each blast.

The video camera was positioned in the middle of the drift roughly 25 metres from the blast-hole collar. Lighting was supplied by a quartz halogen and normal light bulbs located along the drift. An eight foot long drill steel was painted with bars spaced 25 cm apart and was positioned in the middle of the drift near the blast hole collar to provide a scale for distance calculations. The recording was captured in a Hi-8 video mode with a 1/30 second interval between successive images.

The sequence of events for both blasts as observed in the video images is listed in Table 3. It appears that the initial shock wave caused the ejection of small rock fragments along the drift wall shortly after detonation. The peak particle velocities that were measured (see Section 3.4) corresponded to this shock wave.

The velocities of visible blocks and their approximate masses as determined from the video images are summarized in Table 4.

Table 3 Sequence of events during the blasts

Approximate Time (s)	Event
0	A flash of light when the detonator external to the blasthole ignites the non-electric detonation tubing.
1/30	Small rock fragments ejected from drift wall (probable cause: shock wave).
1/30 to 2/30	Ejection of rock starts cutting through wires of velocity probes about 33 msec after detonation
2/30 to 7/30	Ejection of a gas plume from borehole collar (more gases were ejected from the poorly stemmed B-blasthole). Ejection of smaller rock blocks.
6/30	Large volumes of rock start to eject.
8/30 to 15/30	Bulk of the ejected rock in motion
24/30 to 30/30	Largest blocks in motion.
30/30	Most ejected rock on the floor.
30/30 to 48/30	Gravity driven blocks fall to the floor.

The video images showed that the faster blocks were ejected within 1/5 to 2/3 of a second after detonation. The larger, slower moving blocks were ejected near the end of the blast. Velocities of the fastest moving, small blocks could not be measured from the video images. The velocities in Table 4 are therefore biased to the smaller blocks that moved at moderate velocities.

Table 4 Summary of velocities estimated from video records

Block	V_e (m/s)	Mass (kg)	Block	V_e (m/s)	Mass (kg)
A-1	10	< 5	B-1	22	< 5
A-2	6	< 5	B-2	9	< 5
A-3	6	< 5	B-3	7	< 5
A-4	7	5 to 10	B-4	6	< 5
A-5	6	5 to 10	B-5	5	< 5
A-6	5	5 to 10	B-6	4	< 5
A-7	5	5 to 10	B-7	4	< 5
A-8	4	5 to 10	B-8	6	5 to 10
A-9	8	10 to 25	B-9	4	25 to 50
A-10	6	25 to 50			
Gas Plume	40 to 60	-	Gas Plume	40 to 90	-

3.2 Ejection Velocities from Trajectories

The distance that a moving object travels can be used to back-calculate its initial ejection velocity. Based on ballistic trajectories (Fig. 3), the ejection velocity of a mass that travels a horizontal distance, D and falls a vertical distance, H is given by:

$$V_e = D\left(\frac{g}{2H\cos^2\theta + D\sin 2\theta}\right)^{0.5} \quad (1)$$

where θ is the initial angle of motion measured upwards from the horizontal plane.

Each blast resulted in the ejection of hundreds of rock blocks and fragments with a wide range of velocities. The largest rock blocks were found lying close to the ejection site while small blocks were spread over a much wider area. Therefore, the highest ejection velocities were associated with the smallest rocks

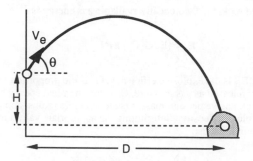

Figure 3 Trajectory of a rock with an initial velocity, V_e and angle, θ

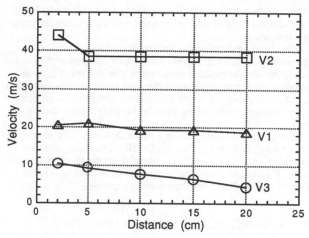

Figure 4 Block velocity versus travel distance for the B-blast

while the lowest velocities were associated with the largest blocks. Rather than estimate the ejection velocity of individual blocks, the overall ejection velocity (in terms of energy) of the entire displaced mass was determined from the position of the muck pile's centre of gravity. The total of kinetic energy in the assembly of ejected blocks can be calculated from the total mass of ejected rock and the trajectory of its centre of gravity.

Examination of the muck piles showed that their approximate centres of gravity had travelled horizontal distances of 2.5 ± 0.5 m and 3.5± 0.5 m respectively for blasts A and B. In each blast, the centre of gravity had also fallen about 1.5 m. The initial trajectory angle for the ejected rock was taken as 0°± 10° based on observation of the video images. This range of angles seems reasonable since the ejected rock is least confined when it initially moves horizontally. These values were used in Equation 1 to predict the velocities listed in Table 2. The approximate mass ejected during each blast is also given in Table 2.

3.3 Ejection Velocities from Velocity Probes

Three velocity probes were used during the B-blast to directly measure ejection velocities. Each velocity probe was made from two telescopic PVC pipes (different diameters). The larger pipe contained six thin timing wires installed through holes in the pipe, perpendicular to its axis. This pipe was fixed in the centre of the drift. A smaller pipe with one end mounted to the drift wall was inserted into the larger stationary pipe containing the wires. When the rock was ejected it forced the smaller pipe through the larger pipe cutting the wires. Knowing the times when each wire was cut (from voltage change measurements) and the spacing between the wires, the ejection velocities were calculated.

Figure 4 presents the velocity versus displacement history for the three probes. The ejection velocities obtained from the velocity probes are high and cover a wide range. It is possible that the highest velocity was associated with a small block (< 20 kg) although this could not be seen in the video record. Figure 4 shows that the ejected rocks reach their maximum velocities before travelling about five centimetres. The velocities then gradually decrease with increasing travel distance. These ejection velocities are comparable to ejection velocities measured with similar velocity probes during two rockbursts simulated with explosives at El Teniente (E. Rojas, pers. comm.). The ejection velocities measured at El Teniente ranged between 1 and 17 m/s.

3.4 Peak Particle Velocities on the Drift Wall

The ppv at the location of the ejected rock was not measured because the geophones installed at these locations were destroyed during the blast and because these geophones, and others nearby, experienced displacements that exceeded the geophones' free travel limit. Good quality waveforms were obtained from geophones that were located further from the blasthole collar.

For example, Figure 5 presents the particle velocities measured at geophones G2 and G5 during the B-Blast. The ppv's listed in Table 2 are the highest values obtained from all non-clipped geophones during each blast and are obviously lower than the ppv's experienced at the sites where rock was ejected.

The particle velocities shown in Figure 5 were generated by the propagation of a shock wave through the rockmass shortly after detonation. This shock wave did not cause the extensive damage seen later during the blast. Only a few dozen small rock pieces were ejected when the shock wave reflected at the drift surface. The shock wave alone did not contain sufficient energy to fracture and eject significant volumes of rock.

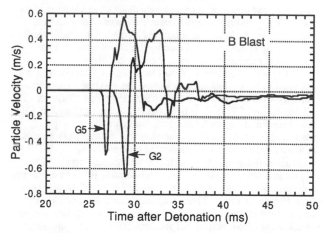

Figure 5 Particle velocities measured on the drift surface for the B-Blast

3.5 Comments on Instrumentation to Measure Velocities

It was possible to quickly position the video camera to record the blast. The video images have an advantage in supplying both estimates of the velocities as well as the block sizes but information is lost when dust or gases fill the drift.. A high speed video camera would have provided more images and hence, more detailed estimates of ejection velocities. However, a standard video camera is a cost effective tool for monitoring blasts and for measuring the ejection velocities of the larger, slower moving rock blocks. Good lighting and camera positioning is essential for blast monitoring.

The velocity probes yielded wire break times that were very accurate and easy to interpret. The cost to build a probe is minimal making them ideal for consumable instrumentation.

However, they took considerable time and effort to set up in the field and the mass of rock driving the piston portion of the probe is unknown.

Post-blast observations of the position of individual rocks in the muck pile provided a good estimate of the range of ejection velocities and the position of the muck pile's centre of gravity yields a robust measurement of the total kinetic energy contained in the ejected rock. It is this total kinetic energy that a ground support system must absorb in order to safely contain the ejected rock.

The geophones were mounted to the drift surface in short (2 cm deep) holes using Plaster of Paris. The Plaster of Paris provided a quick and secure coupling of the geophone to the drift wall. If geophones are used to measure peak particle velocities then it is advisable to use long-travel geophones.

4 INTERPRETATION OF EJECTION VELOCITIES

The energy of the initial explosive is stored as chemical energy and upon detonation it is released and used to drive a number of processes. Explosive energy is released into the surrounding rock in two different forms: detonation pressure and borehole pressure. The detonation, or shock pressure exerts a fragmenting force on the rock. The borehole pressure results from the gas buildup and is slower acting. It may be responsible for some fragmentation, but it is the primary cause of rock displacement.

One fundamental difference between a blast and a rockburst is the absence of gases during a rockburst. Therefore, when using a blast to simulate a rockburst one must interpret the results in terms of the effects caused by the initial shock energy propagation and the effects due to the later gas pressures. The propagation of the shock energy approximates the propagation of seismic energy from a seismic event. This energy may lead to the ejection of rock blocks if the particle velocities are sufficiently high when it reflects from the free surface of the drift. The explosive gas pressure can rapidly displace large volumes of rock and only approximates the ultimate effects of a large seismic event causing rock movement during a rockburst. One issue that arises is the relative roles played by the shock wave and gas pressure in ejecting rock during the two test blasts.

If the rock around the drift had been highly fractured and is assumed to have frictionless contacts, the maximum ejection velocity would be approximately equal to the ppv (Wagner, 1984; Yi and Kaiser, 1993). The peak particle velocities listed in Table 2 and shown in Figure 5 are significantly smaller than the measured ejection velocities showing that the gas pressure was responsible for most of the observed rock ejection.

Empirical damage thresholds (Dowding and Rozen, 1978; Owen and Scholl, 1981; St. John and Zahrah, 1987) based on ppv suggest that damage in a tunnel would be minor if the ppv was less than about 1 m/s. These empirical damage thresholds are applicable for stress waves (seismic waves from earthquakes or shock waves from blasts) and are consistent with the measured peak particle velocities listed in Table 2 as well as observation of small rocks being ejected early in the blasts. Note in Figure 5 that the peaks in particle velocity occur 26 to 29 milliseconds after detonation which corresponds to the first frames of the video records. The first wires in the velocity probe were cut 33 milliseconds after detonation and the ejection may be associated with either the shock wave or gas pressure.

The greater damage observed during the later part of the blasts was caused by the explosive gases. The ejection velocities that were measured using the video camera and travel distances were related to the gas pressures. The average ejection velocity can be predicted by estimating the proportion of explosive chemical energy that is transferred to the kinetic energy of the ejected rock. Blends of ammonium nitrate and fuel oil (AMEX II) are at best 60% to 70% efficient in converting chemical energy into gas and shock energy. In hard rock, only about 5% to 15% of this released energy is shock energy. The larger proportion (85% to 95%) is released as gas energy (Atlas Powder Company, 1987). Taking the length of borehole next to the ejected rock volume (L = 3 m) as a measure of the volume of explosive driving the ejected rock, the theoretically available gas energy is:

$$E_g = \left(E_{ff} G_e C_e\right) \pi r^2 L \rho_e \tag{2}$$

where E_{ff} is the explosive efficiency, G_e is the proportion of energy released as gas pressure, C_e is the chemical energy stored in the explosive per unit mass, r is the borehole radius, and ρ_e is the explosive density. Substituting the appropriate values into Equation 2 yields:

$$E_g = 0.6(0.9)\left(3.8\frac{kJ}{g}\right)\pi(1.9\,cm)^2(300\,cm)\left(1\frac{g}{cm^3}\right) = 7MJ \tag{3}$$

However, the video shows that some of this gas energy escapes out of the poorly stemmed collar of the borehole. Furthermore, some gas energy is used to fracture rock and is lost in frictional processes. Assuming less than half the gas energy is used to propel the rock, the effective velocity of ejection for both muck piles is:

$$E_k \leq 0.5 E_g = \tfrac{1}{2} m V_e^2$$
$$\text{or} \tag{4}$$
$$V_e \leq \sqrt{\frac{E_g}{m}} = 9\,m/s \text{ for A; and } \leq 12\,m/s \text{ for B}$$

These velocities are higher than those calculated from travel distance measurements implying that less than half the gas energy was used to propel the rock.

Ejection velocities calculated from explosive energies or muck pile travel distances are based on general energy considerations. In contrast, the higher velocities measured from the video camera and the velocity probes reflect individual rock blocks that happened to have more than the average kinetic energy per unit mass. Blocks ejected at high velocities are relevant for the design of individual support components.

For the design of support systems to resist blast damage, a measure of the total kinetic energy in the ejected rock is needed. The kinetic energy can be predicted from the available gas energy originating from the explosive. Similarly, the seismic energy generated by a seismic event may provide a basis for predicting the kinetic energy of ejected rock during a rockburst. This energy must be absorbed by the support system including various components.

Further testing at the Bousquet #2 mine is scheduled for early 1993. Single blastholes will be used again to simulate rockburst damage and the response of support systems involving shotcrete and various types of rockbolts will be monitored.

5 CONCLUSION

Two blasts were used to simulate the damage caused by rockbursts. Different methods were applied during each blast to measure the velocities of the ejected rock blocks. The ejected rock contained a range of block sizes and ejection velocities. The higher velocities were associated with the smaller rock blocks while the larger rock blocks had the slowest velocities. Almost all ejected rock was propelled by the explosive gases. The initial shock wave during the blast caused only limited ejections of small rock fragments.

Ejection velocities determined with velocity probes and video camera images were often much higher than the overall velocity back-calculated from the muck pile's centre of gravity. Velocities calculated from consideration of available energy (chemical energy of the explosives) or trajectories of the muck pile's centre of gravity could be used to estimate the total kinetic energy in the ejected rock. Knowledge of this kinetic energy is necessary for energy based design of ground support systems.

ACKNOWLEDGEMENTS

The support and assistance from Serge Lesveque and the staff at the Bousquet #2 mine and Richard Hong (Lac Minerals) is gratefully acknowledged.

REFERENCES

Atlas Powder Company. 1987. *Explosives and Rock Blasting*. Field Technical Operations, Dallas, Texas. Maple Press, 662 p.

Dowding, C.H. and A. Rozen. 1978. Damage to rock tunnels from earthquake shaking. *ASCE Journal of the Geotechnical Engineering Division*, 104(GT2): 175-191.

Owen, G.N. and E. Scholl. 1981. Earthquake Engineering of Large Underground Structures. URS/John Blume & Associates, Engineers, San Francisco, January, Report # FHWA/RD-80/195, 280 p.

Priest, S.D. and Hudson, J.A. 1981. Estimation of discontinuity spacing and trace length using scanline surveys. *International Journal of Rock Mechanics and Mining Sciences and Geomechanics Abstracts*, 18(3): 183-197.

Roberts, M. K. C. and R. K. Brummer. 1988. Support requirements in rockburst conditions. *Journal of South African Institute of Mining and Metallurgy*, 88(3): 97-104.

St. John, C.M. and T.F. Zahrah. 1987. Aseismic Design of Underground Structures. *Tunnelling and Underground Space Technology*, 2(2): 165-197.

Wagner, H. 1984. Support requirements for rockburst conditions. *Proc. 1st International Congress on Rockbursts and Seismicity in Mines*, Johannesburg, 1982, SAIMM. Johannesburg: 209-218.

Yi, X. and Kaiser, P.K. 1993. Mechanisms of rockmass failure and prevention strategies in rockburst conditions. *Proc. 3rd Int. Symp. on Rockbursts and Seismicity in Mines*, Kingston.

Rockbursts and Seismicity in Mines, Young (ed.) © 1993 Balkema, Rotterdam, ISBN 90 5410 320 5

Concentration of rock burst activity and in situ stress at the Lucky Friday Mine

Jeffrey K. Whyatt & Theodore J. Williams
Spokane Research Center, US Bureau of Mines, Wash., USA

Wilson Blake
Hayden Lake, Idaho, USA

ABSTRACT: An unusual concentration of rock burst activity encountered during development of the 5300 level of the Lucky Friday Mine is strongly correlated with in situ stress and geologic anomalies. Three large bursts (local magnitude $M_L \simeq 1$) occurred during and immediately following excavation of three successive 2.5-m rounds in the 101 crosscut on the 5300 level of the mine. The crosscut was being driven through very strong, brittle, sulfide-altered quartzite strata known locally as "blue rock." An overcore stress measurement taken in this crosscut prior to the three bursts indicated a $45°$ rotation in stress direction and a vertical magnitude over twice that of overburden pressure. Subsequent excavation set off the rock bursts, forcing the mine to abandon the crosscut.

The localized nature of the measured stress field has been established by comparing overcore stress measurements, raise bore breakouts, and geologic indications of stress orientations collected throughout the Lucky Friday Mine and the adjacent Star Mine. Evidence of the regional premining stress field orientation was found as close as 60 m to the crosscut stub. The mining induced stress at the stub was estimated with a homogeneous, elastic model to be only about 5% of the natural stress field.

1 INTRODUCTION

The Coeur d'Alene Mining District of northern Idaho has a long history of mining. Several district mines have mined nearly vertical veins of high-grade zinc, lead, and silver to depths of over a mile. Mining to ever-greater depths has been accompanied by an increase in the number of seismic events.** In this report, mining-induced seismic events that have damaged mine openings are referred to as rock bursts.

The Lucky Friday Mine is the most seismically active mine in the Coeur d'Alene District and is generally considered to be one of the most seismically active mines in North America (Jenkins, Williams, and Wideman, 1990). Entire sections of the mine have been severely damaged by large bursts.

Despite the seriousness of the rock burst threat in mines throughout the world, considerable gaps remain in fundamental knowledge of factors that control rock burst violence, location, and timing. The U.S. Bureau of Mines has taken a lead role in investigating both rock burst mechanisms and engineering methods to reduce rock burst hazards in the Coeur d'Alene Mining District. These investigations have included a historical review of rock bursting in the district (McMahon, 1988), case studies of major bursts in the Lucky Friday and Galena mines (Jenkins, Williams, and Wideman, 1990; Swanson, 1993; Swanson and Sines, 1991), experimental numerical modeling of rock burst mechanisms (Whyatt and Board, 1991), rock mass preconditioning techniques (Karwoski, McLaughlin, and Blake, 1979), and mining methods for rock-burst-prone ground (Jenkins and Dorman, 1983; Whyatt, Williams, and Pariseau, 1992). In addition, considerable progress has been made in analysis of seismic data (Riefenburg, 1991) and development of mine-wide (Jenkins, Williams and Wideman, 1990), stope-concentrated (Boler and Swanson, 1990), and

networked real-time (Stebley and Brady, 1990) digital seismic monitoring systems.

This paper looks primarily at the conditions leading to large rock bursts in development openings well ahead of mining at the Lucky Friday Mine. It describes rock-burst activity accompanying development of the mine's 5300 level, in situ stress measurements, and related observations. These data suggest that the tightly concentrated and violent rock bursting encountered during development of the 5300 level was linked to a local geologic structure that was found to be carrying unusually high levels of stress prior to being influenced by mining. This paper was developed as part of an ongoing investigation of the various rock-burst mechanisms and the factors that control their violence, location, and timing.

2 LUCKY FRIDAY MINE GEOLOGY

The Coeur d'Alene Mining District, in which the Lucky Friday Mine is located, lies within the west-central part of a basin that once extended over western Montana, northern Idaho, and southern British Columbia. Sediments filling this basin in late Precambrian times were metamorphosed to become the rocks of the Belt Supergroup, including the quartzites and argillites of the Lucky Friday.

Tectonic activity in the district has led to extensive faulting (fig. 1). Major faults include the Osburn (and its subsidiary faults) and the Dobson Pass; these faults are generally believed to be some of the youngest features in the district. The most recent movement on the Osburn Fault is right lateral, indicating a northwest-trending tectonic stress field (Hobbs et al., 1965).

In the Lucky Friday Mine, two more-localized faults, the North and South Control faults, delineate the ends of the 500-m-long Lucky Friday vein, composed of massive galena, sphalerite, and tetrahedrite. At the present time, this vein has been followed to a depth of approximately 1,550 m below the surface (520 m below sea level). It ranges from several centimeters to 5 m thick and is nearly conformal to an anticline plunging $75°$ to the southeast (fig. 2). Because the vein itself

---

** A seismic event is defined here as a sudden episode of inelastic rock deformation (a shear fracture, tensile fracture, or slip along a discontinuity) that produces seismic waves.

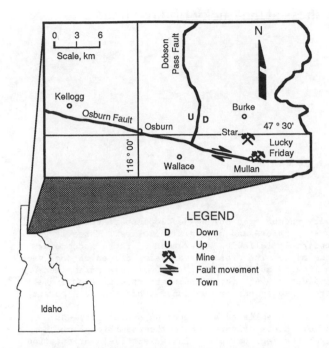

Figure 1. The location of the Lucky Friday Mine, the Star Mine, and major faults in the Coeur d'Alene District of northern Idaho.

dips steeply (70° to 90°) to the south and east, it cuts progressively older rocks with depth.

The rock mass surrounding the vein is made up of vitreous and sericitic quartzite beds (the Revett and St. Regis formations) from 0.3 to 1 m thick, with interbeds of much softer argillite generally less than 2.5 cm thick. Some of the more porous beds of vitreous quartzite have been altered by sulfides in the vicinity of the vein. The resulting rock, known locally as "blue rock", is the strongest, stiffest rock in the mine (elastic modulus $\approx$ 82.7 GPa, uniaxial strength $\approx$ 276 MPa). Scott (1990) provides an in-depth description of geology in the lower portion of the mine.

Studies of rock fabric and fracture patterns in the Lucky Friday Mine and nearby Star Mine have been used to estimate stress field characteristics. A summary of geologic indications of in situ stress orientation is provided in table 1.

3 IN SITU STRESS MEASUREMENTS AND OBSERVATIONS

The rock-burst problems encountered at the Lucky Friday Mine and elsewhere in the district have sparked considerable interest in the district's in situ stress field. Previous efforts at estimating this stress field (e.g., Whyatt, 1986) have concentrated on defining average stress conditions. The unusual concentration of bursting described in this report suggests that stress field variability is at least as important.

Stress field variability, including vertical stresses in excess of overburden estimates, has

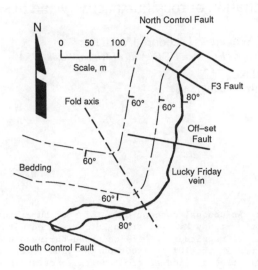

Figure 2. Major geologic features on the 5300 level of the Lucky Friday Mine.

been well documented in the literature. For example, Herget (1988) found that a significant subset of the vertical stress components measured in the Canadian Shield exceeded overburden estimates. Stress variability has also been shown to play an important role in the location of rock bursts in Scandinavian tunnels (Martna and Hansen, 1986) and coal bumps in a French colliery (Gaviglio et al., 1990). In this section, the latest overcore stress measurement data and a number of observations indicating stress field direction are used to investigate whether a pre-existing concentration of in situ stress may account for an unusual concentration of rock burst activity encountered during development in blue rock strata.

3.1 Overcore measurements

Three overcore stress measurements have been conducted in the Lucky Friday Mine and immediate vicinity. The latest measurement (Whyatt, Jenkins and Larson, 1993) was undertaken between June and August of 1986 on the 5300 level of the Lucky Friday Mine. An earlier measurement was conducted on the 4250 level (Allen, 1979; reanalyzed by Whyatt and Beus, 1993). Another measurement of interest was conducted in the now-flooded 7300 level of the nearby Star Mine (Beus and Chan, 1980; reanalyzed by Whyatt, Beus, and Larson, 1993). The results of these stress measurements are summarized in table 2.

The Council for Scientific and Industrial Research (CSIR) biaxial strain cell, commonly known as a doorstopper, was chosen to measure overcoring strains in all three cases. This cell was chosen because it is particularly well suited for sites where core recovery is a problem, since only about 8 cm of overcore are needed for a successful measurement. Jenkins and McKibbin (1986) describe recent overcoring procedures.

Table 1. Estimates of horizontal secondary principal stress orientation based on geologic studies.

Structure	Orientation of σ_{h1}	Reference
Latest movement of the Osburn Fault	NW	Hobbs et al. (1965)
Rock fractures, joints	W-NW	Gresseth (1964)
Geologic fabrics	NW	Gresseth and Reid (1968)
Rock fractures and joints	$N30^{\circ}-40^{\circ}W$	Allen (1979)

Table 2. In situ stress measured at sites in the vicinity sites of the Lucky Friday Mine.

	Magnitude, MPa	Bearing	Plunge
Lucky Friday 4250 level			
σ_1	96	N 37° W	11°
σ_2	57	S 45° W	35°
σ_3	39	N 69° E	53°
Lucky Friday 5300 level			
σ_1	169	S 80° W	34°
σ_2	91	N 4° W	9°
σ_3	86	S 81° E	54°
Star 7300 level			
σ_1	54	N 31° W	15°
σ_2	46	N 82° E	55°
σ_3	37	S 50° W	31°

3.2 Bored raise breakouts

A number of circular raises were bored along with development well ahead of mining. The raises quickly broke out to an elliptical shape, suggesting the direction of the horizontal secondary principal stresses (table 3). The set of raise bore breakouts observed at the Lucky Friday mine are consistent with a northwest orientation of the horizontal maximum principal stress.

Table 3. Horizontal maximum principal stress directions indicated by raisebore breakouts.

Location	Orientation of σ_{h1}
Alimak 3850-168 level	N40°W
Alimak 4050-8 level	N65°W
Alimak 4050-100 level	N75°W
Silver shaft 5100 loading pocket	N45°W
Lucky Friday 5100-107 ore pass	N45°W
Lucky Friday 5100-97 ore pass	N55°W

4 STRESS FIELD VARIABILITY

The goal of many stress measurement programs, including the overcore measurements undertaken in and around the Lucky Friday mine, is to estimate far-field stress conditions for use in rock mechanics analyses. The variability observed in these measurements increases the amount of information needed to provide a reliable picture of the stress field. The significant time and expense incurred in conducting overcore stress measurements requires that the additional information be sought from other sources. Despite the approximate nature of many of these sources, they provide important confirmation of which measurements represent the dominant natural stress field and which are artifacts of local geologic conditions. In this section, stress information from overcore measurements and other sources are examined to determine whether the stress field at the 5300 level site departed significantly from the overall in situ stress field. Two aspects of the stress field, horizontal principal stress direction and vertical stress magnitude, are examined.

4.1 Horizontal stress field orientation

Orientation of the maximum horizontal principal stress (σ_{h1}) was indicated by all the available data (table 4). A graphical representation of σ_{h1} orientations from the entire data set is shown in figure 3. There is a definite clustering of orientations around N45°W with a secondary clustering at N65°-75°W. The S80°W orientation at the 5300 over-

core site is an outlier by at least 25°. The 70° spread in orientation, from N30°W to S80°W, is surprisingly large.

Table 4. Horizontal secondary principal stresses calculated from overcore measurements.

Site	σ_{h1}, MPa	Orientation	σ_{h2}, MPa	Orientation
Lucky Friday 4250	95	N35°W	51	N55°E
Lucky Friday 5300	144	N80°E	91	S10°E
Star 7300	54	N36°W	39	N54°E

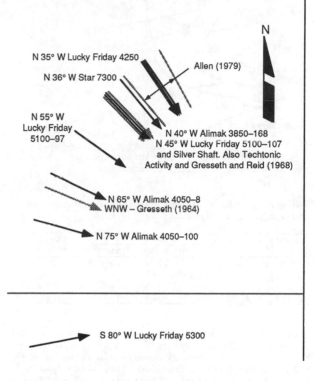

Figure 3. Orientations of σ_{h1} reported in the vicinity of the Lucky Friday Mine.

The locations of the measurements on the 5300 level are summarized in figure 4, which illustrates the localized nature of stress conditions at the 5300 overcore site, particularly with regard to the 97 ore pass. Thus, not only is the 5300 level measurement not representative of the general stress field direction, it also appears that the extent of the stress field is fairly limited.

4.2 Vertical stress magnitude

The equilibrium of a rock mass requires that vertical stress (σ_v) matches overburden weight, at least on the average. Thus, while considerably less information is available on σ_v magnitude than on horizontal stress direction, average σ_v can easily be estimated on the basis of overburden depth. A homogeneous elastic model used to estimate mining-induced stresses at the overcore sites showed the influence of mining to be on the order of only 5% of the stress field. The σ_v magnitudes indicated by the overcore measurements and estimates of overburden pressure at each overcore site are presented in table 5.

While measurements of σ_v at the 4250 level of the Lucky Friday Mine and the 7300 level of the Star Mine do approximate the pressure expected from

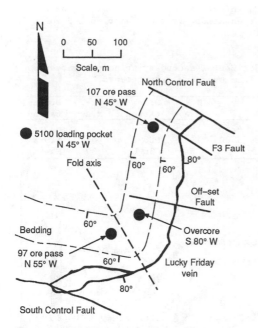

Figure 4. Location of σ_{h1} indications on the 5300 level and their orientations.

Table 5. Vertical stress magnitudes measured in the vicinity of the Lucky Friday Mine.

Component	Lucky Friday		Star
	4250	5300	7300
Depth, m	1,300	1,600	2,200
σ_v, MPa	47	113	44
σ_v, MPa	35	43	60
Deviation from ideal	+36%	+161%	-26%

overburden loading, σ_v measured at the Lucky Friday 5300-level site is over twice the estimated over-burden weight. The deepest site (in terms of ab-solute elevation) is the 5300 site, which lies under the valley floor. The position of the Star Mine under a hill side to the northwest of the Lucky Friday makes the Star deepest in terms of overburden. Thus, the overburden pressure calcu-lation is an overestimate for the Star Mine and an underestimate for the Lucky Friday Mine.

The failure of the Lucky Friday 5300-level meas-urement to approximate overburden pressure is added evidence that it is local in nature.

5 ROCK BURST ACTIVITY IN 101 CROSSCUT

The 5300 stress measurement site was located in the 101 crosscut stub on the 5300 level of the mine. This site was approximately 45 m from the vein, 60 m below active mining, and adjacent to the Offset Fault (fig. 5). The crosscut was developed sufficiently to allow extension without interfering with haulage operations, but was far short of its planned length. The site was particularly attrac-tive for overcoring because the crosscut ended in massive blue rock quartzite that promised good core recovery.

Excavation in the 5300-101 crosscut stub resumed in January of 1990 as miners attempted to gain access to the vein. The first round was taken without incident (approximately 2.5 m of advance), although increased seismic activity was noted. The following round, fired at the end of day shift on January 10, triggered a major rock burst measuring 79 mm on the mine seismograph (a local relative measurement of intensity) with a local magnitude

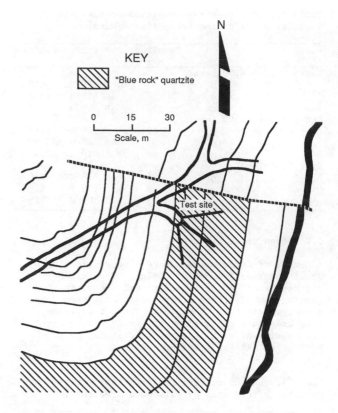

Figure 5. The Lucky Friday 5300 level overcore site and location of blue rock quartzite.

(M_L) of 1.4. Forty-five tons of rock collapsed into the 5300-101 crosscut, creating a 2- by 2- by 6-m void in the roof.

After repairs were completed, a third round was shot on January 17, triggering a burst measuring 31 mm ($M_L \approx 1$). This burst took the right rib of the 5300-101 crosscut and released 72 tons of rock.

Intense seismic activity continued into the fol-lowing day, January 18, causing mine personnel to abandon repair activities. Shortly after the crosscut was evacuated, at 8:09 a.m., a rock burst occurred without warning. This rock burst measured 72 mm ($M_L \approx 1.2$). The burst expelled 150 tons of rock from the right rib and back. This burst was particularly dangerous because it occurred on-shift without warning, whereas the previous bursts had occurred with blasting, when the mine is evacuated. Mine management then decided to abandon the 5300-101 crosscut and develop an alternative accessway in a safer portion of the rock mass.

6 DISCUSSION AND CONCLUSIONS

Observations and measurements suggest that a strong natural stress concentration existed in blue rock in the 101 crosscut on the Lucky Friday Mine's 5300 level. Excavation of the 101 crosscut through this rock produced intense rock bursting, far in excess of the level usually encountered during the driving of development openings, and forced abandonment of the crosscut.

Correlation of natural variations of in situ stress, often described as residual and structural stress, have been reported elsewhere in the liter-ature, often in conjunction with corresponding pat-terns of rock burst or coal bump activity. While the mechanism for stress concentration in many of the reported geologic structures has been apparent, the mechanism and contributing geologic structures at the Lucky Friday Mine are clouded by the complex geologic structure.

However, several geologic features are being analyzed to determine if they can contribute to the observed extreme, but spatially limited, stress concentration. These include the Offset Fault, which places softer rock to the northeast of the blue rock; soft argillite beds that may creep over geologic time; the anticline that dominates the mine; and residual stresses created by injection of the sulfides responsible for alteration of the blue rock. Analysis of these mechanisms will be undertaken in the next phase of this project.

If a clear picture of the controlling factors for this stress concentration can be determined, geologic markers for increased rock-burst hazard could be identified. Furthermore, the concentration of stress in these natural pillars suggests that portions of the rock mass may be naturally destressed by these same mechanisms. Thus, analysis of natural in situ stress variations through geologic structures has the potential to provide a rock-burst-hazard map to guide positioning of development openings. The implications for rock-burst-hazard levels during mining are interesting as well, but complicated by the superposition of mining-induced stresses. These bursts will be examined in the final phase of this work.

REFERENCES

Allen, M.A. 1979. Determining the In-Situ Stress Field on the 4250 Level in the Lucky Friday Mine Using the CSIR Biaxial Strain Gage and Structural Geologic Mapping. M.S. thesis, Univ. of Idaho, 73 pp.

Beus, M.J., and S.S.M. Chan. 1980. Shaft Design in the Coeur d'Alene Mining District, Idaho – Results of In Situ Stress and Physical Property Measurements. U.S. Bur. of Mines, RI 8435, 39 pp.

Boler, F.M., and P.L. Swanson. 1990. Computer-Automated Measurement- and Control-Based Workstation for Microseismic and Acoustic Emission Research. U.S. Bur. of Mines, RI 9262, 9 pp.

Gaviglio, P., R. Revalor, J.P. Piguet, and M. Dejean. 1990. Tectonic Structures, Strata Properties and Rockburst Occurrence in a French Coal Mine. In Rockbursts and Seismicity in Mines: Proceedings of the 2nd International Symposium on Rockbursts and Seismicity in Mines, ed. by C. Fairhurst (Univ. of MN, Minneapolis, MN, June 8-10, 1988). A.A. Balkema, Rotterdam, Netherlands, pp. 289-293.

Gresseth, E.W. 1964. Determination of Principal Stress Directions Through an Analysis of Rock Joint and Fracture Orientations, Star Mine, Burke, Idaho. U.S. Bur. of Mines, RI 6413, 42 pp.

Gresseth, E.W., and R.R. Reid. 1968. A Petrofabric Study of Tectonic and Mining-Induced Deformations in a Deep Mine. U.S. Bur. of Mines, RI 7173, 64 pp.

Herget, G. 1988. Stresses in Rock. A.A. Balkema, Rotterdam, Netherlands, 179 pp.

Hobbs, S.W., A.B. Griggs, R.E. Wallace, and A.B. Campbell. 1965. Geology of the Coeur d'Alene District, Shoshone County, Idaho. U.S. Geol. Survey Prof. Paper 478, 139 pp.

Jenkins, F.M., and R.R. Dorman. 1983. Ch. in Underhand Cut-and-Fill Stoping for Rock Burst Control. U.S. Bur. of Mines, IC 8973, pp. 49-63.

Jenkins, F.M., and R.W. McKibbin. 1986. Practical Considerations of In Situ Stress Determination. International Symposium on Application of Rock Characterization Techniques in Mine Design, ed. by M. Karmis. SME, Littleton, CO, pp. 33-39.

Jenkins, F.M., T.J. Williams, and C.J. Wideman. 1990. Analysis of Four Rockbursts in the Lucky Friday Mine, Mullan, Idaho, USA. In Inter-national Deep Mining Conference: Technical Challenges in Deep Level Mining, ed. by D.A.S. Ross-Watt and P.D.K. Robinson (Johannesburg, S. Afri., Sept. 17-21, 1990). S. Afri. Inst. of Min. and Metall., Johannesburg, S. Afri., pp. 1201-1212.

Karwoski, W.J., W.C. McLaughlin, and W. Blake. 1979. Rock Preconditioning To Prevent Rock Bursts – Report on a Field Demonstration. U.S. Bur. of Mines, RI 8381, 47 pp.

Martna, J., and L. Hansen. 1986. Initial Rock Stresses around the Vietas Headrace Tunnels Nos. 2 and 3, Sweden. In Rock Stress and Rock Stress Measurements: Proceedings of the International Symposium on Rock Stress and Rock Stress Measurements, ed. by O. Stephansson (Stockholm, Sept. 1-3, 1986). Centek, Lulea, Sweden, pp. 605-613.

McMahon, T. 1988. Rock Burst Research and the Coeur d'Alene District. U.S. Bur. of Mines, IC 9186, 49 pp.

Riefenburg, J. 1991. Statistical Evaluation and Time Series Analysis of Microseismicity, Mining, and Rock Bursts in a Hard-Rock Mine. U.S. Bur. of Mines, RI 9379, 15 pp.

Scott, D.F. 1990. Relationship of Geologic Features to Seismic Events, Lucky Friday Mine, Mullan, Idaho. In Rockbursts and Seismicity in Mines: Proceedings of the 2nd International Symposium on Rock Bursts and Seismicity in Mines, ed. by C. Fairhurst (Minneapolis, MN, June 8-10, 1988). A.A. Balkema, Rotterdam, Netherlands.

Stebley, B.J., and B.T. Brady. 1990. Innovative Microseismic Rockburst Monitoring System. In Rockbursts and Seismicity in Mines: Proceedings of the 2nd International Symposium on Rock Bursts and Seismicity in Mines, ed. by C. Fairhurst (Minneapolis, MN, June 8-10, 1988). A.A. Balkema, Rotterdam, Netherlands, pp. 259-262.

Swanson, P.L. 1993. Mining Induced Seismicity in Faulted Geologic Structure: Analysis of Seismicity-Induced Slip Potential. PAGEOPH.

Swanson, P.L., and C.D. Sines. 1991. Characteristics of Mining-Induced Seismicity and Rock Bursting in a Deep Hard Rock Mine. U.S. Bur. of Mines, RI 9393, 12 pp.

Whyatt, J.K. 1986. Geomechanics of the Caladay Shaft. M.S. thesis, University of Idaho, Moscow, ID, 195 pp.

Whyatt, J.K., and M.J. Beus. 1993. In Situ Stress at the Lucky Friday Mine (In Four Parts): 1. Reanalysis of Overcore Measurement on the 4250 Level. U.S. Bur. of Mines, RI in review.

Whyatt, J.K., M.J. Beus, and M.K. Larson. 1993. In Situ Stress at the Lucky Friday Mine (In Four Parts): 3. Reanalysis of Star Mine Overcore Measurement. U.S. Bur. of Mines, RI in review.

Whyatt, J.K., and M.P. Board. 1991. Numerical Exploration of Shear-Fracture-Related Rock Bursts Using a Strain-Softening Constitutive Law. U.S. Bur. of Mines, RI 9350, 16 pp.

Whyatt, J.K., F.M. Jenkins, and M.K. Larson. 1993. In Situ Stress at the Lucky Friday Mine (In Four Parts): 2. Analysis of Overcore Measurement on 5300 Level. U.S. Bur. of Mines, RI in review.

Whyatt, J.K., T.J. Williams, and W.G. Pariseau. 1992. Trial Underhand Longwall Stope Instrumentation and Model Calibration at the Lucky Friday Mine, Mullan, Idaho, USA. In Rock Mechanics: Proceedings of the 33rd Symposium, ed. by J.R. Tillerson and W. R. Wawersik (Santa Fe, NM, June 3-5, 1992). A.A. Balkema, Rotterdam, Netherlands, pp. 511-519.

Rockbursts and Seismicity in Mines, Young (ed.) © 1993 Balkema, Rotterdam, ISBN 90 5410 320 5

Mechanisms of rockmass failure and prevention strategies in rockburst conditions

Xiaoping Yi & Peter K. Kaiser
Geomechanics Research Centre, Laurentian University, Sudbury, Ont., Canada

ABSTRACT: The interaction of plane seismic waves with underground openings is studied by applying two analytical approaches, the dynamics and continuum mechanics approaches. It is found that low frequency waves from a seismic source hundreds of meters away can cause both rock ejections and dynamic stress concentrations at both walls of an opening. These dynamic stress concentrations, superimposed on existing high static stress concentrations, may lead to different types of rock failure including secondary rockbursts (mainly strain bursts). The high frequency, high magnitude seismic waves from nearby rockburst sources may break rock and eject rock blocks from one wall of an excavation whereas the opposite wall is not affected. Formulae for calculating the ejection velocities are developed. Based on this rock ejection model, strategies for mine design, rockmass de-stressing and support design are discussed.

1. INTRODUCTION

Rock failures in mining are referred to as falls of ground induced by gravity or seismicity and rockbursts. Rockfalls are non-violent falls of loose rocks under the influence of gravity, whereas rockbursts are of instantaneous and violent nature (Jaeger and Cook 1984). Two basic mechanisms are involved in rockbursts: (i) instantaneous slip on an existing geological feature (fault-slip) and (ii) instantaneous fracturing of highly-stressed rock. For practical purposes in hard rock mining, the following three types of rockbursts were defined: fault slip, pillar and strain energy rockbursts (Hedley 1987). Fault slip rockbursts involve primarily the mechanism of sliding, strain energy rockbursts the mechanism of fracturing, and pillarbursts often a combination of the two mechanisms.

The Ontario Ministry of Labor (1990) defines a rockburst as the instantaneous failure of rock causing an expulsion of material at the surface of an opening or a seismic disturbance to a surface or underground mine. This definition reflects the concern for the effects of rockbursts rather than their causes. Therefore, as far as support design is concerned, the expulsion of material at the surface of an opening must be prevented or contained, and a rockburst damage model based on the ejection of rock blocks is a logical and realistic basis for support design.

This paper presents results from analytical studies of rock failure mechanisms due to rockbursts. Two approaches, the dynamics and the continuum mechanics approaches, are reviewed and investigated to provide a better understanding of rock ejections into openings during rockbursts. The former approach deals with the dynamic response of discrete rock blocks and the latter deals with dynamic stress concentrations and wave reflections at the openings.

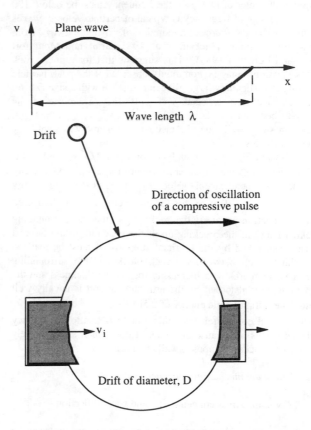

Figure 1. An underground drift in the path of a propagating plane seismic wave.

2. ROCK EJECTION MODELS

2.1 The dynamics approach

When a seismic wave propagates in an elastic continuum, a point or particle of this continuum oscillates around its stationary equilibrium position, and this particle eventually returns to its stationary position some time after the wave has passed by. A particle is prevented from flying off and moving to a different position due to the atomic forces between adjacent particles. The continuum mechanics approach must be employed to deal with this problem (section 2.2). On the other hand, a particle, on the free boundary of an opening, may fly off into the opening if the retaining forces are overcome due to excessive acceleration. Similarly, rock blocks on the free boundary of an opening which are separated from the surrounding rockmass by joints or fractures can be projected into the opening by seismic waves. The dynamics approach to be described in this section can be employed to study this type of problem.

Figure 1 shows a tunnel of diameter D in the path of a plane seismic wave with a wave length λ. This seismic wave is assumed to be sinusoidal with positive (compressive) and negative (tensile) pulses. If the tunnel diameter is much smaller than the wave length or $D\lambda^{-1} \ll 1$, the wave is not disturbed by the tunnel and wave reflection does not occur. Furthermore, a rock block can be considered as a particle if its size is much less than the wave length. If such a rock block is located on the wall of the tunnel facing the source and is separated from the surrounding rockmass by joints or fractures, it can be carried into the tunnel by a positive pulse of a seismic P wave and then left in the tunnel as the wave passes by. On the other hand, if the rock block is on the opposite wall, it may initially remain stationary during the compressive pulse of the P wave and is carried into the tunnel by the negative pulse. Roberts and Brummer (1988) used this approach to analyze rockburst damage mechanisms in South African mines. If the incoming seismic wave is an S wave, rock blocks at the roof and floor of the tunnel may be ejected into the tunnel.

For engineering applications, we may choose $\lambda D^{-1} \geq 10$ as a practical limit to apply this dynamics approach. For this limit, the pulse width is 5 times the opening size, and it will be shown later that the ejection velocity of a rock block is 90 % of the peak particle velocity (section 2.3). In frequency terms, we find $f \leq c(10D)^{-1}$. With a P wave propagation velocity of 6000 m s^{-1} and a tunnel diameter of $D \leq 6$ m, the frequency must be below 100 Hz. This range of frequency is typical of particle velocity waves from rockburst sources hundreds of meters away from measurement points (McGarr et al. 1981, Brink and Mountfort 1984 and Leighton 1984). This suggests that for tunnels, rock block ejections may be commonly observed if the peak particle velocity is sufficiently large. For mined stope with a size $D \leq 60$ m, rock ejection may occur if $f \leq 10$ Hz. Therefore, if the orebody between two adjacent levels is completely mined out, rocks on stope walls may be ejected by seismic waves of very low frequencies.

As shown in Figure 1, a rock block of mass M, isolated by frictionless joints may attain a maximum initial ejection velocity v_i equal to the peak particle velocity v_p. The initial kinetic energy is then $0.5Mv_p^2$. Part of this kinetic energy may be dissipated due to friction between the rock block and the surrounding rockmass and in the breaking-up of the rock block, and the rest must be absorbed by the artificial support. It must be realized that the friction between the rock block and the surrounding rockmass increases with increasing rock block thickness and the kinetic energy absorbed by the artificial support is usually well below the initial kinetic energy of $0.5Mv_p^2$.

The general conclusion from this section is that low frequency seismic waves may cause ejections of both small and relatively large rock blocks on opposed walls of a drift.

2.2 Continuum mechanics approach

2.2.1 Dynamic stress concentration and wave reflection

A propagating planar seismic sine wave encountering a cylindrical underground tunnel (Fig. 1) may cause dynamic stress concentrations or wave reflections at the boundary of the tunnel. Two extreme situations arise depending on the relative size of the tunnel D compared to the wave length λ of the seismic wave.

One extreme occurs if the tunnel is small compared to the wave length, i.e., $D\lambda^{-1}$ approaches zero. In this situation, the loading of the tunnel due to the wave is similar to static loading and may be called quasi-static body force loading. The mathematical proof of this statement can be found in Pao and Mow (1973). The resulting dynamic stress concentrations around the tunnel due to seismic waves can be approximately derived employing the theory of static continuum mechanics.

The other extreme situation occurs if the size of the tunnel is very large compared to the wave length, that is, the ratio $D\lambda^{-1}$ is very large. In this situation, the wall of the tunnel facing the seismic source acts as a free surface to the propagating wave and the theory of wave reflection applies. The wall on the opposite side of the tunnel and the rockmass behind it is shielded from any disturbance. Theoretical treatment of this scenario can be found in Kolsky (1963) and Brady and Brown (1985). The fact that we often observe damage only on one wall of an opening suggests that this damage mechanism may be involved. For engineering applications, we may assume that this extreme situation applies if the opening size is about 6 times the pulse width. The corresponding frequency requirement is therefore $f > 3cD^{-1}$. With $c_p = 6000$ m s^{-1} and $D < 6$ m, the lower frequency limit is of the order of 3 kHz.

For tunnels with diameters D comparable to the wave length λ, the concept of wave diffraction is applicable, but the theoretical treatment is mathematically rather involved (Pao and Mow, 1973). Nevertheless, certain qualitative results can be induced and quantitative results estimated from results for the above two extremes. The concepts of dynamic stress concentration around an opening derived from quasi-static loading and stress wave reflection can be useful in mining and civil engineering, since detailed quantitative analyses are often not feasible due to a lack of detailed data on wave amplitude, wave length and rockmass properties.

2.2.2 Dynamic stress concentration

For the concept of dynamic stress concentration to apply, the relationship between opening size D and wave frequency f proposed in section 2.1 must be satisfied, i.e., $f \leq c(10D)^{-1}$. For a plane wave containing both P and S wave components, propagating in the direction of the x axis (Fig. 1), the dynamic stresses induced in the rockmass are as follows (Brady and Brown 1985):

$$\sigma_x = \rho\, c_p v_x$$
$$\sigma_{xy} = \rho\, c_s v_y$$
$$\sigma_{xz} = \rho\, c_s v_z \qquad\qquad (1)$$
$$\sigma_y = \sigma_z = \sigma_x \frac{\nu}{1-\nu}$$

where σ_x, σ_y and σ_z are normal stresses, σ_{xy} and σ_{xz} shear stresses, v_x, v_y and v_z particle velocities, c_p and c_s longitudinal and shear wave velocities, and ρ and ν the density and Poison's ratio of the rockmass. A typical hard rock has a propagation velocity ratio of $c_p c_s^{-1} = 1.73$ (for a Poison's ratio of $\nu = 0.25$) with $c_p = 6$ km s^{-1} and $\rho = 2600$ kg m^{-3}. For a P wave with a peak particle velocity of $v_x = 1$ m s^{-1}, the peak dynamic stresses are $\sigma_x = 15.6$, $\sigma_y = \sigma_z = 5.2$ MPa, and for an S wave with a peak particle velocity of $v_y = 1$ m s^{-1}, the peak dynamic shear stress is $\sigma_{xy} = 8.9$ MPa. If a positive or compressive P wave pulse is followed by a negative or tensile pulse, for example, the peak tensile stresses for $v_x = 2$ m s^{-1} are $\sigma_x = -31.2$, $\sigma_y = \sigma_z = -10.4$ MPa. The typical tensile strength of intact hard rocks is about 20 MPa (Jaeger and Cook 1979). Taking into account that the strength increases with increasing strain rate, such rock should sustain a peak particle velocity of a few meters per second. This agrees with the suggestion by Brune (1970) that the upper limit of peak particle velocity at the sources of earthquakes is of the order of 1 m s^{-1}. Since a fractured rockmass can not sustain any significant tensile stress, failure is inevitable if the rockmass is in the path of a tensile seismic wave.

It is important to realize that the dynamic stresses due to seismic waves are superimposed on the pre-existing static stresses. Furthermore, there are dynamic stress concentrations

142

around an opening when a seismic wave encounters an underground opening. These dynamic stress concentrations can be obtained by employing the continuum mechanics approach. Since these stresses are functions of time, underground openings are subject to cyclic loadings.

The general conclusion from this sub-section is that low frequency seismic waves can cause dynamic stress concentrations around underground tunnels. If these dynamic stress concentrations are of sufficient magnitudes and are superimposed on high existing static stress concentrations, rockmass failures may occur or secondary rockbursts may be triggered. *but approach gives no indication of Ve*

2.3   Rock block ejection velocities

Ejection velocities for the following two situations will be presented in this section: (i) the opening size is much smaller than the wave length and there is no wave reflections (the wave does not "see" the opening), and (ii) the opening size is much larger than the wave length and the theory for wave reflection applies. If a plane compressive wave with a local peak particle velocity v_{pp} propagates normal to a free surface of the opening, the reflected wave is tensile and the peak particle velocity of the free surface is $2v_{pp}$ (Brady and Brown 1985). A rock block at the free surface may be ejected if the resulting ejection velocity is sufficiently large. The initial ejection velocity of a rock block can be determined by calculating the momentum trapped in the block immediately before the ejection (Wasley 1973 and Kolsky 1963).

Yi and Kaiser (1991) derived the ejection velocities for the two special cases shown in Figures 2a and b. For this purpose, three assumptions were made: (i) a plane sine wave can be approximated by a simpler triangular shape, (ii) the frictional and tensile resistances of the five surfaces of a rock block can be ignored, and (iii) the rock block is in contact with the rockmass such that the incoming seismic wave is transmitted completely across the interface. With the first assumption, the momentum of the wave is underestimated, but linear equations are derived. This underestimation is compensated by the effects of the second assumption which mostly likely introduces the largest error, particularly for thick rock blocks where block boundary friction and energy loss in rock fracturing may dominate. The third assumption is satisfied if the wave peak particle displacement is much larger than the joint aperture. Therefore, the derived ejection velocities represent upper limits which may be used for practical support design.

If a P compressive wave, whose wave length is much smaller than the opening size, approaches the tunnel roof from above (Figure 2a), wave reflection occurs and the reflected wave is tensile. A rock block is ejected when the net stress on the interface, as a result of the incident compressive and reflected tensile waves, becomes tensile. The ratio of rock block initial ejection velocity v_i to the local peak particle velocity v_{pp} was derived by Yi (1993):

$$\frac{v_i}{v_{pp}} = 2 - \frac{4h}{\lambda_p} \qquad h\,\lambda_p^{-1} \le 0.5 \qquad (2)$$

where h is the thickness of the rock block and λ_p the P wave length. The ejection velocity becomes negative for $h\,\lambda_p^{-1} > 0.5$. It can be seen from the above equation that $v_i = 2\,v_{pp}$ if the rock block is reduced to the size of a particle, $v_i = v_{pp}$ if $h = 0.25\lambda_p$, and $v_i = 0$ if $h = 0.5\lambda_p$. The concept of rock block ejection by compressive waves was developed for the Hopkinson Pressure Bar Experiment (Kolsky 1963). This phenomenon was also demonstrated by Watson and Sanderson (1979) who applied

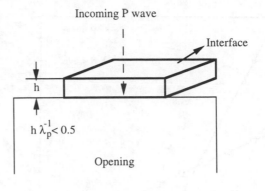

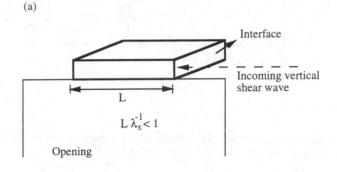

(a)

(b)

Figure 2. Ejection of rock blocks in the roof of an underground opening by (a) a compressive P wave from above and (b) a vertically oscillating S wave from a horizontal direction.

concentrated explosive blasting charges on concrete slabs and observed tension-induced slabbing and spalling.

In the previous section, the wave reflection approach was assumed to be applicable for $\lambda < D/3$. Since $h \le \lambda_p/2$ (Eq. 2), the wave reflection approach is applicable for $h < D/6$ (less than 1 m for a 6 m tunnel) or for $2h \le \lambda_p < D/3$ ($3c_pD^{-1} < f \le 0.5c_ph^{-1}$, for D = 6 m, $1500 < f \le 3000$ Hz). Since seismic waves of frequencies in this range attenuate fast with propagation distance (section 4), only high frequency waves from close-by seismic sources can cause ejections of small rock blocks with dimensions less than one sixth of opening size. For D = 60 m and h = 1 m, we obtain $300 < f \le 3000$ Hz. Frequencies higher than 300 Hz have been obtained for particle velocity waves measured at locations hundreds of meters away from seismic sources (McGarr et al. 1981, Brink and Mountfort 1984 and Leighton 1984). Therefore, if the orebody between two adjacent levels is completely mined out, ejections of fairly large rock blocks (up to 10 m for a 60 m high open stope) from stope walls may be triggered. In reality, this would occur only if the large rock block was already separated from the surrounding rockmass before the arrival of the seismic wave.

If a vertically oscillating S wave propagates horizontally (Figure 2b), no wave reflection occurs and a rock block at the roof may be ejected by the vertical oscillations. The initial ejection velocity can be expressed as (Yi 1993):

$$\frac{v_i}{v_{ps}} = 1 - \frac{L}{\lambda_s} \qquad L\,\lambda_s^{-1} \le 1 \qquad (3)$$

where v_{ps} and λ_s are the peak particle velocity and wave length of the S wave respectively and L is the width of the rock block. Equations (2) and (3) are plotted in Figure 3.

If a P compressive wave, whose wave length is much larger than the opening size, approaches the tunnel roof from above

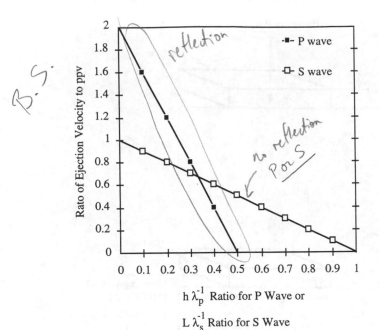

Figure 3. Ratio of ejection velocity to ppv versus ratio of rock block thickness h or length L to wave length relationships.

(Fig. 2a), wave reflection does not occur and a formula similar to equation (3) can be obtained:

$$\frac{v_i}{v_{pp}} = 1 - \frac{h}{\lambda_p} \qquad h\,\lambda_p^{-1} \le 1 \qquad (4)$$

From equations (3) and (4), it follows that the ejection velocity v_i of a rock block is a function of the block size. The maximum value of v_i is equal to the peak particle velocity and is obtained when the block size to wave length ratio approaches zero. The ejection velocity is higher than $0.9\,v_{ps}$ or $0.9\,v_{pp}$ if the opening size $D < 0.1\lambda$ or $f < 0.1\,c_p D^{-1}$ (f < 100 Hz for a 6 m tunnel). For engineering purposes, the rock block ejection velocity may be set equal to the peak particle velocity for frequencies below this limit.

3. ROCK FAILURE MECHANISMS AND STRATEGIES FOR FAILURE PREVENTION IN ROCKBURST CONDITIONS

According to the analyses in the previous section, the rockmass behind an opening may be completely or partially shielded from the seismic waves. While seismic waves with frequency below about 200 Hz are not likely to be affected by tunnels several meters in diameter, the rockmass behind a mined-out stope may be shielded from seismic waves with much higher frequencies. Measurements of seismic waves at locations hundreds of meters away from the source indicate that their dominant frequencies are typically less than 500 Hz (Gibowicz 1984, Fernandez and Heever 1984, McGarr et al. 1981, Brink and Mountfort 1984, and Leighton 1984). The higher frequency waves may have been attenuated or dissipated in the rockmass, or shielded by underground openings located between the sources and the measurement points. High frequency waves, while being blocked, may cause rock ejections at these openings. Seismic waves with frequencies below about 500 Hz can cause ejections of rocks at both small and large openings and induce dynamic stress concentrations around relatively small openings.

For rockburst research purposes, it is unfortunate that strong motion measurements providing peak particle velocities close to the sources have not been measured extensively. For example, the peak particle velocities measured by McGar et al (1981) were around 10 mm s⁻¹. We know neither the amplitudes nor frequencies of these waves at locations very close to seismic sources. However, it may be postulated that frequencies higher than several kilohertz can only be associated with sufficiently large amplitudes at locations very close to a rockburst source. Nevertheless, the following rock failure mechanisms in rockbursts may be suggested:

Seismic waves from seismic sources several hundred meters away can cause both dynamic stress concentrations and ejections of large and small rock blocks at underground openings. These dynamic stress concentrations, superimposed on the existing high static stress concentrations around openings, can cause different types of rock failures including secondary rockbursts near the openings (sub-sources). The higher frequency waves from these nearby rockbursts (say within tens of meters) can cause ejections of relatively small rock blocks.

A comprehensive program for controlling rockburst hazards in a mine must aim at eliminating the occurrences of large-magnitude seismic events and possible sources of secondary sub-events. Should rockbursts occur, damages must be contained employing appropriate support systems. Mine design methodologies aiming at minimizing stress concentrations in the rockmass could reduce the occurrences of primary rockbursts, rock destressing around an opening could reduce the occurrences of secondary rockbursts, and an effective support must help contain ejected rock blocks. Since part of the kinetic energy of an ejected rock block is dissipated through friction, the prime goal of rock support is to maintain the integrity of an opening in order to minimize the energy transferred to artificial support.

4. DISCUSSION

The amplitude A of a seismic wave with wave length λ at a distance x from the source can be expressed as (Kavetsky et al. 1990, Jaeger and Cook 1979, and Dobrin 1976):

$$A = A_o \left(\frac{x_o}{x}\right)^a e^{-\alpha(x - x_o)} \qquad (5)$$

where A_o is the amplitude at distance x_o from the source, and the attenuation coefficient $\alpha = \pi Q^{-1} \lambda^{-1}$. Q is the coefficient of internal friction which characterizes the inelastic absorption by the rockmass. This attenuation coefficient α increases with decreasing wave length. The theoretical value for the exponent a is 0, 0.5 and 1 for a plane (planar source), cylindrical (column source) and spherical (point source) waves respectively. For a fault-slip type rockburst, therefore, a increases from 0 to 1 as the propagation distance x from the geometric center of the source area increases from zero to infinity. From equation (5), the following remarks can be made: (i) higher frequency seismic waves attenuate faster with the propagation distance or lower frequency waves propagate further away from rockburst sources, and (ii) the attenuation of a seismic wave increases non-linearly as it propagates further away from the source (since a increases from 0 to 1).

The effect of frequency on damages of surface structures from blast vibrations was recognized by US Bureau of Mines for the OSM regulation (RI 8507) where, e.g., the peak particle velocity (ppv) limits were 5 and 50 mm s⁻¹ for 1 and 100 Hz frequencies respectively (Atlas Powder 1987). In rockburst applications, the frequency effects of parameters such as the ppv have often been ignored for simplicity reasons or for specific applications (narrow range of dominant frequency). This can be seen from the ppv versus propagation distance relationships developed by McGarr et al. (1981) and Hedley (1990), where the exponential term involving the attenuation coefficient α in equation (5) was ignored. In an underground mine, there exist numerous tunnels and mined-out stopes. Apart from being attenuated in the rockmass, high frequency waves can also be shielded by underground openings. Therefore, only low frequency waves can reach openings beyond stopes and large tunnels. Since large rock blocks can be ejected by these low frequency waves (section 3), large scale ground falls induced by far away seismic sources are often observed in the field, and support design methods must be developed to deal with these extremely dangerous conditions.

5. CONCLUDING REMARKS

Low frequency seismic waves can cause ejections of both small and relatively large rock blocks on opposing walls of underground openings located both close to and far away from seismic sources. The initial ejection velocities can be obtained from equations (3) and (4) if the local peak particle velocity is known. These seismic waves may also cause dynamic stress concentrations around these openings and induce secondary rockbursts (of mainly strain energy type). The dynamic stresses may be calculated from equation (1). High frequency waves from close-by seismic sources may cause ejections of relatively small rock blocks on only one wall of an excavation and the ejection velocity is given by equation (2). The opposite wall should not be affected. Mine design and rockmass de-stressing aiming at reducing or eliminating extraordinary static stress concentrations may reduce the occurrences of large seismic events and secondary rockbursts. Effective artificial support systems must then be applied to contain ejectable rock blocks, and support design must aim at maintaining the integrity of all rock block interfaces in order to facilitate energy dissipation and minimize the energy imparted to artificial support.

ACKNOWLEDGEMENTS

Financial support from MRD (Mining Research Directorate) and NSERC (Natural Science and Engineering Research Council) is gratefully acknowledged. Comments received from Drs R. K. Brummer and D. McCreath were extremely helpful for preparing the manuscript.

REFERENCES

Atlas Powder 1987. *Explosives and Rock Blasting*. Atlas Powder Company, Dallas, Texas. Maple Press, 662p.

Brady, B. G. H. and Brown E. T. 1985. *Rock Mechanics for Underground Mining*. UK: George Allen and Unwin, 527p.

Brink, A. V. Z. and Mountfort P. I. 1984. Feasibility studies on the prediction of rockbursts at Western deep levels. *Proceedings of the 1st International Congress on Rockbursts and Seismicity in Mines*. Johannesburg, 1982, SAIMM, 317-325.

Brune, J. N. 1970. Tectonic stress and the spectra of seismic shear waves from earthquakes. *Journal of Geophysical Research*, 75(26), 4997-5009.

Dobrin, M. B. 1976. *Introduction to Geophysical Prospecting* (3rd edition). New York, London, Sydney, Tokyo and Toronto: McGraw Hill Book Company, 630p.

Fernandez, L. M. and van der Heever P. K. 1984. Ground movement and damage accompanying a large seismic event in the Klerksdrop district. *Proceedings of the 1st International Symposium on Rockbursts and Seismicity in Mine*.

Johannesburg, 1982, SAIMM, 193-198.

Gibowicz, S. J. 1984. The mechanism of large mining tremors in Poland. *Proceedings of the 1st International Symposium on Rockbursts and Seismicity in Mine:* 17-28. Johannesburg, 1982, SAIMM.

Hedley, D. G. F. 1987. *Rockburst Mechanics*. Mining Research Laboratories, CANMET, Canada, 30p.

Hedley, D. G. F. 1987. *Rockbursts in Ontario Mines during 1985*. Mining Research Laboratories, CANMET, SP87-2E, Canada, 20p.

Hedley, D. G. F. 1990. Peak particle velocity for rockbursts in some Ontario mines. *Proceedings of the 2nd International Symposium on Rockbursts and Seismicity in Mine:* 345-348. University of Minnesota, June 8-10, 1988.

Hoek, E. and Brown E. T. 1982. *Underground Excavations in Rock*. The Institution of Mining and Metallurgy, London. England: Stephen Austin and Sons Ltd., 527p.

Jaeger, J. C. and Cook N. G. W. 1984. *Fundamentals of Rock Mechanics* (3rd edition). London and New York: Chapman and Hall, 593p.

Kavetsky, A., Chitombo G. P. F., McKenzie C. K. and Rang R. L. 1990. A model of acoustic pulse propagation and its application to determine Q for a rock mass. *International Journal of Rock Mechanics, Mining Sci. and Geomech. Abstr.*, 21(1): 35-38.

Kolsky, H. 1963. *Stress Wave in Solids*. New York: Dover Publications INC, 213p.

McGar, A., Green R. W. E. and Spottiswood S. M. 1981. Strong ground motion of mine tremors: some implications for near-source motion parameters. *Bulletin of the Seismological Society of America*, 71(1); 295-319.

Ontario Ministry of Labor, 1990. Interpretation 34a of regulation 694. *Occupational Health and Safety Act and Regulations for Mines and Mining Plants*, Occupational Health and Safety Division, 400 University Avenue, Toronto, Canada, M7A 1T7, 350p.

Pao, Y. H. and Mow C. C. 1973. *Diffraction of Elastic Waves and Dynamic Stress Concentrations*. Rand Corporation. New York: Crane, Russak and Company, 693p.

Roberts, M. K. C. and Brummer R. K. 1988. Support requirements in rockburst conditions. *Journal of South African Institute of Mining and Metallurgy*, 88(3), 97-104.

Stillborg, B. 1990. *Rockbolt and Cablebolt Tensile Testing across a Joint*. Unpublished Report, Mining and Geotechnical Consultants, Lulea, Sweden, 19p.

Wagner, H. 1984. Support requirements for rockburst conditions. *Proceedings of the 1st International Symposium on Rockbursts and Seismicity in Mine*. Johannesburg, 1982, SAIMM, 209-218.

Wasley, R. J. 1973. *Stress Wave Propagation in Solids, an Introduction*. New York: Marcel Dekker Inc. 279p.

Watson, A. J. and Sanderson A. J. 1979. The resistance of concrete to explosive shock. *Institute of Physics Conference Series No. 47*: 307-317.

Yi, X. and Kaiser P. K. 1991. *Interaction of Plane Seismic Waves with an Underground Tunnel*. Internal Report No. 91-13-12, Geomechanics Research Center, Laurentian University, Sudbury, Ontario, 24p.

Yi, X. 1993. *Dynamic Response and Design of Support Elements in Rockburst Conditions*. PhD Thesis, Dept. of Mining, Queen's University, Kingston, Ontario, 300p.

2 Mechanics of seismic events and stochastic methods

Rockbursts and Seismicity in Mines, Young (ed.) © 1993 Balkema, Rotterdam, ISBN 90 5410 320 5

Keynote address: Seismic moment tensor and the mechanism of seismic events in mines

S.J.Gibowicz
Institute of Geophysics, Polish Academy of Sciences, Warsaw, Poland

ABSTRACT: There is growing evidence that alternative earthquake mechanisms other than shear failure are possible. The moment tensor approach is in this respect the most general one. A moment tensor can be decomposed into an isotropic part, a compensated linear vector dipole and a double couple. The isotropic component of the source mechanism corresponds to a volumetric change, the compensated linear vector dipole corresponds to a kind of uniaxial compression or tension, and the double couple corresponds to a shear failure. Although moment tensor inversions on a teleseismic scale have been routinely performed for several years, the application of this technique to local events is a relatively recent innovation. A few works only have been published that are related to the use of moment tensor inversion to study the mechanism of seismic events in mines. This approach is mostly used in South Africa and Poland, and to a limited extend in Japan and Canada.

1 INTRODUCTION

Recent results from earthquake focal mechanism studies indicate growing evidence that alternative mechanisms other than shear failure are possible. The most prominent cases of what appear to be anomalous focal mechanisms are reported from seismicity studies in mines. The results from studies mostly based on the first-motion polarity and radiation patterns were reviewed by Gibowicz (1990). Here the results from moment tensor inversions are considered in some detail.

Moment tensors describe completely, in a first order approximation, the equivalent forces of general seismic point sources, the double-couple source being just one of them. In many applications a point source approximation may be quite satisfactory, provided that the source dimensions are small in comparison to the observed wavelenghts of seismic waves. The concentrated force couples and the corresponding formal introduction of the moment tensor and its decomposition into various components are discussed at the beginning of this review. Then various methods of inversion for the moment tensor elements are briefly considered. Although moment tensor inversions on a teleseismic scale have been routinely performed for several years, the application of this technique to local seismic events is a relatively recent innovation. The results of the application of moment tensor inversion technique to study seismicity induced by mining are described at the end of this presentation.

2 CONCENTRATED FORCE COUPLES AND MOMENT TENSOR

It is well known that the displacement field $u_i(\mathbf{x},t)$ generated by a single body force is equal to the convolution of this point force $F(t)$ by the Green's function G_{ij}

$$u_i(\mathbf{x},t) = F(t) * G_{ij}(\mathbf{x},t), \qquad (1)$$

where the asterisk is a commonly used abbreviation for the convolution and the Green's function is simply the medium response to the delta function. If the body force is concentrated but its orientation is arbitrary $\mathbf{f}(\mathbf{x},t) = \mathbf{F}(t)\delta(\mathbf{x}-\xi)$, where $\mathbf{F}(t) = (F_1, F_2, F_3)$, then the total displacement is equal to the sum of the displacements generated by the forces F_1, F_2 and F_3 directed along the $x_1, x_2,$ and x_3 directions (e.g., Pujol and Herrmann, 1990)

$$u_i = F_1 * G_{i1} + F_2 * G_{i2} + F_3 * G_{i3} = F_j * G_{ij}. \qquad (2)$$

Although the single force is one of the simplest models of a seismic source, the supposed external application of a force is unlikely to occur in natural earthquakes. It is more probable that the force action is of self-balancing type, such as a pair of opposite forces acting simultaneously on two adjacent parts of the medium with the resultant force equal to zero.

Let us consider a pair of forces of equal magnitude acting along the positive and negative x_3 directions at a small distance $\in$ apart in the x_2 direction. The two forces are $(0,0,F_3)$ acting at point $(\xi + \in e_2/2)$ and $(0,0,-F_3)$ acting at $(\xi - \in e_2/2)$, where e_2 is a unit vector in the x_2 direction. The total displacement u_i caused by the two forces is the sum of the displacements caused by each force (e.g., Pujol and Herrmann, 1990)

$$u_i = \in F_3 * [G_{i3}(\xi + \in e_2/2) - G_{i3}(\xi - \in e_2/2)] / \in . \qquad (3)$$

Taking the limit of u_i as F_3 tends to infinity and $\in$ tends to zero, in such a way that the product $\in F_3$ remains finite, the following relation is obtained

$$u_i = M_{32} * \frac{\partial G_{i3}}{\partial \xi_2}, \qquad (4)$$

where $M_{32} = \in F_3$. This pair of forces is known in classical mechanics as a couple and the quantity M_{32} as the moment of the couple, which has dimension of force by length and my be a function of time.

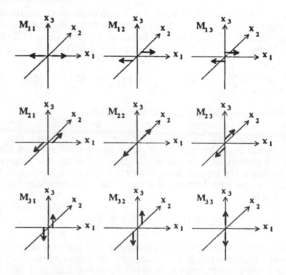

Figure 1. Representation of the nine possible couples M_{ij}. The subindexes i and j denote the directions of the force and the arm of the couple, respectively. (From Aki and Richards, 1980.)

There are nine possible combinations of force and arm directions, shown in Fig. 1, represented by the moment M_{ij} of couple with forces in the x_i direction and arm in the x_j direction. When x_i and x_j are the same, the couple is known as a vector dipol or a couple without moment. All the other couples have nonzero moment, equivalent to torque. If a general body force representing a seismic source can be expressed as a linear combination of couples with moments M_{ij}, then the displacement caused by this force is the sum of the displacements caused by individual couples

$$u_k = M_{ij} * \frac{\partial G_{ki}}{\partial \xi_j} = M_{ij} * G_{ki,j} \quad . \tag{5}$$

The set of nine terms M_{ij} is known as the moment tensor of the source, represented by a matrix $\mathbf{M}$ with elements M_{ij}. The full expression for u_k in (5) can be found, for example, in the book of Aki and Richards (1980).

The displacement u_k generated at a point $\mathbf{x}$ at the time t by a distribution of equivalent body force densities f_i within a source volume V is given by (e.g., Aki and Richards, 1980; Kennett, 1988; Jost and Herrmann, 1989)

$$u_k(\mathbf{x}, t) = \int_{-\infty}^{\infty} \int_V G_{ki}(\mathbf{x}, t; \mathbf{r}, t') f_i(\mathbf{r}, t') \, dV \, dt' \quad , \tag{6}$$

where $G_{ki}(\mathbf{x}, t; \mathbf{r}, t')$ are the Green's functions containing the propagation effects between the source $(\mathbf{r}, t')$ and the receiver $(\mathbf{x}, t)$. The Green's functions may be expanded in a Taylor series (Stump and Johnson, 1977) about the point $\mathbf{r} = \xi$ as follows

$$G_{ki}(\mathbf{x}, t; \mathbf{r}, t') = \sum_{n=0}^{\infty} \frac{1}{n!} (r_{j_1} - \xi_{j_1}) \cdots (r_{j_n} - \xi_{j_n})$$
$$\cdot G_{ki,j_1 \cdots j_n}(\mathbf{x}, t; \xi, t') \quad , \tag{7}$$

where the comma between indices describes partial derivatives with respect to the coordinates after the comma. The location of the reference point would normally be the hypocenter for small sources, whereas for extended faulting an improved representation is obtained by considering the centroid of the source. Defining the time dependent force moment tensor as

$$M_{ij_1 \cdots j_n}(\xi, t') = \int_V (r_{j_1} - \xi_{j_1}) \cdots (r_{j_n} - \xi_{j_n}) f_i(\mathbf{r}, t') \, dV \quad , \tag{8}$$

the displacement can be written as a sum of terms which resolve additional details of the source, known as multiple expansion (e.g., Backus and Mulcahy, 1976; Stump and Johnson, 1977; Aki and Richards, 1980; Jost and Herrmann, 1989)

$$u_k(\mathbf{x}, t) = \sum_{n=1}^{\infty} \frac{1}{n!} G_{ki,j_1 \cdots j_n}(\mathbf{x}, t; \xi, 0) * M_{ij_1 \cdots j_n}(\xi, t) \quad , \tag{9}$$

where $*$ denotes the temporal convolution.

In many applications a point source approximation may be quite satisfactory. Finite sources, on the other hand, may be generated by direct superposition of simple point sources. If the source dimensions are small in comparison to the observed wavelengths of seismic waves, only the first term in relation (9) needs to be considered, and then the displacement can be written as

$$u_k(\mathbf{x}, t) = G_{ki,j}(\mathbf{x}, t; 0, 0) * M_{ij}(0, t) \tag{10}$$

for $\xi = 0$. Assuming that all components of the time dependent seismic moment tensor have the same time dependence $s(t)$, the case known as synchronous source (Silver and Jordan, 1982), the displacement can be expressed as

$$u_k(\mathbf{x}, t) = M_{ij} [G_{ki,j} * s(t)] \quad , \tag{11}$$

where $s(t)$ is often called the source time function. Thus the displacement u_k is a linear function of the moment tensor elements and the terms in the square brackets. If the source time function is a delta function, the only term left in the square brackets is $G_{ki,j}$ describing nine generalized couples (e.g., Jost and Herrmann, 1989), shown in Fig. 1.

In general, the source moment tensor $\mathbf{M}$ of second order, describing a general dipolar source, has nine components M_{ij} and is represented as a 3×3 matrix in a given reference frame. It can be written as (Ben-Menahem and Singh, 1981)

$$\mathbf{M} = M_{ij} \mathbf{e}_i \mathbf{e}_j = \frac{1}{3} (M_{11} + M_{22} + M_{33}) (\mathbf{e}_1 \mathbf{e}_1 + \mathbf{e}_2 \mathbf{e}_2 + \mathbf{e}_3 \mathbf{e}_3)$$
$$+ \frac{1}{3} (2M_{11} - M_{22} - M_{33}) \mathbf{e}_1 \mathbf{e}_1 + \frac{1}{3} (2M_{22} - M_{33} - M_{11}) \mathbf{e}_2 \mathbf{e}_2$$
$$+ \frac{1}{3} (2M_{33} - M_{11} - M_{22}) \mathbf{e}_3 \mathbf{e}_3 + \frac{1}{2} (M_{32} + M_{23}) (\mathbf{e}_3 \mathbf{e}_2 + \mathbf{e}_2 \mathbf{e}_3)$$
$$+ \frac{1}{2} (M_{32} - M_{23}) (\mathbf{e}_3 \mathbf{e}_2 - \mathbf{e}_2 \mathbf{e}_3) + \frac{1}{2} (M_{13} + M_{31}) (\mathbf{e}_1 \mathbf{e}_3 + \mathbf{e}_3 \mathbf{e}_1)$$
$$+ \frac{1}{2} (M_{13} - M_{31}) (\mathbf{e}_1 \mathbf{e}_3 - \mathbf{e}_3 \mathbf{e}_1) + \frac{1}{2} (M_{21} + M_{12}) (\mathbf{e}_2 \mathbf{e}_1 + \mathbf{e}_1 \mathbf{e}_2)$$
$$+ \frac{1}{2} (M_{21} - M_{12}) (\mathbf{e}_2 \mathbf{e}_1 - \mathbf{e}_1 \mathbf{e}_2) \quad , \tag{12}$$

where $\mathbf{e}_i$ and $\mathbf{e}_j$ are the unit vectors along the x_i and x_j directions. The first term on the right hand side of this equation describes a center of compression and the successive terms describe three dipoles along the coordinate axes, three double couples and three torques about the coordinate axes, respectively. This is known as the decomposition theorem. The center of compression comes from the isotropic part of the moment tensor, corresponding to a volume change in the source. The remaining nine sources form the deviatoric part of the moment tensor. This deviatoric part can be further decomposed; a multitude of different decompositions are possible.

The conservation of angular momentum for the equivalent forces leads to the symmetry of the seismic moment tensor (Gilbert, 1970). If the moment tensor is symmetric, then $M_{ij} = M_{ji}$ and the torques in equation (12) vanish. The eigenvalues m_1, m_2, and m_3 of a symmetrical second-order tensor are all real and its eigenvectors $\mathbf{a}_1$, $\mathbf{a}_2$, and $\mathbf{a}_3$ are mutually orthogonal. Then from equation (12) it follows that a moment tensor can be decomposed into an isotropic part and three vector dipoles (Ben-Menahem and Singh, 1981; Jost and Herrmann, 1989)

$$\mathbf{M} = \frac{1}{3} (m_1 + m_2 + m_3) \mathbf{I} + \frac{1}{3} (2m_1 - m_2 - m_3) \mathbf{a}_1 \mathbf{a}_1$$
$$+ \frac{1}{3} (2m_2 - m_3 - m_1) \mathbf{a}_2 \mathbf{a}_2 + \frac{1}{3} (2m_3 - m_1 - m_2) \mathbf{a}_3 \mathbf{a}_3 \quad , \tag{13}$$

where $\mathbf{I} = \delta_{ij}$ is the identity matrix.

Equation (13) may also be written in the form

$$\mathbf{M} = \frac{1}{3} (m_1 + m_2 + m_3) \mathbf{I} + \frac{1}{3} m_1 (2 \mathbf{a}_1 \mathbf{a}_1 - \mathbf{a}_2 \mathbf{a}_2 - \mathbf{a}_3 \mathbf{a}_3)$$
$$+ \frac{1}{3} m_2 (2 \mathbf{a}_2 \mathbf{a}_2 - \mathbf{a}_3 \mathbf{a}_3 - \mathbf{a}_1 \mathbf{a}_1) + \frac{1}{3} m_3 (2 \mathbf{a}_3 \mathbf{a}_3 - \mathbf{a}_1 \mathbf{a}_1 - \mathbf{a}_2 \mathbf{a}_2) \quad , \tag{14}$$

where $2 \mathbf{a}_1 \mathbf{a}_1 - \mathbf{a}_2 \mathbf{a}_2 - \mathbf{a}_3 \mathbf{a}_3$ represents a compressional dipole of strength 2 in the direction of the eigenvector $\mathbf{a}_1$ and two dilatational dipoles each of unit strength along the $\mathbf{a}_2$ and $\mathbf{a}_3$ axes. This type of source is known as a compensated linear vector dipole (CLVD). A general dipolar source with a symmetric moment tensor, therefore, is equivalent to a center of compression and three mutually orthogonal compensated linear vector dipoles. A compensated linear vector dipole is equivalent to two double couples since

$$2 \mathbf{a}_1 \mathbf{a}_1 - \mathbf{a}_2 \mathbf{a}_2 - \mathbf{a}_3 \mathbf{a}_3 = (\mathbf{a}_1 \mathbf{a}_1 - \mathbf{a}_2 \mathbf{a}_2) + (\mathbf{a}_1 \mathbf{a}_1 - \mathbf{a}_3 \mathbf{a}_3); \tag{15}$$

and a double couple is given, for example, by $\mathbf{a}_1 \mathbf{a}_1 - \mathbf{a}_2 \mathbf{a}_2$.

Alternatively, a symmetric moment tensor can be decomposed into an isotropic part and three double couples. Another decomposition of a moment tensor is into an isotropic component and a major and minor double couple introduced by Kanamori and Given (1981). The major couple seems to be the best approximation of a general seismic source by a double couple, since the direction of the principal axes of the moment tensor remain unchanged (Jost and Herrmann, 1989). To construct the major double couple, the eigenvector of the smallest eigenvalue (in the absolute sense) is taken as the null axis, and it is assumed that the purely deviatoric eigenvalues m_i^d of the moment tensor

$$m_i^d = m_i - \frac{m_1 + m_2 + m_3}{3} \qquad (16)$$

are such that $|m_3^d| \geq |m_2^d| \geq |m_1^d|$. Then the complete decomposition can be written as (Jost and Herrmann, 1989)

$$\mathbf{M} = \frac{1}{3}(m_1 + m_2 + m_3)\mathbf{I}$$
$$+ m_3^d(\mathbf{a}_3\mathbf{a}_3 - \mathbf{a}_2\mathbf{a}_2) + m_1^d(\mathbf{a}_1\mathbf{a}_1 - \mathbf{a}_2\mathbf{a}_2), \qquad (17)$$

in which the second term represents the major double couple and the third term represents the minor couple. A best double couple can be constructed in a similar way by replacing m_3^d by the average of the largest (in the absolute sense) two eigenvalues (Giardini, 1984).

Following Knopoff and Randall (1970) and Fitch et al. (1980), a moment tensor can be decomposed into an isotropic part, a compensated linear vector dipole and a double couple. Assuming again that $|m_3^d| \geq |m_2^d| \geq |m_1^d|$ in relation (16) and that the same principal stresses produce the CLVD and the double couple radiation, the following decomposition is obtained (Jost and Herrmann, 1989)

$$\mathbf{M} = \frac{1}{3}(m_1 + m_2 + m_3)\mathbf{I} + m_3^d F(2\mathbf{a}_3\mathbf{a}_3 - \mathbf{a}_2\mathbf{a}_2 - \mathbf{a}_1\mathbf{a}_1)$$
$$+ m_3^d(1 - 2F)(\mathbf{a}_3\mathbf{a}_3 - \mathbf{a}_2\mathbf{a}_2), \qquad (18)$$

where $F = -m_1^d / m_3^d$.

Such a decomposition seems to be the most interesting one for source studies of seismic events induced by mining. The compensated linear vector dipole, corresponding to a uniaxial compression, and the double couple and the corresponding double dipole system that gives the same radiation pattern as the double couple are shown in Fig. 2. The CLVD source was considered as a model for sudden phase transitions in deep earthquakes (Knopoff and Randall, 1970), tensile failure of rock in the presence of high-pressure fluids (Julian and Sipkin, 1985; Foulger, 1988), and source complexity (e.g., Frohlich et al., 1989). The CLVD source corresponding to a uniaxial compression could possibly explain one of the mechanisms of pillar-associated seismic events, observed in situ in deep hard rock mines in South Africa and reported by Lenhardt and Hagan (1990). The other reported mechanism could be explained by a source composed of the CLVD and double couple components. The four possible mechanisms of pillar-associated seismic events at the Western Deep Levels gold mine in South Africa are shown in Fig. 3, reproduced from Lenhardt and Hagan (1990).

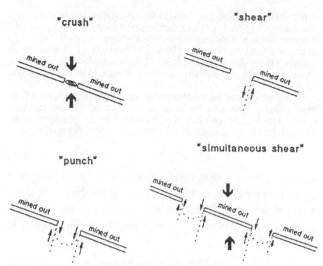

Figure 3. Four possible mechanisms of pillar-associated seismic events at the Western Deep Levels gold mine, South Africa. (From Lenhardt and Hagan, 1990.)

3 MOMENT TENSOR INVERSION

There are various methods of inversion for moment tensor elements. The inversion can be done in the time or frequency domain, and different data can be used separately or in combination. The moment tensor inversion in the time domain can be based on the formulation described by equation (11) (e.g., Gilbert, 1970; Stump and Johnson, 1977; Strelitz, 1978; Fitch et al., 1980). If the source time function is not known or cannot be assessed or the assumption of a synchronous source is not upheld, the frequency domain approach is chosen (e.g., Gilbert, 1973; Stump and Johnson, 1977; Kanamori and Given, 1981). The displacement in the frequency domain, corresponding to the formulation in (11), can be written as

$$u_k(\mathbf{x}, t) = M_{ij}(f)\, G_{ki,j}(f) \qquad (19)$$

for each frequency f. Both approaches (11) and (19) lead to linear inversions in the time or frequency domain, respectively, for which a number of fast computational algorithms are available (e.g., Lawson and Hanson, 1974; Press et al., 1989).

Both equations either (11) or (19) can be written in a matrix form

$$\mathbf{u} = \mathbf{G}\,\mathbf{m} \, . \qquad (20)$$

In the time domain, the vector $\mathbf{u}$ consists of n sampled values of the observed ground displacement at various stations, $\mathbf{G}$ is a $n \times 6$ matrix containing the Green's functions calculated using an appropriate algorithm and earth model, and $\mathbf{m} = (M_{11}, M_{12}, M_{22}, M_{13}, M_{23}, M_{33})$ is a vector containing the six moment tensor elements to be determined. In the frequency domain, equations (20) are written separately for each frequency. The vector $\mathbf{u}$ consists of real and imaginary parts of the displacement spectra, the matrix $\mathbf{G}$ and the vector $\mathbf{m}$ contain real and imaginary parts as well, and $\mathbf{m}$ contains also the transform of the source time function of each moment tensor element. The details of solving equations (20) for $\mathbf{m}$ are given by Aki and Richards (1980). A detailed description of the procedure for regional and local seismograms is given by Oncescu (1986). The application of moment-tensor inversion to microseismic events is described by O'Connell and Johnson (1988).

The main difficulty in the moment tensor inversion is a proper calculation of Green's functions for geologically complex media. The Green's function is in general different for different displacement components and takes different values for particular stations. The simplest approach in the time domain is to use directly the source radiation formulation for P, SV or SH waves. This approach was used by Fitch et al. (1980) and De Natale et al. (1987), and others. A more rigorous approach is based on the method of matrix propagators of Haskell-Thomson (Haskell, 1953), modified by Knopoff (1964) and Dunkin (1965), and used in the frequency domain. Another method of evaluation of Green's functions, especially valuable for highly complex struc-

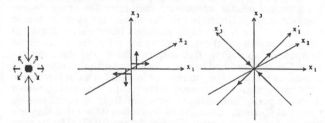

Figure 2. The compensated linear vector dipole (CLVD), corresponding to a uniaxial compression, and the double couple and the corresponding double dipole system that gives the same radiation pattern as the double couple.

tures including possible lateral inhomogeneities, is empirical one using Green's functions determined from observations and a known source. This relative moment tensor determination was first proposed by Strelitz (1980) in a study of subevents of complex deep-focus earthquakes. The method was extended by Oncescu (1986) to individual small events recorded at a few stations.

It is convenient to characterize the moment tensor by its eigenvalues. This can be done by a rotation of the moment tensor from geographical coordinates into its principal axes. Then the moment tensor can be written in the diagonal form

$$M_{ij} = m_i \delta_{ij} , \qquad (21)$$

where m_i are the eigenvalues of $\mathbf{M}$ and the Kronecker delta $\delta_{ij} = 1$ for $i = j$ and $\delta_{ij} = 0$ for $i \neq j$. For a general moment tensor all eigenvalues m_i are different. Seismic sources with no volume change can be obtained by constraining the moment tensor to have zero trace

$$\mathrm{tr}\,\mathbf{M} = m_1 + m_2 + m_3 = 0 , \qquad (22)$$

or in a more general form

$$\mathrm{tr}\,\mathbf{M} = M_{11} + M_{22} + M_{33} = 0 . \qquad (23)$$

The sum of the diagonal elements of the moment tensor divided by 3 is a measure of the volume change associated with the source.

It can be readily shown that for the moment tensor of the double-couple source, one principal value of $\mathbf{M}$ must vanish which means that the determinant det $\mathbf{M}$ must also vanish

$$\det \mathbf{M} = m_1 m_2 m_3 = 0 , \qquad (24)$$

or in a general form

$$\det \mathbf{M} = M_{11} M_{22} M_{33} + 2 M_{12} M_{23} M_{13} - M_{11} M_{23}^2$$

$$- M_{22} M_{13}^2 - M_{33} M_{12}^2 = 0 . \qquad (25)$$

The vanishing of det $\mathbf{M}$ and tr $\mathbf{M}$ are therefore necessary and sufficient conditions for a double-couple source.

The non-isotropic constraint of zero trace on the moment tensor is linear, whereas for double-couple sources the constraint of zero determinant on the moment tensor is nonlinear. To solve the linear system of equations (20) under these constraints, the method of Lagrange multipliers is used (Strelitz, 1980; Oncescu, 1986). The system must be solved iteratively until the determinant det $\mathbf{M}$ and the trace tr $\mathbf{M}$ converge to zero. The scalar seismic moment M_0 can be determined from a given moment tensor, corresponding to a double-couple source, by

$$M_0 = \frac{1}{2} \left(|m_1| + |m_2| \right) , \qquad (26)$$

where m_1 and m_2 are the largest eigenvalues in the absolute sense. The seismic moment can equivalently be estimated by the following relations (Silver and Jordan, 1982)

$$M_0 = \left(\frac{\sum M_{ij}^2}{2} \right)^{1/2} = \left(\frac{\sum m_i^2}{2} \right)^{1/2} . \qquad (27)$$

After the recovery of moment tensor, the deviation of the solution from the pure double-couple model can be evaluated from the ratio (Dziewonski et al., 1981)

$$\in = \frac{|m^d|_{\min}}{|m^d|_{\max}} , \qquad (28)$$

where $|m^d|_{\min}$ is the smallest and $|m^d|_{\max}$ is the largest deviatoric eigenvalue in the absolute sense. The values of this ratio can range from 0 for a pure double-couple source to 0.5 for a pure compensated linear vector dipole. Silver and Jordan (1982)

have developed a method for the estimation of the isotropic and deviatoric components of the moment tensor, introducing the isotropic, deviatoric and total scalar seismic moments. Graphical methods have been recently suggested for identifying non-double-couple moment-tensor components (Pearce et al., 1988; Hudson et al., 1989; Riedesal and Jordan, 1989), and a method for the exact mapping of error bounds on seismic waveforms into bounds on certain moment-tensor properties was presented by Vasco (1990).

In general, moment tensor inversions involve two major assumptions. First, it is assumed that the earthquake may be treated as a point source for a given frequency of seismic waves; second, that the effect of the earth structure on the seismic waves is properly modeled. If the earthquake cannot be represented as a point source or the assumed model of structure is incorrect, the apparent moment tensor may contain a large non-double-couple component, even if the source mechanism is a double couple (Strelitz, 1978; Barker and Langston, 1982). Increasing the complexity of the source structure model, improving the azimuth coverage, and leaving the time function free to compensate for the deficiences of the Greeen's functions decrease the size of the non-double-couple component (Johnston and Langston, 1984).

4 SOURCE MECHANISM OF SEISMIC EVENTS IN MINES

Moment tensor inversions have been routinely performed for several years by the U. S. Geological Survey, and centroid-moment tensor solutions (simultaneous inversion of the waveform data for the hypocentral parameters of the best point source and for the six independent elements of the moment tensor) are regularly published by the Harvard University group for all larger earthquakes recorded at teleseismic distances. The application of a moment tensor inversion technique to local events, however, is a relatively recent innovation (e.g., Saikia and Herrmann, 1985, 1986; Oncescu, 1986; De Natale et al., 1987; O'Connell and Johnson, 1988). Most recently, Ebel and Bonjer (1990) performed moment tensor inversion of small earthquakes in southwestern Germany, and Koch (1991a) examined two methods for moment tensor inversion of waveform data for applicability to high-frequency near-source data. Both these methods, one in the time domain and one in the frequency domain, allow the retrieval of the complete time-dependent moment tensor (Koch, 1991b). Ohtsu (1991) applied moment tensor analysis to acoustic emission recorded during an in-situ hydrofracturing test. He used moment tensor components to classify crack types and to determine crack orientations.

A few works only have been published that are related to the use of moment tensor inversion in studies of the source mechanism of seismic events induced by mining. Spottiswoode (1984) has studied the focal mechanism of 11 mine tremors at Blyvooruitzicht gold mine, South Africa, in the frequency domain, and he found that the data were consistent with zero volume change in the seismic source area and were then interpreted as shear failures on plane striking parallel to the advancing face or to either of two dykes cutting across the face.

Sato and Fujii (1989) have studied the source mechanism of a large-scale gas outburst at Sunagawa coal mine in Japan, which occurred in January 1986. They used a new method to evaluate the moment tensor in the frequency domain and applied it to 15 seismic events recorded by the mine underground seismic network. The procedure consists of two steps. In the first step an iterative least squares method was used to determine the quality factor Q (representing attenuation and scattering effects) and the apparent seismic moment for each record from the P-wave displacement spectrum. In the second step the moment tensor was determined from simple relations between the apparent seismic moment and the moment tensor, taking into account the geometrical spreading, free-surface effect, and direction cosines. Out of 15 studied tremors associated with the outburst, 12 seismic events could be interpreted in terms of a double-couple focal mechanism. In contrast to these results, the moment tensor inversion performed on the observations from two small seismic events at Horonai coal mine in Japan has shown that they are non-double-couple events (Fujii and Sato, 1990). The tremors were associated with longwall mining and were located in the vicinity of the longwall face.

The Integrated Seismic System (ISS) recently introduced into the Welkom seismological network in South Africa (Mendecki,

1990; Mendecki et al., 1990) includes a software package which calculates all components of the higher order moment tensor. Calculations are done in the frequency domain using maximum entropy method for the inversion. The use of higher than second order moment tensors permits to assess a number of source properties, besides those provided by the inversion of a standard second order moment tensor, such as the direction of the rupture propagation, the rupture velocity, duration and size, and the source geometry in terms of the orientation of the plane of rupture and overall shape of the source. Although a few examples of the higher moment tensor inversions for mine tremors were given by Mendecki (1990), no systematic studies of this complex problem have been published so far.

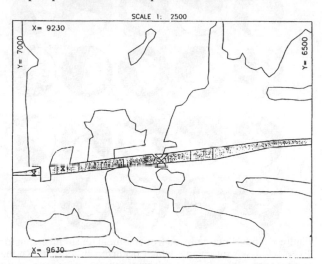

Figure 4. Location of the main seismic event (marked by a large double triangle) and two aftershocks (small double triangles) in a vertical dyke (shaded area), which occurred on April 9, 1991 in a mine in the Orange Free State mining district, South Africa. Their source is characterized by large isotropic components.

Inversions of a standard second order moment tensor are routinely performed within the ISS Welkom system in South Africa, though the results are not readily available in the professional literature. Moment tensor inversions for three seismic events with volume sources, which occurred within a dyke in a mine in the Orange Free State mining district on April 9, 1991, provide good examples to illustrate the kind of information becoming available from moment tensor analyses. The main event with moment magnitude $M = 2.1$ occurred at 07^h09^m and two aftershocks with the same moment magnitude $M = 1.8$ occurred at 09^h24^m and 12^h27^m, respectively. Their location in a dyke is shown in Fig. 5, elaborated by G. van Aswegen from Western Deep Levels mine. The dyke is vertical and strikes EW. The reef being mined is planar and dips 16 degrees easterly. The main event created intense but highly localized damage, and the dyke "exploded". The routine moment tensor inversion provided the following results (in a diagonalized form in coordinate system defined by eigenvectors):

MOMENT TENSOR   ISOTROPIC
$$\begin{bmatrix} 0.14 & 0.00 & 0.00 \\ 0.00 & -0.06 & 0.00 \\ 0.00 & 0.00 & 1.41 \end{bmatrix} = 0.494 \begin{bmatrix} 1 & 0 & 0 \\ 0 & 1 & 0 \\ 0 & 0 & 1 \end{bmatrix}$$

CLVD   DC
$$+ 0.355 \begin{bmatrix} -1 & 0 & 0 \\ 0 & -1 & 0 \\ 0 & 0 & 2 \end{bmatrix} + 0.202 \begin{bmatrix} 0 & 0 & 0 \\ 0 & -1 & 0 \\ 0 & 0 & 1 \end{bmatrix}$$

with the largest isotropic component corresponding to extension and the smallest double-couple (DC) component.

Two aftershocks occurred also in the dyke. Damage during the first aftershock was more wide-spread than that during the main event, but it was less intense. The moment tensor inversion shows more shear than volume change corresponding this time to contraction:

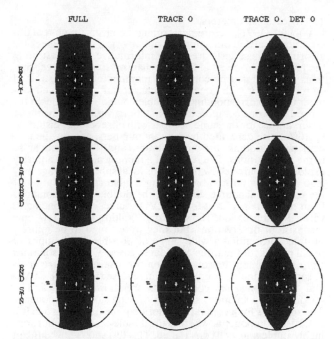

Figure 5. Numerical test of the moment tensor inversion for a seismic source composed of the double-couple (dip-slip reverse fault), 10% CLVD and 10% explosional components; a lower hemisphere equal-area projection is used. Shaded areas represent the regions of up motion for P waves (individual observations marked by +) and unfilled areas represent the regions of down motion (observations marked by −). Three solutions were sought for a fixed network composed of 24 stations with exact synthetic observations, for the same network with observations disturbed by 10% noise generated by random numbers, and for the network with randomly distributed stations and 10% random noise in the observations (solutions are shown in three horizontal rows). These three solutions are: for a general six-free-component moment tensor, for a constrained moment tensor corresponding to sources with no volume changes, and for a constrained moment tensor corresponding to double couple sources; they are shown in three vertical columns. (From Wiejacz, 1991.)

MOMENT TENSOR   ISOTROPIC
$$\begin{bmatrix} -0.44 & 0.00 & 0.00 \\ 0.00 & -1.13 & 0.00 \\ 0.00 & 0.00 & 0.73 \end{bmatrix} = -0.281 \begin{bmatrix} 1 & 0 & 0 \\ 0 & 1 & 0 \\ 0 & 0 & 1 \end{bmatrix}$$

CLVD   DC
$$+ 0.162 \begin{bmatrix} -1 & 0 & 0 \\ 0 & -1 & 0 \\ 0 & 0 & 2 \end{bmatrix} + 0.685 \begin{bmatrix} 0 & 0 & 0 \\ 0 & -1 & 0 \\ 0 & 0 & 1 \end{bmatrix}$$

The third event caused no apparent damage and is characterized by even larger double-couple component and smaller volume change corresponding to extension:

MOMENT TENSOR   ISOTROPIC
$$\begin{bmatrix} 0.15 & 0.00 & 0.00 \\ 0.00 & -0.66 & 0.00 \\ 0.00 & 0.00 & 1.21 \end{bmatrix} = 0.241 \begin{bmatrix} 1 & 0 & 0 \\ 0 & 1 & 0 \\ 0 & 0 & 1 \end{bmatrix}$$

CLVD   DC
$$+ 0.095 \begin{bmatrix} -1 & 0 & 0 \\ 0 & -1 & 0 \\ 0 & 0 & 2 \end{bmatrix} + 0.809 \begin{bmatrix} 0 & 0 & 0 \\ 0 & -1 & 0 \\ 0 & 0 & 1 \end{bmatrix}$$

Wiejacz (1991, 1992) has studied the source mechanism of 60 small seismic events (in the seismic moment range $10^{11} - 10^{12}$ N·m) which occurred in 1990 and 1991 at Rudna copper mine in the Lubin mining district in Poland. He performed the moment tensor inversion in the time domain using the first motion amplitudes and signs of P and the amplitudes of SV waves recorded by the mine underground network composed of over

20 vertical seismometers.

Wiejacz (1991), before performing the inversion on real observations, carried out a number of numerical tests to check his algorithm. These tests are highly informative, well illustrating the problems involved in focal mechanism studies. Three solutions – general six-free-component moment tensor, constrained moment tensor with no volume changes, and constrained moment tensor corresponding to double couple – were sought for a fixed network composed of 24 stations with exact synthetic observations, for the same network with observations disturbed by 10% noise generated by random numbers, and for the network with randomly distributed stations and 10% random noise in the observations. The results of a numerical test for a purely explosional source show that a poor focal sphere coverage may lead even to a not-too-bad-looking double-couple solution of the explosional source. Similarly, the double-couple solutions from the numerical test for a CLVD source, corresponding to vertical compression, show normal faulting with a few only inconsistent observations, but they were found to be entirely unstable presenting a wide range of possible distributions of nodal planes for the three considered cases. The nine solutions from the moment tensor inversion for a seismic source composed of the double couple (dip-slip reverse fault), 10% CLVD and 10% explosional components are shown in Fig. 5. The unconstrained solutions show 8% of explosional components. The solutions for a general deviatoric source have from 5 to 9% of CLVD components, whereas the double-couple solutions are highly stable and well constrained. This test shows how difficult it is to detect the non-double-couple components when they are relatively small and the double-couple sources are dominant.

The horizontal distribution of seismic events and seismic stations at Rudna copper mine, selected by Wiejacz (1991) for the moment tensor inversion, is shown in Fig. 6. The time-independent solutions were obtained for a general six-free-component moment tensor, constrained solutions corresponding to sources without volume changes, and constrained solutions corresponding to double-couple sources. Examples of such solutions for 10 selected events are shown in Fig. 7. In general, the solutions that are well constrained by observations (good coverage of the focal sphere) have dominant shear components, though occasionally the isotropic component (showing both extension and contraction) could be as large as 25 percent of the mechanism. The CLVD component corresponds to uniaxial compression in all cases and is usually larger than the isotropic component.

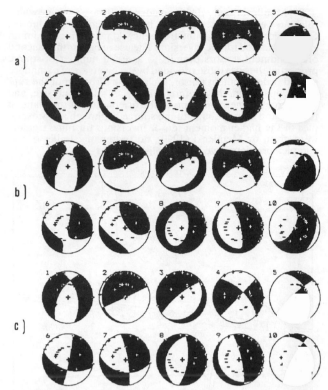

Figure 7. Nodal lines deduced from moment tensor inversion for 10 selected seismic events which occurred in 1991 at Rudna copper mine, Poland. Three solutions are shown for each event: (a) for a general six-free-component moment tensor, (b) for a constrained moment tensor corresponding to sources with no volume changes, and (c) for a constrained moment tensor corresponding to double-couple sources. (From Wiejacz, 1991.)

boratory in Pinawa, Manitoba. From the moment tensor inversion they obtained the ratio of isotropic to deviatoric components and they found a number of explosional and implosional sources. Furthermore, the location of events displaying extensional components corresponded to a breakout observed in the roof of the tunnel. Although they also found a number of purely deviatoric sources, no attempt was made to decompose the deviatoric moment tensors into, for example, the CLVD and double-couple components. Thus the presence of purely shear seismic events could not be detected.

Most recently, McGarr (1992a,b) reported three tremors in the magnitude range from 2.1 to 3.4, recorded in early 1988 on the surface and at an underground station in one of the major gold mines in South Africa, with seismic moment tensors having substantial implosional components. In the mines generating these tremors the subhorizontal tabular ore bodies are offset (typically by several hundred meters) by major faults. Mining in the vicinity of the faults stimulates seismicity resulting in renewed fault slip as well as excavation closure, which manifests as a volumetric contribution to seismic moment tensors.

5 CONCLUSIONS

Moment tensor inversion, as long as the seismic source can be considered as a point source, is probably the best approach to study the mode of failure of seismic events induced by mining. The immediate proximity of openings in underground mines creates favorable conditions for generation of non-shearing seismic events, especially in the stope area. Decomposition of a moment tensor into an isotropic part corresponding to volumetric changes, a compensated linear vector dipole corresponding to a uniaxial compression or tension, and a double couple corresponding to shearing seems to be the most interesting one for source studies in mines, especially where pillar-associated seismic events are observed. It should be noted, however, that relatively small non-shearing components of the source mechanism are difficult to detect; the dominant shear component leading to well-constrained classical double-couple solutions.

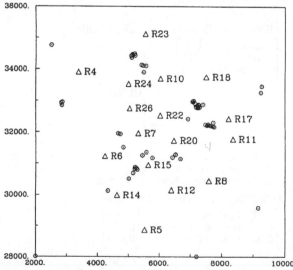

Figure 6. Horizontal distribution of selected seismic events (marked by circles) which occurred in 1990 and 1991 and seismic stations (marked by triangles) used for the moment tensor inversion at Rudna copper mine, Poland. (From Wiejacz, 1991.)

Another case of the application of moment tensor inversion to study the mode of failure of small seismic events induced by mining was recently reported by Feignier and Young (1992). They studied 33 microevents (in the $-4 < M < -2$ moment magnitude range) induced by drilling a tunnel at 420 m depth in the Canadian shield granite at the Underground Research La-

REFERENCES

Aki, K. & P.G. Richards 1980. *Quantitative seismology. Theory and methods.* San Francisco: W. H. Freeman and Company.

Backus, G.E. & M. Mulcahy 1976. Moment tensors and other phenomenological descriptions of seismic sources. I.- Continuous displacements. *Geophys. J. R. Astr. Soc.* 46: 341-371.

Barker, J.S. & C.A. Langston 1982. Moment tensor inversion of complex earthquakes. *Geophys. J. R. Astr. Soc.* 68: 777-803.

Ben-Menahem, A. & S.J. Singh 1981. *Seismic waves and sources.* New York: Springer-Verlag.

De Natale, G., G. Iannaccone, M. Martini & A. Zollo 1987. Seismic sources and attenuation properties at the Campi Flegrei volcanic area. *Pure Appl. Geophys.* 125: 883-917.

Dunkin, J.W. 1965. Computation of modal solutions in layered elastic media at high frequencies. *Bull. Seism. Soc. Am.* 55: 335-358.

Dziewonski, A.M., T.-A. Chou & J.H. Woodhouse 1981. Determination of earthquake source parameters from waveform data for studies of global and regional seismicity. *J. Geophys. Res.* 86: 2825-2852.

Ebel, J.E. & K.-P. Bonjer 1990. Moment tensor inversion of small earthquakes in southwestern Germany for the fault plane solution. *Geophys. J. Int.* 101: 133-146.

Feignier, B. & R.P. Young, 1992. Moment tensor inversion of induced microseismic events: Evidence of non-shear failures in the $-4<M<-2$ moment magnitude range. *Geophys. Res. Lett.* 19: 1503-1506.

Fitch, T.J., D.W. McCowan & M.W. Shields 1980. Estimation of seismic moment tensor from teleseismic body wave data with application to intraplate and mantle earthquakes. *J. Geophys. Res.* 85: 3817-3828.

Foulger, G. R. 1988. Hengill triple junction, SW Iceland. 2. Anomalous earthquake focal mechanisms and implications for process within the geothermal reservoir and at accretionary plate boundaries. *J. Geophys. Res.* 93: 13,507-13,523.

Frohlich, C., M.A. Riedesel & K.D. Apperson 1989. Note concerning possible mechanisms for non-double-couple earthquake sources. *Geophys. Res. Lett.* 16: 523-526.

Fujii, Y. & K. Sato 1990. Difference in seismic moment tensors between microseismic events associated with a gas outburst and those induced by longwall mining activity. *Proc. 2nd Intern. Symp. on Rockbursts and Seismicity in Mines*: 71-75. Rotterdam: Balkema.

Giardini, D. 1984. Systematic analysis of deep seismicity: 200 centroid-moment tensor solutions for earthquakes between 1977 and 1980. *Geophys. J. R. Astr. Soc.* 77: 883-914.

Gibowicz, S.J. 1990. Seismicity induced by mining. *Adv. Geophys.* 32: 1-74.

Gilbert, F. 1970. Excitation of the normal modes of the earth by earthquake sources. *Geophys. J. R. Astr. Soc.* 22: 223-226.

Gilbert, F. 1973. Derivation of source parameters from low-frequency spectra. *Phil. Trans. R. Soc.* A 274: 369-371.

Haskell, N.A. 1953. The dispersion of surface waves in multilayered media. *Bull. Seism. Soc. Am.* 43: 17-34.

Hudson, J.A., R.G. Pearce & R.M. Rogers 1989. Source type plot for inversion of the moment tensor. *J. Geophys. Res.* 94: 765-774.

Johnston, D.E. & C.A. Langston 1984. The effect of assumed source structure on inversion of earthquake source parameters: The eastern Hispaniola earthquake of 14 September 1981. *Bull. Seism. Soc. Am.* 74: 2115-2134.

Jost, M.L. & R.B. Herrmann 1989. A student's guide to and review of moment tensors. *Seism. Res. Lett.* 60: 37-57.

Julian, B.R. & S.A. Sipkin 1985. Earthquake processes in the Long Valley caldera area, California. *J. Geophys. Res.* 90: 11,155-11,169.

Kanamori, H. & J.W. Given 1981. Use of long-period surface waves for rapid determination of earthquake source-parameters. *Phys. Earth Planet. Interiors* 27: 8-31.

Kennett, B.L.N. 1988. Radiation from a moment-tensor source. In D. J. Doornbos (ed.), *Seismological algorithms*, p. 427-441. London: Academic Press.

Knopoff, L. 1964. A matrix method for elastic wave problems. *Bull. Seism. Soc. Am.* 54: 431-438.

Knopoff, L. & M.J. Randall 1970. The compensated linear-vector dipole: A possible mechanism for deep earthquakes. *J. Geophys. Res.* 75: 1957-1963.

Koch, K. 1991a. Moment tensor inversion of local earthquake data – I. Investigation of the method and its numerical stability with model calculations. *Geophys. J. Int.* 106: 305-319.

Koch, K. 1991b. Moment tensor inversion of local earthquake data – II. Application to aftershocks of the May 1980 Mammoth Lakes earthquakes. *Geophys. J. Int.* 106: 321-332.

Lawson, C.L. & R.J. Hanson 1974. *Solving least square problems.* Englewood Cliffs: Prentice-Hall.

Lenhardt, W.A. & T.O. Hagan 1990. Observations and possible mechanisms of pillar-associated seismicity at great depth. In *Technical challenges in deep level mining*, p. 1183-1194. Johannesburg: S. Afric. Inst. Min. Metal.

McGarr, A. 1992a. An implosive component in the seismic moment tensor of a mining-induced tremor. *Geophys. Res. Lett.* 19: 1579-1582.

McGarr, A. 1992b. Seismic moment tensors with well-defined implosional components (abstract). *Papers submitted to Workshop on Induced Seismicity*, p. 116. Santa Fe, New Mexico, June 10, 1992.

Mendecki, A.J. 1990. The Integrated Seismic System (ISS). *Presented at the Seminar on Monitoring and Safety in Civil and Mining Engineering*, Nancy, France, June 1990.

Mendecki, A.J., G. van Aswegen, J.N.R. Brown & P. Hewlett 1990. The Welkom seismological network. *Proc. 2nd Intern. Symp. on Rockbursts and Seismicity in Mines*: p. 237-243. Rotterdam: Balkema.

O'Connell, D.R.H. & L.R. Johnson 1988. Second-order moment tensors of microearthquakes at The Geysers geothermal field, California. *Bull. Seism. Soc. Am.* 78: 1674-1692.

Ohtsu, M. 1991. Simplified moment tensor analysis and unified decomposition of acoustic emission source: Application to in situ hydrofracturing test. *J. Geophys. Res.* 96: 6211-6222.

Oncescu, M.C. 1986. Relative seismic moment tensor determination for Vrancea intermediate depth earthquakes. *Pure Appl. Geophys.* 124: 931-940.

Pearce, R.G., J.A. Hudson & A. Douglas 1988. On the use of P-wave seismograms to identify a double-couple source. *Bull. Seism. Soc. Am.* 78: 651-671.

Press, W.H., B.P. Flannery, S.A. Teukolsky & W.T. Vetterling 1989. *Numerical recipes: The art of scientific computing.* New York: Cambridge University Press.

Pujol, J. & R.B. Herrmann 1990. A student's guide to point sources in homogeneous media. *Seismol. Res. Lett.* 61: 209-224.

Riedesel, M.A. & T.H. Jordan 1989. Display and assessment of seismic moment tensors. *Bull. Seism. Soc. Am.* 79: 85-100.

Saikia, C.K. & R.B. Herrmann 1985. Application of waveform modeling to determine focal mechanisms of four 1982 Miramichi aftershocks. *Bull. Seism. Soc. Am.* 75: 1021-1040.

Saikia, C.K. & R.B. Herrmann 1986. Moment-tensor solutions for three 1982 Arkansas swarm earthquakes by waveform modeling. *Bull. Seism. Soc. Am.* 76: 709-723.

Sato, K. & Y. Fujii 1989. Source mechanism of a large scale gas outburst at Sunagawa coal mine in Japan. *Pure Appl. Geophys.* 129: 325-343.

Silver, P.G. & T.H. Jordan 1982. Optimal estimation of scalar seismic moment. *Geophys. J. R. Astr. Soc.* 70: 755-787.

Spottiswoode, S.M. 1984. Source mechanisms of mine tremors at Blyvooruitzicht gold mine. In N.C. Gay & E.H. Wainwright (eds.), *Rockbursts and Seismicity in Mines*, Symp. Ser. No.6, p. 29-37. Johannesburg: S. Afr. Inst. Min. Metal.

Strelitz, R.A. 1978. Moment tensor inversions and source models. *Geophys. J. R. Astr. Soc.* 52: 359-364.

Strelitz, R.A. 1980. The fate of downgoing slabe: A study of the moment tensors from body waves of complex deep-focus earthquakes. *Phys. Earth Planet. Interiors* 21: 83-96.

Stump, B.W. & L.R. Johnson 1977. The determination of source properties by the linear inversion of seismograms. *Bull. Seism. Soc. Am.* 67: 1489-1502.

Vasco, D.W. 1990. Moment-tensor invariants: Searching for non-double-couple earthquakes. *Bull. Seism. Soc. Am.* 80: 354-371.

Wiejacz, P. 1991. *Investigation of focal mechanisms of mine tremors by the moment tensor inversion.* Ph.D. Theses, Inst. Geophys., Pol. Acad. Sci., Warsaw.

Wiejacz, P. 1992. Calculation of seismic moment tensor for mine tremors from the Legnica-Głogów copper basin. *Acta Geophys. Pol.* 40: 103-122.

Rockbursts and Seismicity in Mines, Young (ed.) © 1993 Balkema, Rotterdam, ISBN 90 5410 320 5

Keynote address: Recent Polish and Czechoslovakian rockburst research and the application of stochastic methods in mine seismology

V. Rudajev

Institute of Geotechnics, Academy of Sciences of Czech Republic, Prague, Czechia

Abstract: Long term cooperation between Czech and Polish specialists in the field of rockburst research resulted in the establishment of a joint scientific meeting with 23 years of tradition. Discussion at the 1992 meeting included common problems, methods, interpretation and new physical approaches, specifically in the following topics: location of foci - in which the contemporary and most sophisticated method utilizes seismic tomography, and foci mechanism - in which a new combined shear/implosive model has been verified. Non-linear processes in sources have been discovered. Prognoses - multichannel site/time/energy of rockbursts occurrence has been investigated.

HISTORICAL REVIEW OF ROCKBURSTS IN POLISH AND CZECH MINES

Although rockbursts were observed in some Czech mines in the second half of the past century (for example 237 rock bumps were macroscopically documented in the Kladno black coal mines within the period from 1880-94, archive documents), their systematic research did not start until the end of the nineteen fifties and has been connected with the construction of permanent local seismic stations. The first mine station (mechanical seismograph Wiechert) in Bohemia was installed in the years of 1903-05 in the ore mine of Příbram (Stauch 1905), unfortunately the records have not been preserved to our time.

A systematic evidence of rockbursts in the Polish Upper-Silesian black-coal basin began with the construction of the local station Raciborz in 1929 (Wierzchowska 1959, 1961, 1963). In the year 1950, the seismic observations were expanded to a further four stations, equipped with photogalvanometric seismographs.

The rockbursts occur in both the coal and the ore mines. The depth of mine workings is not necessarily a factor for the occurrence of rockbursts. This is documented, for example, by the occurrence of seismic events induced by mining in the open-pit mine of Belchatów. During the last years, the following mines were the most endangered by rockbursts:

1. Lubin Cuprum Mine - Poland. The extracted copper ore is situated in the Fore-Sudetic Monocline and the mine region is situated near the Odra Fault Zone. Mining operations started in this region in 1968. The depth of stope is about 850 m. The mine is equipped with underground seismic stations. (Kazimierczyk 1979).

2. Belchatów brown-coal open pit mines - Poland. The mining area is failed by a system of tectonic faults, which are characteristic of the Teisseyre-Tornquist Zone. The mining operations started in 1976 (Belchatów - Mine Field 2), the mined thickness is 100 m, strip width 2 km and its length 1 km (Gibowicz 1985).

3. Polish mines of the Upper Silesian Basin. Mining operations began more than 100 years ago. The black coal seams form a series of levels and are affected by tectonic faults. The mines are equipped with both underground and surface seismic networks. The depth of stope is about 700-1000 m. (Wierzchowska 1974).

4. Ostrava-Karviná basin (OKR), Czech Republic. This region is a part of the Upper-Silesian black-coal basin and its geological and tectonic situation is very complicated. The mining works are situated at depths of 700-1000 m. At these mines there have been constructed:

a) the so-called seismic and seismoacoustic micronetworks (distance of stations a few hundred metres),

b) local networks (combined underground and surface stations at distances of a few kilometres),

c) seismic polygons, in which the stations are mostly situated on the outer border of the basin of Ostrava-Karviná (Slavik 1992).

5. Kladno basin. The region contains a single black coal seam with average thickness of 6 m. It is horizontally deposited at a depth of 450-500 m and is failed by tectonic faults which cause throws of up to 40 m. The hanging wall is formed by sediments (claystones, conglomerates, and sandstones) from which the sandstone banks, some with thicknesses of up to 40 m are most dangerous for the occurrence of rockbursts. A characteristic feature of this region is a tertiary volcanic effusion which causes a significant increase in the horizontal components of the stress tensor (Buben 1962).

6. Příbram Ore Mines. This region contains some of the oldest Czech mines, where the mining of silver, and later of lead and other metal ores started nearly 1000 years ago. The polymetallic ore is contained in igneous rock veins with almost vertical strike. Rockbursts started to occur, when a depth of about 800 m was reached. The mines were closed in 1978, when the stope depth was about 1,500 m. After closing the mines, the rockbursting activity practically ceased, until the time when the water level depth in the flooded mines attained 900 m. At that time, three strong rockbursts were recorded with a local magnitude of about 2.5 and epicentral intensity $I_0 = 4 - 5°$ on the Medvědev-Sponheuer-Kárník 1964 (MSK-64) intensity scale. The reason for these rockbursts can be explained not only by the accumulation of energy in the predisposed parts of the rock mass, but also by the alteration of frictional conditions on the surfaces of the discontinuities. (Rudajev 1985).

7. Příbram Uranium mines. Příbram Uranium deposits belong to the belt of ore mineralization in middle Bohemia and are connected with the course of the old structural line - the so called middle Bohemian suture, which forms a border between the Barrandien and Moldanubical block of the Czech massif. The dimensions of the uranium belt are 25 km in length and 2 km in width. The rock massif is weakened by numerous tectonic disturbances with throws of several metres up to several hundreds of metres. Maximum depths of mining excavations reached almost 2 km (Rudajev 1985).

The strongest rockbursts, which occurred in individual regions, are specified in Table 1. The size of the rockbursts is characterized by local magnitude M_L. The strongest rockbursts were accompanied by aftershocks which lasted several days. These rockbursts affected not only mine openings but were also felt on the surface with intensities reaching five degrees on the MSK-64 intensity scale. The seismic moment of the strongest rockburst (Belchatów) had a value of $M_0 = 250 \times 10^{13}$ Nm.

It can be concluded from the review of rockburst areas outlined in Table 1 that the main causes of the occurrences of these rockbursts are connected neither with the depth of the mine working (geostatic stress), nor with the art of deposition of the extracted raw materials. It would appear that they depend, above all else, on the stress state of the rock mass (magnitude and orientation of main axes of the stress tensor), density and

Table 1. The strongest rockbursts which have occurred in Polish and Czech rockburst regions.

Rockburst region	Date	Time h m s	M_L	Reference
Lubin	24/03/77	07:32:26.1	4.5	Gibowicz 85
Belchatów	29/11/80	20:42:19.1	4.6	Gibowicz 85
Upper Silesian (Szonbierki)	30/09/80 12/07/81	01:02:57.8 11:59:27.5	4.3 4.1	Gibowicz 85 Gibowicz 85
OKR	27/04/83	11:25:01.3	3.7	Travnicek et al. 87
Kladno	23/01/78	10:20:10.0	3.1	Buben et al. 87
Pŕibram	01/12/78	01:17:05.0	2.8	Buben et al. 87

size of fault structures (stress concentrators), physical properties of the rock mass and on the stratigraphy. Technological parameters also contribute to the process and once understood can be used to minimize the risk of rockburst consequences. These include, the form and size of mine workings, the rate of mine advance, and the extent and frequency of blasting.

POLISH CZECH CONFERENCES ON ROCKBURSTS AND SEISMICITY IN MINES

Rockburst problems common to Polish and Czech specialists resulted in cooperation for several measurements, evaluation and also the organization of annual conferences. The meetings have taken place alternately in Poland and Czechoslovakia since 1970. Lectures were published in the national language of the host country in special issues of "Mat. a Prace Pol. Ac. Sci. (1971, 1974) and since 1976 in "Publs. Inst. Geophys. Pol. Ac. Sci.. ser M" (editor Inst. Geophys., Warsaw) and in the review "Acta Montana" published by the Institute of Geotechnics Czech. Ac.

It has been quite evident, since cooperation started, that the mining-induced seismic phenomena require a complex approach to the solution of the main problems. The physical explanation of the process is based on the research of stress distribution and transformation of the deformation energy, their irregularities in time and space, the energetic occurrence and the prediction of rockbursts. Application of an effective rockburst protection, will result in the reduction of the effect of rockbursts and eventually in the control of their occurrence. Intense research, involving both the knowledge of the conditions of rockburst occurrences and technological characteristics, is therefore indispensable.

Thus, the main topics of the meetings referred to above were aimed at the seismic monitoring of local events, determination of the basic parameters of foci (time, site and energy), seismicity of regions, mechanism of the energy release, prognosis of maximum seismic energy and statistical space/time rockbursts prediction. Many papers were aimed at the investigation of precursors, namely at: deformation phenomena; seismoacoustic emission (SAE); waves of ultrasonic frequencies, excited by technical sources to changes of gravity; and to distribution of stress. Some methods have been tested on the basis of laboratory experiments (SAE, variations of the speed and attenuation of the energy of ultrasonic waves, etc.). Results of geophysical research were interpreted and explained on the basis of geology, particularly tectonic characteristics of the regions, and correlated with mining operations.

SEISMIC MONITORING IN POLISH AND CZECH MINES

The first seismic monitoring systems were isolated stations with photogalvanometric recorders. At the present time, most mines are equipped with underground and surface digital stations, which operate in a triggering regime with data transmission into a centrum (by cable or wireless), and record events within the frequency range of 1-100 Hz.

A classic example can be quoted from the OKR region, where networks have been constructed on three levels:
- seismic micronetworks - sensor distance a few hundred metres, frequency band 100-200 Hz and cable connection with the centre,
- local networks - station distance of a few kilometres, stations located underground and on the surface, frequency band 1-100 Hz, and cable connection with the centre,
- seismic polygons - stations are located outside of the active region and their distances are a few tens of kilometres, signals to the recording centre are transmitted by radio, the polygon is interconnected with the regional Czech seismic network, which enables calibration of magnitude scales (Slavik 1992).

Initial interpretation problems were linked with the location of foci and evaluation of the rockburst energy. During the period when isolated photogalvanometric stations were used, the method of the Type Analysis (TA) was developed. This method was based on the similarity of records of seismic waves, (Buben 1964), emitted within a single rockburst area, and mostly connected to a certain working. These sets of rockbursts were used for statistical evaluation, particularly the determination of seismicity parameters. However, the accuracy of the TA location did not allow for either the investigation of source parameters or the investigation of the connection between the tectonic faults and the position of foci.

Seismic networks enabled the development of kinematic location methods, based on time differences of arrival of wave phases into the seismic stations, and also allowed the velocity model of the medium to be accounted for. However, rockburst zones are heterogeneous, anisotropic and discontinuous and their structure is affected by mine openings which cause the variability of all above mentioned properties in time. It follows that the velocities of seismic waves cannot be known with sufficient accuracy. The classic location methods, minimizing the residua (e.g. the Newton's method or the method of generalized inversion), are accompanied with systematic errors.

The reliability of location can be improved by tomographic location methods, determining the velocities of seismic waves as well as the hypocentral coordinates of the foci. A condition for the applicability of these methods is a sufficient amount of input data (i.e. onset times of seismic phases recorded on suitable situated stations). Kijko, 1980 discussed the optimization of the seismic network..

The tomographic method was realized for the local networks of the Kladno and OKR regions (Sileny 1987, 1989, Jech 1990).

The suggested twelve-block seismic model (Sileny 1991) in the Kladno area gives very reliable space distribution of foci, which are clustered into overlying sandstone layers. These more precisely defined focal areas can be studied in detail by micronetworks (or dense networks), where the locations are not affected as much by heterogeneity and discontinuities of mine openings and the environment.

Due to the complexity of the environment at small focal distances, the seismic energy of the rockburst focus cannot be determined with sufficient accuracy. A real, physically measurable quantity, with direct relation to the rockburst energy, is the velocity of the oscillatory motion at the observation site. The spectral analysis of the oscillograph enables the frequency of the signal, carrying the main quantity of energy, to be determined. However, in individual regions, local energetic scales are introduced, which are generally based on the relationship:

$$E = c \int_{t_0}^{T} v(t) \dot{x}^2 dt$$

where c includes the density of rockmass; v(t) is the velocity of seismic waves, which changes by steps in time, t, corresponding to onsets of new phases; $\dot{x}(t)$ is the oscillation velocity of the station's bedrock; and t_0 and T are the beginning and end times of the seismic record of the rockburst. This relationship is simplified into the form:

$$E = c\dot{x}\tau$$

where τ is the duration time of the signal, and the energy, E, is reduced under the application of the absorption law to a sphere with radius R=1 km. At small epicentral distances, the decrease of energy with distance is affected by the spreading of the wave front, i.e. $E = E_0 \cdot d^{-p}$ (e.g. d is the distance and n = 1.7 for the

Kladno region). For the local energy and local rockburst magnitude, the following relation has been found empirically:

$$\log E = 1.9M + Q$$

where $Q = 2.2$ according to Gibowicz (1963) for the Upper-Silesian region, or,
    $Q = 2.0$ according to Buben, et al (1982), for Pribram.
    The local magnitude in the region of the Pribram and Kladno mines is determined on the basis of the relationship:

$$M = \log \dot{x} + 2.1 \log d - 2.33$$
$$\dot{x}/\mu ms^{-1},$$
$$d/km/epicentral\ distance$$

For the assessment of the mechanism of rockburst foci (size and form of the fracture; mode of failure, shearing forces, tensile or implosive; size of displacement in the focus - Burger's vector; failure rate; stress decrease; and seismic moment) the model of shearing failure was initially accepted and the models by Brune (Fucik and Rudajev 1979a) and Madariaga (Gibowicz 1985), respectively, have been applied. The determination of parameters in both quoted models comes from integral characteristics of the emitted seismic waves (shape of their spectra) and thus depends greatly on the properties of the medium through which they are propagated. The formally defined size of foci in the Kladno district ("corner frequency" of spectra) results in unrealistic values (the radius r of foci is comparable with the entire depth of mining workings, r>200 m.

Due to these results and to the fact that the distribution of the direction of the onset of the P-waves in both the Kladno area and the OKR region did not exhibit the typical radiation pattern of distribution of dilatations and compressions (only dilatations were observed in 90% of events) an implosive focus mechanism (Fucik, Rudajev 1979b) has been suggested. This resulted from the conception of the possible deformation of the rock mass towards the free mine space. The focus of such a volume model had a radius comparable with the observed extent of the destructions. However, an explanation of such a mechanism of foci in the overlying rock proved impossible. A combined focus model, consisting of shear and implosive displacement has therefore been applied, which gives fair results from both the geometrical and physical points of view (Rudajev 1985) and is in good agreement with the observed distribution of signs of the phase onsets (dilatations).

However, spectra of seismic shocks also exhibit characteristic features of the modulated frequency, which occur as non-linear dynamic processes in the focus. The non-linearity is connected to the variable rupture rate, the consequence of which is not only a superposition of waves, but also their modulation. Owing to the fact that in such a case the space density of the radiated seismic energy increases (Aksjonov et al. 1992), a small source is able to radiate a considerable quantity of energy. But frequency modulation can also occur in consequence with the non-linear properties of the medium through which the waves are propagated. It is therefore important, for the clarification of the assumed non-linear sources (and also of sources of purely shearing mechanisms), to record shocks at small epicentral distances when higher wave frequencies, which are the main carriers of focal information, are still not absorbed.

APPLICATION OF STOCHASTIC METHODS IN MINE SEISMICITY

Seismic research of individual rockburst regions delimited by the location of foci, is based particularly on the tracing of the flow of seismic energy by the form of cumulative graphs (Benioff graphs) and on the magnitude-frequency distribution. It has been established that:

1. The seismic energy flow is a non-stationary process which depends on the site and progress of mining operations. The deviations from the stationarity are used as precursors of the seismic hazard (Rudajev 1967, Slavik 1992). The paper of Kijko (1985) proposes a possible dependence of the rockburst occurrence on the mining activities, which was also verified in Czech mines (Glowacka et al. 1988, Rudajev 1985, Glowacka et al. 1989),

2. The maximum possible radiated energy can be assessed from the width of the band within which the Benioff graph is

oscillating.
    3. The energy/frequency distribution, in analogy to tectonic earthquakes, has a negative exponential form:

$$N = A \cdot E^{-\gamma} (resp.\ \log N = a - bM)$$

The value of the exponent γ depends on the region and changes with time in dependence of the quantity of released energy. In periods of high rockburst activity (determined from Benioff graphs) its value is lower than during periods of lower activity (Rudajev 1971).
    4. Assuming the stationarity of the energy distribution, the parameters of that distribution can be used for prediction of the occurrence of strong rockbursts. The prognosis of the rockburst occurrence is understood to be the determination of the degree of exposure of the mine to rockbursts during the time interval in question. The degree of exposure has a statistical character, resulting from the space/time distribution of foci and their energy.
    For the probability $P(K_{kr})$ of the occurrence of rockbursts with energy exceeding the critical K_{kr} the following equation holds true (Buben, Rudajev 1973):

$$P(K_{kr}) = \frac{10^{\gamma K_{kr}} - 1}{10^{\gamma K_M} - 1}$$

where K_M is the maximum observed energy within the given period.
    Based upon this relationship, the number of rockbursts N with energy exceeding $10^{4.5}$ J (i.e. rockbursts, which cause the destruction of the mine working) during the ten year period (divided into three-months intervals) for the investigated region of the Kladno mines, has been predicted.
    Results are illustrated in Table 2, which gives the number of bursts, N, predicted values, $\hat{N}$, and the parameter γ.

Table 2.

Year	1965				1966			
Quarter	1	2	3	4	1	2	3	4
γ	0.62	0.62	0.49	0.61	0.46	0.5	0.69	0.46
$\hat{N}$		3	6	10	7	7	6	4
N	5	4	12	6	8	6	4	10

Year	1967				1968			
Quarter	1	2	3	4	1	2	3	4
γ	0.53	0.50	0.44	0.58	0.51	0.38	0.45	0.55
$\hat{N}$	12	6	9	11	7	2	11	9
N	7	11	9	7	3	15	11	10

Year	1969				1970			
Quarter	1	2	3	4	1	2	3	4
γ	0.61	0.69	0.87	0.70	0.75	0.76	0.70	0.60
$\hat{N}$	9	12	8	11	13	21	13	11
N	15	10	15	28	33	20	12	14

Year	1971				1972			
Quarter	1	2	3	4	1	2	3	4
γ	0.74	0.68	0.68	0.90	0.63	0.64	0.74	0.68
$\hat{N}$	14	5	3	3	2	4	3	1
N	4	5	4	1	0	0	1	0

Year	1973				1974			
Quarter	1	2	3	4	1	2	3	4
γ	0.51	0.75	0.68	0.64	0.82	0.78	0.53	0.59
$\hat{N}$	3	8	2	4	4	1	2	4
N	7	3	3	4	1	2	4	6

For the assessment of the probability of occurrence of strong rockbursts the Gumble's distribution of maximum values of the first or third type can be used (Buben, Rudajev 1973).

The mentioned statistical methods are based on the phenomenological approach only. These methods presume a slight stationarity of the release of seismic energy and thus also a certain regularity (uniformity) of mine operations, and are particularly significant for the general assessment of the seismic risk levels in the epicentral area. From this viewpoint, an important empirical relationship has been established for the Ostrava-Karviná mines and its validity has been checked with the conditions of the Kladno mines. This relationship characterizes the relationship between the maximum possible seismic energy, E, and the thickness of the layer, h, where the rockburst focus is situated (Kalenda 1992):

$$\log E = -0.56 + 4.29 \log h$$
(E is in J, h in m)

The direct evaluation of the occurrence of strong rockbursts is performed on the basis of their prediction (i.e. optimum estimation of the site, time, and energy of future rock bursts).

Unlike the local distribution of rockburst foci into individual active areas, the distribution of the seismic activity in time has random character.

During certain periods and on some occasions, deviations from the random distribution in time may be observed, particularly when the rockbursts are affected by only one mine opening. For such a case, an empirical relation has been found in the Kladno mines for the interval between strong bursts, Δt_1, and the magnitude, A_{i-1}, of foregoing bursts, in the form of:

$$\Delta t_i = B \cdot A_{i-1}; \ B \text{ is const.}$$

This relationship agrees with the hypothesis that the cumulation of the deformation energy takes places within a certain isolated area. It has also been established, that in 30% of the events two strong rockbursts occur within a 24 hour interval (analogy with the so-called seismic doublets).

The time prediction of rockbursts is often looked for by studying precursors also found in other observed series. Most frequent are the studies of anomalous distributions in the seismoacoustic emission (Simane and Broz 1980). The method of time clustering of seismoacoustic pulses was in practical use in the Pribram mines. It has been used for imminent application of antiburst measures - stopping of extraction and eventually the application of shock-preventing explosions.

The prediction is generally evaluated by the use of statistical extrapolation methods of random time series. A method of multichannel prediction has been developed, in which the predicted events were formed by time differences between rockbursts and their energies, and as the correlated input series seismoacoustic and convergence data were used(Rudajev 1985).

The extrapolation was performed by modified Wiener filtration, adaptive filtration, and linear regression with the use of Bayes' conditional probability. The prediction's efficiency was tested by the ratio of mean square values of prediction errors to the dispersion of the predicted centred sequence.

$$Q = \frac{\sum\limits_{i=1}^{N-L}(x_i - \hat{x}_i)^2}{N-L} : \frac{\sum\limits_{i}^{N} x_i^2}{N}$$

N is the number of input data x_i,
L is the length of prediction operator,
$\hat{x}_i$ are the predicted values.

The prediction is considered successful, if Q<1. It holds true that, for an absolutely exact prediction, Q = 0.

However, when applying the statistical predictions, which are based on the evaluation of correlation of individual series, only a formal access is involved, which makes use of maximum input information, but is not based on the physical part of the failing process. In principle, the physical access is substituted by various auxiliary input sequences (including technological data).

To improve the rockburst prediction methods a more detailed knowledge of the rockburst-prone areas and their responses to real mining activities is necessary.

REFERENCES

Acta Montana Nos 22(1972), 32(1975), 38(1976) 50(1979),55(1980), 61(1982), 71(1985), 75(1987), 81(1989), 83(1989), 84(1992), Proc. of Mining Geophysical Confer. Inst of Geotechnics, Prague, CR.
Aksenov, V., J. Kozak, T. Lockajicek, V. Rudajev and J. Vilhelm 1993. The seismic source self-organizing, Proc. Acta Montana 88/89, Prague (in press).
Archive documents: Shocks of mining areas and detonations in Kladno black-coal mines (1880-1894). Archiv of Kladno mine company, Czechoslovakia.
Buben, J. 1962. Seismische Untersuchunge von Gebirgsschlägenbei Kladno im Jahre 1960. Freiberger Forschungshefte C 126, p.21-32.
Buben, J. 1964. Seismological research of rockbursts, Ph.D. thesis. Prague (not published).
Buben J., V. Rudajev, Fucik 1981. Seismology in rockburst research. Travaux Geophysiques XXIX, p.103-118. Prague.
Buben J., V. Rudajev and K. Kasak 1982. Parameters of seismic foci from Pribram stations records (in Czech). Proc. "Mining Pribram (Czechoslovakia) in science and technology", p.158-172.
Fucik, P. and V. Rudajev 1979a. Physical parameters of rockburst sources (in Czech). Publs. Inst. Geoph. Pol. Acad. Sci., Warszawa, M-2(123), p.21-35.
Fucik, P. and V. Rudajev 1970b. Modification of the Sharp model for the study of parameters of rockburst foci (in Czech). Acta Montana 50, p.41-54. Prague.
Gibowicz, S.J. 1963, Magnitude and energy of subterrane shocks in Upper Silesia. Studia Geophys. et Geodat, No 1, p.1-19. Prague.
Gibowicz, S.J. 1985. Mechanism of strong rockbursts in Poland (in Polish). Publs. Inst. Geophys. Pol. Ac. Sci., M-6(176), p.21-50. Warszawa.
Glowacka, E., V. Rudajev, and V. Bucha 1988. An attempt of continuous evaluation of seismic hazard induced by deposit extraction for the Robert field in the "Gottwald" mine in Kladno, Czechoslovakia. Publs. Inst. Geoph. Pol. Acad.. Sci., M-10(123), p.311-319. Warszawa.
Glowacka, E., K. Holub, T. Stankiewicz and J. Hajek 1989. An attempt of seismic hazard evaluation based on the extracted deposit amount at the Doubrava colliery (Czechoslovakia). Acta Montana 81, p.43-56. Prague.
Jech, J. 1990. Seismic tomography in Ostrava-Karaviná region. In: Induced seismicity and associated phenomena. Proceeding of conference in Liblice, March 14-18, 1988, p.122-129.
Kazimierczyk, M. 1979. The location of rockbursts' foci in mines with respect to rock-mass properties (in Polish). Publs. Inst. Geoph. Pol. Acad. Sci. M-2(123), p.135-150. Warszawa.
Kijko, A. 1980. Optimum seismic network in Ostrava coal basin, Czechoslovakia, Acta Montana 55, p. 73-96. Prague.
Kijko, A., 1985. Theoretical model for a relationship between mining seismicity and excavation area. Acta Geophysica Polonica 33, p.231-242. Warszawa.
Proc.: Materiazy i Prace Publs. Inst. Geophys. Pol. Ac. Sci., 1971 (N 47), 1974 (N 67).
Proc. Publs. Inst. Geoph. Pol. Ac. Sci., 1976 (M-1/97), 1979 (m-2/123), 1980 (M-3/134), 1981 (M-5/155), 1984 (M-6/176), 1986 (M-8/191), 1988 (M-10/213).
Rudajev, V. 1967, Die Seismizität der Gebirgsschläge bei Kladno. Freiberger Foschungshefte C 225, p.7-19.
Rudajev, V. 1971. On the seismic regime of rockbursts in the Kladno coal mine region (in Polish). Mat. i prace 47, Publs. Inst. Geophys. Pol. Ac. Sci., p.3-11. Warszawa.
Rudajev, V. (ed.) 1985. Physical model of rockburst active region of Uranium mines Pribram (in Czech). Scientific report Inst. Geol. and Geotech. CSAS, Prague.
Rudajev, V., R. Teiseyre, J. Kozak and J. Sileny 1986. New concepts of rockbursts mechanism in coal mines. Pageoph, Vol. 124, Nos 4/5. p.841-855.
Sileny, J. 1987. The tomographic method of location of rockbursts in coal Kladno basin (in Czech), Acta Montana 75, p.65-82. Prague.
Sileny, J. 1989. Effect of the medium parameterization on the tomographic localization of Kladno mining tremors (in Czech). Acta Montana 81, p.71-86. Prague.
Simane, J. and M. Broz 1980. The time distribution of natural rock mass impulses. Publs. Inst. Geoph. Pol. Acad. Sci., M-3(134), Warszawa.

Slavik, J. 1992. Complex processing of seismic, seismoacoustic, geologic and mining engineering databases of active regions of OKR (Czechoslovakia) with aim for rockbursts prognoses (in Czech). Ph.D. thesis, Ostrava, not published.

Stauch, K. 1905. Strong earthquake in North India and its record in Bohemia (in Czech). Hornické a hutnické listy r. 1905.

Travnicek, L., J. Holecko, K. Klima, L. Ruprechtova and A Spicak 1987. Seismological development of progressive failure of rock mass in area prone to anomalous rockbursts. Acta Montana 75, p.21-36. Prague.

Wierzchowska, Z. 1959. The strongest rockbursts in Upper Silesia (in Polish). Reports of GIG Katowice 1959, 1961, 1963.

Wierzchowska, Z. 1974. Development and modernization of microseismological investigations in Upper Silesian coal basin (in Polish). Publs. Inst. Geoph. Pol. Acad. Sci., Vol. 67, p.49-54. Warszawa.

Rockbursts and Seismicity in Mines, Young (ed.) © 1993 Balkema, Rotterdam, ISBN 90 5410 320 5

Rock burst prediction – An empirical approach

A. Shridhar Chavan & N. M. Raju
National Institute of Rock Mechanics, Kolar Gold Fields, India

S. B. Srivastava
Department of Mining Engineering, Indian School of Mines, Dhanbad, India

ABSTRACT: Rock bursts have been associated with the mining of gold at Kolar Gold Fields since many decades. Changes in mining technology were brought about from time to time, to reduce the frequency and intensity, of the occurrence of rock bursts. Monitoring of the mining activity, by way of mine closure measurements and later by the seismic network, have been of great help. In order to make the best use of the data being generated, improvement of the existing method of analysis was warranted. A new concept and method of analysis, for predicting rock bursts has been evolved. The method involves recognition of simple regularity patterns, quantification of closures, delineation of stopes prone to bursting and arriving at the date of occurrence of the event. This technique has been tried a number of times and is found to be very successful.

1 INTRODUCTION

The gold deposit in the green stone belt at Kolar Gold Fields, India, is being mined at ultra depths, associated with a highly stressed environment. Rock bursts are a common feature which occur quite frequently, with varying intensities, in and around the working stopes. Rich ore shoots have been permanently lost due to rock burst occurrence. Damage to machinery and loss of lives, have also been documented.

To understand the nature of ground movement, closure measurements were taken up during the early 1950's in the stopes, drives and cross cuts. In the year 1978 a Seismic network was installed to compute the energy released due to rock bursts and to locate their foci. Thus a large volume of data has been generated. Analysis of the data in the past has helped, in redesigning the mine workings. However rock bursts have only been reduced in frequency and intensity but their occurrence is still a part and parcel of the mining activity. In view of the problems mentioned above, prediction of rock bursts has gained great importance. Analysis of the closure measurements, which is being carried out on a regular basis, has helped only to a certain extent in as far as prediction is concerned. All along, the technique being adopted, is the simple arithmetic mean and standard deviation of the above data for prediction. But this does not give satisfactory results, owing to certain limitations of the method of analysis.

This paper deals with a new concept and method of analysis, which is found to be very satisfactory.

2 EXISTING METHOD OF CLOSURE DATA ANALYSIS

The closure measurements are being taken with the help of tape extensometers in the working stopes, drives and cross cuts on a regular basis. These measurements are repeated on alternate days. To calculate the monthly limits for each measuring station, the closure data from the preceding three months are taken (Krishnamurthy 1972). An arithmetic mean ($\bar{x}$) and the standard deviation (σ) is computed for this data set. Limits of $\bar{x} + 2\sigma$ (warning limit) and $\bar{x} + 3\sigma$ (action limit) are then fixed for that month.

During the course of measurements for the period of a month, if the closure exceeds these limits, the mining personnel are alerted. It is noticed that,

this method of analysis has its own limitations and actual prediction of rock bursts has not been possible. Further more the intensity and possible day of the burst cannot be arrived at.

3 PROPOSED APPROACH FOR ROCK BURST PREDICTION

A study of the ground movement, obtained from the closure measurements, reveal that the movements are not uniform but are erratic in nature. Therefore the use of normal statistical analysis do not give satisfactory results. In the proposed method of prediction, the following stages are adopted.

1. Recognition of patterns of regularity indicative of rock bursts.
2. Quantification of closures and developing a closure index, to understand the significance of the regularity and the intensity of the forthcoming rock burst.
3. Delineation of the stope/stopes, prone to bursting and
4. Predicting the day of the rock burst.

3.1 Recognition of patterns of regularity indicative of rock bursts

From the stope closure measurements taken once in two days, the rate of closure (ROC) is calculated. A simple ratio of the rate of closure (RRC) of any particular day, with respect to the previous day of measurement, is taken and the process is continued. A graph drawn between RRC with each point as a reference and period in days (PD) gives a clear idea of the ground movement for a particular measurement station. It has been noticed that before a rock burst, a certain regularity of the ground movement takes place, where the fluctuations either rise or fall in a controlled manner. This regularity is very much indicative of an on coming event in most of the cases. Figure 1 shows a typical graph where the patterns recognised before rock bursts are indicated.

3.2 Quantification of closure data

Having recognised the patterns, one's interest gets centered to know the intensity of the event to come. This calls for quantification of closure data. The present concept is that, ground movements in

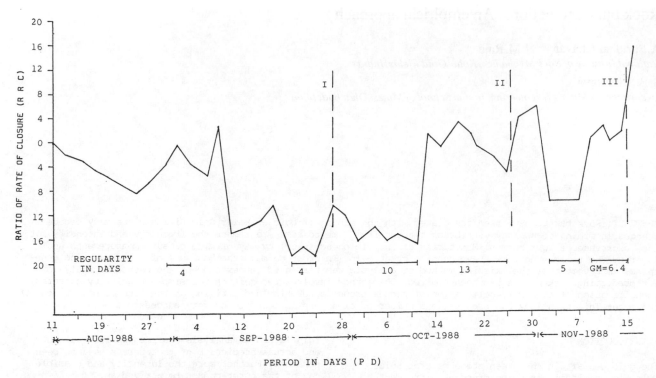

Figure 1. Ratio graph showing ground movements at 100 level, west reef north SD3 station, Northern folds area, Champion Reefs mine, Kolar Gold Fields.

various parts of the mine, do have certain mean levels of fluctuation. If this mean level (background level) is computed and each measurement is reduced with respect to it, further quantification of the closures become relatively easier. The procedure for calculation of background level for a particular case is as given by Claud Lepeltier (1969).

Since the ground movements are erratic, the data treated lognormally provides reliable results. The background level may be obtained from a simplified statistical treatment of the data. For this purpose the cumulative frequency percentage is calculated from logarithmically stretched class intervals, to which the data is subjected. From the date of pattern recognition, the preceding three months of closure data are taken. A plot of the cumulative frequency percentage versus the class interval, on a log-probability paper, would give a straight line, from which the background level is obtained. Having computed the background level, each measurement is to be reduced with respect to it, by simple division. This exercise gives whether the closures are less than, equal to, or greater than the background level. It is seen that when the closures are greater than the background level, there is a build up of stresses around the stope. This reduced data, when multiplied continuously for the data set of three months, would give a Closure Index. This in turn can be related to the intensity of the forthcoming rock burst.

Also as mentioned earlier, when patterns of regularity are noticed, their significance would depend on the magnitude of the Closure Index. If this index is very low, for instance, less than 1×10^6, then the pattern may be taken as insignificant. However, the index of other levels, belonging to the same sequence of stoping operation, is to be examined.

A broad classification of the closure index (table 1) has been arrived at by back analysis of rock burst data. This correlates with the seismic classification of the intensities of rock bursts, followed at Kolar Gold Fields (Srinivasan 1990).

Table 1. Classification of Closure Index.

Seismic classification		Closure
Amplitude	Intensity	Index
Upto 50 mm	Minor burst	1×10^6 to 10^{12}
50 mm – 100 mm	Medium burst	1×10^{12} to 10^{15}
>100 mm	Major burst	1×10^{15}

The computed values of closure index for ten rock bursts in the Osborne's and Northern folds areas of Champion reef mines, Kolar Gold Fields are cited in table 2. It can be seen from the table that most of the cases follow the classification as given in table 1.

3.3 Delineation of stope/stopes prone to bursting

Stoping operation in a high stress environment, usually follow a pattern of leading and lagging of the stoping sequence. In the Champion reef mine of Kolar Gold Fields, for example, the stoping sequence between 93rd to 105th levels are as in table 3. The initial step is to identify the patterns of regularity at all stope closure points. Further, the closure index of these points would give clearly, the state of stress at the various stopes. The stope having the highest closure index would be the one prone to bursting. At times, more than one stope belonging to two different sequences, would have high values of the closure index, thus indicating occurrence of two rock bursts. This is best seen when a number of events occur during an area rock burst.

3.4 Prediction of the date of rock burst

Predicting the date of a rock burst, largely depends

164

Table 2. Values of Closure Index and Amplitude - Duration

Date of Rock burst	Location	Station	Closure Index	Amplitude (mm)	Duration (s)
13.08.86	Osborne's area	102 level SD 21	1.00×10^{16}	100	50
		SD 20	5.40×10^{18}		
15.11.88	Northern folds area	100 level SD 3	7.89×10^{20}	110	08
01.02.89	Osborne's area	94 level SD 2	3.10×10^{22}	100	35
21.03.89	Osborne's area	94 level SD 1	1.50×10^{16}	100	50
		95 level SD 1	1.14×10^{15}		
		95 level SD 2	2.06×10^{19}		
07.06.89	Osborne's area	94 level SD 2	1.50×10^{20}	110	90
		98 level SD 11	3.40×10^{24}		
26.01.91	Osborne's area	98 level SD 12	8.60×10^{17}	110	90
		99 level SD 16	1.90×10^{21}		
07.03.91	Northern folds area	107 level SD 7	1.80×10^{19}	100	10
07.01.89	Osborne's area	102 level SD 22	6.50×10^{23}	90	10
14.12.88	Northern folds area	104 level SD 4	8.60×10^{06}	75	05
19.12.88	Osborne's area	95 level SD 2	5.98×10^{08}	60	08
01.04.86	Osborne's area	101 level SD 17	1.50×10^{12}	110	65

Table 3. Stoping sequence at Glen ore shoot of Champion reef mine.

Level	Stoping sequence
93 to 95	first
95 to 97	second
97 to 100	third
100 to 103	fourth
below 103	fifth

on the ground behaviour during the recent past. The graph shown for pattern recognition, may be conveniently used for understanding the past cycle of movements. An arithmetic/geometric mean calculated from the regularities of the past cycle can be used to predict the rock burst. The arithmetic mean or geometric mean is calculated depending on whether the regularities are uniform or erratic. This method of prediction can be appreciated by the case study presented below.

The rock burst of 15th November 1988 which took place below the 100th level of west reef north stoping, Northern folds area, Champion reef mine is presented. The occurrence of this burst was predicted by the author on 13th November, 1988.

As mentioned earlier, the closure data of the preceding three months, from the date of recognition of regularity patterns, was taken for the purpose of analysis. From Figure 2, it can be seen that the pattern of regularity had commenced from the 9th November 1988, therefore data from the 11th August 1988 was taken for analysis.

The highest ground movement during the span of the three months was 0.107 inch and the lowest was 0.001 inch.

The range of distribution of ground movements = 0.107/0.001 = 107

Logarithm of Range = 2.02938

The log range was divided into a number of classes such that the logarithmic class interval approximated to 0.2. Therefore ten classes were taken in the above case and the cumulative frequency percentage was determined as given in the table below.

Table 4. Statistical analysis of closure data.

Class interval (inch)	Frequency	Cumulative Frequency	Cumulative Frequency %
.001 - .002	6	6	15.40
.002 - .003	4	10	25.60
.003 - .005	1	11	28.20
.005 - .008	4	15	38.50
.008 - .013	5	20	51.30
.013 - .021	7	27	69.20
.021 - .033	5	32	82.10
.033 - .053	4	36	92.30
.053 - .084	2	38	97.40
.084 - .134	1	39	99.99

The cumulative frequency percentage plotted against the class interval, on log probability gave a background level of movement = 0.003 inch as shown in Figure 2. Each movement was then divided by the background level. The resulting values when continuously multiplied gave a closure index of 7.89×10^{20}. This value in turn indicated a major rock burst in the stope below 100 level (SD 3 station West reef North). In figure 1, the observed patterns of regularity are shown. Their period vary between 4 to 13 days. A geometric mean of these gave 6.4 days, by which the date of the forthcoming rock burst was projected from the 9th November 1988, the day of

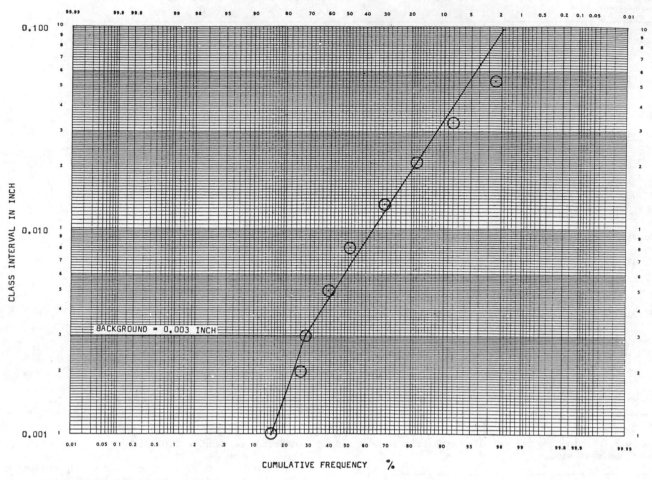

Figure 2. Cumulative frequency plot.

the beginning of regularity. Hence the predicted day of the rock burst was 15th November 1988. Three rock bursts took place in this area on the afternoon of 15th November 1988, causing damage to the stope below 100 level (SD 3 station west reef north). The closure after the rock burst was 0.333 inch. The burst details are given in table 5.

Table 5. Details of rock burst on 15/11/1988

Time (hrs)	Amplitude (mm)	Duration (s)
14:05:52	90	05
14:06:47	110	08
14:06:53	100	10

Along similar lines, past data pertaining to rock burst energy release of Osborne's area, Champion reef mines was studied. Clear indications of area instability are noticed as shown in Figure 3. Here a raise, regularity, a significant drop (indicative of number of events below the background level) and there after an increased seismic activity (area instability) are depicted.

4. CONCLUSION

An alternative theory for the prediction of rock bursts from the mine closure measurements

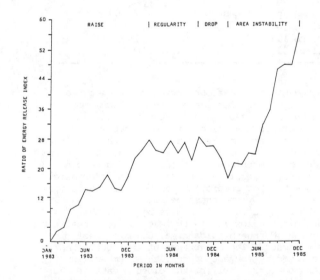

Figure 3. Graph showing energy release index.

presented here, is a new method of predicting rock bursts with respect to location, intensity and day of the burst. By application of this concept for predicting rock bursts, the following conclusions are drawn:

Ground movements caused due to stoping activity

166

are not uniform but erratic in nature. Analysis by way of normal statistical methods do not help in satisfactory prediction of rock bursts. In the method presented here, these limitations have been overcome.

Using mine closure data, the existence of high stresses around stoping areas, can be known from the regularity patterns indicative of rock bursts.

The calculated closure index bring out, the significance of the precursory regularity pattern and indicate the intensity of the expected rock burst.

From the ratio graphs, the cycle of ground movements can be identified and used in predicting the date of the burst.

The closure index at various stopes help in delineating the stope, which has the highest closure index responsible to cause the burst. The advantage of this new approach, as compared to the existing method of analysis are as follows;

The present approach permits an estimation of the intensity of the rock burst being predicted. This has not been possible with the existing method, since warning and action limits do not indicate the intensity.

When precursory regularity patterns are noticed, their significance can be checked with respect to the closure index, where as in the existing method, many a times, the closure exceed the warning and action limits, and no rock bursts occur. At times these limits are hardly reached and bursts are recorded.

Further this approach has also been successfully used in the analysis of rock burst energy release data, where a larger area instability can be identified, from the precursory patterns.

ACKNOWLEDGEMENT

The authors are thankful to the Director, National Institute of Rock Mechanics (NIRM), for permission to present this paper. The cooperation extended by the scientists and scientific staff in the Rock Mechanics Instrumentation and Seismological departments of NIRM is gratefully acknowledged.

REFERENCE

Claud Lepeltier 1969. A simplified statistical treatment of geochemical data by graphical representation. Economic geology vol. 64, pp 538–550.
Krishnamurthy, R. 1972. A review of rock burst research in the Kolar Gold Fields. Proc. of symp. on rockmechanics: Institution of Engineers (India).
Srinivasan, C. & Shrikant B. Shringarputale 1990. Mine induced seismicity in the Kolar Gold Fields. ISSN 0016–8696 Gerlands Beitr Geophysik Leipzig IS 10–20.

Rockbursts and Seismicity in Mines, Young (ed.) © 1993 Balkema, Rotterdam, ISBN 90 5410 320 5

Model of rockburst caused by cracks growing near free surface

A. V. Dyskin
The University of Western Australia, W.A., Australia

L. N. Germanovich
The University of Oklahoma, Okla., USA

ABSTRACT: A mechanism of rock burst caused by crack propagation at the free surface of an excavation is considered. The compressive stress concentration makes the initial (pre-existing) cracks grow parallel to the free surface towards the compression in a stable manner. The interaction with the free surface intensifies the growth and eventually makes it unstable. As the result of the unstable phase of the growth, the cracks suddenly separate thin layers from the rock mass. If these layers become sufficiently long (this depends on the sizes of the stress concentration zone) they buckle and fracture. This process manifests itself as a rock burst. The particular size and magnitude of the rock burst depend on thickness of the layers. A 2-D model is suggested for calculating the stress magnitude for the onset of the unstable crack growth. The model also allows analysing the influence of the confining pressure (the pressure acting perpendicular to the growing cracks) and calculating its value required to eliminate the unstable phase of the crack growth. The initial, stable stage of the crack growth is accompanied by their opening (dilatancy) which results in non-linear increase of the normal component of the surface displacement. The measurements of this displacement can be a basis for the rock burst monitoring.

INTRODUCTION

Rock burst, i.e., a sudden failure of rock near an underground opening accompanied by the relief of significant amounts of accumulated strain energy, is one of the major problems in mining, geotechnical, and petroleum industries. Prediction of rock bursts and safe rock engineering design require understanding and modelling the mechanism of such a phenomenon.

It is now widely accepted that rock bursts occur in zones of high concentration of compressive stresses and are caused by propagation of cracks, parallel to a free surface of the opening (e.g., Fairhurst and Cook, 1966; Gay, 1973; Stacey, 1981; Nemat-Nasser and Horii, 1982; Wittke, 1984; Ewy and Cook, 1990; Talebi and Young, 1992). Such cracks are, in fact, secondary cracks produced by pre-existing defects. Dyskin et al. (1992) have shown that the main role in generating the secondary cracks is played by pre-existing cracks, i.e., voids, the opposite faces of which can contact due to loading.

Dyskin and Salganik (1987), Germanovich and Dyskin (1988), Dyskin et al. (1991), and Germanovich et al. (1990, 1992) have modelled the dilatancy and fracture in unbounded rock due to stable growth of the crack-induced secondary cracks under uniaxial and biaxial compression. However, the influence of a free surface on the crack propagation may be essential. First of all, non-rectilinear free boundaries change the stress state and make the cracks propagate parallel to the boundary, as it has been experimentally shown by Nemat-Nasser and Horii (1982). Second, the presence of the free boundary may change the type of the crack growth: from initially stable to unstable. This assumption follows the fact, that for the crack situated very close to the free boundary, the beam asymptotic method (e.g., Slepyan, 1990) predicts unstable growth. Therefore, the point of starting of unstable crack growth may be treated as the onset of rapid strain energy release and, eventually, as the beginning of rock burst. This paper investigates such a rock burst mechanism.

1. QUALITATIVE DESCRIPTION OF THE MECHANISM OF ROCK BURST

Consider the surface of an opening, in the vicinity of which there exists concentration of compressive stress acting parallel to the surface (fig. 1a). After the stress reaches a certain magnitude, some of the pre-existing cracks start to grow (fig. 1b) towards compression. Initially, the growth will be approximately the same as for an unbounded body under uniaxial compression.

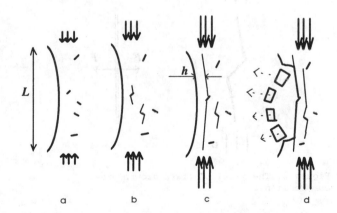

Figure 1. Qualitative description of the mechanism of rock burst.

Experiments (Brace and Bombolakis, 1963; Hoek and Bieniawski, 1964; Nemat-Nasser and Horii, 1982) show that, in this case, the initial pre-existing crack will generate branches (fig. 2) growing towards the maximum compression in a **stable** manner. It means that the crack can increase its length only after corresponding increase in the load. As the crack grows, the influence of the free surface becomes essential. It will be shown below that after reaching a certain crack length this surface makes the crack grow **unstable**. Therefore the crack suddenly increases in length and separates a thin layer from the rock mass (fig. 1c). The thickness h of the layer is determined by the distance between the crack and the free surface. The length of the layer is determined by the length L of the compressive stress concentration zone (fig. 1).

One may imagine two possibilities:

1. If the length L of the stress concentration zone is large, compared to the layer thickness h, the crack may propagate a significant distance; the layer becomes slender and will eventually buckle. This will create a new free surface and the process will be repeated (fig. 1d). Such a mechanism will continue until the length of the zone of stress concentration decreases. The process will manifest itself as a rock burst. The size of the rock burst will, of course, depend on the layer thickness and on the magnitude of stress concentration.

2. If the zone is not large enough, no immediate fracture will occur. However, as the stress level keeps rising, other pre-existing cracks that could not grow under the previous load will now be able to propagate. This will eventually separate the layer into several more thin layers with subsequent buckling. The failure will occur, in this case, by expulsion of small fragments of rock. It may be referred to as spalling.

Some technologies of excavation (e.g., blasting) create a cracked zone around the opening. In this case the size of the cracked zone will control the size of the rock burst.

The following is the 2-D model of the described mechanism of rock burst initiation, including formation of the first thin layer. One has to notice that in plane-strain approximation adopted here the initial pre-existing cracks are assumed to be infinite in one direction (the direction perpendicular to the plane). This assumption is restrictive. Nevertheless, the presented model is the necessary first step in approaching the micromechanical model of rock burst based on the explicit consideration of propagation of pre-existing cracks near the free surface.

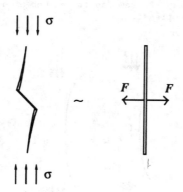

Figure 2. The model of crack growth under compression.

2. CRACK MODEL

Consider the developed stage of the crack growth and let the initial pre-existing crack be of length $2a$ and inclined with respect to the compression direction at an angle α. The simplest model for such a crack is shown in fig. 2. It is a rectilinear crack being opened by a pair of concentrated forces simulating the action of the inclined contact area (the former initial crack). The value of these forces (per unit length of the inclined contact area) is assumed to be equal to the horizontal projection of the shear force tending to displace the opposite faces of the initial crack (see Dyskin et al., 1991):

(2.1) $\quad F = 2\sigma\beta(\alpha)a, \quad \beta(\alpha) = C\sin^2\alpha\cos\alpha$

Here σ is the applied compressive stress acting along the free surface; C is a factor of order unity accounting for the friction and influence of curvature of the secondary cracks.

Using such a model, Dyskin et al. (1991) and Germanovich et al. (1990 and 1992) have found a dependence of the rock dilatancy on the applied compression, which fits in well with all available experimental data and, if normalised, does not require any parameters. Thus, it is reasonable to use this model to simulate crack growth near the free surface as well provided that it is modified to take into account the interaction between the crack and the free surface.

3. INTERACTION BETWEEN CRACK AND FREE SURFACE

Consider the crack growing at a free surface. Since the compressive stress concentration acts parallel to the surface, the developed crack will grow parallel to the surface as well (fig. 3).

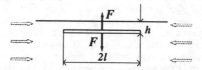

Figure 3. Crack growth parallel to a free surface.

The problem for such a crack has been solved numerically (Germanovich and Grekov, 1993). The values of the stress intensity factors (SIF) when the surface is free are plotted on fig. 4.

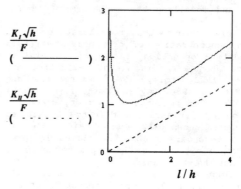

Figure 4. Stress intensity factors vs. the normalised crack length.

The minimum for K_I at $l_{cr} \approx 0.5h$ indicates the transition from the stable crack growth to the unstable one.

If there was no free surface, K_{II} would be zero. Therefore, a finite K_{II} results from the interaction between the crack and the free surface. In principle this should turn the crack trajectory toward the free surface, according to conventional criteria for crack propagation (e.g., Cherepanov, 1979). However, as it has been experimentally shown by Nemat-Nasser and Horii (1982), the crack propagation under compression is strictly parallel to the free boundary (even if the boundary is not rectilinear). A possible explanation for this phenomenon lies into consideration that in reality the fracture of the material at the crack tip occurs in areas that are not infinitesimally small. In such areas, the high compression acting along the free boundary becomes comparable with the crack-induced stress

170

concentration and is able to prevent the crack from changing its direction. Accurate modelling of the crack growth in this case requires the detailed analysis of the fracture mechanism at the crack tip. However, for the proposed simplified model it is sufficient just to take into account the described experimental fact of crack propagation parallel to the free surface and neglect the influence of the shear stress concentration on the crack propagation (Note, according to fig. 4, the values of K_{II} are considerably smaller than K_I up to the moment of the unstable crack growth). This allows using the simple conventional criterion of crack propagation during the stable stage:

(3.1) $K_I = K_{Ic}$

where K_{IC} is rock fracture toughness.

4. DIPOLE ASYMPTOTIC MODELLING OF INTERACTION BETWEEN THE CRACK AND THE FREE BOUNDARY

The fact that the unstable crack growth starts when the crack size is less than the thickness h of the layer between the crack and the free surface, suggests to use the dipole asymptotic method (e.g., Dyskin at al, 1992) to calculate the interaction between the crack and the free boundary. The idea of this method is to use only the remote stress field generated by the crack for modelling the interaction. This stress field being reflected from the free surface creates an additional load which is assumed to be uniform over the crack length. Mathematically it means that only the terms of order of $(l/h)^2$ in the full solution for the problem are calculated. As it has been shown by Dyskin et al. (1992) the dipole asymptotic is accurate enough at least up to distances between interacting objects equal to their sizes. In the problem under consideration, the validity of the dipole asymptotic will be shown by direct comparison with the numerical results.

For the considered case, the dipole asymptotic solution is given by:

(4.1) $K_I = \dfrac{F}{\sqrt{\pi l}} + \dfrac{3Fl^{\frac{1}{2}}}{2h^2\sqrt{\pi}} = K_I^{\infty}\left(1 + \dfrac{3l^2}{2h^2}\right)$

where $K_I^{\infty} = F/\sqrt{\pi l}$ is the SIF for a crack in the infinite plane loaded by a couple of concentrated forces (fig. 2). The unstable crack growth starts at the point of minimum of K_I, i.e., crack propagates unstably when its length

(4.2) $l > l_{cr} = \sqrt{2}/3\, h \approx 0.471h$

Comparison with the numerical solution (table 1) shows the acceptable accuracy of the method up to the critical length and even further. The discussed method has the essential advantage, since the dipole asymptotic method allows simple calculation of rock dilatancy and (in principle) generalisation to the 3-D case.

Taking into account (2.1) and the criterion of crack propagation (3.1), one can obtain the equation governing the stable crack growth

(4.3) $\left(\dfrac{l}{h}\right)^2 - \dfrac{K_{Ic}}{3\beta\sigma}\dfrac{\sqrt{\pi h}}{a}\sqrt{\dfrac{l}{h}} + \dfrac{2}{3} = 0$

The **critical stress** corresponding to the onset of the unstable growth can then be calculated by substituting the critical crack length (4.2) into (4.3):

(4.4) $\sigma_{cr} = \dfrac{2^{\frac{1}{4}}K_{Ic}\sqrt{3\pi h}}{8\beta a}$

Table 1. Comparison of the asymptotic formula (4.1) with numerical results.

Normalised crack size	$k_1 = \dfrac{K_I\sqrt{l}}{F}$		
l/h	Numerical results	(4.1)	Relative error, %
0.05	0.566	0.566	0.06
0.1	0.572	0.573	0.12
0.25	0.615	0.617	0.34
0.5	0.754	0.776	2.9
1	1.17	1.410	20.6

Applied to the cracks closest to the boundary, equation (4.4) gives the lower estimate for the stress of the possible onset of rock burst. The rock burst will actually occur if the size of the zone of compressive stress concentration allows the unstable growing crack to separate a layer sufficiently long for buckling.

The minimum length of the layer required for buckling may be estimated by considering the layer as a plate with fixed edges. In 2-D model (plain strain), the result (e.g., Landau and Lifshitz, 1959) can be written as

(4.5) $l_b = \pi h\sqrt{\dfrac{E}{3\sigma_{cr}(1-\nu^2)}}$

Here E is Young's modulus of the rock, ν is Poisson's ratio.

5. THE GROWING CRACK UNDER CONFINING PRESSURE

Consider now the case of a confining pressure q applied perpendicular to the growing crack. Such a pressure can, for instance, model action of reinforcement of the excavation walls or, if the stability of boreholes is considered, q can model the pressure of fluid in it. The aim of this section is to calculate what magnitude of the confining pressure is required to eliminate the unstable phase of the crack growth.

The influence of the free surface can again be modelled by the dipole asymptotic method. The stress intensity factor for the crack subjected to the combined action of the concentrate forces F and the confining pressure q has the form:

(5.1) $K_I = \dfrac{F}{\sqrt{\pi l}} - q\sqrt{\pi l} + \dfrac{3Fl^{\frac{1}{2}}}{2h^2\sqrt{\pi}} - \dfrac{3Fl^2}{4h^2}q\sqrt{\pi l}$

Here the last two terms account for influence of the free boundary. Fig. 5 shows the dependence of the stress intensity factor on the crack length for different values of the dimensionless confining pressure q_*,

(5.2) $q_* = \dfrac{\pi q h}{F}$

It is seen that there is a critical value of q_*, which eliminates the unstable crack growth. This value is given by

(5.3) $q_* = 0.737$

Using (5.2) and taking into account (2.1) one can calculate the confining pressure required to exclude the unstable growth of the considered crack. This pressure will depend on the angle α of the initial crack inclination. However since the rock burst preparation involves growth of many cracks, one can calculate the average critical value of the confining

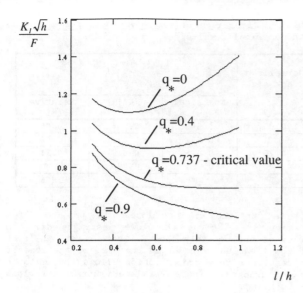

Figure 5. Influence of the confining pressure on the unstable crack growth.

pressure. Averaging β in (2.1) over the pre-existing crack orientations (their distribution is assumed to be isotropic) results in $\beta = 3/(4\pi)$. Substituting this average value into (2.1), one has:

$$(5.4) \qquad q_{cr} = 0.199 \frac{a}{h} \sigma$$

where $2a$ is the initial crack length, h is the distance to the free surface. One can see that even for the extreme case of $h \approx a$ the unstable crack growth can be excluded by the confining pressure of 20% of the compressive stress concentration.

6. DILATANCY AND ROCK BURST MONITORING

Growth of the considered cracks is accompanied by their opening. This appears as **dilatancy** (e.g., Brace et al, 1966), which can serve for prediction of the rock burst.

In the 2-D case contribution of the growing crack into dilatancy is determined by area S of its opening. In the dipole asymptotic method the area can be calculated as a sum of areas of the crack opening due to action of the concentrated forces (2.1) and due to action of additional stresses reflected from the boundary (with the accuracy adopted by the method, the additional stresses are assumed to be uniform; see Dyskin at al, 1992, for details). The result is

$$(6.1) \qquad S = 2 \frac{1 - \nu^2}{E} \beta a l \sigma \left[4 + 3 \left(\frac{l}{h} \right)^2 \right]$$

We consider not a specific crack here, but an average one, since dilatancy is measured as the total opening area (total volume in 3-D case) of the growing cracks per unit area of the considered rock mass section. Dilatancy produced by these cracks appears as an additional (non-elastic) strain in direction perpendicular to the free surface. Let

$$(6.2) \qquad \Omega = 4 N a^2$$

be the dimensionless 2-D concentration (originally introduced by Salganik, 1973) of the pre-existing cracks situated approximately at the distance h from the free surface; N be the number of the cracks per unit area. Then the additional strain has the form

$$(6.3) \qquad \Delta \varepsilon = \frac{1 - \nu^2}{E} \frac{\Omega}{2} \frac{l}{a} \beta \sigma \left[4 + 3 \left(\frac{l}{h} \right)^2 \right]$$

where the averaged value of β is $\beta = 4/(3\pi)$ (see the previous section).

Together with (4.3), equation (6.3) gives the contribution of the cracks to the dilatancy and its dependence of the magnitude of the applied compression and the distance from the free surface. However, in reality, underground openings produce lateral (with respect to the growing crack plane) pressure. This pressure, being zero at the free surface, gradually increases with h. In spite of slow increase, the lateral pressure dramatically affects the ability of the crack for growing (see the previous section and Germanovich and Dyskin, 1988). This makes the zone of significant dilatancy to be narrow. Besides, for technological reasons the biggest pre-existing cracks are usually concentrated near the excavation surface. Therefore, for the first approximation it is sufficient to consider only dilatancy produced by the cracks that are close enough to the boundary.

The average distance h between the boundary and the closest cracks may be estimated as half of the average distance b between the considered pre-existing cracks. It may be expressed thorough the dimensionless crack concentration as follows:

$$(6.4) \qquad h \approx a \Omega^{-\frac{1}{2}}$$

This approximation allows one to simplify the equation of crack growth (4.3). Now this equation can be rewritten, using (4.4), in the following dimensionless form:

$$(6.5) \qquad \left(\frac{l}{h} \right)^2 - \frac{8}{2^{\frac{1}{4}} 3\sqrt{3}} \frac{\sigma_{cr}}{\sigma} \sqrt{\frac{l}{h}} + \frac{2}{3} = 0$$

Then, using (6.4) and (6.5), one can calculate the dilatancy-induced increment Δu of the normal displacement.

Let d be the width of the dilatancy zone. Within this zone, the dilatancy is assumed to be uniform (this rough assumption can be accepted for this simple model. For more detailed calculations, it is necessary to take into account the real distribution of the growing cracks, determined by the full stress state at the opening). As the result, the dilatancy-induced displacement has the following form:

$$(6.6) \qquad \Delta u = \frac{1 - \nu^2}{E} \frac{\sqrt{\Omega}}{2} \frac{l}{h} \beta \sigma d \left[4 + 3 \left(\frac{l}{h} \right)^2 \right]$$

Since $l/h = \sqrt{2}/3$ as $\sigma = \sigma_{cr}$, the equation can be rewritten in the dimensionless form:

$$(6.7) \qquad \frac{\Delta u}{\Delta u^{max}} = \frac{l}{h} \frac{\sigma}{\sigma_{cr}} \left[4 + 3 \left(\frac{l}{h} \right)^2 \right]$$

where

$$(6.8) \qquad \Delta u^{max} = \frac{1 - \nu^2}{E} \frac{7\sqrt{2\Omega}}{9} \beta \sigma_{cr} d$$

and l/h is determined from (6.5). Fig. 6 shows the normalised displacement vs. the normalised compressive stress magnitude.

It is seen that the relationship between the displacement increment and the stress is highly non-linear. This may allow the distinction between this part of the displacement and the total one (which includes the elastic displacement as well) and determine the two parameters Δu^{max}, σ_{cr}. The possible rock burst monitoring could be based on this

172

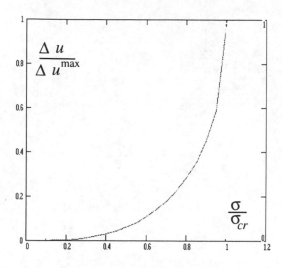

Figure 6. The normalised relationship between the dilatancy-induced inward displacement of the excavation surface and the compressive stress.

recognition. Approaching the end of the curve could indicate close onset of rock burst.

7. CRITICAL STRESS

Critical stress (4.4) can be expressed through rock compressive strength. Considering an average growing crack, one can substitute the corresponding value $\beta=4/(3\pi)$ into (4.4). The fracture toughness K_{Ic} can be expressed through the tensile strength σ_t of a rock "sample" with the same initial cracks (Germanovich and Dyskin, 1988). Using the criterion (3.1) one has $K_{Ic} = \sigma_t \sqrt{\pi a}$. Substituting this into (4.4) and using (6.4), one obtains:

$$(7.1) \qquad \sigma_{cr} = \frac{3^{3/2} \pi^2}{2^{15/16}} \sigma_t \Omega^{-1/4}$$

According to the 2-D model of dilatancy and fracture (Germanovich and Dyskin 1988) the following relation takes place between the tensile and compressive strengths

$$(7.2) \qquad \sigma_c = 5.13 \sigma_t (\Omega)^{-1/4}$$

Substituting (7.2) into (7.1), one finally obtains the following approximate criterion for rock burst:

$$(7.3) \qquad \sigma_{cr} = 0.74 \sigma_c$$

Thus, due to the free surface, rock burst starts when the compressive stress concentration is 26% below the uniaxial compression strength. We have to point out, that (6.5) is a result of the 2-D illustration of 3-D situation. However in the 3-D case, one also can expect that due to interaction with the existing free boundary the critical stress is less than the uniaxial compressive strength.

CONCLUSION

A simple model of rock burst caused by the crack growth parallel to a free surface has been considered. It has been shown that the interaction with the free surface changes the crack propagation from the stable type, at the initial stage of loading, to the unstable one. The unstable crack growth constitutes the initial mechanism responsible for rock burst.

The dipole asymptotic method has been shown to be an appropriate method for modelling interaction between the crack and free surface in the stages of the stable crack growth.

The pre-rock-burst stage is accompanied by dilatancy, which manifests itself by a highly non-linear increase of the normal displacement. This fact can be the basis for rock burst monitoring.

To exclude the unstable crack growth and effectively prevent rock burst it is sufficient to apply the confining pressure of about 20% of the stress concentration.

We have only considered the situation when stresses around an underground opening increase for some reason. The long-term stability of openings is determined by the slow crack growth under fixed compression magnitude. This case of the crack propagation, so-called "delayed fracture" or "fracture corrosion", is not considered in this paper. However in this case, one also might expect that the crack - free surface elastic interaction would eventually cause the unstable crack growth and rock burst.

The authors of this paper are listed alphabetically.

ACKNOWLEDGMENT

The authors thank K.B. Ustinov for discussion of the model of crack propagation. They are grateful to J.-C. Roegiers for the advises which have helped improving this text.

REFERENCES

Brace, W.F. and E.G. Bombolakis 1963. A note of brittle crack growth in compression. J. Geophys. Res. 68, No. 12, 3709-3713.

Brace, W.F., B.W. Paulding and C.H. Scholz 1966. Dilatancy in the fracture of crystalline rocks. J. Geophys. Res. 68, 3709-3713.

Cherepanov, G.P. 1979. Mechanics of Brittle Fracture. McGraw-Hill, New York.

Dyskin, A.V., L.N. Germanovich and R.L. Salganik 1991. A mechanism of deformation and fracture of brittle rocks. In: Rock Mechanics as a Multidisciplinary Science, J.C. Roegiers (ed.). Balkema, Rotterdam, Brookfield, 181 - 190.

Dyskin, A.V., L.N. Germanovich & K.B. Ustinov 1992. On the pore-based mechanism of dilatancy and fracture of rocks under compression. In: Rock Mechanics. Proc. of the 33rd US Symposium, J.T. Tillerson & W.R Wawersik (ed.). Balkema, Rotterdam, Brookfield, 797-806.

Dyskin, A.V. and R.L. Salganik 1987. Model of dilatancy of brittle materials with cracks under compression. Mechanics of Solids 22, No 6, 165-173

Dyskin, A.V., R.L. Salganik and K.B. Ustinov 1992. Multi-scale geomechanical modelling. Western Australian Conference on Mining Geomechanics. T. Szwedzicki, G.R. Baird and T.N. Little (ed.), 235-246.

Ewy, R.T. and N.G.W. Cook 1990. Deformation and fracture around cylindrical openings in rock. Parts I and II. Int. J. Rock Mech. Min. Sci. & Geomech. Abstr. 27, No. 5, 387-407, 409-427.

Fairhurst, C. and N.G.W. Cook 1966. The phenomenon of rock splitting parallel to the direction of maximum compression in the neighborhood of a surface. Proceedings of the First Congress of International Society on Rock Mechanics (Lisbon), 1, 687-692.

Gay, N.C. 1973. Fracture growth around openings in thick-walled cylinders of rock subjected to hydrostatic compression. Int. J. Rock Mech. Min. Sci. & Geomech. Abstr. 10, 209-233.

Germanovich, L.N. and A.V. Dyskin 1988. A model of brittle failure for material with cracks in uniaxial loading. Mechanics of Solids, 23, 111-123.

Germanovich, L.N., A.V. Dyskin and M.N. Tsyrulnikov 1990. Mechanism of dilatancy and column failure of brittle rocks under uniaxial compression. Transaction (Doklady) of the USSR Academy of Sciences: Earth Science Section 313, No. 4, 6-10.

Germanovich, L.N., A.V. Dyskin and M.N. Tsyrulnikov 1992. A model of brittle rock deformation under compression. Izv. AN Rossia, MTT, 27, No. 6. (In Russian. The English translation will be published in Mechanics of Solids 27, No. 6.)

Germanovich, L.N. and M.A. Grekov 1993. Influence of free surface. In: Germanovich L.N., A.P. Dmitriev and S.A. Goncharov, Rock Fracture Thermomechanics. To be published by Gordon and Breach. London-N.Y.

Hoek, E. and Z.T. Bieniawski 1964. Brittle fracture propagation in rock under compression. Int. J. Fracture, 1,137-155.

Landau, L.D. and E.M. Lifshitz 1959. Theory of Elasticity. N.Y. Pergamon Press.

Nemat-Nasser, S. and H. Horii 1982. Compression-induced nonplanar crack extension with application to splitting, exfoliation, and rockburst. J. Geophys. Res. 87, No. B8, 6805-6821.

Salganik, R.L. 1973. Mechanics of bodies with many cracks. Mechanics of Solids 8, No. 4, 135-143.

Slepyan, L.I. 1990. Mechanics of Cracks. Sudostroenie. Leningrad (in Russian).

Stacey T.R. 1981. A simple extension strain criterion for fracture of brittle rock. Int. J. Rock Mech. Min. Sci. & Geomech. Abstr. 18, 469-474.

Talebi, S. and R.P. Young 1992. Microseismic monitoring in highly stressed granite: relation between shaft-wall cracking and in situ stress. Int. J. Rock Mech. Min. Sci. & Geomech. Abstr. 29, No. 3, 25-34.

Wittke, W. 1984. Felsmechanik. Springer-Verlag. Berlin.

Rockbursts and Seismicity in Mines, Young (ed.) © 1993 Balkema, Rotterdam, ISBN 90 5410 320 5

Evaluation of spatial patterns in the distribution of seismic activity in mines: A case study of Creighton Mine, northern Ontario (Canada)

Mariana Eneva
Queen's University, Kingston & University of Toronto, Ont., Canada

R. Paul Young
Engineering Seismology Laboratory, Department of Geological Sciences, Queen's University, Kingston, Ont., Canada

ABSTRACT: The spatial patterns in the distribution of microseismic events in a nickel-copper mine in northern Ontario (Canada) are quantitatively studied using two complimentary techniques, pair analysis and method of correlation dimensions. These techniques allowed us to evaluate the temporal variations of two parameters, the degree of spatial non-randomness and spatial clustering in the distribution of mining-induced events. Increasing and high clustering and degrees of spatial non-randomness appear to be associated with the stress changes preceding some $M \geq 2.0$ events at this mine, while decreasing and low clustering and degrees of non-randomness follow such events. These results agree with previous observations of earthquake distribution in seismically active areas and seem promising in the search for precursory patterns to damaging rockbursts. Continuous monitoring of the parameters used here is suggested to help our understanding of the relationship between microseismic activity and stress changes due to the excavation process.

1 INTRODUCTION

Microseismic events are the single continuous and most abundant source of information about the processes taking place in mines. A common way of examining this type of data is to study the changes of the number of seismic events per unit time (seismicity rates) and to visually associate their locations with geometric and structural features in the mines and where possible, with the advancement of excavation. It is possible, however, to distinguish the development of space-time patterns in the distribution of mining-related activity applying techniques that are significantly different from simple counting of events or visual displays of their locations. Any detection of such patterns would be valuable for the understanding of the way the excavation process changes the existing stresses, although the observation of features preceding damaging rockbursts would be of the greatest interest.

The present study focuses on applying two non-traditional techniques to the study of the space-time distribution of the induced events. Both use interevent parameters, such as distances between events. The first technique, pair analysis, was previously applied to the study of spatial distribution of earthquakes and its changes in time (Eneva and Pavlis, 1988, 1991; Eneva and Hamburger, 1989; Eneva et al., 1992; Eneva, 1992). The second technique is based on the evaluation of the so-called correlation dimensions. It stems from applications in chaos theory and non-linear dynamics (e.g., Grassberger and Procaccia, 1983). This method was used to study the spatial distribution of microfracturing in rocks (Hirata et al., 1987), as well as the spatial distribution of earthquakes (Hirata, 1989; Radulian and Trifu, 1991).

Both techniques are applied here to study the spatial distribution of mining-induced events and its variations in time at Creighton Mine. Nine nickel-copper mines are operated by Inco Ltd. in the Sudbury Basin in northern Ontario (Morrison and MacDonald, 1990). Five of these are seismically active, Creighton Mine being the most active of all. About 10 events of magnitudes $M \geq 2.0$ occur annually in these mines. Sometimes the rate of seismic activity is significantly larger. For example, 35 rockbursts and significant seismic events occurred in 1987 at Creighton Mine, the largest of which was of magnitude M3.6.

The spatial distribution of mining-induced events is studied here through the use of interevent distances. The data consist of event locations supplied by Inco Ltd. The arrival times of the events have been recorded by a system used in many Canadian mines, the so-called MP-250 system. These systems are employed for automatic arrival recognition and arrival time acquisition; i.e., they do not record waveforms. The sensors are installed in arrays encompassing the active mining areas. While the events of magnitudes $M \geq 1.5$ are also detected by the Eastern Canada Teleseismic Network (with three stations on the rim of the Sudbury Basin), smaller events can only be recorded by arrays installed locally in the mines. Whole-waveform systems are also being installed, providing records like the seismograms in earthquake seismology. In this study, however, only MP-250 locations are used.

The algorithm applied at Inco Ltd. to locate the events is described in detail by Ge and Kaiser (1990). Unlike other methods of event location assuming that all arrivals are from P-waves, this algorithm considers S-arrivals and is thought to produce more accurate locations than previous methods.

2 DATA

The data used in this study is from the so-called "dense" array installed at Creighton Mine. Unlike the 32 channels per array used in other mines (Morrison and MacDonald, 1990), this array has a 64-channel capacity, although not all channels have worked continuously. The arrivals from the dense array are used mainly to locate events at depths between 6600 ft (2000 m) and 7200 ft (2182 m). Locations using the arrivals from the so-called "mine-wide" array are preferred at shallower depths.

In addition to the MP-250, we also used information provided by the Geological Survey of Canada (GSC) and CANMET. Both GSC and CANMET supply magnitudes for larger events, while the MP-250 systems cannot be used for this purpose. The CSC magnitudes are chosen here; they are usually somewhat higher than the CANMET magnitudes.

2.1 Selection of data subsets

The MP-250 locations from the dense array used in this work are for the period March 1 - May 31, 1992. 16 of these were of $M \geq 1.5$ and 7 - of $M \geq 2.0$. The total number of MP-250 locations during this period is 46486, of which 441 are blasts. The results reported here are only from a small portion of the data for which we had reasonable grounds to expect the best location accuracy. Three criteria were used to select this data. First, only category A data were considered using the location reliability designations supplied by the location algorithm (A being the highest of four categories; see Ge and Kaiser, 1990). This reduces the number of events to 21299, of which 417 are blasts (i.e., 45% of all non-blast events and 95% of all blasts). Furthermore, we conjectured that the events for which the algorithm assumed only P-arrivals on the first six sensors might be also events which are more accurately located. In general, these would be also the strongest events. Indeed, 90% of them triggered at least 10 phones, and more than 50% - at least 16 phones. This selection further reduced the number of data used to 2398 non-blast events. Finally, to make sure that the data set used consists only of the most accurately located events, we imposed limits in space that outline the greatest cluster of sensors and dropped

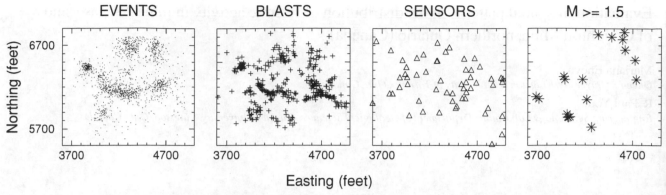

Northing (feet)

Easting (feet)

Figure 1. Maps of the main data set, blasts, sensors, and M≥1.5 events.

data falling outside these limits. The volume thus considered is 1300 x 1300 x 600 ft³ (394 x 394 x 182 m³), the first two dimensions being in the horizontal plane and the last one - in depth. This final reduction of the number of data leaves us with 1654 non-blast events. While representing only 3.6% of the initial data, we believe that these events form a subset of the best locations the MP-250 systems and the Ge and Kaiser's algorithm can produce. It is reasonable to start with this subset and only later proceed to include less accurate locations and smaller events. Figure 1 shows four plots featuring: the final data set of 1654 events within the 1300 x 1300 x 600 ft³ volume; all blasts in the same volume; the sensors of the dense array; and M ≥ 1.5 events that occurred in the same volume from mid-February until mid-June of 1992 (i.e., during the initial period from March to May 1992, ± 2 weeks). These are only part of the 25 M≥1.5 events that occurred in the whole mine during this extended period of time. Figure 1 displays only the horizontal mine co-ordinates of events, blasts and sensors (i.e., easting and northing; in feet).

2.2 Location accuracy

Our ability to evaluate the accuracy of the MP-250 locations is very limited at this time, since no representative reference data is available at present. Reasonable comparisons of MP-250 and whole-waveform locations for the study period are not possible, since the operation of the whole-waveform array was only starting at that time. Yet, the techniques applied here use distances between events and location accuracy is crucial. At the absence of better alternatives, in order to tackle this problem, we compared the actual locations of some blasts with their respective MP-250 locations. Only the actual locations for 44 blasts from the month of March were available to us at this time. All but two are of category A. The discrepancies of the actual and MP-250 locations for the 42 category A blasts vary between 15 and 405 ft (i.e., between 4.5 and 123 m), with median value at 57 ft (17 m) and 67% of the blasts located with accuracy better than 100 ft (30 m). Production blasts (40% of all blasts), as could be expected, are less accurately located than development blasts. Indeed, more than 2/3 of the production blasts are located with accuracy worse than 100 ft (30 m), while the MP-250 locations for most of the non-production blasts (60% of all blasts) are within 50 ft (15 m) from their actual locations. Comparisons of blasts and seismic events, as well as the meaning of "actual blast locations" leave a lot to be argued about, but no other reference frame is available at this time.

3 TECHNIQUES USED

3.1 Pair Analysis

The technique of pair analysis has been developed to evaluate non-randomness in distribution of earthquakes (Eneva and Pavlis, 1988). More meaningful results have been obtained studying the spatial distribution of earthquakes and its changes in time (Eneva and Hamburger, 1989; Eneva and Pavlis, 1990; Eneva et al., 1992; Eneva, 1992), while the temporal distribution of seismic events did not reveal much when studied in this way (Eneva, 1992).

Pair analysis is applied here to quantitatively study the spatial distribution of microseismic events in the volume specified above. The principle used is based on comparison between the observed frequency distribution of interevent distances and the expected distribution of distances between points randomly generated in the same volume. All possible distances are used, so that N events form P=N(N-1)/2 pairs. Thus, the 1654 events described above produce 1367031 pairs. The expected distribution is obtained as an average from many random generations (100 trials of 1654 points each in this study). The random generation used is uniform; i.e., points can occur with the same probability anywhere in the volume specified. Both the observed and the expected distributions are affected by the limited size of the study volume. This dependence is removed for all practical purposes by subtracting the expected distribution from the observed one. The resulting residual distribution for the total data set used here is shown in Figure 2. The relative residual number of pairs in per cent is shown along the vertical axis and distance in feet is shown along the horizontal axis. Positive values indicate excess of pairs at certain distances by comparison with random distribution, while negative values show deficiency of pairs and 0-values show that there is no discrepancy between the real and random distributions. One may think of the excess of pairs at relatively small distances as of clustering of events. Such anomalies can be, however, associated with either too many pairs formed by events occurring spatially close to each other (which coincides with the intuitive perception of clustering) or/and too many pairs formed by points distant from each other (which increases the relative number of pairs with smaller distances). Because clustering is to be understood only in relative sense, we have previously preferred to use the more general term "non-randomness" (see references above) rather than "clustering" and continue to do so here. The dotted lines in Figure 2 designate tolerance limits which cover 90% of the randomly generated frequencies with 95% confidence. Values outside these limits are definitely significant, while values within them are of unknown significance. Further details about the application of pair analysis can be found in the above references.

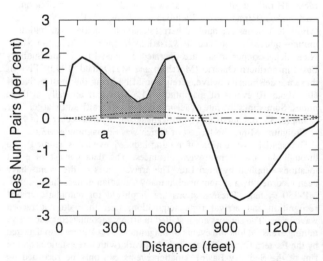

Figure 2. Residual distribution of pairs used in pair analysis.

The residual distribution can be effectively used to quantitatively measure the degree of spatial non-randomness, $c_{[a,b]}$, over certain distance range [a,b]:

$$c_{[a,b]} = \pm \sqrt{\left| \sum_{i_a}^{i_b} \Delta p_i / P \right|} \qquad (1)$$

where the sign depends on whether the anomaly is represented by excess or deficiency of pairs, i_a and i_b designate the discrete distance intervals in which a and b fall, the sum runs from $i = i_a$ to $i = i_b$, Δp_i is the residual number of pairs in interval i, and the total number of pairs is $P = N(N-1)/2$ for N events. This quantity is equivalent to the square root of the stippled area shown in Figure 2 and approximately represents the portion of events involved in the anomaly over distance range [a,b]. Although the parameter thus described measures non-randomness in a non-unique way, it provides a convenient and useful characteristic of the overall spatial distribution of events in a given seismically active volume.

Using the whole data set, we obtain a baseline long-term value for the degree of spatial non-randomness over certain distance range [a,b], further denoted by $c^*_{[a,b]}$. Groups of events are studied in the same manner. Since these groups cover various periods of time, temporal changes in the degree of spatial non-randomness can be studied by comparing these short-term values with the long-term value, $c^*_{[a,b]}$. The short-term values are further denoted by $c^j_{[a,b]}$, where each j designates a specific group. The behaviour of the values of $c^j_{[a,b]}$ during periods associated with larger events such as rockbursts is of utmost interest. While identifying precursory features would be most important, pair analysis can be used in a much more general sense, since it can provide a lot of insight about the inter-relation between the spatial distribution of seismic activity and stress heterogeneity at various stages of the excavation process.

3.2 Correlation Dimensions

The concepts of self-invariance and fractal geometry have found their way in geophysics, more specifically in studying crustal fragmentation and fault geometry (e.g., Aviles and Scholz, 1987), as well as in describing earthquake occurrence in time, space and size (e.g., Hirata, 1989). Only two applications of this type are known to us that use data from mines. Merceron and Velde (1991) carried out fractal analysis of the fractures in a Japanese mine, while Coughlin and Kranz (1991) reported changes in the fractal dimension of the spatial distribution of induced events before and after rockbursts in a silver-copper mine in Idaho, U.S.. The two latter authors applied the method of correlation dimensions introduced by Grassberger and Procaccia (1983). The same method has been also applied to earthquake data (Hirata, 1989; Radulian and Trifu, 1991; Eneva, 1992).

The technique of correlation dimensions is the second technique used in this study. Figure 3 illustrates the so-called correlation integral calculated for the total data set of 1654 events. After some modifications, the correlation integral C(R) at interevent distance R can be calculated from:

$$C(R) = n(r<R)/P = 2n(r<R)/N(N-1) \qquad (2)$$

where C(R) equals the number of all pairs, $n(r<R)$, with interevent distances $r<R$, normalized by the total number of pairs, $P = N(N-1)/2$ for N events. It has been found that for a distribution with a fractal structure and for small values of the distance R, this integral can be represented by a power law, where the power is the so-called correlation dimension d:

$$C(R) \sim R^d \qquad (3)$$

The correlation dimension d is equal to or smaller than the fractal dimension D ($d \leq D$). In the double logarithmic plot shown in Figure 3, the correlation integral, C(R), is shown along the vertical axis, and the distance, R, along the horizontal axis. The curve features a straight segment over certain distance range [A,B], called scaling range. The slope of this line segment equals the correlation dimension d. To designate the scaling range over which the slopes are estimated, we further denote the correlation dimensions by $d_{[A,B]}$. The curve marked by the open circles is derived from the real data, and the filled circles

form a curve obtained from 100 randomly generated data sets, of 1654 events each. Note that the distance ranges [A,B] here and [a,b] in 3.1 above do not have to coincide. Figure 3 shows that the observed curve is above the random curve, and the slope (i.e., the correlation dimension) derived from the straight segment of the observed curve would be smaller than the one derived from the random data. Smaller slopes mean that relatively more pairs are concentrated at the left side of the scaling range (i.e., at smaller distances) than for the cases when the slopes are larger. That is, the smaller the dimensions (slopes) the higher the degree of clustering.

Similar to the degrees of spatial non-randomness, long-term and short-term correlation dimensions are also estimated and are further denoted by $d^*_{[A,B]}$ and $d^j_{[A,B]}$, respectively.

3.3 Pair analysis versus correlation dimensions

The same type of quantities is apparently used in both pair analysis and the correlation dimension estimates. Nonetheless, there are several differences between the two techniques:

1. Pair analysis accounts for the limited size of area and its shape by subtracting expected from observed distributions. No similar procedure has been developed to remove the effect of saturation from the correlation curve; a scaling region [A,B] can be observed only at distances within about not more than 1/3 of the average size of the study volume. For this reason, one can be much more flexible in choosing the distance range [a,b], than the distance range [A,B].

2. The measure of non-randomness from pair analysis over certain range [a,b] characterizes this range as a whole. The correlation dimension, being a slope of a straight segment, is estimated over all the points along this segment and therefore depends on the structure within the scaling range [A,B]. For this reason, identical degrees of spatial non-randomness (from pair analysis) do not imply identical correlation dimensions, even if range [a,b] coincides with [A,B]. This comparison shows that although in both cases distances are used, the pair analysis and the correlation dimension analysis utilize these distances differently.

3. The differences in (1) and (2) above are intimately related to another conceptual difference between pair analysis and the correlation dimension method. Correlation dimensions cannot be estimated unless scaling is present. Pair analysis does not suffer of this limitation.

4 RESULTS FROM CREIGHTON MINE

The long-term values of the degree of spatial non-randomness and the correlation dimension are obtained from the total data set of 1654 events. The short-term values on the other hand, are obtained from overlapping groups of 100 events each, moving forward in time by 20 events. This division results in 78 groups. Figures 4 and 5 show residual distributions and correlation integral for several of the groups. Each curve in Figure 4 spans 1400 ft (424 m) which is the maximum

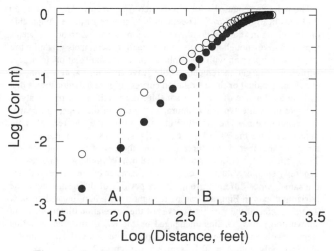

Figure 3. Correlation integral for real and random data.

Groups 9 to 14

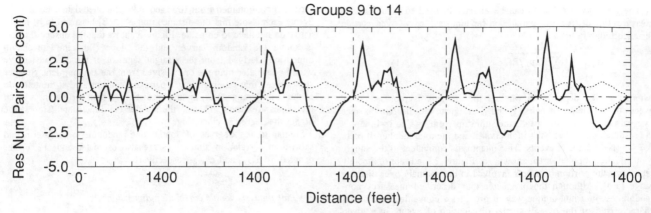

Figure 4. Residual distributions for six groups.

Groups 24 to 28

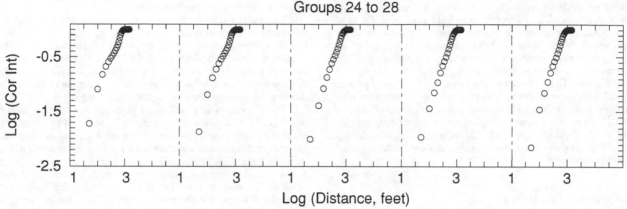

Figure 5. Correlation integral for five groups.

interevent distance, while each curve in Figure 5 spans log(1400 ft)=3.15.

4.1 Choice of parameters

Any distance range can be used to evaluate the degree of spatial non-randomness, but ranges are preferred for which the same type of anomaly is observed in all groups. The distance range [a,b] used here is a=0 ft to b=250 ft (76 m); all group residual curves show significant excess of pairs over this range. Similarly, the range [A,B] is chosen as the range over which scaling is observed for all group correlation curves. In this case, [A,B] seems well defined by A=50 ft (15 m) and B=400 ft (121 m). The spatial degree of non-randomness and the correlation dimension are therefore evaluated over different distance ranges in this work.

The pairs are counted with certain step (increment), Δr, which is 50 ft (15 m) in Figures 2 to 5. The intent is to choose Δr to be comparable or larger than the predominant location error. Δr=50 ft seems representative enough here, given that smaller increments might be too small by comparison with the probable predominant location accuracy, while much larger increments would leave us with too few points in either the residual or the correlation curves. Figure 6 demonstrates that as long as the size of Δr is reasonable, the results from pair analysis indicate the same type of temporal variations in the degree of spatial non-randomness. The short-term values of the degree of non-randomness for the 78 groups, $c^j_{[0,250]}$ (j=1,..., 78), are plotted against the group number, j, for three different choices of the counting step: 25 ft (8 m), 50 ft (15 m), and 125 ft (38 m). While larger Δr's lead to generally smaller values of $c^j_{[0,250]}$, the character of the curves remains the same. Since different groups cover periods of different length, the horizontal axis in Figure 6 can be thought of as a non-linear time axis. The character of the variation curves of the correlation dimensions (not shown here) also remains unchanged by the use of different counting steps in distance. The two next figures therefore show estimates obtained only with Δr=50 ft, same as in Figures 2 to 5.

4.2 Results from pair analysis

The long-term degree of spatial non-randomness, $c^*_{[0,250]}$, estimated from the total data set measures at 26.8 per cent, while only 4.6 per cent can be reached by randomly generated distribution. These estimates are made applying equation (1) to the residual curve and the upper tolerance curve shown in Figure 2. In addition to the anomalies at small distances, anomalies at relatively large distances are also observed for some groups, similar to previous studies of earthquakes (Eneva and Pavlis, 1988, 1991). Their variations from group to group may be very informative for the spatial distribution of mining-induced events, but are not examined here.

The short-term values, estimated from the 78 groups for the range [0,250] are shown in Figure 7 by filled squares connected with a solid

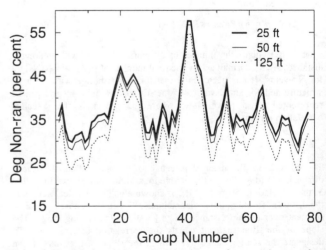

Figure 6. Temporal variations of spatial degree of non-randomness for different counting steps.

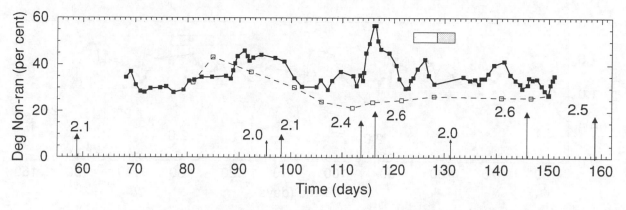

Figure 7. Temporal variations of spatial degree of non-randomness for events and blasts.

line. Note that the short-term degrees of non-randomness, $c^j_{[0,250]}$ ($j=1,...,78$), are all higher than $c^*_{[0,250]}=26.8$ per cent and exceed 40 per cent several times. Unlike the previous figure, the values in Figure 7 are not equally spaced along the horizontal axis, because it features time. Time is given in days from the beginning of 1992 (e.g., day 61 is March 1 and day 152 is May 31). The values of $c^j_{[0,250]}$ are marked at the times when the 100-th event of each group occurs; i.e., at the ends of the groups. Each value, however, is estimated over certain period of time that varies from group to group. The lengths of these periods vary between 2.4 and 10.7 days, with a median length at 4.7 days. These estimates do not include five of the largest periods, because they are artificially long; groups 56 to 60 include days 130 to 132, for which data was missing. The two boxes on the top of Figure 7 mark the missing data. The open box extends over the beginning times of the five affected groups. The box with the stippled pattern shows a gap in the beginning times caused by the missing data. As the short-term values are shown at the ending times of the groups, the values for the five affected groups are the ones immediately following the stippled box.

The variations featured in Figure 7 can be caused by one or more of the following: stress changes due to excavation and blasting; stress changes due to impending larger events and/or rockbursts; stress changes caused by such events that have already occurred; random fluctuations not associated with any of the above.

The role of excavation and blasting is yet to be properly estimated. Information about the excavation process is not available to us at this time. The times and locations of the blasts, however, are known (within certain accuracy). One way to approach this problem, is to apply the same analysis to the blasts. 327 of all blasts were within the volume specified above. Long-term and short term degrees of spatial non-randomness were estimated for these blasts in the same way as before. Only 12 groups could be considered here, since the number of blasts is much smaller than the number of events. The periods covered by these groups vary between 16.2 and 30.9 days with a median at 27 days. The dashed line and the open squares in Figure 7 mark the variation curve for the blasts, which unlike the previous curve, is a direct result of human activity. The comparison of the two variation curves does not suggest any clear relationship.

To estimate the possible relationship with larger events, we compare the times of their occurrence with the changes suggested by the variation curve. The vertical arrows along the horizontal axis in Figure 7 mark the times of M$\geq$2.0 events. Their lengths are proportional to the magnitudes shown next to the arrows. Large arrowheads are used for events that occurred within the study volume. Small arrowheads mark events that occurred outside the volume, but not further than 500 ft (150 m) away. This leaves 8 M$\geq$2.0 events of interest for the study volume within the period between mid-February and mid-June, 1992.

Increased degree of non-randomness seems to precede the larger events. Values higher than 40 per cent are observed prior to most M$\geq$2.0 event. The highest maximum approaching 60 per cent is associated with two closely located events of M2.4 and M2.6 (114-th and 116-th days; i.e., April 23 and 25). The time periods over which the values of $c^j_{[0,250]}$ are observed to increase or remain relatively high vary between 2 and 10 days before the larger events. In at least one representative case, a M2.6 event (day 146, May 25) seems to be preceded by increase and subsequent decrease of the degree of spatial

non-randomness within about 1 week. On the other hand, decreasing and/or low values appear to follow some of the larger events within similar periods of time; a M2.1 event (98-th day, April 7) and especially the M2.4-M2.6 couple later in April are good examples. Similar studies in seismically active areas also revealed increasing degrees of spatial non-randomness preceding M$\geq$5.0 main shocks, as well as decreasing degrees following such events (Eneva and Pavlis, 1988, 1991; Eneva et al., 1992). The continuous search for temporal changes in the degree of spatial non-randomness is therefore well justified for the seismic activity in mines.

4.3 Results from correlation dimensions

The estimate of the correlation dimension using the scaling range between A=50 ft (15 m) and B=400 ft (121 m) for the whole data set of 1654 events yields a long-term value $d^*_{[50,400]}=1.68733 \pm 0.00039$. Because the correlation dimensions are estimated through slopes of straight segments, it is also possible to estimate their standard deviations. Since the dimension of the embedding space is 3, one generally assumes that a correlation dimension between 2 and 3 indicates a distribution intermediate between plane-filling and volume-filling distributions, while a dimension below 2 would reflect some tendency towards distribution along a line. Estimates based on the suggestions given by Nerenberg and Essex (1990) show, however, that the particular combination of number of events and size of volume here may not define a proper scaling range despite the appearance of straight line segments. This is perhaps reflected in the correlation dimension for the randomly generated data estimated at only 2.3265 ± 0.0074, quite below the proper value of 3. Whatever error we make in these estimates, we may assume that it remains approximately the same in time as the study volume does not change. For this reason, we proceed to compare the long-term value of 1.68 with the short-term values $d^j_{[50,400]}$ (Figure 8).

The notations in Figure 8 are the same as in Figure 7, with only two differences: correlation dimensions are shown along the vertical axis instead of degrees of non-randomness and the two curves at the base of the plot show the standard deviations for the groups of events (solid line) and for the groups of blasts (dashed line). This plot indicates that $d^j_{[50,400]}\leq d^*_{[50,400]}$ for all $j=1,...,78$; i.e., the short-term values indicate higher clustering at all times by comparison with the long-term correlation dimension. Using results from laboratory experiments Hirata et al.(1987) have reported decreasing spatial correlation dimension preceding rock failure. Hirata (1989) observed the same change before a large earthquake in Japan and Coughlin and Kranz (1991) reported similar changes associated with at least two rockbursts in a silver-copper mine. In agreement with these observations, our results suggest that decreasing correlation dimensions (increasing clustering) seem to precede some M$\geq$2.0 events at Creighton Mine for about at least 10 days. The M2.1 event on day 98 (April 7) is the most representative example. The M2.0 event just before it might have played some role too, but it occurred outside the study volume. The groups ending on the first few days of another 10 day period preceding the M2.5 event on day 158 (June 6) are also characterised by markedly decreasing correlation dimensions. The observations stop short in this case, however, as the end of data is reached. On the other hand, increasing

179

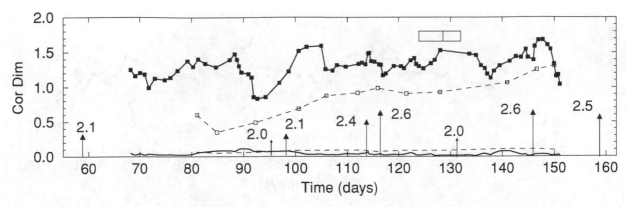

Figure 8. Temporal variations of correlation dimension for events and blasts.

and relatively high correlation dimensions (decreasing and low clustering) appear to follow some of the larger events, most notably the same M2.1 event as above and a M2.6 event on day 146 (May 25). In contrast, the M2.4-M.2.6 couple on days 114 and 116 (April 23 and 25) is not associated with any significant effect on the correlation dimensions. This is very different from the results from pair analysis, where the same couple of events were associated with the greatest anomaly. This demonstrates that pair analysis and the method of correlation dimensions are complimentary techniques and are not necessarily equally sensitive to changes in spatial patterns.

5 CONCLUSIONS

Two complimentary techniques, pair analysis and method of correlation dimensions, were applied to the study of spatial distribution of mining-induced events and its variations in time at Creighton Mine, northern Ontario. Using comparisons with randomly generated points, these techniques allowed us to quantitatively evaluate two parameters: the degree of spatial non-randomness and the spatial clustering in the distribution of microseismic events. Significant temporal variations were observed in these parameters that most probably reflect the stress changes associated with the excavation process. Of particular interest are changes in the spatial pattern possibly caused by stress changes preceding and following $M \geq 2.0$ events. Increasing and high degrees of spatial non-randomness, as well as decreasing and low spatial correlation dimensions (identical with increasing and high clustering) could be observed for up to 10 days prior to some $M \geq 2.0$ events. Conversely, decreasing and relatively low degrees of spatial non-randomness and increasing correlation dimensions (same as decreasing clustering) appear to follow some of these larger events.

The meaning of these results cannot be overstated at this time. The possibility for random fluctuations not related with either the blasts or the larger events, cannot be ruled out. It is necessary to study the time periods surrounding many more larger events. The results presented here, however, are in very good agreement with observations of patterns associated with laboratory rock failure and larger earthquakes. Continuous monitoring of these parameters is therefore strongly recommended not only at Creighton Mine, but in all mines characterized by high microseismic activity and the occurrence of damaging rockbursts.

6 ACKNOWLEDGEMENTS

We thank Inco Ltd., Sudbury (Ontario) for funding this study and providing the data used. At various stages of this work we had numerous helpful discussions with Doug Morrison and Terry Villeneuve from Inco Ltd.; Shawn Maxwell, Ted Urbancic, and Cezar-Ion Trifu from Queen's University; Marc Diederichs from Laurentian University; Robert Wetmiller from GSC; and Michel Plouffe from CANMET.

REFERENCES

Aviles, C.A. and C.H. Scholz, Fractal analysis applied to characteristic segments of the San Andreas fault, *J. Geophys. Res.* 92, 331-344, 1987.

Coughlin, J. and R. Kranz, New approaches to studying rock burst-associated seismicity in mines, in *Proceedings of the 32nd U.S. Symposium, Rock Mechanics as a Multidisciplinary Science*, Ed. J.-C. Roegiers, 491-500, 1991.

Eneva, M., Investigation of the space-time distribution of earthquakes in the Charlevoix seismic zone, Quebec, *Open-File Report, Geological Survey of Canada*, Ottawa, 1992.

Eneva, M. and M. W. Hamburger, Spatial and temporal patterns of earthquake distribution in Soviet Central Asia: Application of pair analysis statistics, *Bull. Seism. Soc. Amer.* 79, 4, 1475-1476, 1989.

Eneva, M., M. W. Hamburger, and G. A. Popandopulo, Spatial distribution of earthquakes in aftershock zones of the Garm region, Soviet Central Asia, *Geophys. J. Int.* 109, 38-53, 1992.

Eneva, M. and G. L. Pavlis, Application of pair analysis statistics to aftershocks of the 1984 Morgan Hill, California, earthquake, *J. Geophys. Res.* 93, 9113-9125, 1988.

Eneva, M. and G. L. Pavlis, Spatial distribution of aftershocks and background seismicity in central California, *Pure Appl. Geophys.* 137, 35-61, 1991.

Ge, M. and P.K. Kaiser, An innovating micro-seismic source location technique, in *Proceedings of the 31st U.S. Symposium, Rock Mechanics Contributions and Challenges*, Eds. W.A. Hustrulid and G.A. Johnson, 43-50, 1990.

Grassberger, P. and I. Procaccia, Measuring the strangeness of strange attractors, *Physica* 9D, 189-208, 1983.

Hirata, T., A correlation between the b value and the fractal dimension of earthquakes, *J. Geophys. Res.* 94, 7507-7514, 1989.

Hirata, T., T. Satoh, and K. Ito, Fractal structure of spatial distribution of microfracturing in rock, *Geophys. J. R. astr. Soc.* 90, 369-374, 1987.

Merceron, T. and B. Velde, Application of Cantor's method for fractal analysis of fractures in the Toyoha mine, Hokkaido, Japan. *J. Geophys. Res.* 96, 16641-16650, 1991.

Morrison, D.M. and P. MacDonald, Rockbursts at INCO mines, in *Proceedings of the 2nd International Symposium on Rockbursts and Seismicity in Mines, Minneapolis 8-10 June 1988*, Ed. C. Fairhurst, 263-267, 1990.

Nerenberg, M.A.H. and C. Essex, Correlation dimension and systematic geometric effects, *Phys. Rev.* A 42, 7065-7074, 1990.

Radulian, M. and C.-I. Trifu, Would it have been possible to predict the 30 August 1986 Vrancea earthquake? *Bull. Seism. Soc. Amer.* 81, 2498-253, 1991.

Rockbursts and Seismicity in Mines, Young (ed.) © 1993 Balkema, Rotterdam, ISBN 90 5410 320 5

Failure mechanisms of microseismic events generated by a breakout development around an underground opening

B. Feignier & R. P. Young
Engineering Seismology Laboratory, Department of Geological Sciences, Queen's University, Kingston, Ont., Canada

ABSTRACT: Thirty seven microseismic events associated with a tunnel excavation are analyzed. Their hypocenters correlate with a zone of damage observed underground. This damage develops in the σ_3 direction similarly to borehole breakouts being observed in the petroleum industry. The goal of this work is to enhance our understanding of the response of the rock mass to an opening in a high-stress environment. A detailed analysis of the source mechanisms is carried out, using moment tensor inversion, to investigate the growth of the breakout in its early stage. We find that as much as 75% of the mechanisms contain a significant negative isotropic component. These events are interpreted as closure of cracks, due to changing stress conditions, that opened during earlier excavation phases.

1 INTRODUCTION

The presence of a volumetric component in the source mechanism of mining-induced seismic events, at various magnitude scales, has been recently discussed in numerous publications. This volumetric component can be either positive in the case of an explosional mechanism (Kozak and Sileny, 1985, Sileny et al., 1986, Gibowicz et al., 1990, Gibowicz et al., 1991, Feignier and Young, 1992) or negative for an implosional mechanism (Kusnir et al., 1980, 1984, Rudajev and Sileny, 1985, Sileny, 1989, Wong et al., 1989, McGarr, 1992, McGarr, 1993). It is clear that shear failures are not the only mechanism observed in mining environments. It is also evident that moment tensor inversion is the most conclusive method to assess and quantify such volumetric source mechanisms (see Jost and Herrmann, 1989 and Gibowicz, 1993, for reviews).

In this paper, we are studying microseismic events generated by a tunnel excavation and associated with the development of a breakout in the roof of the tunnel. Our goal is to better understand the physical process leading to the breakout growth and its relation with the in-situ stresses. We are taking advantage of the sixteen triaxial accelerometers located all around the tunnel to invert for the moment tensor components of the seismic events. We are focusing particularly on the contribution of volumetric sources in this process.

2 THE UNDERGROUND RESEARCH LABORATORY

The Underground Research Laboratory (URL), located in Pinawa, Manitoba, is a research facility operated by Atomic Energy of Canada Limited. Its main objectives are to advance our understanding of the response of a rock mass to excavation and investigate the long term stability of an underground opening. In the "Mine-by" experiment, a 46-meter-long tunnel has been excavated at 420 m depth in the Canadian shield granite. The most interesting feature associated with the excavation is the development of two breakouts in the roof and the floor of the tunnel. Their orientation is parallel to the σ_3 direction, the tunnel itself being oriented perpendicular to σ_1. The in-situ stresses have been acurately defined, based on numerous overcoring tests and hydraulic fracturing experiments (Martin, 1990).

The ratio σ_1/σ_3 is over 3.9 and this highly deviatoric stress field explains the breakout growth. Furthermore, it is worth noting that

no preexisting fractures have been observed in the granite at that depth.

Prior to the excavation of the tunnel, a Queen's Microseismic System (QMS) was installed to monitor the seismic activity. Sixteen triaxial accelerometers were cemented at the bottom of boreholes all around the planned tunnel to optimize for source location and focal coverage (Figure 1). Their frequency response ranges from 50 Hz to 10 kHz (±3 dB) and the sampling frequency is set to 50 kHz. Frequency sweeps from 20 Hz to 20 kHz were used to calibrate the amplitude/frequency response of the acquisition system. An extensive description of the QMS can be found in Young et al. (1993).

Before the Mine-by experiment started, velocity and attenuation surveys were carried out to enhance our knowledge of the rock mass properties. From the velocity survey it was found that the rock mass is roughly homogeneous with $V_P = 5880 \pm 60$ m/s. From the attenuation survey, estimates of Q_P were obtained using the spectral ratio method (Feustel and Young, 1993). They found that $Q_P = 250 \pm 50$ in the 4 to 8 kHz range.

Table 1. In-situ stresses at the Underground Research Laboratory at 420 m depth (after Martin, 1990).

	Magnitude (MPa)	Trend (°)	Plunge (°)
σ_1	55	135	10
σ_2	48	044	05
σ_3	14	290	79

3 MICROSEISMIC DATA

This paper deals with data recorded after the excavation of a one-meter-long section located right at the middle of the tunnel, i.e where source location and focal coverage are optimal. The technique used to remove this piece of rock is called rock breaking. It consisted of drilling one-meter-long holes (45 mm diameter) all around the perimeter of the tunnel. Then, the holes were reamed up to a 100 mm diameter to almost connect each other. After completion of the reaming, the microseismic system was turned on to monitor the seismic activity at that early stage. The system monitored continuously for 4 hours 45 minutes. Then,

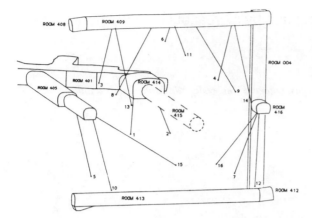

Figure 1. 3D view of the mine-by tunnel (dotted lines) and the instrumentation galleries. Each triaxial accelerometer is indicated by a number at the bottom of a borehole (after Read and Martin, 1991).

hydraulic rock splitters were used to break out the interior of the tunnel plug. Once the rock breaking and the mucking were finished, the seismic system was switched on again. Subsequently, the seismic network monitored continuously for 72 hours and provided data over a 300 hour time window, allowing an extensive analysis of the seismicity related to the breakout growth.

The whole data set has been manually picked and amounts to a total of 359 microseismic events. Source locations have been computed using a homogeneous velocity model giving an average RMS residual of about 0.05 ms (2.5 sample points). The spatial distribution of the events defines two zones of activity in the roof and in the floor of the tunnel (Figure 2a). As shown in Figure 2b, the two clusters perfectly fit the eventual damage zones observed in the tunnel. In the remaining of this paper, we will concentrate on 37 events (Table 2) occuring in the upper notch immediatly after reaming, i.e at the initiation of the breakout. Figures 3a and b display the source location of the data set in longitudinal and cross-section views respectively. One can notice from Table 2 that the first 33 events occur in less than one hour while the last four are spread over four hours. It is also worthwile noting that the last 4 events occur closer to the mining face, indicating a shift in source location with time. On the longitudinal view (Fig. 3a), it is apparent that the seismic events are closer to the tunnel free surface as they are located closer to the advancing face. This is

related to the greater development of the breakout away from the mining face, on average 0.4 m depth, one meter away from the face.

4   MOMENT TENSOR INVERSION

Our approach is similar to the one used by Oncescu (1986) and De Natale and Zollo (1989). According to the representation theorem for seismic sources in general elastodynamic theory and assuming that the Green's functions G_{nk} vary smoothly within the source volume, and expressing the equivalent body forces as the force moment tensor M_{kj}, the displacement d at a point x and a time t is given by:

$$d_n(x,t) = \sum_{m-1}^{\infty} [\frac{1}{m!} G_{nk,j1\ldots jm}(x,t;\xi,\bar{t}) \otimes M_{kj1\ldots jm}(\xi,\bar{t})]$$

where $\otimes$ is the temporal convolution. Using the point source approximation, one can show that only the first term of this series is required (Stump and Johnson, 1977). Since we neglect the higher-order terms, $M_{ki}(\xi,t)$ is a second-order time-dependent moment tensor. If we now assume that all components have the same time history (e.g synchronous source), and that the source time function is a heaviside step function, then the representation theorem becomes:

$$d_n(x,t) = G_{nk,j} \cdot M_{kj}$$

where M_{kj} is a real and symmetric tensor composed of six independent components. To solve this time-independent moment tensor, we used the low-frequency plateau of the displacement spectrum of P-waves corrected for anelastic attenuation and multiplied by the sense of deviation of the first motion. Green's functions (G matrix) are computed for far-field data in a homogeneous, isotropic and unbounded medium. The use of a full-space is justified because our sensors are grouted at the bottom of the bore-holes and the assumption of an homogeneous, isotropic rock mass is supported by the results of the velocity survey. The solution is obtained by computing the Singular Value Decomposition (SVD) of the G matrix. Its robustness is assessed by the condition number, i.e the ratio of the largest to the smallest eigenvalue in the SVD algorithm. A Householder decomposition is then performed to compute the eigenvalues and eigenvectors from the moment tensor matrix, which represent respectively the amplitudes and orientations of the forces acting at the source. The

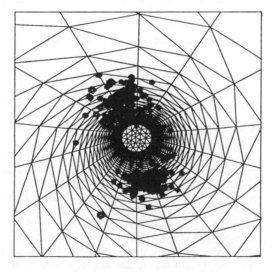
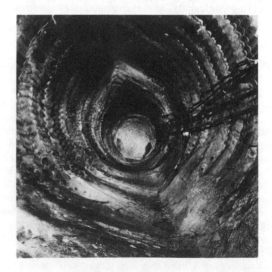

Figure 2. Cross-section of the mine-by tunnel. (a) 3D view of the tunnel with microseismic activity (359 events) related to the excavation of a 1 m section (b) Picture showing the damage observed after excavation.

Table 2. Source locations and moment tensor solutions for the 37 selected events. Err is the location error calculated from the RMS residual, C is the condition number in the inversion, N is the number of phases, M_0 is the seismic moment, λ_1, λ_2, λ_3 are the eigenvalues of the moment tensor, R is the ratio of isotropic/deviatoric component and Sign is the confidence level of having a non-deviatoric solution.

Event #	Date	Time	X (m)	Y (m)	Z (m)	Err (m)	C	N	M_0 (N.m)	λ_1	λ_2	λ_3	R %	Sign %
1	03-05-1992	10:05:22.50	761.06	438.30	123.37	.22	0.5	15	04322.8	-.216	-.388	-1.000	-63.3	99.000
2	03-05-1992	10:05:24.86	761.33	438.09	123.16	.34	1.8	13	02591.6	0.498	-.327	-1.000	-34.8	96.463
3	03-05-1992	10:05:27.66	760.85	437.94	123.57	.25	0.5	15	01122.2	0.141	-.388	-1.000	-51.6	99.000
4	03-05-1992	10:06:25.66	761.09	437.89	123.65	.30	1.5	13	01998.5	1.000	-.489	-0.740	-09.6	00.000
5	03-05-1992	10:07:21.03	761.59	438.25	123.29	.30	1.9	13	01424.9	0.118	-.438	-1.000	-54.1	99.000
6	03-05-1992	10:08:23.81	761.12	436.75	123.84	.47	2.2	9	00912.5	1.000	-.266	-0.582	07.4	59.154
7	03-05-1992	10:08:26.39	761.51	437.59	123.68	.27	0.4	14	01198.6	-.352	-.582	-1.000	-73.1	99.000
8	03-05-1992	10:08:27.38	761.39	437.22	123.71	.27	1.3	13	02467.7	0.575	.512	-1.000	04.1	00.000
9	03-05-1992	10:18:00.53	761.38	438.43	123.23	.22	0.5	15	01751.8	0.338	-.105	-1.000	-34.0	95.855
10	03-05-1992	10:20:01.20	760.87	437.89	123.51	.40	1.9	10	02012.4	-.101	-.350	-1.000	-58.4	99.000
11	03-05-1992	10:20:43.43	760.94	438.14	123.52	.52	1.9	13	04150.6	0.899	.158	-1.000	02.7	00.000
12	03-05-1992	10:20:54.25	760.35	437.69	123.56	.32	1.8	14	04339.7	0.229	-.579	-1.000	-49.9	99.000
13	03-05-1992	10:21:57.58	761.17	438.71	122.55	.47	1.5	12	03679.9	0.353	-.276	-1.000	-40.0	99.000
14	03-05-1992	10:23:44.63	761.79	438.38	123.32	.33	0.5	14	02408.3	0.868	.081	-1.000	-02.5	00.000
15	03-05-1992	10:23:53.31	761.25	437.33	123.16	.51	1.7	11	03859.4	0.585	-.117	-1.000	-24.5	79.976
16	03-05-1992	10:24:45.77	761.49	437.54	123.72	.26	0.4	14	10165.3	-.180	-.455	-1.000	-64.3	99.000
17	03-05-1992	10:24:47.58	761.43	437.61	123.97	.47	0.6	12	02944.5	1.000	-.174	-0.765	03.0	00.000
18	03-05-1992	10:32:55.21	760.50	437.49	123.69	.28	1.8	12	00754.8	0.038	-.468	-1.000	-57.7	99.000
19	03-05-1992	10:32:59.98	761.56	437.89	123.56	.28	0.4	13	03619.9	-.074	-.425	-1.000	-60.0	99.000
20	03-05-1992	10:33:02.73	761.58	438.00	123.60	.31	2.1	14	01769.2	0.006	-.647	-1.000	-59.7	99.000
21	03-05-1992	10:33:06.41	760.73	437.48	123.76	.28	2.4	12	06763.9	0.606	-.350	-1.000	-30.4	94.167
22	03-05-1992	10:33:08.22	760.96	437.57	123.71	.31	1.5	13	00947.7	-.123	-.270	-1.000	-56.5	99.000
23	03-05-1992	10:33:11.02	760.81	437.43	123.67	.29	1.9	10	05787.9	0.126	-.557	-1.000	-54.3	99.000
24	03-05-1992	10:33:14.21	761.63	438.20	123.57	.26	0.5	15	01933.7	0.541	-.284	-1.000	-32.0	99.000
25	03-05-1992	10:33:25.31	761.12	437.27	123.89	.28	1.5	11	01339.9	-.183	-.346	-1.000	-60.9	99.000
26	03-05-1992	10:34:17.87	760.53	436.81	123.78	.24	1.7	13	05188.0	1.000	-.277	-0.468	12.2	00.000
28	03-05-1992	10:45:57.67	760.96	438.08	123.55	.25	1.6	13	01475.1	0.280	-.517	-1.000	-47.2	99.000
30	03-05-1992	10:46:22.56	761.15	438.05	123.72	.35	1.9	13	03562.8	-.135	-.413	-1.000	-61.5	99.000
32	03-05-1992	10:53:24.05	761.39	438.40	123.33	.21	1.4	14	00915.9	-.225	-.466	-1.000	-65.9	99.000
33	03-05-1992	10:56:23.06	761.82	438.37	123.46	.34	1.5	13	01425.2	0.118	-.875	-1.000	-55.5	97.919
34	03-05-1992	11:57:31.75	760.53	437.01	123.84	.23	1.5	14	03215.7	0.442	-.444	-1.000	-39.2	99.000
35	03-05-1992	12:13:26.63	760.19	437.95	123.74	.27	0.5	14	01230.7	0.441	-.393	-1.000	-38.6	99.000
36	03-05-1992	14:23:13.31	760.33	436.95	123.95	.30	1.4	12	00803.4	0.561	-.051	-1.000	-22.6	95.597
37	03-05-1992	15:01:40.23	759.58	436.51	123.73	.23	1.7	14	01518.7	0.361	-.321	-1.000	-41.3	99.000

scalar seismic moment is given by (Silver and Jordan, 1982):

$$M_0 = \left(\frac{1}{2} \sum_{i=1}^{3} |m_i|^2 \right)^{\frac{1}{2}}$$

where m_i are the eigenvalues. The trace of the moment tensor, $\mathrm{tr}(M) = m_1 + m_2 + m_3$, represents the volume change at the source. If $\mathrm{tr}(M) \neq 0$, there is an isotropic part (i.e a volume change) at the source. Then, the moment tensor matrix may be decomposed into isotropic and deviatoric parts (Jost and Hermann, 1989). We quantify the respective percentage of isotropic/deviatoric component as:

$$R = \frac{tr(M) * 100}{(|tr(M)| + \sum |m_i^*|)}$$

where m_i^* are the deviatoric eigenvalues. The ratio R varies from 100 (pure explosion) to -100 (pure implosion), and R = 0 indicates a purely deviatoric source mechanism. However, since the presence of an isotropic component can be due to a poor fit of the data in the inversion, we also computed the pure deviatoric solution by forcing the solution to be traceless. The difference between the 2 distributions (free and deviatoric solutions) is assessed through an F-test that uses the variances between observed and calculated amplitudes for each distribution:

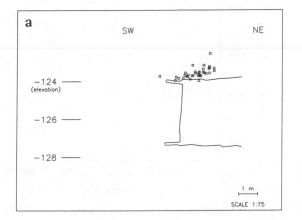

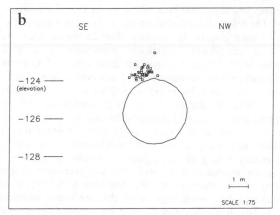

Figure 3. Data set selected (37 events) for the moment tensor inversion. (a) Longitudinal view (b) Cross-section view.

$$F_\chi = \frac{\chi^2(p) - \chi^2(p+1)}{\chi^2(p+1)} * (N - (p+1))$$

where p is the number of degrees of freedom and N the total number of observations. The χ^2 have been computed with 1/amp and 1/sqrt(amp) weighting functions to check the stability of the F-test. However, since the results are extremely similar, we will only give the results using the 1/amp weighting function. Figure 4 shows the significance of the difference between the two distributions as a function of the ratio R. When the difference was not significant at a 50% confidence level, the significance was automatically set to 0. As already pointed out by Feignier and Young (1992), a minimum of 30% of isotropic component is required to produce a significant difference at a 95% confidence level.

We calibrated the program using controlled explosions initially detonated for an attenuation survey. Although not located inside the array, these sources were found to have as much as 85% of isotropic component and were significantly different from a deviatoric solution at a 99% confidence level.

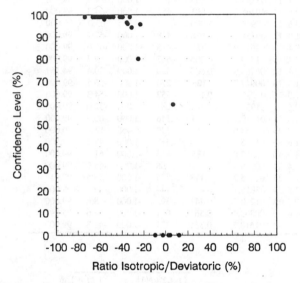

Figure 4. Confidence level in the difference between the free and the deviatoric solution as a function of the ratio R.

5   RESULTS AND INTERPRETATION

From the initial data set of 37 events, stable moment tensor solutions were obtained for 34 events (condition number smaller than 20). Events 27 and 29 had to be dropped because their first motions could not be picked with enough confidence and event 31 was ignored because it is located 5 meters away from the mining face and is probably not related to the rest of the seismic activity. The results from the inversion are given in Table 2. The most important pattern is certainly that 25 events show an implosional component statistically significant at a 95% confidence level. Figure 5 shows the focal sphere of an event containing an implosional component. These events are not purely implosional as shown by either a positive eigenvalue, which indicates a deviatoric component, or a large discrepancy between the largest eigenvalues and the two others, indicating the closure of a crack. The remaining events show mainly deviatoric source mechanisms and no explosional source mechanisms are observed.

Feignier and Young (1992) reported a similar pattern while studying the seismicity associated with the excavation of the beginning of the mine-by tunnel. Although a drill-and-blast technique was then used, they found that the events recorded during the first few hours after the blast were mainly implosional while events recorded much later (over one day) displayed rather

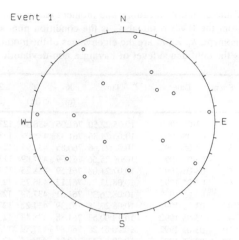

Figure 5. Focal sphere of event 1 (cf Table 2).

shear-explosional source mechanisms. Figure 6 shows the amplitude of the trace of the moment tensor solutions as a function of time. It appears clearly that the events with the largest volumetric components happen in the first half hour of the data set (e.g, within 45 minutes after the end of the reaming).

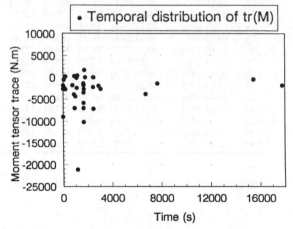

Figure 6. Temporal distribution of the moment tensor traces.

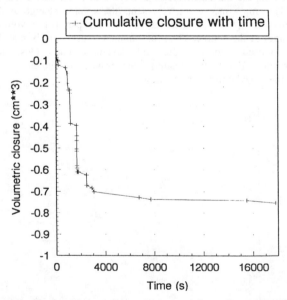

Figure 7. Cumulative closure (computed from the events with an implosional component) with time.

Following McGarr (1992), we computed the volumetric closure associated with these implosional events using:

$$\Delta V = tr(M_{ij})/(3\lambda + 2\mu)$$

(Aki and Richards, 1980). At the URL, we estimated $3\lambda + 2\mu = 1.527\ 10^{11}$ Pa. Figure 7 shows the cumulative closure with time. The total cumulative closure is on the order of 0.75 cm³, with more than 80% of it occuring in the first half hour. We interpret these closures as resulting from the change in the stress field due to the reaming. Small cracks which opened during an earlier round (after rock breaking) are now readjusting to the new stress conditions.

In order to analyze the source mechanisms in greater detail, we will now focus on those providing the best fitting solutions. Our selection criterion is that calculated displacements after inversion must fit the measured displacements within +/- 30%. This criterion was met for 27 events. In order to test the overall quality of the inversion, we computed synthetic seismograms. The original program (Coutant, 1989) was modified to input any kind of moment tensor, namely the solution obtained from the inversion. A homogeneous, isotropic propagation model was used, as in the inversion code. However, a Ricker source function was found to give better results than the Heaviside step function used in the inversion. The source function was arbitrarily centered around 4 kHz to match the frequency content of the real seismograms. As shown in Figure 8, the fit between real and synthetic waveforms is excellent. Figure 9 displays the ratio R for the 27 events. Interestingly, the events are clustered in two distinct groups. The first one includes 6 events with a ratio R centered around 0 (deviatoric sources), the second one includes 21 events with a ratio centered around -50 (sources with implosional component). This behaviour is very similar to the one observed by McGarr (1993) on 10 large rockbursts in South African mines. However, these groupings cannot be simply explained since no clear spatio-temporal differences exist between the 2 groups.

If we now focus only on the events containing an implosional component, we can analyze the orientation of the largest negative eigenvalue. It can be related to the axis normal to the plane on which the closure is happening since in most cases $|\lambda_3| \gg |\lambda_1|$ or

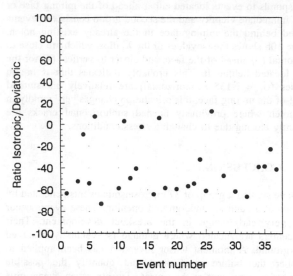

Figure 9. Values of the ratio R for the 27 events with best fitting moment tensor solutions (the event number refers to Table 2).

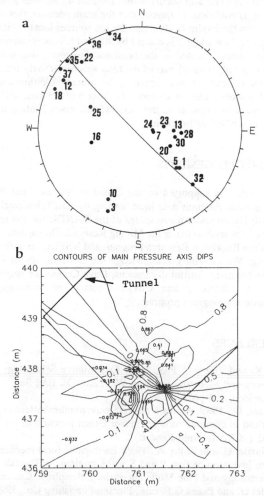

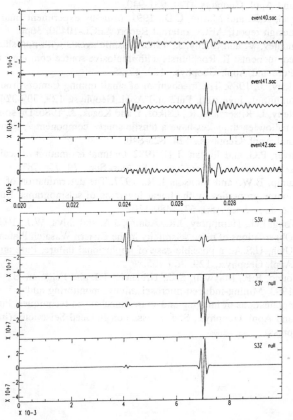

Figure 8. Comparison of real (top) and synthetic (bottom) seismograms for event 1 recorded at triaxial 14 (cf Fig. 1).

Figure 10. Distribution of the largest eigenvalue (λ_3) for the 21 events with an implosional component. (a) stereonet projection (numbers correspond to the event numbers in Table 2). The σ_1-σ_3 plane is indicated (N226 84S). (b) contour map of the λ_3 dips, the tunnel position is shown.

$|\lambda_2|$ (cf Table 2). Figure 10a displays a stereonet with the orientations of λ_3 for all implosional events. Two main clusters are present, one in the NW corner with subhorizontal λ_3, the other in the SE quadrant with strongly dipping λ_3. It appears that these 2 clusters are also distinct spatially, since the first group corresponds to events located either ahead of the mining face or in its immediate vicinity and the second group consists of events located behind the mining face in the already existing notch. Figure 10b shows the isovalues of the λ_3 dips, which are close to horizontal (0) ahead of the face, and closer to vertical (1) for the ones located behind it. This probably indicates that if in-situ stresses (σ_1 = N135 subhorizontal) are relatively undisturbed ahead of the mining face, this orientation changes very rapidly in the notch where previously opened subhorizontal cracks are presently closing due to changing stress conditions.

6   CONCLUSIONS

We have studied a group of 37 microseismic events generated by a breakout around an underground opening. These events occur at a very early stage in the breakout development. Their hypocenters are located where damage is eventually observed underground. A moment tensor inversion has been applied to decipher the failure mechanism and quantify the possible volumetric component at the source. Twenty seven events give very reliable solutions and 21 of them show a significant implosional component. The total cumulative closure is of the order of 0.75 cm^3 and occurs within the first 45 minutes after the reaming is complete. It appears that the main pressure axis at the source fits the in-situ σ_1 direction for the sources located close to the mining face while it is rotated by 90° and becomes subvertical for the sources located in the breakout. However, this data set represents only a small part of the total seismic activity recorded after the removal of this section of the tunnel. Additional work, applying moment tensor inversion to the rest of the data, will give a complete description of the physical processes leading to the development of the breakout.

ACKNOWLEDGEMENTS

This work was supported by the Canadian Nuclear Fuel Waste Management Program with joint funding by AECL Research and Ontario Hydro under the auspices of the CANDU owners group. The authors wish to thank the Mining Research Directorate of the Canadian Rockburst Research Program and NSERC for additional funding. We gratefully acknowledge the contributions of Derek Martin for many fruitful discussions, Dave Collins for helping in the data processing and Olivier Coutant for providing the synthetic seismogram program.

REFERENCES

Aki, K. and Richards, P.G., 1980. Quantitative Seismology: Theory and Methods, Freeman W.H. and Co. Eds, San Francisco, 984 pp.

Coutant, O., 1989. Axitra a discrete wavenumber/reflectivity method to compute moment tensor Green function in layered media, CEA Internal Report.

De Natale, G. and Zollo, A., 1989. Earthquake focal mechanisms from inversion of first P and S wave motions, in Digital Seismology and Fine Modeling of the Lithosphere, Cassinis R., Nolet G. and Panza G.F. Eds., Plenum Publishing Co., 399-419.

Feignier, B. and Young, R.P, 1992. Moment tensor inversion of induced microseismic events: evidence of non-shear failures in the -4 < M < -2 moment magnitude range, Geophys. Res. Lett., 19, 1503-1506.

Feustel, A.J., and Young, R.P., 1993. Attenuation analysis at the AECL Underground Research Laboratory using the spectral ratio method, (in preparation).

Gibowicz, S.J., 1993. Seismic moment tensor and its application in mining seismicity studies: a review, (this issue).

Gibowicz, S.J., Harjes, H.P. and Schafer, M., 1990. Source parameters of seismic events at Heinrich Robert Mine, Ruhr Basin, Republic of Germany: Evidence for nondouble-couple events, Bull. Seismol. Soc. Am., 80, 88-109.

Gibowicz, S.J., Young, R.P., Talebi, S. and Rawlence, D.J., 1991. Source parameters of microseismic events at the Underground Research Laboratory in Manitoba, Canada: Scaling relations for the events with moment magnitude smaller than -2, Bull. Seismol. Soc. Am., 81, 1157-1182.

Jost, M.L. and Herrmann, R.B., 1989. A student's guide to and review of moment tensors, Seismol. Res. Lett., 60, 37-57.

Kozak, J. and Sileny, J., 1985. Seismic events with non-shear components: I. Shallow earthquakes with a possible tensile source component, Pure and Appl. Geophys., 123, 1-15.

Kusznir, N.J., Al-Singh, N.H. and Ashwin, D.P., 1984. Induced-seismicity generated by longwall coal mining in the North Staffordshire coal fiald, U.K., in Rockburst and Seismicity in Mines, Gay N.C and Wainwright E.H Eds., South African Inst. Min. and Metall. Symp. Series, 6, 153-160.

Kusznir, N.J., Ashwin, D.P. and Bradley, A.G., 1980. Mining induced seismicity in the North Staffordshire coal field, England, Int. J. Rock. Mech. Sci. Geomech. Abst., 17, 44-55.

Martin, C.D., 1990. Characterizing in situ stress domains at AECL's Underground Research Laboratory, Can. Geotech. J., 27, 631-646.

McGarr, A., 1992. An implosive component in the seismic moment tensor of a mining-induced tremor, Geophys. Res. Lett., 19, 1579-1582.

McGarr, A., 1993. Moment tensor of ten Witwatersrand mine tremors, Pure and Appl. Geophys., Special Issue on Induced Seismicity (in press).

Oncescu, M.C., 1986. Relative seismic moment tensor determination for Vrancea intermediate depth earthquakes, Pure and Appl. Geophys., 124, 931-940.

Read, R.S., and Martin, C.D., 1991. Mine-by experiment final design report, AECL Internal Report AECL-10430, 36pp.

Rudajev, V. and Sileny, J., 1985. Seismic events with non-shear components: II. Rockbursts with implosive source components, Pure and Appl. Geophys., 123, 17-25.

Sileny, J., 1989. The mechanism of small mining tremors from amplitude inversion, Pure and Appl. Geophys., 129, 309-324.

Sileny, J., Ritsema, A.R., Csikos, I. and Kozak, J., 1986. Do some shallow earthquakes have a tensile source component ?, Pure and Appl. Geophys., 124, 825-840.

Silver, P.G. and Jordan, T.H, 1982. Optimal estimation of scalar seismic moment, Geophys. J. R. Astr. Soc., 70, 755-787.

Stump, B.W., and Johnson, L.R., 1977. The determination of source properties by the linear inversion of seismograms, Bull. Seismol. Soc. Am., 67, 1489-1502.

Wong, I.G., Humphrey, J.R., Adams, J.A. and Silva, W.J., 1989. Observations of mine seismicity in the Eastern Wasatch Plateau, Utah, U.S.A.: a possible case of implosional failure, Pure and Appl. Geophys., 129, 369-405.

Young, R.P., Maxwell, S.C., Urbancic, T.I. and Feignier, B., 1993. Mining-induced microseismicity: monitoring and applications of imaging and source mechanism techniques, Pure and Appl. Geophys., Special Issue on Induced Seismicity (in press).

Rockbursts and Seismicity in Mines, Young (ed.) © 1993 Balkema, Rotterdam, ISBN 90 5410 320 5

Triggering of self-organized system: Implications for the state of the uppermost crust

J.R.Grasso
Observatoire de Grenoble, LGIT/IRIGM, France

ABSTRACT: Earthquakes induced by human activity, in otherwise historically aseismic areas, provide direct evidence that both changes in pore pressure ($\pm\Delta p$) and mass transfers ($\pm\Delta m$) trigger seismic instabilities of the uppermost crust with magnitude ranging from 3.0 to 7.0. The apparently small changes in driving deviatoric stress, close to 1 MPa, regardless of the tectonic setting, confirm that the continental crust everywhere must be nearly at a state of stress near failure. Once the seismicity is triggered, it is sustained by stress changes of at least one order of magnitude less than the trigger stress change. This supports the concept that the induced seismicity belongs to a class of Self-Organized Critical systems (SOC systems) that evolve naturally toward a critical state characterized by power law distributions in space and time. Thus, induced earthquakes define Induced SOC (ISOC) systems that are subsets of the Self-Organized Criticality that are proposed for tectonic earthquakes (Bak and Tang, 1989; Sornette and Sornette, 1989; Scholtz 1991). A break in self-similarity in each ISOC System correlates with the width of the local seismogenic beds. Characteristic dimensions of ISOC phenomena range from the size of mine pillars, in the case of Mining Induced Seismicity, to the thickness of brittle sedimentary beds, for ISOC systems in the vicinity of dams or depleted hydrocarbon reservoirs. These characteristic dimensions mimic the effect that the schizosphere plays on tectonic earthquakes (Scholtz 1991). The earthquake size is locally bounded by the thickness of seismogenic beds, there is no paradox why the major induced earthquakes ($m \geq 6$) occur in so-called "stable" shield provinces where thick shallow seismogenic beds (>10 km) are available. The ISOC systems suggest that the regions of possible seismic hazard are much larger than the areas where earthquakes are common. The study of Induced SOC systems illustrates the basic theoretical ideas of self-similar processes and helps to understand their connections to physical processes of earthquake mechanics. Particularly, the shallow location of most ISOC systems imply that seismic and aseismic instabilities occur within these ISOC systems. There is a source of discrepancy between the fractal geometry of faults and the fractal geometry of shallow earthquakes. Moreover, we can estimate that ISOCS can emerge under low shear stress (~1 MPa) and are sustained by stress changes less than ~0.1 MPa but larger than the ~0.01 MPa of tidal stress.

1 INTRODUCTION

Most of the world's seismically active faults are concentrated along plate boundaries, and slip on them results from relative movements between essentially rigid plates. Stress accumulates continuously as one plate move past another until an earthquake occurs. Yet, many earthquakes occur within the generally aseismic rigid plates, and little is known of the processes causing intraplate seismic instabilities in continental setting. The details of what triggers seismogenic failures well away from plate boundaries, at a particular place and time are generally unclear, despite numerous studies (see for example Byerlee, (1967); Nur (1972); Costain, et al. (1987); Sibson (1989)) that rely, by indirect evidence, on changes in fluid pressure and movements of fluids. An opportunity to test and quantify these possible mechanisms by direct evidence, is offered where changes in hydro-mechanical parameters of the crust are known, i.e. in areas where earthquakes are induced by human intervention.

Six stress indicators are usually used to infer the orientation and the relative magnitude of the contemporary in-situ tectonic stress field in the Earth's lithosphere: earthquake focal mechanisms, well bore breakouts, hydraulic fracturing and overcoring, and young geologic data including fault slips and volcanic alignments (i.e. Zoback and Zoback, 1980). Most of the types of data previously described give an indication of the orientation of some components of the stress tensor without information on the magnitude of the stress components. The hydraulic fracturing stress measurement is the only technique that ideally provides information on both the magnitudes and orientations of horizontal stress (e.g. Haimson and Fairhurst; Zoback and Haimson, 1983). On the other hand, when the mechanics of earthquakes induced by certain types of human activity can be isolated, an estimation of the perturbation of the stress field that drives the faulting is possible in term of driven shear stress. Previous works relate seismic stress parameters (stress drops from induced earthquakes) to the relevant in situ stress and rock properties (Wyss and Molnar, 1972; McGarr et al., 1979; Fletcher 1982; Simpson 1986). In this paper we review major reported cases of induced seismicity including Reservoir Induced Seismicity (RIS) and seismicity that is triggered by subsurface fluid manipulations, with a special emphasis on sampling a variety of so called "stable" tectonic settings, i.e. areas where the deformation rate is slower than in the so called "active" areas where most of the seismicity takes place. The mechanics of triggered earthquakes are used to quantify local critical stress thresholds. Observations reported in Zambia (Gough and Gough 1970), SE China, SW India (Gupta and Rastogi, 1976), western Uzbekistan (Simpson and Leith, 1985), northern Holland, Norway and SW France (Grasso, 1992) show examples of untypical seismicity relative to low historical seismicity and are in the vicinity of either water reservoir impoundments or hydrocarbon extractions. In the second part of this study, each previously low seismicity area is treated as an isolated system, in which external forces, that are applied at the boundaries of the regions, are relatively constant by comparison to the fast fluid induced changes in internal forces that have altered the stress and the stability of the regions. We report observations in which the induced seismicity, once triggered, is sustained over years and obeys power law size distribution. Such a setting argues that induced seismic instabilities, ranging in magnitude from 3.0 to 7.0, define subsets of Self-Organized Criticality. Then local uppermost crustal blocks are moved to criticality as a response to pore pressure changes and mass transfers and define Induced SOC (ISOC) systems.

2 OVERVIEW OF MECHANICS OF FLUID INDUCED EARTHQUAKES

Historically, two types of earthquakes were suggested to be associated with fluid manipulations: those produced by the filling of an artificial reservoir (Carter, 1945; Gupta and Rastogi, 1976; Simpson, 1986), and those associated with fluid injections under pressure greater than hydrostatic (Evans, 1966; Healy et al. 1968; Raleigh et al. 1972). Reservoir impoundments trigger seismicity either by direct loading effects or by coupled poro-elastic effects (Simpson 1986). The direct effects of the load are an elastic effect and an undrained poroelastic effect. The elastic effect will change the radius of the Mohr circle, depending on the tectonic environment, and modify the seismicity accordingly, with respect to the Coulomb failure envelope: induced seismic activity in normal faulting areas and triggered seismic quiescence in thrust faulting area (i.e. Bufe 1976; Jacob et al. 1979). The second direct load effect, the increase of pore pressure immediately below the reservoir, will always favor induced seismicity by moving the Mohr circle towards the origin. Many seismic events of magnitude between 3 and 4 have been reported in the vicinity of subsurface fluid injections, i.e. hydrocarbon fields stimulated by fluid injection or waste storage (Colorado, (Evans, 1966; Healy et al. 1968; Raleigh et al. 1972); N.Y., (Fletcher and Sykes, 1977); Nebraska (Evans and Steeples, 1987); Ohio (Nicholson et al. 1988). Delayed seismic responses to reservoir impoundment correspond with more gradual diffusion of water from the reservoirs to hypocentral depths (Simpson et al. 1988a). Such possible connections on tens of kilometers distance between fluid manipulations and earthquakes were also proposed at Denver and Lacq fields on the basis of pore pressure diffusion associated to deep well activities (Hsieh and Bredehoeft 1981; Grasso et al. 1992a). The delayed mechanisms for induced seismicity are not well constrained due to the lack of knowledge of hydrological properties with depth and will not be used in this work.

Earthquakes induced by fluid loading are understood to result from an increase in pore pressure, which causes the effective normal stress on the fault plane to be reduced. Conversely, there was an apparent paradox when Grasso (1992) reviewed numerous cases where simple hydrocarbon extractions had triggered seismic activity. From the observations of seismic events recorded by local networks in areas surrounding hydrocarbon fields, e.g. in Alberta, Canada (Wetmiller 1986), SW France (Grasso and Wittlinger, 1990; Guyoton et al. 1992) and northern Holland (Haak 1991), it appears that such earthquakes have not occurred in the depleted portions of the reservoir where gas pressure has decreased, but instead that most of the seismicity has occurred above or below the actual gas reservoir. These observations are explained by a poroelastic stress transfer from the depleted reservoir to the levels surrounding the hydrocarbon reservoir (Segall, 1989). The linear relation between the decrease in pore pressure and subsidence near Lacq (Grasso et al. 1992b) supports the poroelastic model of the local strain and stress field. At other sites where the available measurements do not permit unambiguous identification of the rock mass rheology, Grasso (1992) proposes a two step mechanism in which poroelastic stressing occurs before the in-situ faults, which localize displacement, had been reactivated. In these cases seismic slips ($m_{max} \sim 4$) are second order pattern of a main aseismic displacement. In a broader context, massive hydrocarbon recovery reduces the vertical load which can trigger earthquakes in a thrust faulting environment. The same basic relation explains seismic energy released by mine induced earthquakes (McGarr 1976). McGarr (1991) suggests a mechanical connection between the mass withdrawal and the earthquake size by the fact that the seismic deformation (estimated through the seismic moment of the earthquakes) is that required to offset the force imbalance caused by hydrocarbon production (proportional to the mass withdrawal). The other types of seismicity induced by mass withdrawal include those due to quarrying (e.g. Pomeroy et al 1976; Yerkes et al, 1983) and those associated with removal of ice sheets (Stein et al., 1979; Johnston, 1987,1989).

3 INDUCED EARTHQUAKES AS STRESS GAUGES

3.1 Stress changes induced by pore pressure changes

Pore pressure changes that trigger Reservoir Induced Seismicity can be maximized by using the water height of the reservoir. The values stand close to 1-2 MPa for most of $m \geq 4$ RIS (see for example Table 1 of Gupta, 1985). In some areas, fluctuations of the water level by a few meters correlates with seismicity in agreement with.the model of the coupled pore pressure field beneath the reservoir (Simpson and Negmatullaev 1981; Roeloffs, 1988). Roeloffs's analysis argues that a 0.1 MPa size of the stress step is needed to trigger seismicity. Similarly, upon fluid injection performed by washing for salt removal purposes in Dale (NY) a monthly rate of 80 earthquakes was observed, against a background of one natural earthquake per month (Fletcher and Sykes, 1977). Downhole pressures ranged from 5.2 to 5.5 MPa at 450 m (Plitho/Pfluid $\approx$1.9), but for pressures below 5 MPa, there was no seismic activity.

Application of the poroelastic stressing model to the Lacq gas field where hydromechanical parameters are constrained by numerous subsurface data (porosity, bulk and elasticity moduli, Biot coefficient, pore pressure history, and surface subsidence) predicts that a variation in the shear stress of a few bars triggers seismicity above and below the gas reservoir (Segall and Grasso, 1991). At other sites we estimated the maximum induced stress at the onset of seismicity by using either the subsidence and shear modulus of the rocks involved, or a rough value depending on pressure drop and reservoir geometry, as proposed by Segall's model. For all the sites where estimates can be performed, stress changes smaller than 1 MPa appear to trigger seismic instabilities (Grasso 1992). The stress step that triggers seismicity appears to be roughly close to those estimated to trigger seismicity in the surrounding of artificial water reservoir impoundment. Moreover these seismic instabilities occur under a great variety of regional tectonic conditions, e.g. stable in the Netherlands and Texas; stable and/or extensional in the North Sea, Norway and Denmark; and compressional in the Caucasus, France and Canada. Due to the mechanical properties of most of shallow rock matrix involved, this type of seismic slip is a second order process of the main non-brittle deformation (Grasso 1992).

3.2 Stress changes induced by perturbations of the loading system of the upper crust

It is worth mentioning the possible relationship between major earthquakes ($m \geq 6$) and some hydrocarbon production sites. Both in California (McGarr 1991) and in Uzbekistan (Grasso, 1992) major events can be related to upper crustal hydromechanical disturbances produced by subsurface fluid extraction from shallow geological formations. Despite the smaller fluid pressure changes observed at these sites as compared to those previously described to trigger events by pore pressure effect, larger seismic responses and at greater distances to the reservoir are observed (McGarr, 1991; Grasso, 1992). Assuming that mass withdrawal at these sites is related to the major earthquake sequences based on McGarr's (1991) analysis, we estimate a rough stress change for each sequence by multiplying the earthquake stress drop by the ratio of stiffness of the loading system (the vertical force change) to the stiffness of circular crack model. These estimates suggest, although by less direct evidence than the mechanisms involving pore pressure effects, that earthquake faulting occurs under low shear stresses (<1 MPa).

A crude critical mass of a few 10^{11} kg, (Grasso 1992), appears to induce the large thrust events in compressive settings. Such a critical mass raises two remarks. First, the same critical mass was reached on other hydrocarbon fields, where $S_1 \neq S_H$ and the only poroelastic response triggered moderate seismicity ($m<5$) close to the depleted reservoirs (e.g. Lacq field, western Pyrennees foreland; Groningen field, northern Holland; Ekofisk field, central graben Norway). In agreement with admitted standard tectonic settings for the 3 areas given as examples (e.g. Zoback 1992), the lack of thrust faulting in these 3 areas inhibits large seismic failure from occurring in response to mass withdrawal. Only the poroelastic stressing effect, which is independent of the preexisting tectonic stress, can induce earthquakes. On this basis, we propose that the competition between seismicity triggered by poroelastic stress change and the seismic response of the crust to mass transfer is a robust indicator of the local tectonic setting. Second, the crude value of the critical mass withdrawal that induces the thrust events is at the lower limit of the increases in load that trigger a fast normal faulting response to reservoir impoundment (Gupta and Rastogi, 1976; Simpson et al. 1988). Accordingly, normal faulting types of RIS are also indicators that confirm the regional extension setting of the African midplate stress field (Kariba), the local extension state of deformation of different foot hill massifs (Sierra Nevada, Oroville; French subalpine massif, Monteynard) and the western anatolian stress province (Kremasta). These earthquakes are the only 4 normal faulting type of the top 9 $m \geq 5.5$ RIS listed by Gupta (1985). Note that in the Himalayan thrust region, load increases inhibit the occurrence of thrust earthquakes. Accordingly, numerous reservoir impoundments, for water height $\geq$ 100m, have not triggered fast seismic responses (Gupta 1985). For the other fast seismic responses, due to reservoir impoundment, the observed reverse slip or strike slip events are associated with local increases in pore pressure in undrained conditions that are independent of the tectonic setting, as well as the delayed response type observed for large Reservoir Induced Seismicity (Gupta and Rastogi, 1976; Simpson, 1976, 1986; Gupta, 1985). Whatever the mechanism of pore pressure change is (drained or undrained), the water height is the key parameter that bounds the maximum stress changes. Despite that a minimum 100 m water height is admitted to be the critical threshold that triggers RIS (e.g. Rothé, 1970), there is no correlation between water height and the size of the largest RIS (Figure 1). This observation and the unusually high b-values reported for RIS (Gupta and Rastogi, 1976) (i.e. larger number of small shock in the frequency-magnitude diagram), support the hypothesis that the size of the Reservoir Induced Earthquakes is bounded by the size of seismogenic bed of the upper crust. Accordingly, the largest induced earthquakes must occur in areas where the largest seismogenic beds stand, (i.e. shield area). This appears to be verifiable for the top 4 largest induced earthquakes that succeeded to activate fault lengths larger than 10 km in an intraplate area (Gasli, Kariba, Koyna, Hsingfenkiang).

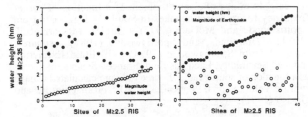

Figure 1 : Size of induced earthquakes as a function of water height of the reservoir. Data are $m \geq 2.5$ RIS from Table 1 of Gupta (1985)

4 INDUCED SEISMICITIES AS SELF-ORGANIZED CRITICAL SYSTEMS (ISOC)

Earthquakes occur in response to the accumulation of the stress pattern imposed by tectonic processes and are controlled by boundary conditions of many sorts. The occurrence of events in space and time in the vicinity of human activities are a direct reflection of the local changes (in space and time) which permit triggering where appropriate conditions are met. A careful examination of the time/space seismicity pattern in previously low seismicity areas where changes in stresses and the mechanical properties of the involved medium can be known, provide unique opportunities to interpret seismic instabilities. The earthquakes, triggered or induced by fluid manipulations, result either from pore pressure changes (effective stress law and poroelastic effects) or from mass transfers (Newtonian effect or isostatic effects). The sites where these shallow earthquakes occur are located in different tectonic settings worldwide.

First, we use the mechanics of induced seismicity to derive the steps in stress that either trigger the seismic failures or sustain the induced seismic instabilities. From the observed cases of induced seismicity and for the 2 proposed mechanisms, pore pressure effect and mass effect, it appears that small stress variations (1-2 MPa), trigger seismic failures in the upper crust in the vicinity of water reservoir impoundments and subsurface fluid manipulations. The observations and models that we present extend the critical small stress threshold previously observed and modeled only where plates interact. It is worth mentioning that small stress transfer (~0.1MPa) has been proposed to explain coseismic slip-triggered or slip-locked by an earthquake on neighbouring faults (Jones et al. 1982; Simpson et al. 1988b) in areas prone to failure,(i.e. naturally seismically activated). Recent models of the coseismic aftereffects of the Landers' California earthquake support that a 0.1-1 MPa stress change affects seismicity 30-100 km apart from the main $m\sim 7.4$ shock (Harris and Simpson 1992, Stein et al. 1992, Jaumé and Sykes, 1992). In the case of induced seismic instabilities, after the principal effect had locally moved the area to criticality, small variations in stress (~0.1 MPA), related to variations either in surface water level storage (Figure 2) or in subsurface fluid pressure drop (Figure 3), control seismic slips on pre-existing discontinuities. Thus, in several areas stress changes of at least one order of magnitude smaller than the trigger stress change can sustain induced seismicity over months or years. These

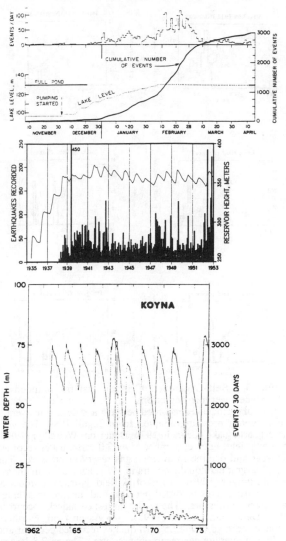

Figure 2 : Self Organized Critical systems induced by reservoir impoundments. -a) Monticello reservoir, USA; number of events per day and cumulative number of events versus lake level (adapted from Talwani et al. 1979; Flecher 1982). -b) Lake Mead, Hoover dam, USA; water level (curve) versus monthly frequency of earthquakes (vertical bars), adapted from Roeloffs (1988). The $m\sim 5$ largest earthquake occurs in 1939. Data from Carder (1970). -c) Koyna, India; same as (b).The $m\sim 6.2$ largest earthquake occurs in 1967. The second largest earthquake, $m\sim 5.2$ occurs in 1973 when the water level was allowed to increase to 1 meter beyond the 1967 maximum, (from Simpson et al. 1988). The time duration of ISOC systems ranges from a few months after the maximum water level had been reached at Monticello, (a), to tens of years at Lake Mead, (b). For the above RIS cases, the earthquakes do not move with an elevated isobaric front. Because local in-situ strengths have unknown magnitude and spatial distribution, events occur both along and behind the preceding lower front pressure. The stress change can be read by transforming the water height tp pressure i.e. 100 m ~ 1 MPa.

observations are intimately related to the concept of the structuration of the whole lithosphere as a Self-Organized critical phenomenon (Bak et al. 1987), which has recently been suggested to be relevant for understanding the processes underlying earthquakes (Bak and Tang, 1989; Sornette and Sornette, 1989; Scholtz 1991). In Figure 4 direct evidence for induced earthquake self-similarity is presented, this being an important criterion for the SOC hypothesis. From Mining Induced Seismicities (MIS), where induced seismic instabilities disappear the no-worked days (see for example Cook (1976)), we can estimate that the shortest lifetime for ISOC systems is a few tens of hours. From seismic nests, that are repetitively activated on volcanoes by moon tides, the minimum stress changes that drive ISOC systems can be estimated to a few

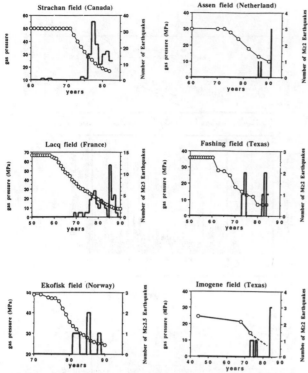

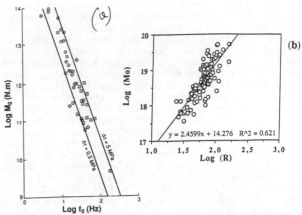

Figure 4 : Example of power law size distribution relation for induced earthquakes. -a) Mining Induced Seismicity, South Africa gold mines. Seismic moment as a function of corner frequency for mine earthquakes in South Africa. Data from Spottiswoode and McGarr (1975) - Circles, McGarr et al. (1981) - Diamonds, and Spottiswoode (1984) - squares. Adapted from Spottiswoode (1984). -b) Fluid Extraction Induced Seismicity, Lacq gas field, France. Seismic moments as a function of source radii. Source radii are calculated using a rupture velocity of 0.9b, R=0.32b /f0. From Feignier and Grasso (1991).

Figure 3 : Self-Organized Critical systems induced by subsurface reservoir depletions. -a) Strachan field, Canada; number of earthquakes recorded per year and decline in average gas reservoir pressure, adapted from Segall 1989. The $m\sim4$ largest earthquake occurs in 1974. Data from Wetmiller (1986). -b) Lacq field, France; number of $m\geq3.0$ earthquakes recorded per year and decline in average gas reservoir pressure. The $m\sim4.2$ largest earthquake occurs in 1981; adapted from Grasso and Wittlinger (1990). -c) Ekofisk field, Norway; number of $m\geq2.5$ earthquakes recorded per year and decline in average gas reservoir pressure. The $m\sim3.8$ largest earthquake occurs in 1988. Data from Grasso (1992). -d) Assen field, Northern Holland; number of $m\geq2.5$ earthquakes recorded per year and decline in average reservoir pressure. The $m\sim2.8$ largest earthquake occurs in 1986. Data for earthquakes (Haak 1991, writt. comm.), Grasso (1992). -e) Fashing field, Southern Texas; number of $m\geq2$ earthquakes recorded per year and decline in average gas reservoir pressure. The $m\sim3.4$ largest earthquake occurs in 1983. Data from Pennington et al (1986). -f) Imogene field, Souththern Texas; number of $m\geq2$ earthquakes recorded per years and decline in average gas reservoir pressure. The $m\sim3.9$ largest earthquake occurs in 1984. Data from Pennington et al (1986). The induced stress at a time t, is roughly related to the gas reservoir pressure as follow (see text for details): $\Delta\sigma\sim3\times10^{-3}$ ($P_{init}-P_t$). Note that if seismicity emerges with $\Delta\sigma\sim1$ MPa, it is further sustained by $P_{init}-P_t \leq10$ MPa, i.e. $\Delta\sigma\leq0.1$ MPa.

hundredths of MPa. Although searches for RIS/tide correlation are weak, due to the poor number of data, tidal-stress triggering of RIS had been proposed as significant for a few RIS sites (Klein, 1976). For the cases reviewed in our study, changes in stress of a few 0.1 MPa change the previous risk assessment based on natural seismicity (e.g. northern Holland; Aquitaine basin, France; SW India, SE China; western Uzbekistan). This implies that the continental crust everywhere must be at a state of stress near failure. This fact greatly expands the region of possible seismic hazards beyond that limited to regions where historic or geologic information reveals evidence of past seismicity.

Second, the size of the induced earthquakes, occurring in previously low seismic area, can be used to assess the size of intraplate seismicity that is difficult to predict using the poor historical seismicity in intraplate areas. The largest induced events ($M_0\geq10^{18}$N.m) triggered by either water reservoir impoundments or hydrocarbon recoveries are located where thick seismogenic beds are reported close to the surface, (i.e. shield area for the top four induced events that occurred in low

seismicity areas: Gasli, central Asia; Hsinfengian, Southeast China; Kariba, Zambia-Zimbabwe; Koyna, Southwest India). In other areas of reported $m>3$ induced seismicity, the thickness of the larger, brittle sedimentary bed (for seismicity induced by impoundment (RIS) or depletion of reservoir (EIS)) or the geomechanical setting (dimension of the mine pillars in the case of MIS) bound the upper size of the earthquakes in agreement with the break in self-similarity (see Figure 5 for example on recurrent break slope in self-similarity). This could be checked more closely by analysis of the common break slope in both b-values and fractal dimensions of induced earthquakes as proposed by Volant and Grasso (1993). This characteristic dimensions for the uppermost crust mimics the concept of characteristic dimension proposed for the whole brittle schizosphere by Scholtz (1982) and observed as the upper boundary of different scaling laws (e.g. McGarr 1986, Shimazaki 1986, Pacheco et al. 1992). On other IS sites, break slopes in either b value or seismic moment release have been proposed to correlate with the local geomechanical setting (Figures 6 and 7). The concept of characteristic earthquake has also been proposed by Jin and Aki (1989) to explain anomalies of b-value that are induced by a large number of earthquakes with the same size. Such subsets of characteristic dimensions would help in determining locally the magnitude and depth of the possible shallow energy release, in which occurrence is critical for damage estimates.

Finally for most induced seismicity cases, the seismicity occurs concurrently both in the pressure (or stress) diffusing front and behind the rising pore pressure (or stress). Only if the distribution of preexisting local strengths has been sufficiently decimates by the induced effective stress change will there be a

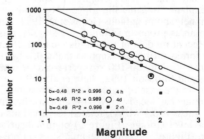

Figure 5 : Time duration of Self-Organized Critical systems induced by subsurface mining. Example of a South Africa gold mine and constant break slope in self-similarity (adapted from McGarr 1976).

190

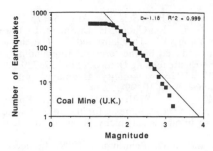

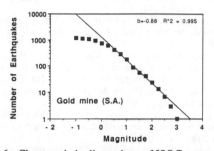

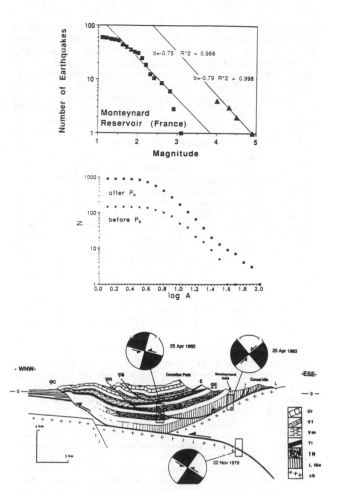

Figure 6 : Characteristic dimensions of ISOC systems: Mining Induced Seismicity. -a) Break slope in b value reported for a deep gold mines induced seismicity from underground seismic observations (data from Deliac and Gay (1984)). -b) Break slope in b value reported for a coal mine induced seismicity. Adapted from Kusznir and Al-Saigh (1984).

simple relationship between the movements of an isobaric front and the location of events near it (Byerlee and Lockner 1977; Krantz et al. 1990). Fractal form of induced seismicity can help to determine the local distribution of geomechanic strengths in both magnitude and position.

5 CONCLUDING REMARKS

Induced seismicity analysis demonstrates that shallow seismic failures occur in a variety of tectonic settings and attest for the key role of the uppermost crust behavior when studying earthquake mechanics. The triggered seismic instabilities obey a subset of scaling laws related to SOC phenomena that mimics the phenomena observed for tectonic earthquakes. Induced seismicity allows us to quantify a 1-2 MPa step in stress that moves isolated stable crustal blocks to criticality. Note that when reported, Extraction Induced Seismicity (EIS) is a regional pattern of the sedimentary basin. ISOC systems emerge simultaneously on neighbouring hydrocarbon fields that share the same extraction pattern , i.e. for 3 Northern Holland fields, two fields located within the central north sea graben, two fields in the Aquitaine Basin and two neighbouring field in south Texas (Grasso, 1992; Pennington et al. 1986). Moreover, with a 0.1 MPa stress step, at least one order of magnitude smaller than the trigger step, induced seismic instabilities naturally evolve toward a critical state and organize themselves (SOC). Such induced SOC (ISOC) systems are sustained for days (MIS) to several years (RIS and EIS). In this paper we show that (1) ISOC systems provide direct evidence that the uppermost continental crust is everywhere at a state of stress near to failure, (i.e. shallow earthquakes may occur everywhere). It confirms the theoretical approach of structuration of the lithosphere as a SOC phenomenon (e.g. Sornette and Sornette 1989; Sornette et al. 1990). (2) We demonstrate that the thickness of the largest local seismogenic bed is the characteristic dimension of each ISOC system. This characteristic dimension mimics the effect that the schizosphere plays for tectonic earthquakes (Scholtz 1991). This dimension is defined by the break slope in self-similarity from small to moderate earthquakes. It ranges from the pillar size in MIS to the thickness of brittle sedimentary beds in the vicinity of RIS and EIS. These dimensions drive the size and the location of the possible shallow earthquakes. Thus the sedimentary basin with soft material which are prone to local destructive amplification of seismic waves are less strongly subject to shallow earthquake. Reversely thick cratonic or shield areas have localized the largest induced earthquakes ($m \geq 6$) and are candidates for future large shallow earthquakes.

Figure 7 : Characteristic dimensions of ISOC systems: Reservoir Induced Seismicity. -a) Break slope in b value reported in RIS, Monteynard Reservoir, France. Data are from the seismic station that operated within the dam site during the 12/1963-12/1967 period (squares). Note that both the m=4.3 shock that occurred in August 1966 and the sequence of 3 $m \geq 4$ earthquakes that occurred in the April 25-27, 1963 period are shifted (diamonds) relative to the number the small shocks ($3.1 \geq m \geq 1$). All the earthquakes are within 7 km distance from the reservoir (data from Plichon et al. 1979). -b) Number of events having an average normalized peak amplitude greater than A (in units of mV/mm) before and after borehole pressurization in rock sample experiments. (from Krantz et al. 1990). -c) Geological cross section of the local structure where brittle calcareous beds and ductile marly beds (Vm and TN levels) are reported. The main shocks and their focal mechanisms are plotted. Adapted from Grasso et al. (1992). Note that the observed pattern of frequency - size relationship of Monteynard earthquakes (a) mimics the laboratory observations (b) of Krantz et al. (1990). Accordingly, the low number of large earthquakes in the 12/1963-12/1967 period could correspond to fractures within a non fluid-pressured area whereas the $m \geq 4$ shocks could be a direct effect of pore pressure -induced seismicity.

Identification and study of such isolated objects within an actual area of low seismicity (e.g. intraplate area) helps both to estimate the size of largest possible earthquake and to understand the behaviour of SOC systems. In terms of seismic risk, the earthquake size estimate, that uses the characteristic dimension of local ISOC system, is complementary to the analysis of Boatwright and Choy (1992) who focused on acceleration source spectra of large cratonic earthquakes. (3) Once triggered ISOC systems are sustained over many years by ~0.1 MPa stress change but larger than the ~0.01 MPa tidal-stress. (4) In the setting of the natural SOC system that is proposed for the whole lithosphere, small stress changes trigger mechanical connections over distances larger than a hundred kilometers (e.g. Sornette et al. 1990). The resulting seismic and aseismic instabilities are those reported as precursory phenomena to earthquakes and these instabilities have been used for several years in China and in the former USSR to calibrate the preparation zone of the major shock (Dobrovolsky et al., 1979). Small stress changes, with the same magnitude order of the ones that sustain ISOC system, can explain large distance anomalies in precursor phenomena to earthquakes when hydrological changes (e.g. Roeloffs, 1989, Silver and Valette-Silver 1992) and electrical changes (e.g. Bernard 1992) are now proposed to be driven by local fluid induced instabilities. (5) Earthquakes induced by human activity, in zones of historically low seismicity, provide direct evidence that both pore pressure changes ($\pm\Delta p$) and mass transfers ($\pm\Delta m$), trigger seismic instabilities of the uppermost crust with magnitude ranging from 3.0 to 7.0. Due to the mechanical properties of the uppermost crust, competition between seismic and aseismic instabilities occur both for large scale precursor processes and main slip behavior (Volant et al. 1992). Aseismic instabilities (slow or quiet earthquakes) are a source of discrepancy when comparing SOC analysis that used fractal geometry of faults and SOC analysis that used fractal geometry of earthquakes. The earthquakes and faults are distant physically and the first attempts to address the possible links are those of Davy et al. (1990). Monitoring ISOC systems provide opportunities to observe and understand how SOC systems emerge and evolve and to characterize the distribution of strengths within geological objects.

Acknowledgments: I am grateful to P. Young who invited me to present this study as an Invited Paper at the 3rd International Workshop on Rockburst and Seismicity in Mines, Kingston, August 1993. I thank D. Simpson for discussions on RIS as well as P. Davy, D. Sornette and P. Volant for fruitful comments. P. Molnar and J.L. Chatelain provided helpful comments on an earlier version of the manuscript. This study benefited from the open collaboration with petroleum companies: Cag, Elf, Nam, Petroland, Phillips, and Shell provided geomechanical data of the fields studied in this work, but the opinions expressed in this study are the author's own. This research was partially supported by the University Joseph Fourier, Grenoble, the Elf company under contract Risk re-assessment within the Lacq Industrial Facilities and the French DBT-INSU Instability Program.

REFERENCES

Bak, P. and C. Tang 1989. Earthquakes as self-organized critical phenomenon, J. Geophys. Res., 94, 15635-15637

Bak, P., C. Tang and K. Weisenfeld 1987. Self-organized criticality : an explanation of 1/f noise, Phys. Rev A 38, 364-374

Bernard, P. 1992. Plausibility of long distance electrotelluric precursors to earthquakes. J. Geophys. Res., 97, 17531-17546.

Besrodny, E.M. 1986. The source mechanism of the Gazly earthquakes of 1976-1984, in Gazly earthquakes of 1976 and 1984, Tashkent, Fan, 94-105.

Blanpied, M.L., D.A. Lockner and J.D. Byerlee 1992. An earthquake mechanism based on rapid sealing of faults, Nature, 13 08 92.

Boatwright, J. and G.L. Choy 1992. Acceleration spectra anticipated for large earthquakes in NE North America, Bull. Seis. Soc. Am., 82, 660-682.

Bufe, C.G.. 1976. The Anderson Reservoir seismic gap - Induced aseismicity? Engineering Geology, 10, 255-262

Byerlee, J.D. 1967. Frictional characteristics of granite under high confining pressure. J. Geophys. Res. 72, 3639-3648.

Byerlee, J.D. and D. Lockner 1977. Acoustic emission during fluid injection into rock. Paper presented at first Conference on acoustic emission/microseismic activity in geologic structures and material, Penn. State Univ., June 1975, eds, H. Hardy and F.W. Leighton, Aedermannsdorf, Switzerland; Trans. Tech. Publications, pp 87-98.

Caloi, P., M. Depanfilis, D. DiFilippo, L. Marcelli and M.C. Spadea 1956. Terremoti della val Padana del 15-16 Maggio 1951: Ann. Geofis., 9, 63-105.

Carder, D.S. 1945. Seismic investigation in the Boulder Dam area, 1940-1945, and the influence of reservoir loading on earthquake activity, Bull. Seis. Soc. Am. 35, 175-192.

Chung-Kang, S., C. Hou-Choun, H. Li-Sheng, L. Tzu-Chiang, Y. Cheng-Yung, W. Ta-Chun and L. Hsueh-hai 1974. Earthquakes induced by the reservoir impounding and their effect on the Hsinfengkiang Dam, Sci Sin., 17, 239-272.

Cook, N.G.W. 1976. Seismicity associated with Mining. Engineering Geology, 10, 99-122.

Costain, J.K., G.A. Bollinger and J.A. Speer 1987. Hydroseismicity : a hypothesis for the role of water in the generation of intraplate seismicity. Seismol. Res. lett. 58, 41-63.

Davis, S.D. and W.D. Pennington 1989. Induced seismic deformation in the Cogdell oil field of West Texas. Bull. Seis. Soc. Am., 79 1477-1494.

Davy, P., A. Sornette and D. Sornette 1990. Some consequences of a proposed fractal nature of continental faulting, Nature, 348, 56-58.

Deliac, E.P. and N.C. Gay 1984. The influence of stabilizing Pillars on seismicity and rockbursts at ERPM, in N.C. Gay and E.H. Wainwright (eds) Proceeding of the 1 Int. Congress on Rockburst and Seismicity in Mines, Johannesburg, 1982, SAIMM, Johannesburg, 1984.

Dobrovolsky, I.P., S.I. Zubkov and V.I. Miachkin 1979. Estimation of the size of the earthquake preparation zones, Pure Appl. Geophys.,117, 1025.

Evans, D.G. and D.W. Steeples 1987. Microearthquakes near the Sleepy Hollow oil field, southwestern Nebraska, Bull. Seis. Soc. Am., 77,132-140.

Eyidogan, H., J. Nabelek and N. Toksoz 1985. The Gazli, USSR, 19 March 1984 earthquake: the mechanism and tectonic implications, Bull. Seis. Soc. Am., 75, 661-675.

Fabre, D., J.R. Grasso and Y. Orengo. 1992. Mechanical behavior of deep rock core samples from a seismically active gas field, Pure Appl. Geophys., 137, 200-220.

Feignier, B.and J.R. Grasso 1991. Relation between seismic source parameters and mechanical properties of rocks: a case study, Pure Appl. Geophys., 137, 175-199.

Fernandez, L.M. 1973. Seismic energy released by the deep mining operations in the Tranvaal and Orange Free State during 1971. S. Afr. Dep. Mines Geol. Survey.

Fletcher, J.B. 1982. A comparison between the tectonic stress measures in situ and stress parameters from induced seismicity at Monticello Reservoir, South Carolina, J. Geophys. Res, 87, 6931-6944.

Fletcher, J.B. and L.R. Sykes 1977. Earthquakes related to hydraulic mining and natural seismic activity in western New-York State, J. Geophys. Res., 82, 3767-3780.

Gough, D.I. and W.I. Gough 1970. Stress and deflection in the lithosphere near lake Kariba. Geophys. Journ. Roy. Astro. Soc., 21, 65-101.

Grasso, J.R. 1992. Mechanics of seismic instabilities induced by the recovery of hydrocarbons, Special Issue Induced Seismicity, Pure Appl. Geophys.. in press.

Grasso, J.R. and G. Wittlinger 1990. Ten years of seismic monitoring over a gas field area, Bull. Seis. Soc. Am.,80, 450-473.

Grasso, J.R. and B. Feignier 1990. Seismicity induced by gas depletion: II Lithology correlated events, induced stresses and deformation, Pure Appl. Geophys, 134, 427-450.

Grasso, J.R., J-P. Gratier, J.F. Gamond and J.-C. Paumier 1992a. Stress diffusion triggering of earthquakes in the upper crust., Special Issue Mechanical Instabilities In Rocks and Tectonics, Journ. Struc. Geol., 14, 915-924.

Grasso, J.R., D. Fourmaintraux et V. Maury 1992b. Le role des fluides dans les instabilités de la croute supérieure: L'exemple des exploitations d'hydrocarbures, in press, Special Issue "Geomechanics", Bull., Soc. Geol. France.

Grasso, J.R., F. Guyoton, J. Frechet and J.F. Gamond 1992c. Triggered earthquakes as stress gauges: Implication for risk re-assessment in the Grenoble area, France, Special Issue Induced Seismicity, Pure Appl. Geophys., in press.

Gupta, H.K. and B.K. Rastogi 1976. Dams and earthquakes, 229 pp., Amsterdam, Elsevier.

Gupta, H.K. 1985. The present status of Reservoir induced seismicity: investigations with a special emphasis on Koyna earthquakes, Tectonophysics, 257-279.

Guyoton, F., J.R. Grasso and P. Volant 1992. Interrelation between induced seismic instabilities and complex geological structures, Geophys. Res. Lett., 19, 7.

Haak, H.W. 1991. Seismiche Analyse van de Aardbeving bij Emmen op 15 februari 1991, Koninklijk Nederlands Meteorologisch Instituut eds., Ministerie van Verkeer en Waterstaat, 14 pp..

Haimson, B.C. and C. Fairhurst 1970. Insitu stress determination at great depth by means of hydraulic fracturing, Proc. U.S. Symp. Rock Mech., 11th, 559-584.

Harris, R.A. and R.W. Simpson 1992. Changes in static stress on southern California faults after the 1992 Landers earthquake. Nature, 360, 251-254.

Healy, J.H., W.W. Rubey, D.T. Griggs and C.B. Raleigh 1968. The Denver earthquakes, Science, vol. 161, 1301-1310.

Hsieh, P.A. and J.D. Bredehoeft 1981. Reservoir analysis of the Denver earthquakes. The case of induced seismicity. J. Geophys. Res., 86, 903-920.

Jacob, K.H., J. Armbuster, L. Seeber and W Pennington 1979. Tarbela reservoir, Pakistan: a region of compressional tectonics with reduced seismicity upon initial reservoir filling, Bull.. Seis. Soc. Am., 69, 1175-1182.

Jaumé, S.C. and L.R. Sykes 1992. Changes in state of stress on the southern San Andreas fault resulting from the California earthquake sequence of April to June 1992, Science 258, 1325-1328.

Johnston, A.C. 1987. Suppression of earthquakes by large continental ice sheets, Nature, 330, 467-469.

Johnston, A.C. 1989. The seismicity of "stable continental interiors", in Earthquakes at North Atlantic passive margins: Neotectonics and post glacial rebound, S. Gregersen and P.W. Basham (eds), Kluwer, Dordrecht, The Netherlands, 740 pp.

Jones, L.M., B. Wang, S. Xu and T.J. Fitch 1982. J. Geophys. Res., 87, 4575-4584.

Krantz, R.L., T. Satoh, O. Nishizawa, K. Kusunose, M. Takahashi, K.Masuda and A. Hirata 1990. Laboratory study of fluid pressure diffusion in rock using acoustic emission. J. Geophys. Res., 95, 21593-21607.

Klein, F.D. 1976. Tidal triggering of reservoir-associated earthquakes. Engineering Geology, 10, 197-210.

Kusznir, N.J. and N.H. Al-Saigh 1984. Some observation on the influence of pillars in Mining-Induced-seismicty, in N.C. Gay and E.H. Wainwright (eds) Proceeding of the 1 Int. Congress on Rockburst and Seismicity in Mines, Johannesburg, 1982, SAIMM, Johannesburg, 1984.

Lukk, A.A. and S.L. Yunga, 1988. In "Geodynamics and stress strain state of the lithosphere of the Central Asia., Ed. Donish, Duschambe.

McGarr, A. 1976. Seismic moments and volume changes. J. Geophys. Res., 81, 1487-1494.

McGarr, A. 1986. Some observations indicating complications in the nature of earthquake scaling, in S.Das, J. Boatwright and C. Scholtz (eds), Earthquake source mechanics, A.G.U. Geophysical Monograph 37, Washington D.C., A.G.U., 217-225.

McGarr, A. 1991. On a possible connection between three major earthquakes in California and oil production. Bull. Seis. Soc. Am, 81, 948-970.

McGarr, A., S.M. Spottiswoode, N.C. Gay and W.D. Ortlepp 1979. Observations relevant to seismic driving stress, stress drops and efficiency, J. Geophys. Res., 84, 2251-2261.

McGarr, A., R.W.E Green and S.M. Spottiswoode 1981. Strong ground motion of mines tremors: some implication for near source ground motion parameters, Bull. Seis. Soc. Am., 71, 295-309.

Nason, R.D., A.K. Copper and D. Tocher 1968. Slippage on the Buena Vista thrust fault, 43rd annual meeting guidebook, AAPG, SEG, SEPM. Am. Ass. of Pet. Geol., Pac. Sect., 100-101.

Nicholson, C., E. Roeloffs and R.L. Wesson 1988. The northeastern Ohio earthquake of 31 January 1986 : Was it induced? Bull. Seis. Soc. Am., 78, 188-217.

Nur, A. (1972). Dilatancy, pore fluids, and premonitory variation of ts/tp travel times. Bull. Seis. Soc. Am. 62, 1217-1222.

Pacheco, J.F., C.H. Scholtz and L.R. Sykes 1992. Changes in frequency-size relationship from small to large earthquake, Nature, 355, 71-73.

Pennington, D.W, S.D. Davis, S.M. Carlson, J. Dupree and T.E. Ewing 1986. The evolution of seismics barriers and asperities caused by the depressuring of fault planes in oil and gas fields of south Texas, Bull. Seis. Soc. Am., 939-948.

Plichon, J.N., P. Gevin, P. Hoang, P. Londe and P. Petterville Q. 51 R. 30, 1979. Sismicité des retenues de grands barrages, *Proceeding of the 13th Int.Comm. Of Large Dams Congress,*

New-Delhi, I. C. O. L. D. editor,151, Bd Haussmann, Paris, vol. 2 , 1347-1362.

Pomeroy, P.W., D.W.Simpson and M.L. Sbar 1976. Earthquakes triggered by surface quarrying: The Wappingers Falls, New-York, sequence of June 1974, Bull. Seis. Soc. Am., 66, 685-700.

Raleigh, C.B., J.H. Healy and J.D. Bredehoeft 1972. Faulting and Crustal stress at Rangely, Colorado, A.G.U., Geophysical Monograph 16, 275-284.

Roeloffs, E. 1988. Fault stability changes induced beneath a reservoir with cyclic variations in water level. J. Geophys. Res. 93, 2107-2124.

Roeloffs, E. 1989. Hydrologic precursors to earthquakes: a review, Pure Appl. Geophys., 126, 177-209.

Rothé, J.P. 1970. Séismes artificiels (man-made earthquakes). Tectonophysics, 9, 215-238.

Segall, P. 1989. Earthquakes triggered by fluid extraction. Geology, 17, 942-946.

Segall, P. and J.R. Grasso 1991. Poroelastic stressing and induced seismicity near the Lacq gas field (France). Eos transactions, Am. Geophys. Union, 72, 44, 293.

Sibson, R.H. 1989. High-angle reverse faulting in northern New-Brunswick, Canada, and its implication for fluid pressure levels. Jour. Struc. Geol. 11, 873-877.

Silver, P.G. and N.J. Valette-Silver 1992. Detection of hydrothermal precursors to large California earthquakes. Science, 257, 1363-1368.

Simpson, D.W. 1986. Triggered earthquakes. Ann. Rev. Earth Planet. Sci., 14, 21-4.

Simpson, D.W. and S.K. Negmatullaev 1981.Induced seismicity at Nurek reservoir, Tajikistan. USSR. Bull. Seis. Soc. Am.,71, 1561-1586.

Simpson, D.W. and W. Leith, "The 1976 and 1984 Gazli, USSR, earthquakes - Were they induced ?", Bull. Seis. Soc. Am., 75, 1465-1468.

Simpson, D.W., W.S. Leith and C.H. Scholtz 1988a. Two types of reservoir induced seismicity, Bull. Seis. Soc. Am., 78, 2025-2040.

Simpson, D.W., S.S. Schulz, L.D. Dietz and R.O. Budford 1988b. The response of creeping parts of the San Andreas fault to earthquakes on nearby faults: two examples. Pure and Applied Geophys., 126, 665-685.

Sornette, A. and D. Sornette 1989. Self-Organized criticality and earthquakes, Europhys. Lett, 9, 197-202.

Sornette, D., P. Davy and A. Sornette 1990. Structuration of the lithosphere in plate tectonics as a Self-Organized Critical phenomena. J. Geophys. Res., 95, 17353-17361.

Scholtz, C.H. 1982. Scaling laws for large earthquakes: consequences for physical models, Bull. Seis. Soc. Am. 72, 1-14.

Scholtz, C.H. 1991. Earthquakes and Faulting: self-organized critical phenomena with characteristic dimension.; in Triste and D. Sherrington (eds), Spontaneous formation of space-time structures and criticality, Kluwer Academic Publishers, Netherlands, 41-56.

Shimazaki, K. 1986. Small and large earthquakes: the effects of the thickness of the seismogenic layer and the free surface, in S.Das, J. Boatwright and C. Scholtz (eds), Earthquake source mechanics, A.G.U. Geophysical Monograph 37, Washington D.C., A.G.U., 209-216.

Spottiswoode, S.M. 1984. Source mechanism of mine tremors at Blyvooruitzicht gold mine. In N.C. Gay and E.H. Wainwright (eds) Proceeding of the 1 Int. Congress on Rockburst and Seismicity in Mines, Johannesburg, 1982, SAIMM, Johannesburg, 1984, pp 29-37.

Spottiswoode, S.M. and A. McGarr 1975. Source parameters of tremors in a deep-level gold mine. Bull. Seis. Soc. Am., 65, 93-112.

Stein, S., N.H. Sleep, R.J. Geller, S.C. Wang and G.C. Kroeger 1979. Earthquakes along the passive margin of eastern Canada, Geophys. Res. Lett., 6, 537-540.

Stein, R.S., G.C.P. King and J. Lin 1992. Change in failure stress on the southern San Andreas fault system caused by the 1992 M=7.4 Landers earthquake, Science, 255, 1687-1690.

Volant, P. and J.R. Grasso 1993. Fractal dimensions and b values of earthquakes: which geomechanical link? Paper presented at E.G.S. XVIII Assembly, Wiesbaden 3-7 May, 1993.

Volant, P., J.R. Grasso, J.C. Chatelain and M. Frogneux 1992. b-value, aseismic deformation and brittle failure within an isolated object: evidence from a dome structure loaded by fluid extraction, Geophys. Res. Lett., 19, 1149-1152.

Wetmiller, R.J. 1986. Earthquakes near Rocky Mountain House, Alberta, and relationship to gas production. Can. Journ. of Earth Sciences, 32, 2,172-181.

Wyss, M. and P. Molnar 1972. Efficiency, stress drops, effective stress and frictional stress of Denver (USA) earthquakes, J. Geophys. Res., 77, 1433-1438.

Yerkes, R.F., W.L. Ellsworth and J.C. Tinsley. Triggered reverse fault and earthquake due to crustal unloading, northwest Tranverse ranges, California, Geology, 11, 287-291.

Zoback, M.D. and B.C. Haimson 1983. Hydraulic fracturing stress measurements, US. National Committee for Rock Mechanics, National Academic press, Washington D.C., 270 pp.

Zoback, M.D.and M. Zoback 1980. State of stress in the conterminous United States. J. Geophys. Res. 85, 6113-6156.

Zoback, M.L.1992. First- and second-order patterns of stress in the lithosphere: The world stress map project, J. Geophys. Res., 97, 11703-11728.

Rockbursts and Seismicity in Mines, Young (ed.) © 1993 Balkema, Rotterdam, ISBN 90 5410 320 5

Fault-plane solutions of microseismicity induced by progressive excavations of a large underground chamber

Tsuyoshi Ishida
Yamaguchi University, Japan

Yasuo Uchita
The Kansai Electric Power Co Inc., Japan

Tadashi Kanagawa
Central Research Institute of Electric Power Industry, Japan

Masaru Urayama
NEWJEC Inc., Japan

ABSTRACT: We monitored microseismicity induced by progressive excavations of a large underground chamber measuring 24.0 m wide, 46.6 m high and 134.5 m long in porphyrite at 280 m below the surface. By superposing P-wave polarity distributions of three located representative microseismic events, a fault-plane solution was obtained. The strike and dip of a nodal plane in the fault-plane solution were similar to those of the most dominant joint surface. The P and T axes also coincided with a stress condition estimated from measured initial stress and a configuration of excavated openings at the time. These two findings indicated that the three events were caused by shear fracturing along the dominant joint surfaces under the influence of the stress condition. We then concluded that fault-plane solutions of microseismic events give us valuable information in understanding rock mass behavior with progressive excavation of an underground chamber.

1 INTRODUCTION

By aiming to use microseismic monitoring in the stability assessment of an underground chamber, we developed a method through small scale in situ experiments (Ishida et al. 1986, 1992) and preliminary monitoring in a real underground chamber (Ishida et al. in press). Recently, we monitored microseismicity with progressive excavations of an underground chamber for electric power generators. Monitoring was successfully carried out and sources of some microseismic events were located. For the located events, fault-plane solutions were obtained and they were interpreted by considering the initial stress condition, directions of predominant joint surfaces and the configuration of excavated openings at the time. This suggests that fault-plane solutions for microseismic events can provide more useful information to assess stability of a chamber than displacement data measured by conventional systems such as extensometers, convergence meters and so on. In this paper, we will introduce our monitoring method and discuss some of the data obtained from monitoring.

2 SITE AND MICROSEISMIC MONITORING

Monitoring was carried out in porphyrite rock masses around a large underground chamber under construction, measuring 24.0 m wide, 46.6 m high and 134.5 m long at 280 m below the surface. The excavations

were started at the upper section and the rock masses were removed in the order of (a), (b) and (c) indicated in Fig. 1. Thereafter, the floor was excavated downward by benches of three meters height

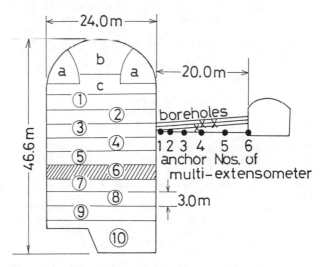

Figure 1. Excavation steps of the chamber, three holes for microseismic monitoring and positions of extensometer anchors in an elevation view. The hole for the extensometer was about 22 m horizontally from the three holes.

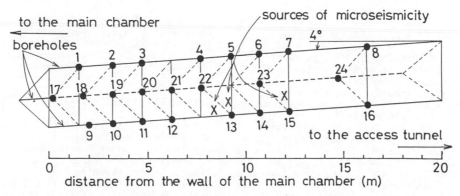

Figure 2. Sensor arrangement for microseismic monitoring. The sensors were positioned in these almost horizontal boreholes.

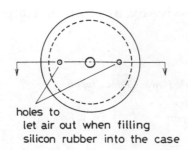

holes to
let air out when filling
silicon rubber into the case

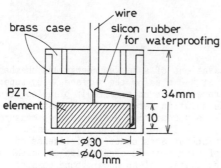

Figure 3. Plane and section view of a sensor.

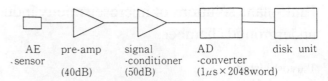

AE
-sensor

pre-amp
(40dB)

signal
-conditioner
(50dB)

AD
-converter
(1μs×2048word)

disk unit

Figure 4. Block diagram of the microseismic
monitoring system.

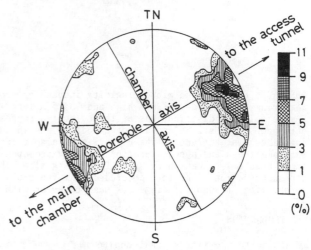

Figure 5. Lower hemisphere equal-area projection of
joint surfaces.

from No. 1 through No. 10.   Monitoring was planned
for excavations of a pillar part in the upper section
(part (b) in Fig. 1) and No.3 and No.6 benches.   For
these excavations, microseismic events were recorded
for two hours just after each explosion near the
boreholes along which sensors were located.   The
three crosses in Fig. 1 indicate sources of microseismic events induced by excavations of No. 6 bench
which is hatched in the figure.   The three events
will be discussed later.

Sensors were set along three holes as shown in Fig.
2.   The three holes were parallel and about 3 m
apart.   The sensor comprised a piezoelectric element
and a brass case as shown in Fig. 3.   The piezo-
electric element was polarized in the axial direction
of the disk and its resonance frequency was 67 kHz.
The system used in monitoring (Fig. 4) was reported
in detail in the previous paper (Ishida et al. 1992),
and the monitoring frequency range was from 5 to 100
kHz.

To obtain a fault-plane solution, it is necessary
to obtain clear traces of P-wave initial motions at
many sensors for a microseismic event (Kasahara
1981).   Microseismic events satisfying the condition
were caused by large rock fractures only.   In
monitoring, although events with small amplitudes
were usually recorded, the events so large amplitudes
as to be able to obtain fault-plane solutions were
sometimes recorded.   Therefore, the total numbers of
events for which fault-plane solutions were obtained
were very small; one for the pillar part, one for No.
3 bench and six for No. 6 bench.

3 DIRECTIONS OF JOINTS AND DISPLACEMENT

Figure 5 shows the directions of joint surfaces
surveyed using a borehole television system in the
three holes before the sensors were placed along
them.   In the figure, the most dominant joint
surface is around a point 68° clockwise from true
north (TN) and 68° from the vertical, meaning that
the strike is N22°W and the dip is 68°SW.   The
strike is almost parallel to the chamber axis, N28°W
as shown in Fig. 5 (the chamber axis corresponds to a
direction through the paper in Fig. 1.), and the dip
indicates that the surface is a steep slope downward
to the chamber.   This suggested that the rock masses
have a tendency to slide easily downward along the
dominant joint surfaces with progressive excavations

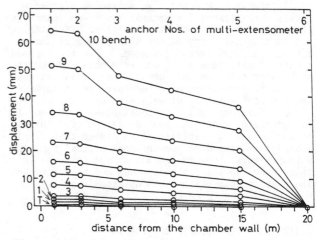

Figure 6. Horizontal displacement arising with
progressive excavations measured using an
extensometer.   The lateral axes indicate distance
from the chamber wall (lower) and anchor positions
(upper).

of lower sections of the chamber.   This tendency was
demonstrated in Fig. 6 showing that excavations at
lower than No. 7 bench caused more than 60 % total
displacement toward the chamber although the extensometer was set at No. 3 bench level.

4 FAULT-PLANE SOLUTION AND DISCUSSION

Three of the six events recorded in excavations of
No. 6 bench showed almost the same patterns in polarity distribution of P-wave initial motions.   Then,
by superposing the three events, we obtained a fault-
plane solution shown in Fig. 7.   (Sources of the
three events are shown in Fig. 1 and Fig. 2.)
Symbol ● indicates stations for compression and
symbol ○ indicates stations for dilatation of P-wave

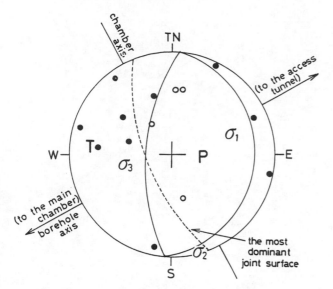

Figure 7. Lower hemisphere equal-area projection of a fault-plane solution obtained by superposing P-wave polarity distributions of the three microseismic events.

initial motions.   Letter positions of P and T indicate pressure and tension axes derived from the solution, respectively.

Before the chamber excavation, the in-situ stress condition was measured by an over-coring method. Letter σ_1, σ_2 and σ_3 in Fig. 7 indicate the directions of the principal stresses.   The magnitudes of σ_1, σ_2 and σ_3 are 10.0, 6.4 and 3.9 MPa, respectively.   Although the P axis is in a similar direction to σ_1 and the the T axis is in that of σ_3, the direction of the P axis is nearer to the vertical than that of σ_1 and the direction of the T axis is nearer to the horizontal than that of σ_3.   This can be understood by considering that the directions of the two principal stresses have turned slightly with progressive excavations of the chamber; one to the vertical and the other to the horizontal from the directions in the initial condition.

The direction of the most dominant joint surface showing a concentration from 9 to 11 % in Fig. 5 is denoted by a broken arc in Fig. 7.   We can see that the strike and dip of one of the nodal planes (N4°E/68°W) is similar to the strike and dip of the most dominant joint surface (N22°W/68°SW).   In addition to this and considering that the three microseismic events were recorded in excavations of No. 6 bench about ten meters lower than No. 3 bench level where the sensors were located, the three events were most likely caused by shear fractures along the dominant joint surfaces under the influence of the in situ stress condition.

At this site, a distinctive feature is that excavations at lower than No. 7 bench caused a large displacement toward the chamber as shown in Fig. 6. As mentioned in the previous section, a large displacement was assumed to be caused by the rock masses sliding downward along the dominant joint surfaces. The shear fractures along the dominant joint surfaces, derived from the fault-plane solution of microseismic events recorded in excavations of No. 6 bench, seemed to be a precursor to the large displacement arising with excavations lower than No. 7 bench. Consequently, if we can obtain fault-plane solutions in real time considering the directions of dominant joint surfaces and in situ stress condition, we can probably predict important rock mass movements around a chamber such as occurrences of large displacements.

5 CONCLUSION

We monitored microseimicity in porphyrite rock masses around an underground chamber with progressive excavations and obtained a fault-plane solution by superposing P-wave polarity distributions of located representative microseismic events.   After discussing the fault-plane solution, we came to the following conclusions:

(1) The strike and dip of one of the nodal planes in the fault-plane solution are similar to the strike and dip of the most dominant joint surface.   The directions of P and T axes in the fault-plane solution agree with the directions of maximum and minimun principal stresses estimated from the in situ measured initial stress condition and the configuration of excavated openings at the time.   From these two findings, the three events were most likely caused by shear fractures along the dominant joint surfaces under the influence of the in situ stress condition.

(2) An extensometer set at No. 3 bench level showed that more than 60 % of the total displacement towards the chamber arose with progressive excavations at lower than No. 7 bench.   The large displacement was assumed to be caused by the rock masses sliding downward along the dominant joint surfaces.   The rock mass movement measured by the extensometer showed that the interpretation mentioned in item (1) is reasonable.

(3) The representative microseismic events for which the fault-plane solution was obtained occurred in excavations of No. 6 bench, while a large displacement arose with excavations at lower than No. 7 bench.   Thus, the shear fractures along the dominant joint surfaces, derived from the fault-plane solution, seemed to be a precursor to large displacement. Consequently, if we can obtain fault-plane solutions in real time considering directions of dominant joint surfaces and in situ stress condition, we may be able to predict important rock mass movements around a chamber such as occurrences of large displacements.

ACKNOWLEDGEMENTS

We would like to thank Mr. T. Nakamura of NEWJEC Inc. (now Toda Construction Co. Ltd) and Mr. A. Yada of the Kansai Electric Power Co., Inc., for their discussion on the obtained data and their kind help in our field works.

REFERENCES

Ishida, T., Kanagawa, T., Sasaki, S. and Urasawa, Y. 1986. AE monitoring during the in-situ direct shear test applied to an underground cavern. Proceedings of the Japan Society of Civil Engineers, 376(III-6): 141-149   (in Japanese)

Ishida, T., Kitano, K., Kinoshita, N. and Wakabayashi, N. 1992. Acoustic emission monitoring during in-situ heater test of granite, Journal of Acoustic Emission. 10: S42-S48

Ishida,T., Kanagawa,T., Tsuchiyama,S and Momose,Y. (in press). High frequency AE monitoring with excavation of a large chamber, Proceedings of The Fifth Conference on Acoustic Emission/Microseismic Activity in Geologic Structures and Materials

Kasahara, K. 1981. Earthquake mechanics: Cambridge University Press.

Rockbursts and Seismicity in Mines, Young (ed.) © 1993 Balkema, Rotterdam, ISBN 90 5410 320 5

Area rockbursts in Kolar Gold Fields: The possible attributes

Prakash Ch. Jha, C. Srinivasan & N. M. Raju
National Institute of Rock Mechanics, Champion Reefs, Kolar Gold Fields, Karnataka, India

ABSTRACT: Area rockburst is a unique feature of Kolar Gold Fields wherein a major burst triggers a series of bursts of same or smaller sizes. During an area rockburst, the entire area under influence remains disturbed over a period of two days or more with a wide scale damage unleashed by the series of rockbursts. This paper examines the origin of this phenomenon. Detailed studies with respect to size and energy distribution, hypocentral distribution, and stress build-up pattern have been made for the associated rockbursts of the three representative area rockbursts of the Champion Reef mine. While examining their possible attributes (causative factors), it is found that they owe their origin to various reasons like the presence of tectonic faults, tectonic stresses, failure of remnants or failure of unfilled (poorly filled) regions in the old workings.

1 INTRODUCTION

Rockbursts in the mines of Kolar Gold Fields are century old phenomena now. In the last 100 years of the mining history of KGF, there are virtually no places left where the incidence of rockbursting has not been reported. A regional seismic network was established in KGF in 1978 for round the clock monitoring of the rockburst activity in the entire mining district. Analysis of seismic data collected from this network helped in a better understanding of the rockburst mechanism and in choosing a safer course of mining. The deeper level burst prone regions of the Champion Reef mine, which has now attained a working depth of below 3.2 km, have been reinforced with PC-based microseismic network. Some new short range microseismic precursors have now been identified (Raju et al, 1991). Round the clock monitoring, analysing the collected data and implementing its fall out are the clues to the sustained mining activity in this mine. But added to the discomfiture of all such efforts is the occurence of area rockburst, a phenomenon in which a series of rockbursts occur in quick succession in a relatively small area, thereby unleashing a devastating effect in the entire area. As per past studies, area rockbursts are reported to recur with a frequency of 10 years (Krishnamurthy and Gupta, 1983). So far area rockbursts have eluded a seismic or any other precursor pattern. This paper attempts to investigate the area rockbursts in detail so as to examine the possible attributes to their origin.

2 GEOLOGY

Geologically the gold bearing hornblende schist of Kolar Gold Fields belongs to the lower Dharwar age. The gold bearing lodes dip towards west at an angle of 40–45° near the surface and gradually become nearly vertical at great depth. Of the many quartz lodes explored on the fields, only two lodes - the Champion and the Oriental lode - are of economic importance. While the Champion lode has been mined extensively in all the three mining districts of KGF, the Oriental lode is being mined on a large scale in the Nundydroog mines only due to economic reasons (Krishnamurthy and Shringarputale, 1990).

The Champion lode is not a continuous fissure filled with quartz, but is a system of lode made up of individual fissures which taper out both along the strike and the dip, the average width being 1.0–1.2m.

In places where the quartz attains a width exceeding 6m, it occurs in a large zig-zag usually called "Folds" which pitch northward (on the Champion lode) in the plane of the lode at an angle similar to the angle of dip (Taylor and sons, 1955). While mining in these folded formations, severe rockbursts were encountered in the past leading to the frequent damages to shafts (Krishnamurthy and Gupta, 1983).

The productive working in these lodes have been interrupted by a series of faults, dykes and pegmatite veins, all involving plane of weaknesses. Of them, two major fault planes i.e. the Mysore North Fault (MNF) and the Tennant's Fault (TF), attach much significance as they either intercept the lode or run parallel to it. The MNF lies only about 8–12m away in the hanging wall of the folded formations of the Champion lode in the Northern Folds area of the Champion Reef mine. The TF runs parallel to the main lode in the shallower levels and lies only 30m away in the hanging wall of the lode in the Champion Reef mine. Major rockbursts including area bursts were reported while mining close to these fault planes and as such their roles in inducing these bursts needs to be examined in detail.

3 AREA ROCKBURST

Area rockburst is unique to the gold mines of KGF mines. Mention of area rockbursts have been made in the past by Miller (1967), Krishnamurthy (1977), Krishnamurthy and Nagarajan (1981), Krishnamurthy and Gupta (1983), Srinivasan and Shringarputale (1990) and Jha (1991). By definition, area rockburst implies a sequence of rockbursts which follow in quick succession with their hypocentres concentrating in a smaller areaof 100–200m radius. It so happens that the occurrence of a major rockburst triggers a chain of rockbursts numbering 20–200 of equivalent or smaller sizes in the same area over a period of few hours to few days. Besides causing wide scale devastation to ore shoots, working shafts, drivages, steel setts etc. in the underground, area rockbursts have their damaging effects on the surface buildings also lying within 2–3 km radius of the epicentral region.

Till date 15 area rockbursts have been reported from the KGF mines, out of which 11 have taken place in the Champion Reef mine itself. However, full details of only 10 area rockbursts that have taken place after the commencement of round the clock seismic monitoring, are available. Some of the area rockbursts have been described by Miller (1967),

Krishnamurthy and Nagarajan (1981) and Krishnamurthy and Gupta (1983). No separate study has been made till date on the area rockbursts as regards their mechanism, periodicity, nature and precursor pattern. Since maximum number of area rockbursts are reported from the Champion Reef mine, three area rockbursts of this mine pertaining to different depth levels and being associated with a large scale damage viz. area burst of 14.3.1982, 19.7.1985 and 22.8.1985 have been analysed in detail in this paper.

4 SEISMIC INVESTIGATION

A new approach for delineation of high stress zones in mines has been developed by Jha and Willy (1991). Based on this approach, it has been possible to divide the longitudinal section of the Champion Reef mine into five seismic zones, as shown in Table 1.

Table 1. Particulars of seismic zones in the Champion Reef mine.

Zone	Limits	Name
I	X = 19750-21000 ft. Z = 8000-10000 ft.	Heathcote's Shaft area
II	X = 21000-23000 ft. Z = 8000-10000 ft.	Osborne's Shaft area
III	X = 17000-19500 ft. Z = 8000-10500 ft.	Northern Folds area
IV	X = 19500-22000 ft. Z = 2000-4000 ft.	Garland's Shaft area
V	X = 16000-18500 ft. Z = 3500-6000 ft.	Bullen's Shaft area

Interestingly, each of these seismic zones veer around some shafts. Almost 70-80% of the rockbursts taking place in the Champion Reef mine confine to these zones. The three area rockbursts chosen for analysis in this paper pertain to the Zone IV, Zone III and Zone V of the Champion Reef mine in the order of occurrence. The seismic investigation of these area rockbursts confine to the zone wise analysis of the following features:
1. energy and event size distribution,
2. hypocentral distribution of associated bursts,
3. stress distribution pattern prior to the area rockburst.

4.1 Energy and event size distribution

In the case of area rockburst in the zone IV of the Champion Reef (CR) mine, a total of 205 bursts were recorded in the 15 days time following a major burst on 14.03.1982. Of them 15 rockbursts were major one accounting for nearly 70% of the energy release whereas other 190 events constituting 92% of the total events, make up only 30% of the energy released. It is interesting to note that in the other two cases also, whereas major events, constitute only 15-30% of total events, they account for 85-95% of the energy release. Table 2 shows the even size and event distribution pattern for the three area rockbursts under study.
   There is reason to believe that for shallower events, where natural state (in-situ) of stress is less as compared to deeper ones, minor events triggered are comparatively much higher and accordingly energy released by these minor events are much smaller as compared to their deeper level counterparts. The triggering effect too gets subdued faster as the centre of activity moves deeper (table 2).

Table 2. energy and event size distribution of associated burts of the three area rockbursts.

Sl. No.	Date of initiation	Active days	Total events	Major events	% of energy released in major events
1	14.03.1982	15	205	15	66
2	19.07.1985	2	47	15	85
3	22.08.1985	10	60	10	93

4.2 Hypocentral distribution pattern

The plot of hypocentral distribution of the series of seismic events associated with the three area rockbursts are shown in figs 1 to 3 where major and minor events are shown by separate symbols. One common observation with the three plots is the fact that the initial event in each of them is a major one and thereafter we get a mix of major and minor events. There is no fixed interval between successive major events and the entire area seems to have become active like a self exciting process of seismic energy release. To examine the influence of fault plane, the hypocentres of seismic events associated with the first area rockburst of March 14, 1982 are plotted on a transverse section, whereas for others, the plot is made on the longitudinal section.

a) Area rockburst of 14.3.1982

The area rockburst of 14th March, 1982 had taken place in the shallow levels old working area (Zone IV) of the Champion Reef mine. It is seen from the plot of hypocentres (fig.1) that majority of the events align parallel to the main reef (lode) near the working shaft. A major fault plane (Tennant's fault) crosses this zone and lies in the hanging wall of the main lode. It is interesting to note that 97% of the total seismic events of this series have taken place in the footwall of the fault plane on the either side of the main lode. This adds weight to the fact that possible attribute to this area burst has to do something with the mining of the main lode. Mining from this zone had ceased in early thirties and even in the past there were reports of severe rockbursts being encountered in this area.

b) Area rockburst of 19.7.1985

This area rockburst had taken place in the deeper level productive region of the Champion Reef mine. The hypocentres of the series of 47 bursts constituting this area rockburst series lie in the zone III of the Champion Reef mine. This zone is historically known to be burst prone and the working shaft in this zone i.e. the Sub Auxiliary Shaft was damaged by major bursts at least 14 times till 1966 (Krishnamurthy and Nagarajan, 1983).
   The hypocentral plot of seismic events (fig.2) shows that major seismic events of this series have taken place very close to the shaft leading even to its damage between 97-103 levels. Other events too cluster very close to this reef and are distributed on the either side of the Sub Auxiliary Shaft. One interesting reature to note here is that unlike in other cases where events have a horizontal scatter, seismic events of this series are more widely distributed in a vertical plane. Surprisingly, though being a working area, there was no indication of any increase in the seismic activity prior to this area rockburst which was active for two days.

c) Area rockburst of 22.8.1985

This area rockburst too had taken place in the old workings (Zone V) where the mining was abandoned in the early forties. The series of 60 events constituting this area rockburst had taken place in the five

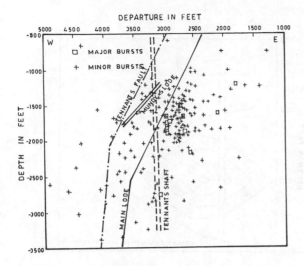

Figure 1. Transverse section plot of the hypocentral distribution of the series of 205 rockbursts associated with the area rockburst of 14.3.1982 in the Zone IV of the Champion Reef mine. Major and minor events are shown by separate symbols. Also shown in the background are the locations of the working shaft, the main lode and the fault plane traversing this area.

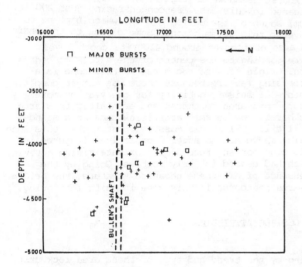

Figure 3. Longitudinal section plot of the hypocentral distribution of the series of 60 rockbursts associated with the area rockburst of 22.8.1985 in the Zone V of the Champion Reef mine. Major and minor events are shown by separate symbols. The location of working shaft is also shown in the background.

days period following major burst on 21st August, 1985. 9 out of the 10 major events of this series are located south of the Bullen's Shaft, whereas minor events are fairly scattered on the either side of the shaft (fig. 3). It is important to note here that unlike the presence of fault planes in the zone of other two area rockbursts described above, there is no such fault plane or any other influencing geological feature in this zone.

4.3 Stress distribution pattern

The longitudinal section of the various seismic zones under the influence of area rockbursts are gridded in 500ft. X 500ft. (150m x 150m) blocks and the cumulative square root of energy released in each block

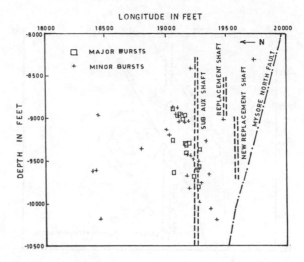

Figure 2. Longitudinal section plot of the hypocentral distribution of the series of 47 rockbursts associated with the area rockburst of 19.7.1985 in the Zone III of the Champion Reef mine. Also shown in the background are the locations of working shaft and the Mysore North Fault traversing this area.

is contoured out in the mining section. The square root of seismic energy release is a measure of the strain build-up prior to the occurrence of that seismic event (Benioff, 1951a). Taking into account the proportionality of stress and strain, these plots are called the section maps of stress for the corresponding period and the contours so drawn are referred as stress contours in arbitrary units. This method of analysing the stress regime has been proved quite successful and is discussed in detail by Jha and Willy (1990) and Jha (1991). The influence of the geological features as also the mining parameters are clearly reflected in such plots. Stress distribution pattern for each of these three area rockbursts is discussed individually.

a) Area rockburst of 14.3.1982

The stress distribution pattern for the area rockburst of March 1982 is shown in fig.4. It is seen that majority of stresses were concentrating between the depth range of 1500-2000 ft. (400-600m) with its focus centered around 200 ft. (70m) east of the main lode. The Tennant's fault, which is further west of the main lode, seems to have no influence on the stress distribution/concentration in this small pocket, nor these stress have any impact on the shaft because the shaft is in the background stress level of 100 value. Going by this feature, the Tennant's fault does not appear to have played any role in the genesis of this area rockburst. Even the hypocentral distribution of associated seismic events do not attach any importance to the presence of this fault plane.

b) Area rockburst of 19.7.1985

Fig. 5 shows the stress distribution pattern during this area rockburst of July 1985. A stress front of nearly 3000 magnitude is concentrated right at the shaft location at around 9250 ft. (3000m) depth level. This explains precisely the reason behind an extensive damage to the Sub Auxiliary Shaft between 97-103 levels which is the area under the influence of stresses of magnitude 2750 and above. A massive stress concentration is seen just at the face of MNF, which is a major geological feature of this zone. The lode occurs here in the folded form and at times, mining on this fold had reached very close to the MNF. The MNF is a massive fault which divides the working

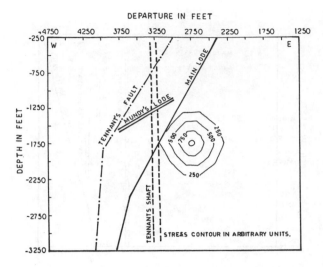

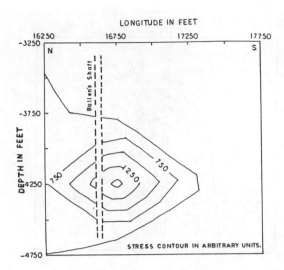

Figure 4. Transverse section of the stress distribution pattern during the area rockburst of 14.3.1982. Major stresses are seen to concentrat east of the main lode.

Figure 6. Longitudinal section of the stress distribution pattern during the area rockburst of 22.8.1985. Major stresses are veering around the Bullen's Shaft.

c) Area rockburst of 22.8.1985

The stress distribution pattern during this area rockburst is shown in fig. 6. Majority of stresses released in this case are concentrating just 200 ft. (60m) south of the Bullen's shaft which happens to traverse this seismic zone. Some damage to the shaft was also noticed at around 4200 ft. depth level corresponding to the centre of the stress concentration. Origin of area rockburst in this zone is a surprising feature because there was almost complete absence of seismic activity for one year or more before this area rockburst and even after this area rockburst, the seismic activities came to a standstill till 1991 when a massive burst rocked this zone again. There are no adverse geological features in this zone nor any record of proximate mining activity which had ceased in early forties. Going by the pressence of all these negative featurs, the origin of area rockburst in this zone is quite intriguing.

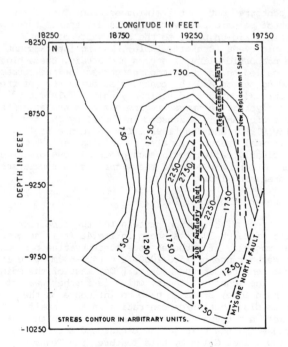

Figure 5. Longitudinal section of the stress distribution pattern during the area rockburst of 19.7.1985. Major stresses are concentrated just ahead of the Mysore North Fault and around the working shaft.

region in the deeper level into two areas viz. the Northern Folds (NF) on the north and the Glen Ore Shoot (GOS) on the south. The fault plane takes a southward trend as it moves shallower and disappears at around 2000 ft. (600m) depth.

The MNF is supposed to be a passive fault with no record of any induced movement along its course. But keeping this area rockburst in view and the history of damaging rockbursts in this zone, the role of this fault plane becomes suspectible.

5 POSSIBLE ATTRIBUTES

Going by the trend and type of these area rockbursts they are certainly not attributable to the proximate mining activity. Knoll and Khunt (1990) have reported a similar type of phenomenon called "tectonic" or "dynamic rockburst" and have termed it as "Type 2 burst" which can only indirectly be connected to the mining activities. The stimulating factor for this type of rockburst is stated to be the presence of high tectonic stresses in the virgin rockmass and the presence of geological fracture planes approximately oriented to the tectonic stress field. Though the characteristic properties of these area rockbursts of KGF are similar to type 2 bursts, they definitely do not owe thier origin to the high tectonic stresses because area rockbursts in KGF have taken place as shallow a depth of 460m in 1952 and 600m in 1982, where there is no evidence of any fracture planes. Hence we have examined the role of various possible attributes with reference to these three area rockbursts.

5.1 Fault Plane

In case of area rockburst on 19.7.1985 in the seismic zone III, where the Mysore North Fault seems to have played a role in the modification of stress distribution pattern (fig. 5). Miller (1965) while commenting on the role of the MNF in the area rockburst of 27.11.1962 has said that the MNF did not behave significantly different from the rest of the country rock and hence ruled out the possibility of any movement along the fault line initiating the collapse. Similar views have been expressed by Krishnamurthy and Nagrajan (1981) and Krishnamurthy and Gupta (1983). The question remains as to wherefrom such large stresses get accumulated to account for these area rockbursts at frequent intervals near this fault plane.

In this context, it is worthwhile to mention that the analysis of the stress regime of the Champion Reef Mine half-yearly and annual basis has revealed that whenever the mining reaches close to the fault plane, the induces stresses fail to redistribute themselves in the surroundings and remain stationary at the face of the fault plane which can contain such high stresses. As a consequence, whenever a critical limit of stress is reached, a massive burst takes place in the working stope (Jha, 1991). This phenomenon is observed in the form of frequent occurrences of violent rockbursts on the either side of the fault plane. In Champion Reef mine, it is normal practice to carry out stoping sequences simultaneously in 3-4 levels. Hence high induced stresses concentrate over a larger area near the fault plane, so that in the event of a critical limit being reached, the entire area erupts into a series of rockbursts of all sizes. This type of area rockburst is observed both in the Northern Folds area and in the Glen Ore Shoot area, which are on the either side of the MNF in the Champion Reef mine. Had there been a movement along the MNF behind the possible attribute to this area rockburst, associated rockbursts would not only have been more severe, but also would have been observed on the either side of the fault plane.

The other fault plane i.e. the Tennant's Fault, which crosses the seismic zone IV, where the area rockburst took place in March 1982, does not seem to have played any role in its genesis. The centre of stress concentration as also the hypocentral distribution of the associated rockbursts of this series do not reflect any correlation with the fault plane.

5.2 Tectonic stresses

Tectonic stresses too influence the genesis of major rockbursts provided the direction of induced stresses is coincident with the maximum principle stress (insitu). We have problem of this type in the Biddick's shaft area of the Northern Folds. This shaft is elliptical in shape with its major axis perpendicular to the maximum principle stress. One major area rockburst on 11.2.1956 in this area severely damaged the Biddick's Shaft in at leasst five consequent levels (Krishnamurthy and Nagrajan, 1981). Even later this shaft was frequently getting damaged by major rockbursts and ultimately a Replacement Shaft had to be sunk for access to the stoping area.

5.3 Mining

Mining has a vital role in the genesis of the area rockburst particularly when the mining is being practiced close to the plane of weaknesses like folds, faults, dykes etc. All these features are present in the Northern folds area (seismic zone III), which has witnessed area rockbursts on varying scale in different regions in 1983, 1985 and 1992. The role of mining in inducing area rockbursts is extensively discussed by Miller (1967), krishnamurthy and Nagrajan (1981) and Krishnamurthy and Gupta (1983). It is stated that for whatever course of mining be adopted, there is a critical area of extraction at which major collapse will occur. However, mining being the remote region (long term effect), the role of geological features needs to be read simultaneously while assigning attributes for area rockbursts of this region (zone III).

5.4 Other attributes

The area rockburst of 14.3.1982 and 22.8.1985 discussed above, do not find their origin to any of these attributes, because they occurred in the old working area free from the influence of mining or any other geological plane of weaknesses. While looking for their possible attributes, the past mining history is critically examined.

It is found that serious rockbursts were encountered in the zone V in the early forties while mining through the Bullen's shaft. This forced to leave a number of large size barrier pillars before abandoning the mining operations. The area rockburst of 22.8.1985, which was active in this zone for five days, came without any precursor. Moreover, out of the 60 bursts so recorded, lo were of major type accounting for the 93% of the energy released. It is opined that due to long term effect of the stress build-up because of ongoing mining activity in the proximate regions, a critical limit is reached when these pillars yield with violent area rockburst. This view is also supported by the fact that only few rockbursts carry the major chunk of the energy released and once again seismicity was practically absent in this zone after this area rockburst of 1985.

The third area rockburst of 14.3.1982 was the shallowest one and one of the largest duration, whherein 214 events were recorded in the 17 days of intense seismic activity. Not much is known about the mining history of this zone, where mining had ceased some sixty years back. The area rockburst in this zone is likely due to failure in the unfilled or poorly filled stoped out areas. The long term effect of stress build-up renders such areas unstable resulting in their failure with a series of bursts of this type. This view is supported by the fact that despite being active for 17 days, only 15 major rockbursts were recorded in this series, while others were of minor to medium size, indicating some wide scale failure of small size.

6 CONCLUSION

The area rockbursts are similarin character to the "Type 2 " rockbursts defined by Knoll and Khunt (1990), but they have a different mechanism of origin. Drawing analogy with the earthquake seismology, the area rockbursts are like earthquake swarms where the number and magnitude of the earthquake increase gradually with time and decrease after some duration. There is no predominant principal earthquake in the swarm, which is the case with the area rockburst also. Even the mechanism of these area rockbursts resembles the mechanism of the earthquake swarm as described by Mogi (1967).

The study of area rockbursts reveals that there is no common attribute to their origin. They take place at all depth levels in both mining and old working areas. They may be attributable to the presence of fault planes or other geological planes of weaknesses, tectonic stresses, failure of pillars or remnants and failure in unfilled or poorly filled regions. Seeing such varied attributes, there cannot be a fixed return period for area rockbursts in Kolar Gold Fields. They may follow at different time intervals in different seismic zones.

However, seeing the past observation of a 10 years of recurrence cycle, some long term precursor needs to be identified for the area rockbursts. Further studies on the seismic behaviour and the analysis of the possible precursor patteerns may throw some more light on this subject.

ACKNOWLEDGEMENTS

The authors are thankful to the management of M/S
Bharat Gold Mines Limited (BGML) for providing the
necessary information about the mining history and
other details. The discussion with Mr. R. Krishnamurthy
ex-chief, R&D unit of BGML, helped to update this
text. We are thankful to Dr. M.V.M.S. Rao for the
useful discussions on this subject. The whole-hearted
support from our scientific assistants of the seismology section, particularly Sri Y.A. Willy and Sri
Y.L.Visweswaraih, is also thankfully acknowledged.

REFERENCES

Benioff, H. (1951a), Earthquake and rock creep, Bull.
    of seis. soc. of America, Vol. 41, pp. 31-62
Jha, P.C. and Willy, Y.A. (1990), Analysis of the
    stress regime of the Champion Reef mine with refere-
    nce to rockbursts, NIRM internal report, November
    1990, pp 1-75 (unpublished).
Jha, P.C. (1991), Seismic risk in mines and rockburst
    prediction, Ph.D. thesis, Indian School of Mines,
    Dhanbad (India), September 1991, (unpublished).
Krishnamurthy, R. (1977), Investigations into the
    stability and design of workings in deep mines, Jl.
    of mines , metals and fuel, Spl. vol. on proceed.
    of BPE seminar   I, pp 80-87.
Krishnamurthy, R. and Nagarajan, K.S. (1981),
    Rockbursts in Kolar Gold Fields, Proceed. of Indo-
    Greman workshop, Hyderabad, pp. 125-140.
Krishnamurthy, R. and Gupta, P.D. (1983), Rock
    mechanics studies on the problem of ground control
    and rockbursts in the Kolar Gold Fields, Paper in
    Rockbursts: prediction and control, IMM, London,
    pp. 67-80.
Krishnamurthy, R. and Shringarputale, S.B. (1990),
    Rockburst hazards in Kolar Gold Fields, Proceed. of
    second symp. on rockburst and seismicity in mines,
    A.A. Balkema pub., Rotterdam, pp. 411-420.
Miller, E. (1967), Notes on rock mechanics research
    in the Kolar Gold Fields, KGF Min. and Met. soc.
    Bull., Vol. XXIII, No. 95, pp. 23-61.
Mogi, K. (1967), Earthquake and fractures, Tectono-
    physics, Vol. 5, No. 1, pp. 35-55.
Raju, N.M., Jha, P.C., Shringarputale, S.B.,
    Srinivasan, C. and Sivakumar, C. (1991), Combating
    the problem of rockbursts at KGF - contribution of
    National Institute of Rock Mechanics, Jl. of mines,
    metals and fuel, Vol. XXXIX, No. 11-12, pp. 370-376.
Srinivasan, C. and Shringarputale, S.B. (1990), Mine
    induced seismicity in Kolar Gold Fields, Gerlands
    Beitr Geophysik, Leipzig, Vol. 94, No. 1, pp. 10-20.
Taylor and sons, J.T.M. (1955), Report of the special
    sub-committee on the occurrence of rockbursts in
    the mine of Kolar Gold Fields, Mysore State, South
    India, pp. 1-30.

Rockbursts and Seismicity in Mines, Young (ed.) © 1993 Balkema, Rotterdam, ISBN 90 5410 320 5

Identification of anomalous patterns in time-dependent mine seismicity

A. Kijko, C.W. Funk & A.v.Z. Brink
ISS International Ltd, Welkom, South Africa

ABSTRACT: A methodology is presented which identifies clusters of anomalous seismicity that may lead to a strong event. Single-link cluster analysis (Frohlich and Davis, 1990; Davis and Frohlich, 1991) and Matsumura's (1984) concept of seismicity description were combined and modified to form a cluster identification algorithm which can operate in real time. Temporal variations in parameters describing seismicity within clusters are used to assign a probability of occurrence of a seismic event which will be above some specified value.

A demonstration of the algorithm is given using seismic data from Western Deep Levels gold mine in South Africa. Probability assessments within clusters prove to be quite successful, indicating that the processes involved in rock mass preparation for large seismic events can be detected. A surprising result from this study is the similarities in probability patterns between clusters of seismicity that are separated by more than 1000 m. This suggests that clusters of seismicity within a mine interact and therefore the preparation process for large events incorporates volumes of rock which could be as large as an individual mine.

1 INTRODUCTION

Many extensive studies have been performed on space-time-magnitude/energy distribution of seismic events, for different regions, and different scales. Even after the elimination of for- and aftershocks, the tendency to form nests, swarms, and clusters is often observed. A variety of statistical methods have been applied in order to quantify the spatial and temporal properties of various catalogs. Some recent investigations include Kagan and Knopoff (1980), Kagan (1981 a,b), Dziewonski and Prozorov (1984), Johnson at al. (1984), Natale and Zollo (1986), Eneva and Pavlins (1988). Fractal formalism to describe the clustering of seismic events was introduced by Smalley at al., (1987). Interesting and conceptually similar approaches were proposed by Matsumura (1984), and recently by Frohlich and Davis (1990), and Davis and Frohlich (1991). In investigations of different kind of seismic activity anomalies, significant attention is given to detect seismic quiescence. Several investigations have found that in some time prior to significant seismic events, a characteristic decrease in seismic activity near the hypocentral region is observed. Comprehensive reviews of such a phenomenon include Habermann (1988), Wyss and Habermann (1988), Reasenberg and Matthews (1988), Wardlaw et al. (1990) and recently Kagan and Jackson (1991).

Such studies relating to the detection of anomalous seismicity in space-time-energy have, to our knowledge, never been applied to mining induced seismicity. Any such approach should be valid in mines. Numerous investigations have proven that in many respects, mine induced seismicity and tectonic seismicity are quite similar and the processes involved with the creation of anomalous seismicity patterns should be scale invariant.

A significant volume of high quality seismic data is required for detection of anomalous seismicity. Perhaps this is one of the reasons why the study of anomalous seismicity has been performed so seldom in mine seismology. Recent developments in quantitative, real-time, digital mine seismology (Mendecki, 1992) have enabled seismologist to collect vast amounts of data without manual processing. A routine quantitative description of a seismic event would include its location, radiated energy, and seismic moment. These parameters are determined automatically, and therefore when combined with a sensitive system, large complete catalogues can be compiled.

This paper introduces a methodology which finds clusters of anomalous seismicity which may be lead to the occurrence of a significant seismic event. The method for cluster detection needs no information about the mine geology or the mining works. Seismic data from the Western Deep Levels (WDL) gold mine is used to demonstrate the techniques. Because of it's great depth, WDL mine has a high rate of seismicity. Real-time mine seismology at WDL mine has provided a large and complete catalogue for analysis.

2 DETERMINATION OF SPACE-TIME CLUSTERING OF SEISMICITY

The formulation of the clustering process is conceptually close to the single-link cluster analysis (SLC) first adapted to seismology by Frohlich and Davis (1990). The difference in the approach used here is the incorporation of a moving time window which is reminiscent of Matsumuras procedure. In SLC (described by Frohlich and Davis, 1990), for each earthquake, a nearest neighbor is found and the earthquakes are linked together. The preliminary set of links forms a collection of so-called trees. At this stage, if there are N events in the catalogue, there is a minimum of $N/2$ links and a maximum of N-1 links. Clusters are formed by linking events in different trees together. The process is iterative: each iteration links together sets of trees to form new, larger trees. The procedure terminates when N-1 links are formed. The links are reordered from the largest link to the smallest link to form a ordered cluster. Clusters with links smaller than some specified maximum length can be formed from this main ordered cluster by eliminating all links greater than a specified length.

For real-time applications in mine seismology, the above SLC procedure is not practical: the procedure requires considerable memory and is also time consuming. We have chosen a less rigid approach for the clustering algorithm. The process is illustrated in Figure 1. Seismic events occurring within a mine are placed into a moving time window of some specified length. For each new event that comes into the window, links are calculated between this most recent event and all the events in the window. The link distance is both a space and time metric. This is a concept similar to Matsumura (1984), Frohlich and Davis (1990), Davis and Frohlich (1991), however we allow the distance between two events to increase exponentially with time rather than linearly. The link distance is a spatial distance which is exponentially weighted according to the product of some time constant and the time difference between the occurrence of the events. If a link is found to be less than the maximum link length, the two events in question are linked together. Only one link is permitted to form between two events which are both within the same cluster; the two events may not necessarily be nearest neighbors. At some stage in the

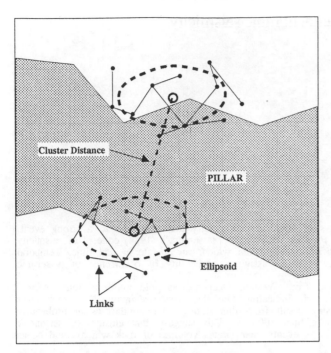

Figure 1. Illustration of the clustering process.

clustering process, an event from one cluster may link to an event belonging to another cluster The algorithm will merge the two clusters together. Occasionally two clusters should physically be linked together, however no events occurred near enough to each other for a link to be formed. This situation can arise from using a time window which is too small. Such a problem is easily amended by periodically testing distances between cluster centroids (Figure 1). If cluster centroids drift near enough to each other, the algorithm will merge them into one.

The choice of the time window length is an important consideration. If the window length is too small, clusters may have no events in the window at some point in time. Such a situation would effectively terminate the cluster. If the window length is too large, many links must be calculated for each new event causing the algorithm to become unnecessarily slow. The average activity rate for the mine must be studied to decide on an appropriate window length.

For each cluster, a number of parameters are monitored so that the seismicity can be characterized. The mean X, Y, and Z along with variances of the coordinates are stored so that the shape of the ellipsoid can be determined via the covariance matrix. The physical cluster volume holds one standard deviation of event locations with respect to the cluster centroid. The minimum, maximum, and cumulative energy, moment, apparent stress, and apparent volume are continuously updated within the clusters. Parameters such as b-slope, activity rate, and the probability of occurrence of a strong event (below) can be determined at any time. Also, the volume of the ellipse relates how concentrated the events are in the area.

3 TIME-DEPENDENT ASSESSMENT OF SEISMICITY

The volumes within a mine where large events are most likely to occur can be identified by their clustering. The time occurrence of seismic events can be determined by statistical analysis. Mine tremors are not strictly a random process, however statistical analysis of seismic events occurrence provides a reasonable basis for future seismic activity assessment. The approach considered in this work is based on continuous evaluation of the certain parameters, and is quite similar to the formalism developed by Lasocki (1990) where certain models of stationary hazard evaluations are used and applied within moving time windows. Thus, the process of seismic events occurrence is time dependent through the time dependence of its parameters.

The idea of temporal variations of seismic activity parameters, especially Gutenberg-Richter parameter b, is not new. Time variations of b-value were found during earthquake sequences in New Zealand (Gibowicz, 1973; Imoto, 1987) and for earthquakes in California (Wyss and Lee, 1973). The variations preceded the 1976 Tangshan earthquake in China (Li et al., 1978) and the 1976 Friuli earthquake in Italy (Cagnetti and Pasquale, 1979). These changes were also observed in laboratory experiments modeling the fault movements (e.g. Main et al., 1989; Meredith et al., 1990). In most cases a decrease in b-values before slips on cut surfaces, followed by an increase after the event, were found. More complex patterns of b-value changes were found by Ma (1978). In a study of variations in b-value before several large earthquakes in northern China, he found that for smaller areas around the major earthquake epicenters the parameter b varies with time from higher to lower values as the earthquake approaches, while for larger areas peak values of b appear immediately before the earthquakes (Reyners, 1981). These observations imply that the parameter b can be used for an assessment of time dependent earthquake hazard and prediction. Characteristic changes of b-vales are considered, amongst the other precursory phenomena, in models of physical processes occurring in the areas of pending earthquakes (e.g., Mjachkin et al., 1975), and variations of $b(t)$ are routinely observed for earthquake prediction in Japan (Shibutani and Oike, 1989).

Precursory phenomena before seismic events induced by mining were also observed. Anomalous seismicity changes (an increase followed by a decrease) were recorded prior to moderate rockbursts in a deep silver mine in northern Idaho, U.S.A. (Brady, 1977), in coal mines in Poland (Gibowicz, 1979), and in gold mines in South Africa (Brink, 1990). Variations of the parameter b from seismic events in mines might provide an indication of stress state in specific areas in underground mines.

Unfortunately, in mining practice there are indications that the frequency-magnitude relation sometimes has curvature, so that the classical Gutenberg-Richter b value is not always adequate. Observations of seismic events induced by mining indicate that the distribution of value x, equal to m, $\lg(M_0)$, $\lg(E)$, or $\lg(M_0 + \alpha E)$, where m is magnitude, M_0 is seismic moment, E is seismic energy, and α is constant, can be expressed as

$$\ln n = A - B(x - x_{min})^C. \tag{1}$$

where n is the number of events, x_{min} is threshold of completeness and A, B and C are the parameters. Relation (1) may be interpreted as being either a cumulative relationship, if n is the number of events equal or larger than x in a given time interval, or as being a density law if n is the number of events in a certain small interval around the value of x. The parameter A is a measure of the level of seismicity. B describes the relative number of small and large events, whereas parameter C controls the curvature of the discussed relation. For $C=1$, the curvature of formula (1) disappears and takes the same form as the classical relation of Gutenberg-Richter.

Time variations of the coefficient A, B and C found during seismic event sequences in mines indicate that they depend on the rheology and structure of the rock and therefore can be considered as parameters controlling the capability to release accumulated stresses and energy. These observations imply that the parameters A, B and C can be used for an assessment of future seismic activity.

If the values of x are assumed to be random variables and distributed according to the relation (1), where parameter C is not necessary equal to 1, its frequency and cumulative probability distribution functions takes the form

$$f(x) = \begin{cases} 0, & \text{for } x < x_{min}, \\ BC(x - x_{min})^{C-1} \exp[-B(x - x_{min})^C], & \text{for } x \geq x_{min}, \end{cases} \tag{2}$$

and

$$F(x) = \begin{cases} 0, & \text{for } x < x_{min}, \\ 1 - \exp[-B(x - x_{min})^C, & \text{for } x \geq x_{min}. \end{cases} \tag{3}$$

Formulas (2)-(3) are known as Weibull distribution functions, (Eadie et al., 1982) and are used mainly in the modeling of times of seismic event occurrence (e.g. Utsu, 1984; Cornell and Winterstein). For any given time t, the the maximum likelihood estimation of parameters B and C are obtained from n events of $(x_1,...x_n)$, recorded in a region during the time period $(t-\Delta T, t]$, according to the relations (Utsu, 1984; Lasocki, 1990)

$$
\begin{cases}
B = \dfrac{1}{<(x-x_{min})^C>}, \\[2ex]
\dfrac{1}{C} + <\ln(x-x_{min})> = \dfrac{<(x-x_{min})^C \ln(x-x_{min})>}{<(x-x_{min})^C>},
\end{cases}
\tag{4}
$$

where $< \cdot >$ mean the sample average. The second equation in (4) is independent of the B parameter and for C can be readily solved by an iterative procedure. Then, parameter B can be estimated from the first equation of (4).

Assuming that the occurrence of seismic events is Poissonian and stationary for a time interval $(t-\Delta T, t]$ where ΔT is long enough to obtain data for a reliable estimation of parameters, the probability of occurrence of at least one event with $x \geq x_{min}$, between t and $t+\Delta t$ is

$$
1 - \exp(-\lambda \Delta t),
\tag{5}
$$

where $\lambda = n/\Delta T$ is the seismic activity rate, n is the number of seismic events with x equal or exceeding the threshold value x_{min}, occurring within time interval $(t-\Delta T, t]$.

Thus, if the process is time dependent through a time dependence of its parameters $B \equiv B(t)$, $C \equiv C(t)$ and $\lambda \equiv \lambda(t)$, but can be considered stationary for any time interval $(t-\Delta T, t]$, the probability of occurrence of a seismic event (PSE) with a value of x, greater than or equal to a given x_0 $(x_0 \geq x_{min})$, within a time interval $(t, t+\Delta t]$, is

$$
P\{x \geq x_0 | (t, t+\Delta t]\} = 1 - \exp[-\lambda_0(t) \cdot \Delta t],
\tag{6}
$$

where $\lambda_0(t)$ is the rate of occurrence of seismic events with $x_0 \geq x_{min}$ and equal to $\lambda(t)[1-F(x_0|t)]$.

In this way, PSE (eq.6) within a cluster can be estimated for any time t and predicted for any time interval $(t, t+\Delta t]$ from the events recorded within the interval $(t-\Delta T, t]$. The length of interval ΔT is selected according to the variability of seismicity within the cluster.

4 WESTERN DEEP LEVELS GOLD MINE AND SEISMIC NETWORK

Western Deep Levels gold mine is situated approximately 70 km west of Johannesburg, South Africa and is recognized as the deepest mine in the world. Two gold bearing reefs are mined according to a longwall concept. The upper reef horizon, the Ventersdorp Contact Reef (VCR), covers a depth range of 1400 m to 2400 m below surface. The deeper reef horizon, the Carbon Leader Reef (CLR), extends from 2300 m to 3500 m below surface. The reefs are separated by approximately 800 m of quartzite rock and are intersected by numerous dykes and some minor faults. The country rock is hard and dry with a uniaxial compressive strength of about 200 MPa. The reefs are extracted to a height of 1 m to 3 m by means of drilling and blasting.

WDL mine is extensively covered with a state-of-the-art digital seismic system. From January 1 to December 21, 1992 more than 14,300 seismic events have been recorded by the Integrated Seismic System (ISS), by a minimum of five seismic stations in the solution, over the local magnitude range -1.0 to 4.7. All the events are recorded and located with a network of triaxially mounted velocity sensor units. Comprehensive seismological processing, in real time, provides all the relevant parameters which are stored in a data base. These parameters include location, seismic radiated energy, seismic moment, and other source parameters. The data on which this study was based was extracted from the WDL mine data base.

5 EXAMPLES OF APPLICATION AT WDL MINE

The ability of the space-time clustering (STC) analysis algorithm to detect anomalous seismicity is demonstrated in Figure 2. This figure shows the seismicity at WDL mine on the CLR for a one year period. The clusters are represented by ellipsoids. A moving time window of 10 days was used with a maximum space-time distance of 250 m, a migration distance of 100 m, and a time decay constant of 5 days. The level of completeness for the data was energy $10^{4.5}$J. Without any knowledge of the mining the STC algorithm was able to detect all the clusters of seismicity in the mine.

By monitoring the PSE for each individual cluster the concentrations of seismicity in the mine can be delineated. Figure 3 shows the PSE, computed with $C=1$ (the classical Gutenberg-Richter relationship), for three clusters over a one year period. The level of completeness for this data is energy $10^{3.5}$J. Each cluster is situated at a stope. Two of these clusters were on the VCR and the other was on the CLR. The depths and volumes of the cluster ellipsoids are given on the figure along with the number of events in the clusters. The vertical axes show the PSE equal and above energy $10^{8.5}$J. The arrows on the PSE curves mark positions of events which released energy equal to or above the energy $10^{8.5}$J threshold. Figure 3 shows that all events above energy $10^{8.5}$J occurred during times of high or rapidly increasing probability.

In some cases it has been found that a more complex approach for PSE assessment is needed. Figure 4 show an example were $C>1$ proved much more effective than $C=1$. In this case, curvature in the Gutenberg-Richter relationship is permitted; the curvature is determined by the nature of the data and can vary with time. For this example, the arrows indicate positions of events which were above energy $10^{9.5}$J. Clearly, in this example curvature in the Gutenberg-Richter relationship must be considered for a more accurate PSE assessment.

Preliminary PSE assessments suggest that if the physical volume of the cluster is small, one must allow for curvature in the Gutenberg-Richter relationship. Larger cluster volumes (encompassing several stopes) show a more classical distribution. These observations indicate that one must be cautious in PSE assessment.

Coming back to Figure 3, this example illustrates that the PSEs have similar patterns in the different areas. The low period of probabilities is quite obvious between July and early September of 1992, followed by a high period in mid-September. Note also that during the period of high probability there were events above energy $10^{8.5}$J in the three clusters between August and September. Also, in November, the cluster on the CLR shows a steady increase in probability through the month, cumulating in two seismic events above energy $10^{8.5}$J. The top PSE curve in Figure 3 (VCR) shows a seismic event that occurred just after the two events on the CLR. The middle PSE curve had no seismic event above energy $10^{8.5}$J, however the PSE increased in a pattern similar to the CLR cluster. A energy $10^{7.9}$J event did occur at the peak in PSE, in this cluster on the VCR, in November.

In general, the PSEs in different clusters show similarities in their patterns regardless of the number of

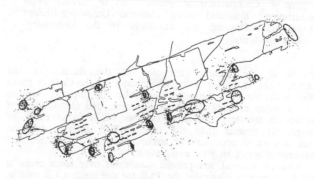

Figure 2. Clusters of seismic events at WDL mine.

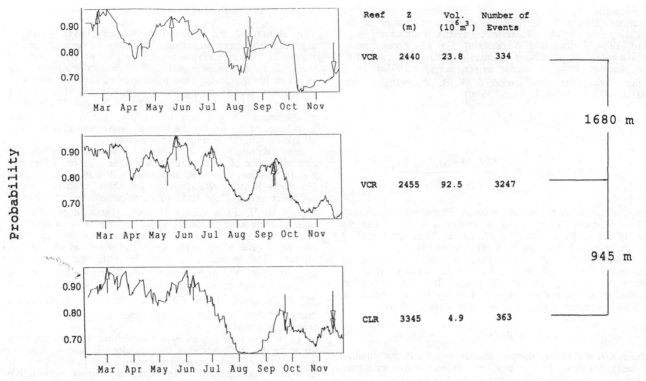

Reef	z (m)	Vol. $(10^6 m^3)$	Number of Events
VCR	2440	23.8	334
VCR	2455	92.5	3247
CLR	3345	4.9	363

1680 m

945 m

Figure 3. PSEs for three clusters at WDL mine.

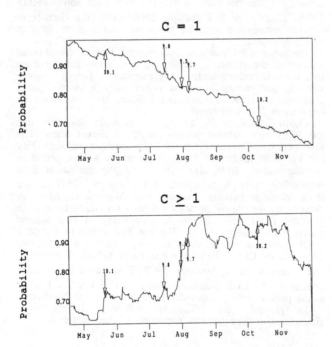

Figure 4. Comparison of probabilities of seismic event occurrence calculated using classical Gutenberg-Richter (top), and allowing for curvature in the Gutenberg-Richter (bottom).

events in the cluster, the size of the cluster, or the depth. Note in Figure 3 that the cluster on the CLR is about 900 m deeper than the other two. Also the physical volumes of the three clusters are quite different, and the number of events in the larger cluster (on the VCR) is an order of magnitude greater than the others. This analysis suggests that one must consider, or make allowance for, interaction between clusters in PSE assessment.

PSE assessment has proved successful over large areas at WDL. Figure 5 illustrates the PSE for the entire CLR mining area between January and November of 1992. Only the larger

seismic events are considered here with energies above $10^{6.5}$J. The energy threshold for the arrows was chosen at $10^{9.8}$J. Even though the events used in this analysis cover a large area, remarkably distinct patterns can be observed. The three large peaks in PSE are separated by about 3.5 months. During times of increasing, or maximum PSE, at least four significant events (above energy $10^{9.8}$J) are observed. Only one large event is observed during times of decreasing or minimum PSE.

The concept of interaction between natural seismicity at different areas is not new. This problem was undertaken by several authors and different approaches are known (e.g. Chiaruttini, et al., 1980; Båth, 1984; Mantovani et al., 1987; Mucciarelli et al., 1988; Alberto et al., 1989). Interaction between two different mining areas was also observed in mines. A simple trial of quantification of such a phenomenon is described by Kijko (1980). Using a different approach such interactions have also been recognized here in a South African gold mine.

6 CONCLUSIONS

A methodology has been introduced which can detect clustering of seismicity in space and time. The spatial coordinates of the clusters indicate where strong seismic events will occur. Using the developed approach, probability of seismic event occurrence as a function of time can be evaluated for any cluster. The algorithm uses only radiated seismic energy and seismic moment and does not rely on magnitude or model dependant seismic source parameters. The applied methods do not require any information related to the local geology and mining activity. It is crucial that enough data is available for analysis by the algorithm. Real-time mine seismology used at WDL mine has provided an excellent catalogue of mine-induced seismic events for analysis. The preliminary results are promising: large events are found to occur during high or rapidly increasing PSE. In some situations, curvature in the Gutenberg-Richter relation must be acknowledged for a proper evaluation of PSE.

An interesting result from this study was the demonstration of interaction between different clusters. The interactions were shown to exist between clusters separated

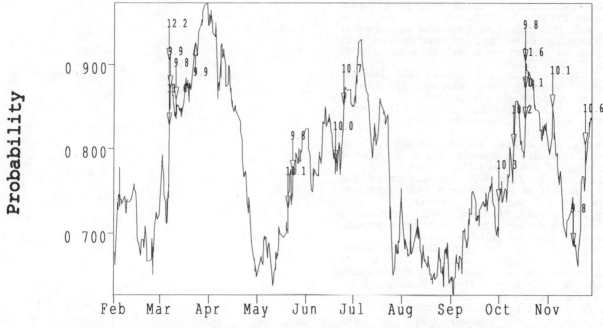

Figure 5. PSE for the Carbon Leader Reef calculated from events with energy greater than $10^{6.5}$ J.

by more than 1000 m. The phenomenon is evident in the similar patterns of PSE within separate clusters. It appears that the interactions between clusters must be considered when evaluating PSE. In the past it has been speculated that the processes involved in the creation of strong seismic events affect large volumes of rock; volumes much larger than any single stope or mining area. This analysis further supports these speculations. Our next target will be the quantification of the degree of interaction between clusters, and its effect on PSE evaluation within clusters.

REFERENCES

Alberto, D., M. Mucciarelli, and E. Mantovani 1989. Use of non-parametric correlation tests for the study of seismic interrelations. *Geophys. J.* 96: 185-188.

Båth, M. 1984. Correlation between Greek and global seismic activity. *Tectonophysics* 109: 345-351.

Brady, B.T. 1977. Anomalous seismicity prior to rock bursts: Implications for earthquake prediction. In *Stress in the Earth*, M. Wyss (ed.), Special Issue, *Pure Appl. Geophys.* 115: 357-374.

Brink, A.v.Z. 1990. Application of microseismic system at Western Deep Levels. *Rockbursts and Seismicity in Mines*, C. Fairhurst (ed.): 355-361. Rotterdam: Balkema.

Cagnetti, V., and V. Pasquale 1979. The earthquake sequence in Friuli, Italy, 1976. *Bull. Seism. Soc. Am.* 69: 1797-1818.

Chiaruttini, C., A. Kijko, and R. Teisseyre 1980. Tectonic discrimination of the Fruli earthquakes. *Bull. Geofis. Ther. Appl.* 22: 295-302.

Cornell, C.A., and S.R. Winterstein,1988. Temporal and magnitude dependence in earthquake recurrence models. *Bull. Seism. Soc. Am.* 78: 1522-1537.

Davis, S.D., and C. Frohlich 1991. Single-link cluster analysis, synthetic earthquake catalogs, and aftershock identification. *Geophys. J. Int.* 104: 289-306.

Dziewonski, A.M. and A.G. Prozorov 1984. Self-similar determination of earthquake clustering, *Computa. Seism.* 16: 7-16.

Eadie, W.T., D. Drijard, F.E. James, B. Sadoulet, and M. Roos 1982. *Statistical Methods in Experimental Physics*, 2nd reprint. Amsterdam: North-Holland Publishing Company.

Eneva, M., and G.L. Pavlins 1988. Application of pair analysis statistics to aftershocks of the 1984 Morgan Hill, California earthquake. *J.Geophys. Res.* 93: 9113-9125.

Frohlich, C., and S.D. Davis 1990. Single-link cluster analysis as a method to evaluate spatial and temporal properties of earthquake catalogs. *Geophys. J. Int.* 100: 19-32.

Gibowicz, S.J. 1973. Variation of the frequency-magnitude relation during earthquake sequences in New Zealand. *Bull. Seism. Soc. Am.* 63: 517-528.

Gibowicz, S.J. 1979. Space and time variations of the frequency-magnitude relation for mining tremors in the Szombierki coal mine in Upper Silesia, Poland. *Acta Geophys.Pol.* 37: 39-49.

Habermann, R.E. 1988. Precursory seismic quiescence: past, present and future. *Pure Appl. Geophys.* 126: 279-318.

Imoto, M. 1987. A Bayesian method for estimating earthquake magnitude distribution and changes in the distribution with time and space in New Zealand. *N. Z. J. Geol. Geophys.* 30: 103-116.

Johnson, C., V.I. Keilis-Borok, R. Lamore, and B. Minister 1984. Swarms, of main shocks in southern California, *Computa. Seism.* 16: 1-6.

Kagan, Y.Y. 1981. Spatial distribution of earthquakes: four-point moment function. *Phys. Earth Planet. Int.* 12: 291-318.

Kagan, Y.Y., and D.D. Jackson 1991. Seismic Gap hypothesis: ten years after. *J.Geophys.Res.* 96: 21419-21431.

Kagan, Y.Y., and L. Knopoff 1980. Spatial distribution of earthquakes: the two point correlation function. *Roy. Astron. Soc.* 62: 303-320.

Kijko, A. 1980. Statistical test of mutual dependence of seismic activities in two adjacent regions. *Publ. Inst. Geophys. Pol. Acad. Sci.* A-10, (142): 125-133.

Lasocki, S. 1990. *Prediction of Strong Mining Tremors*. Zesz. Nauk. Akad. Gorn-Hutn., Geofiz. Stosowana (Cracow), 7: 1-110. (In Pol.; Engl. abstr.)

Li,Q. L., L.Yu. Chen, and B.L. Hao 1978. Time and space scanning of the *b*-value: A method for monitoring the development of catastrophic earthquakes. *Acta Geophys. Sin.* 21: 101-125.

Ma, H. C. 1978. Variations of the *b*-values before several large earthquakes which occurred in north China. *Acta Geophys. Sin.* 21: 126-141.

Matsumura, S. 1984. A one-parameter expression of seismicity patterns in space and time. *Bull. Seismol. Soc. Am.* 74: 2529-2576.

Main, I.G., P.G. Meredith, and C. Jones 1989. A reinterpretation of the precursory seismic *b*-value anomaly from fracture mechanics. *Geophys. J.* 96: 131-138.

Mantovani,E., M. Mucciarelli, and D. Alberto 1987. Evidence of interrelation between the seismicity of the southern

Apennines and southern Dinarides. *Phys. Earth Planet. Interiors* 49: 259-263.

Mendecki, A.J. 1993. Real time quantitative seismology in mines. Proceedings of the 3rd International Symposium on "Rockbursts and Seismicity in Mines", 16-18 August 1993, Kingston, Ontario, Canada.

Meredith, P. G., I.G. Main, and C. Jones 1990. Temporal variations in seismicity during quasi-static and dynamic rock failure. *Tectonophysics* 175: 249-268.

Mjachkin, V.I., W.F. Brace, G.A. Sobolev, J.H. Dietrich 1975. Two models of earthquake forerunners. *Pure Appl. Geophys.* 113: 169-181.

Mucciarelli,M., D. Alberto, and E. Mantovani 1988. Earthquake forecasting in Southern Italy on the basis of logistic models. *Tectonophysics* 152: 153-155.

Natale, G.D., and A. Zollo 1986. Statistical analysis and clustering features of the Phlegraean Fields earthquake sequence. *Bull. Seis. Soc. Am.* 76: 801-814.

Reasenberg, P.A., and M.V. Matthews 1988. Precursory seismic quiescence: a preliminary assessment of the hypothesis. *Pure Appl. Geophys.* 126: 373-406.

Reyners, M. 1981. Long- and intermediate- term seismic precursors to earthquakes - State of the art. In *Earthquake Prediction* D.W. Simpson and P.G. Richards (eds), Maurice Ewing Series 4: 333-347. Washington D.C.: American Geophysical Union.

Shibutani, T., and K. Oike 1989. On features of spatial and temporal variation of seismicity before and after moderate earthquakes. *J. Phys. Earth* 37: 201-224.

Shlien, S. and M.N. Toksöz 1970. A clustering model for earthquake occurrences. *Bull. Seism. Soc. Am.* 60: 1765-1787.

Smalley, R.,Jr., J.-L. Chatelain, D. Turcotte, and R. Prévot 1987. A fractal approach to the clustering of earthquakes: applications to the seismicity of the New Hebrides. *Bull. Seism. Soc. Am.* 77: 1368-1381.

Utsu, T. 1984. Estimation of parameters for recurrence models of earthquakes. *Bull. Earthq. Res. Inst., Univ. Tokyo,* 59: 53-66.

Wardlaw, R.L., C. Frohlich, and S.D. Davis 1990. Evaluation of precursory seismic quiescence in sixteen subduction zones using single-link cluster analysis. *Pure Appl. Geophys.* 134: 57-78.

Wyss, M., and R.E. Habermann 1988. Precursory seismic quiescence. *Pure Appl. Geophys.* 126: 319-332.

Wyss, M., and W.H.K. Lee 1973. Time variations of the average earthquake magnitude in central California. *Stanford Univ. Publ., Geol. Sci.* 13: 24-42.

Rockbursts and Seismicity in Mines, Young (ed.) © 1993 Balkema, Rotterdam, ISBN 90 5410 320 5

Statistical short-term prediction in mining induced seismicity

S. Lasocki
University of Mining & Metallurgy, Krakow, Poland

ABSTRACT: A step-by-step construction of a stochastic model of the local tremor generation process in an area of single excavation has been given. An assumption base for model creation has been obtained from studies of statistical properties of seismic event sequences recorded in such regions. Three variants have been put forward with respect to three different theoretical distributions which can be used to represent the actual distribution of tremor energy. The model is then used in an algorithm of real time estimation of time-varying probability of large tremor occurrence close to mining works. Interpretation of the probability estimates obtained while processing mining data and a practical usefulness of the presented approach have been discussed.

1. INTRODUCTION

Coal extraction in Poland is frequently accompanied by induced seismicity and rockbursts. Even early observations showed, that the majority of mining seismicity is located in a vicinity of active exploitation fronts at the distances up to about 100-200 meters. These tremor swarms move accordingly while the fronts advance (Klonowski & Gerlach 1974, Trombik & Zuberek 1980, Sato & Fujii 1988). Fig. 1 presents a typical frequency histogram of tremors occurring during stope mining, related to the front position.

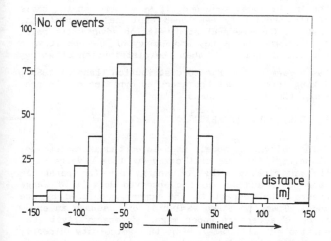

Figure 1. Distribution of seismic events directly connected with mining with respect to a longwall face position.

According to recent opinions (Johnston & Einstein 1990, Gibowicz 1990), these events build up a distinct class of seismicity directly connected to mining works. Both, event rate and energy range for the class depend upon local conditions. In Poland the event rate runs to 20 events daily in one stope and energies observed vary from 10^2, 10^3 J in accordance with quality of recording equipment, to 10^7 and more J. Positions of sources of the events and their energy range distinguish them from other tremors which form a group called seismic events triggered by exploitation. It is presumed that the main controlling influence on generation of events occurring close to mining works has a local state of stress and its redistribution in the stope region. The triggered events are in general stronger, not connected directly with stopes. They usually concentrate around local discontinuities or zones of weaknesses.

Due to the proximity of the class directly connected with mining to mining works the strongest events from this class often cause rockbursts. In fact, a majority of rockbursts is related to seismic events, which may be included in the presented class (Gay & Wainwright 1984, Gibowicz 1990). Because of the possible hazardous effect, any information of known reliability concerning probability of such strong tremors occurrence may be valuable. An idea to predict these events by means of an analysis of the tremor sequence connected with one stope (hence regarded as homogeneous) is based upon an assumption that information contained in the recorded sequence is sufficient for describing a state of tremor generation process in the whole energy range. Similar assumption gave basis for constructing earthquake risk methods.

Contrary to the static earthquake risk methods and the analyses of induced seismicity for constructing regional seismotectonic models, properties of local sequences directly connected with mining imply building time dependent models and performing dynamic analyses. As in the course of mining the local stress field varies and it has a controlling effect on generation of seismic events, it is expectable that all distribution functions describing probability of seismic events occurrence are time-varying. Thus if, on the basis of the observed event sequence, one could infer something about a direction of time changes of defined risk functions, then the real-time analysis would be a helpful tool in a day-by-day decision making process, in rockburst mitigation as well as in evaluating effectiveness of active prevention means (Marcak 1978, Gibowicz 1979).

Studies of properties of observed seismic event sequences with respect to possibilities for constructing dynamic short-term prediction methods have been carried on for years. A significant number of processed seismic data files were acquired from regions of longwall faces and galleries of dozen or so coal mines in Upper Silesia, Poland, in which mining was going on under different geological conditions. The results of these studies, which form assumptions for constructing further presented model, have been supported by a comprehensive and, as far as possible, objective data analysis. Testing groups have been selected so as to enable generalization of results. Because of very wide range of these works it is not possible to present the empirical background of given conclusions. This basis can be found in successively given references.

2 RATIONALE AND MODEL CONSTRUCTION

Standard seismic monitoring networks in mines deliver tremor data in the form of time of event occurrence, its energy at the source and coordinates of hipocentral point. The latter are used to split sequences and to extract event group connected with the particular stope under study. Thus we suppose that the local sequence is represented by two random variables: time of event occurrence and its energy.

Studies have not shown any significant correlation between interoccurrence time and event energy, which supports the assumption that these random variables may be considered as independent (Report on CPBP 1989).

The main argument for introducing dynamic models has been obtained from the studies of stochastic structure of sequence of events coming from the region of a single stope (Lasocki 1990a). The analysis of data has pointed out that the process of tremor generation is non-stationary. Weak events are the principal carriers of process memory because the sequence censored with an energy threshold of 10^4 J usually becomes Poissonian. The analysis showed at the same time that the time variation of process is slow and the process may be regarded as segmental stationary. It was also suggested that it is the generalized Poisson process. The last conclusion has been backed up by Monte-Carlo simulation studies made in order to identify renewal process for sequence of induced seismicity events (Lasocki 1992a). Due to the above it is assumed, within the model, that:
- the process is the generalized Poisson process, for which probability of n events per unit time is:

$$Pr\{n;t\} = [\lambda(t)]^n/n! \; exp[-\lambda(t)] \qquad (1)$$

where $\lambda(t)$ denotes the time-varying process parameter;
- the process has the segmental stationarity property i.e. there exists a finite time interval ΔT, constant and characteristic for local conditions, so that for any time moment t: $\lambda(\tau) \approx const$ when $\tau \in [t, t+\Delta T]$.

As far as tremor energy is concerned, it has been widely observed that the statistical Gutenberg-Richter relation:

$$logN = a + logE \qquad (2)$$

where N is a number of events of energy not less than E and a, b are parameters, is also valid for induced seismicity (e.g. Spottiswood & McGarr 1975, Sato & Fujii 1988, Subbaramu et all 1989, Johnson & Einstein 1990). In local sequences the Gutenberg-Richter relation is also applicable but the parameter b varies in time (Gibowicz 1979, Bath 1984, Johnston & Einstein 1990, Slavik et all 1992). Fig. 2 presents the values of b parameter with standard error bars estimated by means of the maximum likelihood method from tremors

recorded in 15-day periods of mining the longwall 15, Szombierki Mine, Upper Silesia. The points are related to the centres of periods. It may be observed that time changes are significant. This variability certainly is not confined to the Gutenberg-Richter's parameter hence it can be assumed in general that an energy distribution is time-dependent.

The autocorrelation function of sequences of energy of events has not revealed any relation between elements of the sequence, which in connection with the above mentioned studies of the form of renewal process allows to accept an assumption about the mutual independence of events. Hence it will be assumed that a distribution function of energy depends on time only through the time-dependence of its parameters. Additionally it will be supposed that the segmental stationarity is a property of the process, hence the parameters of energy distribution are approximately constant in time intervals of length ΔT.

For this base of assumptions the cumulative distribution function of tremor energy E for any time interval $(t, t+\Delta t)$, $\Delta t \leq \Delta T$, conditional upon event occurrence is given by:

$$F(E|n>0, \Delta t; t) = \{1-exp[-\lambda(t) \cdot \Delta t]\} \cdot F_e(E; t) \qquad (3)$$

where $F_e(E; t)$ is the cumulative distribution function (c.d.f.) of marginal distribution of event energy. The probability that the event having energy greater than or equal to a given value E_p will occurre within the time interval $(t, t+\Delta t)$ is given by:

$$R(\Delta t, E_p; t) = \{1-exp[-\lambda(t) \cdot \Delta t]\} \cdot \{1-F_e(E_p; t)\} \qquad (4)$$

The formula (4) represents the large event risk and is used in prediction analysis.

In order to fully determine the model, one has to state a specific form of the c.d.f. of energy of tremor in the local sequence. If we assume, on the basis of the previously mentioned empirical observations, that the Gutenberg-Richter relation governs the relation between frequency and energy of tremors above a certain threshold E_o then the distribution of energy has a form of the Pareto distribution (Lasocki 1989). Taking into account the accepted dynamics of the process, the c.d.f. becomes:

$$F_e^P(E; t) = 1 - [E_o/E]^{b(t)} \qquad (5)$$

The eqiuvalent model, used in earthquake seismology (Lomnitz 1974), was criticized due to a divergence of the anticipated value of energy for the parameter range $b \leq 1$, and due to the observed deficit of large events with respect to the number predicted by the model. These problems vanish if one assumes an existence of a certain upper magnitude/energy limit (Cosentino et all 1974). For the seismicity directly connected with mining this assumption means that, in a region of works, there exists a maximum of energy which can be discharged as a seismic event. Since it is expectable that this maximum - E_m, if exists, depends upon local failure conditions, it must be considered as time-varying parameter of the process. Introduction of the upper limit modifies the energy distribution to a form of the truncated Pareto distribution (Lasocki 1990b). Its c.d.f. is given by the formula:

$$F_e^T(E; t) = \{1-[E_o/E_m(t)]^{b(t)}\}^{-1} \cdot [1-(E_o/E)^{b(t)}] \qquad (6)$$

Both distributions, (5), (6) positively passed tests of goodness of fit. Because of limited number of strong events in the sequences directly connected with mining and time-variation of the energy distribution in these sequences, an evidence gathered till now does not give any definite answer which one of these two models better represents real data.

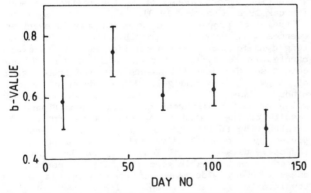

Figure 2. Time variation of the Gutenberg-Richter's b parameter for seismic events recorded in an area of excavation in the course of mining.

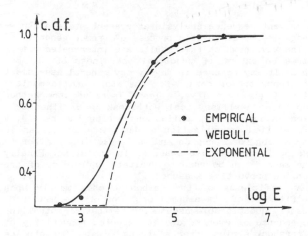

Figure 3. Fits of the cumulative distribution functions of the Pareto distribution model and the Weibull distribution model to the empirical distribution of logarithm of tremor energy.

If energy has the Pareto distribution (the truncated Pareto distribution), the common logarithm of energy has exponential (truncated exponential) distribution. A very interesting model for energy is formed through a generalization of the log-linear shape of the frequency-energy relation. This generalization means that the exponential distribution for the logarithm of energy is replaced with the Weibull distribution. In the time-dependent model the c.d.f. for logarithm of energy will then have a form:

$$F_e^W(x;t) = 1 - \exp[-\alpha(t) \cdot (x-\gamma)^{\beta(t)}] \qquad (7)$$

where $x = \log E \geq \gamma$. Such distribution is applicable for whole range of observed energies and describes, without a necessity of introducing any deterministic lower and upper limits for the range, the nonlinear shape of the frequency-energy relation for low and high energies. The parameter γ represents an actual low energy threshold of recording ability for the observation network, below which no events are recorded. Fits of the c.d.f.-s of the Pareto distribution model and of the Weibull distribution model to the empirical cumulative distribution of logarithm of energy of tremors recorded in the vicinity of single stope are compared in fig. 3. The Weibull distribution gives better fit in whole energy range. The chi-square and the Kolmogorov-Smirnov tests of goodness of fit confirm this conclusion. The details about the Weibull distribution model, reasoning of its capability to represent the distribution of tremor energy and the results of tests of goodness of fit are furnished elsewhere (Lasocki 1990b).

All, the simple and truncated Pareto distribution models as well as the Weibull distribution model showed up their potential ability to estimate strong tremor risk. A number of strong events, estimated by use of these models in selected mining periods were in good, statistically significant agreement with the real number of such events (Lasocki 1991). These results justify application of the models (5), (6), (7) to estimate time-varying probability of occurrence of strong event given by formula (4).

3 ALGORITHM OF ESTIMATION AND ITS ANTICIPATED EFFECTIVENESS

An estimation of the distribution parameters is done with a help of the segmental stationarity property. If the process is stationary then its outcome is a simple statistical sample. Since the segmental stationarity approximation is to be valid for any time period of length ΔT, regardless the position of the period on the absolute time axis, then for n events

recorded in the period, say, $(t-\Delta T, t)$ the set of their energies $\{E_1, .., E_n\}$ constructs the simple sample. This sample can be used to estimate, constant for this period, model parameters. Due to regularity purposes the algorithms used at present make use of the maximum likelihood method. The maximum likelihood estimators for parameters of models (4), (5), (6), (7) and their asymptotic variances have been derived elsewhere (Lasocki 1990b). The resultant estimates are valid for the whole period $(t-\Delta T, t)$ thus, in particular, these values may be assigned to the upper limit t. Hence if Δt small compared to ΔT, the final numeric value of the risk (4), when substituting parameters with estimates, represents the probability of occurrence of the event having energy greater than or equal to E_p estimated for time moment t and predicted for time interval $(t, t+\Delta t)$. Such values are evaluated in real time, every constant short time interval, usually one day.

The length of interval of stationarity ΔT is an external parameter in the model and must be set a'priori according to the event rate and the local variability of conditions controlling tremor generation at the location under study. Two ways of ΔT selection are in use. ΔT may be chosen constant in absolute time units. This does not exclude other estimations with different ΔT values but one sequence of the probability estimates for all time moments is obtained with the use of the same value ΔT. In the other way, more appropriate for excavations where mining operations cause movement of an active front of any kind, ΔT value is continuously updated so that the front advance in ΔT remains constant. This technique is supposed to stabilize variations of tremor generation process corresponding to irregularities of the advance rate. Usual ΔT value amounts 15 to 40 days.

Regardless the way of ΔT selection the method reflects a belief that ΔT is long enough, hence the process is slow enough so that every statistical sample would be made of events recorded in every interval of length ΔT is numerous enough for a reliable estimation of the model parameters. In Polish coal mining this condition is usually fulfilled. It should be stressed, however, that even then the segmental stationarity only approximates the process of continuous dynamics. Within the interval $(t-\Delta T, t)$ more recent observations contain more up-to-date information about the state of tremor generation process at moment t than the data from moments closer to the left-hand-side limit. The estimation does not distinguish among the data hence the estimated probability is delayed with respect to the true probability due to incorporating older data into the estimation procedure, i.e. the data corresponding to the past states of the process. The worse segmental stationary approximation it is, due to longer ΔT selection necessity or due to faster changes of the process, the bigger delay will appear. To ease the problem the estimation procedure has been supplemented with a technique of weighing observations which allows to increase the importance of more recent data in the estimation procedure. Let w_i denote a weight of observation E_i in the sample $\{E_1, \ldots, E_n\}$. The logarithmic likelihood function is then given by:

$$l = \sum_{i=1}^{n} w_i \cdot \ln[f_e(E_i)] \qquad (8)$$

where $f_e(E)$ is the probability density function of energy. At present the exponential shape of the weights is in use. The weight of an event recorded at moment t_i from the point of view of moment t ($t \geq t_i$; $t-t_i \leq \Delta T$) is given by:

$$w_i = \delta \cdot \exp\{-\mu(t-t_i)\} \qquad (9)$$

213

where μ determines a rapidity of 'ageing' of information and δ is a normalization constant. If we assume that the number of events recorded during $(t-\Delta T,t)$, say n, is invariant of the weighing operation then we get:

$$\delta = n \cdot \{ \sum_{i=1}^{n} \exp[-\mu(t-t_i)] \}^{-1} \qquad (10)$$

The weighing technique has been thoroughly discussed elsewhere (Lasocki 1992b).

Not very high event rate for tremors recorded in the region of single stope and the continuity of process dynamics oblige the model parameters to be estimated from poorly populated samples. The feature significantly limits the use of multi parameter models. Some models, which nicely fit real data, cannot be used because estimating procedures diverge while their parameters are being assessed from small samples. In models likely in use the estimator variances can be significant. When an accuracy of energy evaluation is low the error of the probability estimate will still be greater. The weighing of observation, necessary to diminish the mentioned delay of the estimate, due to assigning increased influence to selected data, may on the side magnify the error of a single estimate. Therefore an interpretation of the single value of $R(E_p,\Delta t;t)$ is not very reasonable.

Much more informative is a smoothed series of successive $R(E_p,\Delta t;t)$ values till the moment of inference and this series is interpreted. We may expect the occurrence of strong events (those being the target of interpretation) in periods of a build up of the estimated risk and in periods in which the $R(E_p,\Delta t;t)$ estimates are at about steady high level. Consistently, the strong event should not occur in periods of decreasing trend of estimated $R(E_p,\Delta t;t)$ and when the trend keeps steady, locally low value. Such interpretation is conceptually similar to discerning times of increased probability - the idea introduced by the Moscow research group of earthquake prediction theory (Keilis-Borok & Soloviev 1991).

There is another, purely technical argument for the above given trend interpretation of results of the strong tremor prediction method. The formula (4) determines probability of events from the tail of the distribution of energy, hence rare. Thus a value of this probability in Δt time period (usually one day) is small, usually of several percent. Such probability alone has insignificant value, in terms of expectations, for a decision making process.

The presented outcome of the method is not as could be expected form prediction i.e. an accurate and reliable time of occurrence of predicted event. In author's opinion, however, this sort of prediction is the only result which can be achieved from the seismic event sequence analysis at present state of knowledge of the stochastic process of tremor generation.

The way in which the results are interpreted determines an extent of practical usefulness of the method. It may be used in day-by-day general assessment of strong tremor hazard for the stope and in monitoring of the hazard variations. Therefore the method can be an auxiliary tool while making on line decisions about mining, while considering the necessity of detailed investigation of the state of stress in the mining works region and while deciding whether to use active preventive measures in the stope. Finally the method can be used to evaluate efficiency of the applied preventive measures.

An efficiency of the method depends mainly upon fulfillment of its assumptions. One of the most frequently questionable assumption is that of mutual independence of events, due to expected possibility of aftershock series after a strong tremor. Our studies of tremor sequences belonging to the class of seismicity directly connected with mining operations, recorded after an occurrence of large event do not give indications to contradict this assumption. The similar experience has been gathered in South African deep level mines (Lenhardt 1992). It is clear, however, that in different conditions the accepted model of independent events may not be fully valid.

It is well known that the statistical distribution of energy of tremors provoked by distressing blasting significantly differs from the distribution of energy of spontaneous events. The phenomenon, in turn, increases uncertainty of the probability estimation. Certainly there are some other factors that cause deviations from the model. In the course of their identification the rising a'priori information can be used to complement or rebuild the model and will effect in an increase of reliability of resultant estimates.

4 PRACTICAL EXAMPLE

In order to show results which are usually delivered when the described method is applied to real time processing of seismicity data from mine a fragment of the estimated probability record is shown in figure 4. The data analyzed were recorded in the region of longwall 533 of Katowice coal mine in Upper Silesia, Poland. The exploitation with hydraulic backfill was carried on at the average depth of 600 meters in the footwall slab of the 10 meters thick coal seam. Thick rigid sandstone layers present in this area in the overburden were supposed to be the cause of numerous seismic events accompanying mining works. The event rate was variable, in average 2.6 events daily. Seismicity and particularly strong tremors had significant influence on mining. The face advance was relatively slow and irregular, not greater than 2.5 me-

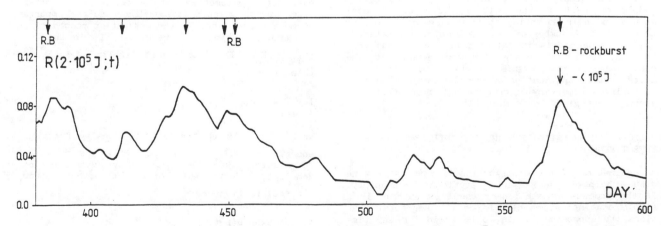

Figure 4. The probability of occurrence of seismic event whose energy exceeds 10^5J, estimated during mining in the longwall 532 in Katowice mine. Arrows mark days of actual occurrence of strong events.

ters a day. Many times works were stopped and active preventive methods, mainly distressing blasting, were applied. There were two rockbursts during the exploitation in the face; one was due to a tremor of $9 \cdot 10^5 J$, the other one was caused by a weaker event of $9 \cdot 10^4 J$.

The analysis was made by use of the Pareto distribution model. Qualitative studies of seismicity from adjacent regions suggested the half of the order $10^3 J$ as the lower limit of applicability of the Pareto distribution (E_o). Because of very irregular front advance the length of interval of stationarity ΔT was kept variable so as to ensure constant 40 meter movement of the face. The weighing constant μ was such that the significance of information decreased by two every ten days. Strong events - the target of analysis were assumed to exceed $10^5 J$. Their actual occurrence time is marked with arrows in the upper part of the figure. The probability was estimated for each day of exploitation.

The figure shows evident relation between the estimated probability build-ups and the strong event occurrences. Also the peaceful period of low probability after the second rockburst followed by the build-up prior to the next strong event is worth noting.

5 CONCLUSIONS

Statistical analysis of induced seismicity data directly connected with mining works showed that the process of tremor generation in the vicinity of a single stope is time-dependent and that information about this feature is stored in its outcomes. Thus attempts can be undertaken to construct time-dependent models of this process, models whose parameters are estimated from recorded event sequences. Then they can be used for state analysis and prediction.

The conclusions gave grounds for building the above presented method of estimating time varying probability of occurrence of strong tremor in mining works region. While it was being constructed attempts were made to use up, to the maximum possible extent, statistical properties of data sequence recorded in the region of single stope. The selected base of assumptions was thoroughly checked in numerous empirical data sets. The accepted assumptions occurred to be at least not contradictory in view of the testing hypothesis theory.

Practical applications of the worked-out algorithm show that even with such limited a'priori information that can be utilized at present, a rate of successful predictions obtained by means of the presented method significantly exceeds 50 percent. As the result the method is useful in general evaluation of strong tremor hazard and may be helpful to the real time decision making process.

An overall effectiveness of the method is determined by precision with which the constructed model of process describes the physical process. Thus the method cannot predict the time of event occurrence with a narrow confidence interval. The present knowledge of mechanisms controlling the tremor generation process certainly does not allow to construct methods which could fulfill these expectations.

It may be expected however, that the efficiency of algorithms similar to the presented in the paper will quickly increase in the course of recognizing and including new a'priori information about state and dynamics of the process. It also seems that efforts should be made to link predictions coming from different observation bases or from different algorithms. This kind of undertakings were initiated by Glowacka (Glowacka 1991) who tried to combine hazard estimates obtained by means of a relationship between mining seismicity and excavation area (Kijko 1985) with other methods of real time risk evaluation, also with the method presented here. The obtained results show that this direction is very promising.

REFERENCES

Bath, M. 1984. Rockburst seismology. *Proc. 1st Int. Congr. on Rockburst and Seismicity in Mines*: 89-94. Johannesburg: SAIMM.

Cosentino, P., V. Ficarra & D. Luzio 1977. Truncated exponential frequency-magnitude relationship in earthquake statistics. *Bull.Seism.Soc.Am.* 67: 1615-1623.

Gay, N.C. & E.H. Wainwright (eds.) 1984. Strategy in mine design - discussion. *Proc. 1st Int.Congr. on Rockburst and Seismicity in Mines*: 265-266. Johannesburg: SAIMM.

Gibowicz, S.J. 1979. Space and time variations of the frequency-magnitude relation for mining tremors in the Szombierki coal mine in Upper Silesia, Poland. *Acta Geophysica Polonica XXVI*: 39-49.

Gibowicz, S.J. 1990. The mechanism of seismic events induced by mining. *Proc. 2nd Int.Symp. on Rockburst and Seismicity in Mines*: 3-27. Rotterdam: Balkema.

Glowacka, E. 1991. *Doctor's dissertation*. Unpublished thesis. Warszawa: Inst. of Geophys. Pol. Acad. Sci. (in Polish)

Johnston, J.C. & H.H. Einstein 1990. A survey of mining associated rockbursts. *Proc. 2nd Int.Symp. on Rockburst and Seismicity in Mines*: 121-128. Rotterdam: Balkema.

Keilis-Borok, V. & A. Soloviev 1991. Introductory lecture. *Workshop on Non-linear Dynamics and Earthquake Prediction*. Trieste: International Centre for Theoretical Physics.

Kijko, A. 1985. Theoretical model for a relationship between mining seismicity and excavation area. *Acta Geophysica Polonica XXXIII*: 231-240.

Klonowski, Z. & Z. Gerlach 1974. The release of seismic energy as a function of dimensions of a ledge remnant. *Publs.Inst.Geophys. Pol.Acad.Sc.* 67: 107-124. (in Polish)

Lasocki, S. 1989. Some estimates of rockburst danger in underground coal mines from energy distribution of MA events. *Proc. 4th Conf. on AE/MA in Geologic Structures and Materials*: 617-633. Clausthal: Trans Tech Publs.

Lasocki, S. 1990a. Non-Poissonian structure of mining induced seismicity. *Proc. XX Czechoslovak-Polish Conf. on Mining Geophysics*. Acta Montana (in print).

Lasocki, S. 1990b. Prediction of strong mining tremors. *Zesz.Nauk. AGH, s. Appl. Geophys.* 7: 1-110. (in Polish)

Lasocki, S. 1991. Predictive ability of models for distribution of tremor energy. *Publs.Inst.Geophys. Pol.Acad.Sc.* M-15(235): 143-152. (in Polish)

Lasocki, S. 1992a. Weibull distribution for time intervals between mining tremors. *Publs.Inst.Geophys. Pol.Acad.Sc.* M-16(245): 241-260.

Lasocki, S. 1992b. Truncated Pareto distribution used in the statistical analysis of rockburst hazard. *Proc. 1st Int.Symp. on Mining Induced Seismicity*. Acta Montana (in print)

Lenhardt, W.A. 1992. Personal communications.

Lomnitz, C. 1974. *Global tectonics and earthquake risk*. Amsterdam: Elsevier.

Marcak, H. 1978. Statistical interpretation of shock series in mining. *Rock Mechanics* 10: 181-186.

Report on CPBP 03.02. 1989. Archivising seismicity data from selected coal mines. Krakow: Inst. of Geophysics. Univ. of Min. & Metall. (in Polish)

Sato, K. & Y. Fujii 1988. Induced seismicity associated with longwall coal mining. *Int.J.Rock Mech.Min. Sci. & Geomech.Abstr.* 25: 253-262.

Slavik, J., P. Kalenda & K. Holub 1992. Statistical analysis of seismic events induced by the underground mining. *Proc. 1st Int. Symp. on Mining Induced Seismicity*. Acta Montana (in print)

Spottiswoode, S.M. & A. McGarr 1975. Source parameters of tremors in a deep-level gold mine. *Bull. Seism.Soc.Am.* 65: 93-112.

Subbaramu, K.R., B.S.S. Rao, R. Krishnamurthy & C. Srinivasan 1989. Seismic investigations of rockbursts in the Kolar Gold Fields. *Proc. 4th Conf. on AE/MA in Geologic Structures and Materials*: 265-274. Clausthal: Trans Tech Publs.

Trombik, M. & W. Zuberek 1980. Location of microseismic sources at the Szombierki coal mine. *Proc. 2nd Conf. on AE/MA in Geologic Structures and Materials*: 179-190. Clausthal: Trans Tech Publs.

Rockbursts and Seismicity in Mines, Young (ed.) © 1993 Balkema, Rotterdam, ISBN 90 5410 320 5

Source mechanisms at the Lucky Friday Mine: Initial results from the North Idaho Seismic Network

P.B. Lourence, S.J. Jung & K.F. Sprenke
College of Mines and Earth Resources, University of Idaho, Moscow, Idaho, USA

ABSTRACT: Since March 1992, the University of Idaho has been collecting seismic data from the Coeur d' Alene (CDA) mining district using the North Idaho Seismic Network (NISN). Unlike in-mine seismic networks, NISN consists of surface geophones with sufficient focal sphere coverage to allow accurate source solutions for events anywhere in the district. Based on the data provided by 262 seismic events recorded in the vicinity of the Lucky Friday Mine between May and September 1992, about half of the events were consistent with double-couple shear and the other half with implosional sources. The preferred failure mode was found to be right-lateral strike-slip movement on the steeply dipping northwest-trending faults, although several other types of failure were identified. The preferred orientation of the σ_1 axis in the mine area is horizontal toward N20W, rotated 20 degrees clockwise from the generally accepted virgin σ_1 axis of horizontal toward N40W. The failure modes of the four largest events studied (M>2.5), which alone accounted for 95% of the seismic energy released during the study period, were consistent with the modes of failure of the entire data set.

1 INTRODUCTION

The study of source mechanisms of seismic events occurring in the Coeur d'Alene (CDA) mining district of northern Idaho can provide useful information for rockburst mitigation (Stickney and Sprenke, 1993; Sprenke et.al., 1991). The research described in this paper is based on seismic events at the Lucky Friday Mine from May 1, 1992 to September 30, 1992. The following questions will be investigated:
a. How many different types of source mechanism can be identified?
b. Does one type of source mechanism appear to predominate and does this vary with the magnitude?
c. What is the general relationship between hypocenter location, source mechanism, geological structures, and stoping geometry.
d. How well do the orientations of the P- and T-axes calculated from the fault-plane solutions compare with the theoretical principal stress orientations calculated by a simple geomechanical model?

2 MINING LAYOUT, GEOLOGY, AND ROCK ENGINEERING DATA

The Lucky Friday Mine produces approximately 12,400 metric tons of ore per month. The average volume of ore removed from stopes is about 3,900 m³ per month. The main ore vein dips steeply (70 to 90 degrees), and consists of massive galena, sphalerite and tetrahedrite. The average stope is mined at a width of two to four metres. Except for some scattered remnants, the main vein has been continuously stoped at depths ranging from 550 metres to 1615 metres below surface (the 5300 level), and for about 400 metres on strike. The current mining method used is mechanized, underhand cut and fill, with extensive ramp development in the footwall. Stopes are advanced horizontally along strike using 3 metre high cuts. After each cut has been backfilled, a new stoping cut resumes beneath the fill.

The country rock consists of quartzite beds folded on a steeply plunging anticline, bounded by nearly vertical faults. The bedding typically dips 60 degrees but with much local variation.

Table 1 shows the estimated virgin principal stress vectors at a depth of 1550 metres. These data are based on overcore stress measurements in the CDA district, mostly performed by the University of Idaho and the U.S. Bureau of Mines in the 1970s.

Table 1. Assumed virgin principal stress vectors

Principal stress	magnitude (MPa)	plunge	bearing
maximum, σ_1	53.1	horiz.	N40W
intermediate, σ_2	40.6	horiz.	N50E
minimum, σ_3	39.7	vert.	-----

3 SEISMIC EVENT MONITORING AT THE LUCKY FRIDAY MINE

The North Idaho Seismic Network (NISN), which began acquiring events in the district in March 1992, is operated by the University of Idaho College of Mines and Earth Resources. The purpose of NISN is to provide seismograph coverage in the CDA mining district and surrounding region. The main NISN research objective is to study the source mechanisms of mining-related seismic events by analyzing their far-field characteristics. Because NISN instruments are not limited to underground stope locations, complete focal sphere coverage is generally possible and highly constrained fault-plane solutions are attainable.

Figure 1 shows the locations of NISN stations in the CDA district. The 21 surface stations and three underground stations are linked by radio telemetry to acquisition computers at the Lucky Friday and Sunshine mines. All NISN surface instruments are vertical component velocity gauges ranging in natural frequencies from 1 Hz to 4½ Hz.

The vast majority of the events detected by the network during this study occurred near the mine workings at the Lucky Friday Mine (Figure 2-3), and very accurate source locations were obtained from the in-mine rockburst monitoring systems. Fault-plane solutions were calculated by the programs FPFIT (Lee and Valdes, 1985). Statistical analyses of the fault-plane solution data to determine overall preferred directions of planes and axes were performed using the program MicroNET (Guth, 1987).

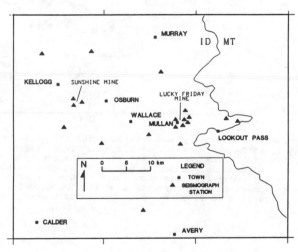

Figure 1. The North Idaho Seismic Network.

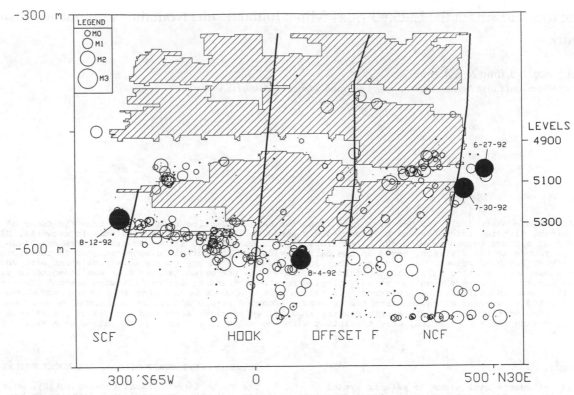

Figure 2. Seismicity (May 1,1992-September 30, 1992) in the Lucky Friday Mine plotted on a longitudinal view of the near vertical vein. The hatched areas are stoped, the unhatched areas are remnants.

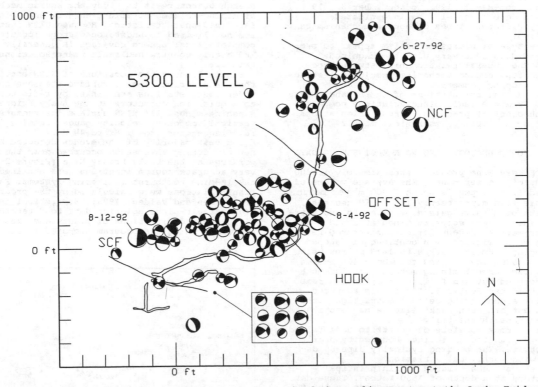

Figure 3. The fault-plane solutions for 104 well-constrained shear-slip events at the Lucky Friday Mine (May 1, 1992-September 30, 1992). The black quadrants are compressional; the white quadrants are dilatational. The radii of the stereonets are proportional to magnitude, which ranges from -1 to 3. The T-axis is located in the center of the white quadrants; the P-axis is located in the center of the black quadrants. Normal faults show white in the center of the nets; reverse faults show black. For strike-slip events, the I-axis (intersection of the two solution planes) is in the center of the net, and the strike-slip direction is from the white quadrants toward the black quadrants.

A source mechanism is a description of how the rock
mass fails at the hypocenter of a seismic event. In
the case of shear slip, which is the common double-
couple source mechanism for natural earthquakes, a
fault-plane solution can be calculated from seismic
network data that yields the orientation of two
mutually perpendicular planes, one of which represents the true failure plane, as well as information
on the type of slip that has occurred. Fault-plane
solutions also yield the P-axis, I-axis, and T-axis,
directions analogous to σ_1, σ_2, and σ_3. The actual
failure plane and exact stress directions can only
be resolved with independent seismological or geological information.

The quality of fault-plane solutions can be ascer-
tained by considering the range of all possible
planes that are consistent with the data. The computer program FPFIT (Lee and Valdes, 1985) is used
to search out all such possible solutions for all
events. For NISN data, we consider a well-
constrained event to be one for which the possible P
and T axes fall in clusters +/- 12 degrees from
their mean position. Thus the strike and dip of
possible solution planes are constrained to this
error range as well.

5 RESULTS

During the study period of May 1, 1992 through September 30, 1992, Lucky Friday personnel compiled
data on 495 events that produced a deflection of at
least 1 mm on their mineyard seismograph. The hypocenters, shown projected on a longitudinal section
view of the active stoping areas in Figure 2, are
generally within 60 metres of an active stope or
tunnel. These seismic events, which ranged in magnitude from -1.0 to 3.0, formed the data base for
our study.

In this 5 month period, the mine averaged 8 seismic events per month with a Richter magnitude of 1.0
or greater. This is nearly one event (M≥1) for every
500 m^3 mined. Simply counting events, however, does
not provide much insight into the seismic energy
released in the mine. The four large events (M>2.5)
that occurred during the study period accounted for
95% of the seismic energy released during the study
period. Nonetheless, the smaller events provide
important data on the stress regime that results in
major seismic events.

5.1 Large Implosional Event

A large implosional event, magnitude 2.5, occurred
on July 30 1992 at the Lucky Friday Mine. A stereonet showing all-dilatational first motions distributed throughout the focal sphere is shown in
Figure 4. The hypocenter coordinates indicate that
the event occurred near a 12 metre remnant of the
vein between the 4900 and 5100 levels (Figure 2).
Although this event was near the North Control Fault
(NCF), there is insufficient seismological evidence
to verify that the NCF was the failure plane. However, because of the all-dilatational first motions,
the source must have involved a stope or tunnel
boundary. This event was the largest implosional
event that occurred during our study period. Using
Ryder's definition (Ryder, 1988), this event could
possibly be called a "crush" event. The mere presence of the NCF near the hypocenter, however, creates some ambiguity. Sileny (1989) showed that an
all-dilatational response can occur when shear dis-
locations make up less than 80% of the total displacement.

5.2 Major Shear Events

The three largest magnitude events that occurred
during our study period at the mine were shear-slip
events. Figure 5 shows the fault-plane solution for
a shear event, magnitude 2.8, which occurred on
June 27, 1992. Hypocenter coordinates indicate that
this event occurred, like the implosional event
described above, near the NCF in the same 5100 level
remnant (Figures 2 -3). In this case, one solution
indicates an 4:1 ratio of left-lateral strike-slip
movement to normal movement on a plane N35E, 50SE.
The auxiliary solution indicates about equal right-
lateral strike slip and normal movement on a plane
N42W, 75SW. The second solution agrees with the
actual strike and dip of the NCF (N60W, 85SW) within
the accuracy of the solution (+/- 12 degrees).

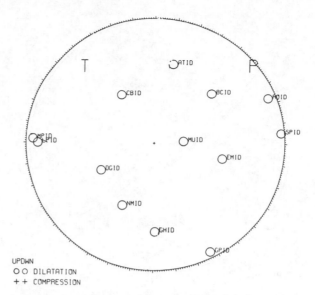

Figure 4. Stereonet showing an implosional mechanism
for the July 30, 1992 M-2.5 event near the North
Control Fault.

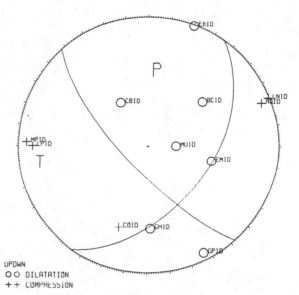

Figure 5. Stereonet solution for June 27 1992 M-2.8
event on the North Control Fault.

Figure 6 shows the fault-plane solution for a
strike-slip shear event, magnitude 3.0, which occurred on August 4, 1992. The hypocenter coordinates
indicate that this event occurred near the Offset
Fault, in the down-dip abutment of the 5400-107
longwall (Figures 2 and 3). In this case, one solution indicates a 7:3 ratio of right-lateral strike-
slip to normal movement on a plane N51W, 81SW. This
solution would agree well with the orientation of
the Offset Fault (N60W, 85SW). The second solution
indicates a 9:1 ratio of left-lateral movement to
normal slip on a plane N35E, 65SE. This solution
could possibly represent strike-slip movement along
the bedding planes.

In contrast to the above events which had geologically plausible solution planes, Figure 7 shows a
fault-plane solution for an event that does not show
correlation with any known geological structure.
This magnitude 3.0 event occurred on August 12, 1992
near the 5210-95 Main Vein stope in the vicinity of
the South Control Fault (SCF) (Figures 2 and 3). The
first solution indicates reverse slip on a near
vertical plane (N5E, 87NW). The auxiliary solution
indicates right-lateral movement along a near horizontal plane (N85E, 15S). Neither fault-plane solution is consistent with the orientation of the SCF,
bedding planes, or any other known structure. One
hypothesis is that the failure plane involves the

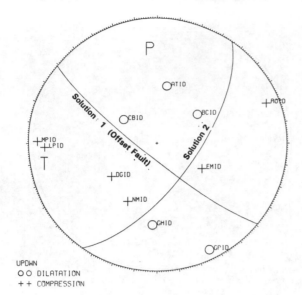

Figure 6. Fault-plane solution for the August 12, 1992 M-3.0 event near the Offset Fault.

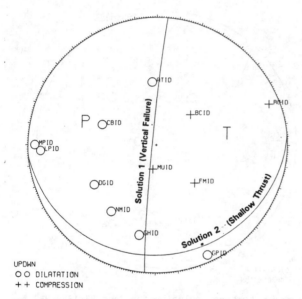

Figure 7. Fault-plane solution for the August 12, 1992 M3 event near the South Control Fault.

shear rupture of intact rock resulting from excess shear stress around the abutments adjacent to the stope.

5.3 Distribution of source mechanisms

Of the 495 seismic events catalogued during the study period, 262 had sufficient energy and signal-to-noise ratio to produce impulsive first motions on enough NISN stations to attempt fault-plane solutions. About half (48%) of the solutions were not consistent with simple double-couple sources, thus pointing out a fundamental difference between mining-induced seismicity and natural earthquakes (Figure 8). No apparent correlation exists between the type of source mechanism and magnitude. Although implosional events occur across a wide magnitude range, the largest detected during our study was magnitude 2.5 while the three largest events were shear events of magnitudes 2.8 to 3.0.

5.4 Overall correlation with geological structures

The preferred orientations of failure planes determined from the seismic data are compared with the average orientations of geological structures in the

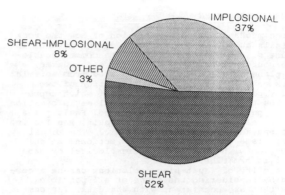

Figure 8. Overall distribution of source mechanisms at the Lucky Friday Mine, June 1, 1992-September 30, 1992.

mine in Figure 9. Only the steeply dipping northwest trending faults correlate with seismic failure planes. The lack of seismic failures along the plane of the vein is expected because the ore is very soft compared to the country rock and cannot store large amounts of strain energy. The lack of seismic failures along the average bedding plane orientations is very surprising. Because bedding is complex in the mine, bedding plane failures may certainly occur in local circumstances, but there is nonetheless, no overall correlation between mean bedding orientations and preferred seismic failure planes.

The orientations of possible failures shown on Figure 9 that do not correlate with the faults are most interesting because some certainly represent geological planes of weakness that are perhaps not yet recognized in the mine. These planes have the following orientations: N5E, 54W; N60W, 55NE; N35E, 63SW; and N50E, 10SE.

The possible solution planes for the three big shear events that occurred in the mine during the study period are also plotted in Figure 9. Each event has a solution plane that falls on the preferred orientations identified from all the events. This indicates a connection between the mechanisms of big events and those of small events in the mine. Two of the large events are consistent with steeply dipping northwest trending faults; the August 12, 1992 event shows a near horizontal solution plane consistent with many other smaller events in the mine, suggesting that this failure plane orientation may be real.

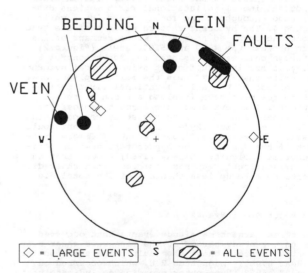

Figure 9. Stereonet projection comparing the preferred orientations of seismic failure planes (hatched areas) with the average orientation planes of known geological structures in the Lucky Friday Mine. All planes are plotted by their poles. Two average orientations for the vein and bedding are shown, one for each limb of the Hook anticline.

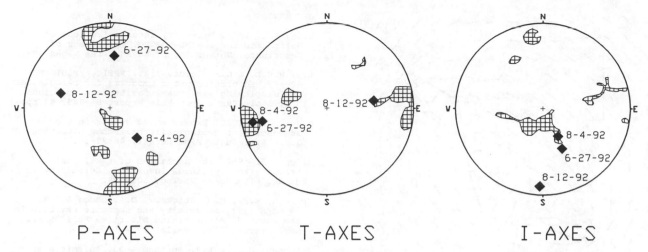

P-AXES   T-AXES   I-AXES

Figure 10. Stereonet projection of the preferred orientations (hatched areas) of the P, T, and I-axes in the Lucky Friday Mine based on fault-plane solutions of 106 well-constrained shear events. The diamonds show the respective axes for the three major seismic events that occurred in the mine during the study period.

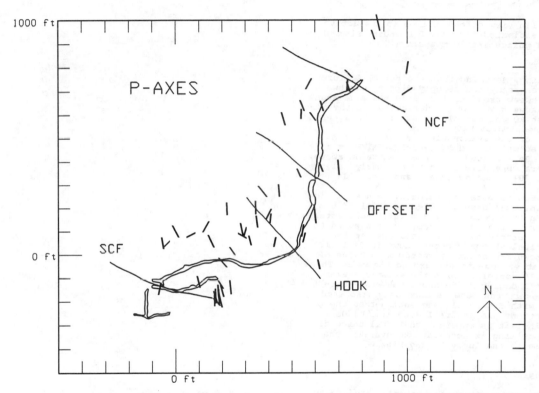

Figure 11. P-axes from seismic events projected on a general plan view of the Lucky Friday Mine.

5.5 Predominant stress orientations and faulting styles.

The preferred orientations of the P-axes, T-axes, and I-axes for seismic failures in the mine are shown in Figure 10 as well as the orientation for the three major shear events that occurred during the study period. The preferred P-axis bearing is strongly north-south. The preferred plunge angle of the P-axis is generally within 30 degrees of horizontal, however a significant number are concentrated in a nearly vertical direction and others at 45 degrees. The preferred T-axis bearing is strongly east-west. The preferred plunge angle is generally within 20 degrees of horizontal, or inclined about 45 degrees, and rarely vertical. The preferred I-axis plunge is nearly vertical although westerly plunges as low as 45 degrees are common; east-north-easterly plunges from 0 to 45 degrees also occur; and low angle northerly plunges are also common.

Based on these preferred orientations, the most common faulting styles in the Lucky Friday Mine are as follows (in order of importance):

a. right-lateral strike-slip movement on the steeply dipping northwest to west-northwest trending faults (2 major events, 30% of shear events).
b. failures along near horizontal planes (or perhaps vertical planes) in intact rock or along unknown geologic zones of weakness (1 major event, 14% of shear events).
c. normal faulting, probably gravity failures, with near-vertical P-axes, in intact rock or perhaps bedding planes along failure planes striking generally north-south (no major events, 24% of shear events).
d. oblique reverse faults with moderately inclined T-axes and horizontal northerly P-axes and with failures in intact rock or bedding planes striking generally NW and NE with dips from 30 to 60 degrees (no major events, 24% of shear events).

A style of faulting that is noticeably absent in the mine area is pure reverse faulting with a vertical T-axis, suggesting that σ_2, not σ_3, should be vertical in geomechanical models of the mine.

The N5W and N85E preferred orientations of the P-axes and T-axes, respectively, as shown in Figure 10

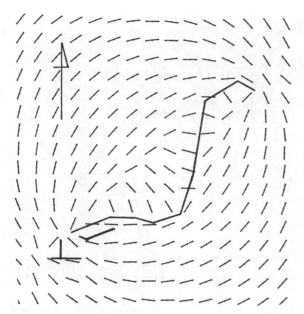

Figure 12. Plan view of the σ_1 axes in the active mining area estimated from a 2-D boundary element model. Field points are 30 m apart.

are interesting. Assuming that the actual failure planes are generally the steeply dipping northwest to west-northwest trending faults in the mine area, then this places σ_1 at N20W, a deflection in the mine area at least 20 degrees clockwise from the accepted virgin σ_1 direction of N40W (Table 1).

About half of the events with well-constrained fault-plane solutions had near-horizontal P-axes (Figure 11). As expected from the discussion above, the P-axes are generally oriented northerly, although about 20% trend northwest and about 10% northeast.

In an attempt to model these data, we used a simple two-dimensional boundary element model of the mine area. In this model, the 3 m thick vein extends perpendicular to the model plane and is assumed to be completely stoped out. The model assumes the virgin principal stress vectors shown in Table 1. The lines shown are constant vector magnitudes of the σ_1 axes at various points around the vein. These vectors should correlate with the seismic P-axes within the ±15 degree differences between P-axes and σ_1 axes. However, the model showed only scattered correlation with Figure 11, suggesting that the model assumptions are generally too simple. Generally, speaking, it is apparent that full three-dimensional modelling is necessary to evaluate the stress regime in the Lucky Friday Mine.

6 CONCLUSIONS

The initial results of source mechanism studies by the the North Idaho Seismic Network at the Lucky Friday Mine indicate that about half of the events are consistent with a double-couple shear-slip mechanism and the other half are consistent with implosional mechanisms. The most common seismic shear failures probably involve right-lateral slip on steeply dipping northwest-trending faults, although several other styles of faulting are also common. The local stress axes orientations suggested by the seismic mechanisms should provide useful constraints on future geomechanical models of the mine.

ACKNOWLEDGEMENTS: Funds for this study were provided by the VPI Generic Mineral Technology Center for Mine Systems Design and Ground Control. The authors wish to thank the management and staff of the Lucky Friday Mine for their complete cooperation in all phases of this project. Valuable technical assistance for this study has been provided by T.J. Williams and J.K. Whyatt, SRC-USBM; M.C. Stickney, Montana Bureau of Mines and Geology; W.R. Hammond, USGS; D.A. Dodge, Stanford University; and G. Stevens, NISN.

REFERENCES

Guth, P.L, 1987. MicroNET, Interactive Schmidt and Wulff nets for the MS-DOS PC. Department of Geoscience, University of Nevada, Las Vegas, NV.

Lee, W.H.K. & C.M. Valdes 1985. FPFIT, FPLOT AND FPPAGE: Fortran computer programs for calculating and displaying earthquake fault-plane solutions. U.S. Geol. Surv. Open-File Report 85-749, 43 pp.

Ryder, J.A. 1988. Excess shear stress in the assessment of geologically hazardous situations. J. S. Afr. Inst. Min. Metall. 88:27-39.

Sileny, J. 1989. The mechanism of small mining tremors from amplitude inversion. Seismicity in Mines. Birkhäuser Verlag, 309-324.

Sprenke, K.F., M.C. Stickney, D.A. Dodge & W.R. Hammond 1991. Seismicity and tectonic stresses in the Coeur d' Alene Mining District. Bull. Seis. Soc. America 81:1145-1156.

Stickney, D.A. & K.F. Sprenke 1993. Seismic events with implosional focal mechanisms in the Coeur d' Alene mining district, northern Idaho. Journal of Geophysical Research (in press).

Rockbursts and Seismicity in Mines, Young (ed.) © 1993 Balkema, Rotterdam, ISBN 90 5410 320 5

The use of pattern recognition method for prediction of the rockbursts

H. Marcak
Institute of Geophysics, Technical University of Mining, Cracow, Poland

ABSTRACT: Fracturing of the rock-mass due to mining exploitation may reveal brittle character described by the plastic-dilatancy model. Such a model, applied to a layered medium, leads to the formation of inclusion zones that dissect the stratigraphically-layered rock body. These zones favour the relaxation of energy, being associated with increased seismic emission from high-energy events. Such a nonuniform mode of seismic emission may be used in the prediction for mining shocks. Application of computer programs based on the method of pattern recognition makes it possible to recognize seismic emission derived from inclusion zones, as well as to predict strong earthquakes.

1 PLASTIC-DILATANCY MODEL OF ROCK FRACTURING

Analysis of deformation of the rock-mass resulting from building up of tectonic and exploitation stresses requires to consider both plastic and brittle types of inelastic strain. The development of microcracks, called dilatation, causes changes in elastic properties of rock medium. Moreover, the cracks are surrounded by zones wherein elastic properties of the rock turn into plastic ones, without producing microcracks (Patton F.D. 1966). Then, movement along a fracture zone may have plastic character, if:

$$Y = \tau - \mu p \qquad 1$$

where: Y - cohesion coefficient,
  τ - shear sresses,
  μ - coefficient of internal friction,
  p - load upon friction surface, and
    friction force: $F = \eta p$; η - coefficient of friction.

To describe interrelationships between stress and strain a dilatancy-plastic model has been introduced (Rudnicki J.W., Rice J.R. 1975), in the form:

$$\gamma^P = 1/h\,(\dot\tau - \mu\dot p) \qquad \gamma^e = 1/G\tau$$
$$\varepsilon^P = \beta\gamma^P \qquad\qquad \varepsilon^e = 1/K\dot p \qquad 2$$

where: γ - rate of shape deformation,
  τ - shear stresses,
  ε - rate of volumetric expansion,
  p - compressional stresses,
  h - coefficient of plasticity,
  β - coefficient of dilatancy,
  G - stiffness coeficient; and
  K - elastic strain coefficient.

Indices "p" denote plastic-related, "e" - elastic-related interrelationships; whereas dot placed above a symbol marks the time derivative.

During sliding of rock-masses, the law of preservation of elastic strain energy, the energetic transformation and dissipation of thermal energy being negligible, leads to the following relationship:

$$-p\varepsilon^P + \tau\gamma^P = \eta p\gamma^P + Y\gamma^P \qquad 3$$

Introducing: $\varepsilon^P = \beta\gamma^P$ we receive:

$$\tau - (\beta + \eta)p = Y \qquad 4$$

meaning that coefficient of internal friction, in the presence of dilatancy, is given by: $\mu = \beta + \eta$
Hence, the condition for plastic flow is fulfilled at much smaller values of this coefficient. Let us take notice of the fact that during plastic flow:

$$\tau = Y + (\beta + \eta)p \qquad 5$$

so, given the presence of dilatancy and a constant coefficient of friction , then the coefficient of plasticity is:

$$h = \frac{\partial \tau}{\partial g^P} = \frac{\partial Y}{\partial g^P} + \frac{\partial \beta}{\partial \varepsilon^P}\,p\beta \qquad 6$$

where: g^P - total plastic strain.
The first coefficient in the formula (6) is a measure of cementation of the medium, whereas the second component - at small loading - is equal to zero.

When the threshold at which dilatancy appears is attained, a rapid change of "h" takes place. This value is of the negative character. The coefficient "h" is a measure of changes of the bonds among rock-forming structural elements.

The inelastic strains described by the above model results in the formation of fracture zones within the rock-mass. Numerous theoretical papers (cf. Rudnicki J.W.,Rice J.R. 1975) consider the origin of fracture zones to be a result of rock-mass instability caused by stress condotions. The condition for instability occuring in a horizontal bed, within which displacement is parallel to its boundaries, is satisfied when we find maximum values of the coefficient "h", satisfying the equation (Nikitin L.W.,Ryzak E.I. 1977, 1984):

$$(\varepsilon - \mu\gamma)(\varepsilon - \beta\gamma) + \frac{h}{K}\,\gamma^2 + \frac{h}{G}\,\varepsilon^2 \leq 0 \qquad 7$$

where: $\varepsilon = \varepsilon^P + \varepsilon^e$, $\gamma = \gamma^P + \gamma^a$

Let us denote maximum values of "h" satisfying this equation as h_o, and the corresponding values μ and β as μ_o and β_o. Maximum values are attained in those zones of the width "a" which intersect zones of fractures, being perpendicular to zone boundaries (cf. Fig. 1), at the angle "ψ".

The conditions of instability can help to determine parameters characterizing fracture zones. Their thickness "a" may be equal either to zero (fracture zone reduced to zero) or to the final thickness.

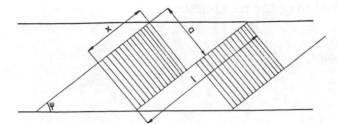

Figure 1. Formation of strain zones in top part of the coal seam.

The angle "Ψ" can be found from:

$$\Psi = \frac{1}{2}\alpha \frac{3K}{G+3K} \quad ; \quad \mu = \alpha \mu_0 \quad ; \quad \beta = \alpha \beta_0 \qquad 8$$

where:

$$\alpha = \frac{\mu - \beta}{2}\left[1 + \frac{h_0}{G}\right]$$

$$x/L = b = \frac{G + \frac{3}{2}K}{3}\left[1 - \sqrt{1 - \frac{3}{4}\frac{G+(3/16)K}{G+(3/4)K}}\right] \qquad 9$$

The above-presented plastic-dilatancy model describes changes of the rock-masses where tangential stresses are transferred through the system of rigid barriers. In fact, such a transfer is not a stationary and boundary conditions change with time. Moreover the rock-mass is a layred which means that mechanical properties of the rock are a function of the distance between the measuring site and the heading.

Differences between assumptions at which model have been constructed and the real distribution of rock-mass deformations may alter only selected properties of the model, leaving its principal characteristics unchanged. Examples of seismological measurements carried out in mines will help to examine the above-described machanism of failure.
A specially established seismologacal network (H. Marcak et al. 1989) has made it possible to localize numerous mining shocks of different energetic levels in the excavation zone of Katowice coal mine (Fig. 2).

These shocks cluster along three parallel lines. Table 1 gives a comparison between measured and calculated parameters.

LEGEND
- 10^4
- 10^5
- 10^6
- 10^7

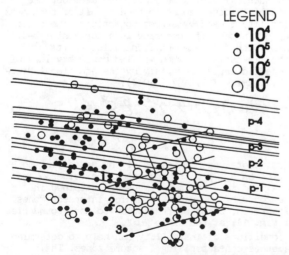

Figure 2. Distribution of shocks generated during exploitation of the sean 510 of Katowice coal mine
1 - sandstones, 2 - coal seam, 3 - inclusion line

Table 1.

	line(1)	line(2)	line(3)
length (m)	178.3	162.1	135.1
length of fracture zones: calculated	72.83	70.35	58.63
length of fracture zones: measured	66	60	50
coefficient b; calculated	0.432	0.432	0.432
coefficient b; measured	0.4	0.55	0.3

The calculated and measured values do not differ significanty. Hence, taking into account differences seen in both real conditions and those assumed in the model formulas, we can accept that this model describes generally the rock-mass behaviour in the case represented by the Katowice mine.

2 APPLICATION OF THE PATTERN RECOGNITION ALGORITHM TO PREDICTION OF MINING SHOCKS

It follows from the previous paragraph that there should exist a possibility of selecting neccesary information from a sequence of measurements preceding the occurrence of high-energy mining shocks. Unfortunately, measuring data are disturbed by the dual character of information sources. Apart from seismic records, resulting from development of the inclusion zone, seismic data concern as well the shocks associated with exploitation stresses. Therefore, in search for suitable prediction algotithms, a method should be find which would require a minimum of initial assumptions, neccesary for extraction of the usefull information.

Such algorithms are given by V.I.Keilis-Borok (1990). We assume that a squence of seismic events in a mine includes large random component; hence, determination of only one variable describing the sequence is insufficient. The algotithms based on the pattern recognition assume that seismic data from the mines are described by the following functions:
- the number of shocks with energy greater than or equal to the threshold E: $N(t|E,S)$, where S denotes the length of time window studied;
- the weighted sum of shocks occuring in the given time:

$$N\left(\Sigma 10^{\beta(lgE-\alpha)}\bigg|m, M', S, \alpha, \beta\right) \qquad 10$$

where: $m = lg\,E$; minimum energy of shocks considered; $M' = lg\,E$; maximum energy of shocks; α and β: parameters of shocks energy distribution;
- the ratio of the number of shocks of different energy:

$$lg\,E_{min} \leq lg\,E \leq lg\,E_{max}$$

$$G(t|m_1, m_2, S) = 1 - \frac{N(t|m_2, S)}{N(t|m_1, S)} \qquad 11$$

- the variance of the number of shocks at two succesive time moments: $V(t|m,s,u) = var\,N(t|m,s) = \Sigma[N(t_{i+1}|m,s) - N(t_i|m,s)]$
- the changes in seismic activity: $q(t|m,s) = \Sigma[a(m)s - N(t_i|m,s)]$, i.e the difference between the average number of shocks with energy exceeding m (the number of shocks falling into the time window of length s, calculated from the average, is $N = a(m)s$) and the real number of these shocks;
- the function of increase and decrease of sctivity: $Q(t,m,s) = \Sigma[N(t_{max}|m,s) - N(t_i|m,s)]$, which represents the time equivalent to maximum activity;
- the deviation of seismic activity from a long-term trend within the time interval t_0 to t:

$$L(t|m,s) = [N(t|m, t-t_0) - N(t-s|t-t_0, s)]\frac{t-t_0}{t-t_0-s} \qquad 12$$

-the increase in activity, i.e. the difference between the number of shocks within two succesive time intervals: (t-s,t) and (t-2s,t-2):

$$K(t|m,s) = N(t|m,s) - N(t-s|m,s) \qquad 13$$

To give a complete estimation of the time of increased probability of shocks occurence, the period of seismic registration should be subdivided into three sets:
- the set of observations preceding a strong shocks of duration t_1;
- the set of observations following a strong shocks of duration t_2, and
- the set of remaining observations.
The problem of recognition of seismic hazard consist in assigning to every seismic observation the probability of falling into individual sets. To define the sets and their limits, the so-called learning process is conducted through discrimination of individual sets on the basis of observation data, subdivided a priori into classes. This process needs to be somewhat simplified, since measurements ought to be discretized. The discretization concerns both the time axis (time moments within which seismic moments are selected) and the function quantity with respect to three levels of small, medium and large function values.

The learning algorithm used in the program of hazard recognition has the following structure:
- examination of distribution of observation vectors within each observation class;
- two classes (i.e. before and after the strong shock) are being assigned probability distributions p_D and p_N.
The second step of the program of pattern recognition ascribes a new observation into relavant class in the following way:

$$w^i \in D \quad if \quad p_D^i - p_N^i \geq \varepsilon$$
$$w^i \in N \quad if \quad p_D^i - p_N^i < -\varepsilon \qquad 14$$
$$w^i \in U \quad if \quad p_D^i - p_N^i < \varepsilon$$

where: i is a multidimensional coordinate in the probability space; and ε is the value defined by the user of the program.
In this manner, we can automatically assign observations to the "before strong shocks" class and give such an assignment the relevant alarm meaning.

Prediction possibilities of offered by the pattern recognition algorithm, which is currently being developed at the International Institute of Earthquake Prediction Theory and Mathematical Geophysics in Moscow, have been tested on the basis of data collected in the Halemba black coal mine in Poland. Seismological observations in this mine, that have been carried out since 1981, concern the exploitation of seams 506 and 507 in two parts: the western F and eastern H. These seams are, respectively, 2m and 3m thick; occur at a depth of 840m to 1010m, and dip 6° - 8° due south. These seams are separated by ca. 30m - thick sandstones. Exploitation started in both parts from the seam 506 with succesive walls. Nearly 7300 shocks have been recorded, six of them being mining tremors.
The procedures of hazard prediction for strong shocks, with the use of a seismological program based on the pattern recognition, have een preceded by introductory processing of all available seismic records of Halemba mine. The analysis consists in adaptation of the data to the real seismic records which are dealt with in seismic hazard prediction (the changes of time and space scale).

3 RESULTS OF CALCULATIONS

Prediction calculations have been performed for

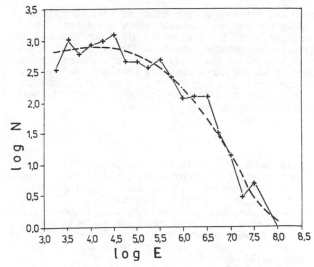

Figure 3. Distribution of shocks recorded in "Halemba" mine; N - number of shocks, E - released energy.

seismic observation data, collected in the part F of the Halemba mine. Figure 3 shows a histogram of the energy of all shocks from this area.

This histogram can be subdivided into three segments, the second of which is a rectilinear one and represents the energy interval from $10^{5.5}$ to 10^7 J. Such a pattern implies that prediction should focus on shocks with energy exceeding 10^7 J. The results obtained are shown in Table 2.

Table 2.
diagnosis of TIPs from 1.1.1904 to 1.1.1994
file with coding pf.rat; file with traits pf. fat
threshold magnitude for strong shocks=7.0
function SIGTH from file pf.pat is used to stop TIP
DELTA for voting=0
threshold for SIGTH to stop TIP=1424 (4.50)
1 region; strog shocks from file pf1.dat

Start of TIP	Strong shocks		End of false	Duration
	Date	M	alarm	month
31.3.1909	31.3.1910	7.0		12.0
31.3.1910	2.9.1910	7.5		5.0
22.3.1918	22.3.1919	7.0		12.0
22.3.1919			22.3.1920	12.0
20.6.1925	20.6.1926	7.3		12.0
	1.12.1934	7.3		failure
14.8.1940	14.8.1940	7.8		0.0
7.7.1950	7.7.1950	7.0		0.0
7.7.1950	30.6.1951	7.3		11.8
	26.10.1953	7.5		failure
1.1.1970			1.1.1973	36.0
1.1.1974			1.1.1975	12.0

total number of strong shocks = 9
7 shocks have been predicted
total duration of TIPs = 112.8 months (10.4% of total time).

Among from 9 strong shocks that occured in the part F, seven have been predicted, alongside with two periods of false alarm, during which shocks failed to occur.

4 CONCLUSIONS

The dilatancy-plastic model makes it possible to recognize both the mechanism of rock-mass deforma-

tion and the mode of formation of inclusion zones.
The latter may have nonstationary character. The
application of the prediction algorithm based on the
pattern recognition method confirmed the presence of
such a model of rock failure and associated development of mining shock sequence that precede the
strong event in a way sufficient enough to make the
model informative for prediction purposes, at least
at the incipient stage of failure.

REFERENCES

Keilis-Borok, V.I.(Editor) 1990. Intermediate-term
   earthquake prediction: models, algorithms, worldwide tests. Physics of the Earth and Planetary
   Interior, No 1-2, vol.61 (special issue).
Marcak, H.,Lasocki, s., Gerlach, Z., Wyrobek, E.
   1989. Ocena zagrożenia tąpaniami na podstawie
   komputerowego przetwarzania wyników obserwacji
   sejsmologicznych w KWK Katowice. Zeszyty Naukowe
   AGH, seria Górnictwo.
Marcak, H. 1992. Zastosowanie modelu dylatancyjno-
   -plastycznego do oceny stanu zagrożeń tąpaniami.
   Zeszyty Naukowe AGH, seria Geofizyka Stosowana,z.9.
Nikitin, L.W.,Ryzak, E.I. 1977. Zakonomiernosti
   rozruszenija gornoj porody z wnutriennim trieniem
   i dilatanciej (in Russian). Fizika Ziemli nr5,
   p.22-38.
Nikitin, L.W., Ryzak, E.I. 1984. O odnoj modeli
   obrazowanija sistiem tektoniczeskich razrywow.
   Geodinam. Issled. nr 7 p.56-76.
Patton, F.D. 1966. Multiple models of shear failure
   in rock. Proc. 1st Congr.Int.Soc. of Rock Mech.,
   Lisbon, 1.
Rudniscki, J.W., Rice, J.R. 1975. Conditions for
   the localization of deformation in pressure -sen-
   sitive dilatant materials. J.Mech. Phys. Solids
   23 no6.

Rockbursts and Seismicity in Mines, Young (ed.) © 1993 Balkema, Rotterdam, ISBN 90 5410 320 5

Source mechanisms, three-dimensional boundary-element modelling, and underground observations at Ansil Mine

R.G. McCreary & D.Grant
Noranda Technology Centre, Pointe-Claire, Que., Canada

V. Falmagne
Minnova Inc., Ansil Mine, Rouyn-Noranda, Que., Canada

ABSTRACT: By using a relatively simple full-waveform monitoring system, an accurate source location algorithm, simplified source-mechanism techniques, a consistent event-magnitude estimation method, and a stress-modelling algorithm suited to the deposit size and geometry, some useful relationships have been developed to understand mining effects at Ansil mine. Comparison of results from three-dimensional boundary-element modelling, source mechanisms, and underground observation, have shown that recorded activity has been induced in areas in which the predicted maximum principal stress magnitude is greater than approximately 55 MPa. Events with a purely shear mode of failure are prevalent in the footwall along geological contacts and the backs of development drifts. Those events without double-couple solutions are scattered throughout the footwall, and are not associated with geological features. Tensile activity, which has been clustered close to footwall development drifts, is associated with the greatest observable underground damage. Implications for further mining are that by modelling various stoping sequences, an "optimum" sequence has been selected based on these "failure criteria".

1 INTRODUCTION

The Ansil deposit, located in the Noranda mining camp in the southern part of the Abitibi greenstone belt, is a deep massive sulphide deposit centered at a depth of 1.3 kilometers (see Figure 1). Striking north-south with an average dip of 50° east, the deposit has a complex geometry in which its thickness may reach up to 35 m, varying significantly both along strike and along dip. Both the upper and lower zones of the deposit are quite thick, connected through the middle by a thinner central portion having a strike length of approximately 100 metres.

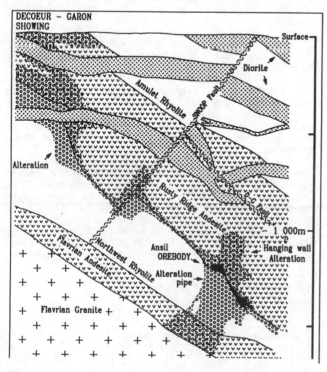

Figure 1. Geological cross-section 46590 N of the Ansil deposit. From Riverin et al., 1990.

As described by Falmagne (1991), rockbursts which were experienced during the sinking of the shaft in 1987 prompted the installation of a 16-channel Queen's University Full-Waveform microseismic system, completed in 1989 (for details concerning the Queen's Full-Waveform system, see Young et al., 1989). Initially consisting of seven hydrophones (7 x 1 channel) and two triaxial accelerometers (2 x 3 channels), the array was later modified to consist of 12 hydrophones, 1 triaxial accelerometer, and 1 uniaxial accelerometer (i.e. 1 channel of a triaxial) to give better coverage.

Since the start of mining at the site, microseismic data has been analyzed on site. Daily data-processing has involved visual sorting of the microseismic data to remove "cultural noise" (blasting and drilling), first-arrival traveltime and polarity picking, and source location.

The success of source-mechanism analyses of data recorded relatively early the mining sequence (see McCreary et al., 1992) initiated a detailed study of mechanisms in relation to results from numerical modelling and underground observation. This work was of special interest due to increasing evidence of the existence of non-double-couple source mechanisms for mine-induced seismicity (e.g. Wong et al., 1989; Potgieter and Roering, 1984; Rorke and Roering, 1984; Kusznir et al. 1984; Gibowicz et al., 1991; Feignier et al., 1992).

With increased extraction of the orebody (ie. following the mining of the thick upper and lower zones), microseismic activity induced by the mining of selected stopes has been analyzed and is presented with implications for sill pillar extraction.

2 NUMERICAL MODELLING OF STRESS EFFECTS

Various numerical models have the potential to be used to predict stress effects as a result of mining. By comparing results from the numerical modelling of different stoping sequences, one hopes to select the optimum sequence which minimizes underground damage and microseismic activity. The ultimate purpose of modelling is to maximize safety and minimize mining costs associated with ground control.

To model stress effects of an underground excavation using two-dimensional boundary element models, the following constraints must be met:

1) the limit of the excavation are "far" from the analysis plane:
2) the excavation cross-section is laterally constant;
3) induced displacements are in the analysis plane.

Due to the complicated shape and relatively small size of the Ansil orebody, none of the above constraints are met making two-dimensional modelling inappropriate. Thus, three-dimensional boundary-element models are most suited for the Ansil deposit.

However, for the results of three-dimensional boundary-element modelling to be properly interpreted, one must understand the assumptions used in its application:
1) the rockmass behaves elastically;
2) the rockmass is homogeneous;
3) the input parameters are accurate;
4) the boundary elements accurately represent the excavation boundaries.

Because one must make these assumptions to apply three-dimensional boundary-element modelling, the only true failure criteria that can be used to interpret the results is to make comparisons with results from visual observations and microseismic monitoring.

3 DATA PROCESSING

3.1 Microseismic data

A modified Geiger algorithm has been used to locate the microseismic events. After obtaining a first estimate of the source location, "poor" arrivals - associated with detrimental path effects - were identified as those in which the residual was greater than 1 ms. By removing the poor traveltimes to leave only "active" traveltimes, and continuing in an iterative approach until all residuals were less than 1 ms, source locations were obtained with an accuracy on the order of 2 m, as determined from calibration tests.

Being limited by a monitoring system with only one triaxial, and an environment in which rapid data-analysis was a priority, simple and efficient procedures for source-mechanism analyses were required for practical application. For these reasons, the source location algorithm was modified to sort events by "simplified" source mechanism (see Figure 2).

If all of the active arrivals used in the source location have positive polarities, we assume that the event has a tensile mode of failure. Similarly, if all are negative, the event is assumed to be implosional. If there are some negative, and some positive (i.e. "mixed" polarities), the event has either a shear mode of failure, or a mode involving a combination of shear,

tensile, and/or implosional components. Events with mixed polarities have been analyzed using a double-couple source-mechanism algorithm described by Brillinger et al. (1980).

Thus, for the purpose of this study, events have been sorted and defined as "tensile", "implosional", "shear with no solution", and "shear with double-couple solution".

In the analysis of data from a source location and velocity calibration exercise using small chemical explosives (see McGaughey, 1992), differences in waveform amplitude were seen to be controlled primarily by the extent of mine development and rock type along the ray path, with intersecting drifts and ore contacts resulting in greatly diminished amplitudes. Plots of waveform amplitude versus ray length revealed no obvious correlation for ray lengths less than 350 m, which was the maximum ray length examined. Obstructions due to mine development dominate other path effects; loss of energy due to scattering dominates spreading and absorption losses. Examination of spectra from the same source as measured by different receivers over similar ray lengths clearly demonstrated the low-pass filtering effect of mine workings along the path. Thus, estimation of event magnitudes by the single triaxial sensor on site was deemed inappropriate since:
1) the general amplitude level as well as the shape of the high-frequency end of the spectrum is highly influenced by propagation effects in the low kilohertz region;
2) the source-to-sensor raypath could be on a low-amplitude azimuth of the source radiation pattern;
3) the triaxial could become saturated (i.e. "clipped") when the source was relatively close to the sensor.

In our study of induced microseismic activity, when each event was recorded by the monitoring system, we assumed that at least one of the hydrophones of the array (presumed to be omni-directional) would record a non-clipped waveform on a maximum-amplitude azimuth of the source radiation pattern, and that the attenuation effects of the mine workings and geology would be minimal. For these reasons, event magnitude for each event was estimated by taking the logarithm of the pressure-distance product for the hydrophone which recorded the maximum voltage (i.e. pressure).

This hydrophone-based method proved to give more consistent magnitude estimates than the triaxial-based method.

3.2 In situ stress measurements and three-dimensional boundary-element modelling

As described by Falmagne (1991), in situ stress measurements were performed using a borehole slotter to determine stress values in the near-field of the orebody before mining was started on site (refer to Kanduth et al., 1991, for a description of the borehole slotter instrument).

The in situ stress measurements indicated that the stress magnitudes were lower than expected for such a great depth (see Table 1 and 2). Furthermore, there seemed to be a rotation of the principal stresses from the upper (i.e. Level 7) to lower zones (i.e. Level 9) of the orebody.

Using the three-dimensional boundary-element modelling algorithm described by Diering (1987), the stress-field was modelled for two different mine-sequence geometries

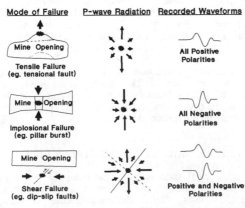

Figure 2. Schematic modes of failure for mine-induced seismicity. In the models (i.e. the first column), solid arrows indicate mine-induced force direction during the induced seismic event (as represented by a star). The second column schematically shows the normalized p-wave radiation patterns denoted by solid arrows. The third column indicates schematic representations of recorded p-waves. Modified after Hasegawa et al., 1989.

Table 1. Level 7 in situ stress measurements.

Level 7 (1220 m depth)			
	Magnitude (MPa)	Azimuth	Dip
σ_1	40.6	231°	1°
σ_2	25.2	141°	49°
σ_3	15.7	324°	41°

Table 2. Level 9 in situ stress measurements.

Level 9 (1340 m depth)			
	Magnitude (MPa)	Azimuth	Dip
σ_1	58.9	281°	4°
σ_2	44.2	13°	29°
σ_3	22.9	183°	61°

on a stope-by-stope basis. The magnitude and orientation of the stress field, as measured in situ, was used as input for the algorithm. Stress magnitude and orientation were linearly interpolated between the Level-7 and Level-9 stress-measurement locations.

4 RESULTS

The results presented in this paper are from the mining of Stopes 8A-26, 8A-28, and 8A-33 from the selected mining sequence. Discussed are the results of modelling and microseismic monitoring, in comparison with results from underground observation. The reaction of the underground support to the mining is also discussed.

Figure 3 shows the three-dimensional boundary-element mesh and the three stopes (8A-26, 8A-28, and 8A-33), including development for Levels 8A and 9.

4.1 Stope 8A-26

A total of 163 events were induced during the mining of Stope 8A-26. The most significant was induced following the first large production blast. Increasingly less-significant microseismicity was induced following the two remaining large blasts.

Most of the activity (160 events) has some component of shear failure, 21 % of which (33 events) has recognizable double-couple solutions. Extremely consistent nodal-plane pole concentrations are evident suggesting failure along pre-existing structure oriented at 150°/84° and/or 038°/17° (strike/dip).

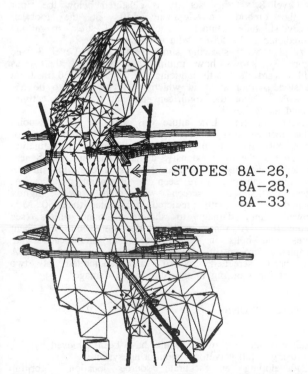

Figure 3. Isometric view looking west of the three-dimensional boundary-element mesh and the three stopes (8A-26, 8A-28, and 8A-33), including development for Levels 8A and 9.

Figure 4 presents the spatial variation of σ_1 and source mechanisms for Stope 8A-26 near Level 8. Activity is clustered to the north and south of the stope within the abutments where σ_1 is predicted to be 55-70 MPa. Within each of the two abutments, large double-couple shear events exist at the massive sulphide contacts where stress is predicted to be the greatest.

Underground observations of Level 8 showed that the footwall area to the west of the stope where stresses were predicted to be low were competent with no damage. This area is "stress-shadowed" by the stope. However, just north of the stope within the abutment where stresses were predicted to be high, 10cm-thick shotcrete over rebar and screen was cracked and showed spalling. Large cracks (5 cm to 8 cm wide) were induced within shotcrete around the drawpoint pillar on Level 8 located at approximately 39100 N, 46550 E near the bottom tip of Stope 8-0 where σ_1 is predicted to be 55-60 MPa. The brow to the north of the stope showed significant damage where shear activity was experienced. The raisebore of Stope 8A-33 - located two stopes to the north - was "dog-eared" with spalling on its eastern and western sides.

4.2 Stope 8A-28

During the mining of Stope 8A-28, 104 microseismic events were recorded by the monitoring system. Of all of the activity, the most significant amount was induced following the final and largest blast of the stope.

Detailed source-mechanism analyses indicates that activity with some component of shear failure is the prominent mode of failure - 16% of which has recognizable double-couple solutions. Around 4% of the events have a tensile mode of failure. Nodal-planes are oriented at 346°/78° and 078°/86° (strike/dip).

A sectional view at 46575 N (Figure 5) shows that all of the activity is in the footwall within regions in which σ_1 is predicted to be 55-100 MPa. Events with a

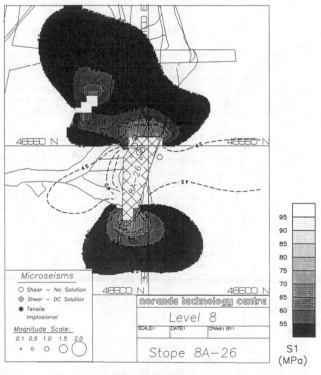

Figure 4. Plan view of Level 8 of the spatial variation of failure modes for activity induced during extraction of Stope 8A-26 compared with predicted predominant principal stress magnitude (σ_1). Failure modes are differentiated by symbol. The plotting size of each event is scaled by its magnitude. Stope 8A-26 (cross-hatched), geological contacts, and development drifts are shown.

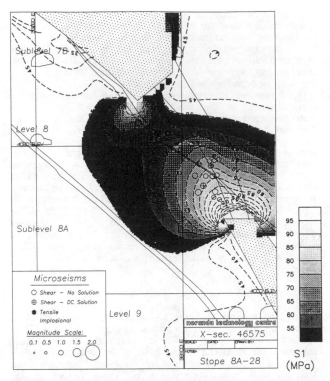

Figure 5. Cross-sectional view at 46575 N (seven metres to the north of the stope) of the spatial variation of failure modes for activity induced during extraction of Stope 8A-28 compared with predicted predominant principal stress magnitude (σ_1). Failure modes are differentiated by symbol. The plotting size of each event is scaled by its magnitude. The upper and lower mined areas (dotted), geological contacts, and intersections of development drifts with the section are shown.

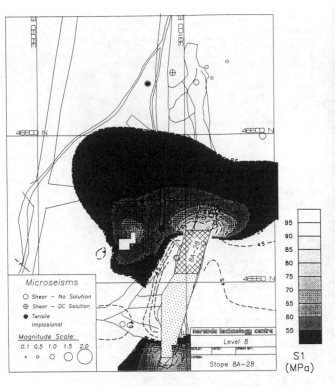

Figure 6. Plan view of Level 8 of the spatial variation of failure modes for activity induced during extraction of Stope 8A-28 compared with predicted predominant principal stress magnitude (σ_1). Failure modes are differentiated by symbol. The plotting size of each event is scaled by its magnitude. Stope 8A-28 (cross-hatched), previously-mined areas (dotted), geological contacts, and development drifts are shown.

purely shear mode of failure are located along footwall porphyry-dyke and massive-sulphide contacts within the footwall. Events with some component of shear failure (ie. positive and negative polarities, no double-couple solution) are scattered throughout the footwall, and are not associated with geological features. One large event is located within the back of the footwall drift on Level 8. The largest events are within high stress areas and near openings.

Figure 6, a plan view showing similar data along Level 8, shows that mining has perturbed the stress field toward the north inducing activity up to 20 m from the stope within areas in which σ_1 is predicted to be 55-80 MPa. Furthermore, anomalous, high-magnitude, shear and tensile activity has been induced as far as 50 m away from the stope in the backs of development drifts where σ_1 is 45-50 MPa. This activity is associated with a fault which passes through this area oriented $070^\circ/75^\circ$.

Underground observation revealed that the raisebore of Stope 8A-33 showed further squeezing. Cracks in the shotcrete of the Level 8 drawpoint pillar located to the east of the bottom "tip" of Stope 8-0 were further opened up to 10-15 cm.

4.3 Stope 8A-33

Induced microseismic activity, 115 events in total, has been directly related to stope extraction with the most significant occurring following the final and largest blast of Stope 8A-33. Activity with some component of shear failure remains the prominent mode of failure at the site - 15% of which having recognizable double-couple solutions. Around 10% of the activity has a tensile mode of failure. Nodal-planes are oriented $328^\circ/42^\circ$ and $085^\circ/76^\circ$ (strike/dip).

Figure 7, a cross-sectional view at 46590 N, shows that events with some component of shear failure are prevalent

in the back of the footwall drifts on Level 8A. Tensile activity is clustered below the floor of footwall drifts on Level 8. Other activity is clustered below the floor of the orebody development drifts at the footwall massive-sulphide contact. The principal stress magnitude is predicted to be 55-85 MPa in these areas.

A plan view showing similar data along Level 8 (see Figure 8), shows how mining has perturbed the stress field toward the north inducing activity up to 30 m from the stope within areas in which σ_1 is predicted to be 55-80 MPa. Note the significant amount of tensile activity located along Level 8.

Near-complete pillar failure was seen for the drawpoint pillar located just to the east of the bottom "tip" of Stope 8-0 (see Figure 9). Underground investigation showed that the most extensive damage was on Level 8 near the tensile activity where many cracks (2 cm wide) in 10-cm thick shotcrete were seen in drawpoint pillars. At these cracks, the screen was quite often slightly sheared. Areas with predicted σ_1 greater than 50 MPa showed minor damage to shotcrete around pillars (see Figure 10). Also seen and recorded were bolts with bent plates, and bolts in which the plates and heads were completely broken-off. Development on Levels 8 and 8A has been screened, bolted, and shotcreted at two different times during mining due to the damage.

5 CONCLUSIONS

Straightforward yet useful results have been achieved by:
1) using a full-waveform monitoring system;
2) developing an accurate source location algorithm which, in a simple manner, accounts for detrimental path effects related to mine geology and, more-importantly, the ever-changing mine geometry;
3) in the analysis of source mechanisms, using only

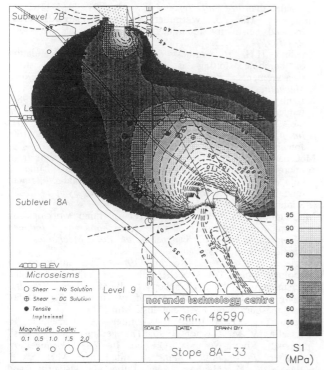

Figure 7. Cross-sectional view at 46590 N (seven metres to the north of the stope) of the spatial variation of failure modes for activity induced during extraction of Stope 8A-33 compared with predicted predominant principal stress magnitude (σ_1). Failure modes are differentiated by symbol. The plotting size of each event is scaled by its magnitude. The upper and lower mined areas (dotted), geological contacts, and intersections of development drifts with the section are shown.

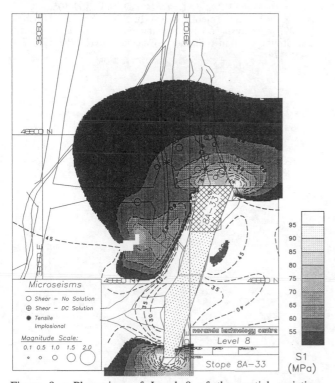

Figure 8. Plan view of Level 8 of the spatial variation of failure modes for activity induced during extraction of Stope 8A-33 compared with predicted predominant principal stress magnitude (σ_1). Failure modes are differentiated by symbol. The plotting size of each event is scaled by its magnitude. Stope 8A-33 (cross-hatched), previously-mined areas (dotted), geological contacts, and development drifts are shown.

Figure 9. Photograph showing pillar damage on Level 8.

Figure 10. Photograph showing spalling of pillar shotcrete on Level 8.

those p-wave polarities of raypaths deemed relatively free of detrimental path effects;

4) simplifying source-mechanism analyses by sorting events by polarity distribution alone;

5) developing a more-consistent array-specific method to estimate relative event-magnitude which accounts for source radiation pattern and path effects;

6) applying a numerical stress-modelling algorithm best-suited for the geometry and size of the mine;

7) using accurate in situ stress measurements as input for the numerical modelling, and corroborating produced results with mapped geology and underground observations.

Results in the comparison of three-dimensional boundary-element modelling, source mechanisms, and underground observation have shown that activity recorded at Ansil in the mining of three different stopes has been induced in areas in which the predicted maximum principal stress magnitude is greater than approximately 55 MPa. The lack of activity at very high predicted stress levels suggests that the rockmass has already failed in these areas and is thus no longer carrying load. This limitation of elastic models suggests that areas experiencing activity may indeed be at stress magnitudes higher than actually predicted.

Events with a purely shear mode of failure have been induced along geological contacts and in the backs of development drifts within the footwall. Events without double-couple solutions - having shear or combination modes of failure - have been scattered throughout the footwall, and are not associated with geological features. Tensile activity, which has been clustered close to footwall development drifts, is associated with the greatest observable underground damage. Predicted maximum principal stress magnitudes over approximately 50 MPa have been associated with shotcrete damage around pillars. Successful in maintaining safe access for ore-haulage with minimum rehabilitation downtime have been shotcrete around pillars and cable bolts in the backs of the drifts.

Significant activity has occurred along two mapped faults oriented $330°/75°$ and $070°/75°$ (strike/dip) respectively. There seems to be two types of activity at Ansil Mine - one type located close to newly-mined stopes associated with high-magnitude induced stresses, and another associated with movement along faults occurring at medium-magnitude stress levels. More work is required to differentiate the causes and effects of the two different types of activity. This work has emphasized the importance of recognizing major structures early on in the life of the mine and monitoring their reaction to mining.

Implications for further mining of the sill pillar are that different stope geometries can be modelled, and areas prone to significant microseismic activity and damage can be delineated as those in which the maximum principal stress magnitude exceeds approximately 55 MPa. Areas prone to minor damage will be those in which the maximum principal stress magnitude is between approximately 50 MPa and 55 MPa. The most significant damage, tensile in nature, will be in the high stress regions around footwall drifts and pillars. Based on these relationships, an "optimum" sill-pillar extraction sequence has been selected in hopes to minimize related damage, maximize recovery (e.g. avoiding small remnant pillars), and facilitate smoother production.

Further relationships between source mechanisms, underground damage, and modelling results is the subject of future work in which other failure criteria will be investigated, such as incremental change in maximum principal stress magnitude ($\sigma_{1_{after}}$ - $\sigma_{1_{before}}$), the difference between the maximum and minimum principal stress magnitudes (σ_1-σ_3), and extension-strain.

REFERENCES

Brillinger. D.R., A. Udias & B.A. Bolt 1980. A probability model for regional focal mechanism solutions. *Bull. Seismol. Soc. Am.* 70: 149-170.

Diering, J.A.C. 1987. *Advanced elastic analysis of complex mine excavations using the three-dimensional boundary element techniques.* PhD. Thesis: Pretoria University, South Africa.

Falmagne, V. 1991. Rock mechanics and microseismic monitoring at Ansil. *CIM Bulletin* 84, 955: 46-51.

Feigner, B., R.P. Young & S. Talebi 1992. Non-double-couple source mechanisms: analysis of excavation-induced microseismic events using moment tensor inversion. *Proc. Workshop on Induced Seismicity, 33rd U.S. Symposium on Rock Mechanics:* 112-113. Menlo Park: USGS.

Gibowicz, S.J., R.P. Young, S. Talebi & D.J. Rawlence 1991. Source parameters of seismic events at the Underground Research Laboratory in Manitoba, Canada: scaling relations for events with moment magnitudes smaller than -2. *Bull. Seismol. Soc. Am.* 81: 1157-1182.

Hasegawa, H.S., J. Wetmiller & D.J. Gendzwill 1989. Induced seismicity in mines in Canada - an overview. *PAGEOPH* 129, 3/4: 423-453.

Kanduth, H.H., R. Corthésy & D.E. Gill 1991. Validation of borehole slotting as a method for in situ stress measurements. *Proc. 7th ISRM Int. Cong. on Rock Mechanics:* 527-532. Rotterdam: Balkema.

Kusznir, N.J., N.H. Al-Singh & D.P. Ashwin 1984. Induced-seismicity generated by longwall coal mining in the North Staffordshire coal field, U.K.. *Proc. 1st Int. Cong. Rockbursts and Seismicity in Mines:* 153-160. Johannesburg: South African Inst. Min. Met.

McCreary, R.G., D. Grant & V. Falmagne 1992. Results from microseismic monitoring and analysis at Ansil Mine: a comparison with three-dimensional boundary-element modelling. *Proc. 16th Can. Rock Mech. Symp.:* 13-24. Sudbury: Laurentian University.

McGaughey, W.J. 1992. The effect of mine workings on seismic waveforms. *Proc. Workshop on Induced Seismicity, 33rd U.S. Symposium on Rock Mechanics:* 13. Menlo Park: USGS.

Potgieter, G.J. & C. Roering 1984. The influence of geology on the mechanisms of mining-associated seismicity in the Klerkshop goldfield. *Proc. 1st Int. Cong. Rockbursts and Seismicity in Mines:* 45-50. Johannesburg: South African Inst. Min. Met..

Riverin G., M. LaBrie, B. Salmon, A. Cazavant, R. Asselin, & M. Gagnon 1990. The geology of the Ansil deposit, Rouyn-Noranda, Québec. In *The Northwestern Québec Polymetallic Belt.* Edited by M.Rive, P. Verpaelst, Y. Gagnon, J.M. Lulin, G. Riverin, and Simard. Canadian Institute of Mining and Metallurgy, Special Volume 43, 143-151. Montréal: CIMM.

Rorke, A.J. & C. Roering 1984. Source mechanism studies of mine-induced seismic events in a deep-level gold mine. *Proc. 1st Int. Cong. Rockbursts and Seismicity in Mines:* 51-55. Johannesburg: South African Inst. Min. Met..

Wong, I.G., J.R. Humphrey, J.A. Adams & W.J. Silva 1989. Observations of mine seismicity in the eastern Wasatch Plateau, Utah, USA. *PAGEOPH* 129, 3/4: 369-405.

Young, R.P., S. Talebi, D.A. Hutchins & T.I. Urbancic 1989. Analysis of mining induced microseismic events at Strathcona mine, Sudbury, Canada. *PAGEOPH* 129, 3/4: 455-474.

Rockbursts and Seismicity in Mines, Young (ed.) © 1993 Balkema, Rotterdam, ISBN 90 5410 320 5

Chaotic behaviour and mining-induced seismicity

Douglas M. Morrison
Inco Limited, Copper Cliff, Ont., Canada

Graham Swan
Falconbridge Limited, Onaping, Ont., Canada

Christopher H. Scholz
Lamont-Doherty Geological Observatory, Columbia University, USA

ABSTRACT: That major mining-induced rockbursts result from stick-slip motion on pre-existing fractures has long been recognised. However, the spatial and temporal distribution of these large seismic events can only loosely be correlated with the location of prominent geological structures and with the timing of production blasting operations and this led recently to consideration of possible chaotic behaviour within a mine system. As a system, the rock environment is naturally complex but often appears to display extraordinary sensitivity to very minor changes - self-organised criticality.

By means of a two-dimensional numerical experiment involving a single frictional surface and a changing mining geometry, spatial aspects of seismic energy release have been studied. Results are presented and discussed which argue for chaotic behaviour to be an intrinsic part of this "simple" representation of mining.

1 INTRODUCTION

There are some 16 mines currently exploiting the nickel-copper sulphide deposits in the Canadian Precambrian Shield, known as the Sudbury Nickel Eruptive (Figure. 1). The mines employ several different mining methods and extend from about 300m below surface to 2400m although the majority of reserves presently being exploited range from 700m to 2000m below surface. Only four of the currently operating mines have a significant problem with rockbursts, defined as seismic or microseismic events which cause any physical damage to underground excavations. Altogether ten mines are currently monitored by a variety of microseismic monitoring systems, detecting events of up to magnitude 4.0. The majority of events are less than magnitude 1.5 but events between magnitude 1.5 and 2.5 are quite frequent. While the problem is often most severe in the largest and deepest of the mines, rockbursts have occurred at less than 600m below surface.

The most seismically active mines in the Sudbury Basin (Figure 1) at the present time are Creighton Mine and Copper Cliff North Mine, operated by Inco Limited, and Strathcona Mine and Lockerby Mine operated by Falconbridge Limited. Much of the most recent research on the problem is being carried out as part of a collaborative programme of research (the Canadian Rockburst Research Programme) with six mining companies and the Universities in Kingston (Queen's) and Sudbury (Laurentian), coordinated by the Mining Research Directorate (MRD).

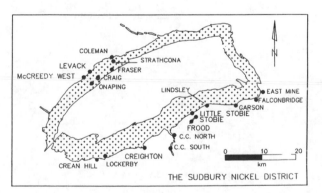

Figure 1. Area of Sudbury Basin showing Operating Mines.

2 THE ROCKBURST PROBLEM IN MINING

Rockbursts are classified as three different types; strain energy bursts, pillar-type bursts and fault-slip bursts. In the case of the first two types, the tactical and strategic approaches adopted by the mining companies have been largely successful in avoiding and controlling these bursts. However, the fault-slip bursts, which are also generally responsible for the largest and most damaging events, have proven to be a more intractable problem.

The orebodies are usually planar ranging from 7m to 100m wide, dipping at 60 to 90 degrees with the principal shaft, ramp and tunnel accesses located in the footwall. Generally, the major principal stress is horizontal and perpendicular to the strike of the ore and is approximately 1.8 times the vertical overburden stress. Locally, the stresses are very variable and of course are strongly influenced by mining operations.

From the industry perspective, the objective of rockburst research is not to predict rockbursts but is to eliminate the cause of the rockbursts wherever possible and, when this is not possible, to manage and control the location, severity and timing of the occurrences to minimise the effect on the mining operations. The principal interest at the present time is the location and severity of the problem rather than its timing and this may be the most important difference between the objectives in mining-induced seismicity and conventional earthquake research.

The mining operations influence the seismicity in several different ways (Morrison, 1991). Some factors are directly related to specific mining activities such as blasting and the backfilling of the mined out excavation with hydraulic tailing fill. However, many events and particularly the largest fault-slip events do not appear to be closely related to specific activities but rather to the excavation of the mine as a whole.

3 EVIDENCE FOR SELF-SIMILARITY

Inco's Creighton Mine is the deepest and most seismically active underground mine currently operating in the Sudbury Basin. Presently, mining extends to 2400m below surface and seismicity is monitored below the 1800m level. In October 1989, a $M_n=3.6$ event occurred on the 6600 level (2200m below surface) in the footwall of the orebody which resulted in severe damage in and around an ore-pass which serves to transfer the broken ore out of the mine. This event, together with a $M_n=1.7$ event, occurred after a great deal of seismicity following a major production crown blast.

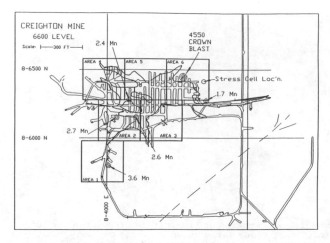

Figure 2. The location of the six Areas on 6600 level and the location of the major events of October and November, 1989.

To statistically analyze all the events associated with this major event, the levels from 6600 to 7000 were arbitrarily divided into six equal-sized areas (Figure 2) and the number of daily events in each area was plotted. Area #6 contains both the crown blast and the 1.7 magnitude event while Area #1 contains the 3.6 magnitude rockburst.

For most of the month of October, all six areas were equally active with about 10 events/day except for Area #3 which consistently had a higher level of daily activity, at 15-20 events/day. The level of seismicity in Area #4, #5 and #6 did not change and are not plotted in Figure 3. As shown in Figure 3, Area #3 showed a particularly sensitive response to the crown blast, to the 1.7 magnitude event (which occurred about 8 hours later) and to the magnitude 3.6 event (which occurred 65 hours after the blast). These first two occurrences are both located within Area #6 but this area did not remain very active for long and even on October 27 and 28, the number of events in Area #3 was always greater than in Area #6. Immediately following the magnitude 3.6 rockburst, Area #1 was the most active area but over the course of the whole day Areas #2 and #3 recorded as many, if not more events than Area #1. After the initial flurry of activity in these three areas, the activity in Area #1 quickly returned to normal, while Areas #2 and #3 continued to show enhanced levels of activity and Area #2 had four other major events.

These results illustrate that the distribution of the small events within the mine is not directly related to the location of very large events. The area which included the 3.6 event was not the most seismically sensitive area and no large events occurred within the most active areas. This tends to suggest that the nature of the large events and the small events is characteristically different.

Evidence of self-organized criticality is shown by the familiar power-law distribution for earthquake magnitude and the power, b, is normally found to be in the range 0.6 to 1.1. Figure 4 shows data for naturally occurring earthquakes in Eastern Canada, 1986-87, as recorded by the Eastern Canada Seismic Network. The same figure shows Creighton Mine data, recorded by the same Network, for the period 1984-1991, Hedley (1992). Both sets of data display a power-law relationship agreeing well in terms of the exponent value and fit well in the normal range.

The very large events typically occur deeper into the orebody wall rocks some distance from the mine workings while the smaller events tend to cluster close the mine openings. The timing of the large events also appears to be somewhat independent of the timing and location of those mining activities such as blasting and back-filling which often promote flurries of small events. Thus, the larger events could be considered to be 'natural events' and the influence of the adjacent mining activity has been to artificially decrease the recurrence times of these natural events.

It has been suggested that the reason for the difference in b values of large and small earthquakes is a mechanical constraint on the faults upon which these events occur. Small earthquakes occur on smaller faults which are contained within the schizosphere (Scholz, 1990) while larger earthquakes occur on faults which extend all the way to the top or bottom of the schizosphere. The b values for large rockbursts and small earthquakes are similar and it may be that the mechanistic constraints on the faults upon which both these sets of events occur are also similar.

In the case of the large rockbursts the dominant mechanism has been shown to be shear while in the case of smaller rockbursts the majority of the events are enriched in P-wave energy and depleted in S-wave energy, implying their failure mechanisms are non-double couple, (Urbancic & Young, 1991). It is possible that the difference in the dominant failure mode of the large and small rockbursts is related to the mechanistic constraints on the faults upon which they occur. Large rockbursts tend to occur deep into the host rocks and are constrained in the same way as faults contained entirely within the schizosphere while the smaller rockburst events occur much closer to the excavations.

4   SPATIAL CONSIDERATIONS

Although the consideration of magnitude recurrence times appears to be the principal interest in earthquake seismology, in the context of mining induced seismicity it is of little practical value.

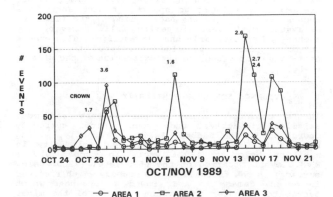

Figure 3. Graph showing the different levels of seismicity in three of the six defined areas in October and November, 1989.

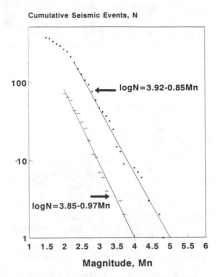

Figure 4. Comparison of b values for regional earthquakes, Eastern Canada and large rockbursts at Creighton Mine.

From the perspective of an operating mine the spatial distribution of large and potentially damaging seismic events are of crucial importance. It is recognised that not all areas of the mine are equally susceptible to damage and spatial variations occur in the severity of seismic events. Once identified, susceptible areas can be protected by means of a variety of tactical support systems and strategic approaches to mine design. Consequently, a major objective of the study of mining induced seismicity is to develop a means of identifying those areas in the mine which are at risk of severe damage.

Because the nature of rock is not well characterized as a continuous, elastic material, a consideration of spacial factors is not limited to excavation geometry. This has led to the development and use of sophisticated two and three dimensional discontinuum models in mine design seismic risk assessment, (Tinucci and Hart, 1990), (Tinucci and Hanson, 1990). With discontinuum model capability has also come the realisation that results can be extremely sensitive to very small changes in initial conditions, (Cundall, 1990). These arise because of the geometrical complexities of discontinuous models in general and in particular, because of the explicit solution of a fault "softening" process used in the UDEC code (Universal Distinct Element Code) (Itasca, 1988).

4.1 Model Description

By means of a simple model the effect of a number of possible geometrical changes on the stability of a nearby fault in terms of the release of seismic energy can be easily shown using the commercially available UDEC code. This code uses an explicit, time-marching solution and was employed to simulate the effect of an excavation sequence on a faulted medium subject to quasi-static loading.

The fault itself was given both cohesion and frictional properties at each contact point described by UDEC's "continuously yielding" constitutive function (Cundall and Lemos, 1990) with the form shown in Figure 5. This function allows the model to simulate progressive damage to the fault during shear and to exhibit the observed phenomena of stress lock-up and stick-slip behaviour. The model could be run say one hundred times with a range of initial conditions in a reasonable time. Below we show the results from only two such conditions; the excavation approaching the fault and retreating from it.

Within the model the fault was divided into nine segments (Figure 6), with contact points at a segment's midpoint where calculations of average normal and shear stress, and shear displacement are made. The difference in stress and displacement between each of eleven excavations in the mining sequence is used to calculate the amount of energy liberated. A simple expression of the form

$$\log E = a \, M_n - b$$

is then used to estimate slip event magnitudes, Ng et al (1992).

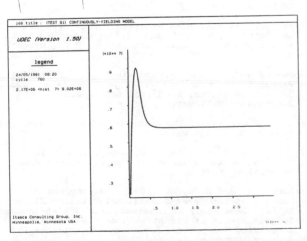

Figure 5. UDEC Continuously Yielding Fault model, shear stress (Pa) as a function of shear displacement, (m).

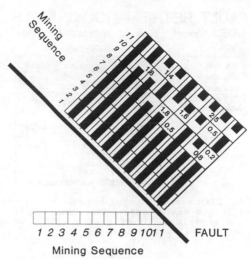

Mining Sequence

Figure 6. Fault Model showing slip events, mining sequence advancing (white=slip, black=no slip).

4.2 Results

At total of 21 slip events were obtained from the two runs. This is too few a number to generate a frequency-magnitude plot and to confirm a power law result from the model. The results do however reveal positional contrasts in slip events along the fault for a given mining sequence, Figure 6. When the sequence is reversed a completely different set of events is observed (Figure 7).

From these data the beginnings of a representation of risk of damage to the excavation can be obtained. A damage criterion based upon peak particle velocity provides the risk map, onto which the slip events are plotted (Figure 8). By running the model repeatedly for a given sequence but with changes to the initial conditions, a frequency of events in the respective damage or risk zone may be found. In this way statements can be made about the probability of seismic events occurring in the vicinity of a mining excavation and causing damage exceeding acceptable limits.

5 DISCUSSION

The use of risk or probabilistic design methods in the geosciences is not new and, as an option, is considered by many to accommodate the observed variations in geological material (Roberds, et al, 1991), (Pine, 1991). All available methods however derive from deterministic models for which the results are only as good as the input distributions used. In

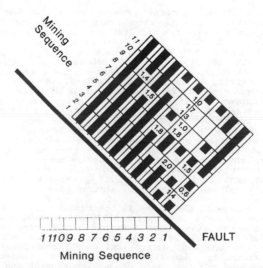

Mining Sequence

Figure 7. Fault model showing slip events, mining sequence retreating.

FAULT RETREAT ROCKBURSTS
UDEC Model - Risk at Shortest Distance

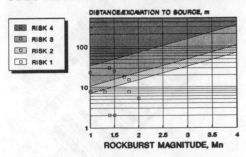

Figure 8. Slip events plotted as apparent magnitudes relative to the mining excavation. The risk level is determined empirically from peak particle velocity measurements, Hedley (1992).

the present application where, aside from a simulation of mining activity, complex earthquake source mechanisms are modelled, questions concerning instability and chaotic behaviour must be investigated more fully. Recently Hobbs (1990) initiated a series of studies using the UDEC code to model the most basic spring-slider system. This model was able to show all the classical behaviour of chaotic dynamic systems; chaotic intervals, periodic windows and period doubling.

It has come to be recognised that the expectation of a definitive and predictable response from a system such as the highly stressed and intensely faulted rockmasses found in the mines of the Sudbury Area is unrealistic and that to continue to develop ever more sophisticated and accurate models will ultimately be fruitless. Nevertheless, it is also clear that very simplistic, chaotic automaton models also have severe limitations in terms of producing real solutions to real problems. Generally, many of the mining seismicity problems are deterministic to an acceptable degree and only a few aspects of the problem display the characteristics of an extremely sensitive chaotic system, namely the fault-slip rockbursts.

Consequently, there appears to be a need for the development of some form of hybrid deterministic model, which incorporates aspects of weakly chaotic systems. In this context the behaviour of some of the discontinuities within the UDEC model would not be driven by a formal constitutive law but by results from an automaton model. The object of this kind of simulation would be to generate the range of likely responses of the discontinuities rather than to calculate a unique solution to a particular set of pre-determined conditions. Such a model could be expected to be less well behaved and more sensitive to minor variations in the input values than typical probabilistic models.

Turning to the general problem of mining induced seismicity, the data from Creighton suggests that there are at least two distinct self-similar groups of events. Another feature of the data is the diffuseness of the activity in time and space which contrasts sharply with the behaviour in natural earthquakes (Scholz, 1991), probably as a result of the proximity of the mine excavations. However, the diffuseness signature for Creighton is different from the Copper Cliff North Mine signature. The categorisation of mine seismicity into shear and non-shear events may also vary from mine to mine. The most obvious mine seismic signature of all is the existence of any significant levels of seismicity. Of the 16 mines currently operating, two are in low stress environments close to surface but of the 14 deep underground operations, only four are subject to significant levels of seismicity, despite striking geometrical and geotechnical similarities.

Indeed, many of the mines in the Sudbury basin which are now, or once were, seismically active have a near neighbouring mine which remained almost completely aseismic. These pairs of mines, where the mining configurations and geotechnical conditions appear to be virtually identical, are often connected underground and the major production excavations are usually only a few hundred metres apart. Finally, the occurrence of seismicity within any particular mine also appears to be episodic. Mines which become

seismically active generally mature from an early aseismic phase, through a seismically active period and then return to a less active phase. Naturally, the energy balance within the mine is dominated by the rate of production and consequently by the rate of convergence of the surrounding host rocks. However, even during the seismically active phase the levels of seismic activity are distinctly episodic, even periodic, despite the fact that a relatively uniform rate of production is maintained throughout the year.

6   SUMMARY

It appears then that for the special application to mining problems, for which at the very least a two-dimensional, spatial fault model is required, ample scope exists to establish rules governing the realism of the results. This is because of the easy access to the underground conditions which govern scale, friction and fault stability. Model development in the context of mining engineering can be seen as an empirical process of calibrating real situations and goes beyond the consistency with the Gutenberg-Richter law displayed by typical cellular automaton models. The output from the kinds of hybrid models proposed here would be an all time, spatially-dependent risk map. Given that researchers in the disciplines of earthquake and mining seismology should be equally concerned with the realism of their models, we suggest that the problems associated with mining induced seismicity merit much greater attention by earthquake seismologists, than they have received in the past.

REFERENCES

Cundall, P.A. (1990). Numerical Modelling of Jointed and Faulted Rock. Proc. Conf. Mechanics of Jointed and Faulted Rock, Vienna (Ed. Rossmanith/Balkema), pp.11 -18.

Cundall, P.A. and Lemos, J.V. (1990). Numerical simulation of fault instabilities with a continuously-yielding joint model. Proc. 2nd. Symp. Rockbursts and Seismicity in Mines, Minneapolis (Ed. Fairhurst/Balkema), pp 147-152.

Hedley, D.G.F. (1992). Rockburst Handbook for Ontario Hardrock Mines. Mining Research Laboratories CANMET, E.M.R.

Hobbs, B.E. (1990). Chaotic behaviour of frictional shear instabilities. Proc. 2nd Symp. Rockbursts and Seismicity in Mines, Minneapolis (Ed. Fairhurst/Balkema), pp.87-91.

Itasca (1988). UDEC - A Universal Distinct Element Code. Itasca Consulting Group Inc., Minneapolis, Minnesota, U.S.A.

Morrison,D.M., Villeneuve, T. and Punkkinen, A. (1991). Factors Influencing Seismicity in Creighton Mine, Proc. 5th Conf. on Acoustic Emissions and Microseismic Activity, Penn. State, U.S.A..

Ng, L., Swan, G. and Hedley, D.G.F. (1992). An Examination of Seismic Energy in Fault Models. In Preparation, Proc. Induced Seismicity Workshop, Santa Fe, June 10.

Pine, R.J. (1991). Risk analysis design applications in mining. Proc. Conf. Research Applications in the Mining Industry, Univ. Nottingham, U.K., October 17.

Roberds, W.J., Iwano, M. and Einstein, H.H. (1990). Probabilistic mapping of rock joint surfaces. Proc. Conf. Rock Joints, Loen (Ed. Barton and Stephansson/Balkema), pp.681 - 691.

Scholz, C.H. (1990). The Mechanics of Earthquakes and Faults. Cambridge Univ. Press.

Scholz, C.H. (1991). Internal Report to M.R.D., Canadian Rockburst Research Project, Laurentian University, Sudbury.

Tinucci, J. and Hart, J. (1990). Development and Testing of the Three-Dimensional Distinct Element Method. Case Study 5: Examination of Mining-Induced Seismic Events at the Strathcona Mine. ITASCA Consulting Group Inc., Report to Falconbridge Ltd., Sudbury, Ontario, January 1990.

Tinucci, J. and Hanson, D.S.G. (1990). Assessment of
Seismic Fault-Slip Potential at the Strathcona Mine.
Proc. 31st U.S. Rock Mechanics Symposium, Colorado
(Ed. Hustrulid & Johnson/Balkema), pp.753 - 760.

Urbancic, T.I. and Young, R.P. (1992). Focal mechanism
and source parameter studies at Strathcona Mine.
Eng. Seismology Lab. Int. Report: MRD-MD001, Queen's
University, Kingston.

Rockbursts and Seismicity in Mines, Young (ed.) © 1993 Balkema, Rotterdam, ISBN 90 5410 320 5

Fracture mechanism of microseisms in Saskatchewan potash mines

Arnfinn F. Prugger
Potash Corporation of Saskatchewan, Sask., Canada

Don J. Gendzwill
University of Saskatchewan, Sask., Canada

ABSTRACT: Over the last decade numerous microseismic monitoring surveys have been conducted at Saskatchewan potash mines. Noticeable earthquakes have occurred at some mines, and small-scale microseismicity is common at all mines. There is no relationship between microseisms and earthquakes. Small events are located within mining level salts, while large events occur in hanging-wall limestones. Some microseismicity is controlled by local geology, but most are a natural rock mechanical result of stress-relief mining. Study of seismic waveforms led to the conclusion that microseisms result from delamination in the roof or floor, failure best modelled by a single force. Slow rupture propagation is indicated.

1 INTRODUCTION AND BACKGROUND

In this paper we present the observational results of monitoring seismicity in potash mines in Saskatchewan, Canada. The primary objective of the study was to gain an understanding of rock failure occurring near potash mining operations through observation of seismic signals caused by these ruptures.

1.1 Geological and Seismological Background

Much of southern Saskatchewan is underlain by the "Prairie Evaporite", a layered sequence of salts and anhydrite which contains the western world's largest reserve of potash. The potassium extracted from the predominantly sylvinite ore has its main use as a fertilizer. The 100-200 m thick Prairie Evaporite is overlain by approximately 500 m of Devonian carbonates, followed by 100 m of Cretaceous sandstone, followed by 400 m of Cretaceous shales and Pleistocene glacial tills to surface, and is underlain by Devonian carbonates, (Fuzesy, 1982). The Phanerozoic stratigraphy of Saskatchewan is remarkable in that units are flat-lying and relatively undisturbed over very large areas. Ten potash mines were brought into production in Saskatchewan in the period 1962 through 1970. Eight of these are conventional underground mines, and two operate using solution mining methods. Mines operate at 900 - 1100 m depth within 15 - 50 m of the top of the Prairie Evaporite. All conventional mine operators utilize some form of stress-relief mining using continuous boring machines. The mining methods employed in the province are discussed in Jones and Prugger (1982).

Saskatchewan has a history of sporadic, low level, shallow seismicity in the south of the province, (the result of movement along a system of north-east and north-west trending mid-continent faults according to Horner, 1983), and in the region near the north edge of the subsurface Prairie Evaporite, (the result of dissolution and collapse in the Prairie Evaporite salts according to Gendzwill, 1988). These events have all been below Richter magnitude +4.0, with the exception of one magnitude +5.5 earthquake which occurred in southern Saskatchewan in 1909.

Noticeable earthquakes have occurred at only five of the ten potash mines in the province. The large events, first noticed in the fall of 1976, have been in the Richter magnitude range +1.5 to +3.7; (Gendzwill et al., 1982; Gendzwill, 1984; Hasegawa et al., 1989; Gendzwill and Prugger, 1990). The proximity of these earthquakes to mines, in areas where previous seismicity was unknown, led to the conclusion that they were induced by mining activity. A failure model has been proposed for the large events (Gendzwill, 1984), but this still needs to be conclusively verified by observation. Seismic signals from events large enough to be called earthquakes now occur only once every 2-3 years, and have yet to be recorded properly. In contrast, microseismic activity (magnitude less than 0) is

common, and has been observed at every mine where monitoring programs have been conducted. Digital recordings of full-wave seismograms have been collected for thousands of microseismic events over the past decade. This discussion is therefore limited to potash microseisms.

1.2 Overview of Methodology

Over the past decade we have conducted short-term seismic monitoring experiments at six potash mines in Saskatchewan and one in New Brunswick (Prugger and Gendzwill, 1991). This discussion will focus on results of the best data set, which was recorded at the Cominco Fertilizers mine at Vanscoy, (which is just west of the city of Saskatoon), in 1988. At Vanscoy, which is the deepest mine in the province at 1050-1100 m, stress-relief mining is realized through sequencing long rooms with narrow yield pillars, as described in Mackintosh (1975 and 1977) and Jones and Prugger (1982). An array of 11 vertical component geophones was set up underground, and seismicity was monitored almost continuously over a six week period. Each station consisted of two Mark Products model L-1 geophones, (1480 Ohm, with a flat frequency response from 4.5 to 250 Hz), mounted to the mine roof using bolts shot into the rock. Unamplified geophone signals were carried up to 1700 m on light (22 gauge), two-pair telephone wire to a set of differential amplifiers run at gain settings ranging from 60 to 78 db. Signals were generally digitized at 0.005 s, (with high-cut filters at 50 Hz), but some events were recorded at higher

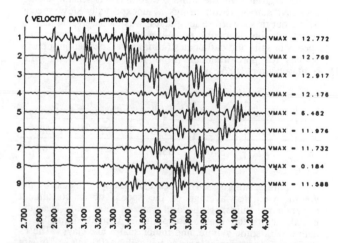

Figure 1. Event seismograms. Time in seconds, all amplitudes normalized to peak values (which are listed at the right of each trace).

sample rates (0.002 s and 0.001 s, with high-cut filters at 100 Hz, 250 Hz, or 500 Hz). Signals were continuously monitored, and seismic events were recognized using an STA/LTA algorithm (Short-Term-Average/Long-Term-Average, as discussed in Lee and Stewart, 1981, for example). Seismograms for each event were plotted, digitally filtered and scaled, as required, and arrival times were interactively picked on a computer screen. Source locations were computed by Simplex optimisation, as described in Prugger and Gendzwill (1988). Representative seismograms are shown in Figure 1.

2 OBSERVATIONS

The first study of seismicity near a Saskatchewan potash mine was carried out at the Potash Corporation of Saskatchewan (PCS) mine at Cory, as discussed in Gendzwill (1984) and Prugger and Gendzwill (1990). Signals were recorded on surface, about 1000 m above the mining level (and the ruptures). There was nothing unusual about the seismograms that were collected, and standard seismic analysis methods were routinely applied. But when recording systems were installed underground in the mine several unexpected characteristics were observed. These included an anomalous magnitude/frequency-of-occurrence relationship, the absence of a recognizable P-wave, remarkable consistency in waveform detail (independent of azimuth or distance), low frequency content of the waveforms, and multiple impulse source functions.

2.1 Magnitude/Frequency of Occurrence

Study of temporal relationships between microseismicity and large earthquakes, and analysis of event magnitudes established that there was NO correlation between the small events and the larger earthquakes at the PCS Cory mine (Prugger and Gendzwill, 1990). Instead there seemed to be two classes of seismicity, low-energy events and high-energy events. The apparently bimodal nature of the magnitude data at Cory was initially treated with scepticism, given that it goes against the commonly held view of what the Gutenberg-Richter (1941) magnitude/frequency distribution is. However, similar bimodal distributions are observed in other mining areas, notably at coal and copper mines in Poland (Kijko et al., 1987). Interpreting the Polish data, Gibowicz (1990) concludes that while it is difficult "to explain their nature" the fact that there are two classes of seismic event generated by two different phenomena is irrefutable. In this light the data from Cory indicate that the mechanism causing micro-earthquakes is not the same as the mechanism that generates the larger events.

2.2 Propagation Speed

As soon as we began to collect data using geophone arrays set up underground in the mine we found that seismic energy travelled across sensor arrays at unexpectedly slow speeds. Move-out times were so long that it was inconceivable that the observed signals could be P-waves. Seismic propagation speeds in salts and carbonates fall into well defined groupings. Values for pure minerals (in m/sec) are as follows (after Schlumberger, 1986):

	Sylvite	Halite	Anhydrite	Calcite	Dolomite
P	4100	4500	6100	6220	6930
S	2310	2540	3380	3450	4230

The bulk seismic speed in rock can be approximated by the volume weighted average of the speeds of its constituent minerals and fluids; (P-speed in water is 1500 m/s). Near the potash layers the Prairie Evaporite is composed mainly of halite, with some sylvite, and minor shales and clays. The P-wave speed on sonic logs is typically 4200 m/s . In-situ measurements of the S-wave speed in salts at the mining level are near 2300 m/s (Gendzwill and Stead, 1992). The carbonates overlying the Prairie Evaporite are typically composed of calcite and dolomite, with halite-filled or brine-filled porosity. Sonic log P-wave speeds are near 5800 m/s, decreasing in proportion to fluid-filled porosity.

The Simplex source location algorithm allows the calculation of seismic wave propagation speed in addition to position coordinates. The average calculated speed for 127 events picked up at Vanscoy was 2400 m/s, close to the shear-wave speed for halite and sylvite, but too slow for any other major rock component; (notably slower than carbonates in the immediate hanging-wall and foot-wall). In microseismic events recorded at the mining level, energy with a P-wave speed was seen on only the rarest occasion. This observation was made at a potash mine in New Brunswick as well as in the Saskatchewan mines.

2.3 Vertical Position

Slow move-out times made it impossible for the source to have been far out of the plane of geophones, which were all at the mining level. Events occurring any distance above or below the mine level would result in faster apparent speeds than were ever observed. When all geophones are at nearly the same elevation, and a microseismic event occurs close to that level, there is a loss of precision in the vertical coordinate of the calculated source location. But the apparent horizontal velocity of the seismic wave is the true velocity when the source and all receivers are at the same elevation. At a different elevation, the apparent velocity is INCREASED by the secant of the angle of the source above the horizontal. The clustering of observed speeds near the S-wave speed of salt indicates that events were generated in the salt, and are close to the mine elevation. If they occurred in hanging-wall carbonates, ray theory requires the waves to propagate at limestone velocity, and the data would cluster around a higher value. The apparent velocity of a head-wave refracted down through the salt to mine-level geophones would also be detected at limestone velocity. So events could only have occurred in the limestones above (or below) the mine level if rock propagation speeds were so low as to be unreasonable. Microseismic events with slow move-out times must have been generated in the salts at, or near, the mining level.

2.4 First Motions

For small events observed near potash mines, coherent seismic phases tended to have the same polarity on all channels: coherent motions were either ALL up or ALL down. This was seen both when geophones were on surface and when arrays were set up underground. In contrast, there were obvious polarity reversals in first-motions for the single large earthquake (mb = +2.4) for which there are data (Gendzwill, 1984).

A double-couple seismic source mechanism causes far-field P- and S-wave radiation patterns with distinct regions of compression and dilatation. As Gibowicz (1990) points out, "so far no systematic differences have been found between mine tremors and natural earthquakes, and most of what has been discovered about the mechanism of earthquakes can be applied to mine tremors". McGarr (1971, 1984) and Wong and McGarr (1990) concur explicitly with this view. Given the strong, (one could almost say overwhelming), evidence that mine-induced seismicity generally involves a double-couple source mechanism, it was disconcerting to find that small microseismic events recorded near potash mines showed NO signs of this type of failure. The observation that coherent seismic phases tended to have the same polarity on all traces is a strong indication that SMALL seismic events in potash mines arise from a non-double-couple fracture mechanism.

2.5 Heterogeneous Failure

Multiple pulses, or apparent "phases", commonly seen on seismograms arrive at the same speed across the array of sensors. Most seismologists analyzing the traces plotted in Figure 1 might presume that the successive pulses, identifiable on all channels, indicate different seismic phases or reflection events in the record. But when each trace is time-shifted by its first-arrival time pick, (to line up all the first-breaks, as shown in Figure 2), it becomes clear that ALL events travelled across the set of geophones at the SAME speed, the salt S-wave velocity. Note also the remarkable phase similarity between traces even to very small features. There is no significant variation in signal character with travel path azimuth; the only alteration to the signal is a decrease of amplitude with distance.

Since they all travel at the same speed, successive pulses seen in these seismograms must be the result of multiple fracture events on the fault, rather than multiple seismic phases (or source to geophone travel paths) from a single fracturing event. Small seismic events in potash mines are a manifestation of a complex, heterogeneous failure process. This is not surprising. While mapping underground fault zones in deep South African gold mines (which generate complex seismicity), Spottiswoode and McGarr (1975) and McGarr et al. (1979) found that shear zones "consisted of a complex series of subparallel shear planes which were sometimes offset in an en-echelon manner, or a network of fractures that bifurcate and join along their lengths." In other words, failure zones are complicated and so, by natural extension, are the ruptures that propagate along them.

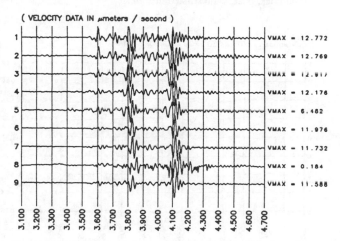

Figure 2. Seismograms shown in Figure 1 time-shifted to align apparent first-breaks.

2.6 Spectral Analysis

Analysis of displacement spectra of seismograms from small events in potash mines was routinely carried out following the method of Brune (1970), as discussed in Bath (1984) and Gibowicz (1990). While calculated values for seismic moment appeared realistic, the computed fault area dimension was consistently much higher than seemed reasonable.

The event shown in Figure 1 radiated 650 J and had a seismic moment of 1260 x 10e+06 Nt-m, for an energy/moment ratio of 5 x 10e-07. The corner frequency of about 43 Hz translates into a Brune fault radius of about 12 to 22 m, or about 33 m if the long-fault model of Haskell (1964) and Savage (1972) is used. Using a representative shear modulus, these values of seismic moment and fault dimension imply displacement on the fault of less than 0.05 mm. The seismic moment seems realistic for a small microseism, but the computed fault dimension seems too large and the computed displacement seems much too small. And would a rupture over tens of metres generate no more than a few hundred J radiated energy? Comparable values were calculated for nearly all recorded events. Similar results were presented by Wong et al. (1989), who found that "harmonic" events associated with a potash solution mine in Utah had calculated source radii of 50 - 100 m for seismic moments in the range 10e+09 - 10e+10 Nt-m: large apparent radii for low-moment events. In potash mines, the scale of observed rock failure does not agree with values calculated using commonly held relationships.

In granite, events of similar radiated energies had seismic moments of about 10e+06 Nt-m and corner frequencies in the 1000 Hz range, which gave calculated fault radii of less than one metre (Gibowicz et al., 1991). In a German coal mine, events with corner frequencies around 10 Hz (giving source radii of over 50 m) radiated over 10000 J energy, and had seismic moments greater than 100,000 x 10e+06 Nt-m

(Gibowicz et al., 1990). For natural earthquakes the ratio of seismic energy to seismic moment is about 10e-05 (Vassiliou and Kanamori, 1992). For potash-related microseisms the ratios of energy to moment, and energy to radius are inconsistent with these examples.

The physical dimension of a failure in the roof or floor of a potash mine room must be controlled by the size of individual openings: at Vanscoy production rooms are up to 11 m wide (and 1200 m long). Small fractures are common near mine openings, with widths of centimetres and displacements as high as many tens of centimetres. Creep displacements, which are considered aseismic, can also be as high as a few tens of centimetres. So at the Vanscoy potash mine a reasonable rupture might involve failure over the width of a room: 11 m, for a radius of about 5.5 m. Entering this fault size into Brune's equation gives expected corner frequencies of 100 - 250 Hz, which is much higher than almost all observations at any potash mine.

The discrepancy between potash mine events and other types of seismicity could be explained if measured frequencies were too low due to insufficient recording bandwidth. Many events were recorded with a high-cut filter of only 50 Hz. But displacement spectra computed for most of these data show corner frequencies below 50 Hz, well within the capability of the system. Furthermore, calibration data recorded using systems with high-cut filters at 100 Hz and 250 Hz showed the same basic results: although some events reached corner frequencies close to 100 Hz, most events were well below 50 Hz. For the great majority of potash microseisms the results shown in Figure 5 are representative.

3 INFERRED FRACTURE MECHANISM

The only model to explain all aspects of the microseismic data recorded at potash mines is the single force directed either up or down. The far field radiation pattern for this type of failure is shown in Figure 3. The biggest constraint on the fracture mechanism of small seismic events in potash mines is the observation that P-waves were absent in the vertical vibration field at the elevation of the microseism: energy seen in the failure plane travelled at a shear wave speed, a result observed time after time. In Figure 3 note the zero amplitude for P-waves, and the maximum amplitude for S-waves in the horizontal plane containing the source, exactly what was seen for nearly all events recorded underground in potash mines.

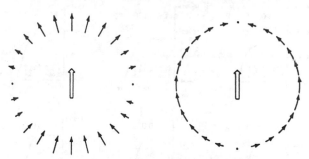

Figure 3. Radiation patterns for far-field P and S waves for a single force directed upwards (from Aki and Richards, 1980).

This mechanism is also consistent with the observation that P-waves WERE seen with sensors installed on surface, well above the failure plane (Prugger and Gendzwill, 1990), with the observation that coherent seismic phases tended to have the same polarity and character on all traces regardless of azimuth, and with the evidence that there is a fundamental difference between small events and large events (which DO involve double-couple shear failure).

The single-force mechanism is rarely invoked to explain seismic observations from naturally occurring earthquakes. The unipolar source has been invoked to explain the highly unusual recording of the seismic signatures of ocean bottom slumping on a flank of Kilauea volcano (Eissler and Kanamori, 1987), and of a turbidity current travelling along the sea-floor at Grand Banks (Hasegawa and Kanamori, 1987). Physically, these two events were submarine "landslides". As well, Kanamori and Given (1982) and Kanamori et al. (1984) argue that the single-force is "the most adequate kinematic representation" of the Mount St. Helens volcanic eruption.

Kanamori and Given (1982) proposed that the single-force fracture mechanism could be modelled by a smooth, half-cosine time function:

$$1/2\ f_o\ [1-\cos(\pi\ t\ /\ \tau)]$$

operating in the interval $0 < t < 2\tau$, and 0 for all other values of t, where "f_o" is the peak force and parameter "τ" is the source-pulse width. This function is plotted for a unit force and various values of τ in Figures 4 (time-domain) and 5 (log-frequency domain). In Figure 5 note the systematic decrease in corner frequency, and increase in the amplitude of the spectral plateau as τ widens. Corner frequency "f_c" is related to pulse-width by:

$$f_c = 1\ /\ 2\tau$$

The pulse-width is proportional to the size of the fracture (radius or length) divided by the rupture propagation speed.

$$D(t) = \left\{ \begin{matrix} 1/2\ [1-\cos(\pi t/\tau)], & 0 < t < 2\tau \\ 0 & ,\ t > 2\tau \end{matrix} \right\}$$

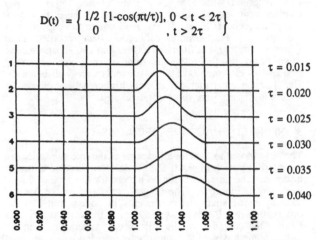

Figure 4. Synthetic displacement seismograms for various pulse-widths, (listed at the right of each trace in seconds).

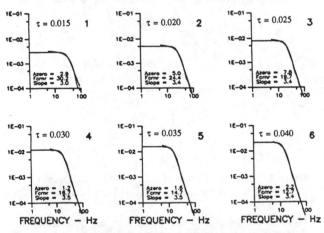

Figure 5. Displacement spectra for synthetic data in Figure 4.

Seismograms for a reasonably complex potash microseism were shown in Figure 1. Data from a simpler record are shown in Figures 6 (original velocity seismograms), 7 (displacement seismograms integrated from velocity data), and 8 (log-displacement spectra). Note the resemblance between the data in Figure 7 and the single-force synthetic shown in Figure 4. Traces remarkably similar to these were recently observed at a South African gold mine (McGarr, 1992, Figures 1 and 4), and interpreted as an indication of "volumetric implosions".

Using standard calculation methods this event radiated 37.2 J and had a seismic moment of 1290 x 10e+06 Nt-m, for an energy to moment ratio of 10e-08. The corner frequency of

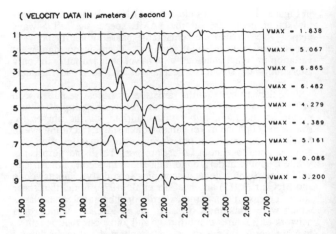

Figure 6. Event seismograms. Time in seconds, true amplitudes, peak values listed at right.

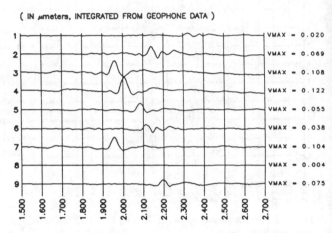

Figure 7. Displacement seismograms computed by integrating the data in Figure 6.

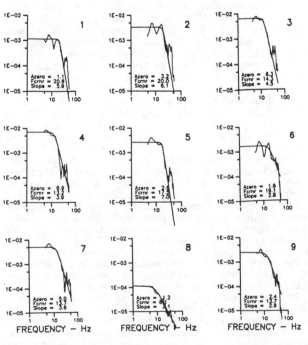

Figure 8. Displacement spectra for data shown in Figures 6 and 7.

about 17 Hz gives a Brune fault radius of 31 to 55 m, (or a Haskell-Savage long-fault length of about 84 m). Using the single-force model, this corner frequency translates into a pulse-width of about 0.030 s. A rupture propagation speed of 90 % the shear wave velocity of salt gives a fault length close to 65 m, which is not realistic. Ruptures this big are hardly ever observed in a potash mine, and a failure this large must surely generate more than 40 J radiated energy. The resulting implication is that the rupture propagation speed must be much slower than the S-wave velocity. But if an 11 m fault-length is assumed, (mining room width), the 0.030 s pulse-width translates into a rupture speed under 400 m/s, about 15 - 20 % the shear-wave speed.

This inference is in good agreement with the findings of Kovach (1974) and Kanamori and Hauksson (1992), who invoke "slow earthquakes" to explain anomalous (subsidence-related) seismic activity near two California oil-fields. Both researchers observed events with "emergent first-arrivals", "a rapid roll-off of high frequencies", "a dominant period much longer than that of ordinary earthquakes with a comparable magnitude", and "smaller energy-to-moment ratios than for ordinary earthquakes". All their descriptions concur with observations of microseismic events near potash mines. Potash microseisms involve slow rupture propagation.

4 SYNTHESIS

At the Vanscoy mine initial "relief" rooms are designed to fail in order to keep subsequent adjacent rooms open (Mackintosh, 1975 and 1977). Failure about potash rooms at Vanscoy is described as follows by Mackintosh (1977): "Opening a room results in immediate stress relief creep of the adjacent rock into the opening. Horizontal stress is transferred to the beds immediately above and below the opening, and because the clay seams act as planes of weakness the beds, unable to withstand the excess loading, buckle or shear. Failure of successive beds continues until a salt bed thick enough to withstand the stress is encountered. Failure stops at this bed which is able to expand plastically at very low rates, and thus provide a stable roof. The beds in the floor are much thinner than the roof beds and continue to heave." Roof-falls and floor-heaves are very well described by the "successive beam deformation" model of Isaacson (1962). Horizontal stress causes rock lamina to slide along the friction-reduced clay partings, eventually deforming, and failing, into the mine opening. Vertical stresses act to accelerate this process. Seismologically, this formation of a horizontal tension crack in the immediate roof or floor, accompanied by movement of rock into the mine room, should appear as an explosion with two opposing force pulses directed up and down (i.e.- the horizontal tension crack discussed earlier). But if the open mine room acts to attenuate one of the pulses (the downgoing pulse for a roof-slab, the upgoing pulse for a floor heave) then this type of failure WOULD appear as

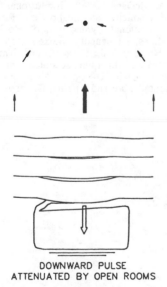

DOWNWARD PULSE
ATTENUATED BY OPEN ROOMS

Figure 9. Single force far-field S wave radiation for proposed roof-slab delamination model.

a single force directed either up or down in the far field. This scenario, depicted in Figure 9, matches both seismological and rock mechanics observation.

Vertical shortening of pillars is common in potash mines as evidenced by rapid room closure rates (Mackintosh, 1975 and 1977; Ong et al., 1983; Herget and Mackintosh, 1987) and by the pattern and amount of surface subsidence. Much of the vertical shortening is taken up by horizontal expansion of the pillar into the room opening along clay seams. Failure along the horizontal clay-seam "shear surfaces" occurs by small-scale plastic flow, not by more violent bursting (Mackintosh, 1977). The most obvious brittle failure seen in pillars is "sloughing" of walls into mine rooms. The mode of rupture is essentially a near-vertical tension crack. No seismic records show any clear evidence of this mode of failure, so it appears that the formation of a wall-slab will not in itself generate enough energy to be seen as a seismic event. It may be, however, that the actual pillar shortening associated with the formation of a wall-slab, is what radiates observed seismic energy. Some events did occur over pillars rather than over mined-out panels. This type of failure would appear as an implosion in potash mine seismic data as depicted in Figure 10.

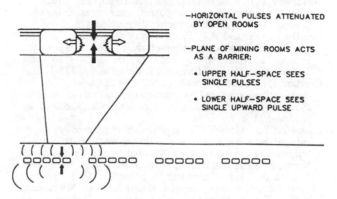

Figure 10. Force pulses due to yield pillar deformation model.

5 CONCLUDING REMARKS

The main conclusion of this study is that seismic signals due to small-scale rock failure about potash mines is best modelled by a single force directed either up or down. The suggested mode of failure is one of roof-slab delamination and/or floor heave in mined-out rooms, and possibly pillar shortening for events that occur over pillars. These types of failure are not usually modelled by a single force; a vertical or horizontal tension crack "explosion", or a pillar shortening "implosion" (both dipoles) seem to be more appropriate models for these types of failure. We suggest that one of the force-pairs is attenuated by the large, open, mined-out area, resulting in the apparent single-force seismic pattern that is so commonly observed. We envision the fracture to form as a horizontal tension crack in the roof or floor of a mine opening. The crack spreads outwards from its point of origin at a speed of about 400 m/s. There is no significant shear displacement in these ruptures.

ACKNOWLEDGEMENT

We would like to thank all of the Canadian potash mining companies who supported our research, both financially and by providing mine access, over the years: Cominco Fertilizers, Denison-Potacan Potash Company, International Mineral and Chemical Company, Potash Company of America, and Potash Corporation of Saskatchewan. Research facilities were provided by the Department of Geological Sciences, University of Saskatchewan. We are especially indebted to Mr. J. Arnold (electronic technician, University of Saskatchewan) and Mr. A. D. Mackintosh (mine engineer, Cominco Fertilizers); we could not have made the observations we did without their assistance. The help of Ms. C. Couch (executive secretary, Potash Corporation of Saskatchewan) in preparing this manuscript is appreciated.

REFERENCES

Aki, K., and P. Richards (1980). Quantitative Seismology, Volumes 1 and 2, W. H. Freeman and Company.

Bath, M. (1984). Rockburst seismology. Rockbursts and Seismicity in Mines, South African Institute of Mining and Metallurgy Symposium Series 6, pp. 7-15.

Brune, J.N. (1970). Tectonic stress and the spectra of seismic shear waves from earthquakes. JGR, 75, pp. 4997-5009; correction: JGR, 76, p. 5002.

Eissler, H.K. and H. Kanamori (1987). A single-force model for the 1975 Kalapana, Hawaii earthquake. JGR, 92, pp. 4827-4836.

Fuzesy, A. (1982). Potash in Saskatchewan. Sask. Energy and Mines Report #181.

Gendzwill, D.J. (1984). Induced seismicity in Saskatchewan potash mines. Rockbursts and Seismicity in Mines, South African Institute of Mining and Metallurgy Symposium Series 6, 131-146.

Gendzwill, D.J. (1988). Seismic patterns in southern Saskatchewar. Abstracts of the Fifteenth Annual Meeting of the Can. Geoph. Union at Saskatoon, Canada, p. 12.

Gendzwill, D.J., R.B. Horner, and H.S. Hasegawa (1982). Induced earthquakes at a potash mine near Saskatoon, Canada. Can. J. Earth Sciences, 19, pp. 466-475.

Gendzwill, D.J. and A.F. Prugger (1990). Seismic activity in a flooded Saskatchewan potash mine. Proceedings of the Second International Symposium of Rockbursts and Seismicity in Mines, Minneapolis, Balkema, pp. 115-120.

Gendzwill, D.J. and D. Stead (1992). Rock mass characterization around Saskatchewan potash mine openings using geophysical techniques: a review. Can. Geotech. J., pp. 666-674.

Gibowicz, S.J. (1990). The mechanism of seismic events induced by mining: a review. Proceedings of the Second International Symposium of Rockbursts and Seismicity in Mines, Minneapolis, Balkema, pp. 3-28.

Gibowicz, S.J., H.P. Harjes and M. Schaefer (1990). Source parameters of seismic events at Heinrich Robert Mine, Ruhr Basin, Federal Republic of Germany: Evidence for nondouble-couple events. BSSA, 80, pp. 85-109.

Gibowicz, S.J., R.P. Young, S. Talebi and D. J. Rawlence (1991). Source parameters of seismic events at the Underground Research Laboratory in Manitoba, Canada: Scaling relations for events with a moment magnitude smaller than -2. BSSA, 81, pp. 1157-1182.

Gutenberg, B. and C.F. Richter (1941). Seismicity of the Earth. GSA Special Paper, 34, pp. 1-134.

Haskell, N.A. (1964). Total energy and energy spectral density of elastic wave radiation from propagating faults. BSSA, 54, pp. 1811-1841.

Hasegawa, H.S. and H. Kanamori (1987). Source mechanism of the magnitude 7.2 Grand Banks earthquake of November 1929: double-couple or submarine landslide. BSSA, 77, pp. 1984-2004.

Hasegawa, H.S., R.J. Wetmiller, and D.J. Gendzwill (1989). Induced seismicity in mines in Canada - an overview. Pure and Applied Geophysics, 129, pp. 423-453.

Herget, G. and A.D. Mackintosh (1987). Mining induced stresses in Saskatchewan potash. Proceedings of the 6th International Congress on Rock Mechanics, Rotterdam, pp. 953-957.

Horner, R.B. (1983). Earthquakes in Saskatchewan: a potential hazard to the potash industry. Proceedings of the first International Potash Technology Conference, Potash '83, pp. 185-191.

Isaacson, E. de St. Q. (1962). Rock Pressure in Mines (2nd Edition). Mining Publications Ltd., London, 260 p.

Jones, P.R., and F.F. Prugger (1982). Underground mining in Saskatchewan potash. Mining Engineering, 34, pp. 1677-1683.

Kanamori, H. and J.W. Given (1982). Analysis of long-period seismic waves excited by the May 18, 1980, eruption of Mount St. Helens - a terrestrial monopole? JGR, 87, pp. 5422-5432.

Kanamori, H., J.W. Given and T. Lay (1984). Analysis of seismic body waves excited by the Mount St. Helens eruption of May 18, 1980. JGR, 89, pp. 1856-1866.

Kanamori, H. and E. Hauksson (1992). A slow earthquake in the Santa Maria Basin, California. BSSA, 82, pp. 2087-2096.

Kijko, A., B. Drzezla and T. Stankiewicz (1987). Bimodal character of the distribution of extreme seismic events in Polish mines. Acta Geophysica Polonica, 35, pp. 491-506.

Kovach, R.L. (1974). Source mechanisms for Wilmington oil field, California, subsidence earthquakes. BSSA, 64, pp. 699-711.

Lee, W.H.K. and S.W. Stewart (1981). Principles and Applications of Microearthquake Networks. Academic Press, New York, 293p.

Mackintosh, A.D. (1975). Applied rock mechanics: the development of safe travelways at the Cominco potash mine. Proceedings of the 10th Canadian Rock Mechanics Symposium, Kingston, Canada, pp. 69-97.

Mackintosh, A.D. (1977). Strata control in a deep Sask. potash mine. Proceedings of the 6th International Rock Mechanics Conference, Banff, Canada, 15 p.

McGarr, A. (1971). Violent deformation of rock near deep-level, tabular excavations - seismic events. BSSA, 61, pp. 1453-1466.

McGarr, A. (1984). Some applications of seismic source mechanism studies to assessing underground hazard. Rockbursts and Seismicity in Mines, South African Institute of Mining and Metallurgy Symposium Series 6, pp. 199-208.

McGarr, A., S.M. Spottiswoode, and N. C. Gay (1979). Observations relevant to seismic driving stress, stress drop, and efficiency. JGR, 84, pp. 2251-2261.

McGarr, A. (1992). An implosive component in the seismic moment tensor of a mining-induced tremor. Geoph. Res. Lett., 19, pp. 1579-1582.

Ong V., P. Mottahed and J. Jones (1983). Measurements of in-situ rock deformation over a mile distance. Proceedings of the First International Potash Technology Conference, Potash '83, pp. 283-290.

Prugger, A.F. and D.J. Gendzwill (1988). Micro-earthquake location: a non-linear approach that makes use of a simplex stepping procedure. BSSA, 78, pp. 799-815.

Prugger, A.F. and D.J. Gendzwill (1990). Results of microseismic monitoring at the Cory mine, 1981-1984. Proceedings of the Second International Symposium of Rockbursts and Seismicity in Mines, Minneapolis, Balkema, pp. 215-220.

Prugger, A.F. and D.J. Gendzwill (1991). Ten years of seismic studies in Saskatchewan potash mines. Presented at the Second International Potash Technology Conference, KALI '91, Hamburg.

Savage, J.C. (1972). Relation of corner frequency to fault dimension. JGR, 77, pp. 3788-3795.

Schlumberger Well Services (1986). Log Interpretation Charts. Houston, 122p.

Spottiswoode, S.M. and A. McGarr (1975). Source parameters of tremors in a deep-level gold mine. BSSA, 65, pp. 93-112.

Vassiliou, M.S. and H. Kanamori (1982). The energy release in earthquakes. BSSA, 72, pp. 371-387.

Wong, I.G. and A. McGarr (1990). Implosional failure in mining-induced seismicity: a critical review. Proceedings of the Second International Symposium of Rockbursts and Seismicity in Mines, Minneapolis, Balkema, pp. 45-52.

Wong, I.G., J.R. Humphrey, J. A. Adams and W. J. Silva (1989). Observations of mine seismicity in the eastern Wasatch Plateau, Utah, U.S.A.: A possible case for implosional failure. Pure and Applied Geophysics, 129, pp. 369-405.

244

Rockbursts and Seismicity in Mines, Young (ed.) © 1993 Balkema, Rotterdam, ISBN 90 5410 320 5

Host structures for slip-induced seismicity at the Lucky Friday Mine

D. F. Scott & T. J. Williams
Spokane Research Center, US Bureau of Mines, Wash., USA

B. C. White
Coeur d'Alene, Idaho, USA

ABSTRACT: The U.S. Bureau of Mines has identified three types of geologic structures that contribute to seismicity at the Lucky Friday Mine, Mullan, ID. They include: (1) steeply dipping, west-northwest-striking faults, (2) argillite interbeds that wrap about a steeply plunging fold axis, and (3) oriented local joints. In some parts of the mine, these features combine to define large blocks. Considerable evidence of slip on these structures has been collected since early work relating geologic features to seismic events (Scott, 1990). This evidence includes: (1) first-motion analyses, (2) detailed geologic observation of slip surfaces, gouge, and offsets, and (3) progressive fracture development in shotcrete. Digital seismic records and damage surveys of major rock bursts collected since 1989 suggest that most bursts originate with slip along one of these three types of geologic structures. Their identification is a key step in defining the complex response of the rock mass to mining of the Lucky Friday vein. Recognition of actively slipping structures in the mine is expected to provide a foundation for models that may identify the mechanisms.

1 INTRODUCTION

The Lucky Friday Mine, Mullan, ID, U.S.A. (fig. 1), is developed primarily along a single tabular vein (the Lucky Friday). This nearly vertical vein ranges from several centimeters to about 4.3 m thick and contains massive galena with secondary sphalerite, tetrahedrite, and quartz-siderite gangue. The vein is sigmoidal in plan view (fig. 2) and is bounded on the north by the North Control Fault and on the south by the South Control Fault. Strike length is about 490 m, and dip is from 70° to 90° to the south, with a southwest rake. Because the vein dips more steeply than the bedding, it contacts increasingly older rocks with depth and intersects the Upper Revett Formation about 550 m below the surface.

Wall rocks are mainly vitreous and sericitic quartzites and argillites of the Precambrian Revett Formation (fig. 3). Vitreous quartzite occurs in flat, laminated to cross-bedded, 50- to 90-cm-thick beds that contain numerous quartz veins. Fractures in the quartzite lie both parallel and at various angles to the bedding. Color is white, light brown, or pale green, with a hardness of >6 on the Moh's scale of hardness. Sericitic quartzite, next in abundance, is softer than vitreous quartzite (about 4 to 4.5 on the Moh's scale) and also occurs in flat, laminated, 30- to 90-cm-thick beds. Color is pale green to light brown. Sericitic quartzite is commonly associated with thinner beds of argillite, which is pale green to light brown, ranges from about 1 to 4 on the Moh's scale, and typically forms 2- to 5-cm-thick interbeds. The argillite beds contain abundant, localized shears, which commonly display gouge and clay-like layers.

Attitudes of fractures and argillite beds differ east and west of the axis of an anticline called the Hook anticline. West of the axis, argillite beds strike about N 80° W, dip from 60° to 75° south or southwest, and are nearly parallel to the vein. However, east of the axis, argillite beds strike about N 4° E and dip from 60° to 75° to the east or northeast. Main fractures (including joints and faults) on both sides of the axis are nearly parallel to the northwest-striking (about N 60° W) North and South Control faults. However, on the east side of the anticline axis, fractures are also perpendicular, forming large blocks of rock that are bounded on all sides by fractures.

2 FAULTS

Various researchers, including Lenhardt (1988); Gay and van der Heever (1982); Gay et al. (1984); Johnston and Einstein (1990); Potgieter and Roering (1984); and Shepherd et al. (1984), have concluded that rock bursts and seismic activity are induced by slip along faults.

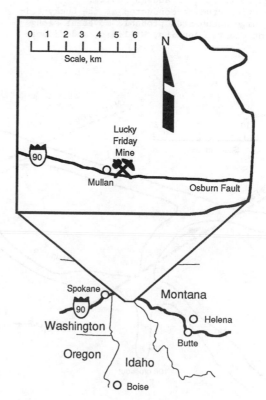

Figure 1. Location of Lucky Friday Mine.

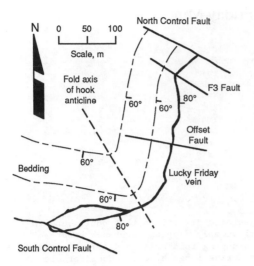

Figure 2. Generalized geology of the Lucky Friday Mine (4900 to 5100 levels).

Faults in the Lucky Friday Mine are silicified; that is, fine-grained quartz has been deposited in the fault zones, resulting in a 0.9- to 1.2 m-thick filling of hard, brittle breccia. Movemet of blocks of rock along these faults should be easy to identify; however, only one case of movement along a fault can be documented. Bureau personnel examined a rock burst site in the Lucky Friday Mine on September 30, 1987. This 2.7-magnitude burst closed two stopes, and upon visual examination, it was found that a block of rock bounded by the North Control and F3 faults exhibited strike-slip movement along the faults (fig. 4), as well as movement upward along argillite beds. In this case, the argillite beds dipped into an opening; the lower friction of the argillite beds was sufficient to initiate failure, resulting in coupled movement along the North Control and F3 faults.

First-motion studies conducted at the Lucky Friday Mine have documented probable movement along the Offset Fault. Jenkins, Williams, and Wideman

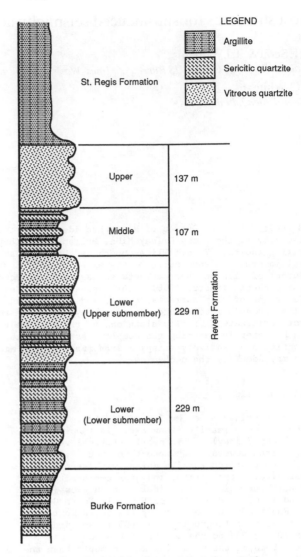

Figure 3. Generalized stratigraphy of the Lucky Friday Mine.

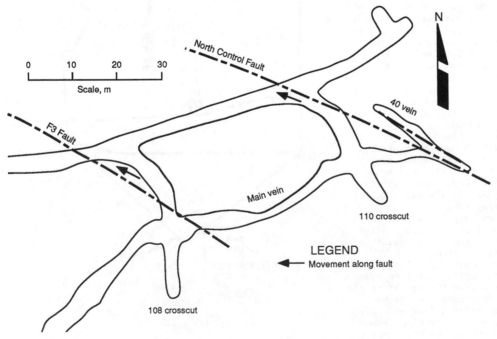

Figure 4. North Control and F3 faults (4660 level).

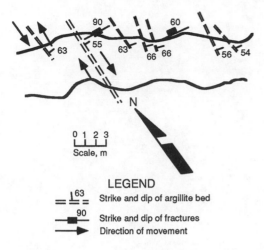

Figure 5. Direction of movement initiated
by rock burst on 5150 sublevel.

(1990) used first-motion studies to calculate
source planes for four rock bursts at the Lucky
Friday Mine. Of the four bursts studied, one burst
had a fault plane solution that almost coincided
with the Offset Fault.

3 ARGILLITE BEDS

Direct evidence indicating that slip has occurred
along argillite beds is difficult to document.
Fallen rock from rock bursts frequently destroys
the point of origin of the event, and mine management often abandons these areas after rock bursts
occur. However, one area where movement along an
argillite bed was verified was in the 5150 sublevel
ramp. No evidence of movement along argillite beds
was apparent in the area prior to a rock burst
registering a magnitude of 2. However, movement
along these beds was evident when the area was re-
examined and mapped. In this case, strike-slip
movement along a 0.3-m-thick argillite bed offset

the rib in the 5150 sublevel ramp (fig. 5) about
2 m. Dip slip during this event was as much as
1 m.

The most obvious indicator of slip between argillite interbeds is seen in cracks formed in
shotcrete applied to the ribs and back of development ramps. These cracks are most evident in
narrow panels of wall rock left exposed near the
floors of the ramps. The argillite beds in these
panels can be traced through to the cracks in the
shotcrete. These cracks begin forming immediately after the shotcrete had hardened. Periodic
application of patching plaster to individual
cracks demonstrates continuous movement as mining
advances. Movement along the argillite beds appears to be mainly dip slip, regardless of the
strike. Most cracks display dip slip toward the
vein (normal faulting). Cracks in the shotcrete
that are parallel to bedding are dilational (dilation in the direction of mined-out vein).

4 SYSTEMS OF LOCAL JOINTS

Detailed geologic mapping of a part of the 5210
ramp has verified slip along preferred joints that
contact argillite beds. At this site, movement of
wall rock was documented by observing fresh rock
fragments along a contact between a joint and an
argillite bed. The strike of the joint was nearly
due east with a dip of 86° (fig. 6), and the strike
of the argillite bed was N 31° W with a dip of 74°.
The joint did not penetrate the argillite, but
sharp, fresh, fine rock fragments (less that 2 mm
in diameter) were found along the contact between
the joint and the argillite bed, as well as along
the joint surface. This type of juncture, between
argillite beds and joints, appears to be a weak
area in the rock mass, providing a third geologic
structure along which slip occurs.

5 CONCLUSIONS

Evidence for seismically induced slip has been
documented along three geologic structures:
(1) steeply dipping, west-northwest-striking

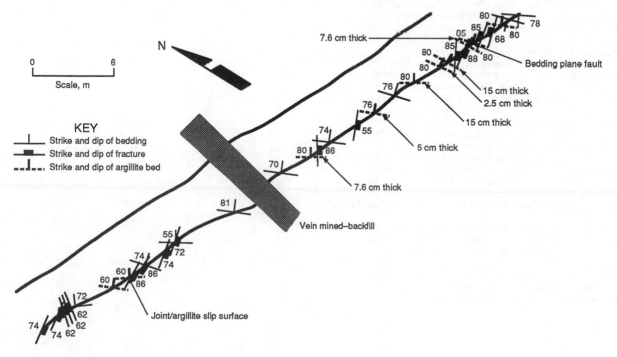

Figure 6. Joint/argillite slip surface in 5210 ramp.

faults, (2) argillite interbeds that wrap about
a steeply plunging fold axis, and (3) oriented
local joints. Mining-induced, as well as tectonic,
stresses in the mine release energy along these
structures. Argillite interbeds and local joints
prove to be weak surfaces within an otherwise hard
and brittle rock mass; friction along these surfaces is much less than in the quartzites, allowing
slip to occur. Slip along the silicified faults is
documented on the basis of fault plane solutions.
Some argillite beds are parallel to these faults,
so it is also possible that argillite beds allow
initial slippage, resulting in coupled failure
between faults.

REFERENCES

Gay, N.C., D. Spencer, J.J. Van Wyk, and P.K. van
    der Heever. 1984. The Control of Geological and
    Mining Parameters in the Klerksdorp Gold Mining
    District. Paper in Rockbursts and Seismicity in
    Mines. S. Afri. Inst. of Min. and Metall., Symp.
    Series No. 6, Johannesburg, S. Afri., pp. 107-
    120.
Gay, N.C., and P.K. van der Heever. 1982. The
    Influence of Geological Structure on Seismicity
    and Rockbursts in the Klerksdorp Gold Mining
    District, South Africa--A Correlation Between
    Geological Structure and Seismicity. Paper in
    Issues in Rock Mechanics, Proceedings, Twenty-
    Third Symposium on Rock Mechanics, ed. by R.R.
    Goodman and F.E. Heuze (Univ. CA, Berkeley, CA,
    Aug. 25-27, 1982). Soc. Min. Eng., pp. 176-182.
Jenkins, F.M., T.J. Williams, and C.J. Wideman.
    1990. Analysis of Four Rockbursts in the Lucky
    Friday Mine, Mullan, Idaho, USA. Paper in the
    International Deep Mining Conference: Technical
    Challenges in Deep Level Mining. S. Afri. Inst.
    of Min. and Metall., Johannesburg, South Africa,
    pp. 1201-1212.
Johnston, J.C., and H.H. Einstein. 1990. A Survey
    of Mining Associated Seismicity. Ch. in Proceed-
    ings of the 2nd International Symposium on Rock-
    bursts and Seismicity in Mines: Rockbursts and
    Seismicity in Mines, ed. by C. Fairhurst (Univ.
    MN, Minneapolis, MN, 1988). A. A. Balkema, pp.
    121-127.
Lenhardt, W.A. 1988. Some Observations Regarding
    the Influence of Geology on Mining-Induced Seis-
    micity at Western Deep Levels, Ltd. Ch. in Rock
    Mechanics in Africa. Int. Soc. for Rock Mech.,
    S. Afri. Natl. Group, pp. 45-48.
Potgieter, G.J., and C. Roering. 1984. The In-
    fluence of Geology on the Mechanisms of Mining-
    Associated Seismicity in the Klerksdorp Gold-
    field. Paper in Rockbursts and Seismicity in
    Mines. S. Afri. Inst. of Min. and Metall., Symp.
    Series No. 6, Johannesburg, S. Afri., pp. 45-50.
Scott, D.F. 1990. Relationship of Geologic Fea-
    tures to Seismic Events, Lucky Friday Mine,
    Mullan, Idaho. Paper in Proceedings of the 2nd
    International Symposium on Rockbursts and Seis-
    micity in Mines: Rockbursts and Seismicity in
    Mines, ed. by C. Fairhurst (University of MN,
    Minneapolis, MN, June 8-10, 1988). A.A. Balkema,
    pp. 401-405.
Shepherd, J., R.L. Blackwood, and L.K. Rixon.
    1984. Instantaneous Outbursts of Coal and Gas
    with Reference to Geological Structures and
    Lateral Stresses in Collieries. Paper in Rock-
    bursts and Seismicity in Mines. S. Afri. Inst.
    of Min. and Metall., Symp. Series No. 6,
    Johannesburg, S. Afri., pp. 97-106.

Rockbursts and Seismicity in Mines, Young (ed.) © 1993 Balkema, Rotterdam, ISBN 90 5410 320 5

Patterns of rockburst seismicity and their relation to heterogeneity of rock materials

C.A.Tang, X.H.Xu & S.Z.Song
Northeast University of Technology, People's Republic of China

ABSTRACT: The patterns of seismicity of rock materials have been studied analytically based on a simple mechanics model for model experiments of rock failure. The results reveal four types of typical seismic events (or acoustic emissions) as the following: 1. main shock; 2. main shock-aftershock; 3. foreshock-main shock-aftershock; 4. swarm, and these led to exactly the same conclusion as obtained from model experiments by Mogi (1985) that they are the results of differences in the structural homogeneity of the rock (medium). According to the theoretical results given in this paper, the type 1 occurs in homogeneous cases, type 2 in Kaiser Effect cases, type 3 in moderately homogeneous cases, and type 4 in heterogeneous or extremely heterogeneous case. The theoretical results agree very well with Mogi's fracture experiments on various types of brittle materials with differing degrees of heterogeneity.

1 INTRODUCTION

Since the occurrence of a rockburst is considered to be a result of unstable failure of rock or rockmass, experimental research on fractures in brittle rock material can provide a powerful indicator in elucidating the mechanism of rock burst. From this viewpoint, experimental research on the fracture characteristics of rocks and similar materials under various conditions is being carried out in laboratories world wide. The most interesting aspect of the laboratory studies is related to seismic activities (or acoustic emissions). For example, during fracture experiments on various types of brittle materials, three types of acoustic emission sequences have been found by Mogi (1985), which led to the conclusion that these types are the result of differences in the structural heterogeneity of the medium.

It is known that the acoustic emissions (AE, hereafter) are transient elastic waves generated by the rapid release of energy within a material which deforms under stress. The AE technique as a tool has been widely used in both laboratory and field investigations of the behaviour of geologic materials and in monitoring the rock (coal) burst in mines. The goal of these investigations and the monitoring has been to (Wilson, 1981):

(1) detect and delineate high stress or potential rock (coal) burst zones;

(2) evaluate control measures to prevent or minimize the occurrence of rock (coal) bursting;

(3) predict the occurrence of an imminent rock (coal) burst so that workers can be evacuated.

Therefore, the study of AE behaviour of geological material is important to mine industry. The existing literature on the generation of AE in geologic material is quite abundant. The proceedings of the three Pennylvania State University conferences on the subject (Hardy and Leighton, 1984) are an indication of the growing interest in the application of AE techniques in evaluation of the behaviour of geologic materials.

However, the theoretical study regarding the AE response (the law of AE) of geologic materials, especially the seismic activities or acoustic emission sequences, is noticeably lacking in the literature. No convincing quantitative explanations have been given about the observed results in this field. Why do acoustic emissions show different patterns for diferent rock types? Furthermore, why do foreshocks sometimes occur prior to an shock and sometimes not? Why do swarms sometimes only occur without a main shock? How can we explain these from the viewpoint of rock mechanics?

With this background, in this paper, the seismic activity (acoustic emission) of a rock specimen is studied based on a simple mechanics model of the testing machine-specimen system. The instability behaviour of this system is approached firstly by determining the deformation rate of the rock specimen. Choosing the variable deformation rate in the theoretical approach is not only because it is observable in practice, but also because it is the independent variable in the laboratory test.

The more important merit of studying deformation rate is that the patterns of acoustic emissions are strongly related to it. By considering the fact that the acoustic emissions are transient elastic waves due to the local damage of rock, the relation between the counts of acoustic emissions and the statistical distribution of the local strength of rock can be established. Then, the seismic events as a function of time can be derived with the help of the deformation rate of rock under certain loading conditions.

2 THE AE COUNTS N OF ROCK

The macroscopic damage phenomenon can be regarded as the synthetic manifestation of many microscopic damages. The so-called acoustic emission is in reality the elastic wave generated by such microscopic damage radiating through the surrounding material. Therefore, the AE activity represents the micro-damages, which is directly associated with the evolution and propagation of the defects within the rock.

We can divide the cross section of a rock specimen into many micro-elements as shown in Fig. 1. For any chosen micro-element V from this section for our discussion, we assume it is big enough to hold many defects (or micro-defects), and, moreover, small enough to be considered as one particle in the continuous media.

Owing to that respective micro-elements hold unequal number of defects, they possess different strength. For simplicity, we have the following two hypothesis:

(1) The micro-elements of rock specimen are Hooke' bodies (the macroscopic integrity of rock may not be a Hooke's body), that is:

$$\sigma_i = e_i \cdot \varepsilon_i = e_i \cdot \varepsilon \quad (i = 1, 2, \cdots) \tag{1}$$

where e_i is the Young's modulus of the ith micro-element.

(2) The strength of micro-elements follows certain distribution density described as the following

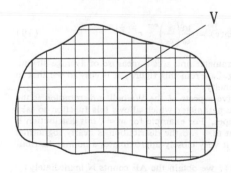

Figure 1. A sketch showing micro-elements in cross-section of rock specimen

$$\varphi = \varphi(\varepsilon_s) \qquad (2)$$

where ε_s is the strength of the individual micro-element (local strength).

So, as the strain of the micro-element increase, this micro-element will fail locally when its strain exceeds the local strength, i. e. when

$$\varepsilon > \varepsilon_s \qquad (3)$$

Suppose the AE rate in one unit area of the damaged micro-element is n, then the damage of the area ΔS will give AE count ΔN

$$\Delta N = n \cdot \Delta S \qquad (4)$$

If the overall cross section has an area S_m, and the AE count when S_m is completely damaged is N_m, equation (4) can be rewritten as

$$\Delta N = \frac{N_m}{S_m} \cdot \Delta S \qquad (5)$$

When the specimen gets a strain increment $\Delta \varepsilon$, the sectional increment of the resultant damage ΔS will be

$$\Delta S = S_m \cdot \varphi(\varepsilon) \cdot \Delta \varepsilon \qquad (6)$$

Inserting equation (6) into equation (5), we have

$$\Delta N = N_m \cdot \varphi(\varepsilon) \cdot \Delta \varepsilon \qquad (7)$$

Therefore, the AE count when the strain in the compressed specimen increases to ε will be

$$N = N_m \int_0^\varepsilon \varphi(x)dx \qquad (8)$$

or

$$\frac{N}{N_m} = \int_0^\varepsilon \varphi(x)dx \qquad (9)$$

This is the AE counts N vs. strain ε when the specimen is loaded. So here $\varphi(\varepsilon)$ is also called acoustic emission rate.

From equation (9) we know that if the probability distribution density $\varphi(\varepsilon)$ is known, then the AE count $N(\varepsilon)$ can be obtained directly. There are many probability distribution density functions can be used, which one is the best depends on the properties of rock. We can choose the appropriate one by comparing it with the experimental results. Hudson and Fairhurst (1969) used Normal Distribution in their simple structure-disintegration model to express the local elemental strength of rock. He et al (1990) used Gaussian distribution for nearly the same purpose. However, the most popular used distribution is Weibull's law (Krajcinovic, 1982; Tang, 1990; etc.). Here we also suppose the micro-element strength of rock follows Weibull's distribution law because it is reasonable to do that. For example, Zhang and Valliappan (1990) found that Weibull's Distribution describes very well the experimental data for distribution of crack length within a specimen of rock mass which is correlated to elemental strength. Other reasons can also be found in Weibull's Theory of the strength of materials (Hudson, 1969).

The Weibull's distribution density can be described as the following:

$$\varphi(\varepsilon) = \frac{m}{\varepsilon_0}\left(\frac{\varepsilon}{\varepsilon_0}\right)^{m-1} e^{-\left(\frac{\varepsilon}{\varepsilon_0}\right)^m} \qquad (10)$$

where m is shape parameter and ε_0 is a measure of average strain. $\varphi(\varepsilon)$ is shown in Fig. 2. From this figure, it is easy to see that m is a index of homogeneity.

One of the attractive aspects of the Weibull's distribution is the presence of the shape parameter which allows this function to take a wide variety of shapes. For example for $m=1$ this distribution is exponential, at about $m=1.5$ the distribution is nearly lognormal and at about $m=4$, it very closely approximates a normal distribution.

From equation (9), we obtain the AE counts N immediately:

$$\frac{N(\varepsilon)}{N_m} = \frac{m}{\varepsilon_0}\int_0^\varepsilon x^{m-1}e^{-\left(\frac{x}{\varepsilon_0}\right)^m}dx = 1 - e^{-\left(\frac{\varepsilon}{\varepsilon_0}\right)^m} \qquad (11)$$

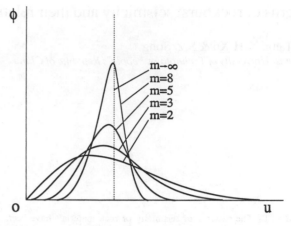

Figure 2.   Distribution of AE counts

3   SIMPLE MECHANICS MODEL FOR MODEL EXPERIMENTS OF ROCK FAILURE

In order to study the patterns of rockburst seismicity, let us consider the following simple mechanism model for a machine-specimen system (as shown in Fig. 3), in which the testing machine and the rock specimen are substituted by an equivalent spring-specimen system loaded in series.

During testing, both the specimen and the testing machine are subjected to forces and, consequently, they both deform. In the process of loading, both ends of the spring move downwards. If the displacement of the upper end (point O_1) is a and that of the lower end (O_2) is u (the deformation of rock specimen), then the force acting on the spring P is given by

$$P = k(a - u) \qquad (12)$$

where k is the spring constant (stiffness of testing machine). This expression describes the loading line of the spring, that is, of the loading machine.

If the load-deformation relation of the rock specimen is described by

$$R = f(u) \qquad (13)$$

then, if the process of loading is viewed quasi-statically, for the system to be in equilibrium, we have $P=R$, that is:

$$k(a - u) = f(u) \qquad (14)$$

or

$$k(u - a) + f(u) = 0 \qquad (15)$$

Since a is the displacement of the upper end (point O_1) related to the loading mode of the testing machine, it must be a function of time, that is:

$$a = a(t) \qquad (16)$$

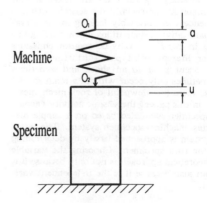

Figure 3.   Simple mechanism model for machine-specimen system

Differentiating equation (15) with respect to t

$$k(\dot{u} - \dot{a}) + f'(u) \cdot \dot{u} = 0 \qquad (17)$$

where $f'(u)$ is the slope of the load-deformation curve of the specimen. This equation can be solved for the deformation rate of the rock specimen

$$\dot{u} = \frac{k \cdot \dot{a}}{k + f'(u)} \qquad (18)$$

Because the upper platen is driven by the flow of the hydraulic liquid or by the motor, the displacement of the upper end (point O_1) will keep increasing constantly if no servo-control or quick manual control is used. So it is reasonable to assume that $\dot{a} = C$ (constant displacement boundary condition). Then equation (18) become

$$\dot{u} = \frac{k \cdot C}{k + f'(u)} \qquad (19)$$

Now we can easily study the behaviour of the specimen-machine system by using equation (19). For example, if $k + f'(u) \to 0$, we immediately obtain the result that $\dot{u} \to \infty$ from equation (19), i. e., the deformation rate of the specimen has a sudden increase (i. e. unstable, or so-called ' soft loading'). And $\dot{u} \approx$ const when $k \gg f'(u)$, this is the so-called ' stiff loading' .

4 PATTERNS OF SEISMICITY AND THEIR RELATION TO HETEROGENEITY OF ROCK

As we know, earthquakes seldom occur as isolated events, but are part of a sequence with variably well-defined characteristics. Foreshock and aftershock sequences are closely associated with a larger event called the mainshock, whereas sequences of earthquakes not associated with a dominant earthquake are called swarms. Mogi in 1963 performed fracutre experiments on various types of brittle materials with differing degrees of heterogeneity, as a model experiment on the occurrence of earthquakes (Mogi, 1985). This produced three types of artifical earthquake (elastic shock or acoustic emissions) sequences: (1) main shock-aftershock types, (2) foreshock-main shock-aftershock type, and (3)swarm type, as shown in Fig. 4. This led to the conclusion that these types are the result of differences in the structural heterogeneity of the medium. That is, type (1) occurs in homogeneous cases, type (2) in moderately heterogeneous cases, and type (3) in extremely heterogeneous cases. When the types of natural earthquake sequences in Japan were examined based on these experimental results, it was discovered that they can indeed be classified into these three types (Mogi, 1985). The same results were also given by Scholz (1990) when he analyzed the earthquake sequences.

Since both rockburst and earthquake are manifestations of the same phenomenon, a very rapid (or ' seismic') release of strain energy that has accumulated in earth' s crust, rockbursts must have the similar patterns of seismic activity to those of earthquakes although generally smaller in magnitude than those of earthquakes.

By a further discussion about equation (19) of deformation rate and equation (10) of the AE counts, the similar theoretical results can be worked out here.

First we should rewrite the equation (10) as the function of t as the following

$$\varphi_t(\varepsilon) = \frac{m}{\varepsilon_0}\left(\frac{\varepsilon(t)}{\varepsilon_0}\right)^{m-1} e^{-\left(\frac{\varepsilon(t)}{\varepsilon_0}\right)^m} \qquad (20)$$

or, considering that $u \propto \varepsilon$

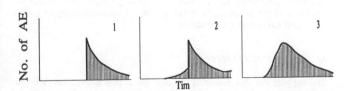

Figure 4. Three ways in which acoustic emissions were found to occur in model experiments on earthquake occurrence (Mogi, 1985)

$$\varphi_t(u) = \frac{m}{u_0}\left(\frac{u(t)}{u_0}\right)^{m-1} e^{-\left(\frac{u(t)}{u_0}\right)^m} \qquad (21)$$

Because equations (10) and (21) is the strength distribution of acoustic emissions, it should be changed into time distribution. This can be done by the following

$$\begin{aligned}\varphi_t(t) &= \frac{1}{N_m}\frac{dN}{dt} \\ &= \frac{1}{N_m} \cdot \frac{dN}{du} \cdot \frac{du}{dt} \\ &= \varphi_t(u) \cdot \frac{k \cdot c}{k + f'(u)}\end{aligned} \qquad (22)$$

where the φ_t expresses the time distribution of AE counts.

To obtain the function $u(t)$, the finite difference method can be used here. In this way, the equation (19) can be rewritten as the following

$$\frac{u_i - u_{i-1}}{t_i - t_{i-1}} = \frac{k \cdot C}{k + f'(u_{i-1})} \qquad (23)$$

Then we obtain u as a function of t as the following

$$u(t) = u_i = \frac{k \cdot C}{k + f'(u_{i-1})}(t_i - t_{i-1}) + u_{i-1} \qquad (24)$$

Suppose the load-deformation (or stress-strain) relation of the

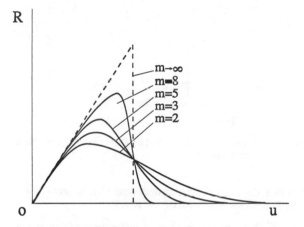

Figure 5. Load-deformation relation of rock (m is homogeneous index)

rock specimen is in forms as shown in Fig. 5, then the curves of AE counts φ vs. time t can be obtained by equations (22) and (24).

For the general boundary condition ($\dot{a} = C$), the curves of AE counts n vs. time t are shown in Fig. 6 (with marks A, B, and C) have been calculated for three values of m (rocks with different homogeneity, as shown in Fig. 2).

A comparison between Fig. 4 and Fig. 6 shows a very good agreement between the theoretical results and the experimental (and observed) results except for the difference between type (1) shown in Fig. 4 and type A shown in Fig. 6, which are considered to be the homogeneous type. The main difference is in that there are almost no aftershocks in the theoretical result. The reason is that because most of the local elements of the rock specimen almost have the same strength due to its homogeneity, the microfracturing will occur almost at the same time, which, obviously will produce a main shock with few aftershocks (shown in Fig. 6A).

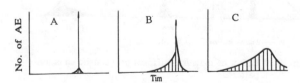

Figure 6. Theoretical results of three types of acoustic emission patterns

Then, if type (1) in Fig. 4 (Mogi, 1985) is not the homogeneous type as shown in Fig. 6A, how can we explain it in terms of homogeneity? According to the idea in this paper, it must be relative to the heterogenous type or the moderately homogeneous type due to its obvious aftershocks. But why does this type have no foreshocks just as those for type B and type C shown in Fig. 6B and C? We reasonably suggest that this special type can be explained by considering the Kaiser Effect.

As we know, if we load a heterogeneous or moderately homogeneous specimen that has previously come under a certain level of stress, according to Kaiser Effect principle, it will keep quiet (i. e. no acoustic emissions) until the level of stress previously applied has been reached. If this stress level previously reached is just the level closing the main shock, another type sequence (main shock-aftershock) occurs (as shown in Fig. 7), which is just the same as type (1) shown in Fig. 4. We call this type the Kaiser Effect type.

So, in summary, there are, at least, four types of acoustic emission (artifical earthquake) sequences, that is

(1) Main shock (homogeneous type, see Fig. 6A);

(2) Main shock-aftershock (Kaiser Effect type, see Fig. 7);

(3) Foreshock-main shock-aftershock (moderately homogeneous type, see Fig. 6B);

(4) Swarm (heterogeneous type, see Fig. 6C),

and these led to the same conclusion as given by Mogi (1985) that they are the results of differences in the structural homogeneity of the rock (medium).

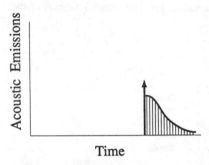

Figure 7.   Theoretical results of Kaiser Effect type of acoustic emission patterns

5   EFFECT OF TESTING MACHINE STIFFNESS ON PATTERNS OF SEISMICITY OF ROCK

In fact, the above results are obtained under the condition of constant stiffness of testing machine. Change the stiffness of testing machine, we can also find its effect on the pattern of seismicity of rock. This is shown clearly in Fig. 8. The same experimental results (Chen et al., 1984) are shown in Fig. 9, which shows a great agreement with the above theoretical results.

Both theoretical and experimental results show that when the stiffness of testing machine is small, the AE rate increases drastically when peak stress is approached, and the spcimen ruptures violently immediately when or soon after peak stress is reached. As the stiffness of testing machine increases, AE rate decreases gradually but develops in a way of swarm. This suggests that microfracturing events may be either very abundant or relatively few, depending on the relative stiffness of the testing machine.

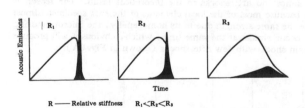

Figure 8.   Effect of machine stiffness on acoustic emission patterns

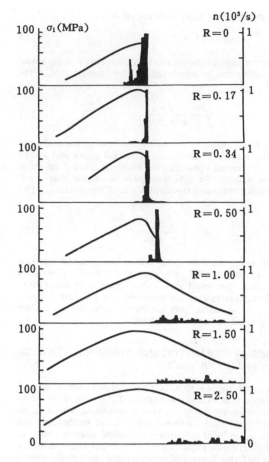

Figure 9.   Change in AE rate in experiments on a marble under different loading conditions (Chen et al. , 1984)

6   CONCLUSIONS

Based on the fact that the acoustic emissions are transient elastic waves due to the local damage of rock, the equations for describing the counts of acoustic emissions (including the Kaiser Effect function) are derived from the statistical distribution of local strength of rock. By using the equation of deformation rate, the patterns of acoustic emissions are studied. It is concluded that there are, at least, four types of acoustic emission sequences depending on whether the rock failure is stable or unstable, that is

(1) Main shock (homogeneous type)

(2) Main shock-aftershock (Kaiser Effect type);

(3) Foreshock-main shock-aftershock (moderately homogeneous type), and

(4) Swarm (heterogeneous type).

From the theoretical results given in this paper, we can see that, just as being pointed out by Mogi (1985), it is not strange for clear precursory phenomena to be observed in some areas and not at all in other areas because they may have different seismic behavious according to their heterogeneity. It is concluded that successful prediction of seismic mainshock depends greatly on the heterogeneity of the area' s structures.

Since the results in this paper give a theoretical explanation agreed very well to Mogi' s fracture experiments on various types of brittle materials with differing degrees of heterogeneity, it is worthy to emphasize that further theoretical and site studies are undoubtedly necessary for anyone concerned with ground control in mining and for geoscientists seeking a better understanding of rockburst and earthquake mechanics.

ACKNOWLEDGEMENT

The support for this research through the National Natural Science Foundation of People' s Republic of China (No. 19002007) is gratefully acknolwdged.

REFERENCES

Chen, Y. , J. S. Hao and W. I. Yan 1984. The effects of machine stiffness on the acoustic properties of marble samples under uniaxial compression. *A Collection of Papers of ISCSEP*: 660-668.

Hardy, H. R. and F. W. Leighton 1984. Acoustic emission/microseismic activity in geologic structures and materials, *Proceedings of the Third Conference*. Clausthal Germany: Trans Tech Publications.

He, C. , S. Okubo and Y. Nishimatsu 1990. A study on the class II behaviour of rock. *Rock Mechanics and Rock Engineering* 23: 261-273.

Hudson, J. A. and C. Fairhust 1969. Tensile strength, Weibull's theory and a general statistical approach to rock failure. *The Proceedings of the Southampton 1969 Civil Engineering Materials Conference (Part II), Edited by M. Te' eni*: 901-914.

Krajcinovic, D. and M. A. G. Silva 1982. Statistical aspects of the continuous damage theory, *Int. I. Solids Structures* 18:551-562.

Mogi, K. 1985. *Earthquake prediction*. Tokyo: Academic Press.

Scholz, C. H. 1990. *The Mechanics of earthquakes and faulting*, Cambridge University Press.

Tang, C. A. and X. H. Xu 1990. Evolution and propagation of material defects and Kaiser Effect function: *J. Seismological Research* 13:203-213.

Wilson, B. 1981. Evaluation of some rock burst precursor phenomena, *Third Conference on Acoustic Emission/Microseismic Acitivity in Geologic Structures and Materials*. Clausthal, Germany: Trans Tech Publication.

Zhang, W. and S. Valliapan 1990. Analysis of random anisotropic damage mechanics problems of rock mass. *Rock Mechanics and Rock Engineering* 23: 91-112.

Rockbursts and Seismicity in Mines, Young (ed.) © 1993 Balkema, Rotterdam, ISBN 90 5410 320 5

Stress release estimates, scaling behaviour, and source complexities of microseismic events

Theodore I. Urbancic, Cezar-Ioan Trifu & R. Paul Young
Engineering Seismology Laboratory, Department of Geological Sciences, Queen's University, Kingston, Ont., Canada

ABSTRACT: We investigate the nature of stress behaviour associated with microseismic events by examining the ratios of different stress release estimates, scaling behaviour, and waveform complexity for 68 excavation related events at Strathcona mine, Sudbury, Ontario. A dependence of static and dynamic stress drop of $\sim M_o^{0.66}$ was found. The majority of dynamic versus static stress drop and source complexity values were below 1.5, suggesting that microseismic sources are simple and without sub-events, and that the observed scaling is therefore related to scale length effects ($\sim$ 2 m). The rupture velocities were found to range from $\sim$ 0.3 to 0.6β, and appear to be responsible for the small values of apparent stress versus static stress drop (below 0.1), indicating that additional complex rupture phenomena such as branching or variable fault roughness may be present. Both static and dynamic stress drops provide equally applicable measurements of stress release for microseismic events.

1 INTRODUCTION

In estimating stress release for excavation related microseismic events (M < 0), it is generally assumed that simple rupture occurs over simple geometries, such as in the calculation of static stress drop based on the spectral approach of either Brune (1970) or Madariaga (1976) for circular ruptures. Taking the static stress drop estimate as an example, an increase in rupture complexity will reduce reliability in source radius estimate and consequently lead to a three-fold increase in static stress drop uncertainty. This suggests that estimates of stress release based on the static stress drop alone could be misleading for complex sources.

Based on the analysis of small to moderately sized induced earthquakes, it has been concluded that the dynamic stress drop provides a more reliable estimate of stress release (Trifu et al., 1992). In routine processing of microseismic events, it has been visually observed that recorded signals can also be complex. However, very few systematic studies of rupture complexity as well as comparison of static and dynamic stress release parameters have been carried out (e.g., Gibowicz et al., 1991). It is the goal of this investigation to examine stress release estimates and waveform character to determine the level of complexity associated with microseismic events, and consequently conclude upon the most appropriate stress release parameter to characterize microseismic data.

To investigate the source complexity of microseismic events, studies have been initiated at Strathcona mine in Sudbury, Ontario. Previous analysis of excavation related events at the mine have found both self-similar and non-similar behaviour in static stress drop estimates (Urbancic et al., 1993), suggesting that a variable complexity linked to site conditions may exist. For this study, the events analyzed (-2.4 $\leq$ M $\leq$ -1.1) were recorded during both active and quiet periods associated with the excavation of ore (10 m x 20 m x 35 m volume) in a highly stressed, moderately jointed/fractured rock mass at about 620 m depth. Unlike the previous studies, we examine ratios of dynamic stress drop, rms stress drop, and apparent stress with static stress drop, as well as measurements of the waveform character to quantify the level of source complexity. We also examine the source scaling behaviour as well as the degree of reliability of the various stress release estimates.

2 DATA ANALYSIS

The data analyzed was collected between September, 1991 and January, 1992, with an underground 64 channel microseismic system operated by Queen's University. The array employed 5 triaxial sensors (dual gain) and additional uniaxial sensors (33) optimally located around the volume of interest. Calibration surveys were performed to determine the system response and analog-to-digital conversion was carried out with 12-bit resolution at a sampling rate of 20 kHz. Only signals with signal-to-noise ratios greater than 4 were used for locating events, allowing for accurate arrival-time picks to better than 0.1 to 0.15 msec. Events were located based on a minimum of 10 P-wave first arrivals using a combined Simplex/Geiger algorithm and incorporating velocity distributions obtained by active velocity surveys (Maxwell et al., 1991). Only the most accurately located events with standard errors of 2 to 3 m, having unobscured ray paths to at least 3 triaxial sensors, and with fault-plane solutions that fit the double-couple model were considered for detailed analysis (Figure 1). This represents a small fraction (68 events) of the entire data set. Worth noting, 25 % of the retained events were recorded during a quiet period of non-excavation.

Source parameters were estimated in both time and spectral domains using the triaxial sensors. These sensors had band-limited flat frequency responses (within ± 3 dB) between 50 Hz and 5 kHz. Typically, the source-sensor separation was on the order of 100 m. The analysis was carried out only on rotated S wave signals. Anelastic attenuation and scattering effects were taken into account by deconvolving signals. Spectra were multiplied with $\exp(\omega t_c^*/2)$, where t^* is $R/\beta Q_s$, R is the source-sensor separation, β is the S-wave velocity (3700 m/s), and Q_s is the average quality factor along the raypath, set to 120 as determined by Feustel et al. (1993). By applying the above Q_s value, the spectral decay was generally well described by a -2 slope (Figure 2). Time domain studies included the analysis of rotated triaxial S-wave acceleration signals and integrated velocity signals. Measurements included peak acceleration and velocity parameters, rms accelerations, S-wave first pulse durations, rise times, and overall signal durations. In all cases, corrections for attenuation have been applied. Peak parameters have been corrected as with the spectral analysis using the derived corner

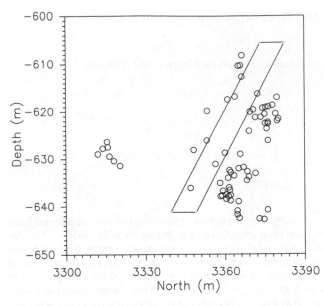

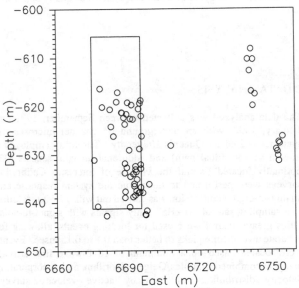

Figure 1. Location of microseismic events relative to the excavation opening created by the advancing mine front from 640 m to 605 m.

frequency whereas rms accelerations, initial pulse durations and rise times, and overall pulse durations were corrected by subtracting t^*, $t^*/2$, and $3t^*/2$, respectively.

3 SOURCE PARAMETERS

Several source parameter estimates were obtained from the spectral analysis. The seismic moment M_o was evaluated by

$$M_o = \frac{4\pi\rho\beta^3 R |\Omega_o|}{F} \quad (1)$$

where Ω_o represents the spectral level vector sum of the components of the S-wave, ρ is the density of the source material (2700 kg/m³), and F accounts for the radiation pattern of the S-waves (0.57). Magnitudes were obtained based on the moment magnitude relationship proposed by Hanks and Kanamori (1979), $M = 2/3\log M_o - 6.0$, where M_o is in N.m. The S-wave seismic energy was calculated from the energy flux J_s following Snoke (1987)

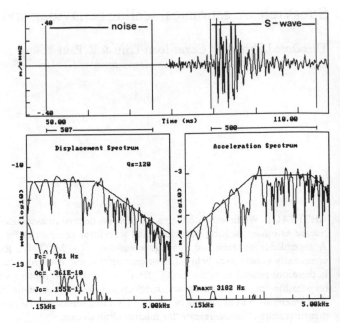

Figure 2. An example of the displacement and acceleration spectra for a rotated S-wave signal and the associated noise spectra.

$$E_s = 4\pi\rho\beta R^2 J_s \quad (2)$$

Estimates of source radius (r_o) were obtained by assuming Brune's (1970) model

$$r_o = \frac{K_s \beta}{2\pi f_c} \quad (3)$$

where K_s is equal to 2.34, and f_c is the corner frequency

$$f_c = (J_s / 2\pi^3 \Omega_o^2)^{1/3} \quad (4)$$

If we assume that f_{max} is related to the source, a radius r_a was obtained by replacing f_c with f_{max} in (3). The static stress drop $\Delta\sigma$ and the apparent stress σ_a were calculated as

$$\Delta\sigma = \frac{7}{16} \frac{M_o}{r_o^3} \quad (5)$$

$$\sigma_a = \frac{\mu E_s}{M_o} \quad (6)$$

where $\mu = \rho\beta^2$. Supplementary source parameters were obtained from time domain analysis. Peak acceleration $\underline{a}$ and velocity $\underline{v}$ values were used to determine the radius r of the most energetic asperity, which gives rise to the maximum values for displacement D, ground velocity adjacent to the fault $\dot{D}/2$, and dynamic stress drop $\Delta\sigma_d$ (McGarr, 1981)

$$r = 2.34 \, \beta \underline{v} / \underline{a}$$
$$D = 8.06 \, R \underline{v} / \beta$$
$$\dot{D}/2 = 1.28 \, (\beta/\mu) \, \rho R \underline{a} \quad (7)$$
$$\Delta\sigma_d = 2.50 \, \rho R \underline{a}$$

An additional estimate of stress drop was calculated from measurements of the rms acceleration a_{rms} (Hanks and McGuire, 1981)

$$\sigma_{rms} = \frac{2.7\rho R}{F}\left(\frac{f_c}{f_{max}}\right)^{1/2}a_{rms} \qquad (8)$$

where a_{rms} values are taken between the S-wave pulse initiation and $1/f_c$.

A measure of the source complexity was obtained by comparing the static stress drop with other estimates of stress release and by comparing the source radius with the asperity radius. An independent estimate of source complexity was obtained by comparing the velocity signal duration of the slip over the entire source (τ) to the pulse duration of the rupture of the first sub-event (τ_r) after Boatwright (1984)

$$c = \frac{\tau/\tau_r + \gamma}{1 + \gamma} \qquad (9)$$

where γ is the waveform compactness, taken to be between 0.5 and 1.0. Examples of signal and pulse duration measurements are given in Figure 3. Based on the measured velocity pulse rise time, $\tau_{1/2}$, estimates of the rupture velocity v_r (Boatwright, 1980) were obtained as

$$v_r = \frac{13r}{16\tau_{1/2} + 12r\sin\theta/\beta} \qquad (10)$$

where θ is the angle between the fault normal and the takeoff direction as obtained from the fault-plane solutions.

4 SCALING RELATIONS

In Figure 4, the observed scaling behaviour clearly suggests a static stress drop dependence on seismic moment, with a 0.70 dependence rather than a 0.33 dependence as suggested by the constant stress drop model. Worth noting is that the same dependence was found for a subset of events associated with the

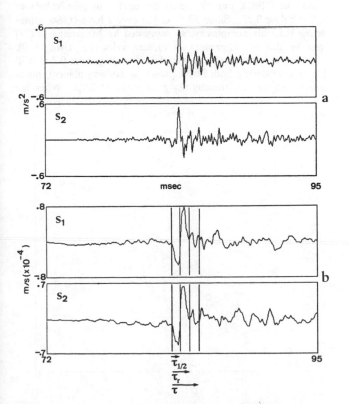

Figure 3. Example of rotated S-wave accelerations (a) and velocities (b), along with complexity measurements.

quiet period of non-excavation. Likewise, a non-similar behaviour was also evident in apparent stress and peak parameters (Figures 5 to 7). Since $E_s \propto M_o^{2.43}$ in our data, from (6) it follows that the apparent stress has a dependence of 1.43 on the seismic moment. In logarithmic scales, R$\underline{v}$ and ρR$\underline{a}$ show dependencies of 0.77 and 0.63 on the seismic moment; these can also be shown to have stress dependencies of $\Delta\sigma \propto M_o^{0.65}$ and $\Delta\sigma_d \propto M_o^{0.63}$, respectively.

A similar static stress drop dependence on seismic moment of about 0.60 to 0.70 was found by Gibowicz et al. (1991) for events with M_o between 4×10^3 and 1×10^6 N.m at the Underground Research Laboratory in Piniwa, Manitoba, and by Urbancic et al. (1993) for events with M_o between 6×10^5 and 1×10^8 N.m in a different area of Strathcona mine. In both studies, these tendencies were related to the presence of characteristic fault lengths and site conditions. Urbancic et al. (1993) also noted that this trend disappeared when multiple data sets were combined, supporting the idea that a constant stress drop model may fit widely dispersed values.

Unlike these reported studies, the non-similar behaviour in our data is strongly supported by independent spectral and time analyses, and the different source models employed. This

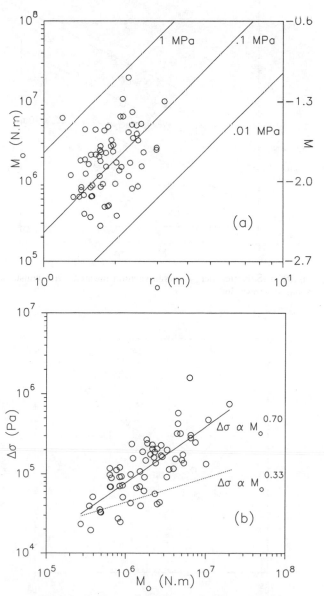

Figure 4. (a) Seismic moment and magnitude versus r_o with constant static stress drop lines; (b) Static stress drop versus seismic moment.

suggests that the stress distribution on the faults within the rock mass is highly inhomogeneous (2 orders of magnitude) for sources with a limited scale length of 1 to 3 m (Figure 4). These results, however, do not allow us to ultimately determine if the observed inhomogeneity is related to the individual seismic sources or to site conditions such as the level of rock fragmentation or the distribution of fractures. To better understand the nature of the observed scaling behaviour, the source complexities of the events need to be examined.

5 STRESS RELEASE AND SOURCE COMPLEXITY

As shown in Figure 8, a well-defined linear relationship of 0.89 was found between the static and dynamic stress drop values. Moreover, the static and dynamic stress drops were similar in magnitude. This is apparent in Figure 9 where 66 % of the analyzed events have $\Delta\sigma_d/\Delta\sigma$ values less than 1.5. Assuming that the errors in the determination of individual stress drop values

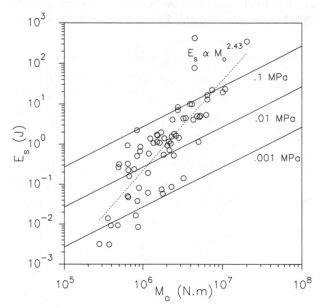

Figure 5. Seismic energy versus seismic moment with constant apparent stress lines.

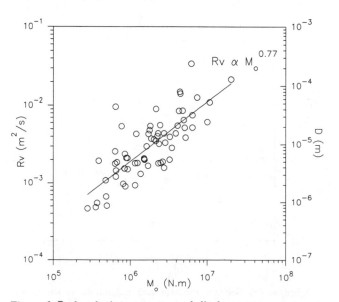

Figure 6. Peak velocity parameter and displacemt at source as a function of seismic moment.

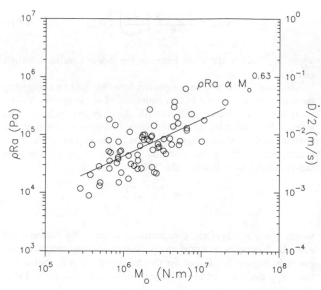

Figure 7. Peak acceleration parameter and slip velocity adjacent to the fault as a function of seismic moment.

range up to 25-50 %, we conclude that the sources are relatively simple. This interpretation is also supported by observed ratios of r/r_o in Figure 10, for which 66 % of the events have values between 0.5 and 0.7, well above the minimum resolvable asperity level of about 0.25 as suggested by r_a/r_o.

Figures 11 and 12 show the ratios of the apparent and rms stress drop to the static stress drop. For $\sigma_a/\Delta\sigma$, about 75 % of the observed values lie between 0 and 0.5, as theoretically predicted (Madariaga, 1977). The $\sigma_{rms}/\Delta\sigma$ values are generally grouped around 0.5, suggesting that the rms and static stress drops are equally suitable measures of the stress release. In terms of complexity, the low apparent stress values suggest that the sources may be somewhat complex. As pointed out by Rudnicki and Kanamori (1981), our r/r_o ratios are unable to give values of $\sigma_a/\Delta\sigma$ below 0.25. Since 33 % of the events have $\sigma_a/\Delta\sigma$ values below 0.1, this complexity, as suggested by Madariaga (1977) may be due to differences in rupture velocity. Using (10), estimates of rupture velocity were found to range from 0.2 to 0.5β (when considering both nodal planes as faulting planes), much lower than the traditionally assigned value of 0.9β. When we

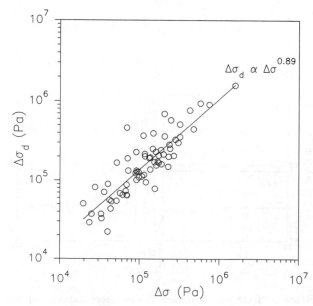

Figure 8. Relationship of dynamic versus static stress drop.

258

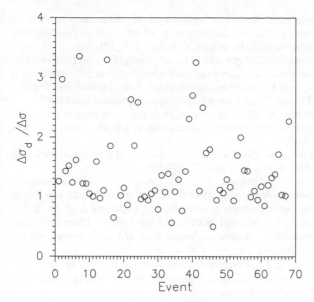

Figure 9. Dynamic to static stress drop values.

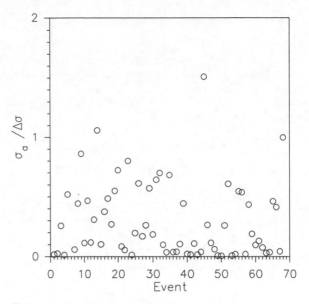

Figure 11. Apparent stress to static stress drop values.

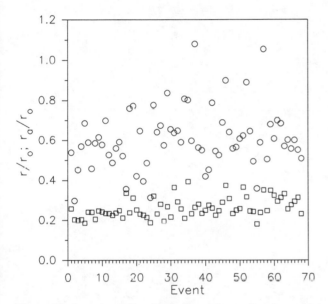

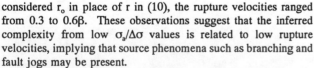

Figure 10. Asperity radius (circles) and f_{max} radius (squares) to source radius.

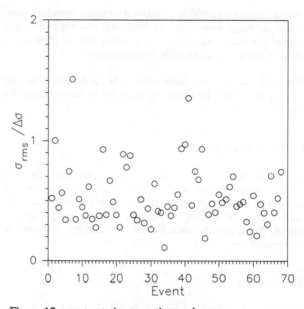

Figure 12. rms to static stress drop values.

considered r_o in place of r in (10), the rupture velocities ranged from 0.3 to 0.6β. These observations suggest that the inferred complexity from low $\sigma_a/\Delta\sigma$ values is related to low rupture velocities, implying that source phenomena such as branching and fault jogs may be present.

Aside from the inferred apparent stress complexities, an independent check on the source complexity was made with (9). In Figure 13, 75 % of the events have c values below 1.5. Since c deals with an inhomogeneous stress distribution and the possible occurrence of sub-events, the above value suggests that the sources are simple, as seen in the $\Delta\sigma_d/\Delta\sigma$ values. It is, however, not possible to derive a meaningful relationship between c and $\Delta\sigma_d/\Delta\sigma$. Based on these analyses, it seems that both the static and dynamic stress drops provide similar values and equally applicable estimates of stress release for microseismic sources. This leads us to believe that the non-similar scaling behaviour observed is not related to an inhomogeneous stress distribution at the source, but is more likely related to characteristic scale length effects.

6 CONCLUSIONS

This study has investigated the nature of stress behaviour associated with microseismic events by examining the ratios of different stress release estimates, scaling behaviour, and waveform complexity for excavation related events at Strathcona mine, Sudbury, Ontario. The findings of this study can be summarized as follows:

• The observed scaling behaviour shows a dependence of static and dynamic stress release estimates of ~ $M_o^{0.66}$. Since about 66 % of the events had $\Delta\sigma_d/\Delta\sigma$ values below 1.5, and 75 % of the events had c values below 1.5, the above scaling dependence does not appear to be related to the source inhomogeneity (in terms asperities and sub-events), but is likely related to scale length site effects (~ 2 m).

• About 66 % of the events had r/r_o values between 0.5 and 0.7. However, 33 % of all events had very small values of $\sigma_a/\Delta\sigma$ (below 0.1), which cannot be explained by a distribution of sub-

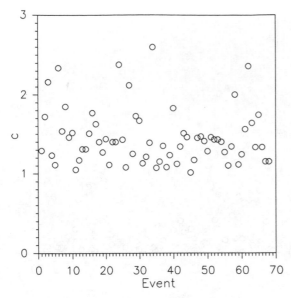

Figure 13. Source complexity values.

events corresponding to r/r_o. Since the rupture velocities were found to range from ~ 0.3 to 0.6β, the low apparent stress values are likely indicative of a complex rupture phenomena such as branching or variable fault roughness.

• The static and dynamic stress drops have similar values and are equally applicable for obtaining estimates of stress release for microseismic data.

ACKNOWLEDGEMENTS

The authors would like to thank Falconbridge Ltd. for their continued support in our research. Funding was provided by the Mining Research Directorate of the Canadian Rockburst Research Program and the Natural Sciences and Engineering Research Council. We gratefully thank Sue Bird for valuable assistance in data preparation.

REFERENCES

Boatwright, J. 1980. A spectral theory for circular seismic sources; simple estimates of source dimension, dynamic stress-drop, and radiated seismic energy. *Bull. Seism. Soc. Am.* 70, 1-27.

Boatwright, J. 1984. The effect of rupture complexity on estimates of source size. *J. Geophys. Res.* 89, 1132-1146.

Brune, J.N. 1970. Tectonic stress and the spectra of seismic shear waves from earthquakes. *J. Geophys. Res.* 75, 4997-5009. Correction in *J. Geophys. Res.* 76, 5002, 1971.

Feustel, A.J., Urbancic, T.I., and Young, R.P. 1993. Estimates of Q using the spectral decay technique for seismic events with M < -1. *in* Proc. 3rd Inter. Symp. on Rockbursts and Seismicity in Mines., in press.

Gibowicz, S.J., Young, R.P., Talebi, S., and Rawlence, D.J. 1991. Source parameters of seismic events at the Underground Research Laboratory in Manitoba, Canada: Scaling relations for events with moment magnitude smaller than -2. *Bull. Seism. Soc. Am.* 81, 1157 - 1182.

Hanks, T., and Kanamori, H. 1979. A moment magnitude scale. *J. Geophys. Res.* 84, 2348-2350.

Hanks, T., and McGuire, 1981. The character of high frequency strong ground motion. *Bull. Seism. Soc. Am.* 71, 2071-2096.

Madariaga, R. 1976. Dynamics of an expanding circular fault. *Bull. Seism. Soc. Am.* 66 639-666.

Madariaga, R. 1977. Implications of stress drop models of earthquakes for the inversion of stress drop from seismic observations. *Pure Appl. Geophys.* 115, 301-316.

Maxwell, S.C., Urbancic, T.I., and Young, R.P. 1991. Final report on three-dimensional seismic tomography at Strathcona mine, Queen's University/Falconbridge Ltd. Technology Transfer Project, 38 p., Queen's University, Kingston, Canada.

McGarr, A. 1981. Analysis of peak ground motion in terms of a model of inhomogeneous faulting. *J. Geophys. Res.* 86, 3901-3912.

Snoke, J.A. 1987. Stable determination of (Brune) stress drops. *Bull. Seism. Soc. Am.* 77, 530-538.

Trifu, C-I., Urbancic, T.I., and Young, R.P. 1992. Estimates of near-source ground motion parameters and source complexity supporting inhomogeneous faulting (abstract). *EOS* 73, 389.

Rudnicki, J.W. and Kanamori, H. 1981. Effects of fault interaction on moment, stress drop, and strain energy release. *J. Geophys. Res.* 86, 1785-1793.

Urbancic, T.I., Feignier, B., and Young, R.P. 1993. Site influence on source scaling relations for M < 0 events. *Pure Appl. Geophys.*, in press.

Rockbursts and Seismicity in Mines, Young (ed.) © 1993 Balkema, Rotterdam, ISBN 90 5410 320 5

Applications of quantitative seismology in South African gold mines

G.van Aswegen & A.G.Butler
ISS International Limited, Welkom, South Africa

ABSTRACT: The relations between radiated seismic energy (E) and seismic moment M_o, expressed as apparent stress or apparent volume (M_o/Apparent stress - Mendecki, 1993) are utilized to detect variation in rock mass behaviour in the Welkom gold mines. Energy Index (Measured E/average for given Mo) is conceptually a sensitive indicator of variation in the state of stress.   The relation E_s/E_p varies with source mechanism and is used to distinguish between fault slip events and dyke/pillar events. Asperities, the source regions of future large tremors, may be delineated by the spatial distribution of small seismic events of relatively high apparent stress. Cumulative apparent volume from a rock volume surrounding a fault correlates with measured creep, consistent with the interpretation that this parameter scales co-seismic non-elastic deformation.

1 INTRODUCTION

Application of basic seismological principles in the quantitative analyses of rock mass behaviour can indicate the variation in space and time of rockburst potential in mines. In this paper, concepts found to be most useful in the Welkom gold field are discussed with reference to example cases. Special emphasis is placed on fault stability since fault slip is responsible for most major tremors in the Welkom and Klerksdorp gold fields.

2 SOME USEFUL SEISMOLOGICAL PARAMETERS

Modern seismic systems offer more than hypocentre locations and local magnitudes. Quality controlled, quantitative seismic source parameter evaluation from digital waveform records (e.g. Mendecki, 1993) provide two robust and largely independent measures of seismic deformation, namely radiated seismic energy [E = energy_P + Energy_S] and seismic moment [M_o = (Moment_P + Moment_S)/2]. The ratio between E and M_o, expressed as apparent stress, is a measure of stress drop (Savage & Wood, 1971 and Snoke et.al 1983).

For fixed M_o, E can vary up to two orders of magnitude (see Figure 1). Variation in apparent stress differentiates seismic source regions of different stress state and/or rock mass properties. This is demonstrated by data from the Welkom gold field where seismic events between magnitude 1.4 and 1.6 associated with the deepest mining, consistently have higher apparent stress than events in the same magnitude range associated with the shallower mining. Empirical relations between apparent stress and other source parameters shows that apparent stress generally increases with magnitude (or moment or energy). To distinguish between variation in apparent stress due to increase in 'size' of the event and that due to local rock mass conditions, apparent stress plots of fixed

local magnitude are useful. In practice it means, however, that the mine seismologist must do a series of plots for different magnitude ranges to get the full picture from available data. To overcome this inconve-

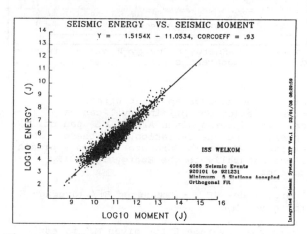

Figure 1a. *The relation Energy vs Moment*

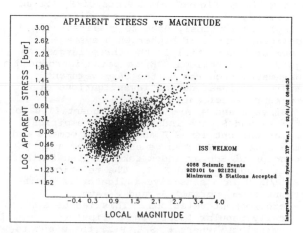

Figure 1b. *Apparent stress increases with magnitude*

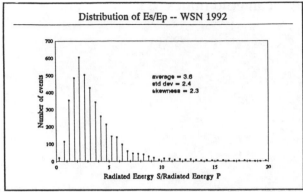

Figure 2a. The distribution of the relation E_S/E_P. Only events recorded by 5 or more three-component seismic stations are used here.

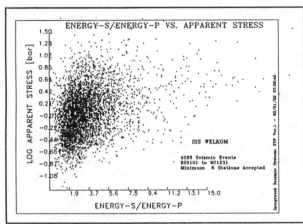

Figure 2b. Energy_S/Energy_P varies independently of apparent stress

nience the average apparent stress per magnitude (or moment) interval can be calculated for a given area/time span of interest and the deviation from these averages used as a parameter. This approach leads naturally to the Energy Index (E_i) concept.

Energy Index (E_i) = (measured E)/ (average E for given M_o)

where the 'average E for given M_o' is estimated through an empirical relation between E and M_o (e.g. Figure 1a). Plots of E_i (or an equivalent 'stress index' - see Glazer (in prep) are routinely used to qualitatively delineate areas of higher than average rockburst potential in the three largest SA gold mining regions. E_i is being introduced a parameter required by the SA government mining engineer during investigations of rockburst accidents.

Since E and M_o are routinely quantified for the P- and S-wave independently, empirical relations between E_P and average moment and the same for E_S can be established and the concept of Energy Index taken a step further to include E_i_P and E_i_S. The relation E_i_P/E_i_S is a sensitive indicator of deviations from 'normal' S- and P-wave energy/moment relations. This led to a locally popular 'pops' and 'slips' terminology: where E_i_S/E_i_P>1, the event is called a 'slip' and in the inverse case a 'pop'. Patterns of the relation E_S/E_P

validates the application of the 'pops' and 'slips' indices (see Figures 2a & b). Variation in this ratio in the case of coal mine related seismic events in Germany was interpreted by Gibowicz et.al (1990) as indicating departure from pure double couple sources.

Another way of expressing the relation between the two basic parameters (M_o and E_i) is 'apparent volume', defined as M_o/apparent stress (Mendecki, 1993). This parameter, with units as m^3 may be totalled over a given area/time span. It potentially scales the volume of non-elastic strain of the seismic source. Deviation from the slope of a cumulative apparent volume vs time plot has been shown to a reflect anomalous rockmass behaviour prior to major tremors.

3 THE ASPERITY MODEL

Characteristics of faults which produced major tremors in the past include an 'optimum' dip (between 50 and 70 degrees), large historical displacement (the larger the fault, the greater the potential coseismic deformation) and, most importantly, extensive mining in the vicinity. The elastic and non-elastic deformation due to the removal of a significant volume of rock mass causes shear deformation of the fault. The way in which this shear deformation is accommodated along the fault surface depends on fault characteristics and mining geometry - some faults slip and others stick, but most do both.

Based on back analyses of major fault slip events (including the measurement of fresh displacements of up to 42cm after the event) and on elementary seismological theory, a working 'asperity model' forms the basis for fault stability evaluation in the Welkom gold field (van Aswegen, 1990). In terms of this model, a major fault slip event is preceded by shear deformation (aseismic creep and/or coseismic slip) over a large area of the structure. At some asperity the shear deformation is arrested, resulting in the accumulation of elastic strain. The eventual violent failure of the asperity causes the major seismic event. A formalisation of the asperity model (see Kostrov and Das, 1988, p.247) relates the potential size of the earthquake source to the size of the asperity and the size of the 'non-asperity' surrounding it i.e. the source size increases with the size of the asperity that breaks as well as with the size of the area around the asperity along which there is freedom to slip. Manifestations of asperities include any one or combinations of: a jog or other geometric aberration, later displacement by a shallowly dipping fault, a local area of high friction (e.g. patches where the fault rock is pseudotachylite) and a mining induced area of high normal stress. The strength of the asperity together with the rate of loading (rate of coseismic and/or aseismic shear deformation) determines the nature of the tremor. If the asperity is too strong, it will never fail and the creep process will be halted - this is an unlikely scenario for the crustal environment where tectonic forces rip continents apart, but it is a practical concept in the mining environment where 'bracket pillars' may be designed to lock faults permanently (at least in terms of the life of the mine).

The recognition of potentially hazardous asperities constitutes prediction in space. The model implies a gradual increase in elastic strain in the environment of the asperity. This should lead to 'rock talk' of a special kind, namely a concentration of small seismic events, mostly characterised by relatively high values of apparent stress. In the case of volume sources (dykes, pillars) situations similar to asperities on faults may be envisaged. Where a part of the rock mass resists deformation while coseismic and aseismic deformation proceeds in the general rock volume, elastic strain can accumulate in the hard patch until eventual violent failure. Seismologically the same precursory concentration of small magnitude, high apparent stress events as in the case of an asperity on a fault can be expected.

4 APPLICATIONS

Four cases where the above concepts are applicable are here described.

A dyke which 'popped'

A dolerite dyke was identified as being 'stressed' by the occurrence of a single event of high apparent stress on it. Subsequently a small reef drive in the dyke was severely damaged by a M_L 2.4 event in the immediate vicinity of the excavation. Some source parameters are: M_o=4.75E12 Nm; E=4.27E7 J; E_P=1.34E7 J; E_S=2.94E7 J; E_S/E_P=2.2; apparent stress = 2.7 bar.

The dyke 'exploded' and completely closed some two metres of development. This was partially opened to recover equipment, providing a unique opportunity to study seismic source damage. Fracturing was intense and fracture spacing bimodal, i.e. zones 20 to 100cm wide display fracture spacing of around 100mm, alternating with zones 5 to 20cm wide in which the fracture spacing is 1 to 10mm. Both types of zones occur as near vertical, small angle conjugate sets sub-parallel to the dyke margins. The maximum lateral displacement along one particular fracture is more than 100mm. Some displacement (<100mm) also occurred along a bedding plane fault (along a shale band above the reef) which cuts through the dyke approximately 1.5m above the damaged drive. The fault was exposed 5m from the drive due to the fall of hangingwall being part of the localised intense damage.

Limits to the size of the source is inferred from the lack of significant damage more than 20m away from the place of intense damage. People working against the dyke, 100m away, felt the event, but, since no rockfall occurred, did not even evacuate. The size estimate in terms of the Brune (1970) model suggests a source radius of 121m and, whether the real source was in the dyke or along the bedding plane fault, these workers should have been virtually within the source. Moment tensor analysis shows a significant deviation from a pure double couple source.

The moment tensor analysis, after being diagonalised through rotation to a new coordinate system refined by eigenvectors, gave the following results:-

MOMENT TENSOR			ISOTROPIC		
0.14	0.00	0.00		1 0 0	
0.00	-.06	0.00	= 0.494*	0 1 0	
0.00	0.00	1.14		0 0 1	

DOUBLE COUPLE			CLVD		
	0 0 0			-1 0 0	
+ 0.202*	0-1 0		+ 0.355*	0-1 0	
	0 0 0			0 0 2	

A major tremor predicted

The asperity model was successfully applied to predict the source region of a major fault slip event (mag. 4.6). A concentration of seismic events of relatively high apparent stress on the fault some months prior to the event was recognised through routine scrutiny of seismic patterns. In June 1990 funds were requested to monitor the phenomenon more closely, but, before instrumentation could be installed, the event occurred. Note that the seismic event pattern viewed in terms of Richter magnitude does not necessarily draw attention to the fault (Figure 3).

Fault/dyke intersection

At the intersection of a major dyke with a fault, a concentration of seismic events of relatively high apparent stress and energy index drew attention (Figures 4a & b). At the intersection, 5 spatial domains can be distinguished, namely the four quadrants caused by the intersection and the dyke itself (Figures 4a,b). All events within the dyke are 'pop' type, and of low apparent stress. The two quadrants on the west side of the fault are dominated by 'slip' type events, events in the northern quadrant being of low apparent stress, those in the southern of high apparent stress. A single event right in the corner of the north-eastern quadrant is of intermediate apparent stress and is a 'pop' type and, lastly, three events in the south-eastern quadrant are of low apparent stress and mixed 'mechanism'. The conclusion is that both structures are seismically active but that the fault has the greater potential to yield a damaging event: it appears to be stuck in the south-western corner of the intersection.

Fault creep, cumulative apparent volume and a geometric asperity

Another major fault (up to 1000m finite dip displacement) which apparently has all the characteristics 'favourable' to yield a major event (e.g. significant modelled 'excess shear stress' induced by extensive mining) is being monitored closely. The area on the fault between the upper and lower cut off of a shale band (associated with the reef) is characterized by soft, clay-rich and generally moist fault rock of obvious low frictional strength. A re-survey of a tunnel transecting the structure indicated some 20cm recent dip displacement. Since no major seismic event was recorded and no damage was found where the tunnel intersects the fault, the displacement had to be relatively slow. The fault transects a shaft near the position where the displacement was measured. The

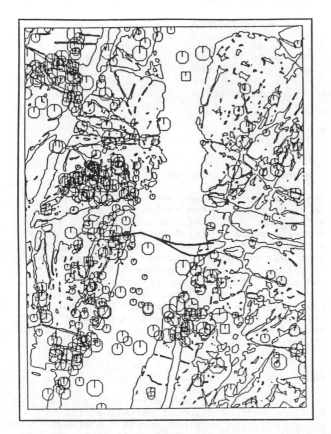 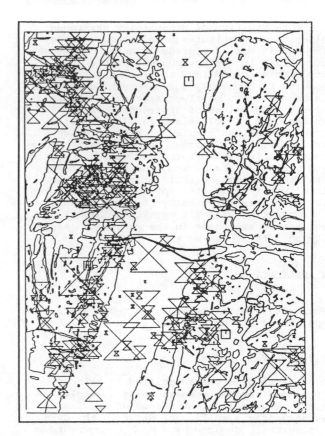

Figure 3a. Seismic events of Richter magnitude 1.0 - 2.5 prior to a major fault slip event. The symbols depict Richter magnitude (A) and apparent stress (B). The westerly dipping normal fault (>800m finite dip slip) causes a 'fault loss' seen as a northerly trending void in the plan view of the planar ore body. The latter dips easterly and is largely mined out to the north and east.

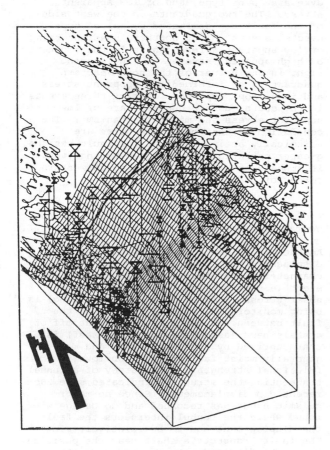

Figure 3b. Orthographic projection of part of the fault shown in Figure 3a. The geometry of the structure was modelled through 'Faultsim' (see Dennison & van Aswegen, 1993). The intersection of a vertical, bifurcating dyke with the fault surface is shown as a bold line over the surface grid. This feature is repeated in Figure 3a for reference.

Excess Shear Stress = $\sigma_s - \sigma_n * \mu$

If the modelled shear stress is greater than the modelled resistance imposed by normal stress and friction coefficient, i.e. ESS is positive, the potential for unstable slip is increased. ESS is widely used in the SA gold mining industry to evaluate fault stability. Short vectors from grid points on the fault surface overemphasizes minor ESS shown by stress modelling: in this case, modelling failed to indicate the likelihood of a major seismic event.

Seismic events, depicted by symbols with vertical tielines to the fault surface, are from the same data set used in Figure 1 but only those within 250m (orthogonally measured) of the fault in the area of interest, are shown. The symbol size variation depicts variation in Energy Index. Values for most events spatially associated with the fault between the two reef horizons are significantly higher than average.

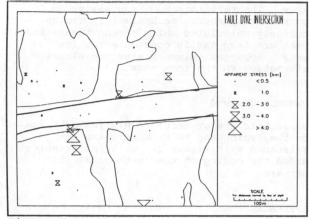

Figure 4a.

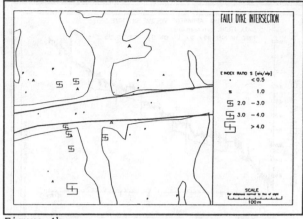

Figure 4b.

Figure 4. Seismicity at the intersection of a vertical dyke (heavy outlines) and a westerly dipping normal fault. (The thin lines are mining faces along the planar ore-body which dips shallowly in the east. Most of the area is mined out except for the fault loss and some irregularly shaped remnant pillars. Events spatially associated with the deeper parts of the fault would, in plan, plot to the west of the fault loss. The symbol size in (b) reflects E_iS while the symbol type reflects categorization according to ES/EP: if this ratio >4.6, the symbol is S, if < 2.6 the symbol is P and the rest are categorized as A. (See Figure 2a.)

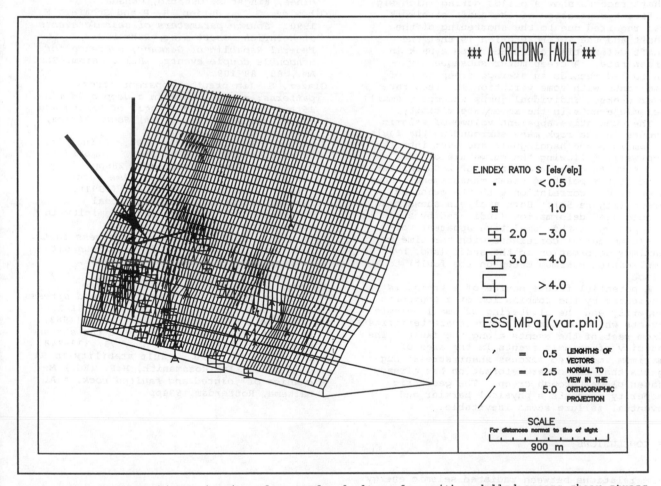

Figure 5a. Orthographic projection of part of a fault surface with modelled excess shear stress (ESS) as vectors and seismic events (with tielines to the surface for perspective). Note the near absence of seismicity positive ESS areas. The symbol size and type as in Figure 3b. Note the change in mechanism in the vicinity of a geometric asperity. The arrow shows the position of a creep meter (see Figure 5b).

265

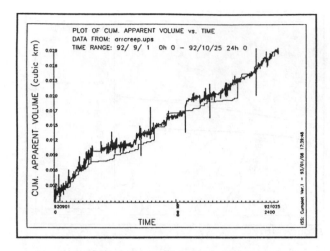

Figure 5b. Plot of cumulative apparent colume from events around the fault shown in Figure 5a, with measured fault creep (darker line) superimposed. The total creep reflected here is 2.5mm.

shaft records show a period during which significant reparation of the shaft steelwork was required due to the shortening of the shaft and the interpretation is that the shaft deformation coincided with a peak in creep rate. A creep meter subsequently installed records an average creep rate of 1mm/month with some variation in creep rate over weeks. Individual jumps accompany small seismic events in the immediate vicinity.

The cumulative apparent volume of seismic events in the rock mass surrounding the fault (500m into the hangingwall and 500m into the footwall, following the curvature of the fault surface for a distance of 2000m either side of the point of creep measurement) shows significant correlation with the measured creep (Figure 5b). Back analysis shows that accelerated deformation (indicated by a steepening of the cumulative apparent volume/time curve) correlates with the time when maximum deformation in the shaft (and, by implication maximum creep on the fault) took place.

A potential future source of a tremor is indicated by the combination of a geometric asperity and the clustering of small seismic events which differ in source characteristics from rest of the events along the fault. The paucity of seismic events in the area of maximum modelled 'excess shear stress' suggests that the shear deformation has already taken place through creep. The geometric asperity presents a physical barrier and eventual failure seems inevitable.

5 CONCLUSIONS

The relations between radiated seismic energy and seismic moment and between S_wave energy and P_wave energy of small seismic events, can be expressed in several ways useful for the mine seismologist to evaluate rock mass behaviour. The asperity model is a valid basis for fault stability evaluation in mines.

Areas yielding consistently high apparent stress events are considered to reflect locally high stress conditions and vice

versa. The relations E_s/E_p can assist in understanding source mechanism, it can be routinely calculated while moment tensor analyses are less easily done on a routine basis. Apparent volume scales non-elastic deformation in the rock mass.

ACKNOWLEDGEMENTS

The authors are grateful to Mrs Ida de Lange for the typing and to Mr R Ferreira for assistance with Figure 2. Dr T. Stankiewicz guided the coding of the 'orthogonal fit' software.

REFERENCES

Brune, J.N. (1970). *Tectonic stress and the spectra of seismic shear waves from earthquakes.* J. Geophys. Res., I5, 4997-5009.

Dennison, P. and van Aswegen, G. (1993). *Stress modelling and seismicity on the Tanton fault: a detail case study in a deep gold mine.* Submitted to: 3rd International Conference on Rockburst and seismicity in mines, Kingston, Ontario, Canada.

Gibowicz, S.J., Harjes, H.-P and Schäfer, M. 1990. *Source parameters of seismic events at Heinrich Robert mine, Ruhr basin, Federal Republic of Germany: evidence for nondouble-couple events.* Bull. Seism. Soc. Am., 80, 88-109.

Glazer, S. (in prep). *Apparent stress patterns: a case study in a deep gold mine.* 15th Congress of the Council of Mining and Metallurgical Institutions, South Africa, 1994.

Kostrov, B.V. and Das, S. 1988. *Principles of seismic source mechanisms.* Cambridge University Press, Cambridge. 286pp.

Mendecki, A.J. (1993). *Real time quantitative seismology in mines.* (1993). Keynote lecture: 3rd International Conference on Rockburst and seismicity in mines, Kingston, Ontario, Canada.

Ryder, J.A. 1988. *Excess shear stress in the assessment of geologically hazardous situations.* J.S.Afr. Inst. Min. Metall., 88,1,27-39.

Savage, J.C. & M.D. Wood (1971). *The relation between apparent stress and stress drop.* Bull. Seism. Soc. Am. 61, 1381-1388.

Snoke, J.A., A.T. Linde & I.S. Sacks (1983). *Apparent stress: an estimate of the stress drop.* Bull. Seism. Soc. Am., 73, 339-348.

van Aswegen, G. 1990. *Fault stability in SA gold mines.* In: Rossmanith, H.P. (Ed.) *Mechanics of jointed and faulted rock.* A.A. Balkema, Rotterdam, 994pp.

Rockbursts and Seismicity in Mines, Young (ed.) © 1993 Balkema, Rotterdam, ISBN 90 5410 320 5

Investigation of a 3.5 M_L rockburst

T.J.Williams & J.M.Girard
Spokane Research Center, US Bureau of Mines, Wash., USA

C.J.Wideman
Spokane Research Center, US Bureau of Mines, Wash. & Montana College of Mineral Science and Technology, Butte, Mont., USA

ABSTRACT: On September 19, 1991, a 3.5 magnitude (M_L) seismic event caused extensive damage to the northeast part of the Lucky Friday Mine near Mullan, ID, where mining had created a region of ore pillars and backfilled stopes. Measurements taken since 1987 indicated that a block of rock between the North Control Fault and the F3 Fault was experiencing closure. Inspection of damage on the 4660 level suggest left-lateral movement east of the F3 Fault, right-lateral movement west of the North Control Fault, and up-dip movement along bedding. Maximum damage in the 5100-106 stope occurred between the two faults and appeared to be related to convergence of the southwest wall on the vein.

First-motion analyses of a full waveform record of the event showed good correlation between the calculated source mechanism orientation and the strike and dip of hanging wall bedding, indicating that movement along bedding was the source of the event. Analysis of damage with respect to geologic structures, data from long-term wall closure monitoring, and first-motion analyses of digital full waveform data indicated that this event resulted from block movement along bedding between the two parallel faults.

1 INTRODUCTION

Hecla Mining Co. and the U.S. Bureau of Mines have been conducting research on seismicity and rock-bursting at the Lucky Friday Mine near Mullan, ID, in an attempt to understand the mechanisms associated with rockbursting at the mine. One successful research effort influenced mine management's decision in the mine switching to mechanized underhand mining with a single stair-stepped face that eliminated pillars below the 5100 level (Williams and Cuvelier, 1990; Noyes et al., 1988). This method is also now being used to recover remnant ore pillars left by overhand mining above the 5100 level. The rockburst reported upon in this paper was associated with mining one of these remnant pillars.

The Bureau installed a digital seismic monitoring system based on the U.S. Geological Survey's (USGS) PC Quake system (Lee et al., 1988), and by 1989, a three-dimensional array of triaxial velocity gauges was surrounding the mining area (Jenkins et al., 1990). The purpose for installing this system was to record full waveform data from seismic events for first-motion studies and other standard earthquake engineering investigations (Aki and Richards, 1980). A double-couple, first-motion model was chosen to represent the data because over 90% of the seismic events recorded contained both dilatational and compressive first arrivals. Events that have all dilatational first arrivals are implosional events (Wong and McGarr, 1990). These events are thought to be volumetric closure of the mine walls along mined-out areas.

In-mine studies involving detailed geologic mapping (Scott, 1988) and manually read closure monitoring stations have been instituted in areas of special interest. Blake and Cuvelier (1988) identified a seismic sequence associated with one of these areas (4660 northeast lateral) and suggested that block movement between two bedding-plane faults along argillite beds might have been the source mechanism for one of the rockbursts. The closure instruments on the 4660 level were designed to identify the blocks and show if they were moving.

The damage from the rockburst, local geology, and nearby mining patterns were evaluated to compare with first-motion data gathered by the digital seismic system and confirm the use of a double-couple model for studying these events. Hypotheses as to why deformation was confined within the block boundaries were developed.

2 ANALYSIS OF GEOLOGY, MINING, AND DAMAGE FOR THE 3.5 M_L ROCKBURST

On September 19, 1991, at 9:20:28 AM, a rockburst with an approximate local Richter magnitude (M_L) of 3.5 damaged the northeast side of the Lucky Friday Mine on two levels. The event resulted in approximately 2,000 tons of rock being expelled into several mine openings. The 5100-106 underhand stope, cut 12, at an elevation of 500 m below sea level, and the northeast lateral on the 4660 level at an elevation of 400 m below sea level sustained most of the damage. Lesser damage occurred in the 4900-105 and 107 stope at the same elevation as the 4660 level.

Figure 1 is a map of the damage in the 5100-106 stope. Here, a 20-m-high remnant ore pillar is being mined for 120 m along the Main vein and 45 m along the 40 vein (North Control Fault). As a result of the event, extensive damage closed almost 80% of the stope, from the access ramp at the approximate location of the F3 Fault and east to the point where the Main vein intersects the North Control Fault. Along this stretch, the fill collapsed on top of flyrock from the walls, indicating that the fill failed after the walls failed. The cemented backfill was of high quality so failure did not result from poor-quality fill. Visible cracks in the fill indicated that fill failure resulted from shearing of the fill, shock loading, or rapid stope wall convergence.

Damage to the stope walls in this zone appeared mostly along the hanging wall, indicating that, if failure was not the result of pillar collapse, the event most likely originated out in the hanging wall. The 3-m-thick layer of debris in the stope made it impossible to determine if the floor heaved. No movement along bedding, which dips at 60° to southeast, was visible in this area.

Miners reported damage to both walls in the mined portion of the 40 vein but no fill failures.

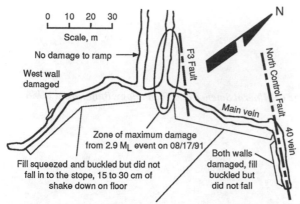

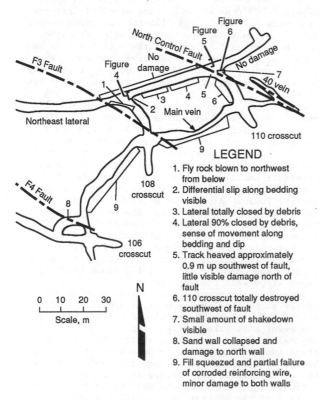

Figure 1. 5100-106 stope plan with damage locations.

Damage to the west side of the stope consisted of 15 to 30 cm of shakedown from the walls for the entire distance to the face and two areas of moderate damage to the foot wall.

The second access ramp to the 106 stope in the footwall was undamaged. The original access ramp was destroyed by a 2.9 M_L burst on August 17, 1991, and had been abandoned. Maximum stope damage occurred from this access, which is just west of the F3 Fault, to the east.

On the 4660 level, there was approximately 1,000 tons of displaced wall rock in the northeast lateral, 108 cross-cut, and 110 cross-cut between the North Control Fault and the F3 Fault. Figure 2 shows locations of the damage. There was no damage in the lateral west of the F3 Fault. The floor had heaved and the southeast wall had been displaced northwest immediately to the east of the fault, indicating left-lateral movement along the fault and up-dip movement along the bedding.

Figure 2. 4660 level with damage locations.

Figure 3 shows the southeast corner of this intersection. The effects of upward movement along the bedding are definitely visible at this location, with the mass of the beds being pushed up into the lateral. Movement along bedding was footwall up and hanging wall down (reverse faulting). The bedding dips 55° to 60° to the southeast and is characterized by a series of 0.6-m-thick vitreous quartzite beds separated by 8-mm-thick argillite beds. The soft, talc-like argillite acts as a lubricant to the hard quartzite beds, allowing movement along the bedding planes.

Figure 3. Dip slip on bedding at 108 cross-cut corner.

In the northeast lateral east of the 108 cross-cut, the southeast wall and floor moved up and into the opening, completely closing the lateral. The damage in the 108 cross-cut and northeast lateral is remarkably similar to the damage at this same location from a 2.7-M_L event on September 30, 1987, which suggested the block movement hypothesis.

The area of the 4660-110 cross-cut was accessed by climbing down a vent raise from the 4450 level. Figure 4 shows the lateral looking west while Figure 5 shows a view into the 110 cross-cut. West of the North Control Fault, the floor of the lateral had heaved and had been pushed to the northwest, indicating right lateral movement with a vertical component on the fault and normal movement along the bedding. Railroad track in the 110 cross-cut was heaved up on the southwest side of the fault, and the vent tube had been crushed against the back. The south wall of the lateral was also heaved up-dip to the northwest along bedding. Immediately to the east of the fault, there was only minor shakedown damage. Figure 6 is a vertical cross section along the 110 cross-cut west of the fault showing mining, geology, and damage.

Figure 4. Dip slip on bedding bounded by the North Control Fault looking along the northeast lateral.

Figure 5. Vertical motion along North Control Fault in 110 cross-cut.

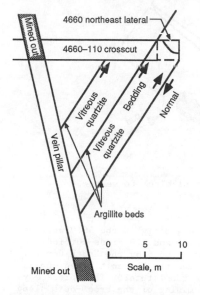

Figure 6. Idealized section along the North Control Fault showing dip-slip movement along bedding to the northeast lateral.

There was also some damage in the 4660-106 ramp where it crossed the F4 Fault; however, a definite sense of movement was not identified.

The 4900-107-109 stope had one unfilled cut at the 4660 level. The stope was accessible from the 4660-106 ramp to within 6 m of the 4660-110 cross-cut. Damage in this area was limited to the collapse of some rotten timber and minor amounts of fill related to corroded reinforcement wire. While the area between the F3 Fault and the 110 cross-cut is in the same block as the heavily damaged lateral, it was barely damaged.

3 NORTHEAST LATERAL CLOSURE DATA

Manually read closure monitoring stations were established (Figure 7) across the northeast lateral in 1987 to identify whether blocks were moving. Figure 8A and B show the closure data collected for 3-1/2 years. Prior to the burst, the readings indicated that the block between the North Control Fault and the F3 Fault was experiencing creep-type failure while the rocks on either side of the faults were stable. Stations 1 and 1A, west of the F3 Fault, showed no closure from the burst, and stations 7 and 8 east of the North Control

Fault had less than 2.5 mm of change. The readings across the corner of the block at the North Control Fault, station 6, showed no change from before the burst. The four closure measurement stations between the North Control Fault and the F3 Fault (2 through 5) were destroyed by movement along the south wall during the burst.

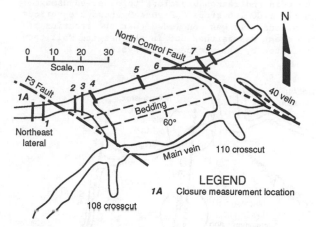

Figure 7. Closure measurement locations on the 4660 level.

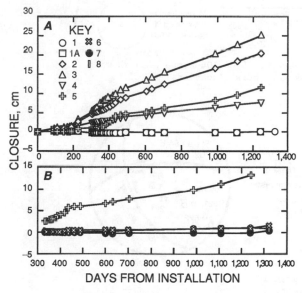

Figure 8. Closure measurements in the northeast lateral. A and B.

4 DIGITAL SEISMIC DATA ANALYSIS

Waveforms were analyzed for the event recorded as 91091902 on the macroseismic system. Several unusual features were noted. First, there was uncertainty in picking first-motion polarity and arrival time because there was a set of low-amplitude first motions followed by a set of very large amplitude arrivals a few milliseconds later. Because of the magnitude of the damage associated with this event, such small first motions were unusual at the mine. Small P-waves are associated with shear energy release, however, so the small P-waves indicate this was a shear type event.

Second, there was uncertainty in determining the location of the hypocenter. The mine's microseismic system located an event near the vein at 20714, 20678, -1615 (mine coordinate system), but this event may have been a microseismic event that took place before the large one. Initially the Bureau's macroseismic system, using a constant velocity model, placed the event at 20964, 20619, -1501 in the hanging wall near the North Control Fault. A second hypocenter location of 20820, 20614, -1600 was obtained by recalculating the macroseismic system data using the CALCX Lotus program and adjusting velocities to correspond with the latest blast hole calibration data. This location is close to the vein and reasonably close to observed damage in the 5100-106 stope. Figure 9 shows the various locations. It was concluded that the location of the event was farther out in the hanging wall than indicated by the microseismic system.

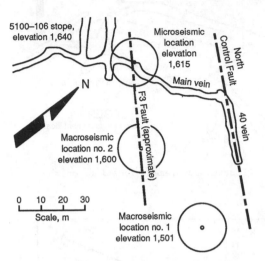

Figure 9. 5100-106 stope plan showing possible burst locations.

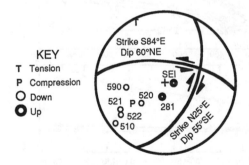

Figure 10. Double-couple solution for event.

The source mechanism was plotted for event 91091902 (Figure 10). All arrivals were dilatational, but a double-couple solution can fit the data. The N25°E solution is within 10° of the strike and dip of bedding in this area.

The University of Idaho's North Idaho Seismic Network is a backup to the Bureau's array for determining source mechanisms, but the network was not operating at the time because of transmitter problems.

5 DATA ANALYSIS AND INTERPRETATION

The source mechanism for this event may have been a single slip on the North Control Fault, the collapse of a highly stressed pillar or pillars, block movement between two faults with differential slip along bedding, movement along the F3 Fault, block movement associated with the creation of a shear in

the hanging wall, or a combination of any or all of these mechanisms.

The damage pattern in the 5100-106 stope indicates the event involved movement of the southeast wall between the North Control Fault and the F3 Fault into the stope. This is suggested because mining along the North Control Fault would reduce the normal forces across it and the F3 Fault, allowing movement toward the mined-out veins. A soft argillite bed could be the fourth side of a prism-shaped block that failed, creating a new shear fracture across bedding toward the top of the ore pillar. Figure 11 is an idealized view of the displacement of this block into the mined-out area as a result of the high horizontal load created by mining-induced stresses on the ore pillar.

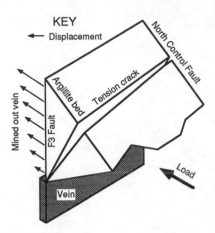

Figure 11. Movement of block to vein with formation of shear.

Damage patterns and geology in the 4660 northeast lateral, 108 cross-cut, and 110 cross-cut indicate that a block of rock bounded by the North Control Fault, the lateral, and the F3 Fault moved up-dip along bedding toward the lateral. This block may be unstable because mining of the cross-cuts along the faults has weakened boundaries of the block. Positioning the lateral close to the vein (approximately 20 m in the footwall) may have created a pillar too small to support the loads transferred to it and provided the block a place to move. The movement on the 4660 level could be visualized as a deck of cards sliding out of a box.

There are two possible explanations for the fact that the mode of failure within the block changed from creep to instantaneous seismic failure, as evidenced by the destruction of closure readings. The first is that loading from the 5100-106 area was transferred by the burst to the 4660 area, which then failed immediately. The second possibility is that the shockwave from the 5100-106 area caused the damage on the 4660. The authors prefer the load transfer rather than the burst theory because of the magnitude of damage on the 4660 level. A contributing factor may have been increasing the number of stopes on this limb of the vein from two to three only months before the event, thus increasing the rate at which load was being transferred to the pillars.

Regardless of which mechanism caused the damage on the 4660 level, the important observation was that 90% of the damage occurred within a well defined block of ground between the North Control Fault and the F3 Fault.

First motion analysis of the digital macroseismic waveform data for the event indicates that movement along bedding in the hanging wall of the 5100-106 stope was the source mechanism because one component of the solution agrees very closely, within 10°, of the strike and dip of the local bedding.

It is possible that if more geophones had been available, they would have sensed dilatational motion. If so, then a double-couple solution may not be correct. It would then be possible to associate the event with pillar failure, though damage observations and the source location did not support this mechanism.

At the present time, it is believed that the most probable interpretation of the data is that shear failure in the hanging wall of the 5100-106 stope allowed a prism-shaped block of beds between the North Control Fault and the F3 Fault to move into the mined-out portion of the vein above the stope. The load from the 5100-106 stope was transferred to the 4660 pillar, which failed simultaneously.

6 CONCLUSIONS

Using geologic mapping observations of mine damage and closure measurements, the movement of blocks of wall rock was defined as the source mechanism for a large rockburst. This information was combined with source location and first-motion studies using digital seismic records to develop an understanding of the failure process. Understanding the relationship between geologic features, mining, mining-induced stresses, and rockburst source mechanisms will result in improved mine designs. This research concluded that blocks of competent quartzite created by mining could fail along weak argillite beds and were the source of seismic activity.

REFERENCES

Aki, K., and P.G. Richards. 1980. Quantitative Seismology: Theory and Methods. Freeman, Cooper, San Francisco, CA.

Blake, W., and D.J. Cuvelier. 1990. Developing Reinforcement Requirements for Rockburst Conditions at Hecla's Lucky Friday Mine. In Rockbursts and Seismicity in Mines: Proceedings of the 2nd International Symposium on Rockbursts and Seismicity in Mines, ed. by C. Fairhurst (Minneapolis, MN, June 8-10, 1988). A.A. Balkema, Rotterdam, Netherlands, pp. 407-409.

Jenkins, F.M., T.J. Williams, and C.J. Wideman. 1990. Analysis of Four Rockbursts in the Lucky Friday Mine, Mullan, Idaho, USA. Paper in the International Deep Mining Conference: Technical Challenges in Deep Level Mining, ed. by D.A.S. Ross-Watt and P.D.K. Robinson (Johannesburg, South Africa, Sept 17-21, 1990). South Africa Inst. of Min. and Metall., Johannesburg, South Africa, pp. 1201-1212.

Lee, W.H.K., D.M. Tottingham, and O.J. Ellis. 1988. A PC-Based Seismic Data Acquisition and Processing System. U.S. Geol. Surv. Open File Report 88-751, p. 31.

Noyes, R.R., G.R. Johnson, and S.D. Lautenschlaeger. 1988. Underhand Stoping at the Lucky Friday Mine in Idaho. Paper No. 12, 94th Annual Northwest Mining Assoc., Spokane, WA, December.

Scott, D.F. 1990. Relationship of Geologic Features to Seismic Events, Lucky Friday Mine, Mullan, Idaho. In Rockbursts and Seismicity in Mines: Proceedings of the 2nd International Symposium on Rockbursts and Seismicity in Mines, ed. by C. Fairhurst (Minneapolis, MN, June 8-10, 1988). A.A. Balkema, Rotterdam, Netherlands, pp. 401-405.

Williams, T.J. and D.J. Cuvelier. 1990. Report on a Field Trial of an Underhand Longwall Mining Method to Alleviate Rockburst Hazards. In Rockbursts and Seismicity in Mines: Proceedings of the 2nd International Symposium on Rockbursts and Seismicity in Mines, ed. by C. Fairhurst (Minneapolis, MN, June 8-10, 1988). A.A. Balkema, Rotterdam, Netherlands, pp. 349-353.

Wong, I., and A. McGarr. 1990. Implosional Failure in Mining Induced Seismicity: A Critical Review. In Rockbursts and Seismicity in Mines: Proceedings of the 2nd International Symposium on Rockbursts and Seismicity in Mines, ed. by C. Fairhurst (Minneapolis, MN, June 8-10, 1988). A.A. Balkema, Rotterdam, Netherlands, pp. 45-52.

Tectonic stresses in mine seismicity: Are they significant?

Ivan G.Wong
Woodward-Clyde Federal Services, Oakland, Calif., USA

ABSTRACT: Lithostatic stresses and their redistribution around mine excavations are thought to be the dominant driving force in the generation of most mine tremors and rockbursts. In some cases, however, it has been suggested that tectonic stresses also play a significant role. These cases have often involved the largest mine tremors, M 4 to 5+. I have reviewed the available focal mechanisms from cases of mining-induced seismicity in an attempt to evaluate the effects of tectonic stresses. Data from the World Stress Map Project were used as a basis for characterizing the broad-scale tectonic stresses operative in the mining districts which were investigated. Based on the focal mechanism data, compressive tectonic stresses appear to play a major role in the generation of mine seismicity worldwide. In general, mine tremors and rockbursts occur in regions being subjected to moderate to high horizontal compressive tectonic stresses and which often possess moderate and sometimes high levels of natural seismicity. The larger events, for which focal mechanisms are most often available, are generally the result of reverse and sometimes strike-slip faulting. This mode of deformation probably reflects the increases in shear stress along pre-existing zones of weakness i.e., faults, due to unloading. In cases where tectonic stresses may not be the primary driving force, they could act to trigger mine tremors where the principal source of strain energy is due to the redistributed lithostatic stresses.

1 INTRODUCTION

Since the early investigations in the 1950's in the gold mines of South Africa, rockbursts and mine seismicity have been considered to be the mechanisms by which redistributed and concentrated lithostatic stresses around mine excavations are violently released. Other components of the in situ stress field, such as tectonic and residual stresses, have also been recognized by several investigators as having a possible effect on rockbursts and mine tremors (e.g., McGarr et al. 1975; Gibowicz 1984; Knoll and Kuhnt 1990; Wong 1993). In particular, cases influenced by tectonic stresses have often involved some of the largest mine tremors ever observed (magnitude [M] 4 to 5+), events occurring at distances beyond a few tens of meters or beneath the mine workings, or events resulting from seismic slip along large faults. In addition to the in situ stresses, the role of pre-existing faults has been critical in many cases of mine seismicity as described by numerous investigators (e.g., Gay et al. 1984).

To understand completely the role of tectonic stresses in mine seismicity, the other components of the in situ stress field must be characterized. Unfortunately, it has been difficult to distinguish and quantify these components in mines. Overcoring, which is by far the most common method for characterizing the in situ stresses in mines, appears to suffer from a number of problems and hence is not considered to be wholly reliable (Zoback 1992). The determination of mine tremor and rockburst focal mechanisms, which have been increasingly utilized in mine seismicity research, has proven to be an extremely useful tool. This is especially the case if the characteristics of the tectonic stress field are known based on other types of data such as earthquake focal mechanisms and other in situ measurements (e.g., borehole breakouts and hydraulic fractures).

In 1986, the World Stress Map Project of the International Lithosphere Program was initiated with the objective of compiling a global database of contemporary in situ stress measurements in the earth's crust using a variety of geophysical and geological techniques (Zoback 1992). This data has allowed for the first time, a comprehensive view of crustal stresses on a global scale. In this paper, I have reviewed, in the context of these tectonic stresses, much of the available published focal mechanism data from cases of mining-induced seismicity in an attempt to evaluate their role in the generation of mine tremors and rockbursts.

2 CRITERIA FOR SEISMIC SLIP

In general, most mine tremors and rockbursts are no different than tectonic earthquakes in their source processes; both are the result of shear failure in brittle rock (McGarr 1971; Spottiswoode and McGarr 1975). (See Wong and McGarr [1990] for a discussion of one form of non-shear failure in mine seismicity.) Brace and Byerlee (1966) suggested that the primary cause of earthquakes is frictional sliding (stick-slip motion) on pre-existing faults although the fracturing of intact rock on a small scale may also be involved. McKenzie (1969) stated that pre-existing faults may slip even at very low resolved shear stresses before the failure of intact rock and the formation of new fractures. Furthermore, Raleigh et al. (1972) suggested that seismic slip along pre-existing zones of weakness is favored over fracture of intact rock if the zone is oriented within 10° to 50° of the maximum principal stress (Figure 1). This observation was in reference to a case of induced seismicity due to fluid-injection at the Rangely oil field in Colorado.

Based on an analysis of heterogeneous fault strength in a pervasively fractured crust, Hill and Thatcher (1992) suggested that: (1) slip will be energetically favored on faults oriented 45° to 50° to the maximum principal stress if the coefficient of friction along these faults is only 20% to 25% lower than along faults at the optimum Coulomb angle of 25° to 30° with typical intrinsic coefficients of 0.70 to 0.75; (2) for very small values of frictional strength and low ambient shear stress, the 45° angle for optimal fault slip is only weakly favored over a variety of fault orientations centered on 45°; and (3) poorly oriented faults with angles greater than 80° can possibly slip if the effective coefficient of friction along such faults is less than 0.2.

Because shear failure can occur along faults or fractures even when they are oriented rather unfavorably to the maximum principal stress, advancing such pre-existing zones of weakness towards failure may not require large increases in shear stress. Observations of other forms of induced seismicity (e.g. reservoir induced) suggest that low levels of induced stresses can trigger seismic slip on faults and other pre-existing zones of weakness if they are already stressed to a state of near failure.

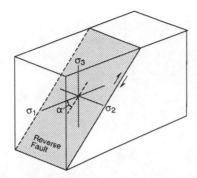

Figure 1. Illustrated is a block diagram of a reverse fault in a compressional stress regime where the maximum (σ_1) and intermediate (σ_2) principal stresses are horizontal. The minimum principal stress (σ_3) due to the lithostatic load is vertical. Seismic slip will occur on a pre-existing zone of weakness e.g., fault rather than the fracture of intact rock if it is oriented favorably to σ_1. This orientation, as indicated by the angle α, is generally 10° to 50° although the angle is dependent upon the effective coefficient of friction along the fault plane. The maximum shear stress acting on the fault plane is equal to $1/2(\sigma_1 - \sigma_3)$ acting in a direction which bisects the angle between σ_1 and σ_3.

A specific example of a small triggering stress was associated with the seismicity induced by surface mining in a Wappinger Falls, New York quarry. A sequence of earthquakes with a mainshock of body-wave magnitude (m_b) 3.3 occurred at depths of less than 1.5 km beneath an active limestone quarry (Pomeroy et al. 1976). The mainshock focal mechanism indicated thrust faulting and a northeast-southwest trending P-axis. Pomeroy et al. (1976) suggested that crustal unloading due to the quarry operations in the highly compressive stress regime in this part of the eastern U.S. triggered the earthquake activity. The triggering or unloading stress was estimated to be only 7 bars.

3 MINE SEISMICITY AND TECTONIC STRESSES

With these criteria and examples in mind, let us examine some of the most significant cases of mine seismicity worldwide to assess whether tectonic stresses are a factor in their generation. In the following, selected references are cited which I believe are notable and representative of the available published focal mechanism data.

3.1 United States

Within the U.S., the best documented cases of mine seismicity occur in the silver, lead, and zinc mining district of the Coeur d'Alene in northern Idaho and the coal mining areas of the eastern Wasatch Plateau and Book Cliffs in central Utah (Figure 2). Both cases occur within the western U.S., where several tectonic stress regimes exist (Zoback and Zoback 1989), in large part because of the proximity to a major plate boundary along the Pacific Coast and the high level of associated tectonism.

Focal mechanisms determined by Lourence et al. (1993) for 106 tremors in the Lucky Friday mine in the Coeur d'Alene district exhibit predominantly strike-slip and oblique-reverse faulting with generally north-south oriented P-axes (Figure 2). The strike-slip faulting focal mechanism of a M 4.1 earthquake which occurred at a depth of about 10 km, 15 km northeast of the Lucky Friday mine indicates a northwest-southeast trending P-axis (Sprenke et al. 1991). These mechanisms are all consistent with the Coeur d'Alene district being located within the Pacific Northwest compressional stress province as defined by Zoback and Zoback (1989) (Figure 2) rather than an extensional regime as suggested by Sprenke et al. (1991). Overcoring measurements in the vicinity of the Lucky Friday mine also indicate a compressional stress field with the northwest-southeast trending maximum principal stress about 1.3

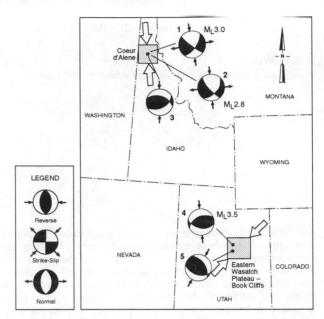

Figure 2. Typical focal mechanisms of mine seismicity in the intermountain U.S. Inward arrows on focal mechanisms indicate horizontal projections of the maximum principal stress and outward arrows represent the minimum principal stress. Mechanisms 1 – 3 are from Lourence et al. (1993) and 4 – 5 from Wong et al. (1989). Magnitudes of largest tremors are shown. Large arrows indicate the regional maximum principal stress direction.

times greater than the lithostatic or minimum principal stress (Lourence et al. 1993).

The mine tremors in the Coeur d'Alene appear to be similar to tectonic earthquakes based on the focal mechanism data; both are the result of generally strike-slip or reverse faulting in a compressive stress regime. This suggests that these mine tremors can be viewed as triggered earthquakes and that tectonic stresses play a major role in their generation.

In the eastern Wasatch Plateau, focal mechanisms of mine tremors of up to Richter magnitude M_L 3.5, exhibit reverse faulting on generally west-to-northwest-trending planes (Wong et al. 1989) (Figure 2). These events occur as deep as 2 to 3 km beneath the mines which are located only a few hundred meters beneath the ground surface. Williams and Arabasz (1989) determined 10 focal mechanisms for mine tremors beneath the East Mountain mines in the eastern Wasatch Plateau; all but one mechanism exhibited reverse faulting in response to a northwest to northeast-oriented maximum principal stress. As is the case for the Coeur d'Alene, the eastern Wasatch Plateau is located adjacent to the Intermountain seismic belt, a major zone of seismicity in the western U.S. A composite focal mechanism of three small earthquakes, which occurred about 5 km away from the coal mines, indicates reverse faulting and a northeast-southwest P-axis (Wong et al. 1989). The latter is consistent with the suggestion that this area occurs within a stress field which is characterized by northeast-southwest compressive stresses. In contrast, the tectonic stress field of the Intermountain U.S. surrounding the eastern Wasatch Plateau and Book Cliffs is extensional in nature (Zoback and Zoback 1989).

Based on two-dimensional finite element modeling, Wong (1985) estimated that the stresses induced by coal mining could be as large as several hundred bars in the vicinity of the mine workings and still a few bars at depths of 2 km beneath the mines. These observations suggest that at least some of the mine seismicity in the eastern Wasatch Plateau, including the larger events, are probably due to triggering by unloading of tectonically pre-stressed faults in a compressive stress regime (Wong 1985; Wong et al. 1989).

Although no known focal mechanism data are available, some coal mining areas in the eastern U.S. have been subjected to mine tremors. One such area is Buchanan County in western Virginia (Bollinger 1989). The eastern U.S. is a region characterized by generally northeast-southwest oriented compressive tectonic stresses and a low to moderate level of earthquake activity (Zoback and Zoback 1989).

3.2 Eastern Canada

The deep metalliferous mines of eastern Canada, particularly in the Sudbury Basin of northern Ontario, have undergone significant mine seismicity and rockbursts in the past decade. This portion of eastern Canada is also seismically active with earthquakes resulting from reverse, and sometimes strike-slip faulting in a compressive tectonic stress field (Adams and Bell 1991). The maximum principal stress in eastern Canada is generally oriented northeast to east although there appears to be a wide range of local variability (Adams and Bell 1991) (Figure 3). These horizontal compressive stresses are very high, often being at least two times larger than the lithostatic stress.

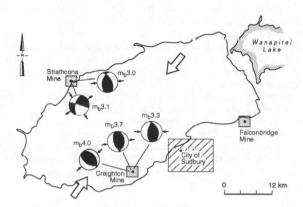

Figure 3. Typical focal mechanisms of mine seismicity in the Sudbury Basin. Focal mechanisms and base map (modified) are from Wetmiller et al. (1990).

Wetmiller et al. (1990) determined focal mechanisms for five large mine tremors in the Sudbury Basin between 1984 and 1987 (Figure 3). The mechanisms of two events in the Strathcona mine, m_b 3.1 and 3.0, exhibit strike-slip and reverse faulting with northwest-southeast and east-west-trending P-axes, respectively. The focal mechanisms of three Creighton mine tremors, m_b 4.0, 3.3, and 3.7, exhibit reverse faulting in response to a northeast-southwest to east-west-oriented compressive stress field. As noted by Wetmiller et al. (1990), four of these large mine tremors appear to be very similar to tectonic earthquakes in that they are the result of generally reverse faulting in the compressive tectonic stress field of eastern Canada (Figure 3).

Urbancic and Young (1993) also report reverse faulting in the Strathcona mine based on 796 microseismic (M < 0) focal mechanisms. Most of the mechanisms exhibit northeast-striking reverse faulting in response to a northwest-southeast oriented maximum principal stress similar to the m_b 3.1 mine tremor mechanism determined by Wetmiller et al. (1990). This stress orientation appears to correspond to the local stress field as indicated by overcoring tests. A smaller number of reverse faulting focal mechanisms indicate a northeast-southwest compressive stress field similar to the regional tectonic stress field (Urbancic and Young 1992) (Figure 3). These data suggest that both the local and regional tectonic stress fields influence the generation of seismicity in this mine. The local stress field may represent a distortion of the regional stress field due to the mine openings. It may also be the case that northwest-southeast compression actually represents a local variation of the regional stress field on the northwestern side of the Sudbury Basin where the Strathcona mine is located (Figure 3). Although the data of

Urbancic and Young (1993) are for much smaller events than the other examples cited in this study, the observation that even such small tremors are the result of reverse and not normal faulting is noteworthy (see Section 4).

3.3 Europe

Mine seismicity is fairly widespread in Europe involving a wide range of mineral deposits. Some of the earliest studies were of mine tremors in Polish mines. Gibowicz (1984) computed focal mechanisms for four large mine tremors which occurred in three different mining districts in Poland. Two events were the result of normal faulting although their focal mechanisms exhibit different T-axes orientations: northwest compared to northeast (Figure 4). The tremors were a M_L 4.5 event on 24 June 1977 located between two copper mines in the Lubin Basin and a M_L 4.3 event on 30 September 1980 in the Szombierki coal mine near Bytom in the Upper Silesian Basin.

A M_L 4.6 event in the Belchatow surface coal mine, which occurred on 29 November 1980, resulted from reverse or possibly strike-slip faulting in response to generally northeast-southwest compressive stresses (Figure 4). The focal mechanism of a second large Szombierki mine tremor, M_L 4.1, on 12 July 1981, exhibits oblique-reverse slip and also a northeast-southwest P-axis similar to the Belchatow event.

According to Grünthal and Stromeyer (1992), the tectonic stress field in eastern Europe is characterized by northeast-southwest compressive stresses (Figure 4). Thus the two reverse faulting focal mechanisms (No. 1 and 4; Figure 4) suggest that the Belchatow and 1981 Szombierki tremors were directly influenced by the tectonic stress field. In contrast, the normal faulting mechanisms suggest that for events No. 2 and 3 (Figure 4), the lithostatic stresses were the most important (Gibowicz 1984). In all cases, however, because underground mining is performed at depths of less than 1 km, Gibowicz (1984) suggested that tectonic stresses must be involved to some degree in the generation of these large mine tremors since the lithostatic stresses are relatively small. All four mine tremors had focal depths of 2 km or less and thus the interaction between lithostatic and tectonic stresses may be complex with either being dominant; specifically, the principal stresses may be nearly equivalent such that the orientations of the maximum and minimum principal stresses are interchangeable.

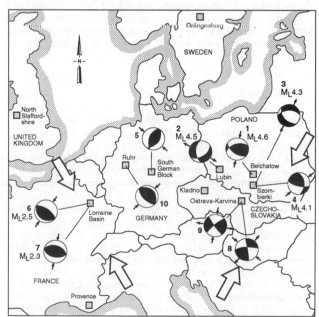

Figure 4. Significant occurrences of mine seismicity in Europe and typical focal mechanisms of mine tremors. Focal mechanisms are: 1–4, Gibowicz (1984); 5, Knoll and Kuhnt (1990); 6 – 7, Hoang-Trong et al. (1988); 8 – 9, Sileny (1989); and 10, Hinzen (1982).

Other significant European cases of mine seismicity include (Figure 4): (1) the potash mines in the seismically-active South German Block in Germany; (2) coal mines in the Ruhr district, West Germany; (3) coal mines in the Provence field and the Lorraine Basin in France; (4) iron ore mines near Grängesberg in central Sweden; (5) coal mines in North Staffordshire, United Kingdom; and (6) coal mines in the Kladno district near Prague and in the Ostrava Karvina district in northern Morovia (and the Upper Silesian Basin), both in Czechoslovakia. All cases occur within the compressional regimes of either western Europe, which is characterized rather uniformly by a northwest-southeast maximum principal stress (Müller et al. 1992) or eastern Europe (Figure 4).

In the South German Block, a focal mechanism of a damaging rockburst exhibits reverse faulting and a northwest-southeast P-axis (Knoll and Kuhnt 1990) consistent with the orientation of the regional maximum principal stress (Figure 4). A composite focal mechanism of 21 small tremors ($M_L \leq 2.3$) in the Ruhr district also reveals reverse faulting but in response to a northeast-southwest maximum principal stress (Hinzen 1982) (Figure 4). Reverse faulting focal mechanisms are typically observed for rockbursts in the Provence coal mines (Revalor et al. 1990), which may also reflect an influence of the tectonic stress field. In situ stress measurements in the Provence coal field indicate that the maximum principal stress is about 1.5 times greater than the lithostatic stress (Gaviglio et al. 1990). Most of the mine tremors observed in the Lorraine coal mines are also the result of reverse faulting (Hoang-Trong et al. 1988). Their focal mechanisms exhibit northeast-southwest trending P-axes, almost orthogonal to the northeast-southwest compressive stress direction for western Europe (Figure 4). Unfortunately, no focal mechanisms are available for the Grängesburg rockbursts (Båth 1984) and the mechanisms for the North Staffordshire mines (Westbrook et al. 1980) are not well constrained.

Extensive seismological studies have been performed in the Czechoslovakian coal mines in the past decade. An analysis by Sileny (1989) indicates that Kladno tremors are the result of predominantly normal faulting with varying components of strike-slip faulting and for the Ostrava Karvina events, strike-slip and normal faulting. The Kladno mine seismicity does not apparently reflect a dominant tectonic influence since the normal faulting indicates an extensional stress field rather than the compressive tectonic regime of eastern Europe. The relatively shallow depths of these events (less than 1 km) may account for the lack of a significant tectonic stress influence. Although the Ostrava Karvina strike-slip focal mechanisms show considerable variability, they all exhibit a horizontal maximum principal stress which may reflect some tectonic influence on the mine tremors (Figure 4).

3.4 South Africa

The first studies of mine seismicity and certainly the most significant in terms of advancing our understanding of the source processes of mine tremors were performed in the deep gold mines of South Africa. Seismic events thought to be associated with the deep gold mining were first observed in 1908 in the Witwatersrand district.

In one of the earliest seismological studies, Spottiswoode et al. (1971) observed that most large seismic events in the Witwatersrand were the result of shear failure. They determined focal mechanisms for 11 events (M_L 1 to 3.5) in the East Rand Proprietary Mines and found them to be consistent with a source mechanism appropriate for tectonic earthquakes associated with normal faulting. This type of mechanism is compatible with underground observations of burst fractures and with the violent failure predicted by McGarr (1971) on the basis of theoretical stress calculations around a stope. According to McGarr's model, the maximum principal stress is oriented nearly vertical in the hanging wall and footwall in the region of highly stressed rock extending typically 10 m ahead of the face. The focal

mechanisms, as expected for normal faulting, all exhibited a vertical maximum principal stress; however, the direction of the minimum principal stress varied considerably, suggesting that a uniform stress field of tectonic origin was not present (Spottiswoode et al. 1971).

From 1971, when a seismic network was established covering the four major mines in the Klerksdorp district, more than 6,000 events ranging from M_L 0.2 to 5.4 were recorded through 1981 (Gay et al. 1984). The mine tremors, including the largest events, appeared to concentrate near faults, especially those with significant lithologic contrasts across them. In situ stress measurements in some areas showed high deviatoric stresses, suggesting that residual stresses (originating from past tectonic deformation) may play a dominant role in the generation of some events (Gay et al. 1984).

Potgeiter and Roering (1984) determined focal mechanisms for 10 tremors, M_L 1.8 to 3.8, that occurred during 1980 and 1981 in the Klerksdorp district. Focal mechanisms of five events located near a major fault and within 25 m of an advancing stope exhibited normal faulting with the maximum principal stress oriented vertically, similar to the observations of Spottiswoode et al. (1971) in the Witwatersrand district. All mechanisms exhibited a nodal plane parallel to the fault, strongly suggesting that the fault had been reactivated by the mining. The remaining events occurred near an intrusive diabase dyke and an advancing stope. Somewhat surprisingly, their focal mechanisms exhibited reverse faulting with the maximum principal stress oriented horizontally and to the northwest. One nodal plane of each mechanism was parallel to the dyke, suggesting shear failure along this structure. According to Potgeiter and Roering (1984), the contrast in faulting and associated stress fields of the 10 closely-spaced events suggests that (1) localized residual stresses were principally responsible for the generation of the reverse faulting tremors along pre-existing geologic structures and (2) the normal faulting events were due solely to mining-induced stresses acting on the major fault.

Brummer and Rorke (1990) analyzed five large South African rockbursts, M 3.6 to 5.2, which occurred from 1982 to 1988. All events were the result of slip on pre-existing faults and four of the five tremors, for which the actual fault displacements could be observed, displayed normal slip. In addition, two of the events exhibited both normal and reverse slip. According to Brummer and Rorke (1990), reverse slip is possible when two stopes are involved (see their Figure 10). Such movement will occur in the hanging wall of the upper stope and the footwall of the lower stope. Unfortunately, focal mechanisms are apparently not available for these events; thus the coseismic rupture process involving both normal and reverse slip cannot be characterized.

Based on rather sparse stress data for South Africa compiled and quality ranked as part of the World Stress Map Project, the region appears to be characterized by a strike-slip faulting stress field (vertical intermediate principal stress) at its southern end and an extensional regime (intermediate principal stress generally trending north to northwest) in the rest of the country (Zoback 1992). If indeed the mining districts are located in an extensional regime, this would be unlike other major occurrences of mine seismicity which are located in compressional tectonic regimes.

3.5 Other Countries

One of the most significant occurrences of mine seismicity and rockbursts in terms of human and economic losses is in the Kolar gold fields of southern India. Mining has been conducted for more than a century reaching to depths of 3.2 km (Srinivasan and Shringarputale 1990). Although no focal mechanisms for mine tremors have been determined to date, Srinivasan and Shringarputale (1990) classify the mine tremors into: (1) events in close proximity to the active stopes resulting from stress redistribution around the mines and (2) events located away from

276

the stopes due to the redistribution of regional stresses. According to Gowd et al. (1992), the Kolar gold fields are located in a seismically active portion of India which is characterized by northwest-southeast compressive tectonic stresses.

Other important cases of mine seismicity have occurred in: (1) Japanese coal mines including the Horonai mine in the Ikushunbetsu Basin (Sato et al. 1989); (2) southwestern Australia e.g., the Mt. Charlotte gold mine (Lee et al. 1990); (3) metalliferous and coal mines in China such as the Chengzi coal mining district (Mei and Lu 1987); and (4) mines in coal and metallic ores in the Soviet Union including the Tkibuli-Shaorsk coal field (Petukhov 1987). For the Horonai mine, well-constrained focal mechanism data are not available although the solutions that have been determined suggest normal faulting. To my knowledge, no focal mechanisms have been calculated for the Mt. Charlotte, Chinese or Soviet Union mines. Japan and China are characterized by strike-slip faulting stress regimes (northeast to southeast maximum principal stress) and southwestern Australia by east-west compressive tectonic stresses (Zoback 1992). Unfortunately the state of stress is poorly known in the Soviet Union (Zoback 1992) although it is likely that most of the country is subjected to compressional tectonic stresses.

4 DISCUSSION AND CONCLUSIONS

In several occurrences, such as the Sudbury Basin, the eastern Wasatch Plateau, the Coeur d'Alene, the South German Block, and some of the Polish mines, mine tremors, especially the larger events, are indistinguishable from natural tectonic earthquakes which also occur in these regions. Both are the result of reverse and to a lesser extent, strike-slip faulting in response to compressive stresses characteristic of the tectonic stress fields. In some cases, such as the Lorraine Basin, the Ruhr district, the Strathcona mine in the Sudbury Basin, and possibly the Ostrava Karvina district, the P-axes of the focal mechanisms deviate from the regional compressive stress direction suggesting that the tectonic stresses are a significant but not necessarily dominant factor in the generation of the mine tremors. This assumes of course, the uniformity of regional stress fields which is not always the case. In particular, variations of the regional stress field occur in Europe due to localized tectonic sources (Grünthal and Stromeyer 1992). Thus some of the above cases may not actually deviate significantly from the local stress fields. The uncertainties in the P-axes of focal mechanisms, the maximum principal stress direction, and the correlation between them also needs to be considered in such assessments.

Although focal mechanism data are not available for the mine tremors occurring in the Grängesburg, Kolar, eastern U.S., Mt. Charlotte, and Chenghzi mines, they are located in regions characterized by compressive tectonic stresses. Thus based on a review of available focal mechanism data and our current understanding of the state of stress worldwide, it appears that for the most part, the significant occurrences of mine seismicity are located in compressive tectonic stress regimes. Such areas are also often characterized by moderate and sometimes high levels of earthquake activity attesting to active tectonism and significant tectonic stresses.

The observation that cases of mine seismicity generally occur in compressive regimes may not be all that surprising given the fact that much of the earth's crust is being subjected to compressive rather than extensional tectonic stresses. However, mining has been and is being performed, for example, in many areas of the extensional western U.S. with no significant cases of mine seismicity having occurred.

The mine seismicity and rockbursts taking place in the deep gold mines of South Africa may be the major exception. Historically, South Africa has been characterized by only minor earthquake activity and thus tectonic stresses do not appear to be significant as is often the case in such intraplate settings. Yet

because of the predominance of normal faulting in South African mine seismicity, extensional tectonic stresses should not be discounted as a factor in influencing the generation of mine tremors and rockbursts in these deeps mines. Given this possibility, the great depth of mining and hence high lithostatic stresses may be the most important factor in the generation of seismicity in these mines although as noted by several investigators, residual tectonic stresses in some cases may also be significant.

It is commonly believed that normal faulting due to stope closure (McGarr 1971) is the mechanism which generates most mine tremors. Examples may be the mine seismicity in the Kladno, North Staffordshire and some of the Polish mines. Thus it is somewhat surprising that reverse and to some extent, strike-slip faulting appears to be the primary mode of deformation in most regions of mine seismicity based on the available focal mechanism data.

These observations suggest that compressive tectonic stresses play a greater role in the generation of mine seismicity worldwide than was previously thought. As has been noted by a number of early researchers (e.g., Smith et al. 1974), the effect of mining in the earth's crust, in a relatively global sense, is to decrease lithostatic stresses through unloading by the removal of mass. The horizontal stresses will also be slightly increased or decreased due to the unloading and Poisson's effect (Wong 1985). In a compressive stress regime, this effect will advance pre-existing zones of weakness such as faults toward failure by decreasing the vertically oriented minimum principal stress and increasing any shear stresses acting on the faults (Figure 1). (Note in a strike-slip stress regime, unloading mainly decreases the intermediate principal stress although the horizontal maximum or minimum principal stresses will also be altered possibly resulting in increased shear stresses along steeply dipping faults.) As a result, coseismic rupture may occur depending on the coefficients of friction along such faults and their orientation with respect to the maximum principal stress. This effect is illustrated by the familiar Mohr circle diagram (see Figure 6 in Wong [1985]). In mines where extraction is being conducted at multiple levels, the unloading effect can occur throughout the workings.

In addition to this unloading effect, mining obviously leads to stress redistribution around the mine openings where both the in situ vertical and horizontal stresses can be altered. Again in a compressive stress regime, a decrease in the lithostatic stress (minimum principal stress) or increase in the maximum horizontal stress can lead to failure and mine seismicity.

ACKNOWLEDGMENTS

The preparation of this paper was partially supported by Woodward-Clyde Professional Development funds. My thanks to Fumiko Goss for her assistance. My appreciation to Markus Båth, Ken Sprenke, Mike Stickney, Ted Urbancic, Paul Young, Mary Lou Zoback, John Adams, C. Srinivasan, and Katsuhiko Sugawara for reprints. The paper benefitted from critical reviews by Jackie Bott, Paul Young and an anonymous reviewer.

REFERENCES

Adams, J. and J.S. Bell 1991. Crustal stresses in Canada. *In* D.B. Slemmons, E.R. Engdahl, M.D. Zoback, and D.D. Blackwell (eds.), Neotectonics of North America, Geol. Soc. Am. Decade Map Volume, p. 367-386.

Båth, M. 1984. Rockburst seismology. *In* N.C. Gay and E.H. Wainwright (eds.), Rockbursts and Seismicity in Mines, S. Afr. Inst. Min. Metall. Symp. Series No. 6, p. 7-15.

Bollinger, G.A. 1989. Microearthquake activity associated with underground coal-mining in Buchanan County, Virginia, U.S.A. PAGEOPH 129:407-413.

Brace, W.D. and J.D. Byerlee 1966. Stick-slip as a mechanism for earthquakes. Science 153:990-992.

Brummer, R.K. and A.J. Rorke 1990. Case studies on large rockbursts in South African mines. *In* C. Fairhurst (ed.), Proc. 2nd Int. Symp. Rockbursts and Seismicity in Mines, Balkema Publishers, Rotterdam, p. 323-329.

Gaviglio, P., R. Revalor, J.P. Piguet, and M. Dejean 1990. Tectonic structures, strata properties and rockbursts occurrence in a French coal mine. *In* C. Fairhurst (ed.), Proc. 2nd Int. Symp. Rockbursts and Seismicity in Mines, Balkema Publishers, Rotterdam, p. 289-293.

Gay, N.C., D. Spencer, J.J. van Wyk, and P.K. van der Heever 1984. The control of geological and mining parameters in the Klerksdorp gold mining district. *In* N.C. Gay and E.H. Wainwright (eds.), Rockbursts and Seismicity in Mines, S. Afr. Inst. Min. Metall. Symp. Series No. 6, p. 29-37.

Gibowicz, S.J. 1984. The mechanism of large mining tremors in Poland. *In* N.C. Gay and E.H. Wainwright (eds.), Rockbursts and Seismicity in Mines, S. Afr. Inst. Min. Metall. Symp. Series No. 6, p. 17-29.

Gowd, T.N., S.V. Srirama Rao, and V.K. Gaur 1992. Tectonic stress field in the Indian subcontinent. J. Geophys. Res. 97:11,879-11,888.

Grünthal, G. and D. Stromeyer 1992. The recent crustal stress field in central Europe: Trajectories and finite element modeling. J. Geophys. Res. 97:11,805-11,820.

Hasegawa, H.S., R.J. Wetmiller, and D.J. Gendzwill 1989. Induced seismicity in mines in Canada - an overview. PAGEOPH 129:423-453.

Hill, D.P. and W. Thatcher 1992. An energy constraint for frictional slip on misoriented faults. Bull. Seismol. Soc. Am. 82:883-897.

Hinzen, K.G. 1982. Source parameters of mine tremors in the eastern part of the Ruhr-District (West Germany) J. Geophys. 51:105-112.

Hoang-Trang, P., J.F. Gueguen, J.M. Holl, and P. Schroeter 1988. Near field seismological observations in the Lorraine coal mine (France): Preliminary results. *In* Proc. Workshop Induced Seismicity and Associated Phenomena, Liblice, Czechoslovak Academy of Sciences, v. 1, p. 64-74.

Knoll, P. and W. Kuhnt 1990. Seismological and technical investigations of the mechanics of rockbursts. *In* C. Fairhurst (ed.), Proc. 2nd Int. Symp. Rockbursts and Seismicity in Mines, Balkema Publishers, Rotterdam, p. 129-138.

Lee, M.F., G. Beer, and C.R. Windsor 1990. Interaction of stopes, stresses, and geologic structure at the Mount Charlotte mine, western Australia. *In* C. Fairhurst (ed.), Proc. 2nd Int. Symp. Rockbursts and Seismicity in Mines, Balkema Publishers, Rotterdam, p. 337-343.

Lourence, P.B., S.J. Jung, and K.F. Sprenke 1993. Source mechanisms at the Lucky Friday mine: Initial results from the North Idaho Seismic Network (this volume).

McGarr, A. 1971. Violent deformation of rock near deep-level, tabular excavations--seismic events. Bull. Seismol. Soc. Am. 61:1453-1466.

McGarr, A., S.M. Spottiswoode, and N.C. Gay 1975. Relationship of mine tremors to induced stresses and to rock properties in the focal region. Bull. Seismol. Soc. Am. 65:981-993.

McKenzie, D.P. 1969. The relation between fault plane solutions for earthquakes and the directions of the principal stresses. Bull. Seismol. Soc. Am. 59:591-601.

Mei, J. and J. Lu 1987. The phenomenon, prediction and control of rockbursts in China. *In* G. Herget and S. Vongpaisal (eds.), Proc. 6th Int. Congress Rock Mech., v. 2, p. 1135-1140.

Müller, B., M.L. Zoback, K. Fuchs, L. Mastin, S. Gregersen, N. Pavoni, O. Stephanson, and C. Ljunggren 1992. Regional patterns of tectonic stress in Europe. J. Geophys. Res. 97:11,783-11,803.

Petukhov, I.M. 1987. Forecasting and combating rockbursts: Recent developments. In G. Herget and S. Vongpaisal (eds.), Proc. 6th Int. Congress Rock Mech., v. 2, p. 1207-1210.

Pomeroy, P.W., D.W. Simpson, and M.L. Sbar 1976. Earthquakes triggered by surface quarrying - the Wappinger Falls, New York sequence of June 1974. Bull. Seismol. Soc. Am. 66:685-700.

Potgeiter, G.J. and C. Roering 1984. The influence of geology on the mechanisms of mining associated seismicity in the Klerksdorp gold field. *In* N.C. Gay and E.H. Wainwright (eds.), Rockbursts and Seismicity in Mines, S. Afr. Inst. Min. Metall. Symp. Series No. 6, p. 45-50.

Raleigh, C.B., J.H. Healey, and J.D. Bredehoeft 1972. Faulting and crustal stress at Rangely, Colorado. *In* H.C. Heard, I.Y. Borg, N.L. Carter and C.B. Raleigh (eds.), Flow and Fracture of Rocks, Am. Geophys. Union Monograph 16, p. 275-284.

Revalor, R., J.P. Josien, J.C. Besson, and A. Magron 1990. Seismic and seismoacoustics experiments applied to the prediction of rockbursts in French coal mines. *In* C. Fairhurst (ed.), Proc. 2nd Int. Symp. Rockbursts and Seismicity in Mines, Balkema Publishers, Rotterdam, p. 301-306.

Sato, K., Y. Fujii, Y. Ishijima, and S. Kinoshita 1989. Microseismic activity induced by longwall coal mining. *In* H.R. Hardy (ed.), Proc. Fourth Conf. Acoustic Emissions/ Microseismic Activity in Geologic Structures and Materials, Trans Tech Publications, p. 249-263.

Sileny, J. 1989. The mechanisms of small mining tremors from amplitude inversion. PAGEOPH 129:309-324.

Smith, R.B., P.L. Winkler, J.G. Anderson, and C.H. Scholz 1974. Source mechanisms of microearthquakes associated with underground mines in eastern Utah. Bull. Seismol. Soc. Am. 64:1295-1317.

Spottiswoode, S.M. and A. McGarr 1975. Source parameters of tremors in a deep level gold mine. Bull. Seismol. Soc. Am. 65:93-112.

Spottiswoode, S.M., A. McGarr, and R.W.E. Green 1971. Focal mechanisms of some large mine tremors on the Witwatersrand. Chamber of Mines Circular 3/71, 16 p.

Sprenke, K.F., M.C. Stickney, D.A. Dodge, and W.R. Hammond 1991. Seismicity and tectonic stress in the Coeur d'Alene mining district. Bull. Seismol. Soc. Am. 81:1145-1156.

Srinivasan, C. and S.B. Shringarputale 1990. Mine-induced seismicity in the Kolar Gold Fields. Gerlands Beitr. Geophysik Leipzig 1:10-20.

Urbancic, T.I. and R.P. Young 1993. Structural characterization of highly stressed rock masses using microseismic fault-plane solutions. *In* Proc. Int. Soc. Rock Mech. Conf. on Fractured and Jointed Rock Masses (in press).

Westbrook, G.K., N.J. Kusznir, C.W.A. Browitt, and B.K. Holdsworth 1980. Seismicity induced by coal mining in Stoke-on-Trent (U.K.). Eng. Geol. 16:225-241.

Wetmiller, R.J., M. Plouffe, M.G. Cajka, and H.S. Hasegawa 1990. Investigation of natural and mining-related seismic activity in northern Ontario. *In* C. Fairhurst (ed.), Proc. 2nd Int. Symp. Rockbursts and Seismicity in Mines, Balkema Publishers, Rotterdam, p. 29-37.

Williams, D.J. and W.J. Arabasz 1989. Mining-related and tectonic seismicity in the East Mountain area, Wasatch Plateau, U.S.A. PAGEOPH 129:345-368.

Wong, I.G. 1985. Mining-induced earthquakes in the Book Cliffs and eastern Wasatch Plateau, Utah, U.S.A. Int. J. Rock Mech. Min. Sci., Geomech. Abstr. 22:263-270.

Wong, I.G. 1993. The role of geologic discontinuities and tectonic stresses in mine seismicity. *In* J.A. Hudson and E. Hoeks (eds.), Comprehensive Rock Engineering, Principles, Practice and Projects, v. 6, Underground Project Case Histories, Pergamon Press, Oxford (in press).

Wong, I.G., J.R. Humphrey, J.A. Adams, and W.J. Silva 1989. Observations of mine seismicity in the eastern Wasatch Plateau, Utah, U.S.A. A possible case of implosional failure. PAGEOPH 129:369-405.

Wong, I.G. and A. McGarr. 1990. Implosional failure in mining-induced seismicity: A critical review. *In* C. Fairhurst (ed.), Proc. 2nd Int. Symp. Rockbursts and Seismicity in Mines, Balkema Publishers, Rotterdam, p. 45-51.

Zoback, M.L. 1992. First-and-second-order patterns of stress in the lithosphere: The World Stress Map Project. J. Geophys. Res. 97:11,703-11,728.

Zoback, M.L. and M.D. Zoback 1989. Tectonic stress field of the conterminous United States. *In* L.C. Pakiser and W.D. Mooney (eds.), Geophysical Framework of the Continental United States, Geol. Soc. Am. Memoir 172, p. 523-539.

Rockbursts and Seismicity in Mines, Young (ed.) © 1993 Balkema, Rotterdam, ISBN 90 5410 320 5

Expert judgement in making decisions on mining hazards

G.Woo
BEQE Ltd, Sunbury-on-Thames, UK

W. P. Aspinall
Aspinall & Associates, Sunbury-on-Thames, UK

ABSTRACT: Modern decision analysis techniques and probabilistic assessments entail the optimal use of both sparse observational data and subjective degrees of belief to provide a basis for evaluating a hazard or outcome. In this context, the use of expert opinion requires the formal appraisal of elicited judgements to quantify their calibration and informativeness; poorly informed individuals or those who are not impartial may distort an assessment if their contributions are not identified as such. A method for calibrating expert judgement is described which is appropriate for mining hazard analysis and illustrated by a simulated exercise in forecasting energy release in a mining seismicity sequence.

INTRODUCTION

Geophysical monitoring of mines has the important safety function of providing a basis for a real-time appraisal of the risks of rockbursts and induced tremors, which potentially pose a socio-economic threat to mining operations. The past occurrence of severe accidents caused by such events highlights the need for formal decision-making procedures to curtail mining operations in circumstances where danger is judged to be imminent. Decisions of this kind impose on seismologists a major burden of responsibility which is increased by awareness of the serious commercial repercussions of false warnings.

Any decision as to the current level of threat is clouded by the profound degree of uncertainty in the interpretation of conditions which may give rise to sequences of seismic events. At the purely scientific level, certain categories of rockbursts may be associated with a prior build-up of activity, but others may not give any obvious premonitory signal of their occurrence; knowledge of the stress state, seismic properties, fracturation and geology of the country rock can help but their contributions to a prognosis are inexact. Given the present state-of-the-art, the academic Earth scientist might well prefer to reserve his judgement. Yet the practical demands of decision-making on mining safety requires that this expert judgement be exercised. As is often the case in such work, use of expert judgement is traditionally accepted as implicit, and not elicited in any structured way. Because of the absence of any formal procedure for the elicitation of expert judgement, results are susceptible to bias, particularly if the responsibility for making decisions falls on one or two individuals. On the other hand, the common practice of simply averaging the judgements of several individual experts fails to discriminate between them on any grounds, even though reason may exist to suspect the value of some of them. Concern over simplistic methods of aggregating expert judgements leads on to the consideration of methods which involve some calibration of experts.

The issue of decision-making in the presence of uncertainty is a common problem in engineering risk assessment: cancelling a space flight, for example, is as onerous a decision as closing a mine. Techniques for the elicitation of expert judgement in situations of great uncertainty have recently been developed for the European Space Agency. These are based on the concept of an optimal decision-maker, constructed out of the weighted judgements of a selected set of specialist individuals with expertise in a variety of relevant disciplines. In the case of mine safety, seismologists, mining and rock mechanics engineers, and geologists would be among those selected. The weights are assigned through a calibration procedure to test the informativeness and reliability of individuals in making their judgements. A computer program EXCALIBR has been written to implement the construction of the optimal decision-maker.

This aid to decision-making would be very useful in the context of mine safety, for example, because experts would be encouraged to give their true opinions, which would not be quoted directly, but would be incorporated into judgements of the optimal decision-maker. Without this type of formal procedure, traditional methods of gauging opinions may succumb to external socio-economic pressures, which may tend to bias individual decisions in an over-conservative manner. The algorithm is also suitable for application to other aspects of decision-making in an industrial enterprise.

MATHEMATICAL FORMALISM

A salutary lesson learned from the elicitation of expert judgement in the context of risk analysis is that, however well qualified they may be, experts differ widely in their abilities to estimate parameter values and associated confidence bands. Some experts may be poorly calibrated, in the sense that their median estimates may be consistently biased and discordant with reality. Other experts may be over-confident in their views, in the sense that they assign far too narrow uncertainty bands on their median estimates; still other experts may be uninformative, in the sense that they assign such broad, cautious uncertainty bands that little useful information is conveyed.

Recently, an elegant mathematical approach has been devised (Cooke, 1991) to calibrate expert judgements. This method overcomes many of the technical difficulties which have beset previous attempts, while being convenient and efficient for practical use. In Cooke's approach, each expert is asked to assess the values of a range of quantiles for a given set of calibration parameters, as well as for the parameters of actual interest. An optimal decision-maker is then constructed on the basis of the performance of the experts on the calibration parameters. This construction is based on the mathematical theory of scoring rules to gauge the quality of experts. Although first conceived as a way of eliciting probabilities, scoring is a systematic way of rewarding the positive aspects of expert judgement, and is implicit in any performance-based human assessment.

The mathematical formalism is as follows:

Let q_e $(e=1,2,...E)$ be the distribution of the E experts whose judgements are elicited for an uncertain parameter. Then if p_A denotes the analyst's distribution:

$$p_A = \sum w_e \, q_e \; ; \text{ where } \sum w_e = 1, \, w_e \geq 0. \qquad [1]$$

For the discrete case, where a quantity can take n values, the entropy $H(p)$ is defined to indicate the lack of information in the distribution:

$$H(p) = -p_i \, \ln(p_i)$$

If s is a probability distribution over the same n values, the relative information of s with respect to p can be defined as:

$$I(s,p) = s_i \, \ln(s \, / \, p_i)$$

$I(s,p)$ is always non-negative, and $I(s,p) = 0$, if and only if s is identically equal to p. $I(s,p)$ is commonly taken as an index of the information learned (sometimes referred to as the index of surprise) if one initially believes p, and later learns s is correct.

Let X_1, X_2,...X_M be the set of calibration variables; let X_{M+1}, X_{M+2},....X_N be the variables of actual interest; let f_1, f_2,....f_R be the quantile probabilities elicited and let G_{me} be the minimum information cumulative distribution for X_m, defined as the distribution for which the entropy is as large as possible, and which satisfies the constraint that the f_i quantile values agree with the assessments of expert e.

Indices of calibration and informativeness of expert e's assessments, respectively $C(e)$ and $H(e)$, can then be defined (Cooke, 1991) in terms of $I(s,p)$ and G_{me}, and a corresponding weight w_e can be attributed to expert e according to the formula:

$$w_e \; \alpha \; I_\alpha \, [C(e)] \, . \, \{C(e)/H(e)\},$$

where $I_\alpha[C] = 0$ if $C < \alpha$
$\qquad\qquad = 1$ if $C \geq \alpha$

The parameter α may be chosen to optimise the analyst's distribution p_A in equation [1]. With this definition, the weight assigned to expert e increases with the calibration index $C(e)$, and decreases with $H(e)$.

The availability of the free parameter α to optimise the analyst's distribution in a manifestly impartial manner, avoids critical judgement of the quality of individual experts, who might otherwise be self-conscious over public accountability. In practical applications, the calibration term $C(e)$ can vary over three orders of magnitude for a reasonable large group of experts, whereas the informativeness index $H(e)$ rarely varies by a factor greater than about three. There is a natural trade-off between calibration and informativeness: an expert can achieve good calibration by being uninformative. However, high informativeness is not a substitute for poor calibration and, in Cooke's scheme, informativeness modulates the calibration

term, providing a quantitative means of distinguishing between experts of similar calibration.

HYPOTHETICAL MINING APPLICATION

Suppose there to be a major mining company, Molar Extractors Inc., with properties in various parts of the world, and that this company has an immediate concern about the hazards of rockburst in one particular, economically-marginal mine, Deep Larynx. Suppose, also, that Molar has on its staff three full-time professional Earth scientists, say, a structural geologist, a seismologist and a mining engineer, and that the company is prepared to supplement this expertise by three external specialists, who have been used in previous cases elsewhere. Of these six experts, perhaps one has real knowledge of the conditions pertaining in Deep Larynx; the others, however, can bring extensive experience, in their own different disciplines, from other environments such as the Polish coalfields or the South African gold mines.

The issue for the Molar management is how to extract a rational, unbiased assessment of the hazard at Deep Larynx from their group of experts, using as inputs the various and varied opinions, without placing an undue onus on the local man to shoulder responsibility for the decision. For the purposes of this illustrative example, the experts are required to anticipate the total seismic energy release in a specified volume of rock ahead of the working face for succeeding twelve hour intervals, up to 72 hours ahead. They could equally as well be asked to predict the maximum event magnitude, or the number of seismic events, or the location of maximum energy release.

The formal elicitation of the experts' opinions would proceed as follows. An independent facilitator is needed to conduct the elicitation and perform the calibration. He must have a detailed understanding of the proper procedures for eliciting expert opinion and must be sufficiently cognisant with the particular subject that he can understand all the technical issues, even though he himself has very limited experience in the matter; for rockburst hazard assessment, an earthquake seismologist trained in elicitation techniques might be appropriate, for instance.

The experts would be provided with datasets from a number of case histories from which empirical results are available, and with data from the current situation which is causing concern. The datasets could comprise seismic, geological and mining information on previous rockburst episodes, stress drops, focal mechanisms, tomographic velocities, in situ stress measurements, bedding plane geometry, rock type, extraction rates, and so on. It would not matter unduly if some individual experts were personally familiar with the outcomes of some of the earlier cases, or if the datasets were incomplete in some aspects. For each case, the expert would be invited to provide his lower 5%, 50% (median) and upper 95% values for seismic energy release in the six time intervals, the quantile estimates being a convenient way of representing a probability distribution for the parameter being considered. Where an individual's degree of belief is vague, his probability distributions will tend to be broad and hence uninformative; where his degree of belief is full of conviction, his probability distributions will be peaked and thereby informative. If, in the latter case, the expert's median value is close to the empirical result he will be well calibrated, if not, then his opinion is over-confident and may be significantly in error. Examples of inter-expert variation in the estimation of one normalised parameter are shown in Figure 1; the parameter has an expected value of 0.5.

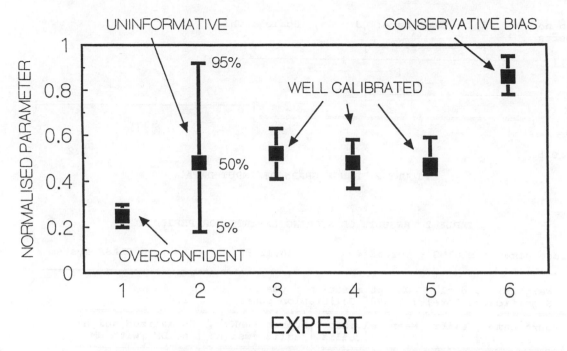

FIGURE 1   TYPICAL INTER-EXPERT VARIATION IN 5%, 50% AND 95% ESTIMATES OF A PARAMETER

The experts' estimates for the test cases are assembled and input to program EXCALIBR, for comparison with the known results. Their indices of calibration and informativeness are quantified and these indices are then used to construct the optimal decision-maker. In the simulated example which follows, the parameter being assessed is total seismic energy release in Joules (expressed as a logarithm) for two future time periods: Period 2 is 12-24 hours ahead and Period 4 is 36-48 hours ahead. The scoring results of the experts are given in Table 1. For Period 2, Experts 2, 3 and 5 are equally well calibrated with Expert 2 providing the most information and, hence, gaining the highest weighting in constructing the optimal Decision-Maker (DM) for this case. The DM is better calibrated than any individual expert but provides less information content than Expert 2.

The ranges of the experts' input values for one test case for Period 2, the empirical realization ("Real") and with the estimate given by the optimal Decision-Maker ("DM") are depicted in Figure 2. In most trial cases, Expert 1 overestimates seismic energy release, while Expert 6 provides

estimates of such variability as to undermine any confidence in his judgement.

There is some change in the scoring of the experts when they are asked to forecast further ahead in Period 4; the scoring results are summarised in Table 2. Now, Experts 2, 4 and 5 are well calibrated with Expert 4 providing most information and thus gaining the highest weighting in the construction of the Decision-Maker. Note that Expert 1 appears to provide more information than the others but he is so poorly calibrated than he achieves a near-zero weighting in the decision-making process. The optimal Decision-Maker (DM) is better calibrated than any individual expert but has to bow to Expert 4 when it comes to assessing seismic energy release in Period 4.

The calibre of the inputs provided by the experts, as exemplified by their ranges for one test case on Figure 2, can be summarised: Experts 2, 4 and 5 produce helpful guidance on the problem; Expert 3 is too cautious in his opinion for the earlier period but contributes to the assessment for the later interval; Expert 1 almost invariably underestimates energy

TABLE 1   RESULTS OF SCORING EXPERTS FOR PERIOD 2.

Case name   : MINING - PERIOD 2   28.12.92   CLASS system

Weights : global.   DM optimisation : yes.
Significance level : 0.390   Calibration power :   1.0

Expert name	Calibr.	Mean rel.infor. total	realiz.	Number realiz.	UnNorm. weight	Normalized weight no DM	with DM
1 1	0.00100	1.005	0.942	5	0	0	0
2 2	0.39000	0.925	0.964	5	0.37608	0.46054	0.33120
3 3	0.39000	0.586	0.550	5	0.21450	0.26267	0.18891
4 4	0.06000	0.674	0.613	5	0	0	0
5 5	0.39000	0.662	0.580	5	0.22603	0.27679	0.19906
6 6	0.00100	0.554	0.587	5	0	0	0
DM	0.73000	0.469	0.437	5	0.31888		0.28083

Item no. :   3   Item name: 3   Scale : UNI

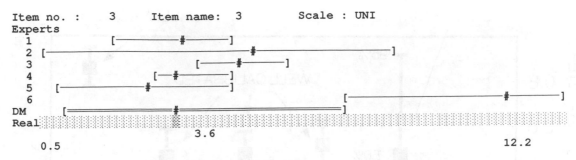

```
Experts
  1              [————#————]
  2    [————————————————#——————————————]
  3              [————#————]
  4         [———#———]
  5    [————————#————]
  6                                      [————————————#————]
 DM    [=========#=========]
Real
              3.6
   0.5                                                          12.2
```

FIGURE 2   RANGE GRAPH OF INPUT DATA.

TABLE 2   RESULTS OF SCORING EXPERTS FOR PERIOD 4.

Case name : MINING - period 4   28.12.92   CLASS system

Weights : global.   DM optimisation : yes.
Significance level : 0.390   Calibration power :   1.0

Expert name	Calibr.	Mean rel.infor. total	realiz.	Number realiz.	UnNorm. weight	Normalized weight no DM	with DM
1 1	0.00500	1.330	1.413	5	0	0	0
2 2	0.39000	0.148	0.163	5	0.06366	0.08184	0.06029
3 3	0.00100	0.984	1.022	5	0	0	0
4 4	0.39000	1.061	1.140	5	0.44479	0.57181	0.42123
5 5	0.39000	0.707	0.691	5	0.26941	0.34635	0.25514
6 6	0.01000	0.544	0.583	5	0	0	0
DM	0.73000	0.375	0.381	5	0.27806		0.26334

release, and Expert 6 gives highly variable predictions throughout. On the strength of these scorings, it might be questioned whether mythical Experts 1 and 6 are in the right jobs!

For Period 2, the median quantile evaluated by the optimal Decision-Maker indicates seismic energy release of 2.2×10^5 J, with lower (5%) and upper (95%) quantiles of 3.1×10^2 J and 1.6×10^8 J; the comparable figures for Period 4 are 6.5×10^6 J, with lower and upper quantiles of 1.9×10^3 J and 8.0×10^{10} J respectively. (Note the enlarged uncertainty manifest in the longer term forecast, reflecting the experts' increased reservations). If the expert opinions were simply aggregated with equal weights (a traditional but now discredited approach), the means for the two cases would be 3.9×10^5 J and 5.2×10^7 J; both are more pessimistic than the Decision-Maker, in the second instance by almost an order of magnitude. The solutions computed by the Decision-Maker would be the optimal values to set against any criterion established for halting or proceeding with mining operations.

DISCUSSION

Induced seismicity, and rockbursts in particular, are considered to be one of the most serious hazards associated with mining in highly-stressed rock. The severity of the problem is expected to increase as active mining is extended to ever greater depths. Recently, there has been a marked growth in applicable techniques for measuring all manner of related phenomena and a burgeoning of theories seeking to explain such seismic events (see, for example, Fairchild, 1990; Knoll, 1992). Yet, the fundamental processes are still unclear; some seismic episodes pose grave threats to safety whilst others cause little or no damage. In these circumstances, a probabilistic procedure is required which can incorporate both observational data of many kinds and the subjective degrees of belief of experts. The issue is one which is one which is increasingly being encountered in engineering aspects of earthquake hazard assessments (Woo, 1992).

The proper elicitation of expert judgement should be an integral component of any seismic hazard assessment. Because the elimination of bias and the treatment of uncertainty are both critical to the outcome, attention must be paid to the inherent tendencies of individual experts to exaggerate or under-estimate values, or to be over-confident or self-effacing in their responses. Calibration exercises allow these traits to be identified and quantified. Methods now exist for improved decision making, through the optimal fusion of expert judgements, taking into account the variability and informativeness of the selected experts.

The introduction of these optimal decision analysis techniques can systematically rationalize the important element of expert judgement present in any seismic hazard assessment or other evaluation made in the face of uncertainty, improving its accuracy and reliability.

REFERENCES

Cooke R.M. (1990) "Experts in Uncertainty: Expert Opinion and Subjective Probability in Science". Oxford University Press, 321pp.

Fairchild C. (ed.) (1990) "Rockbursts and Seismicity in Mines". Proc. 2nd Int. Symp. on Rockbursts and Seismicity in Mines, Minneapolis; A.A. Balkema, Rotterdam, 439pp.

Knoll P. (1992) "Induced Seismicity". A.A. Balkema, Rotterdam, 469pp.

Woo G. (1992) Calibrated expert judgement in seismic hazard analysis. In: Proc. 10th World Conf. Earthq. Engng, Madrid, 333-338.

3 Monitoring of seismicity and geomechanical modelling

Rockbursts and Seismicity in Mines, Young (ed.) © 1993 Balkema, Rotterdam, ISBN 90 5410 320 5

Keynote address: Real time quantitative seismology in mines

Aleksander J. Mendecki
ISS International Limited, Welkom, South Africa

ABSTRACT: The paper describes the concept of quantitative, real time seismology in mines and an example of a specific implementation. A seismic event is considered to be described quantitatively when, apart from its timing and location, at least two independent parameters pertaining to the seismic source, e.g. seismic moment and radiated seismic energy or seismic moment and stress drop are determined reliably. Real time seismology implies immediate, quality controlled automatic seismological processing with built-in functions to respond instantaneously when certain conditions are met. The basic requirements for the data acquisition and control system are presented and the main aspects of quantitative seismological processing and its quality control when run in automatic mode are described. General guidelines for interpretation are offered and examples are presented to demonstrate the feasibility of quantitative real-time seismology and the role it can play as a management tool in mines and in providing valuable insight into understanding of rock mass response to mining. Directions of future research and development are also discussed.

1 QUANTITATIVE AND REAL TIME

1.1 Why quantitative?

It has been recognised that a one-parameter description of the seismic source, whether it is body wave magnitude or seismic moment or seismic energy is for all practical purposes inadequate. Seismic events of similar magnitude can differ significantly in terms of radiated seismic energy or seismic moment, reflecting difference in stress and strain regime at the source. Dziewonski and Woodhouse (1983) described the body wave magnitude as "... a hopelessly inadequate measure of the size of an earthquake" and they give an example of two seismic events with the same magnitude M=5.9, which differ in seismic moment by factor of 400.

Let us see how the situation changes when seismic source is described by both radiated energy and seismic moment.

Strain accumulated in the rock mass can be divided into elastic and inelastic components. No potential energy can be associated with inelastic strain thus a seismic event occurs as a result of elastic strain drop. The energy released during a seismic event is due to the transformation of elastic strain to inelastic strain. This transformation may occur at different rates, ranging from slow creep (stable strain) to a very fast dynamic seismic event (unstable strain). Because the slow type events have a long time duration at the source they radiate predominantly lower frequency waves, as opposed to dynamic sources of the same size. Since excitation of seismic energy can be represented in terms of temporal derivatives of the source function one may infer that a slower source process implies less seismic radiation. In terms of fracture mechanics the slower the rupture velocity, the less energy is radiated; the quasi-static rupture would radiate practically no energy.

The most classical approach to representing seismic sources by seismologists is finding a distribution of forces or moments that is equivalent to the inelastic deformation at the source. The seismic moment density or stress glut, introduced by Backus and Mulcahy (1976a,b), represents the internal stress necessary to cancel the strain produced by the internal non-linear process. The total seismic moment density integrated over the source region is the seismic moment tensor that measures the inelastic deformation of the seismic source. The physical source region is precisely the volume where the seismic moment density (or stress glut) is non-zero.

Let us imagine now the source of a seismic event associated with a relatively weak geological feature or with a soft patch in the rock mass. Such a source will yield slowly under lower differential stress resulting in a low apparent strain event - the ratio of seismic energy (*E*) over seismic moment (*M*). The opposite applies to the source associated with a strong geological feature or hard patch in the rock mass. Apparent stress σ_A

$$\sigma_A = \mu E/M, \qquad (1.1)$$

where μ is rigidity, has been recognised as a reliable, model independent measure of dynamic stress release at the source (Madariaga 1976, Snoke et al, 1983, Boatwright, 1984). Figure 1.1 shows the apparent stress associated with seismic events of magnitude between 1.3 and 1.5 in the Welkom area, RSA. The ratio of the largest to smallest apparent stress for seismic events of similar magnitudes in this case is well over 100.

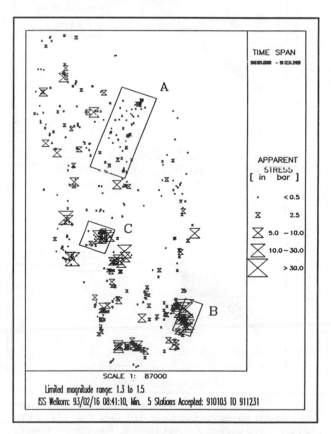

Figure 1.1. Apparent stress of seismic events within the restricted magnitude range of 1.3 to 1.5 in Welkom area, RSA for the period 3 January to 31 December 1991. A(2200m), B(3000m) and C(2500m) indicate examples of areas (and depth below the surface) of consistently low, consistently high or large variations in the apparent stress.

Figure 1.2 shows energy vs moment plots for thousands of seismic events from two reef horizons at Western Deep Levels (WDL) and in the Welkom area. Again we see a consistently higher ratio of radiated seismic energy for the same seismic moment associated with events occurring at greater depth and within a more competent and less faulted rock mass. Gibowicz et al. (1990) consider apparent stress an independent parameter of stress release in the case when P and S wave contribution to the seismic energy is included. In general the apparent stress expresses the amount of radiated seismic energy per unit volume of inelastic coseismic deformation.

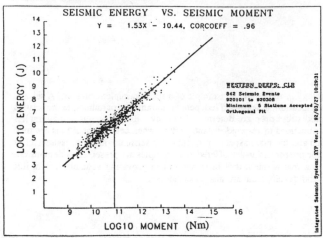

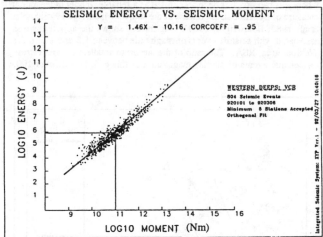

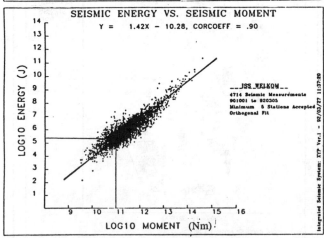

Figure 1.2. Seismic moments and radiated seismic energies of events associated with two mining horizons at WDL Gold Mine: CLR(3100m) and VCR(2100m) and with Welkom Goldfields. Please note consistently higher energies for the same seismic moments associated with the deeper and more competent rock. The greater scatter in the energies at Welkom, RSA could be explained by varying depth of mining between 1500 and 3000m, and numerous faults and/or dykes of different (soft and hard) character, and displacement of up to 600m.

Although the seismic waveform does not provide direct information about the absolute stress (for bit of hope in inferring absolute stress from seismic radiation see Spudich, 1992), but merely about the dynamic stress drop at the source, a number of seismological studies and numerous underground observations suggest that a reliable estimate of stress drop, or preferably the apparent stress, can be used as indicator of the local level of stress. However, in the case of a so called complex event or multiple events, the rapid deformation process at the initial source can push the stresses in the adjacent volume to a level much higher that could normally be maintained by the rock, producing higher apparent stress sub-event(s) that need not be an indication of a generally high ambient stress prior to the event. Although the estimate of apparent stress does not depend on the rupture complexity (Hanks and Thatcher, 1972), the complexity of the event should be tested before the meaningful interpretation in terms of stress can be given.

The unstable strain can usually be localized in a zone which is narrow in comparison with the wave length as seen in the far field: a planar source. In some cases, seismic events which occur close to the underground excavations exhibit a volume rather than a planar character with many zones of permanent deformation and with complex geometry accompanied by a local volume change.

If we may interpret the source volume V, as the region with large inelastic strain ϵ_P (of the order of stress drop $\Delta\sigma$ over rigidity μ), the seismic moment may then be defined as the product of the estimated stress drop and source volume (Madriaga, 1979).

$$M = \mu\epsilon_p V = \Delta\sigma V => V = M/\Delta\sigma. \qquad (1.2)$$

Since apparent stress σ_A scales with stress drop and since there is less model dependence in determining the apparent stress than there is in corner frequency related stress drop, and because in general $\sigma_A \leq 2\Delta\sigma$ (Savage and Wood, 1971), one can define the apparent volume as

$$V_A = M/(2\sigma_A) = M^2/(2\mu E). \qquad (1.3)$$

Apparent volume, like apparent stress depends only on energy and moment, but because of its scalar nature (m^3), can easily be manipulated in the form of cumulative or contour plots etc., providing an insight into the rate and the distribution of coseismic deformation and/or stress transfer in the rock mass. Figure 1.3 presents an interesting example of the cumulative seismic energy, seismic moment and apparent volume before a major seismic event; only the cumulative apparent volume shows any significant trend before the event. Van Aswegen and Butler (1993) showed a case where the apparent volume associated with fault related seismic events correlates with measured creep on the fault.

I would propose the following general guidelines for qualitative assessment of the stress and strain regime and its impact on rock mass stability, based on quantitative description of microseismicity:

(i) Consistently higher (or lower) values of apparent stress associated with seismic events of similar moment in the area of interest would indicate a higher (or lower) level of stress and/or rock strength (e.g. figure 1.1 areas A and B).

(ii) Big variations in apparent stress associated with seismic events of similar moments in given area would be an indication of inhomogeneous stress and/or rock strength (e.g. figure 1.1 area C).

(iii) An area with a high gradient of cumulative coseismic inelastic deformation would be a potential source of high shear stress where a large event or swarm of events could develop. This interface, if accessible, could also be a target for triggering or preconditioning (soft triggering) action.

(iv) A significant increase in the rate of coseismic inelastic deformation increases the potential for a larger seismic event (figure 1.3).

(v) An increase in the rate of coseismic inelastic deformation while changes in average apparent stress or in stress drop for events of similar seismic moment are fairly static would be an indication of potential instability in the form of a large seismic event or a swarm of smaller events (see figures 1.4 and 1.5). One would expect an increase in the rate of microseismicity at the same time.

The notion in (v) could be explained by the classical instability criterion

$$\int_V \dot{\sigma}\, \dot{\epsilon}_p\, dV < 0 \qquad (1.4)$$

where $\dot{\sigma}$ and $\dot{\epsilon}_p$ are rate of change of stress and inelastic strain respectively. V in (1.4) would include so-called source nucleation zone where slip-weakening occurs, shear stress decreases and seismic activity increases (Ohnaka, 1992). Occasionally seismic quiescence may occur due to the dilatancy hardening or inhomogenieties within the nucleation zone (Frank 1965; Ohnaka 1992). Stress rate in (1.4) would be inferred from changes in average apparent stress for similar moments, and strain rate would be measured by changes in cumulative seismic moment or by changes in cumulative apparent volume.

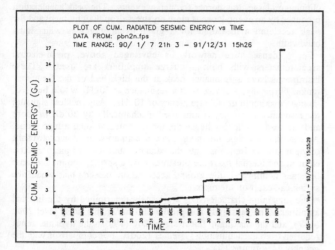

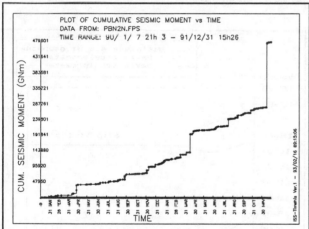

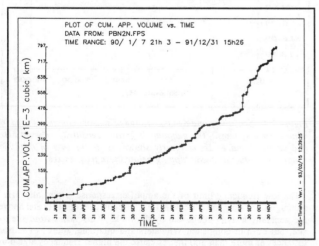

Figure 1.3. Cumulative plots of radiated seismic energy, seismic moment and apparent volume for the area surrounding the source of a seismic event of energy $E = 2 \times 10^{10}$ J

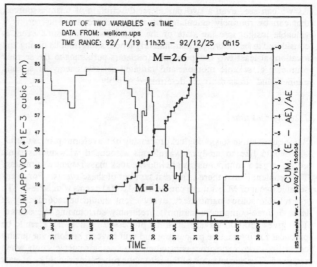

Figure 1.4 Cumulative apparent volume V_A (thick line with dots indicating seismic events) and cumulative energy index EI (thin line) versus time for selected area in Welkom Goldfields. Two largest seismic events are indicated by vertical lines. EI = (E-AE)/AE, where E is the seismic energy for a given event and AE is the average seismic energy for the seismic moment of the event, as derived from a statistical relation for the given region (e.g. see figure 1.2).

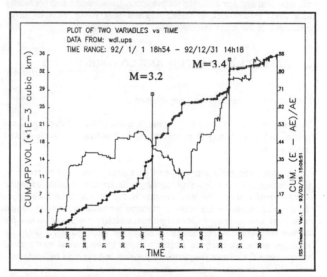

Figure 1.5. As figure 1.4, but for a selected area at Western Deep Levels mine. A few days before the second event of M = 3.4 the rate of cumulative apparent volume increased but energy index fluctuated.

The negativeness of (1.4) is a necessary but not a sufficient condition for rock mass softening behaviour since in that case both total inelastic and elastic strain rate should be included. However, near the end of the hardening regime the elastic strain rate is minimal and almost all further strain increments are of an inelastic nature. An inelastic strain rate can be divided into the rate of inelastic volumetric strain and the rate of inelastic distortion, the ratio of which is called the dilatancy angle. The dilatancy angle is found to be constant near and at the peak strength of the material. The coseismic expression of the dilatancy angle would then be the ratio of the rates of change in the isotropic and deviatoric components of the moment tensor or the rates of change P energy over S energy of micro events.

The basic assumptions made in formulating the guidelines are:
- positive correlation between coseismic and aseismic deformation
- rupture nucleates within or at the interface of an intensively cracked or softened region which, in case of quasistatic load, is large enough to be identified by the microseismic network
- the critical size of the nucleation zone decreases with an increase in the strain rate (Kato et al. 1992).

As we can see, even a two parameter description of seismic sources, that can be routinely calculated by mine seismic networks, offers a valuable insight into the state of the rock mass. A seismic event is considered to be described quantitatively when, apart from its timing and location, at least two independent parameters pertaining to the seismic source, e.g. seismic moment and radiated seismic energy or seismic moment and stress drop are determined reliably.

1.2 Why real time?

Since quantitative seismological processing of waveforms proves useful, one would like to analyze waveforms associated with microseismic events to get a continuous insight into the rock mass response to mining. The problem is that there is a great number of these events. For every seismic event of $M \geq 3$ there can be over 10 000 events with $M \geq -1$. To get reliable source parameters, every event should be recorded by at least 5 three-component stations, preferably surrounding the source. That gives 150 000 waveforms to be processed. It does not seem to be practical for interactive processing and will have to be done via quality controlled, automatic processing where the operator interacts only when required. The success rate of the automatic processing should be at least 70% but the success rate of the quality control must be close to 100% to exclude unreliable data from the interpretation. Another feature of the system or preferably even of the underground stations is control/alarm functions, activated immediately when a preset value (e.g. strong ground motion, creep and/or closure rate etc.) or certain conditions (e.g. extreme rate of increase in apparent volume and/or rate of change of pattern in apparent stress in the area) are met.

So why quantitative and real time? - to make seismology a 24 hour management tool for assisting in preventing, controlling and predicting rock mass instabilities in mines and not only for a "nice to have" research tool for mine seismologists.

2 THE DATA ACQUISITION AND CONTROL SYSTEM

2.1 Basic features and basic requirements:

In many respects a seismic system is a specialised data acquisition system and as such may be characterised by bandwidth, noise level and dynamic range.

Bandwidth: The minimum range of frequencies that must be recorded for meaningful seismological processing is determined from the expected corner frequencies of events occurring in the volume to be monitored. To correctly measure the seismic moment we need frequencies down to at least an octave or 5 spectral points, whichever is lower, below the corner frequency of the largest event to be analyzed. To correctly measure the radiated seismic energy we need frequencies at least 5 to 10 times above the corner frequency of the smallest event to be analyzed.

Sensitivity and dynamic range: The signal bandwidth determines the sensor type, which in turn dictates the capabilities of the seismic system.

Geophones: If the frequencies of interest lie between 4Hz and 500Hz, as could be the case for magnitudes between $M_L=0.5$ and $M_L=3$, and especially if a large area is to be covered, so that the average spacing between sensor sites is about 1000m, then miniature 4.5Hz geophones fulfil the sensor requirements.

Their strong point is their sensitivity over this frequency range (typically more than 20 V/m/s when damped) which means that, when used in combination with good amplifiers, the ambient ground noise determines the system noise level of 10^7 m/s at the quietest sites in mines. Because the output reflects ground velocity, the kinetic energy may be calculated directly, and only one integration is necessary to obtain displacement for the seismic moment. They are inexpensive and relatively easy to install in boreholes long enough to reach intact rock, as long as some care is taken to ensure they are precisely vertical or horizontal (within 2 degrees).

The weak points of these geophones are distortion and clipping introduced at larger displacements (1mm to 3mm), and spurious resonances at frequencies above 150Hz if driven transversely. Because the geophone output reflects ground velocity, the displacement clipping is not immediately apparent on inspection of the seismograms.

The theoretical dynamic range of these sensors is very large, e.g. they can measure a velocity of 1 m/s at 100 Hz, which yields a range of 140dB relative to the noise level of 10^7 m/s. In seismic monitoring the high velocities are produced by large events at much lower frequencies (below 20Hz), where the displacement clipping mentioned above reduces the maximum velocity by an order of magnitude, and the dynamic range required from the data acquisition system to a manageable 120dB.

Accelerometers: Where frequencies above 500Hz are of interest and/or the network is small and dense then accelerometers are the sensors of choice. The sensitivity of a given accelerometer to ground velocity or seismic energy increases with frequency, allowing smaller events to be detected than by a comparable geophone, and without the danger of clipping on large, low corner frequency events. The peak acceleration produced by an event depends more on stress drop than moment and the peak acceleration measured for a given event increases with signal bandwidth.

For a dense mine network as envisaged above, piezo-electric accelerometers with built-in charge amplifiers are used. These instruments have a resonance peak at the high end of the frequency range. Typically, a device with a resonance at 20kHz would be used, giving a maximum useable frequency of 10kHz. Any incident signal at the resonance frequency is amplified mechanically by 30 dB or more and can lead to clipping hence the usual choice of such a high value. An insufficiently rigid mounting can cause an apparent downward shift of the resonance frequency into the band of interest. The piezo-electric element and inertial mass are prestressed by a spring, and the response becomes nonlinear if the ground acceleration exceeds that which the spring can exert on the mass.

The remaining characteristics are largely determined by the internal amplifier. The sensitivity is typically 200 mV/g, the noise level $20\mu g$ ($4\mu V$), and the maximum linear signal 20g (4V) so the dynamic range required from the acquisition system is 120dB. For comparison with the geophone see figure 2.1.

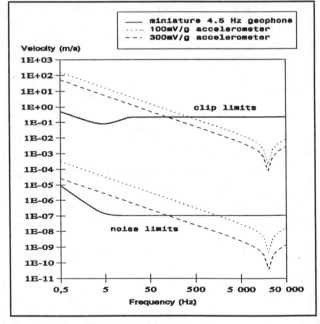

Figure 2.1. Sensitivity and dynamic range of sensors commonly used in mine seismic systems. The geophone's greater sensitivity between 5Hz and several hundred Hz is clearly shown, as is the loss of dynamic range due to displacement clipping below these frequencies.

The low frequency limit of the amplifier response is typically 3Hz, although some only roll off below 1Hz. Frequencies of 5kHz and above are generally heavily attenuated by the rock. This varies from site to site, but even in competent Witwatersrand quartzite, source to sensor distances of more than 300 metres cause the higher frequency signal to be lost. The advantage of this is that the ground noise at these frequencies is very low.

2.2 Data Transmission

The seismic information is generated as a voltage at the outputs of sensors and must be transmitted to a central site. Basically four transmission methods are feasible:

Analogue - Base Band: The sensor signal is simply amplified and transmitted to the central site for digitising and recording. Advantages: simple, low bandwidth requirement, allows centralised triggering, timing. Disadvantages: noise introduced during transmission even with expensive instrumentation cable reduces dynamic range (need several transmission channels per sensor at different gains), channel must be checked manually and often to see whether interference and attenuation are within specification

Analogue - Frequency Modulation: The sensor signal is used to vary the frequency of a carrier signal which is transmitted to the central site where the original signal is recovered, for digitising and recording. Advantages: signal may be shifted in frequency away from known interference, signal quality not directly related to channel attenuation, carrier strength at receiver may be automatically monitored for transmission problems, more than one signal may be multiplexed onto a single channel, allows centralised triggering, timing. Disadvantages: modulation - demodulation process limits the dynamic range and must be manually calibrated, requires wide bandwidth (10 times signal bandwidth), interference may still corrupt signal.

Digital - Continuous: The signal is digitised at the sensor site and the digital values are transmitted to the central site. Advantages: no constraints on dynamic range, transmission errors may be automatically detected with a very high degree of probability, status and calibration data may be interleaved with signal data, multiple signals may be multiplexed onto one channel, allows centralised (digital) triggering, timing. Disadvantages: requires very wide bandwidth (100 times signal bandwidth), interference causes loss of data.

Digital - Store and forward: The signal is digitised at the sensor site and the values are stored temporarily and fed to a trigger algorithm. When a trigger is detected, the central site is informed and the significant data saved for transmission on request to the central site for recording. Advantages: no constraints on dynamic range, no constraints on signal bandwidth, retransmission on transmission error, many sites may be multiplexed onto one channel, status and calibration on command. Disadvantages: each site triggers independently, each site must maintain network time, bi-directional transmission is required.

2.3 An example - The Integrated Seismic System (ISS)

The ISS is comprised of transducers, remote stations, a communication system and a central computer. Following the above classification, this is a digital store and forward system. The exception is the Cluster remote station which forms a miniature analogue base-band subnetwork that may report to a central site or analyse its own events. The dynamic range of 120dB is preserved in the Cluster by limiting cable lengths to 300m, the range of the accelerometer internal amplifier.

Each remote station acquires data from one or more triaxial seismic transducers and a number of nonseismic sensors. Remote stations are Intelligent Seismometers (IS), Processing Seismometers (PS) or Clusters (CL). Depending on the rate of occurrence of seismic events and the communication bandwidth available, seismic waveforms may be processed at the sensor site (PS), transmitted directly to the central computer (IS and/or PS) or processed at some communication nodes. The communication nodes: Intelligent Multiplexers (IM), Processing Multiplexers (PM) and Network Multiplexers (NM), are used to ensure maximal use of resources in bringing the required data and/or results to the central computer as quickly as possible (figure 2.2). Communication is performed at standard baud rates via modem on 2 or 4 wire cable, or radio channels, or via local area network over short distances.

The ISS is digital and intelligent. The advantage of a digital system lies primarily in maintaining the integrity of acquired seismological data. With conversion to a digital format as close as possible to the sensors, maximum dynamic range can be ensured. The seismometers allow for a dynamic range of >120dB with a resolution of 12 bits. Accurate calibration of the seismic waveform data is easy to maintain. Digital communication between the remote sites and a central computer allows for transmission of waveforms with no amplitude or phase distortion.

The remote units are intelligent, allowing for immediate processing at the site, effective use of low cost communication lines and handling of

nonseismic measurements. If the communication bandwidth is too low to permit all data to reach the central computer, then the decision as to which data to discard must be made at the remote site. Both types of seismometer discard smaller amplitude waveforms if larger ones do not have buffer space. The Processing Seismometer (PS) yields a higher throughput of events, because processing is generally quicker than digital waveform transmission. Procedures for handling certain control functions are incorporated in the software and hardware. The control functions include NScram (nonseismic scram) and SScram (strong ground motion scram), which switch on alarms whenever a preset value is exceeded on nonseismic or seismic input channels respectively. A special control function RScram (rockburst scram) is being developed for the cluster unit that could be activated when an extreme rate of increase or extreme rate of change of pattern in measured parameters is detected.

The functions performed by the different stations are described below. In addition hardware configuration diagrams for seismometers are shown.

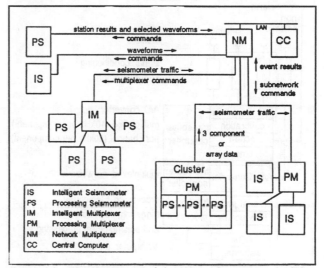

Figure 2.2. Various options of the ISS configuration. An ISS is composed of INTELLIGENT and/or PROCESSING SEISMOMETERS and/or CLUSTERS connected to a central computer.

An Intelligent Seismometer:

- calibrates and monitors a triaxial set of geophones or accelerometers and nonseismic sensors
- keeps network time
- triggers on seismic events
- describes all triggered events to the central computer in terms of time, amplitude and duration, and nonseismic values
- sends waveforms on request
- reads and sends values of nonseismic sensors on request
- keeps largest events if triggers are faster than transmission
- optionally switches on an alarm if seismic or nonseismic values exceed set limits

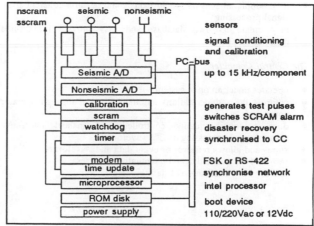

Figure 2.3. Intelligent Seismometer hardware

An Intelligent Multiplexer

- combines messages from several remote stations onto one communication line to the central computer
- rebroadcasts signals from the central computer including time synchronisation

A Processing Seismometer

- performs all the functions of an IS
- may use higher sampling rates
- does automatic quality controlled seismological processing on 3C waveforms
- makes waveforms available periodically or on magnitude or when automatic processing, as indicated by quality indices, fails.

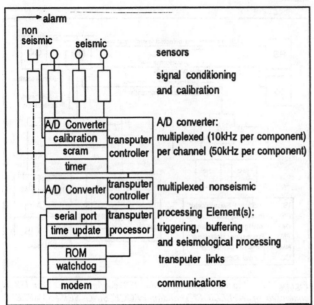

Figure 2.4. Processing Seismometer hardware. Transputer based modules may be paralleled for higher sampling rates or faster processing.

A Processing Multiplexer:

- performs all the functions of an IM.
- may calculate seismological parameters from 3C waveforms
- may perform event processing, turning seismometers into a sub-network

A Cluster:

- performs all the functions of several processing seismometers and a processing multiplexer for a small sub-network
- may sample all sensors synchronously and perform coherent array signal processing
- may create its own database and may provide for on-line quantitative analyses

The Central Computer

- operates under an open system
- controls the remote stations automatically and through operator commands
- does automatic quality controlled and manual interactive seismological processing on 3C waveforms
- stores and allows retrieval of event data and waveforms
- provides for qualitative and quantitative analyses and for 3D visualization of data

3   QUALITY CONTROLLED SEISMOLOGICAL PROCESSING

The following description gives only the most important aspects of quality controlled seismological processing of three-component waveforms developed to run in an automatic mode. All quality indices, whether indicated or not, are normalised to the range $(0,1)$.

3.1 P-picker

- dc offset removal
- short term/long term average => P-pick
- test for spikes or short signal duration events

3.2 Direction and Rotation

- max amplitude Amax
- dominant period = T(Amax)
- signal to noise ratio, SNR
- polarization windows = T(Amax)/4, T(Amax)/2
- eigenvectors (Ei) and eigenvalues (λ_i) of covariance matrix
- degree of polarization for P, DPP = $1 - \lambda_2/\lambda_1$
- direction of P-wave, DIR = E_1
- quality index of P-pick, QI_P = DPP·SNR
- quality index for direction:
  QI_DIR = DIR(T(Amax)/4)* DIR(T(Amax)/2)* QI_P
- rotation to P, SH, SV

3.3 S-picker

- polarization filter PF on P, SH, SV
  PF = $F_1 \cdot F_2 \cdot F_3 \cdot F_4 \cdot F_5$, normalised to (0,1)
  F_1 = deflection angle
  F_2 = degree of polarization for S

$$DPS = \frac{(1-\lambda_2/\lambda_1)^2 + (1-\lambda_3/\lambda_1)^2 + (\lambda_2/\lambda_1 - \lambda_3/\lambda_1)^2}{2(1+\lambda_2/\lambda_1 + \lambda_3/\lambda_1)^2}$$

  F_3 = EnergyS/Energy P
  F_4 = Energy S/Energy Po
  Po - initial pulse on P-wave
  F_5 = SNR on SH and SV
- S-pick => PF > threshold
- quality index of Spick QI_S = $F_1 \cdot F_2 \cdot F_5$
- quality index of rotation QI_ROT = QI_P·QI_S

3.4 Location

- single station locations SSL from DIR and S-P
  quality index QI_SSL = QI_DIR · QI_ROT
- multiple station location, NS ≥ 2, NS - number of stations

$$LOC(xyz) = \min \sum_{j=1}^{NA} ||NQIj * Vj(NOAj - NCAj)||^{L1/L2}$$

$$+ \sum_{j=1}^{ND} ||NQIj * ||LOC(xyz) - DIRj)||^{L1/L2}$$

NA,	-	number of observed P and S arrivals
ND	-	number of nearest stations with quality directions
NQIj	-	normalized quality index for the j-th arrival and/or directions, Σ NQIj = 1
Vj	-	average wave(s) velocity for the j-th station
NOAj	-	normalized observed arrivals at the j-th station

$$NOA_j = \frac{1}{NA} \sum (OAj) - OAj$$

NCAj	-	normalized calculated arrivals at the j-th station
‖LOC(xyz)-DIRj‖	-	distance between the location in current iteration and the direction from the j-th station
L1/L2	-	first or second norm

min - downhill simplex method with the three starting points:-
1- coordinates of the nearest station
2- average coordinates of triggered stations
3- location based on directions from the nearest ND stations

- quality index for location, for $4 \leq NS \leq 12$

QI_LOC = QI_CONF*QI_RES
QI_CONF - quality index of configuration of stations
QI_CONF = dat(A)/(NS/3)^3 · NS/12

$$A = \begin{vmatrix} \sum \cos^2 \alpha_i & \sum \cos\alpha_i \cdot \cos\beta_i & \sum \cos\alpha_i \cdot \cos\gamma_i \\ \sum \cos\alpha_i \cdot \cos\beta_i & \sum \cos^2\beta_i & \sum \cos\beta_i \cdot \cos\gamma_i \\ \sum \cos\alpha_i \cdot \cos\gamma_i & \sum \cos\beta_i \cdot \cos\gamma_i & \sum \cos^2\gamma_i \end{vmatrix}, \quad \sum = \sum_{i=1}^{NS}$$

α_i, β_i, γ_i - directional angles between the j-th station and the hypocentre

$$QI\_RES = \begin{cases} \dfrac{5\%AHD - Res[m]}{5\%AHD} & if\ Res[m] < 5\%AHD \\ 0 & otherwise \end{cases}$$

5%AHD - 5 percent of average hypocentral distance to the stations taking part in the location procedure
Res[m] - residium in metres from the location procedure

3.5 *Spectral parameters per component: P, SH, SV*

- window for P SH and SV
  P - 10 cycles and < S-P
  S - 15 cycles
- filtering between f_1 and f_2
  f_1 - lowest frequency of sensor
  f_2 - sampling frequency/5
- taper - cosine 5-10%
  if f_2-f_1 is large then multitaper
- FFT
- instrument correction
  spectrum (f)/system response (f)
- attenuation correction:

$$A = \exp\left(\frac{R \cdot \pi \cdot f}{(Q + Q_o f) \cdot V}\right)$$

R - distance
Q_o=0 if no info on Q(f)
Q - initial value of Q from an envelope of coda (scatter) and average wave velocity (inelastic attenuation), e.g. low scatter but low velocity => lower initial Q
- displacement spectrum DS(f):
  from velocity - DS(f) = VS(f)/($2\pi f$)
  from acceleration - DS(f) = AS(f)/($2\pi f$)^2
- cepstral filtering:
  C(q) = FFT^{-1} Ln(DS(f))
  filtering - zeroing after the n zero crossing
  n - optimized during spectral fitting
  FDS(f) - filtered displacement spectra
  FDS(f) = exp(FFT(C(q)))
- spectral fitting for P,SH,SV - via downhill simplex for n=3,4,5,6

$$SPECT(\Omega, fo, Q) = \min \sum_f \| wf(f) \cdot (FDS(f) - MDS(f) \|^{L1/L2}$$

MDS(f) = model displacement spectra
MDS(f) = A Ω/(1 + (f/fo)^2)
wf(f) ~ SNR(f)
constrains: 0.25 Q $\leq Q_p \leq$ 2.5 Q
0.2 $Q_p \leq Q_{SH} \leq Q_p$,
0.5 $Q_{SH} \leq Q_{SV} \leq$ 2 Q_{SH}
0.5 T(Amax) $\leq$ fo^P $\leq$ 2T(Amax)
0.25 fo^P $\leq$ fo^{SH} $\leq$ 2.5 fo^P
0.5 fo^{SH} $\leq$ fo^{SV} $\leq$ 2 fo^{SH}

starting point: Ω - average from first 5 spectral point
fo - dominant period
Q - initial value of Q

- integral of velocity power spectrum - IVPS

$$IVPS = \int_{f1}^{f2} FVS^2 (f)\, df$$

FVS(f) - cepstrum filtered velocity spectrum
- quality indices for spectral parameters

$$QI\_\Omega = 1 - \frac{\sum_{f1}^{fo} \||FDSj - MDSj\||^{L1/L2}}{\sum_{f1}^{fo} |FDSj| + |MDSj|}$$

$$QI\_fo = 1 - \frac{\sum_{fo1}^{fo2} \||FDSj - MDSj\||^{L1/L2}}{\sum_{fo1}^{fo2} |FDSj| + |MDSj|}$$

$$QI\_IVPS = 1 - \frac{|IVPS - IMVPS|}{IVPS + IMVPS}$$

$$f_{o1} = (f_1 + f_o)/2 \qquad f_{o2} = (f_o + f_2)/2$$

IMVPS - integral of model velocity power spectrum (spectral fit)
- bandwidth corrections for power spectrum for (0,f1) and (f2,∞)
- quality index for spectrum

$$QI\_SPECT = QI\_\Omega * QI\_fo * .QI\_IVPS$$

3.6 *Moment Tensor*

- polarity of P_i, i = 1, NS
- quality indices for polarities of QI_PP$_i$, see QI_P
- quality index for moment tensor spectrum

QI_MTS$_j$=QI_Ωj*QI_DIRi*COSi(< P-direction and hypodirection)
j=1,...,NC, NC - number of valid components

- initial moment tensor components M_K^o by solving system of linear equations via Singular Value Decomposition

$A_{kj} M_K^o = wf_j \Omega_j$, for wf_j~QI_MTS$_j$ > threshold

$M_K^o = $ (M11, M21, M22, M31, M32, M33)

- final moment tensor M_K^T via downhill simplex

$$M_K^T = \min \sum_{j=1}^{NC} \| wf(j)(\pm\Omega j - A_{kj} M_k)\|^{L1/L2}$$

$$\sum_{j=1}^{NC} \| wf(j)\,(|\Omega j| - |A_{kj} M_k|)\|^{L1/L2}$$

- eigen-vectors E and eigen-values λ of moment tensor
- diagonalization of moment tensor, M_D^T

$$M_D^T = E \cdot M^T \cdot E^T = \begin{bmatrix} \lambda_1 & 0 & 0 \\ 0 & \lambda_2 & 0 \\ 0 & 0 & \lambda_3 \end{bmatrix}$$

- decomposition of moment tensor into isotropic component M_V^T and deviatoric component M_D^T + M_{CLVD}^T

$$M_D^T = M_V^T + M_{DC}^T + M_{CLVD}^T$$

$$M_V^T = M_V \begin{bmatrix} 1 & 0 & 0 \\ 0 & 1 & 0 \\ 0 & 0 & 1 \end{bmatrix}, \quad M_{DC}^T = M_{DC} \begin{bmatrix} 0 & 0 & 0 \\ 0 & -1 & 0 \\ 0 & 0 & 1 \end{bmatrix}, \quad M_{CLVD}^T = M_{CLVD} \begin{bmatrix} -1 & 0 & 0 \\ 0 & -1 & 0 \\ 0 & 0 & 2 \end{bmatrix},$$

- scalar moment, M

$$M = \|M_V^T + M_{DC}^T + M_{CLVD}^T\|_E$$

- radiation pattern
  R_i^P, R_i^{SH}, R_i^{SV}; $R_i^S = [(R_i^{SH})^2 + (R_i^{SV})^2]^{\frac{1}{2}}$
  if MT not resolved - average values
- quality index for moment tensor

$$QI\_MT = QI\_LOC \cdot (\Sigma \; QI\_MTS_j)/NC$$

3.7 Source parameters - per station and per event

- weighted averages of source parameters per station
  and per event from spectral parameters:
  seismic moment [Nm]
  radiated seismic energy [J]
  size [m]
  stress drop [Pa] etc.
- apparent stress and apparent volume:

$$\sigma_A = \frac{\mu(E_P + E_{SH} + E_{SV})}{\|M_V^T + M_{DC}^T + M_{CLVD}^T\|_E} \quad \text{if moment tensor resolved}$$

$$\sigma_A = \frac{\mu(E_P + E_{SH} + E_{SV})}{\frac{1}{2}[M_P + (M_{SH}^2 + M_{SV}^2)^{\frac{1}{2}}]} \quad \text{otherwise}$$

$$V_A = \frac{\|M_V^T + M_{DC}^T + M_{CLVD}^T\|_E^2}{2\,\mu(E_P + E_{SH} + E_{SV})} \quad \text{if moment tensor resolved}$$

$$V_A = \frac{(\frac{1}{2}[M_P + (M_{SH}^2 + M_{SV}^2)^{\frac{1}{2}}])^2}{2\mu(E_P + E_{SH} + E_{SV})} \quad \text{otherwise}$$

- global quality indices for source parameters

4 FUTURE DEVELOPMENT

4.1 Quantitative Seismology

Quantitative description of seismic sources can offer far more than has been done so far, but considerable effort will have to be made before we can apply it routinely. More detailed analysis of high quality near-field waveforms could provide the elements of the higher order moments of the moment distribution, that would estimate the time duration of inelastic deformation at the source, the rupture velocity, and the size, shape and orientation of the source region (Bachus and Mulcahy, 1976a,b; Bachus, 1977). Our attempts to calculate the higher order moments for mine tremors using the maximum entropy scheme developed by Brown (1979) have been only partly successful to date and are still far from routine application.

To improve the success rate of quantitative seimology as it stands at the moment, we need to incorporate information about the velocities and attenuation in the rock mass into the seismological processing. A number of 3D ray tracing routines exist (e.g. Vidale and Ammon 1989, van Tier and Symes, 1991) but one will have to satisfy here two conflicting demands, the accuracy of calculation and computational time. Because mining can change rock mass properties considerably over time, one will need to update this model by inverting for velocities and amplitudes (seismic tomography) using data from natural and/or artificial sources (Maxwell and Young, 1993). Since wave amplitudes are sensitive to first order pertubation in geometrical spreading the joint inversion of amplitudes and travel times can resolve sharper gradients, significantly improving on the results from inverting travel times alone. (Nowack and Lutter, 1988; Neele et al. 1993).

Finally, one would like to be able to model routinely the generation of seismic ground motion by realistic seismic sources for different rock mass conditions - synthetic seismograms - and then invert for the source and the medium simultaneously.

4.2 Real-time systems

Centralised vs distributed processing power: Quantitative seismology puts an obvious demand on the processing power of real time systems. One can concentrate that processing power at the central site or distribute it instead to the individual seismic stations. The centralized system although simpler suffers from lack of flexibility and in the case of longer communication lines puts an additional cost on the system. The cost of running cable to each sensor site is not decreasing in real terms, whereas the cost of processing at each site is decreasing.

Parallel processing: The advent of the single chip microprocessor has broken Amdahl's Law, which applies to traditional systems and says that the power of a processor increases as the square of its cost. Provided sufficient processing power can be incorporated in a single chip, a multi-processor system can be far more cost effective than a large single processor. For seismological processing, "sufficient" power became available with the Inmos T800, the Intel 80486 and the Motorola 68040 which incorporate floating point and integer processors on a single chip. Of them, the T800, while the oldest and least powerful, was designed for parallel processing with high speed communication links and a multitasking kernel in hardware, and is ideal for embedded systems. In this way comprehensive seismological processing may be performed cost effectively at site with consequent reduction of 40:1 in data transmission per event. The same technology allows the power of the central site to be increased incrementally with the number of stations.

Wireless communication: Experiments with Very Low Frequency (VLF) wireless voice communication in underground mines are showing promising results, and combined with at site processing may serve to eliminate the cost of cabling from the seismic system.

4.3 Quantitative interpretation

Although the quantitative seismology is slowly becoming a reality in the mining environment, interpretation of the results is still more of a qualitative rather than quantitative nature. While testing the different concepts of qualitative evaluation of the changes in the strain and stress regime from seismic records, one needs to quantify the significance of these changes for different types of rock, geological conditions and mining methods. We should be able to replace words such as "consistently higher", "big variations", "high gradient", "significant increase" with numbers, paving the way for the automatic interpretation (Kijko et al. 1993).

REFERENCES

Backus, G.E. & M. Mulcahy 1976a. Moment tensor and other phenomenological descriptions of seismic sources. I - Continuous displacements. Geophys. J.R. Astr. Soc. 46: 341-361.

Backus, G.E. & M. Mulcahy 1976b. Moment tensor and other phenomenological descriptions of seismic sources. II - Discontinuous displacements. Geophys. J.R. Astr. Soc. 46: 301-329.

Backus, G.E. 1977. Interpreting the seismic glut moments of total degree two or less. Geophys. J.R. Astr. Soc. 51: 1-25.

Boatwright, J. 1984. Seismic estimates of stress release. J. Geophys. Res. 89: 6961-6968.

Bolt, B.A. ed. 1987. Seismic Strong Motion Synthetics. New York: Academic Press.

Brown, D.R. 1989. A maximum entropy approach to under-constraint and inconsistency on the seismic source inverse problems. Finding and interpreting seismic source moments. Ph.D Thesis, University Witwatersrand, Johannesburg.

Cichowicz, A. 1993. An automatic S-phase picker. Bull. Seism. Soc.

Am. 83: 180-189.

Dziewonski, A. & J.H. Woodhouse 1983. *Studies of the seismic source using normal-mode theory.* In "Earthquakes: Observations, Theory and Interpretation". H Kanamori & E. Boschi, ed. North-Holland, 45-137.

Frank, F.C. 1965. *On dilitancy in relation to seismic sources.* Rev. Geophys. 3: 484-503.

Gibowicz, S.J., H.P. Harjes & M. Schäfer 1990. *Source parameters of seismic events at Heinrich Robert Mine, Ruhr Basin, Federal Republic of Germany: Evidence for non double-couple events.* Bull. Seism. Soc. Am. 80:88-109.

Gibowicz, S.J. 1993. *Keynote lecture: Seismic moment tensor and the mechanism of seismic events in mines.* Proc. 3rd Intern. Symp. on Rockbursts and Seismicity in Mines. Rotterdam: Balkema.

Hanks, T.C. & M. Thatcher, 1972. *A graphical representation of seismic source parameters.* J Geophys. Res. 23: 4393-4405.

Kato, N., K. Yamamoto, H. Yamamoto & T. Hirasawa. 1991. *Strain-rate effect on frictional strength and the slip nucleation process.* Tectonophysics 211: 269-282.

Kijko, A., C.W. Funk and A van Zyl Brink, 1993. *Identification of anomalous patterns in time-dependent mine seismicity.* Proc. 3rd Intern. Symp. on Rockbursts and Seismicity in Mines. Rotterdam: Balkema.

Madariaga, R. 1976. *Dynamics of an expanding circular fault.* Bull. Seism. Soc. Am. 66: 639-666.

Madariaga, R. 1979. *On the relation between seismic moment and stress drop in the presence of stress and strength hetrogenity.* J. Geophys. Res. 84: 2243-2250.

Maxwell, S.C. and R.P. Young, 1993. *Verification of a 3D passive image and further corroborative evidence of the seismicity-high velocity correlation.* J. Geophys. Res., (in press).

Neele, F., J.C. Vandecar and R. Snieder, 1993. *A formalism for including amplitude data in tomographic inversions.* Geophys. J. Int., (in press).

Nowack, R.L. and W.J. Lutter, 1988. *Linearized rays, amplitude and inversion.* Pure Appl. Geophys. 128: 401-421.

Ohnaka, M 1992. *Earthquake source nucleation: a physical model for short-term precursors.* Tectonophysics 211: 149-178.

Park, J., C.R. Lindberg & F.L. Vernon III. 1987. *Multitaper spectral analysis of high-frequency seismograms.* J. Geophys. Res. 92: 12675-12684.

Savage, J.C. & M.D. Wood 1971. *The relation between apparent stress and stress drop.* Bull. Seism. Soc. Am. 61: 1381-1388.

Snoke, T.A., A.T. Linde & I.S. Sacks 1983. *Apparent stress: An estimate of the stress drop.* Bull. Seism. Soc. Am, 73: 339-348.

Spudich, P.K.P. 1992. *On the inference of absolute stress levels from seismic radiation.* Tectonophysics 211: 99-106.

van Aswegen, G. & A. Butler 1993. *Applications of quantitative seismology in SA gold mines.* Proc. 3rd Intern. Symp. on Rockbursts and Seismicity in Mines. Rotterdam: Balkema.

van Tier, J. and W.W. Symes, 1991. *Upwind finite-difference calculation of traveltimes.* Geophys. Vol. 6: 812-821.

Vidale, J.E. and C. Ammon., 1989. *Efficient seismic traveltime and amplitude calculations and application to velocity inversion and migration: Presented at the 1989 Fall Meeting of the AGU, San Francisco.*

Rockbursts and Seismicity in Mines, Young (ed.) © 1993 Balkema, Rotterdam, ISBN 90 5410 320 5

Keynote address: Some applications of geomechanical modelling in rockburst and related research

M. D. G. Salamon
Colorado School of Mines, Golden, Colo., USA

ABSTRACT: The goal of rockburst research is to alleviate the hazard arising from seismicity. To achieve this it is essential to find ways of linking the mechanism of seismic event initiation to mining activities. Geomechanical modeling may provide this critical connection. The first example discussed involves the simulation of seismicity caused by the sudden slip of preexisting random flaws in hard rocks. The instability is created by the passing stress wave induced by the extraction of a tabular excavation at great depth. The obtained data resemble surprisingly closely the seismicity observed in South African gold mines. The second example relates to the creation of instabilities in coal mining and the relationship of these conditions to coal bumps. Here seismic studies may provide the critical data required to evaluate the proposed model and may even yield the numerical value of the strain softening parameter.

1 INTRODUCTION

Underground mines are exposed to various dangers. The most feared of these threats have always been those which occur unexpectedly and arc associated with some form of violence. Explosions or the disturbance of unstable equilibria (rockbursts, coal bumps, etc.) are examples of this type of events.

Historically, the greatest losses of lives resulting from mining disasters have probably been associated with methane and coal dust explosions. These tragedies have been part of underground coal mining from the earliest days. Much research effort has been devoted to the solution of this problem. The research has been highly successful. Today, responsible mine management, working in collaboration with a well trained and competent mine crew, can look forward to operating without methane and/or coal dust explosions. Thus, theoretically, the explosion problem has been solved. Although tragic disasters continue to occur occasionally, but these are almost entirely due to human fallibility.

Unfortunately, the research conducted so far to combat the consequences of instabilities in the rock mass, have not progressed anywhere near to the same level of success. The only way mining engineers can be assured to avoid future rockburst casualties is by ceasing to operate rockburst prone mines. It is not even possible to predict with any certainty whether a new mine will be predisposed to endure rockbursts. At present, the human control over the occurrence of most rockburst-like events is, at best, tenuous. Although progress has been significant, nevertheless much remains to be done to achieve a satisfactory level of control over the rockburst hazard.

The main aim of this paper is to demonstrate the usefulness of geomechanical modeling in rockburst research. The most convincing way of doing this is to show the power of modeling through examples. It is somewhat unusual for a non-specialist to give a keynote address at a specialist meeting. This keynote falls into this category. The aim here is to highlight, by bringing in an outsider's point of view, an aspect of the rockburst problem which has received insufficient attention in recent years. Rockbursts and coal bumps (outbursts of gas in coal and other mines also fall into this group, but this problem is not dealt with here) are probably the most adverse manifestations of mining induced seismicity. Unless the relationship between mining activities and the seismicity induced by them is well understood, it is unlikely that miners will ever achieve satisfactory control of this menace.

The time has arrived to strive for quantitative developments. There is an urgent need for progress of a kind which will enable mining engineers to implement improvements with reasonable confidence in their success. At the current state of development of rock mechanics such quantitative advances are usually made through the judicious joint application of numerical modeling and field observations. Such approach, which is usually referred to as back-calculation, is needed to calibrate and then validate the model.

Numerical modeling can be applied usefully in many aspects of rockburst research. For example, the following aspects of the problem would certainly benefit from geomechanical modeling:

* relationship between mining activities and the related seismicity,
* source mechanisms and
* effects of seismic waves on mining excavations.

The objective of the modeling in the first problem area is to establish a quantifiable linkage between mining operations and the seismic hazard. Success in this area may provide the basis for the development of the mining engineers' primary defenses against the devastations of rockbursts. Most of the remainder of this paper is devoted to the demonstration of some applications of this type of modeling.

Much of the early work in mine seismology was guided by the approaches taken in the study of natural earthquakes. Consequently, it has been assumed that the source of an event is remote from the point of observation. The presumption of such remoteness permits significant simplifications in the study. In mining, however, this assumption is frequently not valid. Furthermore, it is essential to the better understanding of the initiation of events that the circumstances at and near the source are studied in detail. Hence, there is an important field of modeling in relation to the source mechanism. Promising work in this area is being done by S.L.Crouch and his associates (Mack and Crouch 1990).

It is surprising that the interaction between the approaching stress waves and the effected mining excavations has not been studied more intensively. Furthermore, rockbursts represent only a small subset of the large set of seismic events which occur in seismically active mines. The following statement was made a decade ago (Salamon 1983): "Virtually no systematic research has been done to elucidate the basis of setting apart those seismic events which become rockbursts from those which do not."

Gibowicz (1990) lamented the lack of progress in the same area in his keynote address to the last Symposium. A start has been made, again by Crouch and his coworkers (Mack and Crouch 1990), but much remains to be done.

Also, little is known about the mechanism(s) through which the dynamic effects, caused by the seismic events falling into the rockburst category, damage mine openings. This is a significant area of research where modeling can play a major role, since the support of mine cavities against the shock loads of rockbursts is the ultimate mode of defence of mine workers and equipment.

2 INSTABILITIES IN MINING

It is generally accepted that seismic events arise from the disturbance of *unstable* equilibria (Cook 1965; Salamon 1983; Brady 1990). Due to the recent visit to the U.S.A. by A.M. Linkov of St. Petersburg, a long delayed insight was gained into the work of Russian scientists in the field of mining induced instabilities. In his notes, which have achieved reasonably wide distribution, he makes two fundamental observations (Linkov 1992). *First*, he argues that "Instability is always induced by nonlinearity." Later, after remarking "...that stability investigations are always based on some definition of stability", he suggests that the essence of dynamic phenomena in mines is the acquisition of kinetic energy by the surrounding rock mass. Thus, his *second* basic observation is that the criterion for instability can be put in the form of the following inequality:

$$\delta W_k \geq 0 \qquad (1)$$

where δW_k is an increment in kinetic energy.

Mining is a progressive activity, that is, mining openings *are made to grow* in size, usually in steps. These steps may be small or can be large. Several metalliferous mining methods involve the simultaneous blasting of a large volume of ore (e.g. sublevel caving, sublevel open stoping). The large and sudden change in the boundary stresses associated with this type of extraction always results in the release of kinetic energy (in addition to that arising from the conversion of the chemical energy of the explosives) into the surrounding rocks (Salamon 1984). This will happen regardless whether or not the rock mass is fully elastic and can be avoided only if mining advances in *small* steps. Thus, kinetic energy can be released even if the constitutive equations are linear.

This deduction indicates that Linkov's criteria require a minor rephrasing. It may be said that instability will occur if:

* nonlinearity exists in the system and
* the imposition of *small* virtual displacements induces kinetic energy in the rock mass.

Hasegawa *et al.* (1989) defined six possible ways in which mine tremors can occur. This is a very useful aid for the visualization of instabilities. It indicates that the creation of discontinuities must be added to the list of possible system nonlinearities. Later it will become apparent that the classification of Hasegawa *et al.*(1989) is not exhaustive.

3 SEISMIC EVENTS IN HARD ROCK MINES

3.1 Preamble

Deep hard rock mines have been battling with the rockburst hazard for many years. The gold mines of South Africa and the Kolar Gold Field in India are perhaps the best known examples of mining districts which have tried to combat this menace for over many decades. While these are the best known examples of the problem, there are many other mining regions, spread over most continents, where rockbursts are a serious threat.

It would appear that the South African gold industry has been in the forefront of rockburst research since the late 1950's. Field observations of displacements, numerical computation of mining induced displacements and stresses (Cook *et al.* 1966) and the numerous seismic investigations, following on the pioneering work of Cook (1964), have led to a broad appreciation of the rockburst problem.

However, as Gibowicz (1990) remarked as recently as three years ago, while it is known qualitatively that factors such as mining layout, geological structure, local tectonics, etc. play an important role, none of these influences have been quantified.

Two measures of proneness to rockbursting have been proposed. These are:

* Energy Release Rate (ERR) introduced by Cook *et al.* (1966), and the
* Excess Shear Stress (ESS) proposed by Ryder (1988).

Salamon (1983) pointed out that ERR is a measure of the elastic stress concentration at the face of tabular excavations such as those resulting from the extraction of narrow gold reefs in South Africa. Linkov (1992) indicated that ERR, at least in two dimensions, can be expressed in terms of the stress intensity factors well known in fracture mechanics.

Although ERR has gained wide acceptance in South Africa, it can be only of limited value in combating the rockburst hazard. Its limitation arises from the narrowness of its scope. The magnitude of ERR depends only on the virgin stress field, the elastic properties of the rocks and the layout of the mining excavations. ERR is independent of the geological structure, the presence or otherwise of flaws (discontinuities) in the rock mass and the potential instability of these flaws. Another important shortcoming of ERR is its inability to recognize failure. This weakness may lend itself, for example, to stabilizing pillar designs where the possibility of pillar or pillar foundation failure is ignored. In spite of the limited value of ERR, it is frequently used in layout design, because of the ease of its computation and its ability of reflecting the adverseness of the mining geometry. Clearly, however, it must be used with caution.

Ryder (1988) in introducing the notion of Excess Shear Stress (ESS) observes that "ERR has been found to correlate well with depth of face fracturing, hangingwall conditions, and the decrease in seismicity observed after the introduction of stabilizing pillars in a number of deep mines; it is the preferred criterion used in current designs of stabilizing pillar and backfill systems." He goes on to say that, according to seismological observations, "...mine seismicity falls into at least two classes: one associated with the crushing of highly stressed volumes of rock ..., and the other with slip or rupture along planes in the rock mass...". Interestingly Linkov (1992) also talks about *volume* and *contact* related instabilities. On the basis of the existence of the second type of events, Ryder then defines the Excess Shear Stress as the difference between the "Prevailing shear stress prior to slip, minus the dynamic strength of the plane", where the plane's dynamic strength is the product of the dynamic coefficient of friction and the normal stress acting on the contact surfaces.

The main advantage of this approach to the assessment of the rockburst hazard is that it takes into account the dominant mode of instability occurring in these mines. It will be seen later that the large majority of seismic events in hard rock mines belongs to the family of shear instabilities. The technique is particularly effective when the stability or otherwise of a known geological feature (e.g. a fault) is to be evaluated. The practical disadvantage of the method is that, while it identifies regions of potential high seismicity, the relationship between ESS and the mining activities is obscured.

To make progress in the fight for the alleviation of the rockburst threat, it is essential to evolve a tool which provides the opportunity of relating seismicity to the changing mining layout and to the geological environment. The objective of the method to be described later, is to furnish the basis of such a tool.

3.2 Features of seismicity in South African deep mines

Over the decades a great deal has been learned about seismicity in the South African mines. Unfortunately, the observations collated in earlier years refer to rockbursts, while most of the more modern data relate to seismological measurements, that is, to seismic events. As it was said earlier, rockbursts are only a small subset of the large set of seismic events, consequently, strictly speaking, conclusions drawn from the host set need not be valid for the subset and vice versa. Since the method to be introduced purports to simulate the generation of seismic events, it will be more appropriate to discuss seismological observations first.

Shear failure: Wong and McGarr (1990) say in the first sentence of the abstract of their paper that: "To date, in-depth studies of rockbursts in the deep South Africa gold mines have revealed no clear evidence indicting a fundamental distinction in the source mechanism between tectonic earthquakes and mine tremors; both appear to be the result of shear failure." Nevertheless, in the main body of the paper they go on to propound that a few examples of seismic events have been detected which suggest the possibility of a failure mechanism other than pure shear. Most mine tremors in South Africa appear to be similar to natural earthquakes with respect to stress drops, source dimensions, magnitudes and the relationship between magnitudes and seismic moments (Spottiswoode and McGarr 1975). Thus, the accumulated evidences convey the conclusion that among South African mine tremors shear failures, possibly stick-slip type slides along preexisting planar or near planar discontinuities or fresh fractures, play a dominant role (Gane *at al.* 1946; Ryder 1988).

These mines produce gold from narrow reefs which extend in their own plane over large distances. Obviously, the working of such tabular ore bodies results in the creation of mainly tabular mine excavations. The unfortunate consequence of this geometrical constraint is an unusually high concentration of stresses in the vicinity of the edges of the mine openings. Clearly, this feature of the mining induced stress field has a significant influence on the seismic activities.

Concentration of events near to active mining: Overwhelming evidence suggests that most mining induced seismic events in South Africa are closely associated with active mining and inoperative parts of the mines are relatively free of tremors. This means that a large proportion of events locate in the vicinity, say within 100-200 m, of advancing faces. This is so even when the mine geometry is symmetric, so the stress distributions around the moving and stationary faces are virtually identical (Cook 1976).

Concentration of events near the reef: Another feature of seismicity in these mines is the conglomeration of events near the reef plane. Most locations occur at perpendicular distances not greater than 150-200 m from the ore body. Furthermore, there appears to be a bias towards the hangingwall, that is, more then half of the events are located above the reef (Cook *et al.* 1966; Cook 1976; Deliac and Gay 1984).

Stochastic nature of seismicity: A decade ago it was observed (Salamon 1983) that:

"No mining geometry or layout appears to exist which unfailingly leads to a seismic occurrence. At best, the likelihood of a seismic event occurring is grater in one situation than in another. Thus, seismicity is a statistically predictable rather than deterministically foreseeable phenomenon."

Unless this stochastic element of mine-induced seismicity is reproduced by the modeling of the phenomenon, the results are unlikely to be of lasting value.

The earlier and far less reliable data was obtained through the statistical analysis of rockburst observations provided by mine personnel in the course of a systematic survey performed in six mines. The survey was initiated in 1963. The results arising from the analysis of the survey data were reported by Cook *et al.* (1966) and repeated in a much abbreviated form by Salamon (1983). The frequency of occurrence of rockbursts, which was the basis of the analysis, was measured in terms of number of events per unit area. The main conclusions of this study can be summarized as follows:

Excavation size: The rockburst incidence increased with span until about 180 m was reached. As mining progressed further, the incidence decreased and at a span of about 275 m the incidence dropped to about 65% of the maximum value.

Abutment size: The rockburst incidence increased with a decrease of abutment or remnant size (i.e. unmined portion of the reef which is fully or partially surrounded by mined-out areas) until a remnant area of 150 m^2 was reached; it then decreased until the abutment was extracted.

Depth below surface: A significant positive linear correlation was found between rockburst incidence and the depth below surface, provided the data were so categorized that the influence of other parameters was eliminated. Similarly, at island remnants, where the abutment areas could be regarded as being comparable, a linear increase in the rockburst incidence was found with an increase in depth. Interestingly, Gibowicz (1990) also remarks that increasing depth appears to result in a growing number of large events.

Stoping width: It was found that in comparable stopes an increase in stoping width was, in most cases, associated with an enhanced incidence of violent occurrences. However, it was difficult to determine whether the relationship was causal; a rockburst could result in a fall of hangingwall, thereby increasing the apparent stoping width. The analysts were handicapped also by the relatively narrow range of stoping widths in the sample.

Dykes and faults: Both of these types of geological features were shown to have highly significant influence on the rockburst incidence, reinforcing indirectly the earlier conclusion that many mine tremors are due to slip on preexisting discontinuities.

It would appear from this review that a significant body of observations exists to support an attempt to model seismic event generation. The tremendous volume of data accumulated during recent years through the seismic installations at various mines provide a background for 'calibration runs' in conjunction with layouts with well documented seismic history. The aim of such work should be to explore the relationship between seismicity and the mining activities responsible for their initiation.

4 MODELING OF SEISMICITY IN HARD ROCK MINES

4.1 Goals for and specifications of modeling

Apart from the earlier mentioned applications of ERR and ESS, no systematic attempt appears to have been made to relate seismicity to mining activity. Unless such linkage can be achieved in a reasonably effective manner, there is little hope for mining engineers to combat the rockburst problem which may menace their mines. Any successful modeling, of course, must mimic reasonably closely the features of seismicity described in the previous section. If this requirement can be fulfilled, the modeling should also have the flexibility to handle realistic mining situations. Hence, the goals of the modeling exercise should be:

* To incorporate reasonably realistic approximations of the dominant source mechanism, that is, shear failure; with the view to test against each other the seismological findings and the results of the proposed approach to modeling.

* To provide, if the modeling methodology proves effective, a tool for mining engineers and researchers for the evaluation of mine plans, the assessment of the relative hazardousness of various geological environments and for the appraisal of possible measures for combating the problem.

The goals stated above are quite ambitious. Clearly, to simulate the generation of seismic events it is necessary to incorporate into the model discontinuities along which sudden slip can occur. At the same time, to cater for the fulfillment of the second goal, it is essential to formulate a model which can handle complex, three-dimensional mining geometries with relative ease.

Fortunately, a large volume of evidence was collated during the 1960's to suggest that the rock mass surrounding the tabular excavations of the gold mines behaves as a linearly elastic medium, except in the very near vicinity of the openings themselves (Ortlepp and Cook 1964; Ortlepp and Nicol 1964; Ryder and Officer 1964; Cook *et al.* 1966). This feature of the conditions in these deep level hard rock mines permits the formulation of a trial model with the required properties.

4.2 *Some Basic Results*

It is accepted that most tremors are initiated when the resistance which binds the opposite faces of a discontinuity are overcome and these surfaces slip suddenly relative to each other. As such a slip is associated with a sudden drop in the shear stress acting on the faces in contact, some energy, W_r, will be released. A part of this energy is consumed during the slide, heating the rock surfaces, W_h, and the reminder is transformed into kinetic or seismic energy, W_k (Salamon 1984). Thus,

$$W_k = W_r - W_h \tag{2}$$

Shear type seismic events are created by slip along either fresh fractures or preexisting flaws. It is convenient and reasonable, at least in the first approximation, to assume that these discontinuities are planar in both cases. A fundamental difference between a preexisting flaw and a fresh fracture is that the former has a location and orientation which is independent of the prevailing stress field, while these properties of the latter are determined by the stresses. It is assumed in this study that the source discontinuities are of *geological origin* (i.e. predate mining) and, in the pre-mining state, the virgin stress field has frozen together their opposing surfaces.

Postulate that the rock mass is linearly elastic around a flaw and the magnitudes of the shear stress prior to and after the slip are τ and $\mu\sigma$, respectively. Salamon (1974) has shown that the estimates of the released energy and the energy consumed (heat energy) during the slide are as follows:

$$W_r = \frac{1}{2} \int_A (s_t \, \tau + \mu\sigma) \, dA \tag{3}$$

$$W_h = \mu \int_A s_t \, \sigma \, dA \tag{4}$$

Consequently, the generated kinetic energy is given by:

$$W_k = \frac{1}{2} \int_A s_t(\tau - \mu\sigma) \, dA \tag{5}$$

In these expressions A is the area of the flaw and s_t is the slide or ride. Clearly, for these results to be sensible, the normal stress over the surfaces in contact must be compressive ($\sigma > 0$) and $\tau - \mu\sigma > 0$, with $\tau = |\tau_t|$, where τ_t is the actual shear stress. Also, μ denotes the *dynamic* coefficient of friction.

The frictional resistance to slip between the contact surfaces of an existing flaw, τ_r, is commonly found to increase linearly with the normal stress and can be expressed as:

$$\tau_r = \mu_s\sigma + c \tag{6}$$

where c is the cohesion and μ_s is the *static* coefficient of friction. Slip will occur if $\tau - \tau_r > 0$, that is, if

$$\tau - (\mu_s\sigma + c) > 0 \tag{7}$$

The stress drop during the slide is:

$$\Delta\tau = \tau - \mu\sigma \tag{8}$$

If mining is performed in small steps then τ exceeds τ_r only slightly. In this case the stress drop can be approximated by the following expression:

$$\Delta\tau \simeq (\mu_s - \mu) \, \sigma + c \tag{9}$$

When slip occurs the contact surfaces are mobilized without a change in the value of the normal stress (Salamon 1964). The assumptions adopted here for the control of slide are identical to those employed by Ryder (1988) when introducing his ESS concept.

Further simplifications occur if the flaws are assumed to be circular with radius R and the normal and shear stresses are taken to be constant over the flaw area:

$$W_r = \frac{1}{2} (\tau + \mu\sigma)A \, s_{av} \qquad W_h = \mu\sigma A \, s_{av} \tag{10}$$

and

$$W_k = \frac{1}{2} \Delta\tau A \, s_{av} \tag{11}$$

Here $A = \pi R^2$ is the area of the flaw and s_{av} is the *average slip* or ride of the contact surfaces on each other. Since the ride distribution is given by (Salamon 1964):

$$s_t = \frac{4(1-\nu)\Delta\tau}{\pi(1-\nu/2)G} \sqrt{R^2 - r^2} \tag{12}$$

the average ride is:

$$s_{av} = \frac{1}{A} \int_A s_t \, dA = \frac{8(1-\nu)\Delta\tau R}{3\pi(1-\nu/2)G} \tag{13}$$

In these formulae G is the shear modulus and ν is the Poisson's ratio.

The seismic moment, M_o, is defined as the product of the average slip, shear modulus and the area of the flaw. It was noted by Hedley (1985) that the expression of mean slip facilitates the computation of M_o. After correcting some minor arithmetical errors of Hedley, the following result is obtained:

$$M_o = \frac{8(1-\nu)\Delta\tau R^3}{3(1-\nu/2)}$$ (14)

This expression is identical to that published by Brune (1970,1971) for a circular fault, if $\nu = 1/4$. Hedley noted also that the results in (10) and (11) permit the evaluation of seismic efficiency, η, which is defined (McGarr 1984) to be the ratio of the seismic energy to the total released energy. Therefore,

$$\eta = \frac{\Delta\tau}{\tau + \mu\sigma} \approx \frac{(\mu_s - \mu) + c/\sigma}{(\mu_s + \mu) + c/\sigma}$$ (15)

The seismic efficiency is generally reported to be in the order of a few percent at depths ranging from 1.5-2.0 km. If it is accepted that the normal stress at the flaw is likely to increase with depth, then the result in (15) confirms, provided that $c \neq 0$, that the seismic efficiency reduces with increasing depth (McGarr 1984).

According to some data published by Dieterich (1977) and also by Jaeger and Cook (1979), the dynamic coefficient of friction is some 90-95% of its static counterpart. Byerlee (1977) found that for a wide range of rock surfaces the cohesion c is nearly zero and the static coefficient of friction μ_s clusters in the range from 0.5 to 1.0. South African testing of quartzite, the dominant rock formation in many gold mines, corroborates this finding since it has yielded a μ_s value of 0.6 (Ryder 1988). Ryder, in the same publication, also suggests that the value of c is unlikely to be greater than 10 MPa.

The results in (11) and (14) reveal that a simple relationship exists between the seismic moment M_o and the kinetic energy W_k:

$$W_k = \frac{\Delta\tau}{2G} M_o$$ (16)

If the logarithm is taken of both sides of this equality the following expression is obtained:

$$\log W_k = \log M_o + K$$ (17)

where

$$K = \log\frac{\Delta\tau}{2G} \approx \log\frac{(\mu_s - \mu)\sigma + c}{2G}$$ (18)

Spottiswoode and McGarr (1975) established, using data recorded at ERPM, the following empirical relationships:

$$\log W_k = 1.5 M_L - 1.2 \qquad \log M_o = 1.2 M_L + 4.7$$ (19)

Here W_k and M_o are measured in MJ and M_L is the magnitude of the events. Of these expressions probably the latter can be regarded more robust, since several other authors have reported similar results (Ryder 1988; Gibowicz 1990). This pair of formulae allows the derivation of an empirical function relating kinetic energy and seismic moment:

$$\log W_k = 1.25 \log M_o - 7.08$$ (20)

where -7.08 corresponds to K in (17). It is interesting to note that the coefficient of $\log M_o$ is unity in (17), while it is 1.25 in (20). These values seem remarkably close.

For both earthquakes and mine tremors the relationship between the frequency of occurrence and magnitude has been found to be of the form:

$$\log N = a - b M_L$$ (21)

where N is the number of events of magnitude greater than or equal to M_L. Parameters a and b are determined empirically. McGarr and Wiebols (1977) has computed three sets of values for these constants, corresponding to three different parts of ERPM:

(i) a = 1.68, b = 0.59;   (ii) a = 1.66, b = 0.61
                     (iii) a = 2.19, b = 0.63 (22)

The relationship in (21) can form the basis of comparisons if the value of a is normalized in some manner. In the case of narrow reefs, where the mineralized width is smaller then the working height or the stoping width, it will be advantageous to use an areal measure of extraction. In the case of massive ore bodies, it may be preferable to employ volume or tonnage for normalization.

Since the reef is narrow at ERPM, it is logical to use an area, for example, 100,000 m^2. The numerical values in (22) were related to the extraction of 5,280 m^3, 5,420 m^3 and 7,150 m^3 in the three areas. Assuming that the stoping width was approximately 1 m, the frequency-magnitude relations of McGarr and Weibols are:

(i) $\log N_o = 2.96 - 0.59 M_L$
(ii) $\log N_o = 2.93 - 0.61 M_L$ (23)
(iii) $\log N_o = 3.34 - 0.63 M_L$

where now N_o is the number of events of magnitude greater or equal to M_L per 100,000 m^2 mined.

4.3 *Framework of the model*

The relationship between mining activities and seismicity is modeled by simulation. The simulator consists of two parts. The first part generates the environment or the rock mass containing flaws which will be subjected to the stresses induced by mining. The second part mimics mining and exposes the flaws to enhanced stresses and in the process triggers some mine tremors.

Flow generation. First a fictitious rectangular parallelepiped is established in the rock mass. The block is of sufficient size to contain the relevant part of the ore body, together with such a large portion of the surrounding rock mass that outside it the stresses induced by mining are negligible.

The stochastic nature of the mine seismicity is introduced in two ways. Firstly, it is recognized that geological discontinuities are not distributed in a regular manner in the ground, but they are random variables, which are often grouped into systematic joint sets. However, neither the orientation nor the frequency of the joints, or flaws, in a set are completely regular. Also, not *all* discontinuities belong to one of the sets. Thus, a flaw distribution must be generated in the parallelepiped to simulate, as realistically as possible, the distribution of the weaknesses in the prototype. The flaw distribution is generated employing a suitably designed Monte-Carlo technique. If no information is available with respect to the distribution of the discontinuities then, as a fall back assumption, it may be postulated that both the locations and orientations of the flaws are randomly distributed with an uniform probability density function.

Secondly, it is known that the size of the flaws is also a random variable. As the discontinuities are taken to be circular, a radius is attributed randomly to each of the flaws, by sampling, for example, a Weibull probability density distribution (Scheaffer and McClave 1990):

$$f(R) = \frac{\gamma_o}{R_o} \left(\frac{R}{R_o}\right)^{\gamma_o-1} \exp\left[-(R/R_o)^{\gamma_o}\right] \qquad (24)$$

The expected or mean value of this distribution is:

$$R_m = \Gamma(1 + 1/\gamma_o) R_o \qquad (25)$$

where $\Gamma(.)$ is the gamma function, γ_o and R_o are positive parameters. For $\gamma_o = 1$ the above density function reduces to the well known exponential distribution. The cumulative distribution function of $f(R)$ in (24) is as follows:

$$F(R) = 1 - \exp\left[-(R/R_o)^{\gamma_o}\right] \qquad (26)$$

and the flaw radius corresponding to random number r_j is given by:

$$R_j = \frac{R_m}{\Gamma(1+1/\gamma_o)} \left[-\ln(1-r_j)\right]^{1/\gamma_o} \qquad (27)$$

It would appear that the useful range of the parameter of the distribution is $0 < \gamma_o \leq 1$.

Conceptually several other parameters, such as the cohesion, the coefficients of friction, elastic properties, etc., could be treated as random variables. It appears at this stage that to do this would result in unnecessary complications; the required stochastic behavior of the simulated seismicity is achieved adequately by regarding the distribution, orientation and size of the flaw as random variables.

The total number of discontinuities, $\bar{N}$, is computed from a predetermined flaw density per unit volume, δ_o, and the volume of the parallelepiped. Next, on the basis of the accepted stochastic model, the following parameters are computed for each of the $\bar{N}$ flaws: the coordinates of the centroid, two angles defining the direction of the normal and the radius. These data are then stored in an appropriate computer file for further processing.

Generation of seismic events. Once the hypothetical block containing the appropriately distributed discontinuities have been prepared 'mining' can commence. One of two situations could arise; either the block has not been mined before or the intention is to investigate the rockburst hazard in an existing mine. As the former case is simpler, it is assumed first that the block is free of excavations.

Mining is assumed to take place in steps. This is a reasonable assumption, since virtually all hard rock mining is still performed by blasting. The consecutive steps of mining are designed to simulate the actual mining progress as closely as practicable. Ideally, the advances in extraction should resemble closely the actual mining activities, both in terms of the sizes of advances and their locations.

Each mining step is associated with a cycle of computation during which all flaws are visited and the driving shear stress τ and the frictional resistance $(\mu_s \sigma + c)$ are computed at the centroids of the flaws in turn. The computation can be carried out, assuming that the rock mass is an elastic continuum, by any numerical technique which can handle the problem effectively. Once the driving and resisting stresses corresponding to a particular mining stage are available for one of the flaws the test prescribed by the inequality in (7) can be performed. If the inequality is satisfied, then an instability occurs, or in other

words, a seismic event is triggered. Since the radius of the flaw and the normal and shear stresses at its centroid are known, the relationships in (11), (13), (14), (15) and (18) allow the computation of the seismic moment, M_o, kinetic energy, W_k, average slip, s_{av}, seismic efficiency, η, and the parameter K. Finally, an empirical estimate of the Richter magnitude can be obtained from one of the formulae in (19). The choice will depend on a judgement concerning the relative reliability of these empirical expressions. Of course, if in a mining area a locally established relationship is available, it will be preferable to employ such expression.

If the test prescribed in (7) suggests that a particular flaw is still stable, then this discontinuity is not triggered, at least not during this mining step.

To maintain the simplicity of the model four, somewhat arbitrary rules are accepted: first, the possible interaction between neighboring flaws are ignored; second, flaws which reveal tensile stress at a stage of mining are taken out of further consideration; third, a discontinuity which has been triggered once will not be triggered again and last, flaws, the location of which are mined out before they have slipped, will not be triggered. The last of these criteria is obvious. No attempt is made to provide scientific justification for the introduction of the others. They are dictated by pragmatism arising from experience which suggest that a modeling project should start with the simplest plausible model and this should be made more complex only if it fails to reproduce, with acceptable accuracy, some vital feature of the prototype.

The simple rules stated in this section are sufficient to construct the flow chart for the main component of the seismic event generator. Assume that mining takes place in $\bar{M}$ steps. In the i-th step (i = 1,2,...,$\bar{M}$) all $\bar{N}$ flaws are examined. The analysis concerning the j-th flaw (j = 1,2,...,$\bar{N}$) during the i-th step of mining is shown in Figure 1. It is assumed that the data concerning the $\bar{N}$ discontinuities are stored, for example, in a matrix with $\bar{N}$ rows where each row corresponds to a flaw. In the same matrix, columns are provided for the storage of normal stress, NSTi, shear stress, SSTi and for a mining step identifier, i. Initially these three columns contain zeros, which are replaced with nonzero quantities only when a flaw is triggered, mined out or when the normal stress at a centroid is found to be tensile. 'n the first test of the chart, shown near to its top, 'NST0' denotes the normal stress value already stored in the matrix.

In principle, the analysis shown here is performed for $\bar{N}$ flaws during all $\bar{M}$ steps of extraction, that is, $\bar{N} \times \bar{M}$ times. If the problem to be analyzed is large, then this simple scheme of calculation may become too burdensome. In such instances ingenuity can be applied to reduce the computational task.

If the problem to be tackled involves the analysis of an existing mine, then the scheme outlined above requires modification. First, generate the flaws in the earlier defined parallelepiped, ignoring the mine cavities. Next, simulate the previous mining in a manner analogous to that explained before, but using much larger steps of extraction in order to minimize the task. The purpose of this coarse mining is to attempt to trigger the events which would have been activated by the early operations. While this shortcut does not ensure perfect simulation, it does reduce the burden of preparation and computation.

4.4 An example: single panel in a narrow reef

Definition of the problem. Naturally, the best method for testing the concepts outlined in the previous sections is to simulate mining and the corresponding seismicity in situations where good seismic data are available. However, there has not been an opportunity to do this so far. The next best step is to carry out extensive simulation of a simple idealized mining geometry,

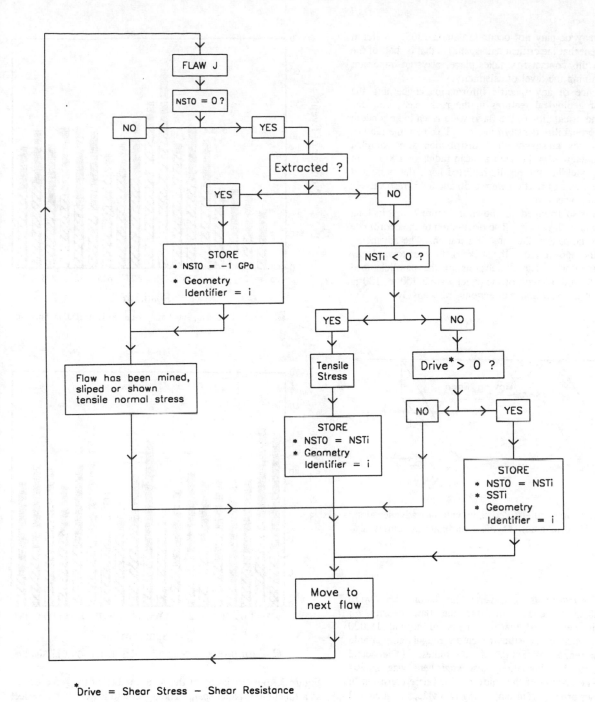

*Drive = Shear Stress – Shear Resistance

Figure 1.   Evaluation of flaw j during the i-th mining step.

which resembles reasonably closely the method of mining at some of the deepest mines in the Witwatersrand basin. This is the mining method to which the description of seismicity in Section 3.2 corresponds. This method is the longwall system which can be approximated in two dimensions by the widening of an isolated, parallel-sided panel.

To check whether the proposed model, which can be described with justification as a 'seismicity simulator', can reproduce some or all of the main features of South African seismic activity described in Section 3.2, it was decided to perform simulations at 2,000 m and 3,000 m depths, in a reef which dips at 20 degrees. The stoping width was taken to be equivalent to a maximum convergence, s_m, of 1.0 m, which in some simulations

was reduced by backfill to an effective maximum convergence of 0.5 m. Further, it was arranged that only one of the faces (edges) of the panel was advanced, the other was left stationary. In the course of simulations the span of the panel was increased from zero to 2,000 m in 2.5 m steps, which compares reasonably well with the normal face advance of about 1 m. The Young's modulus, Poisson's ratio and the specific weight of the rocks, γ, were taken to be 70 GPa, 0.2 and 25 kPa/m. The virgin principal vertical and horizontal stresses were postulated to be γH and $k\gamma H$, respectively, where H is the depth of mining. The value of k was either 0.6 or 1.0. The stresses induced by mining were computed from the well known analytical solution for an isolated panel where contact between the hangingwall and the footwall may or may not occur (Salamon 1968). Later it became apparent that the critical half-span, L_c, that is, half of the span at which this contact first takes place, plays an important role in determining the level of seismicity.

In the absence of any specific information concerning the distribution of geological features in the rock mass, both the distribution and orientation of the flaws were taken to be random with uniform probability density function. To obtain the radii of the discontinuities an exponential distribution was sampled (Weibull parameter $\gamma_o = 1$), with a mean radius of 100 m. In all simulations yielding the results reported here, the static and dynamic coefficients of friction were 0.50 and 0.45, respectively and the cohesion was zero.

The panel was so arranged, as shown in Figure 2, that its long axis was parallel with the dip. The orientation of the coordinate system is also indicted in the same illustration. The advancing face of the panel moved along the strike of the reef towards the positive x direction. The dimensions of the parallelepiped enclosing a 100 m dip-length of the panel were 2,700 m, 100 m and 1,600 m in the x, y and z directions, respectively.

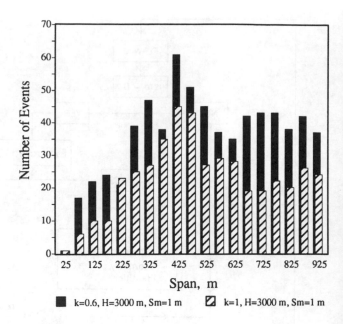

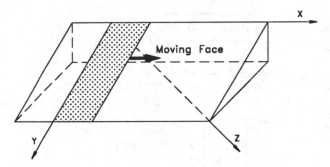

Figure 2. Down dip parallel-sided panel with face advance along the strike. The orientation of the coordinate system is also shown.

Results of simulations. Several simulation runs were performed using various values for the flaw density δ_o. Eventually, it was found that a density of around 16,000 flaw/km^3 gave seismic intensities which are roughly comparable to those experienced by South African mines. (The actual density used in the simulations discussed here was 16,204 flaws/km^3.) The portion of the study reported here consisted of eight base cases arising from employing two values for depth H and two values for the horizontal stress factor k, giving four cases for each of the two values of maximum convergence ($s_m = 0.5$ m, 1.0 m). The first impression from the basic results is that the frequency of tremors reduces as the virgin shear stress is diminished by changing k from 0.6 to 1.0. Similarly, a reduction in event frequency was recorded if the critical half-span was decreased by either an increase in depth or a reduction in the maximum convergence. However, it became apparent later that reliance on event frequency might lead to misleading conclusions.

Excavation size. In Figure 3 the histograms of the total event frequency rate, as a function of span (the data is plotted separately for two sections of this span), are depicted for H = 3,000 m using the two values of k. It is obvious that the rate increases rapidly first, reaching a maximum at a span which is approximately equal to $2L_c$. As the span is increased further the rate of triggering gradually diminishes, as the width of the open region reduces from $2L_c$ at contact, to a value somewhat less then L_c ($2L_c/\pi$; Salamon 1968). It seems that most of the tremors are activated in a region scaled in some manner by the

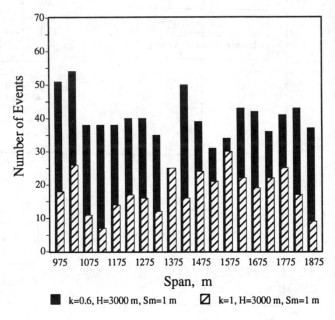

Figure 3. The variation of the total number of triggered events as a function of panel span. The upper and lower charts reflect the data from 0 - 950 m and 950 - 1,900 m, respectively.

width of the open zone behind the moving face. It is interesting to observe that the picture depicted in this illustration is similar to that discussed in Section 3.2 concerning the effect of excavation size on rockburst intensity.

Concentration of events near to the moving face. Figures 4(a) and (b) illustrate dramatically that the vast majority of tremors occur near the active face. Here positive distances represent locations ahead of the advancing face. The former diagram depicts the frequency of events, while the latter shows the distribution of the generated kinetic energy. These results are analogous to the field observations reported by great many field investigators, as it was discussed in Section 3.2. This is especially exciting result since it was computed in a situation where the stress distribution is obviously symmetric with respect

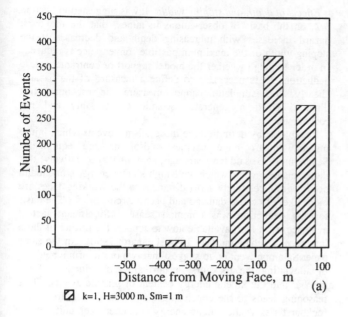

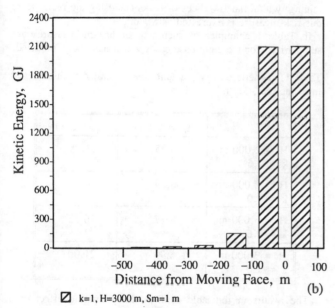

Figure 4. Distributions of the number of events (a) and the corresponding kinetic energy (b), relative to the advancing face. Positive distances signify locations ahead of the moving face.

to the central axis of the panel. It is obtained because near the moving face the stress changes induced by each mining advance are greater than those in the vicinity of a stationary face. Also, the advancing workings sweep new sets of flaws at each step of mining (Salamon 1983).

Concentration of events near the reef. Figures 5(a) and (b) illustrate the distribution of tremors relative to the reef plane. A positive distance in these illustrations conveys a position in the footwall. Diagrams (a) and (b) show the distributions of events and the kinetic energy generated by them, respectively. Clearly the events concentrate closely to the ore body as was found experimentally in the field. Also the calculated distributions, similarly to field observations, are biased towards the hangingwall. It seems, however, that the spread of the computed data, especially for k = 0.6, is greater than that reported by seismologists. It is unclear at this stage whether this discrepancy is a weakness of the model, or merely due to erroneous assumptions concerning, for example, the virgin stress

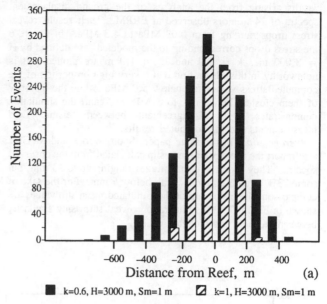

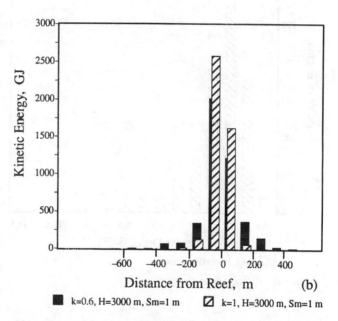

Figure 5. Distributions of the total number of events (a) and the corresponding kinetic energy (b), relative to the reef. Positive distances place points in the footwall.

distribution, or the values of the coefficients of friction, and/or that of the cohesion.

Frequency and magnitude relationship. It has been possible to fit to the results of the simulation the classical frequency-magnitude relationship given in (21). The results for $s_m = 1$ m, when normalized for an extracted area of 100,000 m^2, are given next:

$$
\begin{aligned}
H = 2,000 \text{ m}; \quad k &= 0.6: \quad \log N_o = 3.34 - 0.61\, M_L \\
H = 2,000 \text{ m}; \quad k &= 1.0: \quad \log N_o = 3.22 - 0.69\, M_L \\
H = 3,000 \text{ m}; \quad k &= 0.6: \quad \log N_o = 3.18 - 0.60\, M_L \\
H = 3,000 \text{ m}; \quad k &= 1.0: \quad \log N_o = 2.87 - 0.53\, M_L
\end{aligned}
\tag{28}
$$

Although the linear fit of the data have not been particularly good, the constants in these relationships compare remarkably well with the results of McGarr and Wiebols (1977) as given in (23).

Stress drop. Spottiswoode and McGarr (1975) published the

results arising from the analyses of the ground displacement spectra of 24 tremors observed at ERPM. Their results reveal stress drops ranging from 0.02 MPa to 4.3 MPa. In Figure 6 the stress drops corresponding to the modeled case defined by H = 3,000 m, k = 1.0 and s_m = 1.0 m are plotted. It is noteworthy in this illustration that a very high proportion of the computed stress values fall below 5.0 MPa, while the majority of them clusters between 1 to 5 MPa. Again, the simulation demonstrates a broad agreement between seismic field measurements and the computed results.

Average slip. In the same paper, Spottiswoode and McGarr also report the average or mean slip calculated from the observed spectra. They report slip magnitudes ranging up to 13 mm, but some 73% of their slip values fall below 5 mm. For the sake of comparison, the histogram of the simulated mean slip values are shown in Figure 7. This illustration reveal surprising similarity between the field and modeled values.

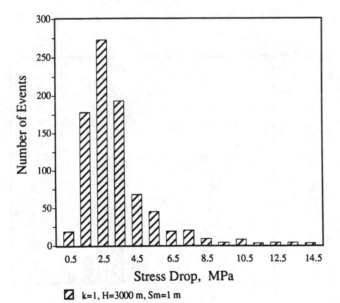

k=1, H=3000 m, Sm=1 m

Figure 6. Histogram of stress drops.

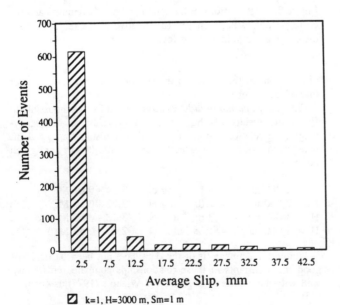

k=1, H=3000 m, Sm=1 m

Figure 7. Histogram of average slip.

Effects of depth and stoping width. It was suggested in Section 3.2, on the basis of observations in mines, that the rockburst hazard increases with increasing depth and decreases if the stoping width or the maximum possible convergence is reduced. In order to judge whether the model support or contradicts these deductions, it is necessary to define a measure of the hazard. Clearly, this simulator cannot measure the rockburst risk directly, since it generates seismic events only and not rockbursts.

Rockbursts by definition are those seismic events which injure miners, impair mine cavities and/or damage equipment. Salamon (1983) said ten years ago, somewhat speculatively, that those seismic events which have high kinetic energy content and occur within relatively short distances to the working faces are more likely to cause damage and become rockbursts. The basis for his deduction was common sense. Significant factual evidence is, in fact, available now to support the former of these inferences (Deliac and Gay 1984; Lenhardt 1990). In the same paper Salamon went on to suggest that two of the prime methods of alleviating the rockburst hazard are the reduction of the number and the kinetic energy content of seismic events. This reasoning leads to the conclusion that a useful measure of the rockburst risk is the kinetic energy generated, per unit area of mining, within narrow slices of rocks enclosing the reef. Here this risk measure is expressed as MJ/m^2.

In Table 1 examples of such risk or hazard measures are summarized for the eight base cases which have been modeled.

Table 1. Kinetic energy per unit area generated within ± 100 m of the reef, MJ/m^2.

	s_m = 0.5 m	s_m = 1.0 m
H = 2,000 m k = 0.6	6.35	11.05
H = 2,000 m k = 1.0	3.89	8.18
H = 3,000 m k = 0.6	8.17	16.22
H = 3,000 m k = 1.0	6.99	21.00

The results in the table confirm the mitigating effect of a reduction in the stoping width or in the maximum possible convergence. This deduction supports the use of backfill as a possible rockburst control measure. The adverse effect of increasing depth is also obvious from the tabulated data. In all instances the risk measures are higher at the greater depth, if all other parameters are kept unchanged. It is interesting to note that a simple count of seismic event frequency suggests the opposite conclusion. The influence of the horizontal stress factor k is not quite so clear-cut. In three of the possible four comparisons the hazard appears to be less if k = 1, but not so in the fourth case.

Deductions. It can be inferred from the discussion in this section that the described model appears to have a remarkable ability to mimic the available seismic data, at least in a general sense. It is readily conceded that the results have turned out to be better than expected. After all, the flaws in the rock mass are treated as preexisting features and the fresh fractures are ignored. While this simplified approach obviously does not represent the true situation, but perhaps the model's behavior is better than expected because the similarity of the slip condition in (7) to the Coulomb failure criterion.

Only limited numerical experimentation has been performed with the proposed method. It is surprising, therefore, that even this restricted work has resulted in the selection of a set of parameters which appears to yield apparently acceptable results. This set of parameters is as follows:

Weibull parameters: $\gamma_o = 1$, $R_m = 100$ m,
Flaw density: $\delta_o = 16{,}000$ flaw/km³,
Static coefficient of friction: $\mu_s = 0.50$,
Dynamic coefficient of friction: $\mu = 0.45$,
Cohesion: $c = 0$.

While the seismicity provided by this set of constants compares well with that experienced by ERPM and other similar mines, much more will have to be done to prove conclusively the utility of the proposed model. Future work must include the comparisons of observed and simulated seismic activities. Nevertheless, it seems already clear that the preliminary results reported here establish a strong case for considerable follow-up work. This is especially so because the basic structure of the model lends itself to generalization and could be incorporated readily into an already established powerful boundary element system such as MINSIM. In that environment it would facilitate the wide spread application of the seismic simulator in the fight for the alleviation of the rockburst hazard.

5 COAL BUMPS AND RELATED PROBLEMS

5.1 Basic approach

Coal mines experience a variety seismic events, some of which are similar in origin to the shear instabilities discussed in relation to hard rock mines (Gibowicz 1990). Others appear to be quite different such as pillar bumps, which have represented a persistent problem, for example, in some mines of the Appalachian Basin of the United States (Iannacchione 1990). As Gibowicz (1990) points out this type of event is one of those in the classification of Hasegawa et al. (1990) which does not lead to a double-couple type focal mechanism. Such failures do not only happen in pillars, but have been observed to occur, for example, on longwall faces or at the edges of large extracted areas. Lippmann (1990) notes that such bumps "...seem to occur only where the rock in the roof and floor adjacent to the seam ... are about 10 times stiffer and stronger than the coal." Even if not all cases where coal bumps occur conform to this description, the observation does justify the restriction of the discussion here to cases in which nonlinearity, an essential prerequisite to instability, is restricted to the coal seam.

Far less seems to be known seismically about coal bumps than shear type instabilities. Therefore, the discussion here differs from that employed in the previous section. First a brief presentation is given of some modeling results which appear to suggest the occurrence of some instability in certain mining situations. Next, it is argued that only seismic observations can determine whether the instabilities predicted by the model are real or, in other words, whether the proposed model describes the behavior of the prototype acceptably well.

5.2 Description of the model

It has been found in recent years that a linear layered model describes reasonably well the displacement field induced by the extraction of a coal seam. Encouraged by this finding, Salamon (1992) has proposed that this simple model could be used to load a coal seam which contains a system of mining excavations. He suggested also that the coal itself should have strain softening properties. This conclusion arose from the analysis presented in the paper which revealed that the currently most favored method for the estimation of pillar strength, that is, the confined core

method, has a fatal flaw. *Every* time when the coal strength at the edge of a pillar is overcome, the method produces an instability. Since such instabilities, fortunately, do not occur regularly in mines, this shortcoming must be cured. The most obvious remedy appears to be to sophisticate the modeled behavior of coal and allow strain softening.

Here only a much abbreviated version of the model proposed for the investigation of instabilities and, ultimately, the strength and yield characteristics of pillars in coal mines, is presented. Its detailed discussion is given in the earlier quoted paper. It consists of a set of three equations. The first of them is derived from the layered model of the stratified roof and floor sediments.

This equation relates the induced vertical stress, $\sigma_z^{(i)}$ (only horizontal stratification is considered) and the convergence of the roof and floor of the seam, s:

$$\frac{d^2 s}{d x^2} = \frac{2 \sigma_z^{(i)}}{\lambda E} \tag{29}$$

where E is the Young's modulus of the surrounding rocks and λ is a parameter proportional to the effective lamination thickness of the strata (Salamon 1991). The second equation expresses the equilibrium of forces in the seam:

$$t \frac{d \bar{\sigma}_x}{dx} + 2 \tau_{xz} (x, t/2) = 0 \tag{30}$$

In this relationship $\bar{\sigma}_x$ is the mean horizontal stress, t is the seam thickness and τ_{xz} is the contact shear stress between the surrounding rocks and the coal. The third relationship is the constitutive law of the coal which is taken to be of the form shown in Figure 8, where κ is the Rankine constant and σ_c is the uniaxial strength. The applied stress-strain relationship is different in the different parts of the seam, depending whether the coal is still elastic or yielding. The descending branch of the curve in Figure 8 is defined by:

$$\sigma_z = \frac{1}{1 - \phi} \left[(1 + \delta)(1 - \sin\rho) \ \sigma_c + (1 + \phi) \ \sigma_x - 4\delta G_s \varepsilon_z \right] \tag{31}$$

where

$$\phi = (1 - 2\nu_s) \ \delta + (1 + \delta) \ \sin\rho \tag{32}$$

In these expressions ρ and ν_s are the angle of internal friction and the Poisson's ratio of the coal. The dimensionless parameter δ is proportional to the slope of the descending branch of the stress-strain relationship. In obtaining solutions of this system, it has been assumed that a yielding edge of a pillar or ribside is restrained by friction only.

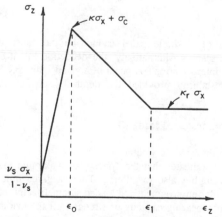

Figure 8. Complete stress-strain curve of coal with strain softening.

5.3 *Some preliminary results*

Since the investigations using the model outlined in the previous section have not reached an advanced stage, the results reported here are merely meant to underline the importance of linking seismic research and modeling studies.

The purpose of the investigation featured in this section is to provide practical guidance for layout design in coal mines. More specifically, the ultimate aim is to approach the prediction of coal pillar strength from a mechanistic point of view and to attempt to lay down the foundations for the design of yielding coal pillars and fenders (narrow pillars). However, the occurrence of coal bumps in certain mines and the research which has been done by various investigators provide a warning that instabilities may occur under some circumstances and these could result in hazardous mining conditions. Two types of instability could arise. First, the pillar edge might fail suddenly and violently and second, uncontrolled collapse of one or several pillars may take place.

Linkov (1992) reports that analytical studies carried out in Russia have indicated that in the course of widening of an isolated panel, eventually, a critical span is reached at which instability at the ribsides is created. The instability occurs when the vertical stress at the faces becomes zero. The results reported by Linkov are based on the use of the homogeneous isotropic elastic substance. This medium is known to be inadequate when the behavior of a stratified sedimentary system is to be modeled. In spite of the deficiency of the model employed in the study, the warning in Linkov's report cannot be disregarded.

In analyzing the problem involving an isolated panel, first the stress and convergence distributions were derived. It was noted that with increasing span, S, the elastic ribsides soon start to yield and, as a result, the vertical stress at the pillar edges begins to diminish. This process causes the maximum stress to migrate away from the exposed face to the point where the strength of the coal is overcome and the seam starts to yield. Let the distance from the edge to the yield point be a_y. First this yield distance grows gradually in accordance with the increase in span. Later a_y commences to accelerate and, in fact, at a particular span the slope of the $a_y = F(S)$ function has a singularity, revealing that the system at this stage is *unstable*.

Further analysis indicates that at this instability the total induced load transferred to a ribside has its *maximum* and this load can be increased *only* by providing some support to the exposed face. This eventuality is not considered here. If there is no contact between the roof and floor, this deduction has a very interesting consequence. At a given depth H, let the span at which the maximum load bearing capacity of the ribsides is reached be S_m and the critical span at which roof to floor contact first occurs be $S_c = 2L_c$. If

$$S_m \geq S_c \qquad (33)$$

then the mining of a single panel cannot be the cause of instability in the seam. Alternatively, the deduction might be given a different interpretation. A panel of a given span, say S, cannot induce instability provided the depth H satisfies the inequality:

$$H \leq (\text{Maximum transferred load})/\gamma S \qquad (34)$$

where γ is again the specific weight of the rocks.

Little reflection indicates that the concept of maximum load bearing capacity applies also to pillars. Thus, a given pillar, placed in a particular environment and loaded symmetrically, has a well defined load bearing capacity or strength. If the loading is asymmetric, the strength is still defined, but now as a function of the asymmetry, and it is less than that corresponding to the ideal loading condition.

Whether the model allows the deformation of the pillars to grow beyond the convergence corresponding to their strength, still remains to be investigated. Such pillars have an important support role in pillar extraction and in the control of longwall chain pillar systems. They are commonly referred to as 'yield pillars'.

To perform the analyses involved in this important area of mining rock mechanics requires the use of a large number of properties. Reasonable estimates can be made of the numerical values of most of these parameters. The notable exception is the magnitude of the strain softening parameter δ. Virtually no acceptable information exists concerning the value of this property. Data obtained from laboratory experiments cannot be regarded reliable since the descending branch of the deformation curve of a specimen is usually controlled by one or, at most, a few slip surfaces.

Both span S_m and the corresponding a_y are highly sensitive to the value of parameter δ. The application of seismic instrumentation in these circumstances could resolve two fundamental issues:

* Does, in fact, the predicted instability (coal bump) occur?
* What is the span at the instability?

Apart from seismic measurements, there does not appear to be a simple and convincing method answering these questions. If, in fact, a span corresponding to a coal bump can be measured seismically then it becomes possible to estimate the corresponding value of δ. Such success would advance the understanding of the basic problems of the rock mechanics of coal mining.

6 CONCLUSIONS

There can be little doubt that the success of the mining engineers' fight for the alleviation of the rockburst and coal bump hazards hinges to a considerable extent on the ability of researchers to provide tools to relate seismicity, or the occurrence of instabilities, to mining operations. The key to achieving this linkage appears to be geomechanical modeling. The opportunities for modeling for this specific purpose appears to be numerous. Two particular applications are outlined in the main body of the paper. These examples probably represent the opposite extremes of the applications of modeling techniques.

The first example involves the modeling or, perhaps more accurately, the simulation of the generation of seismic events in hard rock mines. The model was designed to mimic as closely as possible the main features of observed seismicity. The approach is presented in a general form, but its application is illustrated through the simulated triggering of seismic events by the extraction of a tabular deposit at some considerable depths. The actual example involves a single tabular excavation, a face of which is advanced in small steps to correspond to the actual mining process as closely as practicable. It is postulated that all seismic events are due to a sudden slip of the contact surfaces of preexisting circular flaws. The locations, orientations and radii of the flaws are selected using Monte-Carlo techniques.

The simulation has resulted in a surprisingly realistic general picture of the seismicity in deep level South African gold mines, which use the longwall method of mining. It has been possible to quantify, at least in a preliminary sense, the parameters involved in the modeling. These encouraging results suggest that the initial work should be followed by more specific case studies, comprising of the comparisons of observed and simulated seismic data. It would appear that if this follow-up work is reasonably successful also, then the proposed method can readily be converted into a tool which could aid mine design directly.

The second example involves the discussion of some early results of modeling in coal mining. The final aim of this research is to develop an understanding of the behavior of coal ribs and pillars when their edges are in a strain softening state. The model consists of two layered linear media within which a coal seam with elastic or strain softening properties is sandwiched. The preliminary results of the modeling indicate that coal ribsides and pillars have a definite strength, but also, that instabilities may develop in the course of mining. These instabilities may turn out to be the mathematical manifestations of coal bumps which disrupt and sometimes even endanger the operation of some coal mines.

It is suggested in conjunction with this example that seismic measurements may be the most appropriate tools to resolve whether a mathematical instability is in fact a seismic event and, if it is, the same observations may be employed to back-calculate the strain softening parameter of the model. Thus, in this instance seismology is proposed as a device of research which could aid the development of effective models for coal mine design.

Acknowledgments: It is acknowledged that the seismic simulation project was partially funded by the Research Organization of the Chamber of Mines of South Africa. The results presented in the illustrations were computed by Ms. Maria T. Tchonkova-Parashkevova. The modeling work concerning coal mining is part of an ongoing research project performed at the University of New South Wales and sponsored by the Joint Coal Board, New South Wales, Australia. The author is grateful for all help and support.

REFERENCES

Brady, B.H.G. 1990. Keynote lecture: Rock stress structure and mine design. In C. Fairhurst (ed.), *Rockbursts and Seismicity in Mines*. p. 311-321. Rotterdam, Balkema.

Byerlee, J. 1977. Friction of rocks. *Proc. 2nd Conf. on Experimental Studies of Rock Friction with Application to Earthquake Predication*. p. 55-79. Menlo Park, U.S. Dept. of Interior, Office of Earthquake Studies.

Cook, N.G.W. 1964. The application of seismic techniques to problems in rock mechanics. *Int. J. Rock Mech. Mining Sci.* v. 1, p. 169-179.

Cook, N.G.W. 1965. A note on rockburst considered as a problem of stability. *J.S. Afr. Inst. Min. Metall.*, v. 65, p. 437-445.

Cook, N.G.W. 1975. Seismicity associated with mining. In W.G. Milne (ed.), *Induced Seismicity*. Eng. Geolog. v. 10, p. 99-122.

Cook, N.G.W., E. Hoek, J.P.G. Pretorious, W.D. Ortlepp & M.D.G. Salamon 1966. Rock mechanics applied to the study of rockbursts. *J. S. Afr. Inst. Min. Metall.* v. 66, p. 435-528.

Deliac, E.P. and N.C. Gay 1984. The influence of stabilizing pillars on seismicity and rockbursts at ERPM. In N.C. Gay & E. H. Wainwright (eds.), *Rockbursts and Seismicity in Mines*. p. 257-263. Johannesburg, S. Afr. Inst. Min. Metall.

Gibowicz, S.J. 1990. Keynote lecture: The mechanism of seismic events induced by mining. In C. Fairhurst (ed.), *Rockbursts and Seismicity in Mines*. p. 3-27. Rotterdam, Balkema.

Hasegawa, H.S., R.J. Wetmiller & D.J. Gendzwill 1989. Induced Seismicity in mines in Canada-An overview. In S.J. Gibowicz (ed.), *Seismicity in Mines, Pure Appl. Geophys.*, v. 129, p. 423-453.

Hedley, D.G.F. 1985. Personal communication.

Iannacchione, A. 1990. Behavior of a coal pillar prone to burst in the Southern Appalachian Basin. In C. Fairhurst (ed.), *Rockbursts and Seismicity in Mines*. p. 295-300. Rotterdam, Balkema.

Jaegar, J.C. & N.G.W. Cook 1979. *Fundamentals of Rock Mechanics*. 3rd Ed. London, Chapman and Hall.

Linkov, A.V. 1992. Dynamic phenomena in mines and the problem of stability. *Notes from a course presented at the University of Minnesota*. Distributed by MTS Systems Corp.

Lippmann, H. 1990. Keynote lecture: Mechanical considerations of bumps in coal mines. In C. Fairhurst (ed.), *Rockbursts and Seismicity in Mines*. p. 279-284. Rotterdam, Balkema.

Mack, M.G. & S.L. Crouch 1990. A dynamic boundary element method for modeling rockbursts. In C. Fairhurst (ed.), *Rockbursts and Seismicity in Mines*. p. 93-99. Rotterdam, Balkema.

McGarr, A. 1984. Some applications of seismic source mechanism studies to assessing underground hazard. In N.C. Gay and E.H. Wainwright (eds.), *Rockbursts and Seismicity in Mines*. p. 199-208. Johannesburg. S. Afr. Inst. Min. Metall.

McGarr, A. and G.A. Wiebols 1977. Influence of mine geometry and closure volume on seismicity in a deep-level mine. *Int. J. Rock Mech. Min. Sci.* v. 14, p. 139-145.

Ortlepp, W.D. & N.G.W. Cook 1964. The measurement and analysis of the deformation around deep, hard-rock excavations. *Proc. 4th Int. Conf. on Strata Control and Rock Mechanics*. p. 140-152. New York, Henry Crumb School of Mines, Columbia University.

Ortlepp, W.D. & A. Nicoll 1964. The elastic analysis of observed strata movement by means of an electrical analogue. *J.S. Afr. Inst. Min. Metall.* v. 65, p. 214-235.

Ryder, J.A. 1988. Excess shear stress in the assessment of geologically hazardous situations. *J. S. Afr. Inst. Min. Metall.* v. 88, p. 27-39.

Ryder, J.A. & N.C. Officer 1964. An elastic analysis of strata movement observed in the vicinity of inclined excavations. *J. S. Afr. Inst. Min. Metall.* v. 64, p. 219-244.

Salamon, M.D.G. 1964. Elastic analysis of displacements and stresses induced by mining of seam or reef deposits - Part IV: Inclined reef. *J. S. Afr. Inst. Min. Metall.* v. 65, p. 319-338.

Salamon, M.D.G. 1968. Two-dimensional treatment of problems arising from mining tabular deposits in isotropic or transversely isotropic ground. *Int. J. Rock Mech. Min. Sci.* v. 5, p. 159-185.

Salamon, M.D.G. 1983. Rockburst hazard and the fight for its alleviation in South African gold mines. In *Rockbursts: prediction and control*. p. 11-52. London. The Institution of Min. Metall.

Salamon, M.D.G. 1984. Energy considerations in rock mechanics: fundamental results. *J. S. Afr. Inst. Min. Metall.* v. 84, p. 237-246.

Salamon, M.D.G. 1991. Deformation of stratified rock masses: A laminated model. *J. S. Afr. Inst. Min. Metall.* v 91, p. 9-25.

Salamon, M.D.G. 1992. Strength and stability of coal pillars. *Proc. of the Workshop on Coal Pillar Mechanics and Design*. p. 94-121, U.S. Bureau of Mines, Information Circular/1992, IC 9315.

Spottiswoode, S.M. & A. McGarr 1975. Source parameters of tremors in a deep-level gold mine. *Bull. Seismol. Soc. Am.* v. 65, p. 93-112.

Wong, G.I. & A. McGarr 1990. Implosional failure in mining-induced seismicity: A critical review. In C. Fairhurst (ed.), *Rockbursts and Seismicity in Mines*. p. 45-51. Rotterdam, Balkema.

Characteristics of acoustic emission during tests of Chilean rocks

P. M. Acevedo & S. A. Medrano
IDIEM, University of Chile, Santiago, Chile

ABSTRACT: As is already well-known Chile is geographically located along the meeting zone of the Nazca and South American Plates. Consequently, tectonic activities are causing not only seisms along the whole country but also tectonic phenomena such as the rockbursts that have been affecting the deepest mines. This work introduce results concerning analysis of the behavior of core samples from rocks that are exhibiting problems. Studies have been effected in the laboratory, as concerns both index properties and rock classification using acoustic emission tests.

1 INTRODUCTION

The samples taken from the tunnels where rockburst occurred were subjected to ultrasound tests measured with acoustic parameters (counts number). This allowed to plot a special curve termed the Curve of Characteristic State or, more briefly, the State Curve. It was found that the curve profiles of the diverse types of rocks tested are similar in shape and that they can be employed for the classification of rocks.

Strength parameters determination as well as stresses measurement and assessing are fundamental problems in the construction and design of tunnels and caverns in rocks. This work describes an useful testing method based on ultrasound detection employing acoustic parameters as counts number and pulses number. These laboratory tests allow to plot a special purpose curve termed the Characteristic Curve.

Subsequently the acoustic parameters constituted by the counts number and the pulses number can be measured by monitoring the acoustic emission in the site. Then a weighted pulses index can be obtained through the numerical handling of these acoustic parameters, as explained below. Thereafter, using said index together with characteristic curves, the in situ failure load percent can be estimated.

Figure 1 shows the sequence of steps and/or processes involved in the assessment of in situ stresses.

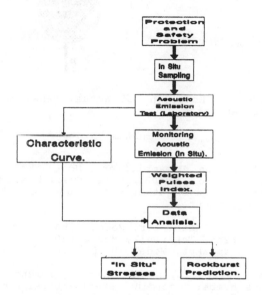

Figure 1. Sequence of steps and/or processes involved in assessing in situ stresses.

2 EQUIPMENT AND TESTING METHOD

The program of tests was carried out using brazilian test, as shown in Figure 2.

This type of tests was selected because behavior in border of tunnels is well represented by a tension test (Fig.3), and taking account of other advantages such as simplicity. Rupture is well defined, it does not hazard integrity of acoustic pulses detectors devices (transducers) and failure specimen is reached with relatively small load, and for this reason elasticity effects in loader device are reduced.

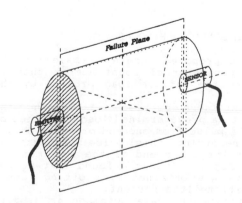

Figure 2. Setting of transducers during brazilian test. Sensors are settled longitudinally to failure plane.

The acoustic parameters were measured using the commercial modular equipment Dunegan Enderco Serie 3000, including the following basic modules:

1. Rack
2. Signal conditioner dual 302A
3. Counter dual 303
4. Timer 402
5. Audio monitor 701
6. Distribution analyzer 920A
7. Amplitude detector 921

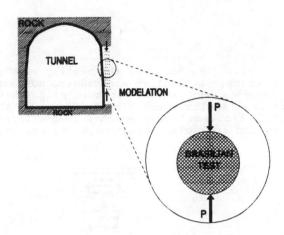

Figure 3. Modelling of actual behavior in brazilian tests.

In addition, Preamplifiers, Model 1801, and Sensor, Model S140B, were available. The information was gathered through the following equipment:

8. x, y Inscriptor, Model 115A
9. Screen, Tektronik, Model 620
10. Loudspeaker, Model 701 S

The capabilities of this equipment are counts-number recording, events number recording and amplitude distribution discerning.

The frequency response of the piezoelectric transducer S140B ranges 100-400 Khz, transducer signal is filtered and amplified trough a preamplifier 1801-170B, with a fixed gain of 40 db.

3 CLASSIFICATION OF ROCKS

The method proposed herein is based on the acoustic parameter of count number that is related to the failure load produce by the specimen.

The test were carried out loading the specimen to a predetermined load, and then by emitting a pulse from one end of the specimen and by recording this pulse through the sensor at the other end of the specimen, is shown in Fig.2. It was a found that a special curve was obtained by plotting counts versus failure load percent.

The experimental data obtained in diorite specimen exhibit are exhibiting a pointed behavior as noticeable in Fig.4.

The parameters proposed for the classification of rock are: the elastic zone, the elastic limit, and the beginning of macrofracturing. These parameters are shown in Fig.5.

This proposed method is useful because it was found that the profiles of the diverse curves of characteristic states remain invariant and hence they can be employ as a pattern.

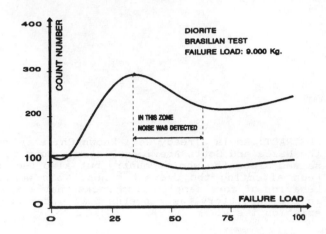

Figure 4. State Curve (S.C.). The residuals Counts (R.C.) are measured by emitting and receiving pulses in the unstressed specimen.

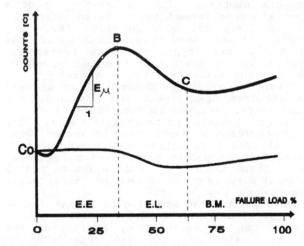

Figure 5. State Curve and its classification Zones

4 ASSESSING OF IN SITU STRESSES

As already pointed out, and empirical method whose steps are shown in Fig.1 is being proposed herein in order to compute in an approximate fashion the stress level present in rocks. The appertaining process uses acoustic-emission tests for getting a characteristic curve by plotting (accumulated-counts)/pulses versus failure-load percent. The experimental data are referred to a pure acoustic emission, that is to say, to the recording of sound emitted in rocks stressed through microfracturing propagation.

Upon plotting the characteristic curve, the next step is as follows. By monitoring the acoustic emission, this curve can be used to assess the stress level. There is an indetermination that is assessed by employing a weighted pulses index as explained below.

312

4.1. Characteristic Curve

This Curve is obtained by measurements of "accumulated counts" (by pulse). In sundry types of rocks it was observed a similar behavior (Andesite, Diorite, Dacite). The acoustic behavior is representing the evolutive process is representing the evolutive process of rocks with internal stresses. Fig. 6 shows the characteristic curve whose profile remains invariant for the several classes of rocks tested. The following aspects are readily observable in Fig.6.

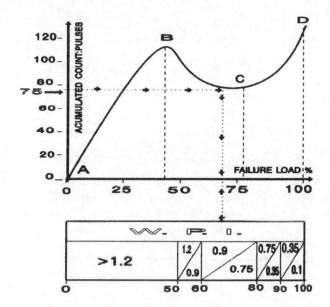

Fig.6: Characteristic Curve obtained by sensing acoustic emission due to microfracturing n the rock stressed.

a. Increasing acoustic activity till reaching a maximum value.
b. Thereafter, decreasing activity down to a minimum value.
c. Then, new increase in acoustic activity till reaching rock failure.

The characteristic curve was empirically obtained through the testing of sundry samples of rocks and the corresponding classification is shown in Table 1.

Table 1: Failure load percents associated to curve segments in characteristic

CURVE SEGMENT	FAILURE LOAD (%)
A - B	0 - 50
B	50 - 60
B - C	60 - 75
C	75 - 85
C - D	85 - 100

In situ stresses in rocks may be successfully determined by using a numerical procedure that permits to assess the indetermination when employing the characteristic curve. Figure 3 allows to see that if having a known value for the ratio accumulated counts/pulses, then there is no certainty at all on which is actually the corresponding load-failure percent.

The empirical results concerning the acoustic behavior of stressed rocks have shown that the method suggested in the present work is useful. This method selects the pulses in the band ranging from 10 and 300 counts, for 90 db amplification.

The Weighted Pulses Index (WPI) is computed as a weighted mean between counts and pulses, as follows:

$$WPI = \frac{\sum_{i=2}^{N} (X_{i-1}-X_i)*C_i}{\sum_{i=2}^{N} [(X_{i-1}-X_i)][X_i-X_1]} = 1$$

C_i = counts threshold
X_i = over-threshold pulses,
N = pulses number

Several experiences have enable the authors to define the ranging values shown in Table 2.

Table 2: Failure load percent associated to the weighted pulses index (WPI).

WPI	FAILURE LOAD (%)
1 - 2	> 50
1.2 - 0.9	50 - 60
0.9 - 0.75	60 - 80
0.75 - 0.35	80 - 90
0.35 - 0.1	< 90

This ranges are graphically shown in Figure 6.

5. CONCLUSIONS

1. The stressed rocks are exhibiting a characteristics acoustic behavior shown through the Characteristic Curve (C.C.) and the Curve of Characteristic States.
2. The method of the "petite sismique" --as termed in french-- in acoustic emission applied to stressed rocks allows to assess Young's Modulus as well as the elastic and plastic limits.
3. Using the "petite sismique" technique in acoustic emission, the stress induced in rocks can be sensed and the associated maximum stress position can b recognized theoretically. Thus it would be possible to predict rockbursts.
4. Acoustic emission tests provide other way for assessing dynamic parameter such as Poisson's Ratio.
5. Qualitatively, it was observed that the acoustic parameter counts number is directly increasing with the discontinuities that are present in rocks.
Finally, acoustic emission tests provide a non-destructive means to characterize rocks. Thus, in combination with in situ monitoring, a useful method is provide herein to duly assess in situ stresses.

ACKNOWLEDGMENTS

The authors would like to express their gratitude to the Chilean National Council for

Science and Technology, and to the DIB-IDIEM of the University of Chile. The testing program was carried out in Geotechnical Section of the IDIEM Institute.

REFERENCES

Arjona,R. 1986.Thesis of Civil Engineering, University of Chile.
Rezowalli, J., King, M., Myer, L. 1983. Cross-hole acoustic surveying in Basalt. Int.Journal Rock Mech. Min. Sci. & Geomech.
Rivera, Gino (1982) Diseño Estructural de Túneles (Tunnels Estructural Design.) Thesis of Civil Engineering, University of Chile.

Rockbursts and Seismicity in Mines, Young (ed.) © 1993 Balkema, Rotterdam, ISBN 90 5410 320 5

3-Dimensional modelling of fault-slip rockbursting

P. Bigarre & K. Ben Slimane
INERIS, Nancy, France

J. Tinucci
ITASCA Consult. Gr., Minneapolis, Minn., USA

Abstract: A research program has been carried out at INERIS aiming to quantify rockburst potential from mining-induced fault-slip. As a part of the research, numerical modeling of fractured rock mass has been undertaken, using the three-dimensional distinct element code 3DEC. Results presented in this paper demonstrate a very good agreement between calculated deformations of modeled faults and the experienced rockburst sequence of the Estaque-sud district of the colliery.

1. INTRODUCTION AND GENERAL SETTING

At the Provence colliery, coal is mined at a depth reaching 1100 m. The seam thickness is around 3 meters, while strata dips westward around 10°. The longwall face method with caving process is used, involving high mechanization and self-advancing support for faces of 200 m of span. Rate of production has increased steadily all along the past, reaching now the value of 11 tonnes per shift and per day, with a average, daily rate advance of 6 meters per day and per working faces.

Nowadays, the coal mine experiences a daily average of 20 seismic events of magnitude 1.5 and greater, 15% of which are magnitude 2 and more. Most of these events are attributed to the goaffing process associated with the longwall mining operation. However, on an annual basis, many of these events result in serious rockbursting damages at the advancing face and along haulage gateways. As regards the southern part of the colliery and the mining of Estaque-sud district, which began in 1987, many major tectonic faults have been suspected to play a major part in dynamic loading of the coal seam through fault-slip induced by mining.

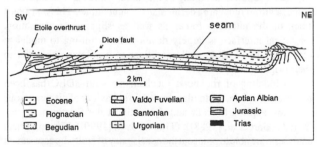

fig. 1- Geological cross-section of the basin

The general geological setting of the basin is quite simple. After Caviglio & al [1988], the structure is overridden by a major thrust sheet overthrusting northward, with a average dip of 25° (figure 1). Strike-slip, sub-vertical faulting is present all over the area, with lengths of several hundreds of meters. Two major zones have been distinguished in the coal field, with regard to the direction of the strata and stress measurements. In the zone we are concerned with, strong anisotropy of the principal stress components has been measured many times, characterized by high horizontal tectonic stresses and a sublithostatic vertical stress. Both have been explained by the regional, geological history (Piguet & Georges [1981], Revalor [1986], Gaviglio [1985]). Associated strata are made of limestones, qualified as hard and brittle (Josien [1981]).

2. ROCKBURST MECHANISMS AT THE ESTAQUE-SUD DISTRICT

For the last fifteen years, stimulated by the steady increase in the daily mine tremors and annually rockburst occurences, a research program has been undertaken at INERIS, aiming first to understand the mechanisms involved and then to improve prevention. Classified with regard to rockbursts locations and effects at the Provence coal mine, three main types of bursts have been recognized, as (Revalor [1988]) (figure 2):

- type 1: ends of the faces, especially on the old panel side. These bursts are now largely controled by means of destressing holes (figure 2a), although this method lacks of accuracy,

- type 2: coal bumps, buckling of the floor, more current at the present time, over length sometimes greater than a hundred meters, can affect the gateways either ahead of the face (old panel side) or behind the working face, at a distance ranging from 50 to 150 meters (figure 2b)

- type 3: strain bursts in unmined, overloaded stiff pillars (figure 2c)

Seimic energy associated with rockbursts varies around 10^8-10^9 Joules, with an associated Richter magnitude ranging from 2.2 to 3. At the Estaque-sud district of the mine (figures 5-6), mining started in 1987 with longwall T13. During the 3 years following period, with a span of one panel wide (200 m), then two (400 m) and three (600 m), 22 rockbursts[1] were recorded, starting with the mining of the second panel T14, most of the events being of type 2. A schematic description of the larger damages is suggested in figure 3, with following characteristics:

- violent expulsion of the coal in the gateway,

- no significant fracturation or convergence of the immediate hangingwall,

- quite often accompanied by floor heavage reaching 1 meter (whether due to buckling mechanism or deeper shear failure is still not clear (Mathieu [1989]). It is worthy noting that this kind of damage has been controlled for the last year by floor slotting ahead of the face, although the efficiency of this method has not been accurately estimated, due to the lack of data since its implementation.

---

1) We include here all significant dynamic events recorded, ranging from dynamic spalling to large, underground damages described here-after.

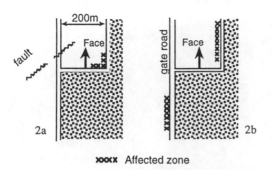

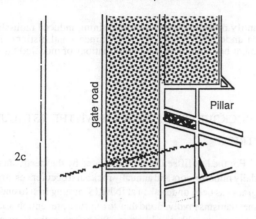

xxxx Affected zone

d - Location in a pillar

fig. 2- kinds of rockbursts

The main hypothesis for type 2 events tends to classify them more precisely as rockbursts triggered by dynamic loading generated by large mine tremors, induced by tectonic fault-slip or sudden failure of stiff bedding planes in the high roof. Naturally, potential means of confirming this kind of mechanism are very few because of the difficulties to get data. Extensions of the mining areas, poor access to faulted areas (one mined seam) and poor understanding of the roof behaviour do not permit to get valuable information. Two types of investigations have therefore been undertaken:

- developing a mine-scale seismic network, able to give location of each mine tremors with good accuracy as well as its energy and seismic moment. This should permit to relate the located focus of the mine tremor and the underground damaged areas and thus assert which mechanism of rupture may be considered. This has been undertaken two years ago (Ben Slimane & al [1990]). INERIS is currently improving the network to get accurate locations and better focus parameters,

- analysing with all avalaible data the major, suspected faults respectively with mining geometry and scenario to get a better understanding of potential fault-slip behaviour. This has been undertaken recently and use of numerical methods is presented in this paper.

fig. 3- schematic description of the damages

3. NUMERICAL MODELING

Numerical modeling has been carried out aiming to quantify rockburst potential for the seismic triggering mechanism from fault-slip along major, pre-existing geologic structures. Because of both the mining configuration and orientation of the faults of the Estaque-sud area to be modeled, it was chosen to undertake three-dimensional numerical analysis. Eventually, the strongly discontinuous nature of the problem conducted us to the choice of the distinct element method.

Due to the lack of seismic data over the period of mining of the Estaque-sud district and the insitu conditions for mined areas of such extents, the aim of this study was to:

- examine the ability of the three dimensional distinct element method (3DEC, Itasca) to study fault-slip assessment for a complex system of discrete, deformable blocs,

- examine the fault-slip potential for large-scale faults lying in the mined area and correlate in space the modeled mining process and the incremental plastic deformations with the insitu recorded rockburst sequence,

- to bring forward a methodology of modeling closely associated with geological survey and above all with data from the newly settled seismic network available.

4. 3DEC SOFTWARE

3DEC is a PC-based computer program using the distinct element method and a central finite difference scheme to simulate the mechanical response of three-dimensional blocky systems. Handling either rigid or deformable blocs, the formulation used permits to simulate large displacements and rotations of the blocs relative to one another, including detection of new contacts, while the solution scheme is explicit in time. During each increment of time, Newton's law of motion is used to obtain velocities and displacements from the unbalanced forces. Mechanical calculations may be described as in figure 4, which shows the importance of the contact logic implemented. A complete description of this and of the calculation cycle are given by Cundall [1988] and Hart & al [1988] respectively. Note that when deformable blocs are used, modeled joints are sudivided in subcontacts corresponding to the finite difference tetrahedral zoning of the faces, while each surface node is the centroid of an area defined as the subcontact. This one keeps track of the interface forces as well as slipping or separation. Graphical interface is largely developed, permitting to model as efficiently as possible well-conditionned problems compared to their original complexity.

Modeling of rockburst mechanisms with 3DEC has been undertaken before, to simulate fault-slip behaviour of discontinuous medium, applied to fault and dyke slip at the Strathcona mine, Canada (Hart & al, [1988]), (Tinucci & al, [1990]). The numerical analyses were able to point out the consistency of fault-slip assessment in mine-induced seismicity and rockbursting.

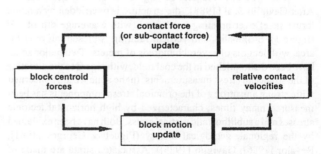

fig. 4 - 3DEC scheme calculations (after Hart, [1988])

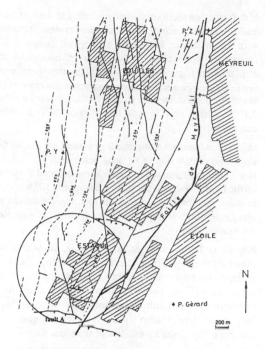

fig. 5 - map of the most recent mined districts

5. NUMERICAL PARAMETERS

Our 2800x1600x2400m model of the rock matrix consits of 188 convex blocks formed by 7 structural features, comprising 5 large-scale faults (500 to 2000 m of length, figures 5-6)) and two bedding planes (dipping east-west 10°) located at 150 m in the upper roof (representing coarsely the Fuvelian strata), and below the seam, and in the footwall respectively. Four longwalls (T13, T14, T25 and T15) are simulated, with a mining scenario reproducing the in-situ excavation process geometry. The four longwalls were then excavated in 15 incremental steps, made of deletion of blocs, (figure 6) with equilibrium reached at each step, providing a quasi-static analysis able to put forward the influence of the incremental mined areas on the plastic deformations along the modeled faults. Each step represents an excavated volume equivalent to two months of mining at a rate of 100 meters per month.

Deformable blocs, zoned by around 70.000 finite-difference zones, are assumed to behave elastically, while all structural features follow a perfect, elastoplastic behaviour, based on a Mohr-Coulomb yield condition (table 1).

table 1 - elasto-plastic parameters of the model

table 1	structural features	rock matrix MPa
stiffnesses	kn,ks=10000 MPa / m	K,G=13333,8000
M.C. parameters	Fric=35°, Coh= 0, Rt= 0 MPa	

table 2 - input stresses

table 2	σ1	σ2	σ3
value MPa	-40	-20	-17
dip °	120	30	0
dip dir. °	0	0	90

Initial pre-mining state of stress is chosen to be very close to available field measurements obtained in the Etoile-sud district, closest to the one modeled. Values and orientations are noted in table 2.

Two parameters are quantified in order to relate plastic deformation along each feature to each sequential excavation:

- $$M = \sum_{SubC.} A_s D_\tau$$ where A_S is the subcontact area and D_τ

its tangential displacement. M may be interpreted coarsely as the seismic moment of the fault divided by its shear modulus. It characterizes the mechanical moment acting on the structure while new equilibrium is reached.

- $$\Delta E = \frac{1}{2} \sum_{SubC.} F_s D_\tau$$ where F_S is the tangential force acting

at the subcontact location. ΔE is the non-recoverable, released energy dissipated by the excess shear force induced at each step.

These two parameters are related, in such a multi-step, quasi-static analysis by the relation: $$M = \frac{\Delta E}{\tau}$$ where τ is the

tangential stress acting at the subcontact location.

6. RESULTS

Figures 7 and 8 indicate the energy dissipated through plastic strain and the parameter M at each step both with the rockburst sequence plotted on the right, vertical axis versus the step number of the simulation. Results show that only the simulated overthrusting fault (fault A) and upper bedding plane (feature I) show large plastic deformations. These deformations appear above all from starting of longwall 14 (mining step 8-9). Summing up briefly, we can do the following comments :

- the qualitative correlation between M for both strutures and rockburst sequence shows a good agreement
- rupture mechanism along fault A is due to shear failure, induced by both decrease in normal stress and increase in shear stress, coupled with stress tensor rotation. The amount

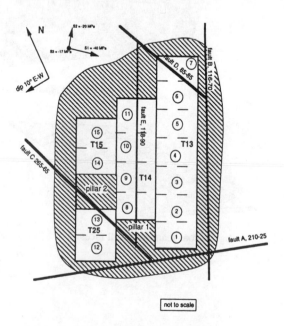

fig. 6 - modeling of the Estaque sud district
faults are indicated with dip direction and dip

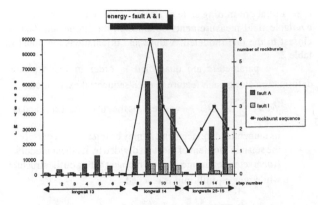

fig. 7 : energy dissipated at each step for joints A & I

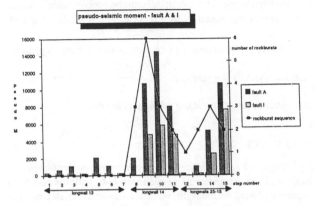

fig. 8 : parameter M at each step for structures A & I

of deformation seems essentially sensible to the width of the mined out area, i.e. extension from east to west. It decreases with extension in length of a longwall (steps 6-7-11). Geometric projections of locations of maximum shear displacement (figure 9) on the seam plane are approximately, vertically plumbed with the step excavation. Energy dissipated at step 9, if calculated on a daily basis, provides a value of 3.10^9J, which corresponds to an event with a order of magnitude of 3.

- failure mechanisms along bedding plane I are of two types: shear and tensile due to flexural behaviour. The relatively small energy dissipated is due to a low induced shear stress on the fault, parallel to the seam, and dipping 10° with regard to original principal stresses. In fact, tensile failure takes place essentially at steps 10, 11 and 15 with the widening of the mining area.

- Slip along fault A is essentially oblique, i.e. with a mixed offset of reverse and right-lateral strike faulting. This motion corresponds to the natural, thrust faulting of the structure. Interaction of the two structures A and I close to their intersection is difficult to estimate. It is worth noting that they most likely amplify each other because of the compatibility of the kinematics of the blocs.

- Along other features, there is no noticeable plastic deformation although induced stresses are unfavourable for most of them, i.e. ratio τ/σ increases, except significant deformation along fault C. However, if large strain was to be obtained on this modeled joint, it would conduct to reduce friction to unrealistic values (20-22°).

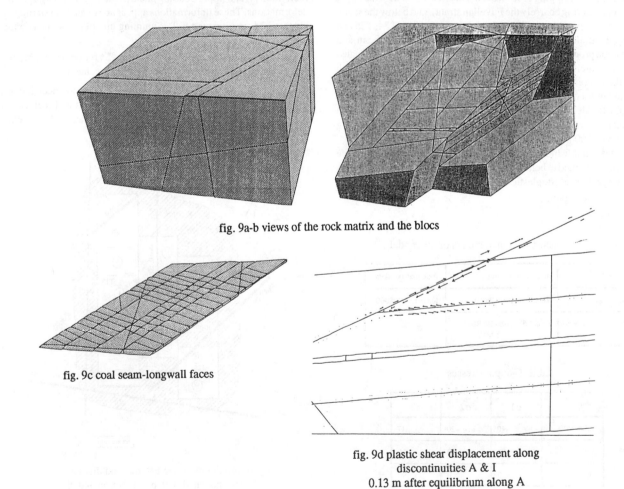

fig. 9a-b views of the rock matrix and the blocs

fig. 9c coal seam-longwall faces

fig. 9d plastic shear displacement along
discontinuities A & I
0.13 m after equilibrium along A

7. SUMMARY AND CONCLUSIONS

In fact, the aim of this study is to evaluate whether using modeling of typical, discontinuous problems in prediction of fault-slip rockbursting might be useful. This study shows an good agreement between the rockburst sequence and the response of some of the discontinuous features lying in the mined areas. Mechanisms involved in the response of the system are clearly identified, critical geometries and spans are pointed out, fault-rupture locations can be calculated, fundamental parameters as dissipated energy and seismic moment are estimated and seem realistic. The three-dimensional aspect in modeling is pointed out as very critical.

However, at the present time, no calibration can be demonstrated. As well, impact of failure along the structures on mining areas are impossible to estimate on a modeling point of view. Therefore, this methodology of three-dimensional modeling by the distinct element method appear as a method to be calibrated as closely as possible (figure 10):

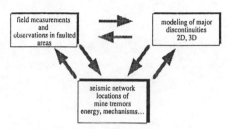

fig. 10 interacting fields to exchange data

- with insitu observations, i.e. detailed, geological survey of the faults and their characteristics, of the local disturbances of the coal seam, and others particular underground conditions.

- above all with seismic results from the settled network in 1991, giving presently locations and daily seismic energy distributions. Back analysis of selected tremors (location, energy, radiation pattern,...) compared with modeling could provide very interesting results. The seismic network is still undergoing improvements to locate more accurately, particularly in depth, and to gain understanding of the focal mechanisms involved in rockburst occurences. Association of seismic analysis and dynamic modeling is expected to come out as very promising.

In every case, this research directly benefits the mine by providing a method able to quantify numerous potential fault-slip problems, or qualify rockburst mechanism.

ACKNOWLEDGEMENT

we greatly appreciate the help of Dr Fairhurst and Dr Piguet, who made that work possible.

REFERENCES

Ben Slimane K., Besson J.C., Mandereau G., Chambon C., 1990 "La surveillance sismique: un outil d'aide à la planification des chantiers miniers sujets aux phénomènes dynamiques". 3ième colloque Polono-Français, Wroclaw. Studio Geotechnica et Mechanica Vol XI n°1 et 2, pp 51-68.

Ben Slimane K., 1990 "Sismicité et exploitation minière", Rapport final INERIS.

Cundall P. A., 1988 "Formulation of a three-dimensional distinct element model - Part I a sheme to detect and represent contacts in a system composed of many polyhedral blocks", Int. J. Rock Mech, Min. Sci. & Geomech. Abstr. pp 107-116

Gaviglio P., Revalor R., Piguet J.P., Dejean M, 1988 "Tectonic struture, strata properties and rockburst occurence in a french coal mine", Proc. 2nd Int. Symp. "Rockbursts and Seismicity in Mines", Minneapolis, C. Fairhurst, Balkema.

Gaviglio P., 1985 "La déformation cassante dans les calcaires fuvéliens du bassin de l'Arc. Comportement des terrains et exploitation minière" Thèse Doctorat d'Etat, Marseille.

Hart R., Cundall P.A., Lemos J., 1988 "Formulation of a three-dimensional distinct element model - Part II Mechanical calculations of motion and interaction of a system composed of many polyhedral blocks", Int. J. Rock Mech, Min. Sci. & Geomech. Abstr. pp 117-12.

Hart R., Board M., Brady B., O'Hearn B., Allan G., 1988 "Examination of Fault-slip induced Rockbursting at the strathcona Mine", in Key questions in Rock Mechanics, Proc. 29th U.S. Rock Mechanics Symp., Minneapolis, Cundall & al, Balkema.

Itasca Consulting Group, Inc, 3DEC version 1.3, Minneapolis, Minnesota.

Josien J.P., 1981 "Lutte contre les coups de couche", rapport final EUR 7869, convention d'étude CECA 7220-AC/307.

Mathieu E., 1989, "Apport de l'écoute sismoacoustique pour la surveillance des chantiers miniers affectés de coups de terrains", Doctorat, Institut Polytechnique de Lorraine.

Piguet J.P., Georges G., 1981 "Influence de la profondeur et des facteurs naturels sur le comportement des ouvrages miniers, rapport final EUR 7848, convention CECA 7220-AC/304.

Revalor R., 1986 "U.E. Provence, stress measurements by hydraulic fracturing, Etoile, Etoile-sud", Internal report Cerchar.

Revalor R., Josien J.P., Besson J. C., Magron A. 1985 "Seismic and seismo-acoustic experiments applied to the prediction of rockbursts in French coal mines", Proc. 2nd Int. Symp. "Rockbursts and seismicity in Mines", Minneapolis, C. Fairhurst, Balkema.

Revalor R., 1991 "La maîtrise des coups de terrains dans les exploitations minières", Doctorat, Institut Polytechnique de Lorraine.

Tinucci J.P., Hanson D.S.G, 1990 "Assessment of seismic fault-slip potential at the Strathcona mine", in Rock Mechanics contri. & Chal., Hustrilid & Johnson, Balkema.

Rockbursts and Seismicity in Mines, Young (ed.) © 1993 Balkema, Rotterdam, ISBN 90 5410 320 5

A comparative study of seismic source location methods in underground mines

D. P. Blair
CSIRO Division of Geomechanics, Melbourne, Vic., Australia

ABSTRACT: Five field trials were conducted each using four seismic source location methods to estimate the coordinates of a point explosive source in an underground rock mass. Method 1 used a least squares solution, Method 2 used an iterative improvement to this solution and Methods 3 and 4 used, respectively, a 5-point and 8-point simplex optimization scheme. The experimental results showed that there was little difference between Methods 2 to 4 and Method 1 was inferior to all other methods. Synthetic models for all trials showed results consistent with the experimental data provided that a small amount of travel-time scatter was assumed. However, when the scatter was increased, Method 1 was found to be the most robust. The numerous openings that are present in some mine environments could produce a large uncertainty in travel-times; under such conditions Method 1 might well be the superior method. Thus caution must be exercised when making claims regarding the superiority of any one method for seismic source location in real environments.

1 INTRODUCTION

Significant rockbursts within mines are often related to the interaction of the varying stress field with the local geology. Thus a knowledge of the location of the resulting seismic event, with a knowledge of the mine geology, may yield an insight to possible causes of the rockburst. It is for this reason that many mines in regions prone to rockbursting require the accurate location of any major local seismic event.

The literature is replete with the comparison of various algorithms for the location of seismic sources (Eccles and Ryder, 1984; Kat and Hassani, 1989). In this regard it is well known that the experimental uncertainties associated with the travel-time measurements will cause an uncertainty in the estimated location of the source. This uncertainty is also modified by the geometry of the detector array. Three standard techniques are usually used in order to investigate the dependence of source location upon the travel-time uncertainty for a given array geometry. The first technique involves the use of experimental travel-time data for the location of production blasts within a mine (Swanson et al 1992). The second technique involves source location within small-scale models (Collins and Belchamber, 1990), and the third technique involves synthetic modelling (Kat and Hassani, 1989).

In the first and second techniques, the natural scatter inherent in the observations produces a corresponding uncertainty in the location of the real source. The uncertainties can be calculated since the actual location of the source is known. In the third technique exact travel-times are calculated for an assumed source location. A component of random scatter is then added to the travel-times and the source location is calculated for each set of randomly perturbed travel-times.

Thus, to date, most comparative studies of source location algorithms involve the location of blast sources within underground environments, or the location of point sources within small-scale or synthetic models. Such studies suffer from some disadvantages. Firstly, blast sources are often distributed (non-point) sources producing seismic waves whose travel path is intersected by discontinuities such as mine openings etc. The uncertainty in source origin combined with unknown travel paths may result in poor comparative tests of source location algorithms. Secondly, the material properties and nearby location of many surfaces of small-scale models may yield travel-times that are not representative of mine environments. Thirdly, synthetic models assume a certain distribution for the "natural" scatter in travel-times that also may not represent the real environment.

The main aim of the present work is to overcome such difficulties associated with standard comparative tests. In this regard, various algorithms are to be used to locate a point seismic source within a drill-hole in an underground mine. Although the in-hole technique was primarily developed to locate the end of a drill-hole that had deviated from its planned path, it does provide a highly localised source in a real environment that is generally free of discontinuities

between source and detector. Furthermore, since the drill-holes were in mine regions which were largely undeveloped, access was limited and the source was often well outside the volume spanned by the detector array. Thus the technique should provide a rigourous test of location algorithms in real environments for which limited access may exist and the source may be well outside the detector array.

2 EXPERIMENTAL TECHNIQUE

Studies by Blair and Spathis (1982) showed that a small detonator (i.e. No 8 STAR detonator) when fired in a water-filled borehole produced an excellent high frequency signal which could be transmitted up to 100 m in competent rock. Furthermore, repeated firings of the detonator did not significantly alter the seismic source injected into the surrounding rock mass. In the present study a No 8 STAR detonator was carefully lowered down the borehole to a fixed depth (usually the end of the hole).

A number of Piczotronics PCB 308B accelerometers were attached to small aluminium bases, which, in turn, were bonded (using PLASTIBOND) directly to the rock surface. The detectors were strategically placed throughout available openings in order to yield the best possible coverage of the expected location of the source.

The output of the accelerometers was fed into a signal conditioner-amplifier and then into a RACAL STORE DS 7-channel instrumentation tape recorder. The recorder was operated in the Frequency-Modulation mode, with a tape speed of 60 ips; this operation yielded an upper (3 dB) frequency response of 40 kHz. However, the recorded signal still had some useful energy above 40 kHz, and resulted in a travel-time resolution of approximately 5 microsec. Assuming a typical velocity of 6000 m/s, this travel-time resolution results in a distance resolution of 3 cm, and is adequate for the present purposes. The tape was subsequently re-played through a NICOLET digital oscilloscope in order to obtain the travel-time differences for all detectors. These travel-time differences were then inserted into the various source location algorithms in order to predict the location of the detonator within the drill-hole.

A total of 5 trials was conducted, and at the end of each trial, the mine operators excavated towards the predicted location. Once found, the actual location of the source was surveyed in order to perform a comparison between the predicted and actual locations.

3 THE SOURCE LOCATION METHODS

Four seismic source location methods are used in the present study. Method 1 is a least squares matrix method (Blake et al, 1974), Method 2 is an iterative improvement to the least squares solution (Salamon and Wiebols, 1974), Method 3 is a simplex optimization using an (N+1)-point simplex (Prugger and Gendzwill, 1988) and Method 4 is a simplex optimization using a

2N-point simplex (Box, 1965); N is the number of variables of interest. In the present case N=4 since the velocity as well as the 3 spatial coordinates are required.

The present implementations of Methods 1 and 2 differ from most standard treatments. The least squares solution for Method 1 is achieved by singular value decomposition (SVD) of the system matrix, and also provides the initial "guess" for Method 2. SVD is the preferred method of equation solution since it is known to be more robust than the standard methods of Gaussian elimination or LU decomposition (Press et al, 1989). The corrections applied at each iteration step within Method 2 are also evaluated in a least squares sense using SVD. In the present study the SVD algorithm given by Press et al (1989) is employed.

The Newton-Raphson iteration technique of Salamon and Wiebols (1974) assumed a known rock velocity, and iterations ceased when all 3 spatial coordinates of the source had converged to within a specified tolerance. However, the present implementations of Methods 1 and 2 do not rely on a known velocity. The 4 unknowns estimated by Method 1 are the 3 spatial coordinates of the source and the rock p-wave velocity. These estimations provide the initial values for Method 2, in which the iterations are terminated when all spatial coordinates and the velocity have converged to within a specified tolerance (0.1 percent in the present work).

The present implementations of simplex optimization also do not employ a known input velocity. The input 5-point simplex for Method 3 is chosen to be a pyramid with a square base and a height equal to the base side. The initial "guesses" for the source coordinates and velocity are specified at each of the 5 points, and span a likely range of values over the simplex figure. For example, a typical simplex may have a base of side 50 m centred on the first-hit detector, and cover a velocity range of 3000 m/s to 7000 m/s. However, in practise it was found that the estimated source location was very insensitive to the starting size chosen for the simplex, provided it was large enough to encompass the source region. The simplex algorithm, AMOEBA, of Press et al (1989) is used to provide the estimation of the source location and rock velocity. The input 8-point simplex for Method 4 is chosen to be a cube of typical side 50 m, and whose centre is the first-hit detector. The velocity range of 3000 m/s to 7000 m/s is also used. The 8-point simplex is used in an attempt to maintain the full dimensionality of the simplex figure as it converges towards the location of the source. It is well known (Box, 1965) that an N+1 simplex is more liable than a 2N simplex to collapse into a subspace. Thus the use of the 2N simplex is simply a guarantee against premature collapse. It should also be appreciated that the 2N simplex also allows a finer increment than does the N-1 simplex for all 4 variables throughout the simplex volume. In the simplex optimization for Methods 3 and 4, the source location is obtained by minimizing the L2 norm of the error function described by Prugger and Gendzwill (1988). The simplex optimization is also constrained by the standard technique of including a penalty (cost) function within the error function to be minimised.

4 RESULTS

Five in-hole trials were conducted in the present investigation. Figures 1 and 2 show the results in plan view and cross-section, respectively, for Trial 1. The location of each detector is shown as well as the estimated source locations using all 4 methods; the location of the actual source is also shown.

Figures 3 to 10 show the results for the remaining trials. It is quite obvious that most sources were located well outside the volume spanned by the detector array. As mentioned earlier, this was due to the limited underground access in the region. The results, as expected, show that the estimated location of the source depends not only upon the method used, but also on the actual location of the source with respect to the array geometry. The further the source from the array volume, the poorer the estimate in source location.

It is worthwhile noting that, in Trials 1 to 4, the source was found ultimately by excavating towards the estimated coordinates. Trial 5 was conducted in order to track the position of the pilot hole for a large raisebore. This trial was conducted in order to verify that the pilot hole was on course in plan view (Easting and Northing); at the outset it was realised that a large uncertainty would exist in the estimated reduced level (RL). This fact is shown clearly in Figures 9 and 10 for which the estimated location in plan view is good, whereas the estimated location in RL is poor.

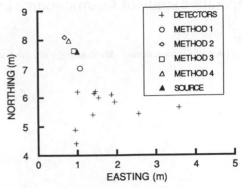

Figure 1. The results for Trial 1 shown in plan view.

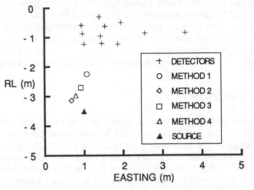

Figure 2. The results for Trial 1 shown in cross-section.

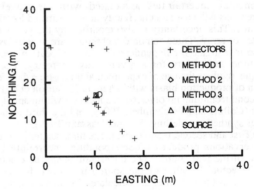

Figure 3. The results for Trial 2 shown in plan view.

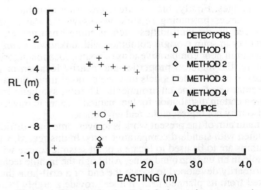

Figure 4. The results for Trial 2 shown in cross-section..

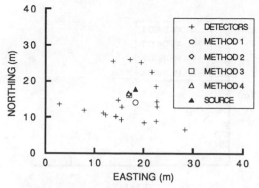

Figure 5. The results for Trial 3 shown in plan view.

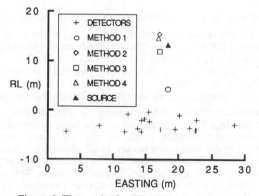

Figure 6. The results for Trial 3 shown in cross-section.

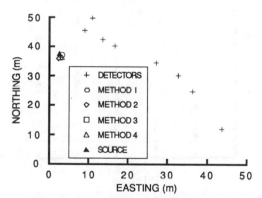

Figure 7. The results for Trial 4 shown in plan view.

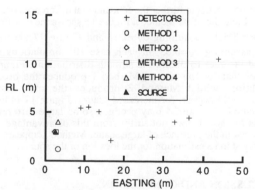

Figure 8. The results for Trial 4 shown in cross-section.

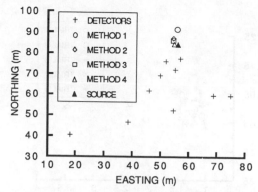

Figure 9. The results for Trial 5 shown in plan view.

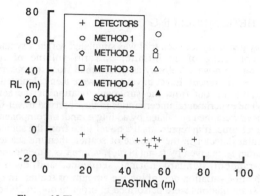

Figure 10. The results for Trial 5 shown in cross-section.

Figure 11 shows the difference, Δd (in m), between the actual source and the estimated source for all trials. It is quite obvious that Method 1 is inferior to all other methods for all trials. Furthermore, there appears little difference in the average performance of the remaining methods.

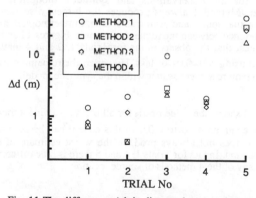

Fig. 11 The difference, Δd, in distance between the actual and calculated sources for the four methods of source location.

Figure 12 shows the scatter observed in the travel-times for all trials. The results are plotted in a scale-independent form in which v_p is the least squares estimate of the velocity, Δt is the difference between the theoretical travel-time (based upon v_p) and the experimental travel-time, d_0 is the distance between the source and the closest detector and d is the distance to any detector. Thus $\Delta t v_p / d_0$ is the scale-independent scatter in the observed travel-times. The results suggest that the scatter never exceeds 25 percent, and that it is typically 5 percent or less.

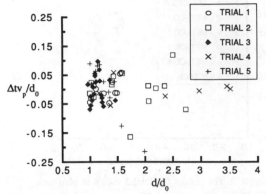

Figure 12. The scatter in experimental traveltimes for all trials.

5 SYNTHETIC MODELLING

The array geometry for each trial was simulated within a synthetic model consisting of an infinite, uniform volume of rock possessing the p-wave velocity as determined from the observed travel-time differences. Exact travel-times were then calculated for waves radiating out from the surveyed location of the actual source. The experimental uncertainties in the observed travel-time differences were then simulated by adding a random component to each travel-time. If t_0 represents the travel-time from the source to a particular detector in the absence of scatter, then the scattered travel-time, t_s, is given by $t_s=t_0(1+rc_v)$, where r is a member of a pseudo-random standard normal sequence, and c_v is the coefficient of variation describing the amount of scatter. In each trial, 200 simulations were performed in order to obtain 200 data sets for the travel-times. If μ represents the mean of all travel-times, t_s, and σ the associated standard deviation of the set, then $c_v =\sigma/\mu$. The difference, Δd, in distance between the actual source and the estimated source (in the presence of scatter) was determined for each of the 200 simulations; the mean difference μ_D, was then evaluated. Figure 13 shows the value of μ_D for all trials assuming a travel-time scatter given by $\sigma/\mu=0.01$.

A comparison of Figures 11 and 13 shows that there are similar features for the observations and synthetic modelling. In particular, Method 1 always produces the largest discrepancy between actual source and estimated source. Furthermore, there is little difference between the remaining models. Figures 11 and 13 also suggest that the observations for Trials 1, 2 and 4 may be simulated using $\sigma/\mu=0.01$ or less. However, the remaining trials appear to require a larger scatter within the synthetic model.

Figure 14 shows the value of μ_D for all trials assuming a travel-time scatter given by $\sigma/\mu=0.03$. In this case it can be seen that Method 1 does not always produce the worst estimate of the source location. In fact for Trials 1, 4 and 5 there is little difference between any of the 4 methods of source location.

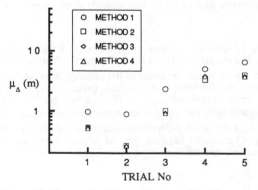

Figure 13. The results of the synthetic model using $\sigma/\mu=0.01$

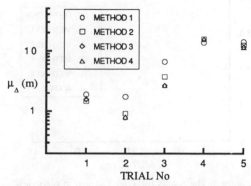

Figure 14. The results of the synthetic model for $\sigma/\mu=0.03$.

In order to obtain an insight to the spread in the values of Δd, Figure 15 shows the standard deviation, σ_Δ, of the set of Δd values for the 200 simulations. The 4 methods are described by the same symbol shapes as used in Figures 13 and 14. The empty symbols of Figure 15 show the results for $c_v=0.01$ and the solid symbols show the results for $c_v=0.03$.

The results indicate, with the exception of Trial 2, that the spread in Δd values increases with increasing c_v. This finding is not unexpected since Trial 2 possessed the best detector coverage of the source, and so only produces minimal source scatter irrespective of the travel-time scatter. In all other trials the source volume is not optimally covered by the detector array and so produces significant uncertainties in source location. This result clearly demonstrates that source location uncertainties are dependent upon both the inherent scatter in travel-times as well as the detector array geometry with respect to the source.

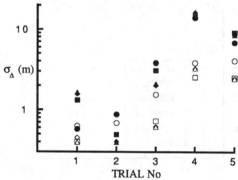

Figure 15. The standard deviation, σ_Δ, for all trials.
Empty symbols - $c_v=0.01$, solid - $c_v=0.03$.

In order to investigate further the fundamental differences between Method 1 and the other methods, Figure 16 shows the distribution in Δd for Trial 1 assuming $c_v=0.01$ and Figure 17 shows the results assuming $c_v=0.03$; in each case 10^5 simulations were required in order to obtain the smooth distributions. Figure 16 indicates that for low scatter Method 1 produces the broadest distribution, whilst Method 2 produces the most narrow distribution. However, for increased scatter, Figure 17 indicates that Methods 2, 3 and 4 may produce some large differences, whereas Method 1 does not suffer from this disadvantage. This implies that in the presence of large scatter Method 1 appears to be more robust in its estimation for the location of the source.

6 DISCUSSION AND CONCLUSIONS

It should be appreciated that the pseudo-random normal distribution with $\sigma/\mu=0.03$ implies that nearly all of the travel-times will have a scatter less than 9 percent of the mean. Although Figure 12 indicates that some data have scatter greater than 9 percent, the majority have less scatter. Thus the synthetic model

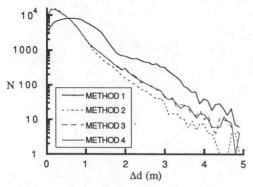

Figure 16. The distribution of Δd assuming $\sigma/\mu=0.01$.

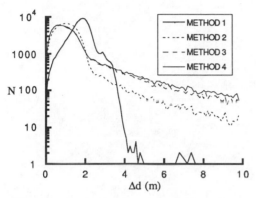

Figure 17. The distribution of Δd assuming $\sigma/\mu=0.03$.

with $\sigma/\mu=0.03$ is a reasonable first approximation to the maximum scatter.

The experimental trials showed that Method 1 was inferior to all other methods for source location since this method produced the largest difference between the locations of the estimated source and the actual source. The experimental results also showed that there was little difference between the performance of the remaining methods for source location (Figure 11). The synthetic models also showed results consistent with the experimental data provided that the travel-time scatter (c_V) was assumed to be small. However, when the scatter was increased, it was found that only Method 1 remained robust in the sense that it did not estimate significant differences between the estimated source and the actual source. In all other trials a reasonable number of simulations predicted large uncertainties in the source (Figure 17). All the results, especially those of Figures 16 and 17 indicated that Methods 3 and 4 produced an insignificant difference in source locations. This finding indicates that there was no premature collapse of the 5-point simplex figure (Method 3) as it converged to the source coordinates. The simplex methods (Method 3 and Method 4) were also used to minimise the L1 norm of the error function for selected trials. However, it was found that minimising the L1 norm always produced a source location estimate that was worse than that obtained by minimising the L2 norm.

The numerous openings that are present within most operating mine regions would be expected to produce a large scatter in observed travel-times as the seismic wave diffracts around the openings. The present investigation suggests that Method 1 might well be the method of choice under such conditions. Thus caution must be exercised when making claims regarding the superiority of any one method for seismic source location in real environments. In practice it is best to have a suite of location algorithms; if all algorithms estimate similar locations for a particular source then it is probable that the estimated source is close to the actual source. On the other hand, any large discrepancies in predictions should be further investigated. Any non-optimal geometry of the detector array as well as the level of uncertainty in measured travel-times could well be the cause of the discrepancy.

ACKNOWLEDGEMENTS

A number of people were involved with the author in collecting the underground travel-time data over an extended period. In this regard the assistance of L. Hunt (CSIRO) as well as C. Woodall, J. Jiang, G. Baird and N. Wemyss (Western Australian School of Mines) is gratefully acknowledged. The assistance of mine personnel at each site is also appreciated.

REFERENCES

Blair, D.P. and A.T. Spathis 1982. Attenuation of explosion-generated pulses in rock masses. J. Geophys. Res., 87, B5, 3885-3892.

Blake, W., Leighton, F. and W.L. Duvall 1974. Microseismic techniques for monitoring the behaviour of rock structures. U.S.B.M. Bulletin 665.

Box, M.J. 1965. A new method of constrained optimization and a comparison with other methods. Computer J, 8, 42-52.

Collins, M.P. and R.M. Belchamber 1990. Acoustic emission source location using simplex optimization. J. Acoust. Emission, 4, 271-276.

Eccles, C.D. and J.A. Ryder 1984. Seismic location algorithms: a comparative evaluation. Proc. First Int. Congress on Rockburst and Seismicity in Mines, Johannesburg, 89-93.

Kat, M., and F.P. Hassani 1989. Application of acoustic emission for the evaluation of microseismic source location techniques. J. Acoust. Emission. v8, 99-106

Press, W.H., Flannery, B.P., Teukolsky, S.A. and W.T. Vetterling 1989. Numerical recipes in FORTRAN. Cambridge University Press.

Prugger, A.F., and D.J. Gendzwill 1988. Microearthquake location: A nonlinear approach that makes use of a simplex stepping procedure. Bull. Seism. Soc. Am. 78, 799-815.

Salamon, M.D.G. and G.A. Wiebols 1974. Digital location of seismic events by an underground network of seismometers using the arrival times of compressional waves. Rock Mechanics, 6, 141-166.

Swanson, P.L., Estey, L.H., Boler, F.M. and S. Billington 1992. Accuracy and precision of microseismic event locations in rock burst research studies. U.S.B.M. Rept. of Investigations 9395.

Rockbursts and Seismicity in Mines, Young (ed.) © 1993 Balkema, Rotterdam, ISBN 90 5410 320 5

Stress modelling and seismicity on the Tanton fault: A case study in a South African Gold Mine

P.J.G. Dennison
Steyn Gold Mine, Freegold, AAC (Research conducted under the supervision of the University of the Witwatersrand, Johannesburg), South Africa

G. van Aswegen
ISS International Limited, Welkom, South Africa

ABSTRACT: Stress modelling, taking into account the geometrical and frictional properties of the Tanton fault (finite normal displacement ±120m) was correlated with its dynamic behaviour during extensive mining of a tabular orebody at ± 2400m depth. Stress measurements proved the absence of tectonic stress. Stress change monitoring helped to calibrate the numeric models. Zones of instability develop adjacent to fault/reef intersections, and advance with mining. Small seismic events locate around these zones where deformation was arrested due to increased normal stress and/or roughness and/or changes in geometry. Sources of larger tremors (M_L>3) coincided with areas of locally higher normal stress and were preceded by coseismic and non-seismic (inferred from modelling) shear deformation in adjacent areas. A mining method which prevents the formation of isolated pillars against the fault and which stiffens the fault/mining system includes initial mining at minimum extracted span against the fault first and the leaving of regional support parallel to the fault. Such methods should minimize fault related rockbursting.

1 INTRODUCTION

The paper outlines the research methods and the results obtained from a stability analysis of the Tanton fault, under the influence of mining operations in its immediate vicinity. The Tanton fault is one of several, westerly dipping, normal faults affecting the easterly dipping planar orebody. Finite displacement is 120m. The dip ranges from 30 to 50 degrees. The strike length of the fault defining the area of interest is 4km. Mining of the tabular ore body in the vicinity of the fault, starts at a depth of 1800m in the west and extends to a depth of 2600m below surface, towards the east. The integration of numerical modelling with stress change, fault creep and seismicity, has indicated the importance of the character of the fault, (geometry, roughness and friction), the mining layout and mining sequence as well as the nature of the regional support upon the behaviour of this fault. Interpretation of this information provides a basis for specifying mining methods to be applied in future.

A more general aim of this project was to outline particular methods for fault stability evaluation (in mines where planar ore bodies are excavated), in terms of financial and manpower input and practically useful output. Work on this project started in September 1988 and data collected up to March 1992 is considered.

2 METHODS

Stress measurements

Stress measurements were deemed necessary as no information was available regarding the in-situ stress tensor. Comprehensive in-situ stress measurements were completed at four different sites in the immediate footwall of the fault. The doorstopper method was chosen for practical reasons, having considered the CSIRO triaxial and hydro-fracturing methods as possible alternatives. Prior to measurement, the field stresses at suitable measurement sites were modelled to ensure that mining influences were minimal.

Stress-change monitoring

This was undertaken with two aims in mind. Firstly, to measure the two-dimensional stress change, in a plane normal to the fault, thereby monitoring the actual change in the stability of the fault as a function of shear and normal stress. Secondly, the results provided a control function for numerical modelling. Three of the four in-situ stress measurement sites were instrumented, using GEOCON and IRAD CAGE vibrating wire instrumentation. This form of instrumentation was chosen for its long term stability characteristics. At each site three gauges were installed, forming a rectangular rosette, in a bore hole oriented parallel to the fault, the rosette lying in a plane normal to the fault. In June 1990, a further three such monitoring sites were added to this network.

Fault creep monitoring

In order to monitor the distribution of fault creep, so as to compare the distribution of produced slip as modelled and interpreted using the EXCESS SHEAR STRESS (ESS) concept (see Ryder 1988) and to monitor co-seismic slip, fault intersections were instrumented using displacement monitoring instrumentation. Two methods were employed. Mechanical gauges were installed at ten sites, five in excess of 50 metres in the footwall of the reef on the down thrown side of the fault and the rest in the fault loss pillar between the displaced reef horizons. Four electronic tiltmeters, three in the fault loss and one 50m in the footwall of the reef on the down thrown side, were installed in March 1991 (see Figure 4).

Geological observations

Geological observations provided information

327

regarding the character of the fault. In particular, the geometry of the fault and the variable roughness, as well as a qualitative interpretation of variable friction, are three parameters which were measured, interpolated and extrapolated across the surface and incorporated in the numerical modelling. The information was also used in the calculation and interpretation of results. The geometry of the fault surface was modelled using Faultsim, a package of computer programmes which prepares benchmark data for Minsim-D input, as well as post processing and graphical presentation of results being part of the ISS 'application software' in Welkom. The variable roughness was quantified by measuring the amplitude and wavelengths of undulations as mapped at fault intersections. This was used to interpret the local angle of friction according to a method as proposed by Ida, (1978). Roughness expressed as the angle of friction was then interpolated and extrapolated using the Faultsim package.

Numerical modelling

To model the stress distribution and ESS upon the fault, as a result of actual and idealised mining layouts incorporating different regional support, extensive use was made of Minsim-D, a 3D boundary element programme. This programme, developed specifically to calculate field stresses in the region of tabular excavations, was developed by the Chamber of Mines Research Organisation. The boundary element solution, assumes the rockmass to be continuous, isotropic, linearly elastic and infinitely strong. The model has been widely used to evaluate the stability of faults, using the ESS concept (Ryder, 1988; Napier, 1988; Webber, 1990).

In addition, to simulate the non-elastic behaviour of the rockmass, along a particular section of interest, the programme UDEC (Itasca Consulting Group Inc., 1989), which makes use of the distinct element method, was used. This programme allows the modelling of the rockmass as an assemblage of blocks with variable shape, size and material properties, separated by discontinuities having variable properties.

Seismic monitoring

Seismicity was monitored over a two and a half year period from the beginning of 1990, using two seismic networks. The regional ISS network provided seismic data for events with local magnitude greater than 0.7. A COMRO PSS micro-seismic network, installed in the region of the Tanton fault, provided data for seismic events with local magnitude in the range -2 to 1.0. For each mining step, seismic events locating in the vicinity of the fault were selected from the seismic data base (compiled from the above two systems) and plotted together with the modelled stresses on the fault surface (using further functions of the Faultsim package). In this way the seismicity could be correlated with the stress distribution on the fault, being a function of sequence of mining, nature of regional support, fault geometry, roughness and friction.

3 RESULTS

Stress measurements

The three dimensional stress tensor as measured at a depth of 2350m, is

Sx(east-west)	=	37.217
Sy(north-south)	=	34.691
Sz(vertical)	=	73.747
Txy	=	- 0.125
Tyz	=	0.762
Txz	=	- 5.851

This agrees with magnitude as predicted from overburden weight. This tensor indicates a lack of tectonic influence.

Stress change Monitoring

The stress meters successfully monitored the actual stress changes as mining progressed in the vicinity of the fault. The results for one of the sites is shown in Figure 1. The correlation between monitored Minsim-D modelled stress, is shown in Figure 2 and the correlation between measured results and that modelled through UDEC, is shown in Figure 3. These results are discussed in greater detail under numerical modelling.

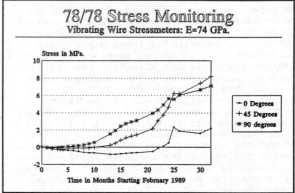

Figure 1. The 0° gauge is horizontal and the 90° gauge is vertical. The 45° gauge completes a rectangular rosette.

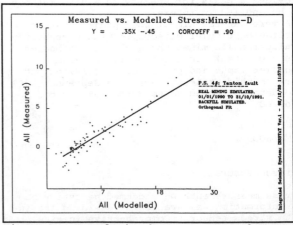

Figure 2. Correlation between measured stress changes and modelled values as predicted using Minsim-D.

All mechanical gauge sites located in the fault loss reflected a shortening of base length, or squeezing, from a fraction of a millimetre to a maximum of 2mm, as a result of increased normal stress. One of the sites 75/82 (see Figure 5) situated across a reef intersection in the immediate footwall of the reef, where shear deformation was predicted by numerical modelling, reflected considerable a-seismic creep. 78/82 (see Figure 6) also situated in a region where shear deformation reflected some creep. These were sites situated in the immediate footwall of the mining horizon on one particular side of

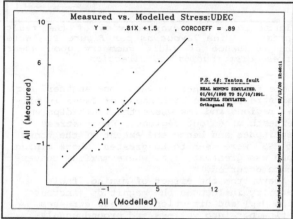

Figure 3. Correlation between measured stress changes and modelled values, as modelled using the discontinuum method (UDEC)

the fault. (These sites occurred within zones of positive ESS). The electronic tiltmeters accurately recorded the co-seismic slip, up to 5mm, which resulted during the largest event to effect the fault and which located in the fault loss close to where these instruments were positioned. The fault creep as monitored at 75/82 and 78/82 fault intersections, (see Figure 4), both being sites in the immediate footwall of the respective up thrown and down thrown reef horizons, are shown in Figures 5 and 6. Other sites, showed little or no dis- placement, locating beyond the zone of shear displacement, as predicted by numerical modelling. Unfortunately there was no access to areas of the fault where maximum slip was predicted by modelling.

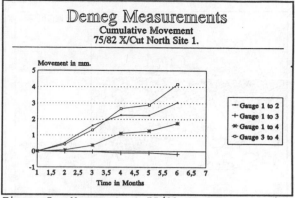

Figure 5. Movement at 75/82 crosscut, October 1988 to March 1989, indicating normal slip and dilation. Modelling predicted shear displacement here.

Geological investigations

The fitted fault surface appears in Figure 7. A plan distribution of roughness, as calcu- lated using the method as proposed by Ida (1978) and interpolating using the Faultsim package, is shown in Figure 8.

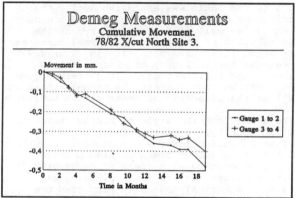

Figure 6. Movement 78/82 crosscut, January 1990 to October 1991. Decrease in base length indicates squeezing and shear displacement. Modelling predicted shear displacement.

Interpolation of the semi-quantitative esti- mates of roughness, as shown by Figure 8, allowed a unique value for the angle of fric- tion at each of the grid points to be used in the calculation of ESS and for seismic data interpretation.

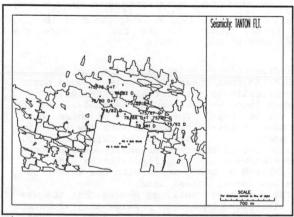

Figure 4. Mine plan (looking East) of President Steyn No.4 Shaft, showing creep monitoring sites. D=demeg, T=tiltmeter.

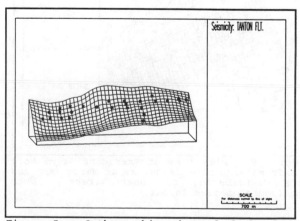

Figure 7. Orthographic view of fault from (looking south-East), showing principle mapped intersections used to fit the fault surface. Note the undulatory geometry.

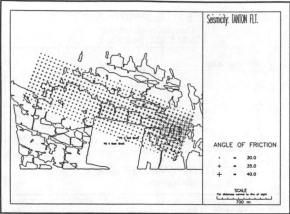

Figure 8. Plan view (looking East) of President Steyn No.4 Shaft showing variation in fault surface roughness. Note increased roughness towards the South.

Numerical modelling

Minsim-D modelling indicated the existence and growth of areas of positive ESS, as mining advanced. This was most significant upon the northern half of the fault, where mining adjacent to the fault was extensive during the monitoring period. The influence of mining geometry as well as fault geometry, on the distribution of the shear and normal stresses, are shown in Figures 9 to 12. In Figures 9-12 P denotes the influence of mining while G denotes the influence of fault geometry, upon the distribution of shear and normal stress as well as the direction of shear stress trajectory.

These show how the style of mining, which by method locally, involves extraction on two fronts advancing towards one another, results in the formation of areas of high normal stress coinciding with areas of converging shear stresses, as the final pillars are formed. Numerical modelling predicted the existence of positive ESS zones below the reef intersections (see Figures 13 & 15). Modelled shear stress changed direction ahead of mining faces, resulting in the formation of zones of convergent shear stresses. As pillars formed

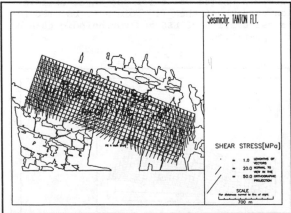

Figure 9. Plan view of President Steyn No.4 Shaft showing mine layout as at March 1992 and the distribution of shear stress. Note influence of mine layout.

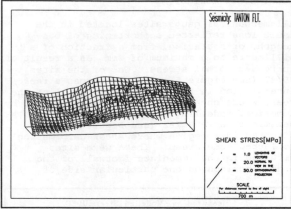

Figure 10. Orthographic view of the fault surface. Mine layout as per Figure 9. Note the influence of fault geometry upon shear stress distribution and direction.

where faces advanced towards one another, normal stresses were high ahead of faces in pillar regions and low where the fault dips underneath mined out regions. Shear stress magnitudes and hence the extent of the zones of ESS, were seen to be greatest where mining span was greatest, i.e. where maximum convergence occurred.

Apart from the effect of mining, fault geometry influences the magnitude (increase with dip) and direction of shear stresses ("flowing" into valleys and around undulations). Normal stress increases where the fault flattens and decreases where it steepens.

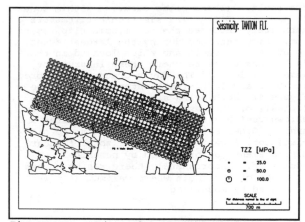

Figure 11. Mine plan President Steyn No.4 Shaft as at March 1992 showing the distribution of normal stress on the fault. Note the influence of mine layout.

The discontinuum model, according to UDEC, indicated that shear displacement would occur specifically in the footwall of the down thrown stope as well as the hangingwall of the up thrown stope. The location of these zones of shear deformation agrees with the location of positive ESS, as predicted using Minsim-D. The extent of the zone of shear deformation and the areas of 'limiting friction', as calculated by UDEC, agree with the extent of the low normal stress zone.

The numerical modelling showed that the use of widespread backfill, particularly where the mining span was large, would reduce the shear

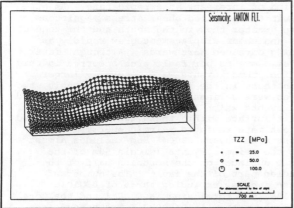

Figure 12. *Orthographic view of the fault surface. Mine layout as per Figures 9 & 11. Note the influence of fault geometry upon normal stress distribution.*

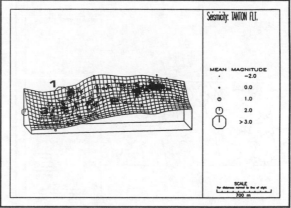

Figure 14. *Orthographic view of Tanton fault showing all seismic events from January 1990 to May 1992. Number indicate the sequence of three large events.*

stress, increase the normal stress and thereby reduce the areas of positive ESS and thus enhance the stability of the fault. Stiff fill ribs parallel to the fault would further enhance the stability and ultimately systematic pillars parallel to the fault would best stabilise the structure.

The boundary element method, overestimated the induced stress at the stress monitoring sites. A better correlation was proven between the discontinuum model and monitored stress (Refer to Figures 2 and 3). The fundamental reason for this is the assumption, as made by the boundary element solution, of an infinitely strong, non yielding, continuous rockmass. Therefore, no stress relief processes (due to fracturing as well as movement of discontinuities) are considered by the model, resulting in a poor correlation with actual measurements. However, the predicted location and interpreted extent of slip agrees with the slip as predicted by the discontinuum model. This model in turn, predicts stress change results which agree with monitored values. Both forms of modelling predict zones of slip, which agree with the distribution of monitored seismicity.

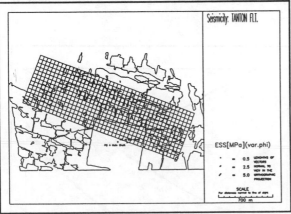

Figure 13. *Plan view of President Steyn No.4 Shaft showing the distribution of ESS in relation to mine layout as at March 1992.*

Therefore, although the Minsim-D model is inaccurate in predicting sites specific results, it can predict zones of instability on a fault provided that accurate and realistic input data is supplied, (fault and geometry), and results are interpreted in a logical, realistic and correct manner. For these reasons, Minsim-D will remain an important tool to model fault stability, until three dimensional discontinuum or hybrid methods become commercially available and affordable.

Seismic monitoring

In excess of 700 events were located upon the fault (selected in a zone 100m wide, 50m either side of the fault) in the period January 1990 to May 1992. Three relatively large (M>3.0) and damaging events occurred on the fault during the study period. Figure 14 shows the distribution of this seismicity.

4 INTERPRETATION OF SEISMICITY

Space does not allow the presentation of all the large number of seismic distribution plots analyzed. Emphasis here is on the interpretation of seismicity relative to modelled stress and other parameters.

Figures 15,16 and 17 show the respective distribution of ESS, normal and shear stresses for June 1991, as modelled by the boundary element solution. Superimposed upon this are the seismic events which were recorded in the corresponding three month mining step. Figure 18 shows the shear displacement, as modelled by the discontinuum model for a particular raise line. Superimposed on this are the relevant seismic events.

The fault can be split into two seismogenic regions. In the south, a large number of small events with local magnitude varying from -2.0 to 1.0, clustered in the fault loss pillar towards the South. Mining in the vicinity is associated with the removal of a remnant pillar abutting the fault. Here the fault is relatively flat, roughness is high and the mined out span is small. Therefore, shear loading is relatively low, normal stress is relatively high and thus the local shear

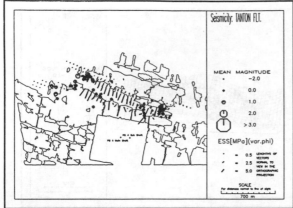

Figure 15. Plan of mine layout as at June 1991 showing seismic events and ESS as at June 1991, calculated using an average angle of friction of 20°.

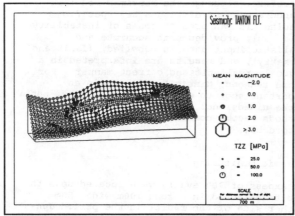

Figure 16. Orthographic view of Tanton fault showing normal stress distribution as at June 1991 and seismic event distribution associqated with this mining step.

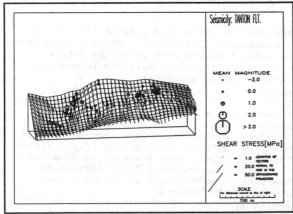

Figure 17. Orthographic view of Tanton fault showing shear stress distribution as at June 1991 and seismic events associated with this mining step.

strength is high. In the north, the general dip of the fault is steeper, mapping of fault intersections indicated lower roughness, mining spans are large and unbroken and considerable extraction took place, during the

period of interest, at a rapid rate. As a result the modelled shear stress magnitudes were greater than in the south and the zone of ESS and shear displacement grew rapidly as mining advanced northwards. Although, fewer seismic events per fault area occurred in this region, these events were generally larger than those in the south. To qualify this, two areas were delineated on the fault surface and all events which located within 50m of the fault surface within these two areas were used to estimate radiated seismic energy/m². The first area covers 6300m² and includes almost all the seismic events (within 50m of the fault surface) on the northern half of the studied part of the fault. The number of these events, excluding three of LOG10(e) > 7, is 196, the total area is 6300m² and the total radiated energy is 21600kJ. The second area in the south similarly includes >90% of the events along the southern half. Here the area is only 2615m², the total number of events (excluding 1 of LOG10(E) >7) is 528 and the total radiated seismic energy is 6740kJ. Thus, in the northern area, .03 recorded events/m² yielded 3.4 kJ/m² of energy while in the southern area, 2.0 recorded events/m² yielded 2.5 kJ/m² of energy. The difference in seismic behaviour, may be interpreted as being a function of the difference in stiffness between the north and the south. The system was stiffer in the south than in the north.

In general, seismic events occurred adjacent to areas of positive ESS, particularly where there was a transition from low to high normal stress, or where there were significant changes in fault geometry. This could be explained by a-seismic shear deformation, originating where shear stress exceeds frictional strength, continuing through regions of low normal stress and being partially arrested in areas of relatively higher normal stress, or where shear deformation was arrested as a function of geometry. As resistance to deformation increases, the proportion of the co-seismic component thereof increases. The shear deformation ·continues to load stronger asperities, which fail more violently resulting in larger seismic events.

Seismic events cluster where shear stress trajectories change direction and/or converge,. This occurs around pillars and ahead of active mining faces, as well as where fault geometry

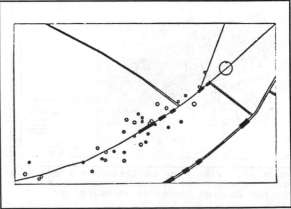

Figure 18. Section looking north through fault/mining, showing the extent of shear deformation as predicted using UDEC.

changes. Some concentration of events were observed where the fault undulates. One particular cluster is associated with flattening of the fault at depth.

Of the three large damaging events, none occurred in areas of positive ESS. The first of these events was preceded by a concentration of small events over a period of nine months. The apparent stress of many of these small events was higher than average, indicating the existence of an asperity of unknown physical nature (the nearest fault exposure was 70m from the main shock hypocentre). The third event was also preceded by similar foreshock clustering. In this case, geometric modelling indicated the presence of a minor undulation. More importantly, a remnant pillar was formed directly over the source region, which increased the normal stress acting upon this asperity. Therefore, mining produced or at least enhanced the asperity, which eventually failed.

The second seismic event was somewhat different to the other two damaging events. It occurred in a seismic gap region between the southern and northern seismogenic regions. The only precursory activity nearby was a cluster of small events more than 100m south of the hypo-centre. The source area was characterised by:- (i) virtual absence of seismic events, (ii) high normal stress and (iii) a significant normal stress gradient (high normal stress in the fault loss, above the fault reef intersection and very low normal stress immediately below reef).

In general, events in regions where shear stress was relatively high and normal stress low, were characterised by relatively large seismic moment and proportionally lower energy release and therefore significant co-seismic deformation. Conversely high normal stress under similar shear stress conditions enhanced co-seismic energy release. Increase in roughness also enhances energy release per unit deformation (moment). Therefore stress distribution, if controlled, can limit the effect of co-seismic deformation upon the fault and the surrounding rockmass.

Summarising, the distribution of seismic events upon the fault correlates with the distribution of shear and normal stress (being a function of mining and fault geometry), the predicted shear displacement, the distribution of ESS and the geometry and roughness of the fault surface. Interpretation of modelled results provides an explanation, consistent with the asperity model of crustal seismology, in terms of which major tremors result from the failure of strong patches on a fault following co-seismic and a-seismic creep in its proximity.

Stress modelling can assist in recognising a potentially unstable 'seismic gaps' i.e. where the modelling predicts significant instability adjacent to a gap and seismic monitoring confirms significant co-seismic deformation adjacent to the gap.

5 TOWARDS SAFER MINING

The seismic behaviour of a fault depends on the stiffness of the system made up of the fault, the rock mass and the adjacent mining. Fault dip, roughness and the elastic/ plastic nature are natural controls of the stiffness. Mine layouts and regional support can, however, have significant local influence. To reduce the potential of a major fault slip event affecting the excavations, shear deformation along the fault should be controlled. In terms of presently applied technology, the general aim should be to stiffen the system as much as possible to reduce shear deformation along the fault. The layout can be stiffened by making use of pillars and/or backfill to reduce the effective span of the extracted reef horizon which abuts the fault. By reducing the effective span, the volumetric closure across the extracted span, as well as the maximum closure at the centre of the mined span is reduced. Abutment stresses are therefore also reduced in magnitude. Pillars must be systematic, running parallel to the fault/reef inter-section. The spacing of the pillars will depend on the degree of required stiffening. Stiff fill ribs can be used as a substitute for pillars, allowing a greater percentage extraction. Soft backfill should be placed between the pillars/stiff fill ribs. This will further reduce closure and hence shear deformation along the fault. The soft fill will also assist to maintain the integrity of the pillars/stiff-rib system.

The layout and mining sequence should be planned so that no scattered pillars (remnants) are formed which will abut the fault/reef intersection. Scattered pillars form regions of relatively higher normal stress around which the shear stress converge. This results in a localised increase in resistance to shear deformation. Stiffening pillars should, therefore, be systematic and not random. A regular grid of parallel pillars, rather than one large pillar abutting the fault surface, is preferred. Regular pillars limit the maximum closure across the mined span more efficiently than a single large pillar. The first pillar, within the system of parallel pillars, may be left bracketing the fault, if this is considered beneficial from a ground control point of view.

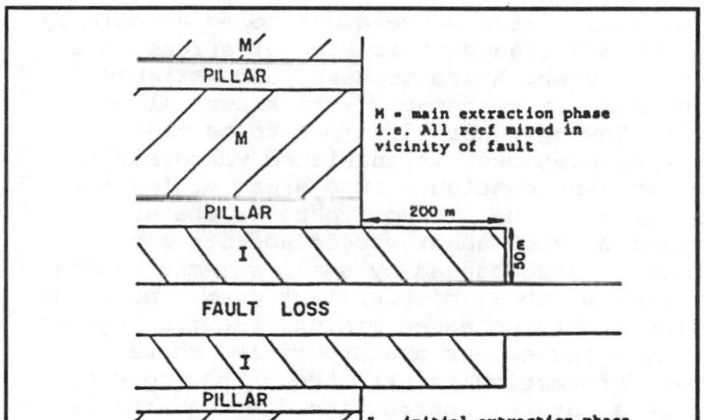

Figure 19. Idealised schematic layout of preferred mining sequence and support, to reduce the potential for violent fault slip.

A schematic of the proposed layout incorporating factors as discussed above is shown in Figure 19. The reef immediately abutting the fault should be extracted first, at minimum span to ensure that minimal shear deformation occurs whilst mining in the immediate vicinity of the fault. Adjacent to this a pillar/stiff-fill rib is left. The principle extraction phase then continues beyond this pillar/rib. Regular parallel pillars/ribs are then left as one moves away from the fault. The primary slot against the fault should be far enough ahead of the principle extraction phase, to ensure that the shear deformation on the fault in the region of this mining face is minimal and to place this face far enough ahead from a possible large seismic event, which may occur during the main extraction phase. Numerical modelling and ground motion/seismic induced damage investigations, will allow this distance to be quantified. The strip should always be backfilled to enhance stability and to form a seismic damping buffer. Where two such strips meet, this remnant should be mined without delay. In this way the formation of scattered pillars abutting the fault is avoided.

An alternative approach which could be applicable where a fault which is 'soft' and difficult to clamp, is not to stiffen the system at all, but to simply mine away from the fault leaving no remnant pillars. The fault is thus allowed to yield from the first onset of mining and, if it should slip violently when the inevitable asperity breaks after substantial mining induced deformation, the mining faces are already some distance away from the source of strong ground motions and/or co-seismic closure. Before such a strategy would be viable, however, technology should be available to either (or both) predict the timing of the larger events or to trigger them at the mine's convenience.

6 CONCLUSIONS

Possibly the most important conclusion to be reached from the above, is confirmation that the asperity model, to describe seismic behaviour of faults which deform under induced load, is applicable to the mining situation. This establishes a basis for planning mine layout and extraction sequences when mining in faulted ground.

The correlation between monitored seismicity and the modelled distribution of stress on a fault, for which the geometry and variable roughness are reasonably well known, allows the following generalisations to be made:-
Shear displacement is initiated where ESS is positive and continues into areas of low normal stress. The 'creep front' is characterised by high shear stress and its propagation is accompanied by small seismic events. The area of shear displacement grows under the influence of increased mining. Source regions for damaging events are asperities where the shear deformation is arrested. The asperity can be a natural obstruction due to variable geometry or increased roughness. Alternatively, it could be an area of mining induced high normal stress and converging shear stress. An asperity may be characterised by the clustering of small seismic events of relatively high apparent stress. Alternatively, the asperity could be a characterised by seismic quiescence - numerical modelling can assist to distinguish between a 'good' and a 'bad' seismic gap.

Of all the different techniques which have been applied for the studying of the behaviour of the Tanton fault, three methods have proven to be of most value, in terms of input and practically useful output. These are seismic monitoring, fault mapping and the modelling of stress in a stepwise fashion, accurately simulating the actual mining. Physical measurements of in-situ stress and the monitoring of stress change and fault creep provide results which are very site specific and consequently are not very useful to evaluate and understand the behaviour of the rockmass under the influence of advancing mining. Site specific measurements are quite sensitive to over- or under estimations due to local rockmass behaviour not representative of the larger rock volume of interest. In contrast, seismic waves travel through the rockmass and bring information about dynamic processes controlled by the entire rockmass. Numerical models are oblivious of minor local anomalies and thus also gives an overall view of potential rock mass behaviour. Information about fault geometry and frictional characteristics came a long way to 'integrate' seismic information with model predictions. It also showed, however, serious shortcomings of presently available modelling software to accommodate input of deformation data from actual seismic and/or non-seismic measurements.

Mining layouts and sequences should aim to limit shear deformation and to avoid inducing localised areas of high normal stress gradient surrounded by shear stress convergence upon a fault. A mining layout incorporating systematic pillars and backfill should be used to reduce the magnitude of induced shear stress and increase the stabilising normal stress over the whole fault surface of interest. Reef adjacent to a fault should be mined first when the general mine span is still small. This will also ensure that no remnant pillars, which induce locally higher normal stress and thereby form asperities, are left to be removed at the final stage of mining.

ACKNOWLEDGEMENTS

The authors are grateful to President Steyn Gold Mine for sponsoring the project and to the management, particularly Mr André Wilkens, for active support. Mrs. Ida de Lange did all the typing and Mr M.F. Handley assisted with numerical modelling.

REFERENCES

Dennison P.J.G. (1990) *Detailed progress report, outlining all work up until November 1989, towards the thesis 'An investigation into the effects of adjacent mining and fault character upon fault stability and its associated seismic beha-iour.* Ph.D thesis progress report, University of the Witwatersrand, Johannesburg.

Dennison, P.J.G. (in prep). *An investigation into the effects of adjacent mining and fault character upon fault stability and its associated seismic behaviour.* Ph.D thesis University of the Witwatersrand, 1993.

Ida, Y. (1978). *Propagation of slip along frictional surfaces.* Pageoph, 116, 931-963.

MINSIM-D (1992). *Instruction Manual, Chamber of Mines*, Johannesburg South Africa.

Napier, J.A.L. 1988. *The application of excess shear stress to the design of mine layouts*, Mining & Metallurgy, 87, 397-405.

Ryder, J.A. (1988). *Excess shear stress in the assessment of geologically hazardous situations*: J.A. Ryder, J S Afr. Inst. Min. & Metall. 88, 27-39.

UDEC. *Universal Distinct Element Code*. Instruction Manual, ITASCA Consulting Group.

Webber, S.J. (1989). *Seismic Moments and Volume Changes in the Klerksdorp Goldfields*. COMRO research report, No.6/89, Chamber of Mines, Jhb., S. Afr.

Webber, S.J. *Numerical Modelling of repeated fault slip*. Chamber of Mines informatory circular no. 28/89.

Webber, S.J. 1990. *Numerical modelling of repeated fault slip*. J S Afr. Inst. Min. & Metall. 90, 133-140.

Rockbursts and Seismicity in Mines, Young (ed.) © 1993 Balkema, Rotterdam, ISBN 90 5410 320 5

Estimates of Q using the spectral decay technique for seismic events with M <-1.0

A. J. Feustel, T. I. Urbancic & R. P. Young
Engineering Seismology Laboratory, Department of Geological Sciences, Queen's University, Kingston, Ont., Canada

ABSTRACT: Seismic data recorded with triaxial accelerometers bandlimited between .05 and 5.0 kHz were used to estimate Q for events with M<-1.0 associated with an excavation at 620 meters depth. The spectral decay slope analysis technique, which assumes an ω^{-2} dependence of a non-attenuated source displacement function, was used to determine Q. The results indicate Qp and Qs to be between 100 and 150 in a moderately fractured rockmass. Large deviations in Q (>250) from average values were observed which may indicate zones of low and high attenuation associated with the excavation. Errors in the technique are relatively low (~20%) suggesting it is a viable method to passively estimate Q using high frequency seismic data.

1 INTRODUCTION

The loss or absorption of energy in a seismic wave propagating through a rock mass is attributed to many mechanisms including: geometrical spreading, scattering, dispersion, and energy loss due to heat or internal friction. Aside from geometrical spreading, each of these mechanisms are considered to be attenuating properties of a rock mass. Investigations of the quantity of seismic wave attenuation/absorption (α) are important in geological environments for which detailed seismic information is desired. Accurate seismic rock mass assessments are dependent upon knowing the attenuating or energy absorbing properties that effect seismic waves. Typically, α of the rockmass is described by the quality factor Q (proportional to the inverse of α), where Q is assumed to be frequency independent and represents the ratio of stored energy to dissipated energy. Determining the attenuation characteristics of in-situ rock is critical in order to accurately correct received seismic signals for path energy loss; thus facilitating reliable source parameter calculations. Strong ground motion concerns and seismic hazzard assesments are also highly dependent on signal attenuation. In addition, with dense raypath coverage and a large sensor array it is possible to image Q and use it to complement velocity structure for rockmass characterization; attenuation has been shown to be more sensitive to changes in rockmass condition than velocity (DaGamma, 1971; Young et al., 1979). However, reliable in-situ measurements of Q are very difficult to obtain given the sensitive nature of processing methods required for its quantification.

The most common techniques used for Q determination include the spectral ratio method, rise time method, Q-coda method, and displacement spectra decay analysis. The use of each technique requires that specific criteria be met by the data in terms of source function assumptions, sensor array geometry, signal quality, attenuation mechanisms, and whether the assessment is being carried out in-situ or in the laboratory. Furthermore, three of the four methods can be used in-situ on both passive (seismic source) and active (blast source) data while the spectral decay method is typically a passive technique.

Applications of the aforementioned techniques for in-situ analysis are almost exclusively used for investigations of regional or global seismic attenuation. More recently, Evans and Zucca (1988), Scherbaum (1990), and Zucca and Evans (1992) used inversion techniques to study Q tomography in seismically active regions of the earth's crust. Boatwright et al. (1991) studied attenuation characteristics from the Loma Prieta earthquake data using the spectral decay technique. Whitman et al. (1993) also used the spectral decay technique for estimates of Q in regions of the Central Andean Plateau. Very little work has been presented which investigates Q in mining environments. Cichowicz et al. (1990) and Cichowicz and Green (1989) represent the most prominent work in this area. Assuming a single scattering model, they used Q-coda methods for attenuation studies of mining induced seismicity in South African

Mines believing this method to give reliable in-situ estimates. In the 1989 study, measured values for Qp ranged from 30 to 100 in a fractured region of an excavation area with an upper limit of 300 outside of this area. Values for Qs ranged from 20 to 78 near the excavation.

Although the studies mentioned above likely represent a fraction of the work being done on in-situ attenuation, they help support the fact that the majority of published studies do not address the problem of seismic attenuation in mines. In this study, we apply the spectral decay analysis technique of Q detection normally used on data with moment magnitude (M) > 2.0 and corner frequencies between 0 and 8 Hz to mining-induced microseismic data with M between -1.0 and -2.4 and corner frequencies typically between 0.5 and 1.0 kHz. The goals of the study are:

1) Determine if the technique is applicable and adaptable to mining environments and microseismicity.

2) Provide a measurement of Q that can be used for source parameter calculations.

3) Discuss the effect of variations in Q on source parameters.

2 METHOD

Many attenuation analysis techniques attempt to compare a non-attenuated reference signal to an attenuated signal and quantify the energy loss. The principle difference between techniques is the reference waveform. In the case of spectral ratios and rise time methods, the reference waveform is the same signal recorded at a smaller distance from the source. The displacement spectral decay technique is based on the assumption that failure for a seismic event is in a pure-shear sense and that the non-attenuated displacement spectra is well represented by an ω^{-2} dependence past the corner frequency as described by Aki (1967) and Brune (1970). The ω^{-2} model appears to be satisfied over a wide range of magnitudes and frequencies (e.g., Hanks, 1979; Hanks and McGuire, 1981; Chael, 1987; Urbancic et al., 1993).

Calculation of Q requires that the user specify the corner frequency (f_c) and maximum frequency (f_2) or system frequency band limit. The slope of the signal is then determined within this frequency band and compared to a slope of -2. Equations 1 and 2 describe the relationship between the exponential decay constant of attenuation α for plane wave propagation, and the assumed frequency independent quantity Q (Aki and Richards, 1980; Johnston and Toksoz, 1981).

$$A(x) = A_o e^{-\alpha x} \qquad (1)$$

$$Q = \frac{\pi f}{C \alpha} \qquad (2)$$

Where x represents distance of propagation, Ao and A(x) represent the wave amplitude at the source and after propagation distance x, respectively. Based on the above equations, values of Q are calculated automatically by equation 3.

$$Q = \frac{\pi(R)(\log_e)\left(f_{(i+1)} - f_{(i)}\right)}{C\left[\left(2\log\dfrac{f_{(i)}}{f_{(i+1)}}\right) + \left(\log\dfrac{A_{2^{(i)}}}{A_{2^{(i+1)}}}\right)\right]} \qquad (3)$$

Where R is source-sensor distance, $f_{(i)}$ is f_c, $f_{(i+1)}$ is f_2, C is Vp or Vs (set to 6095 m/s and 3700 m/s, respectively), $A_2(i)$ is the signal amplitude at f_c, $A_2(i+1)$ is the signal amplitude at f_2, and $\log_e$ is a constant conversion factor from log to ln. The left hand term in the denominator represents the -2 decay slope of the reference signal whereas the right hand term in the denominator represents the initial slope of the received signal.

3 DATA SELECTION AND PROCESSING

Data was selected from the Strathcona Mine, Ontario that was collected on the Queen's Microseismic System (Young et al., 1993) during periods of active mining following production blasts, during non-mining quiet periods, and following the completion of excavation of an approximately 7000 m³ volume of rock at 620 meters depth. Monitoring was carried out with 5 triaxial accelerometers and 33 uniaxial accelerometers that lie within 100 meters of the excavation site. Events were located using a combined Simplex-Geiger method and incorporating an isotropic velocity structure. The triaxial accelerometers were band limited between .05 and 5.0 kHz with a flat frequency response between .05 and 3.5 kHz. The system sensitivity was set at approximately 102 V/g and signals were sampled at a rate of 20 kHz. A total of 41 events were processed from those located in 4 subsets: FW1, FW2, and FW3 from the excavation footwall, and 7 events from cluster 2 (C2) as shown in Figure 1. Fault plane solutions were computed for all recorded events to identify those which fit a double couple shear source mechanism of failure; thus allowing the supposition that the events satisfy the shear model for the spectral decay method. Specifically, 48 events from these clusters were chosen as they represented the highest spatial event density which matched the fault plane solution criteria. In addition, triaxial #2 records are used because of the low signal to noise ratio and the unobstructed path (i.e., stope and drifts) between the event clusters and receiver; as was the case with sensors #4 and #5. Where discrepancies from the mean Q values were noted on triaxial sensor #2, estimates from triaxial sensors #1 and #3 were used to investigate the variances; tests indicated that the anomalous values were consistent on all sensors considered.

All of the signals were rotated by eigenvalue/eigenvector decomposition according to the method described by Matsumura (1981) such that the first channel of the accelerometer pointed in the direction of the P-wave, while the 2nd and 3rd components represent the S-waves (s1 and s2) orthogonal to the P-wave orientation. A time window of 10 msec with a 10 point cosine taper filter was applied to both the P- and S-waveforms. The window was of sufficient size for a representative frequency spectrum of the main P- and S-wave groups, while not too long as to contain significant scattered energy within the wave coda.

Following data rotation and windowing, displacement spectra were calculated and Q values were determined for each channel and event. For each calculation, several (≥ 8) manual iterations were performed while varying f_c and f_2 until the Q consistently converged to a similar value. f_2 was typically ≥ 2.8 kHz and f_c was determined manually by visual fit as well as theoretically according to Snoke (1987) from the energy flux Jc and spectral level Ωo as:

$$f_c = \left(\frac{J_c}{2\pi^3\Omega_o^{\,2}}\right)$$

Once appropriate values were set for f_c and f_2, the Q values would typically converge to within ± 20 for values between 50 and 150, and ± 100 for values between 350 and 500; where 500 represents the maximum value detectable in our applications and can be assumed to represent infinite Q. As the slope of the displacement decay approaches -2, subtle changes in f_c and f_2 caused larger variations in Q values. Although the errors for this method are not readily quantifiable, repetitive processing proved the aforementioned ΔQ values to be consistent. These values of ΔQ represent an average error of approximately $\pm 20\%$ for the full range of values measurable (Q between 20 and 500). The error levels are similar to those found by Whitman et al. (1993) of $\pm 20\%$ for the spectral decay technique applied to seismic events with much lower corner frequencies (i.e., 2-6 Hz).

4 RESULTS

Qp, Qs1, and Qs2 estimates from triaxial #2 records are summarized in Table 1 for the events analyzed. The data is ordered chronologically according to period, date, and time. Most of the FW3 events occur during period 3, while FW1, FW2, and C2 events were dispersed throughout the other periods. Histograms were made of the data in order that trends in the Q values and event clusters could be visualized.

In Figures 2a and 2b, both Qp and Qs values are presented for all FW events recorded on triaxial #2. The plots show similar asymmetric relationships with a maximum ranging from 50 to

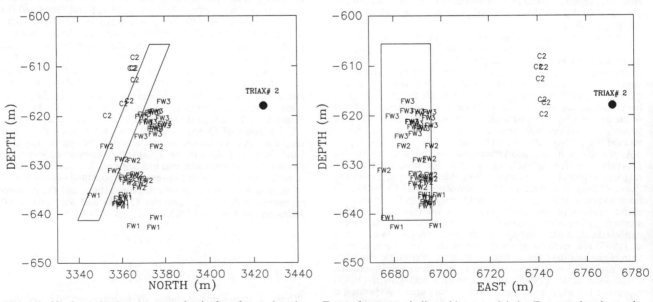

Figure 1. North vs. depth and east vs. depth plot of event locations. Event clusters are indicated by event labels. Excavated rockmass is outlined in both figures by thin solid line. Triaxial accelerometer #2 is identified with filled circle.

Table 1. Qp, Qs1, and Qs2 calculated for each event and ordered according to date, time, and location cluster. P1-P4 correspond to period 1 through period 4.

Event		DATE	TIME	CLUSTER	Qp	Qs1	Qs2
P1	1	10/25/91	17:08:37	C2	342	270	300
	2	10/25/91	17:18:48	FW1	500	500	500
	3	10/25/91	17:47:57	C2	72	105	67
	4	10/25/91	17:51:07	FW2	100	129	138
	5	10/25/91	19:07:12	FW2	78	88	121
P2	6	1/6/92	23:04:09	FW1	76	66	72
	7	1/7/92	2:54:25	FW3	175	168	139
	8	1/7/92	23:06:22	FW1	59	91	70
	9	1/7/92	23:06:50	FW2	248	250	227
	10	1/7/92	23:25:36	FW1	150	135	147
	11	1/7/92	23:27:40	FW1	332	202	500
	12	1/7/92	23:38:37	FW2	129	105	90
	13	1/8/92	0:38:12	C2	50	55	60
	14	1/8/92	2:50:56	FW2	115	72	97
	15	1/8/92	3:11:33	C2	115	154	82
	16	1/8/92	3:20:54	FW1	157	180	181
	17	1/8/92	4:28:01	FW1	163	159	162
	18	1/8/92	5:51:12	FW1	79	287	134
P3	19	1/17/92	15:08:01	FW3	60	48	49
	20	1/17/92	15:11:31	FW1	50	96	81
	21	1/17/92	15:25:25	FW2	56	123	109
	22	1/17/92	15:36:53	FW1	64	90	77
	23	1/17/92	15:38:39	FW1	363	392	500
	24	1/17/92	15:45:42	FW1	56	70	71
	25	1/17/92	15:57:03	FW1	53	80	84
	26	1/17/92	15:57:11	FW3	500	500	500
	27	1/17/92	16:03:12	FW1	113	247	140
	28	1/17/92	18:37:16	C2	53	60	74
	29	1/17/92	18:43:21	FW2	422	337	425
	30	1/20/92	23:07:38	FW3	98	100	107
	31	1/20/92	23:28:43	FW2	67	76	81
	32	1/20/92	23:41:51	FW2	54	106	139
	33	1/20/92	4:02:33	FW3	102	152	82
	34	1/21/92	7:10:54	FW3	133	81	126
	35	1/23/92	22:35:38	FW2	50	66	98
	36	1/24/92	23:32:06	FW3	207	231	194
P4	37	1/24/92	23:35:37	FW3	286	261	215
	38	1/24/92	23:37:52	FW3	369	352	500
	39	1/25/92	23:59:57	C2	39	75	59
	40	1/25/92	0:44:03	FW3	225	150	175
	41	1/25/92	0:47:12	C2	130	166	121
	42	1/25/92	1:05:10	FW3	93	94	126
	43	1/25/92	1:54:52	FW3	184	146	241
	44	1/25/92	2:25:36	FW3	155	116	101
	45	1/25/92	2:56:49	FW3	296	160	144
	46	1/25/92	3:10:54	FW3	500	238	500
	47	1/25/92	4:15:46	FW3	167	189	221
	48	1/25/92	22:46:35	FW2	61	146	64

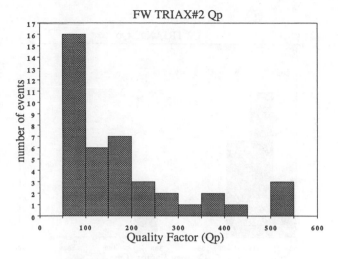

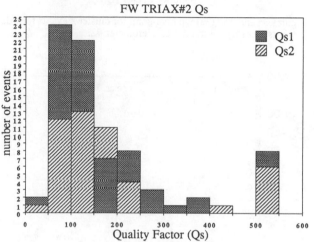

Figures 2a and 2b. (2a) Qp histogram for footwall events recorded on triaxial accelerometer #2. (2b) Qs1 and Qs2 histogram for footwall events recorded on triaxial accelerometer #2.

Qp=9/4Qs. He assumes, based on previous work by Anderson et al. (1965) and Anderson and Hart (1978), that compressional energy losses in the Earth are negligible, and the lack of attenuation of pure compressional motion indicates that shear mechansims such as grain boundary sliding dominate the attenuation processes. If we consider that the wavelengths of seismic signals from this study for both P- and S-waves are on the order of the joint spacing in the rockmass (3-6 meters) and that the sensor is spaced to within approximately 16 wavelenghts from the source, then it is possible that the dominant mechanism of attenuation observed at the sensor is scattering due to joints rather than intrinsic attenuation due to grain boundary energy losses. If this is the case then the P- and S-waves are both affected by the attenuating mechanism, rather than mainly the S-wave as in Burdick's relationship. Similar to the results from this study, Cichowicz and Green (1989) reported a Qp/Qs ratio of aproximately 1.45 for a fractured rockmass near an underground opening. Their ratio is also significantly less than that proposed by Burdick.

Figures 4a and 4b show a plot of Qp, Qs1, and Qs2 as a function of time. The events are chronologically ordered from period 1 to period 4. The trends of this plot reinforce the suggestion from the histogram analysis that the average values are between 50 and 150. Worth noting is the slight increase in Q values for the FW3 events during period 4.

The fact that there are outlying points (Q between 250 and 500) raises concern about the method of calculation as well as the mechanism of signal attenuation. Figure 5 is a schematic diagram of the excavation around which the FW and C2 events occur. The small circles represent low Qp (<250) while the larger circles represent high Qp (>250). If the values are valid, then it appears that high Qp events occur in very close proximity

150 (considering errors mentioned in the previous section). Above this maximum the values tend to fall exponentially to one. The peak at 500 is an artifact of the limitations of the applied method as previously described.

Figures 3a and 3b show Qp and Qs values calculated for all C2 events recorded on triaxial #2. These figures show the same asymmetric trend, with respect to one another, with values falling exponentially to zero above 150 (this trend is less delimited compared to the FW data due to a limited number of events). In comparison to the FW data, it seems that the C2 event region also has a Qp and Qs maximum between 50 and 150. In addition, a few outlying points exist which suggest an anomalous low attenuation region either along the path or at the source volume for these events .

5 DISCUSSION

By observing the trend of the Qp histograms compared to that of the Qs histograms it is concluded that Qs is equal to or greater than Qp in most cases (see also Table 1). Neglecting values >250 and considering errors, the overall Qp/Qs ratio from this study is approximately 0.86. This is not consistent with the previously suggested relationship of Burdick (1978) that

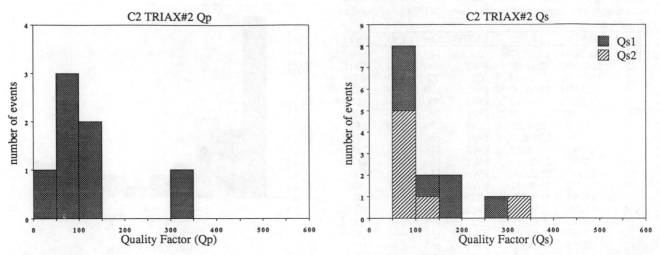

Figures 3a and 3b. (3a) Qp histogram for cluster 2 events recorded on triaxial accelerometer #2. (3b) Qs1 and Qs2 histogram for cluster 2 events recorded on triaxial accelerometer #2.

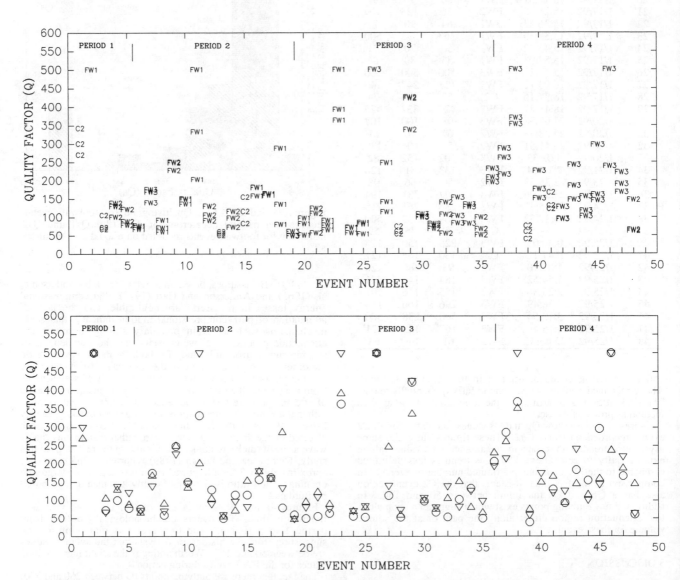

Figures 4a and 4b. Q values calculated for period 1 through period 4 (indicated at top of figure). Event number corresponds to event numbers in Table 1. (4a) Values are labeled according to location cluster. (4b) Open circles represent Qp, up triangles represent Qs1, down triangles represent Qs2.

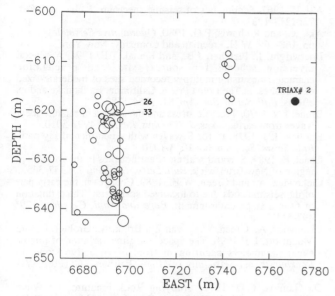

Figure 5. East vs. depth plot of event locations according to Qp estimate. Large open circles represent Qp>250, small open circles represent Qp<250. Locations of events 26 and 33 are identified in figure. Excavated rockmass is outlined by thin solid line. Triaxial accelerometer #2 is identified with filled circle.

to low Qp events. Figures 6a and 6b illustrate the difference between the energy decay of these two events. Figure 6a is event 26 with Qp=500 and Figure 6b is event 33 with Qp=102. In comparison, event 33 has a much greater energy decay characteristic than event 26, resulting in a much lower Qp value; the solid straight line represents a -2 slope for reference. It is presumed that because the events plot near one another their path to the sensor is nearly identical. Thus, differences in Q for side-by-side events appear to be related to the source function. It is possible that because events 26 and 33 are separated by 3 days, we may be measuring a time dependent attenuation that is adjusting to the changing stress conditions due to excavation. If this were the case it should be evident in Figure 4, yet it is difficult to draw this conclusion based on the figure. Considering the errors on event locations, relative to one another, to be from 5 to as much as 13 meters (Maxwell and Young, 1993), we can increase the separation between high and low Q events. At the maximum location error limits, this re-orientation may be enough to group all high Q events and all low Q events into separate clusters (or sub-clusters), thereby identifying zones of high and low Q. In any case, it is likely that these spatial Q relationships do exist, at least to some degree, and the values are constrained enough to suggest that the method can identify the energy decay characteristics for a given event volume.

The accuracy of the spectral decay method for this data is hinged upon the most important assumption of this technique: that these events should have spectral decay slopes of nearly -2 for a non-attenuated signal. This assumption alone controls the overall validity of the values as well having implications in explaining the outlying points (Q>250). There are conditions which may cause the non-attenuated spectra not to follow the ω^{-2} dependence. These include:

1). source rupture complexity
2). possible non-shear mechanisms of failure
3). sensor orientation with respect to source
4). varying rupture velocities
5). local source scattering effects

Although these and other mechanisms may be responsible for variations in the ω^{-2} dependence, it is not within the scope of this study to consider all possibilities in the detail that would be required to resolve the problem. However, we can suggest reasons why issues 1, 2, and 3 above may be eliminated as concerns for error.

Based on the work of Urbancic et al. (1993) on the same data set, these events can be considered as simple, rather than complex. Considering that first motion polarities from 33 uniaxial sensors and 5 triaxial sensors were used to constrain the fault plane solutions, we are confident that shearing is the major

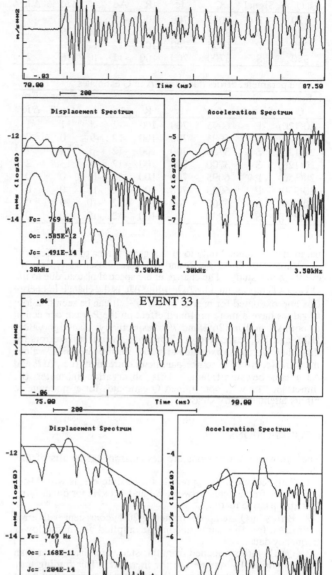

Figures 6a and 6b. (6a) Recorded P-wave signal, acceleration spectrum, and displacement spectrum for event 26 (Qp=500) and (6b) event 33 (Qp=102). Signals have a 10 msec (200 point) window applied to calculate upper signal in the displacement and acceleration spectrums. Lower signal in spectra is noise as calculated by a 500 point window applied to the signal before the P-wave arrival. Fc is corner frequency, Oc is spectral level, and Jc is energy flux as indicated in the displacement spectrum. Solid line in displacement spectra has -2 slope past the corner frequency for reference to decay characteristic in text.

component of failure for these events. Finally, as mentioned previously, values calculated from sensor #2 were confirmed by sensors #1 and #3, at approximately 45° on either side of sensor #2.

Given that we accept the method assumptions, the results from this study indicate that with sufficient data and familiarity with that data, this method can identify trends in Q values for a given rockmass that are related to attenuation either at the source or along the propagation path between the source and sensor.

The significance of the average Q estimates as well as the errors in the measurements need to be addressed in terms of their impact or corrective effect on the signal spectra. Table 2 illustrates the effect of Q values as well as ΔQ values on typical source parameters for this data set. The table is based on equations 1 and 2 and is a very loose approximation of these

Table 2. Effect of Q on source parameters. Variables are described in METHOD section of text.

Q	Signal	C	fc	R	Ao	A	$\Delta\Omega$o %	ΔJc%
120	P	6095	700	100	-12	-9	26	7
120	S	3700	700	100	-12	-7	39	15
350	P	6095	700	100	-12	-11	10	1
350	S	3700	700	100	-12	-10	16	2

Source parameter errors due to errors in Q estimates.

Error Q	Signal	C	fc	R	Ao	A	Error $\Delta\Omega$o %	Error ΔJc%
-.20(120)	P	6095	700	100	-12	-8	11	1
+.20(120)	P	6095	700	100	-12	-9	0	0
-.20(120)	S	3700	700	100	-12	-6	14	2
+.20(120)	S	3700	700	100	-12	-8	13	2
-.20(350)	P	6095	700	100	-12	-11	0	0
+.20(350)	P	6095	700	100	-12	-11	0	0
-.20(350)	S	3700	700	100	-12	-10	0	0
+.20(350)	S	3700	700	100	-12	-10	0	0

parameters, intended only to give a general idea of the effects of Q. Values for Q, C, f_c, R, and Ao were chosen as typical for events in this study. The change in the spectral plateau due to the Q values is represented by $\Delta\Omega$o while ΔJc is the change in energy flux approximated for brevity by $(\Delta\Omega o)^2$. It can be seen that low Q values have a more profound effect on the P-wave correction to both Ωo and Jc (26% and 7%, respectively) than high values (10% and 1%, respectively). Similarly, given that the errors for this study were approximated at ±20% for the range of Q values measurable, errors in source parameter corrections (i.e., $\Delta\Omega$o and ΔJc) can be significant. This observation indicates the importance of well constrained Q estimates for a moderate to highly attenuating rockmass.

6 CONCLUSIONS

The following conclusions are drawn from the results of this study:

1. This technique is applicable for seismic events with M<-1 and corner frequencies between 0.5 and 1.0 kHz for quantifying Q with a passive method.

2. Errors in Q are approximately ±20%; consistent with error estimates for the same technique applied to much lower frequency data.

3. Qp/Qs ratios obtained from this study are much lower than those traditionally reported in the literature. This finding may indicate that significant energy losses occur in compression when the scale length of the heterogeneities (i.e., fractures and joint sets) approximate the dominant wavelengths.

4. The displacement spectra decay slope technique estimates Qp and Qs to range from 100 to 150 near the excavation area studied at the Strathcona Mine, Ontario.

5. Some areas of high Q (>250) may exist which could be better defined with a larger data set and optimization of source location errors.

6. It is possible that large variations in Q are a result of source function or source type rather than path effects.

ACKNOWLEDGMENTS

The authors wish to thank Falconbridge, the Mining Research Directorate, the Natural Sciences and Engineering Research Council, and Queen's University for their financial support. We would also like to thank Cezar Trifu and the Queen's Engineering Seismology Lab personnel for their contributions and Sue Bird for her extensive pre-processing of the data.

REFERENCES

Anderson, D.L., Ben-Menahem, A., and Archambeau, C.B. 1965. Attenuation of seismic energy in the upper mantle. *J. Geophys. Res.* 70: 1441-1448.

Anderson, D.L., and Hart, R.S. 1978. Attenuation models for the earth. *Phys. Earth and Planet Interiors* 16: 289-306.

Aki, K. 1967. Scaling law of seismic spectrum. *J. Geophys. Res.* 72: 1217-1231.

Aki, K., and Richards, P.G. 1980. *Quantitative Seismology,* v.1. pp. 168-169. W.H. Freeman and Company: New York.

Boatwright, J., Fletcher, J.B., and Fumal, T.E. 1991. A general inversion scheme for source, site, and propagation characteristics using multiply recorded sites of moderate-sized earthquakes. in The Loma Prieta, Califonria, Earthquake and its effects. *Bull. Seism. Soc. Am.* 81: 1754-1782.

Brune, J.N. 1970. Tectonic stress and the spectra of seismic shear waves from earthquakes. *J. Geophys. Res.* 75: 4997-5010.

Burdick, L.J. 1978. t* for S waves with a continental ray path. *Bull. Seism. Soc. Am.* 68: 1013-1030.

Chael, E. 1987. Spectral scaling of earthquakes in the Miramichi region of New Brunswick. *Bull. Seism. Soc. Am.* 77: 347-365.

Cichowicz, A., and Green, W.E., 1989. Changes in the early part of the seismic coda due to localized scatterers: The estimation of Q in a stope environment. *Pure and Appl. Geophys.* 129: 497-511.

Cichowicz, A., Green, W.E., van Zyl Brink, A., Grobler, P., and Mountfort, P.I. 1990. The space and time variation of micro-event parameters occuring in front of an active stope. in *Rockbursts and Seismicity in Mines,* Fairhurst, C. (Ed.): 171-175.

Da Gamma, C.D. 1971. Studying Rock Fractures by Wave Attenuation Methods. Symp. Soc. Intern. Mecanique des Roches, Nancy:1-2.

Evans, J.R., and Zucca, J.J. 1988. Active high-resolution seismic tomography of compressional wave velocity and attenuation structure at Medicine Lake Volcano, Northern California Cascade Range. *J. Geophys. Res.* 93: 15016-15036.

Hanks, T. 1979. b values and $\omega^{-\gamma}$ seismic source models: Implications for tectonic stress variations along active crustal fault zones and the estimation of high-frequency strong ground motion. *J. Geophys. Res.* 84: 2235-2242.

Hanks, T., and McGuire, R. 1981. The character of high-frequency strong ground motion. *Bull. Seism. Soc. Am.* 71: 2071-2095.

Johnston, D.H., and Toksoz, M.N. 1981. Definitions and terminology, in Seismic Wave Attenuation, SEG Geophys. Reprint Ser. 2, D.H. Johnston and M.N. Toksoz (Eds.). pp. 1-5. Society of Exploration Geophysicists: Tulsa.

Matsumura, S. 1981. Three-dimensional expression of seismic particle motions by the trajectory ellipsoid and its applications to the seismic data observed in the Kanto District, Japan. *J. Phys. Earth* 29: 221-239.

Maxwell, S.C., and Young, R.P. 1993. Stress change monitoring using induced microseismicity for sequential passive velocity imaging. same issue.

Scherbaum, F. 1990. Combined inversion for the three-dimensional Q structure and source parameters using microearthquake spectra. *J. Geophys. Res.* 95: 12423-12438.

Snoke, J.A. 1987. Stable determination of Brune stress drops. *Bull. Seism. Soc. Am.* 77: 530-538.

Urbancic, T.I, Trifu, C-I., and Young, R.P. 1993. Stress release estimates, scaling behavior and source complexities of microseismic events. same issue.

Whitman, D., Isacks, B.L., Chatelain, J-L., Chiu, J-M., and Perez, A. 1992. Attenuation of high-frequency seismic waves beneath the Central Andean Plateau. *J. Geophys. Res.* 97: 19929-19947.

Young, R.P., Coffey, J.R., and Hill, J.J. 1979. The Application of Spectral Analysis to Rock Quality Evaluation for Mapping Purposes. *Bull. Int. Assc. Engng. Geol.* 19: 268-274.

Young, R.P., Maxwell, S.C., Urbancic, T.I., and Feignier, B. 1993. Mining-induced microseismicity: Monitoring and applications of imaging and source mechanism techniques. *Pure and Appl. Geophysics, special issue: Induced Seismicity.* in press.

Zucca, J.J, and Evans, J.R. 1992. Active high-resolution compressional wave attenuation tomography at Newberry Volcano, Central Cascade Range. *J. Geophys. Res.* 97: 11047-11055.

Rockbursts and Seismicity in Mines, Young (ed.) © 1993 Balkema, Rotterdam, ISBN 90 5410 320 5

An automatic data analysis and source location system (ADASLS)

M. Ge & P. Mottahed
Elliot Lake Laboratory, CANMET, Energy, Mines and Resources Canada, Ont., Canada

ABSTRACT: An Automatic Data Analysis and Source Location System (ADASLS) was recently developed at CANMET. It is designed for general application in the mining industry where a rapid, scientific and comprehensive analysis of the automatically determined microseismic event data is required. ADASLS is a manual driven code in that it is controlled by approximately 130 adjustable parameters, which allows the code to adopt the local source locations and to optimize the analysis procedure. The code has been tested extensively and its performance has been excellent. It has been adopted by two major Ontario mines, the Creighton mine at Sudbury and the Kidd Creek mine at Timmins.

1 INTRODUCTION

Over the past 10 years MP250 systems have been widely used by the Canadian mining industry to monitor mining induced seismicity, and have become an essential element in mine design and operation in rockburst prone mines. Currently, this system is used by about 20 mines across Canada. The MP250 system is designed for automatic event detection and source location. An event is defined if there are at least a specified number of arrivals with signal voltages exceeding the threshold level during a prescribed time period. Threshold and time window are the two basic mechanisms used for automatic event recognition. After an event is identified, its location is subsequently calculated.

Unlike other types of seismic data, which are primarily used for research, MP250 data directly serve the mining industry on a daily basis. These data are vital for mine safety as they provide real-time locations of rockbursts and related seismic activities. Very often they become the only means to assess ground conditions. MP250 data are also extensively used for mine design. Therefore, it is of primary importance to improve MP250 location data.

Recently, a computer code, Automatic Data Analysis and Source Location System (ADASLS), was developed at CANMET. The purpose is to provide the mining industry with a reliable tool, with which automatically determined source location data, such as those from MP250 systems, can be analyzed on a scientific basis. The code is designed for general application, especially for those rockburst-prone mines where the microseismic monitoring program has to be carried out on a daily basis.

A source location data analysis code must have two qualifications if it is to be used for general purposes. First, it must be able to solve a wide range of problems which may be encountered in the daily data analysis. Secondly, it can rapidly adopt the local source location conditions and optimize the analysis procedure accordingly. In this paper we are going to address these problems as well as the general approach to use the code. An evaluation of the code performance is also presented.

2 THEORETICAL BACKGROUND

There are three key elements in the analysis of the automatically determined event data. These elements, listed in order of importance, are:

- raw data analysis;
- reliability analysis; and
- source location strategy.

2.1 Raw data analysis

The arrivals picked by MP250 systems are complex in nature. In addition to P-wave picks there are many S-wave and noise picks. The primary object of the raw data analysis is to identify these arrival picks, which is the key for successful use of MP250 data.

Based on traditional theory, it is necessary to have waveforms in order to identify arrival picks. However, MP250 systems do not record waveforms, they only record arrival times. Therefore, the study of raw data with the conventional source location methods becomes impossible and an assumption of P-wave arrival has to be made for all picks. As a result the source location accuracy is poor for most of the events and the data utility rate is low.

The analysis of the raw data without full waveform capture was developed, a part of this approach is given by Ge (1988), Ge and Hardy (1988), and Ge and et al (1990). A full and comprehensive discussion of the related theory is under preparation by CANMET.

The raw data analysis consists of two parts, namely arrival time difference analysis and residual analysis. The arrival time difference analysis deals with the theory of consistency of arrival times. According to the theory, the arrival time differences associated with an event must be consistent with the array size, the array density, the distance between the transducers, and the relative position of the source to the array. The key concept is the theoretical limit of arrival time difference which states: For a given velocity model there exists a theoretical limit of arrival time difference for each pair of transducers. Thus, the channel status can be interpreted in terms of the consistency of the observed arrival time differences relative to the associated limits.

Residual analysis is the theory about channel residual constitution, the distribution of channel residuals, and the interpretation of channel residuals in terms of their associated physical phenomena. This analysis is developed to serve two purposes. First, to cross-check results from the arrival time analysis, and second, to execute several refinement tasks that cannot be carried out by the arrival time analysis. The residual analysis supplements the arrival time analysis and renders the microseismic data more meaningful for source location.

The final product of the raw data analysis is the event based velocity mode which specifies the physical status of each channel and assign P- and S-wave velocities to P- and S-wave channels, respectively. Channels with erroneous triggering will be dropped. In comparison with the traditional P-wave velocity model, the event based velocity model has two distinctive features. It is not a pre-assumed model, but is constructed based on the analysis of the physical status of the arrival picks. Furthermore, this velocity model recognizes the variations of types of arrival picks, therefore it is event oriented. The event based velocity model allows the source location to be carried out on a realistic basis for MP250 data.

2.2 Reliability analysis

The objective of reliability analysis is to provide a rational assessment of raw source location data and source location solutions. The need for this analysis is based on two distinct reasons: (a) The microseismic event data as determined by an automatic event recognition machine are very complicated. Many of them are associated with significant errors that make meaningful solutions almost impossible. The ability to identify this type of event is important for confident use of the location data. (b) The source location accuracy is affected by many factors, especially the array geometry. A small initial error may cause a major location problem with a poor array geometry. It is, therefore, necessary to assess the stability of a solution.

The reliability analysis is carried out in two stages. The first stage is associated with the raw data analysis. At this stage the raw event data are systematically analyzed. The channels with significant errors, or identified as outliers, are dropped from further source location calculations. If an event is so noisy that the number of remaining channels is not sufficient to obtain an analytical solution, it will be dropped. This type of event accounts for at least 20% of the recorded events based on the investigations conducted at several mine sites. These events, in general, are very small, typically with only five channels, and full of outliers. These events would significantly contaminate the source location picture if they were not detected and subsequently dropped.

The second stage is an analysis of the source location solutions. This analysis involves a number of aspects, of which the primary ones are event residual, sensitivity, and hit sequence. In addition, head residual, energy level, and number of channels are also considered for special applications. The final product of the reliability analysis is the event rank which is a rational assessment of raw event data or source location solutions. The event rank ranges from A to E, Table 1 depicts the event rank with the assigned meaning.

Table 1. Event rank.

Rank	Meaning
A(1)	Very reliable
B(2)	Reliable
C(3)	Acceptable
D(4)	Not acceptable
E(5)	Noise - no solution provided

2.3 Source location strategy

The basic source location method used by ADASLS is called the Hybrid method, utilizing both the Simplex and USBM algorithms.

The Simplex source location method was introduced by Prugger and Gendzwill. The details of the method have been given in their two representative papers (Prugger and Gendzwill, 1988; Gendzwill and Prugger, 1989). However, these papers are severely defective in the area of error estimation both mathematically and conceptually. These problems have been corrected in a recent paper by Ge (1991).

The Simplex method is an iterative method, utilizing a curve fitting technique, known as the Simplex algorithm (Caceci and Cacheris, 1984), to search for a source location solution. The method, as the other iterative method, is very flexible in the use of different velocity models as well as optimization methods. The main advantage of the Simplex method over other iterative methods is that it is less vulnerable. The calculation speed of the method is relatively slow.

The USBM method was developed in the early 1970s at the U.S. Bureau of Mines. The development of this source location method is part of the Bureau of Mines' effort to make the acoustic emission/microseismic (AE/MS) technique an effective engineering tool for determining the stability of rock structures. The method was first published in 1970 and was further modified in 1972 (Leighton et al., 1970, 1972). Since then the method has become the major mine oriented AE/MS source location method used in North America.

The USBM method is a non-iterative method; it finds the solution directly. It is easy to use, and the users do not have to worry about choosing the guess solution, setting the convergency criterion, and especially the divergent problem. The optimization procedure, either the least-squares-method or the absolute value method, can be easily incorporated into this source location method. The method offers the fastest solution among several popularly used methods. The main disadvantage of the method is that it can only handle the single velocity model.

In comparing the two methods the Simplex method is efficient in handling the event based velocity model. It almost always provides a better solution when there are a substantial number of S-channels. On the other hand, it has been found from calibration studies (analysis based on rockburst or blast events which location are precisely known) that the USBM method offers compatible or even better accuracy for main events. An important advantage of the USBM method is that it can effectively handle the erroneous channels when the number of such channels is limited, whereas the Simplex method may be very sensitive to them.

The Hybrid method is designed to take the advantage of the two described methods. It determines which solution is more reliable based on both theoretical and empirical considerations. The analysis is carried out in two stages. The first stage analyzes the relevance of the original data to the source location method. As a result an algorithm may not be used, or may receive more weight. The second stage is the comparison of the solutions given by the two methods. It is mainly based on three criteria, namely, the sensitivity, the event residual, and the head residual.

The head residual is a critical concept used in the Hybrid method. It examines how well the observed arrival times of the first several channels have been matched by the calculated arrival times. The emphasis of the residuals associated with the first several channels is based on the fact that arrival time errors are not randomly distributed. In general, the channels triggered earlier should have the smaller errors and this is not difficult to understand from practical reasons. The key here is the distance. The earlier triggered transducers are closer to the source, which will effectively reduce the uncertainties associated with the velocity model. The shorter distance also means a higher energy level and thus the sharper arrival, which reduces the timing errors. The calibration study shows that solutions with smaller head residuals are more reliable.

3 MAIN TECHNICAL AND OPERATIONAL FEATURES

ADASLS offers a broad range of data analysis and source location techniques as well as many convenient features. A complete discussion of these features is beyond the scope of this paper. In this section only those which have the major impact on general use of the code are discussed.

Manual-driven ADASLS is a user-friendly, manual-driven code in that it is controlled by over 130 adjustable parameters. With this feature, users have almost complete control of the data analysis and source location procedure. In particular, users are able to specify the local source location conditions and to adjust the analysis procedure accordingly which is critical for the successful data analysis.

Digital filter One of the main tasks of ADASLS is to identify S-wave channels and outliers. A wide range of techniques have been developed to deal with this difficult problem. In ADASLS all these techniques are written in the form of digital filters to allow them to be conveniently and flexibly used for general purposes. A digital filter has two general characteristics:
a) the function of each technique is precisely defined and can be used independently;
b) the function criteria are in the control of users.

Wide technical choices In addition to many unique features which cannot be found from conventional methods, such as raw data analysis, the code also provides wide choices of the data analysis and source location techniques. In terms of source location one may use any of three methods: Simplex method, USBM method, or Hybrid method. For each method one may use either of two principal optimization approaches, namely the least-squares method, and the absolute value method. Furthermore, there is the option of the velocity back calculation.

Multiple applications ADASLS can be used as either a daily data processor or a research tool. It offers three operational modes, namely continuous analysis mode, detailed analysis mode, and user controlled mode. The continuous analysis mode is used for a quick processing of daily data. It only stores very concise results. The detailed analysis mode provides a comprehensive result of the analysis. This mode is essential for a detailed study of individual events, for the assessment of the performance of transducers, and for the evaluation of the array effect. The user control mode accepts the velocity model provided by users. This mode is mainly used for the study of very complicated events.

Finally, it has to be emphasized that ADASLS was written based on a broad range of first-hand source location experience obtained from the Campbell Red Lake Mine, the North Mine (Inco), the Kidd Creek Mine (Falconbridge), and the Creighton Mine (Inco). These experiences are essential for the detection of real problems, for the development of ideas to solve problems, and for the evaluation of the data analysis and source location techniques. There would be no ADASLS without these experiences.

4 HOW TO USE ADASLS

Although ADASLS is a powerful tool for the analysis of microseismic source location data, it is not a black box. The effectiveness of the code largely depends on how it will be used. There are three key elements which should be emphasized, which are data analysis, array optimization and staff training.

One of the basic principles of ADASLS is to take local conditions into account. This is essential for automatic analysis of MP250 data as the source location environment varies from mine to mine. The principle is implemented by evaluating the microseismic data from the mine site and customizing the code for the mine site. The code customization includes the parameter setting and the possible routine development work to deal with special problems. Therefore, data evaluation as well as code customization is an essential step for the successful application of ADASLS. It would be a disaster if the parameter setting used for the Dense Array at Creighton mine were applied to the Kidd Creek Mine, or even to the Mine-Wide Array at the same mine.

Reliable source location depends on many factors. The most important one is the transducer array geometry which fundamentally determines the achievable source location accuracy. When the array geometry is attributed to poor source location accuracy there is no other cure except the optimization of the array. Therefore, the major array problems should be identified and corrected before the application of ADASLS.

The theory and technique utilized in ADASLS are new as a whole, and it would be difficult for an average mining engineer to comprehend the theory and methodology of ADASLS system without on job training, when the theory and technique could be understood in terms of real data from mine sites.

With these considerations, as well as the experience gained from work conducted to date, it has been concluded that the transfer of ADASLS is a three-step process:
1. Analysis of the characteristics of microseismic source location data, which is an essential step for code customization and staff training.
2. Customization of the computer code, including parameter setting, development of special routines if necessary, calibration study, and the evaluation of the performance of the customized code.
3. Staff training, which will provide engineers with the basic knowledge of the principle, theory and technique used by the code, the necessary skills in the interpretation of the source location data, and the ability to reset some parameters.

We believe that this procedure serves the best interests of industry. This procedure has been followed exactly at Campbell Red Lake Mine, Creighton Mine, and Kidd Creek Mine.

5 CODE PERFORMANCE

ADASLS was tested extensively at two Ontario mine sites. More than thirty sets of calibration data were used in the test. Calibration data here refer to either rockburst or blasting related events, of which the precise locations are known for the main events. The number of events in each data set varies, ranging from one to about fifty. The merit of using the calibration data is that both the source location accuracy and the effect of the data analysis techniques used in the code can be objectively evaluated.

The performance of ADASLS has been excellent. The main advantage of ADASLS over the conventional methods is its robustness in dealing with complicated microseismic events. The ability of the code to discriminate S-wave channels and outliers plays a major role in its success. The code is effecting in handling both large and small events. The seismicity pattern as determined by ADASLS is therefore reliable and meaningful. In comparison, the conventional methods are ineffective for smaller events. Since the majority of the daily recorded events are small, and the large ones constitute only several percent, the conventional methods often generate false patterns of seismicity. Figure 1 shows such an example.

In this figure the locations of a rockburst and 20 aftershock events as determined by ADASLS and a conventional method are compared. The locations of aftershock events calculated by ADASLS are all, except one, in the immediate vicinity of the

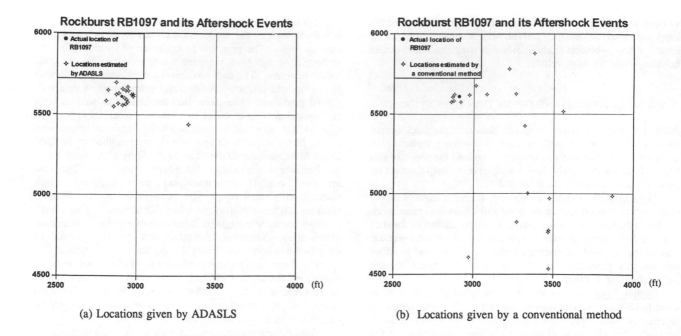

(a) Locations given by ADASLS

(b) Locations given by a conventional method

Figure 1   Locations of RB1097 and its aftershock events as determined by ADASLS and a conventional method

Table 2. A comparison of locations by ADASLS and a conventional method for rockburst RB1097 and its aftershock events

Event No.	Total Channel	Velocity Model Given By ADASLS	ADASLS			Mine		
			X	Y	Z	X	Y	Z
		Actual Location	2907	5615	5252	2907	5615	5252
36	16	PPPPPPPPPPPPPPPP	2886	5635	5293	2879	5624	5307
37	8	PSSSPSSS	2929	5650	5273	3319	5426	5250
38	16	PPPPPPPDPPPDPPPD	2884	5632	5309	2868	5608	5309
39	11	PSSPSPSSPDS	2928	5566	5320	3469	4759	5318
40	13	PSSSPPPSPSDSS	2920	5604	5297	3476	4771	5330
41	13	PPPPSSPPSPSDS	2854	5550	5417	3475	4532	5318
46	16	PPPSPPPPSSPSPPSS	2869	5700	5287	3081	5624	5337
48	11	DPPPSPSPPPP	3325	5441	5439	3270	4822	5629
49	13	PSSSPPPSPSSDS	2945	5649	5299	3264	5630	5428
50	16	PPPPPPPPDPPPPPPP	2819	5652	5399	2875	5584	5376
52	11	PPSSPPSPDSS	2978	5614	5304	3338	5003	5172
53	16	PPPPPPPPPSPSPSPS	2880	5570	5314	2859	5577	5322
54	9	PPSSPSPSD	2943	5672	5279	3379	5880	5027
55	16	PPPPPPPPPPPPPPPP	2975	5626	5278	3015	5679	5236
56	7	PSPSPSS	2965	5625	5283	3222	5784	5288
57	12	PPPSSPSDSPDS	2941	5578	5299	3560	5519	5632
58	7	PPSSPSPDS	2869	5624	5303	3875	4985	5268
59	16	PPPPPPPPPPPPPPPP	2935	5595	5284	2920	5577	5306
60	13	PPPSSPPSPPSSS	2910	5560	5352	2976	4605	5276
61	13	PPPPSSPSPDSSD	2808	5687	5361	3478	4972	5267
62	16	PPPPPPPPPDPPPPPP	2909	5660	5324	2976	5623	5284

346

main event. In contrast those given by the conventional method scatter over a very large area. Thirteen of these are shifted significantly to the eastern direction, while eight have their locations flopped from hanging wall to the footwall. The cause of large errors associated with the conventional method may be seen from Table 2, which contains somewhat more detailed information about rockburst RB1097 and its aftershock events. The numbers in the first column are the sequence numbers of the events, as appeared in the original data file. Number 36 is the main event. The numbers given in the second column are the total channel numbers. In the third column are the velocity models as determined by ADASLS, where P, S and D stand for P-, S- and dropped channel, respectively. The locations determined by ADASLS are given in the next three columns. The mine's solutions are shown in the last three columns. The actual location of the main event is located directly above the main event (No. 36). The solutions with major errors are given shaded background for easy distinction. A very interesting pattern which can be observed from the table is that the poor solutions given by the conventional method are always associated with those events which have a number of S-wave channels. This explains the cause of failure with the conventional method.

6 CONCLUSION

Automatic Data Analysis and Source Location Systems (ADASLS) is a source location code recently developed at CANMET. It is designed for general application in the mining industry where a rapid, scientific and comprehensive analysis of MP250 data is required.

ADASLS offers advanced data analysis and source location technology. The most attractive feature of this code is its ability to identify the type of arrival picks (i.e., P-wave, S-wave and outliers) without waveforms. This feature makes it possible to analyze MP250 data on a scientific basis, and cannot be found from any conventional methods. The source location algorithm used in the code is also unique. It is called the Hybrid method. The Hybrid method analyzes the result of the Simplex and USBM methods and searches for the most reliable one. The other distinctive feature of the code is the reliability analysis. With this capability, the code provides not only the source location, but also the confidence associated with it.

ADASLS is a user-friendly, manual-driven code, controlled by over 130 adjustable parameters. This feature allows users to specify the local source location conditions and to optimize the analysis procedure. This feature, together with the sophisticated data analysis and source location techniques, makes the code very flexible in dealing with a wide range of source location problems.

The code has been tested extensively. It has been adopted by two major Ontario mines, the Creighton Mine at Sudbury and the Kidd Creek Mine at Timmins. The performance of the code has been excellent. The code is planned to be introduced to other Canadian mines where MP250 systems are in use.

7 ACKNOWLEDGEMENT

The authors would like to thank D. Morrison (Inco) and Thiann R. Yu (formerly Falconbridge) for their support of the industrial application of ADASLS. S. Zou and N. Disley from the Kidd Creek Mine, and R. Tan, T. Villeneuve and A. Punkkinen from Inco assisted the calibration study.

REFERENCES

Caseci, M.S. & W.P. Casheris 1984. Fitting curves to data (the Simplex algorithm is the answer). Byte 9/5:340-362.

Ge, M. 1991. Error estimation for the Simplex source location method. Submitted to Bull. Seismol. Soc. Am.

Ge, M. & P.K. Kaiser 1990. Interpretation of physical status of arrival picks for microseismic source location. Bull. Seismol. Soc. Am. 80:1643-1660.

Ge, M. & H.R. Hardy Jr. 1988. The mechanism of array geometry in the control of AE/MS source location accuracy. Proc. 29th U.S. Symp. on Rock Mechanics, Minneapolis, MN, pp. 597-605.

Ge, M. 1988. Optimization of transducer array geometry for acoustic emission/microseismic source location. Ph.D. Thesis, Dept. of Mineral Engineering, Pennsylvania State University, 237p.

Gendzwill, D. & A. Prugger 1989. Algorithms for micro-earthquake location. Proc. 4th Conf. on Acoustic Emission/ Microseismic Activity in Geologic Structures. Trans. Tech. Publ. pp. 601-615.

Leighton, F. & W. Blake 1980. Rock noise source location techniques. U.S. Bureau of Mines Report of Investigation, RI 7432.

Leighton, F. & W.I. Duvall 1972. A least squares method for improving rock noise source location techniques. U.S. Bureau of Mines Report of Investigation, RI 7626.

Prugger, A. & D. Gendzwill 1988. Microearthquake location: a non-linear approach that makes use of a simplex stepping procedure. Bull. Seismol. Soc. Am. 78:799-815.

Rockbursts and Seismicity in Mines, Young (ed.) © 1993 Balkema, Rotterdam, ISBN 90 5410 320 5

Acoustic emission/microseismic observations of laboratory shearing tests on rock joints

C. Li & E. Nordlund
Luleå University of Technology, Sweden

ABSTRACT: Shearing tests were carried out on rock joints to examine the characteristics of acoustic emission/micro-seismic (AE/MS) activities during shearing. Two rock specimens, one with a rough artificial joint and another a smooth natural joint, were sheared in a direct shear test machine. The tests revealed that AE was more actively generated on the rough joint than on the smooth joint. Large AE bursts coincided with the shear stress peak. The b-values of the amplitude, duration and rise time of AE events reached the minimum corresponding to the stress peaks. An increase in the normal stress brought about more AE activities. The roughness of the rock joint, the strength of asperities on the joint and the normal stress imposed on it are crucial to the intensity of potential rockbursts and AE/MS activities.

1 INTRODUCTION

Unstable or stick-slip motion on preexisting rock joints or mine-induced fractures is one of the reasons which causes rockburst and seismicity. One of the methods for monitoring this kind of motion, and some other stability problems of underground rock structures, is the acoustic emission/microseismic (AE/MS) technique (Hardy 1981). AE/MS activities are associated with structural instabilities in underground mines (Hardy and Ersavci 1988). Laboratory shearing tests are expected to provide some hints to explain AE/MS monitoring data and reveal the mechanism of rockburst caused by the sliding motion of rock discontinuities.

This paper introduces the AE characteristics of shearing tests conducted on rock joints. The purpose of these tests was to understand the frictional sliding of pre-existing joints by using the high frequency AE/MS monitoring technique. The AE examinations could supply some fresh information for monitoring rockbursts and seismicity. A Stripa granite specimen with an artificial joint and an Offerdal slate specimen with a natural joint were tested using a direct shearing test machine. AE was recorded during shearing. The AE records demonstrated that the normal stress imposed on the joint surfaces influenced the AE intensity. The distributions of the amplitude, duration and rise time of AE events were obviously changed when the shear stress passed the peak point.

2 SPECIMENS AND TEST FACILITIES

The test results of two specimens will be reported in this paper. The first specimen was a Stripa granite with an artificial joint surface (196x130 mm) which was made by splitting an intact rock core. The joint surface, therefore, was very rough with JRC (Joint Roughness Coefficient) (Barton & Choubey 1977) being 12. Its profile is shown in Figure 1. The second specimen was an Offerdal slate with a natural joint surface (132x122 mm) with JRC=2. Its profile is shown in Figure 2.

The specimens were sheared using a servo-controlled shear test machine the capacity of which was 500 KN (Figure 3). Constant normal stresses were applied to the specimens. The shearing displacement control was used with a shearing speed of 8.25 µm/second.

AE was detected by a AE-transducer attached to the sheared specimen, recorded and analyzed by the 8900 LOCAN AT — a computerized AE instrument. The frequency band of the AE-transducer was in the range of 0.1 to 1 MHz. The total gain of the AE measurement system was chosen to be 60 dB.

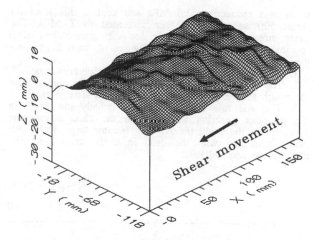

Figure 1. Profile of the artificial joint surface for the Stripa granite specimen.

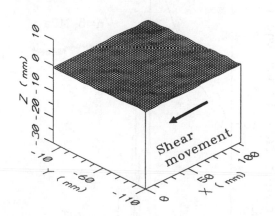

Figure 2. Profile of the natural joint surface for the Offerdal slate specimen.

3 TEST RESULTS AND DISCUSSION

The normal stress applied to the tested specimen was 5 MPa in the beginning of shearing. The shear stress dropped to the residual level after it passed the peak point. During shearing at the residual level, the normal

Figure 3. Servo-controlled direct shear test machine.
1. machine frame
2. vertical actuator
3. shear block
4. horizontal actuator

stress was reduced to 2.5 MPa and kept at this level for about 500 seconds, and then increased to 7 MPa. The shear stress and AE records for the granite and slate specimens are illustrated in Figures 4 and 5, respectively.

The AE records prompted the following observations:
1. The AE count rate was much larger for shearing on the rough granite joint than on the smooth slate joint.
2. The AE rate remained at a relatively low level in the pre-peak shearing stage. When the shear stress was approaching the peak, the AE rate became large. Large AE bursts coincided with the drop in shear stress from the peak to the residual level.

3. The AE rate was reduced when the shearing movement increased at the residual level.
4. A decrease in the normal stress brought about a decrease in the AE rate, while an increase in the normal stress induced more AE activities.
5. When the normal stress was increased from 2.5 to 7 MPa, the subsequent shearing process was similar to that in the pre-peak stage. The AE rate started to increase with an increase in the shear stress until a new stress peak (for the granite specimen) or a new residual stress level (for the slate specimen) was reached. Large AE bursts occurred again at this new peak stress.
6. The shear stress was built up rapidly within a short distance of shearing, both at the beginning of shearing and when the normal stress was increased from a low level to a high level at the residual shearing stage.

The shear stress was clearly oscillating during shearing at the residual level for the granite specimen. The magnitude of the stress oscillation tended to become larger when the normal stress was increased. The joint surface of the granite specimen after being sheared (Figure 6) showed that it was not sheared all over the surface but only at a few contact points. The slate specimen, however, was sheared uniformly all over the asperities on the joint surface. The stress oscillation was closely associated with the shearing process. If the shearing occurs at only a few asperities, stress oscillation would occur. Otherwise, the shearing would be smooth. The oscillation occurs not only on rough artificial joints but also on natural joints (Li and Nordlund, 1990) which have strong asperities and a relatively rough joint surface. The shear stress oscillation depends on both the roughness of the joint surface and the strength of the asperities on the surface.

A new shear stress peak occurred when the normal stress was increased during the residual shearing process for the granite specimen. For the slate specimen, however, the shear stress directly entered a new residual level of shearing. It seems that whether a new

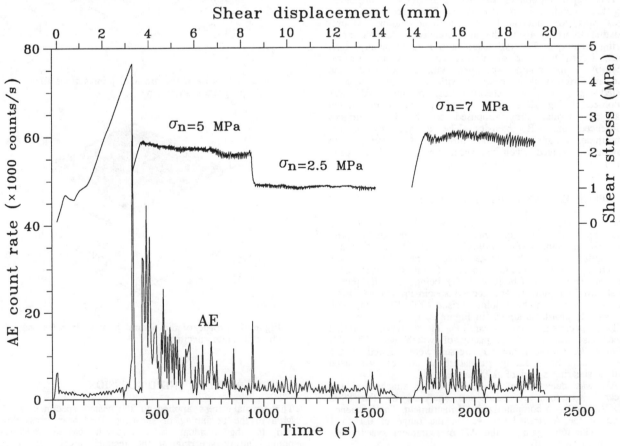

Figure 4. Shear stress curve and the corresponding AE records for the granite specimen.

350

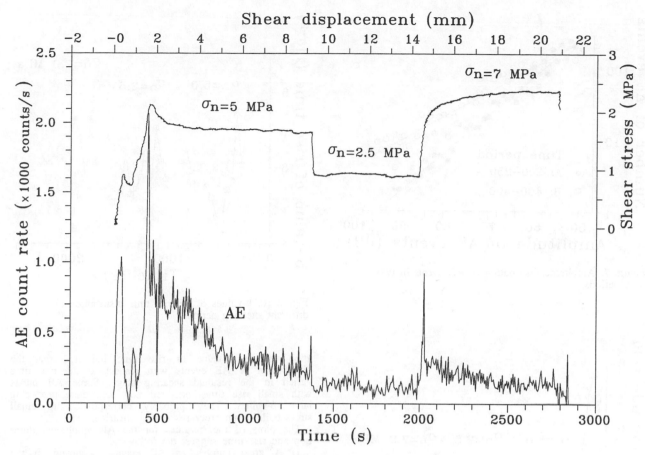

Figure 5. Shear stress curve and the corresponding AE records for the slate specimen.

shear stress peak occurs or not after an increase in the normal stress depends on the properties of joint surface. It is inferred from the test results that an increase in the normal stress during the residual shearing would result in a new shear stress peak, if the joint surface is rough enough and the joint asperities are strong. Otherwise, the shear stress would directly enter the new residual shearing without occurrence of a shear stress peak.

AE activities can be characterized by a number of measured parameters, such as the AE count rate, and the amplitude, duration and rise time of AE events. Taking the AE amplitude as an example, we examined the variation of its distribution in different shearing stages. Figure 7 shows the amplitude distributions of AE events in two shearing stages for the granite specimen. Referring to Figure 4, we can see that the shearing between 200 and 250 seconds (period A) was in the pre-peak stage, while the shearing between 400 and 450 seconds

Figure 6. Joint surface of the granite specimen after sheared. The white gouge on the top block locates the sheared asperities.

(period B) was just after the shear stress peak and at the residual shearing level. The amplitude distribution can be expressed by a regression equation LOG(N)=a-b×(AM), where N is the number of AE events and AM the AE amplitude in dB. The slope of the regression line, the so-called b-value, is a quantitative measure of the amplitude distribution. A small b-value implies that a large number of AE events have high amplitudes. It is seen from Figure 7 that more AE events with high amplitudes occurred in period B than in period A. Consequently, the b-value of AE amplitude in period B was smaller than in period A. In addition, a large number of AE events with high amplitudes between 95 and 100 dB occurred in period B, but not in period A.

The b-values of AE amplitude calculated over the whole shearing process are plotted in Figure 8. For the normal stress of 5 MPa, the b-value was high in the pre-peak shearing stage. It dropped to its lowest level when the shear stress peak was reached. It was then increased gradually when the shearing entered the residual stage. When the normal stress was reduced from 5 MPa to 2.5 MPa, the shearing continued at a relatively low residual shear stress level, and the b-value of AE amplitude remained relatively unchanged. When the normal stress was increased from 2.5 to 7 MPa, the shear stress was built up rapidly (Figure 4), and the corresponding b-value of AE amplitude during this period was relatively high. The b-value dropped to a low level when the new shear stress peak was reached, and then it was increased to a relatively high level during the period of a new level of residual shearing.

The b-values of AE duration and rise time varied in a similar way to the b-value of the AE amplitude, especially during the shearing periods under the normal stresses of 5 and 7 MPa, see Figures 9 and 10. The b-value of rise time under the normal stress of 2.5 MPa oscillated very much. The shearing under this normal stress was completely in the residual stage. The oscill-

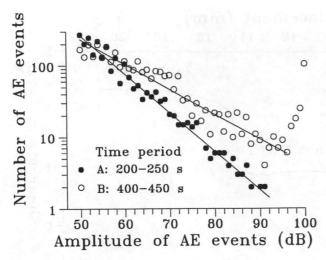

Figure 7. Amplitude distribution of AE events in two time periods.

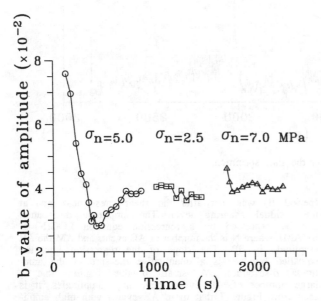

Figure 8. b-values of the amplitude distribution in different shearing periods.

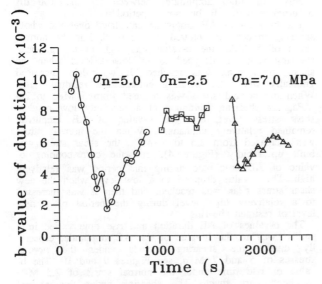

Figure 9. b-values of the duration distribution in different shearing periods.

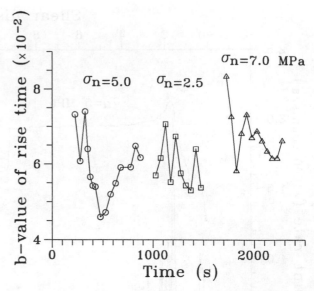

Figure 10. b-values of the rise time distribution in different shearing periods.

ation of the b-value of rise time indicated that the distribution of AE events with respect to the rise time varied in the residual shearing stage. Some AE bursts with small rise times occurred now and then, producing large b-values. They probably corresponded to small stress building-up processes during shearing.

The plots of the b-values for the AE amplitude, duration and rise time suggest the following:

1. A great number of AE events, occurring in the period when the shear stress was built up, had small amplitudes, short durations and short rise times.

2. The number of AE events with high amplitudes, long durations and long rise times dramatically increased when the shear stress was approaching the peak point.

3. At the residual shearing stage, the number of AE events with high amplitudes, long durations and long rise times obviously decreased.

4. A large number of AE events which had very high amplitudes (95-100 dB) occurred now and then due to the so-called stick-slips, especially for the shearing on a rough joint surface, see Figure 7.

Based on the laboratory observations, we infer that if a rockburst could potentially be caused by sliding on a rock discontinuity, the strength of the rock, the roughness of the discontinuity and the normal stress imposed on it are crucial. Not only the first shear stress peak could induce a big rockburst, but an increase in the normal stress during the shearing period could also generate a rockburst. All of these processes could be monitored by a proper AE/MS monitoring system and be understood by examining variations in the amplitude, duration and rise time of AE events.

4 CONCLUSIONS

1. The drop of shear stress after the peak point was larger and more AE events were generated on the rough granite joint than on the smooth slate joint.

2. Large AE bursts coincided with the shear stress peaks. An increase in the normal stress brought about more AE activities during the shearing movement.

3. The b-values of the amplitude, duration and rise time of AE events were high at the stage when the shear stress was built up. They dropped to their lowest levels at shear stress peaks.

4. A large number of AE events occurring during the residual shearing had relatively small amplitudes, but a certain number of AE events which had quite high amplitudes (95-100 dB) occurred now and then due to the shearing stick-slips.

5. The roughness of a preexisting rock discontinuity, the strength of asperities on the discontinuity and the normal stress imposed on it are crucial to the intensity of AE/MS bursts and potential rockbursts.

ACKNOWLEDGEMENT

This work was supported by the Swedish National Science Research Council (NFR) and the Swedish Research Council for Engineering Sciences (TFR) The authors would like to thank Josef Forslund, Ulf Mattila, Kjell Havnesköld and Mats Holmberg for their assistance during the laboratory tests.

REFERENCES

Barton, N.R. & V.Choubey 1977. The shear strength of rock joints in theory and practice. *Rock Mech.* **10**:1-54.

Hardy, H.R. Jr 1981. Applications of acoustic emission techniques to rock and rock structures: A state-of-the-art review. *Acoustic Emissions in Geotechnical Engineering Practice,* ASTM STP 750, 4-92. Baltimore.

Hardy, H.R. Jr & M.N. Ersavci 1988. High frequency acoustic emission/microseismic studies associated with structural instabilities in underground mines. *Proc. 2nd Int. Symp. of Rockburst and Seismicity in Mines*: 199-204. Rotterdam: Balkema.

Li, C. & E.Nordlund 1990. Characteristics of acoustic emissions during shearing of rock joints. *Rock Joints — Proc. of Int. Symp. on Rock Joints*: 251-258. Rotterdam: Balkema.

Rockbursts and Seismicity in Mines, Young (ed.) © 1993 Balkema, Rotterdam, ISBN 90 5410 320 5

The use of numerical modelling in rockburst control

N. Lightfoot
Chamber of Mines Research Organisation of South Africa, Johannesburg, South Africa

ABSTRACT: Both linear elastic and non-linear inelastic codes are being used at COMRO to investigate aspects related to rockburst control. Two methods of rockburst control, fluid injection and preconditioning, are being investigated. Some of the questions that have arisen during the course of this research have been addressed by means of numerical, as well as analytical, modelling. Case studies relating to applications centred on three rockburst control field experiments are outlined. A number of insights into the applicability of the codes themselves and the application of rockburst control techniques have come to light as a result of the modelling exercises.

1 ROCKBURST CONTROL ON DEEP LEVEL GOLD MINES

COMRO is currently undertaking research into two methods of pro-active rockburst control with a view to future implementation on seismically active mines. Each method addresses a different type of rockburst mechanism.

The first method, known as controlled fault slip, is an attempt to use fluid injection to cause incremental slip on faults at predetermined positions and times. By preventing the build up of large strains on fault surfaces it may be possible to prevent the large single occurrences of slip on faults that result in large, damaging rockbursts.

The second method, known as preconditioning, uses high gas pressure explosives set off in the fracture zone ahead of a deep level stope face to reduce the strain energy stored in the fracture zone (Adams, et al, 1993). Preconditioning attempts to reduce rockbursts associated with violent failure of the rock mass in the immediate vicinity of the stope face.

Mathematical modelling, both numerical and analytical, has been used extensively to address issues that have arisen during the course of the large scale field experiments that are being conducted in this area of research.

2 THE NUMERICAL MODELLING PROGRAMS

Stress analysis programs designed specifically for rock engineering purposes are available in many shapes and forms. The work described here concentrates on two types of programs, linear elastic and discontinuum, but two finite element plasticity programs are also mentioned briefly. The choice of a specific program is dependent upon the models that are built when addressing specific problems. Indeed, it is not unusual to use a number of programs to address different aspects of individual problems. The six programs that are mentioned in the bulk of this paper are briefly described here.

2.1 Linear elastic programs

The two dimensional displacement discontinuity program MINAP (Crouch, 1976) has a number of special purpose constant strength boundary elements that include stope and Mohr-Coulomb fault behaviour. MINSIM-D (Napier and Stephensen, 1987) is a three-dimensional displacement discontinuity program that is well suited to modelling narrow, tabular ore bodies.

2.2 Discontinuum programs

Much of the work that is described in this paper places a large emphasis on the discontinuous nature of the rock mass that surrounds a deep level gold mining stope to a distance of some metres. For this reason the Universal Distinct Element Code UDEC developed by

Cundall (Cundall, et al, 1978, De Lemos, 1987) and marketed by Itasca Consulting Group (Itasca, 1992) has proved invaluable. UDEC is a hybrid code that allows the representation of complex blocky rock assemblages. Individual rock blocks can be discretised into multiple finite difference zones to allow internal deformations. The program uses an explicit time marching algorithm based on dynamic relaxation that enables the simulation of both quasi-static and fully dynamic behaviour.

3DEC (Cundall, 1988, and Hart, et al, 1988) is a three dimensional version of UDEC that incorporates many of the facilities of the two dimensional code.

2.3 Non-linear inelastic continuum programs

Generally, finite element programs have been avoided because of the difficulty in setting up simulations for such programs. However, two finite element programs have been used when it has proven difficult to solve problems in any other way. The two programs are LUSAS (FEA, 1990) and the hybrid boundary element-finite element program BEFE (Beer, 1990).

3 CONTROLLED FAULT SLIP BY FLUID INJECTION

The controlled fault slip project involves pumping high pressure fluids (up to 30 MPa) onto fault surfaces by means of wellbores that are drilled to intersect the fault surface in the vicinity of concurrent mining. Both analytical and numerical modelling have been used to help identify the optimum position or positions at which the wellbores should intersect the fault surface and to predict the fluid pressures and volumes that should be used.

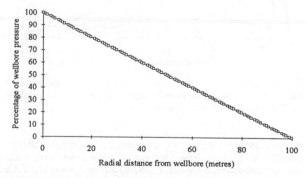

Figure 1. The UDEC solution for the pressure distribution in a parallel plate fracture around a single wellbore.

The fluid flow logic in the program UDEC (Lemos and Lorig, 1990) was used to investigate the amount of slip that would result from pumping fluids at various pressures over different areas of a fault surface. A typical wellbore-fault fluid pressurisation profile is shown in figure 1. The UDEC fluid flow logic produces a linear decay from the well pressure to the boundary at which the fluid is allowed to escape from the system (the fluid boundary condition is zero pressure). This linear decay is an artifact produced by the two dimensional nature of the code. The wellbore is seen, implicitly, as an infinite slot running perpendicular to the plane of the UDEC mesh, rather than as a single point source of fluid injection. This results in over-estimated fluid pressures in the fault or joint away from the wellbore-fault intersection.

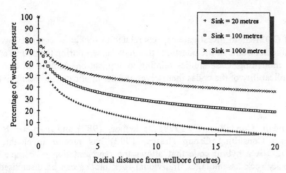

Figure 2. The analytic solution using the bounded parallel plate model for the pressure distribution around a single wellbore.

The analytic solution for the pressure field around a wellbore intersecting an infinite parallel plate fracture (Stratford, et al, 1990) has been used to develop a bounded parallel plate fluid injection model that describes the steady state pressure around the wellbore when boundaries of zero fluid pressure are prescribed at given radial distances from the wellbore. The pressure field around a single wellbore for different boundary distances is shown in figure 2. For a zero pressure boundary prescribed at twenty metres from the wellbore the pressure drops to below twenty five percent of the wellbore pressure within a five metre radial distance. Single wellbores cannot achieve high pressures over large enough areas of a fault surface to affect the stability of the fault. This has resulted in considerable attention being given to practical procedures for pumping onto multiple wellbores in order to obtain high pressures over reasonably large areas of the fault surface. Figure 3 illustrates the pressure distribution on a straight line between two and three wellbores. The pressure will drop off rapidly away from this line of maximum value just as it drops off away from individual wellbores. This hints at the truly three dimensional nature of the fluid injection problem.

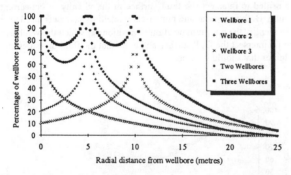

Figure 3. The analytic solution using superposition of the bounded parallel plate model for the pressure distribution around two and three wellbores spaced at 5 metre distances.

Work is currently under way to incorporate the bounded parallel plate model as an effective stress constraint in a two dimensional, static, displacement discontinuity code (MINAP) that incorporates fault slip capabilities. This will be used to investigate the optimum injection points for intersecting fault-stope scenarios. It will also be used to investigate the coupled behaviour of multiple fault systems: what

happens to neighbouring faults when fault slip is initiated on one fault by fluid injection?

Work in the field of joint surface and fluid percolation modelling (Pyrak-Nolte, et al, 1988 and Stratford, et al, 1990) has identified the strong dependence of inter-fracture fluid flow on the surface asperity-aperture distribution. The joint surface roughness and void distribution must be accounted for in fluid injection modelling. This can only be achieved by considering the fault surface as a two dimensional plane rather than a one dimensional slit. It is proposed that the bounded parallel plate model be incorporated in a three dimensional displacement discontinuity code to achieve this.

4 STRESS DISTRIBUTION AHEAD OF SQUARE TUNNELS

Preconditioning of a tunnel at approximately 2500 metres depth is being undertaken to determine the types of explosives that may be most suitable for preconditioning stopes in rockburst prone areas. The tunnel is a 4 metre square development end. To get the most benefit of the space available a 4 metre long preconditioning hole is drilled in the bottom corner of the face and instrumentation holes are then drilled on a pattern that fans away from this corner. It is assumed that the hole runs along the centre line of an axi-symmetric stress distribution and that the tunnel has little effect on the stress field except in the immediate vicinity of the face. Numerical modelling is being undertaken to test the validity of this assumption and the possible influence of the tunnel on the experimental results. Furthermore, during the experiment it was observed that fracturing produced around the tunnel face was preferentially oriented perpendicular to the direction of bedding, suggesting that this direction corresponded to that of the maximum principal stress which was reoriented from the vertical as a result of slip on the weak bedding planes.

4.1 The influence of bedding on stress orientation

Because the problem involves multiple discontinuities in the form of bedding planes, UDEC was used to investigate the existence of a mechanism for the development of principal stress directions being oriented perpendicular to bedding. The strike of the bedding at the preconditioning site is parallel to the tunnel and the dip is approximately 20 degrees through the tunnel. The vertical stress is known to be about twice the horizontal (i.e. the k ratio or the ratio of horizontal to vertical stress is 0.5). Under these conditions the bedding planes can only slip if they have a very low angle of friction. Zero tensile and cohesive strengths were assumed based on field observations of bedding plane discontinuities. To allow bedding plane slip a coefficient of friction of 0.05 was used, this equates to an angle of friction of around 6 degrees. No attempt is made here to explain the physical existence of such a low angle of friction, it is simply a way of assuring that the bedding in the model will slip.

The stress tensor components after the simulation are shown in figure 4. It is clear that an applied vertical and horizontal principal stress direction results in a tilted stress tensor after slip is allowed on the joints and as was predicted, the major principal stress is reoriented towards a direction that is perpendicular to the dip of the bedding. As can be anticipated from the applied stress field all joint slip is sinistral.

4.2 The stress distribution ahead of a square tunnel

The state of stress ahead of a finitely dimensioned tunnel is a three dimensional problem. In reality it involves non-linear inelastic behaviour in the immediate vicinity of the tunnel skin. As the non-linear behaviour is probably confined to a small zone around the tunnel it is ignored in this analysis and the rock mass is approximated as wholly elastic. The program 3DEC was used, initially, due to its availability and (for a three dimensional code) simplicity of use. A single square tunnel was modelled in a large rectangular block. Cross-sections were taken ahead of the tunnel face at an angle perpendicular to the tunnel axis to investigate the magnitude of the major principal stress in this region. High stress concentrations were observed to develop immediately ahead of the face in the area just above the tunnel footwall and just below the hangingwall (figure 5).

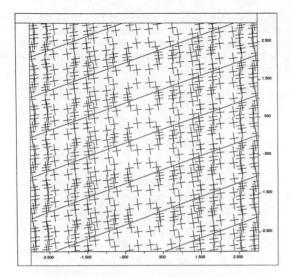

Figure 4. The stress tensor orientation after low angled bedding has been allowed to slip under the influence of a vertically oriented stress field with an initial k ratio of 0.5. The final k ratio is 0.76.

These stress concentrations are a function of the square shape of the tunnel and its orientation within the virgin stress tensor.

The identification of these areas of high stress concentration reflect field observations on a deep level gold mine. Strain bursting occurred in a tunnel that was being developed using longer than normal rounds (Leach, 1992) at a point in the face that corresponds to the hangingwall stress concentration. Such strain bursting is not observed in the majority of tunnels that are developed under the same conditions but where the rounds are shorter. It is postulated here that the strain bursts were a result of the combination of the elastic stress concentrations and the long rounds.

Under normal circumstances the tunnel face remains in the naturally developed fracture zone. High stress concentrations cannot develop in this fracture zone and the face is stable. However, when the tunnel advance is large enough with a single blast, the new face is formed in unfractured rock which is capable of supporting the high stresses. Immediately after the blast the face at the points of high stress concentration is loaded to the point of near fracture. Any disturbance to the face will push it beyond the point of equilibrium and it can fail violently as a strain burst.

The practical implications of this hypothesis could be that the judicious use of preconditioning under such circumstances should alleviate the possibility of face bursting associated with rapid tunnel

advance at great depth. The hypothesis has not yet been tested in practice.

The 3DEC modelling is inherently flawed as the solution is non-symmetric for a symmetric, elastic problem. This is a result of the tetrahedral finite difference mesh that is used in 3DEC. It is not a problem with the program but a problem resulting from the use of the program for solving a problem for which it is ill suited. More in-depth analysis in this area is being pursued using both the finite element program LUSAS and BEFE. The effects of the reorientation of the stress tensor and variations in the ratios of the principal stress components are to be investigated.

5 PRECONDITIONING OF STOPES

It has been recognised that the most suitable orientation for the large diameter blast holes that are used for preconditioning deep level gold mine stope faces is parallel to the face. However, questions have arisen as to where to place the blast holes. Should they be on the plane of the reef or some distance into the footwall or the hangingwall? Should they be placed several metres ahead of the face so that they are in solid rock which will be fractured by the blast thus generating new slip surfaces to release inherent strain energy? Or, should the holes be drilled close to the stope face in the zone of natural fracturing where gas from the blast can cause pre-existing fractures to slip and release any stored strain energy?

Numerical modelling has been used to investigate the behaviour of the fractured rock ahead of the stope face. Blasts have been simulated to determine the types of mechanism that might be induced to precondition the rock mass. As an understanding of the behaviour of fractured rock under high stress and the effect of blasting has developed the question of positioning the holes to achieve the maximum benefit is being addressed.

5.1 The stope model and boundary conditions

The UDEC simulation uses a mesh that represents the rock around the stope as a 12 metre wide by 16 metre high block assembly that is, initially, composed of eight individual blocks (figure 6). The reef horizon is represented as a 2 metre high zone that cuts through the middle of the block. Horizontal joints exist symmetrically about the reef horizon, the first is 1 metre vertically from the reef contact and the second is 2 metres from the first. A vertical joint cuts through the reef horizon two metres from the right hand side of the block. This facilitates the removal of a 2 metre high by 2 metre wide block of rock to crudely simulate mining at the stope face.

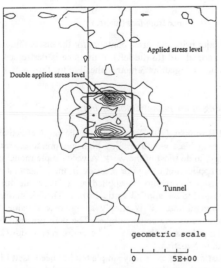

geometric scale

0   5E+00

Figure 5. Contours of the stress distribution ahead of a square tunnel.

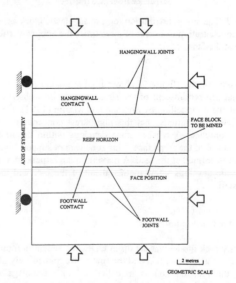

Figure 6. The basic UDEC block assembly and the applied boundary conditions. The arrows represent stress boundaries where the vertical stress is twice the horizontal stress and there is no shear component of stress.

The blocks are internally discretised into finite difference zones to allow deformability and the generation of stresses internal to individual blocks. The constitutive behaviour applied to each block is simply that of linear elasticity with elastic constants that reflect a typical Witwatersrand Basin quartzite (a Young's modulus of 70GPa and a Poisson's ratio of 0.2). The interaction between the blocks is governed by a Mohr-Coulomb constitutive law that can be described in terms of a friction angle only. For the sake of simplicity the joint surfaces have no tensile or cohesive strength.

The left hand side of the block is fixed in the horizontal direction to act as an axis of symmetry. The other vertical boundary and the two horizontal boundaries are prescribed forces in terms of applied stresses. The vertical stress is twice the horizontal stress and no shear component is prescribed on the boundary. Stresses are applied internally in the block prior to any mining to enable rapid convergence of the initial mesh.

5.2 An elastic control model

The block assembly was used in its initial form to generate a quasi-elastic solution for the stope mining problem that could be used as a reference for the discontinuum modelling. A 2 by 2 metre right hand block was removed to simulate mining and the mesh was cycled to a new state of equilibrium.

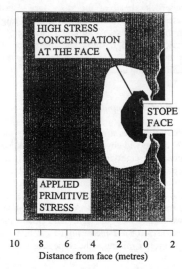

Figure 7. The stress distribution for a quasi-elastic block assembly. A stress concentration in excess of three times the applied vertical stress occurs at the face.

The elastic model (figure 7) shows high stress concentrations at the face that are reminiscent of the stress singularities observed during finite element modelling by Snyman and Martin (1992). It is known, from field observations, that this high stress concentration does not exist in practice. A crushed zone develops within the first 2 to 3 metres ahead of the stope face. This zone cannot support high stresses as it is unconfined at the face. A maximum stress peak of some finite magnitude develops a few metres ahead of the face and the face is destressed.

5.3 The discontinuum model

A blocky rock model was set up in UDEC as shown in figure 8. The two metre thickness of reef and the first metre immediately above and below the reef horizon is modelled as an assemblage of small triangular blocks. Each block is discretised to a maximum of two finite difference zones using the automatic mesh generator in UDEC. The discretisation was not done to provide deformability of individual blocks but, rather, to enable the code to report the stress state in individual blocks.

The discontinuum model develops a zone of load shedding in the failed material at the face. High stresses build up some three metres

behind the face but these do not exceed four times the applied primitive stress and are not sufficient to fail intact rock. The discontinuum model exhibits a stress distribution that is in accordance with underground observations. No stress singularity develops at the face (figure 9).

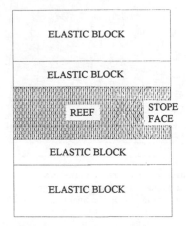

Figure 8. The UDEC discontinuum model with the reef plane and the area 1 metre above and below represented by an assembly of triangular blocks with 60°, 90° and 120° jointing.

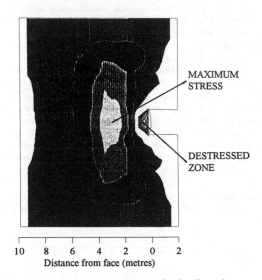

Figure 9. Major principal stress contours for the discontinuum model. A 2 metre zone ahead (to the left) of the face is destressed with the peak stress occurring about 3 metres ahead of the face.

5.4 Modelling of the preconditioning blast

Rorke and Brummer (1990) suggest that the most effective type of preconditioning blast would employ high gas pressures, rather than shock energy, to do most of the work. A very simple model of such a blast is the application of a pore pressure in the joints that are to be blasted. This pore pressure is applied as a Heaviside function of magnitude equal to the applied vertical stress. There is nothing magic about this magnitude, it is simply a convenient value to apply without recourse to any better number. The pore pressure is applied over a rectangular area that covers the entire reef width (2 metres) and a length of 5 metres.

The blast model is extremely simple but has been useful for initial comparative analyses. Improvements to this model would incorporate the transient nature of an explosive blast - possibly a bilinear ramp function with both a climb and decay based upon field measurements of blasting. The area of the blocky system over which the pore pressure is applied should be elliptical with a higher applied pressure at the centre (blast-hole) and a step-wise decay towards some reasonably spaced perimeter. The magnitudes of the applied pressure should be selected to have some bearing on reality.

5.5 Computer simulations of preconditioning

From the steady state situation shown in figure 9, the pore pressure model of preconditioning was applied to the reef horizon in a zone from 5 metres to 10 metres ahead of the face. The block assembly was then brought to a new state of equilibrium. The new stress state is shown in figure 10. The application of the simulated preconditioning blast so far ahead of the face had no significant effect on the state of stress in the block assembly

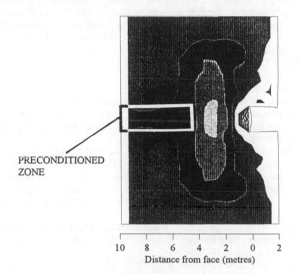

PRECONDITIONED
ZONE

Distance from face (metres)

Figure 10. The major principal stress distribution after preconditioning in the confined rock ahead of the face.

The same process was repeated from the same initial conditions but this time the pore pressure was applied in the zone from the face to 5 metres ahead. The resultant stress state is shown in figure 11. The new stress state is significantly altered from that prior to the blast. The area immediately ahead of the face has been significantly destressed and the peak stress has been moved to a position that is some 6 to 8 metres ahead of the face.

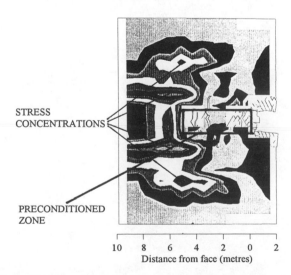

STRESS
CONCENTRATIONS

PRECONDITIONED
ZONE

Distance from face (metres)

Figure 11. The major principal stress distribution contours after preconditioning in the unconfined rock up to 5 metres ahead of the face.

This initial analysis suggests that the application of fluid (blast) pressures too far ahead of the face has no significant effect as the rock mass is too highly confined to enable it to mobilise and destress. Blasting should be located in areas of the rock that are relatively free to move, allowing the dissipation of stored strain by slip along pre-existing fractures.

The state of stress after preconditioning also exhibits zones of high stress concentrations at the contacts between the discontinuous rock mass and the elastic rock mass 1 metre above and below the reef horizon. Further stress concentrations occur at the horizontal fractures cutting through the elastic rock. It was postulated that the stress concentrations 1 metre above and below the reef are associated with the transition from discontinuous to elastic rock and would not occur if the transition zone was more gradual or further from the reef horizon. To test this a second set of simulations was performed with the discontinuous rock mass extending to 3 metres above and below the reef horizon (figure 12). All other conditions were identical to the previous simulations. The principal stress state prior to preconditioning was similar to that observed in the previous simulations.

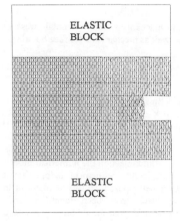

ELASTIC
BLOCK

ELASTIC
BLOCK

Figure 12. The second discontinuum block assembly with the discontinuous rock extending to 3 metres above and below the reef horizon.

Again, preconditioning in the zone 5 to 10 metres ahead of the stope face had a very limited effect. Preconditioning in the zone up to 5 metres ahead of the face produced distinct changes to the stress field (figure 13). In this case the zone of maximum stress is centred at about 7 metres ahead of the face. The stress concentrations observed 1 metre above and below the reef horizon in the previous simulation are not seen at all indicating that they were a result of the elastic to discontinuous transition.

There is a much stronger break in symmetry about the horizontal centre line of the reef than was observed in the previous simulation. This is a result of the increase in the number of degrees of freedom available to the discontinuous rock. The more of the rock mass that is represented as a discontinuum the more degrees of freedom there are available to the system as a whole. The more degrees of freedom in the system the more it is able to behave in an unconstrained manner.

Two points of high stress exist about 2 metres ahead of the face in line with the footwall and hangingwall reef contacts. These are caused by lock up on blocks at these points resulting in high elastic strain in the individual blocks. Such stress concentrations may well be common ahead of stope faces and, under extreme circumstances, could represent locally unstable points that could form the initiation points for face bursts.

This type of discontinuum modelling is being continued. In particular, large blocky rock assemblages are being used to test some of the assumptions (such as the boundary conditions) that were made in the preconditioning modelling. Furthermore similar assemblages are being used to investigate fundamental rock mass behaviour to

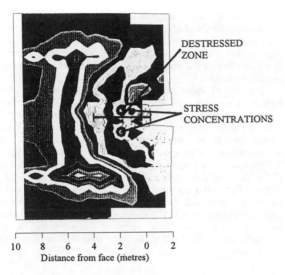

DESTRESSED ZONE

STRESS CONCENTRATIONS

| 10 | 8 | 6 | 4 | 2 | 0 | 2 |

Distance from face (metres)

Figure 13. The major principal stress distribution contours after preconditioning in the unconfined rock up to 5 metres ahead of the face in the second discontinuum model.

help understand mechanisms for horizontal stress generation in the hangingwall as well as mechanisms for face bursting.

6 CONCLUSIONS

Numerical modelling can provide answers to practical problems in many fields of rock engineering including rockburst control provided a rational approach to problem solving is adopted. The modelling is not necessarily quantitative but, more usually, provides qualitative answers, often on a comparative basis.

In the field of fluid injection the numerical models that have been used require considerable refinement before they can answer the questions that are being posed. However, many of the limitations of the current approach have been identified and work is now progressing in a direction that should provide the necessary answers.

Three dimensional modelling of the stress state ahead of a square tunnel has identified a non-intuitive stress distribution in the immediate face. This stress distribution may lead to face bursting in tunnels under exceptional circumstances but further numerical analysis may result in approaches to preconditioning that could ameliorate this problem.

Distinct element modelling of the rock mass around deep level stopes exhibits behaviour that is not captured by more traditional elastic or plastic continuum approaches. The discontinuum modelling has shown that a mechanism that involves that mobilisation of existing fractures by the injection of high gas pressures from a blast hole situated in the fractured rock close to the stope face is most likely to provide optimum preconditioning results.

Acknowledgments

This work forms part of the Rockburst Control research programme of the Rock Engineering Division of COMRO.

REFERENCES

Adams D.J., Gay N.C. and Cross M. 1993. Preconditioning - a technique for controlling rockbursts. *Proceedings of the 3rd International Symposium on Rockbursts and Seismicity in Mines*, Balkema. Rotterdam.

Beer G. 1990. Application of numerical modelling to the analysis of excavations in jointed rock. *Mechanics of Jointed and Faulted Rock*, Rossmanith (ed.) Balkema, Rotterdam.

Crouch S.L. 1976. Analysis of stresses and displacements around underground excavations: an application of the displacement discontinuity method. *University of Minnesota geomechanics report MPLS Mn 55455.*

Cundall P.A. 1988. Formulation of a three-dimensional distinct element model - Part I. A scheme to detect and represent contacts in a system composed of many polyhedral blocks. *Int. J. Rock mech. Min. Sci. & Geomech. Abstr.* Vol. 25, No. 3, pp 117-125.

Cundall P.A., Marti J., Beresford P. and Asgian M. 1978. Computer modelling of jointed rock masses. *Technical report N-78-4 U.S. Army engineers Waterways experiment Station.*

De Lemos J.V. 1987. A distinct element model for dynamic analysis of jointed rock with application to dam foundations and fault motion. *Ph.D. Thesis*, University of Minnesota.

FEA Finite Element Analysis Ltd. 1990. LUSAS User Manual.

Hart R., Cundall P.A. and Lemos J. 1988. Formulation of a three-dimensional distinct element model - Part II. Mechanical calculations for the motion and interaction of a system composed of many polyhedral blocks. *Int. J. Rock mech. Min. Sci. & Geomech. Abstr.* Vol. 25, No. 3, pp 107-116.

Itasca Consulting Group, 1992. UDEC Universal Distinct Element Code: Version 1.8. User's Manual.

Leach A.R. 1992. Pers Comms.

Lemos J.V. and Lorig L.J. 1990. Hydromechanical modelling of jointed rock masses using the Distinct Element Method. *Mechanics of Jointed and Faulted Rock*, Rossmanith (ed.), Balkema, Rotterdam.

Napier J.A.L. and Stephensen S.J. 1987. Analysis of deep-level mine design problems using the MINSIM-D boundary element program. *APCOM 87. Proceedings of the Twentieth International Symposium on the Application of Computers and Mathematics in the Mineral Industries. Volume 1: Mining.* Johannesburg, SAIMM.

Pyrak-Nolte L.J., Cook G.W. and Nolte D.D. 1088. Fluid percolation through single fractures. *Geophysical Research Letters,* Vol. 15 No. 11, pp 1247-1250.

Rorke A.J. and Brummer R.K. 1990. The use of explosives in rockburst control techniques. *Rockbursts and seismicity in mines,* Fairhurst (ed.), Balkema, Rotterdam.

Snyman M.F. and Martin J.B. 1992. The influence of horizontal and vertical discontinuities near a deep tabular excavation in rock. *COMRO reference report No. 4/92, Project GR2A.*

Stratford R.G., Herbert A.W. and Jackson C.P. 1990. A parameter study of the influence of aoerture variation on fracture flow and the consequences in a fracture network. *Rock Joints,* Barton & Stephansson (eds), Balkema, Rotterdam.

Rockbursts and Seismicity in Mines, Young (ed.) © 1993 Balkema, Rotterdam, ISBN 90 5410 320 5

Applications of numerical methods in rockburst prediction and control

Lu Jiayou, Wang Bing & Wang Changming
Institute of Water Conservancy and Hydroelectric Power Research, Beijing, People's Republic of China

Zhang Jinsheng & Wang Zhenliang
China An Neng Construction Corporation, Beijing, People's Republic of China

ABSTRACT: Numerical methods have been widely used to analyse the stresses in the surrounding rock of underground caverns. However, their use in the analysis of rockburst in still seldom seen, mainly because of the complex mechanism of rockburst and the difficulties in constructing a mathematical model for it. In this paper, the results of some researches in connection with rockbursts in the Tianshengqiao Hydropower Station underground project will be presented. Based upon the characteristics of brittle fracture of rocks and rockburst phenomena in the Tianshengaiao tunnel and through the verification by physical modelling and rockburst analyses by theorise of strength and failure instability, a mathematical model has been established and prediction and controll of rockburst have been studied by means of numerical simulation.

1. GENERAL DESCRIPTION

Rockburst is a kind of dynamic instability caused by the suddn release of strain energy during the brittle fracture of the surrounding rock of underground caverns, in which the fragments of fractured rock are projected out. In slight rockbursts, the rock is fractured into small fragments, whereas serious rockbursts may project big rock blocks out and thus threaten the safety of constructors and machinery.

In some other countries, the study of rockburst has already had a history of several tens of years and has attracted special attention in the recent ten–odd years. The study of rockburst in China was only started ten years ago. In the 1980's, coal mining departments started systematical studies of bumps. In the mid–1980's, the hydroelectric power system began to study the problem of rockburst systematically in combinatio with the problem encountered in the diversion tunnel of the Tianshengqiao Two–Stage Hydropower Station. Both of these studies have attained some achievements. However, breakthrough in this respect has not been achieved yet, and mature theory for guiding the prediction vention of rockburst and powerful measures for rockburst control are still lacking.

The main reasons for the slow development of theoretical research on rockburst are that the mechanism of rockburst is very complicated and the establishment of a mathematical model is very difficult. Regardless of these, an effective method for numerical analysis should be developed in order to raise the ability of rockburst prediction and control in the design and construction of underground works. Fortunately, it is possible to get some information on the regularity of rockbursts because they are associated with brittle rocks and the brittle failure of the surrounding rock of caverns is directly observable. Based upon the characteristics of brittle rock failure and the failure phenomena in the tunnel of the Tianshengqiao Two–Stage Hydropower Station, an attempt has been made by the authors and co–workers, in which a mathematical model has been established through theoretical analysis and has been verified by comparison with the result of physical modelling and actual situation of rockburst events. Then, a computer program for FEM analysis has been compiled and researches on the prediction and prevention of rockburst have been performed.

2. FAILURE MECHANISM AND STRENGTH CRITERIA OF ROCKBURST

From the analysis of rockbursts in tunnel of the Tianshengqiao Two–stage Hydropower Station and the description of rockburst cases in other countries, rockbursts mostly took place in two forms: splitting failure and shear failure.

2.1 The splitting failure mechanism of rockburst

Near the surface of a tunnel, the radial stress σ_r in rockmass is relatively lower. On the periphery, the minor principal stress $\sigma_3 = \sigma_r = 0$, and the major principal stress $\sigma_1 = \sigma_\theta$. Because the differential stress $\sigma_\theta - \sigma_r$ is greater, it is possible that the surroundign rock fails even when σ_θ is less than the compressive strength of rock if the rock is brittle. Furthermore, because σ_θ is the major principal stress and trends parallel to the periphery of the opening, the surfaces of brittle fracture at the periphery of the tunnel must also run parallel to the periphery of the tunnel and appear as splitting fracture. Obviously, the splitting failure of rockburst in tunnels belongs to brittle fracture. An answer to the questions of why the rock fragments can break away from the rock mass and how can they be thrown out may be as follows. Though macroscopically the surrounding rock has entered only a fracture state but not its ultimate state at the moment when rockburst occurs, the microcracks or cracked zone in it may have already attained an unstable state of stress and therefore an unstable fracture surface may be formed by the propagation of a single fracture surface or the coalescence of a number of fracture surfaces. Such a local instability of rockmass may result in the splitting and peeling–off of rock or rockburst. Because the elastic energy can only be released through fracture surfaces, a part of it is dissipated in the propagation of cracks and a part of it is transmitted to the adjacent rock mass (including the energy transmitted outward in the form of stress waves), only a small portion of the released energy transforms into dynamic energy for rock fragments project. Rockburst events of the splitting fracture type are relatively weaker, though there exists the possibility of injury to constructors caused by individual small rock blocks which are projected out violently. This kind of rockbursts may take place in rockes of high compressive strength under relatively lower stress in the surrounding rock. Theoretically, the brittle failure of the surrounding rock is an instantaneous phenomenon. However, the actualities of the diversion tunnel of the Tianshengqiao project have demonstrated that rockburst events may be quite active within 24 kours after excavation and can continue for a month or so in some cases. In the period of rockburst continuation, it is usual that the working face has been advanced for quite a long distance and therefore the disturbance of static stresses in the surrounding rock can no longer give influence to the place of rockburst occurrence. So, the sustained rockburst events should be considered as the rheological fracture instability of microcracks caused by the rheological behavior of rock.

The rheological behavior of rocks is rather complex and their energy release rates include two parts as mentioned above. For the state of low stress, there would be only the elastic aftereffect; while steady—state viscous flow takes place only when the stress has exceeded a certain critical value. In underground works, the rheological behavior of soft rocks is often reflected by the macroscopic deformation of the surrounding rock of caverns; whereas that of hard rocks is mainfested by the retarded splitting failure of caverns rather than by an obvious macroscopic rheological deformation. It is just such an effect which results in retarded rockburst.

2.2 The shear failure mechanism of rockburst

According to the Coulomb—Navier criterion of failure, the analysis of a circular tunnel by means of the theory of limiting equilibrium thows that the fracture near the periphery of the tunnel is of a logarithmic spiral shape. Theoretical analysis also shows that the fracture of an opening which has straight walls would be wedge—shaped.

From the failure traces of the surrounding rock left by rockburst in the adit No. 1 and the main tunnel No. 1 near it, it can be found that the fracture surfaces run in the longitudinal direction of the tunnel, make an angle of $18° - 40°$ with the wall and are arranged in parallel with one another. No metter in the main tunnel or in the adit, the fracture surface always extends in a backward direction from the working face (see Fig. 1).

These "agnail—shaped" traces of shear failure remaining in the surrounding rock also reveal that the stress in the surroundign rock has reached an ultimate state (or a peak value). This kind of fracture surfaces, which run along the longitudinal direction of the tunnel and parallel to one another and obliquely cut the tunnel wall, reflects the effect of the longitudinal stress on shear failure. According to theoretical analysis, when the vertical and horizontal components of initial geostresses are respectively the major principal stress σ_1 and minor

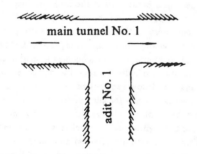

main tunnel No. 1

adit No. 1

Fig. 1   Effects of the direction of excavation on the pattern of fracture surfaces

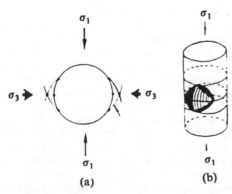

(a)   (b)

Fig. 2   Effects of the intermediate stress σ_2 on the failure of circular tunnels

(a) σ_2 parallel to the longitudinal axis of tunnel
(b) σ_1 parellel to the longitudinal axis of tunnel

principal stress σ_3, the shape of fracture surfaces of the surrounding rock would be a logarithmic spiral as shown in Fig. 2(a). Otherwise, when the maximum initial geostress acts in the axial direction of the tunnel, the shape of fracture on the tunnel wall would be as shown in Fig. 2(b). Since the state of stress on the newly excavated rock surface in the process of tunnel excavation is the worst one where $\sigma_r = 0$, the fracture surfaces develop towards the free surface exposed by excavation and no conjugate fracture surfaces exist. When the ultimate strength of rock is reached and instability occurs, not only the rock in the fractured zone will release more energy than at the beginning of failure, but also a part of the rock outside the fractured zone will undergo a stress drop and release some energy. Therefore, a larger amount of elastic energy will transform into dynamic energy and thus will lead to the violent projection of rock fragments.

Because of the existence of microcracks, crystals, grains as well as joints, fissures, etc., which affect the zone of shear fracture, the fracture surface sometimes shows an irregular shape, not necessarily a logarithmic spiral or wedge shape. The rock may be thrown out in blocks or in bulky volume. It is also possible that the fracture zone is logarithmic spiral or wedge—shaped but the fragments in the fractured zone are still of miscellaneous shapes. What is interesting is that regardless of the attitudes of microcracks and cracks, rockbursts under lower stresses are always splitting failure parallel to the surface of the opening's periphery, which is determined by the brittle nature of failure.

2.3 The failure criteria of rockburst

In so far as stated above, it can be seen that the two mechanisms of rockburst, splitting failure and shear failure, are two types of failure at different stress levels. The stress states for the brittle failure and the ultimte rupture of rocks are determined respectively by the Griffith criterion and the Coulomb—Navier criterion. Of course, that the stress state in the surrounding rock of a tunnel satisfies these two criteria is a necessary but not sufficinet condition for judging the possibility of rockburst and its intensity.

The occurrence criterion of retarded rockburst induced by the rheological fracture of rocks is a much more complicated problem needs to be studied further.

3. THE INSTABILITY CRITERION OF ROCKBURST

A rockburst event is the consequency of instability. Therefore, it is necessary not only to discuss the failure criterion but also to discuss the instability criterion. For a deformation system in equilibrium, the potential energy must have a stationary value. Whether the state of equilirbium is stable or not depends on whether the stationary value is a maximum or a minimum. If the potential energy π of a deformation system has a stationary value, the first order variation of π must be zero, i.e.,

$$\delta\pi = 0 \tag{1}$$

When the stationary value is a minimum, the second order variation of the potential energy of the system must be greater than zero and therefore the system is stable. When the stationary value is a maximum, the second order variation of π must be less than zero, i.e.,

$$\delta^2\pi < 0 \tag{2}$$

and thus the system is in a state of unstable equilirbium. This is the criterion for judging whether the equilirbium state of a deformation system is stable or not. The incremental form of eq. (2) is

$$\delta^2\pi = \int_V \delta\{d\varepsilon\}^T [D_{ep}]\delta\{d\varepsilon\}dV < 0 \tag{3}$$

The negative value of eq. (3) can occur only when the elastoplastic stiffness matrix (D_{ep}) is a negative definite one, i.e., only when the peak of the $\sigma - \varepsilon$ curve is exceeded and the surrounding rock begins to soften.

Similar result can also be obtained from the criterion of rock stability given by Drucker which states that the rock will be unstable when the relation

$$\int d\sigma \cdot d\varepsilon < 0 \tag{4}$$

is satisfied. Eq. (4) can only be satisfied when the peak of the $\sigma - \varepsilon$ curve is exceeded and strain softening begins.

In the previous analysis, the failure criterion for the occurrence of rockburst is the Griffith criterion which corresponds to the commencement of microcrack propagation in rock. In a compressive stress field, it reflects the stable propagation of cracks. According to linear elastic fracture mechanics, the condition for crack propagation in terms of energy is

$$\frac{d}{da}(U - W) = 0 \tag{5}$$

where U —— energy released due to cracking;
       W —— work needed to form the newly created fracture surface;
       a —— half length of the crack.

$$\frac{du}{da} = G \tag{6}$$

where G —— rate of energy release at the crack tip; it is the driving force which reflects the energy source for crack propagation.

$$\frac{dw}{da} = R \tag{7}$$

where R —— resistance to crack propagation.
The critical rate of energy release G_c givenly the Griffith criterion is

$$G_c = \pi \sigma_c^2 a / E \tag{8}$$

It is known from eq. (2) that eq. (5) reflects the stable propagation of crack and therefore is equivalent to the Griffith criterion. Obviously, $G = R$ or $G = G_c$ is the condition for stable crack propagation.

The above analysis shows that the Griffith criterion only reflects that the rock begins to fail but is unable to reflect the possibility of rockburst. For the occurrence of rockburst, the condition for unstable rock fracturing must be satisfied, i.e.,

$$G > G_c \tag{9}$$

when rock fractures unstably. In that case, the excessive energy released by the surrounding rock

$$\pi = U - W = \int_{a_1}^{a} (G - R)\, da \times thickness \tag{10}$$

becomes the energy source for the high–speed crack propagation, the propagation of stress waves and the projection of fractured rock fragments away from the rockmass. From the analysis in a previous paragraph, it is known that the Coulomb–Navier criterion is the failure criterion for rockbursts of the shear fracture type. In addition to the failure criterion, another necessry condition for the occurrence of rockburst that the microcrack zone in the rock deformation system should be in a strain softening state as the criterion of rock instability must also be satisfied simultaneously.

The stiffness theory of rockburst is a reflection of he instability theory from this aspect. Using the ratio F_{CF} between the loading stiffness K_m and the unloading stiffness K_s obtained from the complete stress–strain curve, it is possible to judge the possibility of rockburst occurrence. According to the instability theory, rockburst may possi-

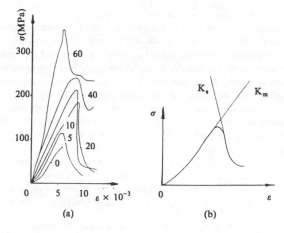

Fig. 3   (a) Complete stress–strain curve (b) and the stiffness of rock before and after the peak stress

bly occur when the stiffness ratio F_{CF} is less than 1, i.e.,

$$F_{CF} = K_m / |K_s| < 1 \tag{11}$$

Fig. 3 gives the result of triaxial compression test on the Tianshengqiao limestone using an MTS servo–controlled stiff testing machine. For the two steps of $\sigma_3 = 10$ and 20 MPa, $F_{CF} < 1$; while for two other steps of σ_3 the value of $F_{CF} \approx 1$. For a same kind of rock under different confining pressures, the value of F_{CF} may sometimes be less than 1 and sometimes be approximately equal to 1 owing to the inhomogeneity of rock, resulting in a certain discrepancy among the test results of a same group of rock samples. However, a general understanding obtained from the test results is that the likelihood of rockburst occurrence at Tianshengqiao is rather high. Before the establishment of the instability theory, to judge the likelihood of rockburst by using the stiffness criterion and the strength criterion jointly is an effective method. So far the possibility of rockburst occurrence in the limestone at Tianshengqiao has been demonstrated. Whether rockbursts will occur or not depends on the stress state and strength of the surrounding rock. This is just the subject to be studied by the mumerical method.

4. DETERMINATION OF INITIAL STRESSES IN THE ROCKMASS BY ROCKBURST BACK–ANALYSIS

To know the initial stresses in the rockmass is an important prerequisite for numerical analysis. If there are no measured data of initial stresses but rockbursts have occurred in the tunnel, it is suggested here that the initial stresses in rockmass can be determined from rockburst back–analysis. According to the stress concentration around an opening, it can be inferred that the vartical geostress σ_v and horizontal geostress σ_H in the rockmass should be principal stresses σ'_1 and σ'_3 or σ'_3 and σ'_1 respectively when rockburst occurs on the wall or roof and floor. It is first assumed that σ_v is due to gravity which is roughly true as verified by actually measured data. Based on this assumption, computation can be made for a series of λ values where $\lambda = \sigma_H / \sigma_v$. The λ value which gives a result resembling the actual state of failure is then used to determine the geostress.

At the cross–section 0 + 905 m in the adit No. 2, rockbursts occurred on the wall. It can therefore be determined that σ_H is σ'_1 and σ_v is σ'_3. Because the phenomenon of rockburst was splitting failure and the intensity was judged to be class 1, computation can be made for $\sigma_H = \lambda\sigma_v(\lambda = 1.1, 1.2, 1.3, ...)$ by use of the Griffith criterion. The tensile strength of rock is taken to be $\sigma_t = 1.5$ MPa ~ 3.7 MPa. It has been decided to take $\sigma_H = 1.3\ \sigma_v$ through the analysis of results. In numerical predictions of unexcavated rockmass, the initial geostresses can thus be taken as $\sigma_v = \gamma H$ and $\sigma_H = 1.3\,\sigma_v$.

The computational results are not necessarily the two components of the real geostress in a plane normal to the tunnel axis. Besides, the error accumulated in various links may make the computed geostress differ much from the actually measured geostress. However, some of the errors produced in back–analysis may vanish automatically. Therefore, the approach of predicting rockburst from previous rockburst, i.e., to predict rockburst by computing σ_H in unexcavated rockmass in a longer tunnel by use of the stress ratio λ iven by back–analysis, is of practical value, especially when there are no actually measured data.

Taking $\lambda = 1.3$ and the vertical geostress $\sigma_v = \gamma H$, numerical rockburst predictions for two cross–sections, 5 + 550 m and 6 + 550 m, can be made as follows. The buried depths of the two cross–sections are 580 m and 440 m respectively. So the vertical geostresses are

$$\sigma_v = \gamma H = 16 MPa, \quad \textit{for cross–section} \quad 5+550m$$

and

$$\sigma_v = \gamma H = 12.1 MPa, \quad \textit{for cross–section} \quad 6+550m$$

Letting the horizontal geostress be $\sigma_H = 1.3\,\sigma_v$, then

$$\sigma_H = 1.3 \times 16 MPa = 20.8 MPa, \quad \textit{for cross–section} \quad 5 + 550m$$

and

$$\sigma_H = 1.3 \times 12.1 MPa = 15.7 MPa, \quad \textit{for cross–section} \quad 6 + 550m$$

There are no data of measured geostresses near the two cross–sections. However, the measured geostress of two measuring points at a certain distance from them can b used for comparison, as shown in Table 1.

Table 1   Measured geostresses at two points in the adit No. 2. and the main tunnel No. II

Test site		Adit No. 2 0 + 792.2 m	Main tunnel No.II 6 + 805 m
Buried depth of measuring point (m)		230	405
σ_1	Stress value (MPa)	25.86	31.60
	Azimuth	S19°E	S57.60°E
	Dip / dip angle	NW $\angle$ 40.5°	NW $\angle$ 0.96°
σ_2	Stress value (MPa)	16.15	22.31
	Azimuth	S12°E	S32.34°E
	Dip / dip angle	NW $\angle$ 45.3°	NW $\angle$ 25.8°
σ_3	Stress value (MPa)	7.23	15.07
	Azimuth	S85.10°E	S34.84°E
	Dip / dip angle	NW $\angle$ 15.7°	NW $\angle$ 64.17°

Given by back–analysis are the stress components acting in a cross–section which should have the relation

$$\sigma_H < \sigma_1$$

and

$$\sigma_v > \sigma_3$$

with the principal stresses in space.

In the results of back–analysis, the σ_H values in both cross–sections are less than 25.86 MPa, the smaller one among the σ_1 values; the σ_v value in cross–section 5 + 550 m is greater than 15.0 MPa, the greater one among the σ_3 values; only the σ_v value in cross–section 6 + 550 m is less than 15.07 Mpa, but still greater than 7.23 Mpa. It seems that the geostress values given by back–analysis essentially tally with the normal regularity.

5. NUMERICAL PREDICTION OF ROCKBURST

Numerical calculation had been made before the excavation of the main tunnel reached cross–sections 5 + 550 m and 6 + 550 m to prediction possibility of rockburst occurrence and the possible intensity and failure range of rockburst. The deduction that σ_H is σ'_1 is equivalent to a predition that the roof and floor of the two cross–sections would be the places of rockburst.

The geostresses from back–analysis mentioned above and the mechanical properties of the rock given in Table 2 were used for calculation.

Table 2   Physico–mechanical parmeters of the surrounding rock

Compressive strength (MPa)	80	Elastic modulus (MPa)	3×10^4
Tensile strength (MPa)	3	Poisson's ratio	0.3
Cohesion (MPa)	7.75	Volume weight (g / cm³)	2.76
Angle of internal friction (°)	68		

The ranges of rockburst failure of the two cross–section are indicated by the hatched areas in Figs. 4 and 6 respectively. The maixmum tangential stress in cross–section 6 + 550 m appears at the top and bottom and is 32 MPa, accounting for 40% of the uniaxial compressive strength of the surrounding rock. The stress state only reaches the value indicated by the Griffith criterion and therefore the rockburst wuld be of intensity I. The maximum depth of fracture surface is 40 cm. The maximum tangential stress in section 5 + 550m also appears at the top and bottom; its magnitude is 42.69 MPa which is also only the value given by the Griffith criterion and therefore the rockburst would also be of intensity I. The maximum depth of fracture surface is 80 cm.

After excavation, rockbursts occurred in the segment from 6 + 624 m to 6 + 674 m but not in the cross–section 6 +550 m because of the existence of karst there. Rockbursts occurred at a place a little to the left of the apex of arch roof and a symmetrical location of the arch floor, associated with a sound like the cracking of ice. The depth of rockburst failure in this segment was generally 20 − 40 cm, with a maximum of 50 cm. Fig. 5 shows the actuality of failure in cross–section 6 + 635 m where the floor heaved and cracked and agnail–shaped fracture surfaces inclined at an angle of 15° − 20° to the boundary of opening occurred. Boreholes showed that cores longer than 10 cm could not be obtained within 5 m depth and radar detection indicated that a zone of rock relaxasion existed in a range of 0 − 4.6 m beyond which the rock became intact rapidly. Comparison between the computational result and the actuality of rockburst shows that the calculated fialure depth is close to the maximum depth of actual fracture surfaces of rockburst. However, shear failure occurred in the actual failure mechanism in cross–section 6 + 635. At present, it is still not known how deep the range of shear failure extended. It is only known that the calculated result is close to the maximum visible depth of fracture surfaces. Besides, microcracks (relaxation zone) in the surroudign rock of cross–section 6 + 635 m have extended to a greater depth. These two pieces of information both suggest that the actual rockburst is of higher intensity and the fracture surface is of greater depth than the prediction.

In the cross–section 5 + 550 m, rockburst also occurred at a place about 25° to the left of the apex of arch roof but it was weaker, about of intensity I. But the width and depth of failure were much smaller than the prediction, the maximum depth was 15 cm or so (see Fig. 7).

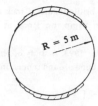

Fig. 4   Predicted range of failure in cross—section 6 + 550 m by FEM

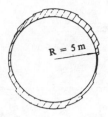

Fig. 6   Predicted range of failure in cross—section 5 + 550 m by FEM

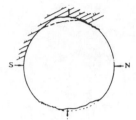

Fig. 5   Actul range of rockburst failure in surrounding rock of cross—section 6 + 550 m

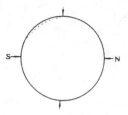

Fig. 7   Actual range of rockburst failurein surroundign rock of cross—section 5 + 550 m

The discrepancy between the computational result and the actualities comes from a number of sources, including the inexact simulation of geostress, the deviation of mechanical parameters of rock, the inhomogeneity of rock, the effects of pre—existing microcracks in the surrounding rock, etc. In addition to these, another important reason is that the criterion used in computation for judging rockburst occurrence is the Griffith criterion; it corresponds to the situation that the energy release rate of surrounding rock, G, reaches its critical value G_c. In fact, the condition for rockburst occurrence is $G > G_c$ which corresponds to an energy release larger than that required for rock failure, in which the excessive energy becomes dynamic energy to produce stress waves propagating outwards in the surrounding rock. In the present computation, the dynamic effect of stress waves cannot be reflected. For the above reasons, it is only possible to compare the rockburst—induced failure range and failure pit depth visible by naked—eyes with the computational result but impossible to reflect the relaxation zone outside the failure zone.

6. APPLICATION OF NUMERICAL METHODS IN ROCKBURST PREVENTION AND CONTROLL

6.1 Choice of opening shape

Calculation was made for the case of $\sigma_v = \gamma H = 11.9$ MPa, $\lambda = 1.1$ and $\sigma_H = 13.1$ MPa. For comparison, three rock strength were used. The tensile strength σ_T of rock was taken to be 2.5 MPa, 2.9 MPa and 3.3 MPa in turn.

The results of computation are shown in Fig. 8 Areas hatched with oblique lines are areas of splitting failure while areas hatched with cross—lines are areas of shear failure (the same below).

It can be seen from the figure that there occurs only splitting failure in circular openings when the rock is of lower strength. Because the initial stresses in rockmass approximates to hydrostatic stress field, the range of fracture is distributed all around in the cross—section and the fractured depth is uniform. The failure ranges for the other two opening shapes are a bit smaller, and the smallest is that for the horse—shoe—shaped opening. For the latter two shapes, localized intense rockburst would occur at the turning of the bottom arch. When the rock strength is higher, the circular corss—section would be the best and free from any fracture, and then is the circular—rectangular cross—section which is superior than the horse—shoe—shaped one.

It can thus be seen from the above that the optimal choice of opening shape is not only related to the initial stress state in rockmass but also depends on the strength of rock.

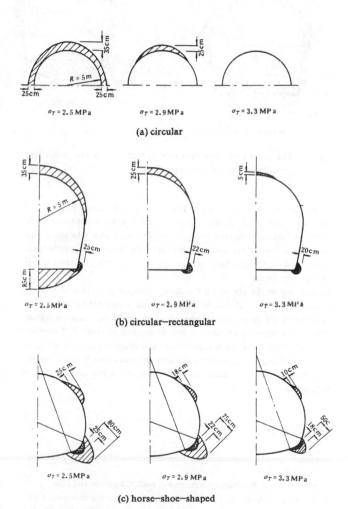

(a) circular

(b) circular—rectangular

(c) horse—shoe—shaped

Fig. 8   The range of surrounding rock failure in tunnels of different shapes

6.2 Choice of excavation scheme

Construction schemes of excavating the tunnel into the final shape in three steps and in two steps were computed and compared. The tensile strength of rock used in computation is $\sigma_T = 2.5$ MPa.

The result of computation showed that the failure ranges for the two

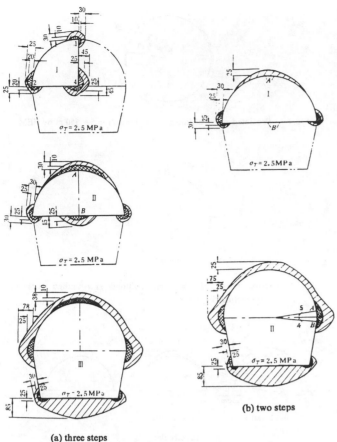

(a) three steps

(b) two steps

Fig. 9   The states of surroundign rock failure in each excavation step for different excavation schemes

shemes are roughly the same (see Fig. 9), except that the enlargement of the upper stage after the first excavation in the three—step scheme would result in intense rockburst to occur on the roof arch whereas in the two—step scheme there would be only splitting failure type rockburst to occur correspondingly. Both schemes would have entirely the same failure at the lower corner.

In addition to the above, all the three corners would fail and produce rockburst when the left half of the upper stage is excavated in the first excavation of the three—step scheme. However, if the two—step excavation scheme is adopted, rockbursts at four places can be avoided. Therefore, when choosing excavation schemes for a rockburst—prone tunnel, it is worthy noticing that to increase the steps of excavation is not favorable because the more the steps of excavation, the more the chances of rockbursts.

REFERENCES

Herget, G., 1974. Ground stress determinations in Canada. Rock Mechanics, 6.

Farmer, I., 1982. Engineering behavior of rock. Chapman and Hall.

Broch, E. and Sorheim, S., 1984. Experiences from the planning, Construction and supporting of a road tunnel subjected to heavy rockbursting. Rock Mechanic and Rock Engineering. 17.

Maury, V., 1987. Observations, researchs et resultats recents sur les mechanismes de ruptures autour de galeries isoless. Proc. 6TH Inter. Cong. on Rock mech. Montreal, Canada.

Lu Jiayou., 1987. Study on mechanism of rockburst in a headrace tunnel. Proc. of the Intern. Conf. on Hydropower, Oslo, Norway.

Lu Jiayou, Wang Changming and Huai Jun., 1988. FEM analysis for rockburst and its back analysis for determining in—situ stress. Proc. of the Sixth Inter. Conf. on Numerical Methods in Geom. Innsbruck, Austria.

Lu Jiayou et al., 1989. The brittle failure of rock around under ground openings. Proc. of Inter. Symp. on Rock at Great Depth. PAU, Franch.

Yang Shuqing and Lu Jiayon., 1993. A study of physically simulating on the mechanism of rockburst around tunnel. In this Proceedings.

Rockbursts and Seismicity in Mines, Young (ed.) © 1993 Balkema, Rotterdam, ISBN 90 5410 320 5

The effect of excavation-induced seismicity on the strength of Lac du Bonnet granite

C. Derek Martin
AECL Research, Whiteshell Laboratories, Pinawa, Canada

R. Paul Young
Engineering Seismology Laboratory, Department of Geological Sciences, Queen's University, Kingston, Ont., Canada

ABSTRACT: A 46-m-long circular test tunnel at the 420 Level of the Underground Research Laboratory was monitored during excavation with 16 triaxial accelerometers. The accelerometers were grouted in place at the end of specially drilled boreholes and provided optimal focal sphere coverage during test-tunnel excavation. During excavation of the tunnel, spalling in the roof and floor resulted in typical well-bore breakout geometry. The location of the induced seismicity associated with the excavation generally fell within one radius of the tunnel wall and appeared as random events. However, within the well-bore breakout geometry the microseismic events were clustered suggesting crack damage and coalesence. Laboratory experiments were carried out to investigate the effect of crack damage on rock strength. The preliminary results suggest that a small amount of crack damage can reduce intact rock strength by 50% or more. It appears that the induced seismicity may be responsible for the damage and the associated reduction in rock strength observed around the underground openings at the 420 Level.

INTRODUCTION

Excavation of an opening at depth in a geological medium commonly results in cracking in the zones of maximum compressive stress concentration around the opening. The general form of this cracking in brittle rocks such as granite is spalling and slabbing which often leads to well-bore breakout geometry. The energy released by this cracking process is referred to as excavation-induced seismicity. A unique experiment in Atomic Energy of Canada Limited's Underground Research Laboratory (URL), called the Mine-by Experiment, was carried out to investigate the source mechanisms associated with excavation-induced seismicity around a 3.5-m-diameter test tunnel excavated at a depth of 420 m in massive unfractured granite. In addition to the seismic component of the experiment, laboratory work was also carried out to investigate the effect of crack damage on laboratory strength. This paper provides preliminary results from the experiment and examines the effect of excavation-induced crack damage on rock strength.

SITE DESCRIPTION

Geology

The URL is located within the Lac du Bonnet granite batholith, which is considered to be representative of many granitic intrusions of the Canadian Shield. The batholith trends east-northeast and its elongated body is about 75 by 25 km in surface area and extends to a depth of about 10 km. The batholith, dated as Late Kenoran age (2680 ±81 Ma), lies in the Winnipeg River plutonic complex of the English River gneiss belt of the western Superior Province. The batholith is a relatively undifferentiated massive porphyritic granite-granodiorite. The massive, medium- to coarse-grained porphyritic granite is relatively uniform in texture and composition over the batholith, although locally it displays subhorizontal gneissic banding (Everitt et al 1990).

A section through the URL reveals that jointing essentially stops at a depth of about 220 m at the location of the URL shaft (Figure 1). The test tunnel was excavated at a depth of 420 m, approximately 150 m below the nearest fault (Fracture Zone 2) and about 200 m below any regular joint patterns (Figure 1). Boreholes drilled to depths of over 1000 m in the vicinity of the URL indicate the massive granite persists with depth. Excavation

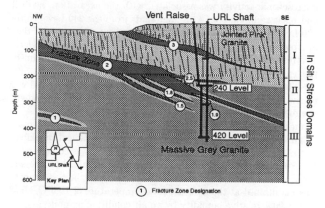

Figure 1: Geological section through the URL showing the location of the 420 Level and the in situ stress domains at the URL. Below Fracture Zone 2 the rock is massive unfractured gneissic grey granite.

of the 420 Level access tunnels and the instrumentation galleries associated with the Mine-by Experiment also did not encounter any joints or fractures on the scale of the excavations (Figure 2). Hence, one can conclude that the Mine-by Experiment was carried out in massive unfractured granite.

In Situ Stress

The in situ stresses at the URL have been extensively investigated using traditional methods, such as overcoring, hydraulic fracturing and borehole breakouts, and by non-traditional methods such as microseismic monitoring, convergence monitoring, and under-excavation techniques (Martin 1990). This extensive characterization program has defined three distinctive stress domains at the URL (Martin in prep(a)). At the 420 Level the in situ stresses are part of Stress Domain III, which extends from about 300 m to a depth of at least 512 m (Figure 1). Within this domain, at the depth of the test tunnel, the in situ stress magnitudes are $\sigma_1 = 55$ MPa, $\sigma_2 = 48$ MPa and $\sigma_3 = 14$ MPa. The test tunnel was excavated in the azimuth direction of 225°, approximately parallel to σ_2 to maximize the stress concentration around the tunnel (Figure 2). Note that the orientation of σ_1 looking into the test tunnel is about 14° from the horizontal, plunging to the southeast.

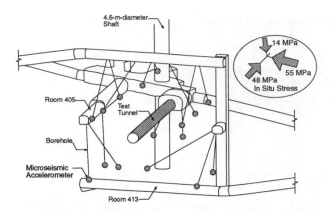

Figure 2: Location of the Mine-by test tunnel and the microseismic triaxial accelerometers on the 420 Level.

TEST TUNNEL EXCAVATION

It is well known that cracking occurs around an opening excavated in a highly stressed brittle medium. Two examples, of extreme scales, are the process zone around the tip of an advancing crack (Pollard and Aydin 1988, Labuz et al 1987) and the cracking associated with deep-level mining (Joughin and Jager 1983). Earlier work at the URL had established that considerable microseismic activity was associated with the excavation, by drill and blast techniques, of a circular shaft from the 240 Level to the 420 Level (Talebi and Young 1992). In order to determine if the microseismic activity was caused by the blasting or simply related to stress redistribution, a test tunnel for the Mine-by Experiment was excavated without the use of blasting.

The 3.5-m-diameter test tunnel had a circular profile and was excavated parallel to σ_2 (Figure 2). This configuration provided the maximum stress concentration in the roof and floor of the tunnel. The tunnel was excavated in 1-m and 0.5-m increments (Figure 3) using perimeter line drilling and mechanical breaking of the rock stub (Onagi et al 1992). Excavation was completed in two 8-h shifts, but experimental activities constrained progress to 1 round about every 3 days. The temperature of the test tunnel was maintained at the ambient rock temperature of 10.5°C ±0.5° and > 90% relative humidity by an air conditioning unit. Extensive state-of-the-art geomechanical instrumentation was installed prior to the start of the excavation, and was used to monitor the complete mechanical response of the rock mass around the tunnel (Read and Martin 1992).

Prior to the construction of the test tunnel, access tunnels off the 420 Level and the instrumentation galleries were excavated using drill-and-blast techniques. The stress magnitudes were sufficient to induce spalling of the roof and floor of the tunnels oriented approximately parallel to σ_2. The tunnels excavated approximately parallel to σ_1 were essentially unaffected by the stress concentration. The spalling resulted in typical well-bore breakout (notch) geometry (Figure 4).

MICROSEISMIC EVENTS AND OBSERVATIONS

In addition to the mechanical instrumentation around the test tunnel, an array of 16 triaxial accelerometers was also installed to monitor the microseismic events associated with the excavation (see Figures 2 and 3). The accelerometers, with a frequency response from 50 Hz to 10 kHz (±3 dB), were grouted in place at the end of diamond-drilled boreholes (Figure 2). The accelerometer array was designed for focal sphere coverage and a source location accuracy of about ±0.25 m near the centre of the tunnel. The sampling rate was set to 50 kHz, allowing the study of seismic events with moment magnitudes as small as -6. The sequencing of the construction schedule for the test tunnel provided about 12 h of quiet time for monitoring after the initial perimeter drilling and about 12 h of quiet time for monitoring

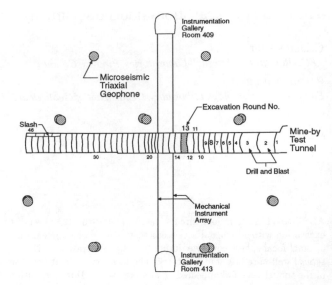

Figure 3: Vertical cross section through the test tunnel showing the location microseismic triaxial accelerometers and the location of Round 13.

after mechanical breaking of the rock stub. This provided a total of about 24 h of monitoring per round of tunnel advance.

Preliminary processing of the microseismic data was carried out in the field using automated source location computer software developed at Queen's University (Collins and Young 1992). Some 25,000 events were source located. Inspection of all 46 rounds showed similar trends. For this paper, a detailed analysis of Round 13, located near the centre of the tunnel, has been carried out by manual picking of the first P and S wave arrivals. In addition to the microseismic data, detailed survey information on the development of the breakout geometry is also available. The location of Round 13 is shown on Figure 3.

The perimeter drilling was carried out in a pattern similar to that shown in Figure 5 for Round 13. During the perimeter drilling, cracking was commonly observed as the drilling approached the roof where the maximum stresses were concentrated. Figure 5 shows the 47 microseismic events recorded over a 10-h period immediately after the perimeter drilling was completed. At this point the events do not show strong clustering, although there is a slight grouping of events where the breakout is eventually first observed. Note that the first breakout is near vertical, yet the stress orientation would suggest that the breakout should be off-centre by 14° (Figure 5). It appears that the direction of the perimeter drilling plays some role in defining the initiation of the breakout.

After the initial microseismic monitoring period the rock stub

Figure 4: Example of the well-bore breakout (notch) geometry observed in a tunnel excavated parallel to σ_2.

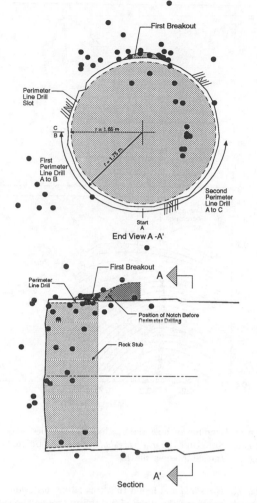

Figure 5: Location of microseismic events at the end of perimeter drilling of Round 13. Note the slight clustering of events in the roof where the first breakout eventually appears.

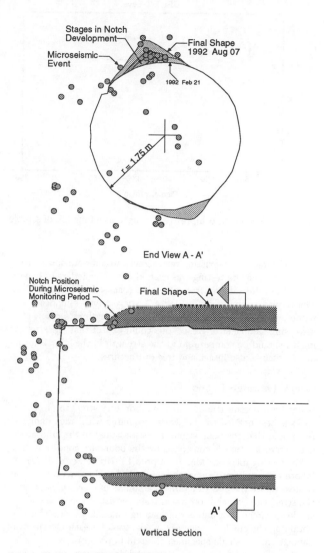

Figure 6: Location of microseismic events at the end of Round 13. Note the strong clustering of events in the roof where the notch will eventually develop. Also note the cracking occurring ahead of the face.

shown in Figure 5 was broken out, from the bottom to the top, using mechanical hydraulic rock splitters. This process took about 6 h. After the rock stub was removed, 52 new microseismic events were recorded during a 16-h monitoring period (Figure 6). These events show strong clustering in the roof, particularly where the breakout eventually occurred. Feignier and Young (1992) analyzed the microseismic events from Round 3 of the test tunnel and found, using a moment tensor inversion technique, that all of the events located where the breakout eventually occurred were dilational (see Figure 3 for location of Round 3).

It would appear that the concentration of induced seismic events is defining the region where the breakout geometry will appear. This concentration of events could be similar to the process zone ahead of an advancing crack tip (Labuz et al 1987). Visual observations during breakout development indicate a process zone does develop at the apex of the notch. Further investigations are under way to determine if the concentration of seismic events does occur at the notch process zone. Note that the dates in Figure 6 do not reflect the actual times required for the notch to develop but the dates of the actual notch survey. The development of the notch was related to the advancement of the tunnel and to the scaling carried out for safety reasons. It should also be noted that the formation of the notch is not evident at the tunnel face because of the 3D geometry, but starts forming about 0.75 to 1 m back from the tunnel face, and is fully developed 2 m back from the face.

It should be noted that in the preceeding discussion and in Figures 5 and 6, microseismic events are only concentrated in the roof of the tunnel. In fact micorseismic events also show the same

clustering in the floor but it occurs after the tunnel has advanced another round. Thus the clustering in the roof and floor is offset by about 1 m along the tunnel axis. This time lag for the floor events is due to the confining stress provided by the weight of the tunnel muck in the floor. Another feature of the excavation rounds investigated to date is the induced seismicity occuring ahead of the tunnel face (Figure 6). Presumably this damage is occurring because of the stress concentrations caused by the flat face. The moment magnitudes for the events associated with Figure 6 ranged from -6 to -4. Similar cracking has been observed in the deep mines of South Africa (Joughin and Jager 1983), although this damage is small by comparison. In the next section we investigate the effect of damage on brittle rock strength.

ROCK STRENGTH

One of the most common reasons put forth for the reduction in strength from the laboratory to in situ conditions is the effect of scale. Brown (1971) reviewed published data and found that laboratory strength was influenced by such factors as specimen geometry, end restraint, stress gradients, surface conditions and the initial degree of microcracking present. Brown concluded that, under uniform stress conditions, intact rock that was

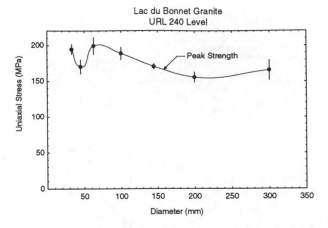

Figure 7: Unconfined compressive strength as a function of sample size.

relatively free of pre-existing microcracks would not show a size effect. Scale-effects testing was carried out on cylinders of Lac du Bonnet granite samples from the URL (Jackson and Lau 1990). The results, summarized in Figure 7, suggest only a modest reduction in the unconfined compressive strength as a result of increasing sample size. Hence, one could expect that the strength in situ should be comparable to the strength in the lab, since the in situ rock is also intact and free of fractures.

Crack Damage Locus

It has been demonstrated that the laboratory strength of brittle rocks is best defined by the stress associated with crack damage (σ_{cd}), and that the peak strength, as measured in the laboratory, is a parameter that is controlled by the boundary conditions used in the testing method (Martin in prep(b), Hudson et al 1972). A laboratory study using damage-controlled testing was carried out to investigate the effect of various amounts of crack damage on the strength. Damage-controlled tests consist of subjecting a single specimen to a number of loading-unloading cycles and obtaining the crack damage and peak stress for each increment of damage from a single specimen (Martin in prep(c)).

A total of 6 uniaxial and 31 triaxial damage-control tests were conducted on samples of Lac du Bonnet granite collected from the 420 Level of the URL. The results from one sample, MB124205 is presented, although all test results showed similar trends (Martin in prep(b)). MB124205 was tested at a confining stress of 2 MPa and was subjected to 63 damage increments, 20 of which occurred in the prepeak region of the stress-strain curve. Figure 8 shows the axial stress/axial strain/volumetric strain response for MB124205. The σ_{cd} curve, in Figure 8, is the locus of points corresponding to the stress level where the crack damage is detected, i.e., where the volumetric strain reverses for each cycle. The migration of the σ_{cd} locus in the prepeak region is illustrated by the volumetric strain versus axial strain for damage increments 3 to 25 (Figure 8). Note that the rapid reduction in σ_{cd} occurs by increment 10 before the peak strength of the sample is reached. We suggest that the results reflect the accumulated damage to the sample from crack growth with each new increment. Martin and Lajtai (in prep) also investigated the crack damage locus using Indiana Limestone and Rocanville Potash and concluded that the more brittle the material the smaller the amount of crack damage required for the strength to decrease to the damage threshold.

Martin and Read (1992) carried out extensive analysis of the tunnels on the 420 Level of the URL using $Examine^{2D}$ (Curran and Corkum 1991). When the laboratory peak strength or the peak strength adjusted for the 10% scale effects was used, all the analyses indicated that the tunnels had a strength to stress ratio greater than 1, i.e., indicating no failure. These analyses were clearly in conflict with the observations (Figure 4). Martin and Read (1992) repeated the analyses using a failure envelope based

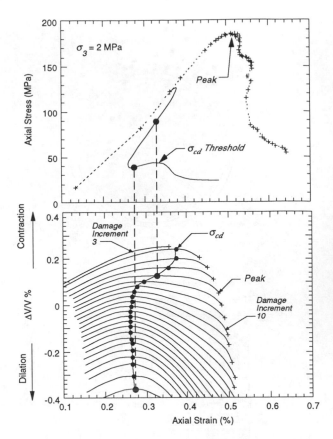

Figure 8: Example of how the σ_{cd} locus is mapped and the σ_{cd} Threshold. Note that the σ_{cd} Threshold is reached before the peak strength of the sample is achieved.

on the σ_{cd} Threshold and found that this failure criterion adequately predicted the zones of instability around the tunnels. In order to develop the σ_{cd} Threshold, damage must be occurring to the rock mass. It would appear that the cracking, i.e., the microseismic events, in Figures 6 and 5 may be sufficient to cause the damage that reduces the in situ strength to the crack damage threshold. It should be noted that the microseismic events in Figures 6 and 5 are not the only crack related events taking place around the test tunnel, but merely the events recorded by the 16 triaxial accelerometers. For example, Carlson and Young (1992) recorded, using 1 MHz transducers, over 720 microseismic events in the sidewall of the test tunnel during a 13 h monitoring period. Hence, the damage is probably associated with microseismic events that have a wide range of moment magnitudes.

CONCLUSIONS

The results of our studies thus far indicate that damage in the form of excavation-induced microseismicity is occurring ahead of the advancing Mine-by test tunnel. The microseismicity occurs in the areas ahead of the tunnel where breakouts eventually appear. Although random events are recorded within one radius of the test tunnel it appears that damage to the rock occurs only where seismic events cluster. Our findings also suggest that this damage ahead of the tunnel face in the location of the maximum stress concentration is necessary in order for the well-bore breakout geometry to develop. The induced seismic events are caused by the stress redistribution near the tunnel face as the test tunnel is exacavated and range in moment magnitude from -6 to -4.

Laboratory testing of intact samples has also shown that small amounts of damage can significantly reduce the stress associated with crack damage to the crack damage threshold. Analysis of the tunnels on the 420 Level, using traditional laboratory peak strength, and laboratory peak strength adjusted for scale effects, indicated that all openings should have been stable. Observations, however, revealed that failure was occurring by spalling and slabbing. Re-analysis of the tunnels using a failure envelope based on the crack damage threshold matched the observations. The evidence suggests that the low rock strength, ≈ 40% of the peak strength, found around the tunnels excavated at the 420-m depth is, in part, caused by the induced microseismicity near and ahead of the advancing tunnel face.

ACKNOWLEDGEMENTS

This work was, in part, jointly funded by AECL and Ontario Hydro under the auspices of the Candu Owners Group. The balance of the support for this work was provided by a grant from the Natural Sciences and Engineering Research Council of Canada to the second author.

REFERENCES

Brown, E.T. Strength-size effects in rock material. In *In Proc. ISRM Symp. Rock Fracture*, pages 2–11, Nancy, 1971.

Carlson, S.R. and R.P. Young. Acoustic emission and ultrasonic velocity study of excavation induced microcrack damage in the Mine-by tunnel at the Underground Research Laboratory. Atomic Energy of Canada Ltd. RP015AECL, Engineering Seismology Laboratory, Queen's University, Kingston, Canada, 1992.

Curran, J.H. & B.T. Corkum. *Examine^{2D}–A 2D boundary element program for calculating stresses around underground excavations in rock, Version 3.1*. Data Visualization Laboratory, University of Toronto, Toronto, Canada, 1991.

Collins, D.S. & R.P. Young. Monitoring and source location of microseismicity induced by excavation of the Mine-by Test Tunnel: Preliminary analysis. Atomic Energy of Canada Ltd. RP013AECL, Engineering Seismology Laboratory, Queen's University, Kingston, Canada, 1992.

Everitt, R.A. A. Brown, C.C. Davison, M. Gascoyne, and C.D. Martin. Regional and local setting of the Underground Research Laboratory. In R. S. Sinha, editor, *Proc. Int. Symp. on Unique Underground Structures, Denver*, volume 2, pages 64:1–23. CSM Press, Denver, 1990.

Feignier, B. & R.P. Young. Moment tensor inversion of induced microseismic events: Evidence of non-shear failures in the $-4 < m < -2$ moment magnitude range. *Geophysical Research Letters*, 19(14):1503–1506, July 24 1992.

Hudson, J.A. E.T Brown, and C. Fairhurst. Shape of the complete stress-strain curve for rock. In E.J. Cording, editor, *Proc. 13th Symp. Rock Mechanics*, pages 773–795, Urbana, Illinois, 1972. American Society of Civil Engineers, New York.

Jackson, R. and J.S.O. Lau. The effect of specimen size on the mechanical properties of Lac du Bonnet grey granite. In A. Pinto da Cunha, editor, *Proc. 1st. Int. Workshop on Scale Effects in Rock Masses, Loen, Norway*, pages 165–174. A.A.Balkema, Rotterdam, 1990.

Joughin, N.C. & A.J. Jager. Fracture of rock at stope faces in South African gold mines. In *Proc. Rockbursts: Prediction and Control*, pages 53–67, London, 1983. The Institution of Mining and Metallurgy, London.

Labuz, J.F. S.P. Shah, and C.H. Dowding. The fracture process zone in granite: Evidence and effect. *Int. J. Rock Mech. Min. Sci. & Geomech. Abstr.*, 24(4):235–246, 1987.

Martin, C.D. Characterizing in situ stress domains at the AECL Underground Research Laboratory. *Can. Geotech. J.*, 27:631–646, 1990.

Martin, C.D. Stress heterogenity and geological structures. In B. Haimson, editor, *Proc. 35th U.S. Rock Mechanics Symposium*, Madison, Wisconsin, in prep(a).

Martin, C.D. *Strength of massive granite around underground openings*. PhD thesis, Civil Engineering Department, University of Manitoba, Winnipeg, Manitoba, Canada, in prep(b).

Martin, C.D. & E.Z. Lajtai. Crack damage and rock strength. *Submitted to Int. J. Rock Mech. Min. Sci.*, in prep(c).

Martin, C.D. & R.S. Read. The in situ strength of massive granite around excavations. In P.K. Kaiser and D. McCreath, editors, *Proc. 16th Canadian Rock Mechanics Conference, Sudbury*, pages 1–10, 1992.

Onagi, D.P. S.G. Keith, and G.W. Kuzyk. Non-explosive excavation technique developed for the excavtion of AECL's Mine-by Experiment Test Tunnel at the Underground Research Laboratory. In *Proc. 10th Annual Canadian Tunnelling Conference, Banff*, pages 393–403. The Tunnelling Association of Canada, 1992.

Pollard, D.D. & A. Aydin. Progress in understanding jointing over the past century. *Geological Society of America Bulletin*, 100:1118–1204, August 1988.

Read, R.S. & C.D. Martin. Monitoring the excavation-induced response of granite. In J.R. Tillerson and W.R. Wawersik, editors, *Proc. 33rd U.S. Symposium on Rock Mechanics, Santa Fe*, pages 201–210. A.A. Balkema, Rotterdam, 1992.

Talebi, S. & R.P. Young. Microseismic monitoring in highly stressed granite: Relation between shaft-wall cracking and in situ stress. *Int. J. Rock Mech. Min. Sci. & Geomech Abst.*, 29(1):25–34, 1992.

Rockbursts and Seismicity in Mines, Young (ed.) © 1993 Balkema, Rotterdam, ISBN 90 5410 320 5

Stress change monitoring using induced microseismicity for sequential passive velocity imaging

S.C. Maxwell & R. P. Young
Engineering Seismology Laboratory, Department of Geological Sciences, Queen's University, Kingston, Ont., Canada

ABSTRACT: A new technique to image temporal velocity changes, with passively monitored induced microseismicity, is proposed and applied to data recorded at Falconbridge's Strathcona Mine, Sudbury, Canada. The motivation for this work is that the method offers the ability to exploit historical seismic data, to back analyze the temporal changes in the velocity structure. The sequential passive source velocity imaging method involves cross-correlating the spectra of event doublets, recorded during different periods, to accurately measure the arrival time delay between the events. The relative event location is then computed from the time delays and used to correct for the temporal travel time delay for the differences in locations. Corrected time delays are then used to compute images of velocity differences. Controlled blast data was used to examine the accuracy of relative locations, which was found to be 2.8 m compared to 13.5 m for the calculation of absolute locations. Corrected blast doublet time delays were inverted to test the sensitivity of the method, which confirmed the occurrence of no significant velocity changes. A case study of imaging velocity differences with event doublets recorded at different times, demonstrated a correlation between velocity decreases and both the location of induced microseismic events and a m_N 2.6 tremor. The velocity decreases were attributed to fracture unclamping due to stress field rotations resulting from mining. The microseismic data was also used to compute a static image of the velocity structure, representative of the average velocity over time. The events were found to locate in zones of anomalously high velocity in the static velocity image, corroborating similar observations made in both mining-induced and natural seismicity studies.

1 INTRODUCTION

Recent studies have shown an association between zones of high P-wave seismic velocity and the hypocentral location of seismic events and between low velocity and aseismic zones, for both induced seismicity (e.g., Young and Maxwell 1992, and Maxwell and Young 1992) and naturally occurring seismicity in seismogenic zones (e.g., Ishida and Hasemi 1988, Zhao 1990, Lees 1990, Lees and Malin 1990, Popandopulo 1990, Michael and Eberhart-Phillips 1991, Amato et al. 1991, Ogata 1991, and Nicholson and Lees 1992). Furthermore, correlations have been observed between high and low velocity and decreased and increased *b*-values respectively (Ogata 1991), increased radiated seismic moment (Michael and Eberhart-Phillips 1991) and coseismic displacement (Nicholson and Lees 1992). The mechanism explaining this velocity-seismicity correlation is not well understood, but is believed to be either due to strength (e.g., Michael and Eberhart-Phillips 1991) or stress (e.g., Popandopulo 1990 and Young and Maxwell 1992) effects, or a combination of the two. The strength mechanism postulates that the velocities are mapping natural stiffness heterogeneities, where high velocities correspond to competent rock capable of storing sufficient strain energy to be the site of a seismic event. Alternatively, the stress mechanism postulates that the velocity anomalies map stiffness heterogeneities produced by stress changes, where high velocities correspond to highly stressed zones. The high stresses will preferentially close fractures in the rock, increasing the rock mass stiffness and capability to store strain energy. In mining applications, several researchers have reported stress effects accounting for observed velocity heterogeneities (e.g., Mason 1981, Kormendi et al. 1986, Young et al. 1990, McGaughey 1990). Furthermore, direct observation of core discing has been used to confirm the presence of highly stressed ground in a seismically active-high velocity anomaly (Young and Maxwell 1992). Regardless of the mechanism, this seismicity-velocity structure correlation offers a practical tool to probe the seismic potential of a rock mass, particularly in mining-induced seismicity applications.

In order to isolate temporal stress variations from static strength and/or lithologic effects, temporal sequential velocity imaging techniques using controlled source data have been used (e.g., Kormendi et al. 1986, Young et al. 1989, McGaughey 1990, and Maxwell and Young 1992, which will be hereafter referred to as Paper I). In Paper I, an association was also observed between zones of velocity decrease and the location of the induced seismicity, corresponding to zones of high velocity in the static velocity structure. The velocity decrease was attributed to stress reductions reducing the clamping stress on fractures, facilitating the release of stored strain energy.

The objective of this paper is to evaluate the potential of a sequential imaging velocity difference from passive source data. The technique involves cross correlation of event doublets recorded during different periods, computing relative event locations to correct the measured time differences for differences in hypocentral coordinates, and then computing images of velocity difference. In Paper I, evaluation of sequential controlled source imaging techniques demonstrated the improved accuracy of computing velocity difference images from time delays along common ray paths over computing differences in two successive velocity images. The computation of the difference between two successive passive source images may be more inaccurate, since differences in the event locations of the two sets may cause different ray sampling and associated spatial resolution limitations in the two images. Alternatively, inversion of passive source delay times will ensure identical sampling if the event doublets are located close together. Furthermore, cross-correlation is a more accurate method of measuring time delays, compared to subtracting two visually picked times. Controlled source data will be used to test the accuracy of relative event locations and subsequent time difference corrections. Induced seismicity recorded at Falconbridge's Strathcona Mine, Sudbury, Canada before and after a m_N 2.6 tremor will be used to further analyze the relationship between seismicity and temporal velocity changes.

2 RELATIVE EVENT LOCATION

Relative event locations may be computed by inverting travel time differences between two events (e.g., Evernden 1969, Ito 1990). The technique determines the relative location of a secondary event relative to a "master event", which can be computed with relatively greater precision than the absolute location determined by the arrival times of the event. Relative event locations may be calculated by applying Geiger's method (e.g., Lee and Stewart 1982) to the measured travel time differences, which are inverted instead of the travel time residuals. Multiple iterations of the linearized technique can be applied by correcting the measured delays for the relative location computed in the previous iteration.

In order to examine the accuracy of relative event locations, P-wave arrival time data recorded from a series of controlled explosions at known locations was used. The blasts were part of a 3D controlled source imaging experiment, performed during June 1991 in an area of Strathcona Mine known as the "abutment pillar" (Figure 1). The blasts were recorded by a microseismic monitoring system (Young et al. 1989) comprised of 37 sensors, 31 high gain single axis accelerometers, and 6 triaxial accelerometers recorded on dual gain. Arrival time differences of a total of 17 blast "doublets", at an approximate spacing of 30 m, were calculated using a cross-correlation algorithm (Paper I) to an accuracy of 0.05 ms. Furthermore, absolute arrival times of each explosion were visually measured for signals with a clear arrival, with an average estimated uncertainty of 0.2 ms or four sample points. Absolute arrival times were used to compute absolute locations for the 17 secondary events of the doublets. Figure 2 shows the errors in the absolute locations, which were computed by comparison with the known blast locations. On average the locations were in error by 13.5 m. Relative locations were then computed using the measured arrival time differences, giving an average accuracy of 2.8 m (Figure 2).

3 SEQUENTIAL PASSIVE IMAGING

The application of sequential imaging techniques to corrected arrival time delays was first tested using the blast doublet data discussed above. The blasts were recorded within a 2 day period, during which no temporal velocity changes are likely to have occurred. Therefore, the corrected arrival time data should indicate no velocity change. Figure 3 shows a histogram of the corrected arrival time delays. The standard deviation is 0.24 ms, which is believed to represent the data uncertainty corresponding to both the measurement error and correction errors due to the location inaccuracies. The arrival time delays were then scaled using an average velocity of 6000 m/s, in order to ensure similar scaling of velocity perturbations as found in a static image (Paper I). No coupling was used between the velocity and hypocentral components of the passive source image, following studies of the importance of the coupling performed on Strathcona data which indicated little effect in resulting velocity images (Maxwell and Young, in press). Damped least squares images were computed (see Paper I for a description of the 3d imaging methodology) for a velocity model consisting of 150 cells (30m x 30m x 25m), 5 in the mine easting direction, 6 in the mine northing direction and 5 in depth. Stochastic error estimates were also computed using a data error of 0.28 ms, which was determined using the variance of the inversion residuals corrected for the degrees-of-freedom equal to the number of residuals less the number of cells. The resulting image contained no cells where the velocity difference was larger then the corresponding error estimate.

Sequential passive source velocity images were also computed for a region of Strathcona Mine known as the "remnant pillar" (Figure 1). The remnant pillar was the site of

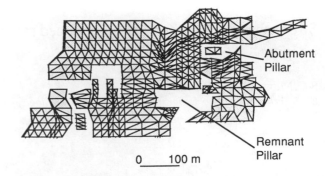

Figure 1. Longitudinal section of Strathcona Mine (shaded areas have been mined-out and backfilled).

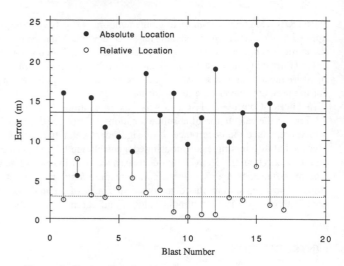

Figure 2. Source location errors.

significant induced seismic tremors and associated ground falls, which were attributed to mining in an adjacent cut-and-fill stope. Event doublets were selected between microseismic event catalogues recorded during October 1990 and April 1991, recorded with the instrumentation discussed above. A total of 35 pairs were selected such that both events contained similar number of arrival times on coincident sensors, had relatively close absolute hypocentral locations, had low estimated location errors, and similar focal mechanisms. Arrival time delays between the pairs were measured, and used to compute relative locations. Figure 4 is a histogram of the resulting corrected arrival time differences, with a standard deviation of 0.31 ms. Comparison with the blast data show a significantly different distribution (0.00038% significance level from a F-test), indicating a possible temporal change in the velocity structure between the recording periods. A velocity model was constructed of 96 rectangular cells (30m x 30m x 23m) 6 in the mine easting direction, 4 in the mine northing direction, and 6 in depth. A damped least squares image was computed with corresponding stochastic error estimates. The data residuals were again used to estimate the data error, 0.27 ms in this case. The resulting image show zones of significant velocity change in relationship to the error estimates. Presentation and interpretation of the images will be included in the next section.

4 RELATIONSHIP BETWEEN SEQUENTIAL IMAGE AND INDUCED SEISMICITY

A static passive source velocity image of the remnant pillar was calculated using a sequence of 110 microseismic events recorded

374

at Strathcona Mine between May and August 1990, corresponding to 3,380 arrival time measurements. The static image velocity model was parametrized using the 96 cell sequential image model, and was constrained on the boundaries by an initial coarse mine-wide static velocity model. The error estimates of the static velocity image are larger then the observed temporal velocity changes. Therefore, the image is representative of the average velocity structure over the period of the sequential image. Figure 5a and 6a shows a perspective view of the passive velocity image results on a vertical cutting plane, along with the geometry of the active mining stope. Also shown is an isosurface of 3D contouring of volumetric event density, which is a 3D contour surface essentially enclosing a cluster or swarm of microseismic events. The event density was calculated using 1200 microseismic events recorded between the event catalogues. Figure 5b and 6b show an orthogonal view of the cutting planes superimposed with the location of the events located within 30 m of each plane, and the hypocentre of a m_N 2.6 tremor recorded during March 1991. Figure 5c and 6c show the the stochastic velocity error estimates. It is interesting to note the correlation between concentrated microseismic activity and regions of high seismic velocity greater than 6400 m/s, which corroborates the seismicity-high velocity relationship.

Figures 7 and 8 are the velocity difference images corresponding to the planes plotted in Figures 5 and 6, respectively. Velocity anomalies are only plotted in the areas where the velocity change is greater than the error estimate.

The velocity difference images show velocity decreases in the zones of concentrated microseismic activity corresponding to high velocity zones in the static image. Similar results were obtained during a sequential controlled source velocity imaging case study at Strathcona (Paper I). Furthermore, Figure 7b shows that the m_N 2.6 tremor is also associated with a zone of velocity decrease. The mechanism of the association between the seismicity and velocity decrease may be due to unclamping

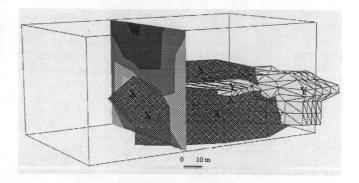

Figure 5a. Perspective view of the remnant pillar showing the active mining stope (marked "Y"), an isosurface of event density (marked "X") and the static velocity image on a vertical north-south cutting plane.

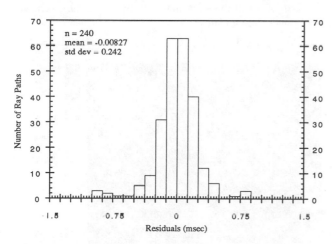

Figure 3. Histogram of location corrected arrival time delays for the blast doublets.

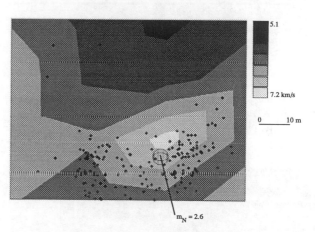

Figure 5b. Orthogonal view of the north-south static velocity cutting plane, superimposed with the location of events located within 30 m of the plane.

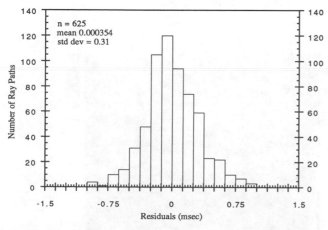

Figure 4. Histogram of location corrected arrival time delays for the event doublets.

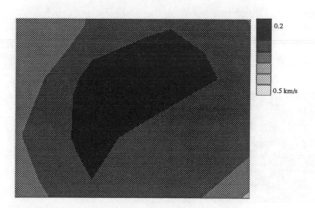

Figure 5c. Velocity image error estimates.

of the fractures, possibly due to rotations of the stress field resulting from the concurrent mining activity. Such an unclamping would facilitate the release of stored strain energy. Damage from the induced tremor may have also contributed to the observed velocity decrease, and could be isolated by applying the sequential imaging technique to the foreshock/aftershock sequences.

5 CONCLUSION

A technique to image temporal velocity variations has been proposed. Cross-correlated arrival time delays, between the waveforms of event doublets from different time periods, are used to compute relative locations. Arrival time delays are corrected for the relative location and inverted for the velocity change. A controlled source experiment was used to illustrate the increased accuracy of relative locations (average error of 2.8 m), compared to absolute locations (average error of 13.5 m). Inversions of the arrival time residuals confirmed that no significant velocity changes, relative to error estimates, were

obtained. A case study of velocity differences, imaged with event doublets recorded in October 1990 and April 1991, showed significant scatter in the corrected arrival time delays relative to the controlled source data. The resulting velocity difference images demonstrated a correlation between velocity decreases and both the location of induced microseismic events and a m_N 2.6 tremor. The locations of the events corresponded to zones of anomalously high velocity in a static velocity image, computed with passively recorded microseismic events. The observed velocity decreases were attributed to fracture unclamping due to stress field rotations resulting from mining.

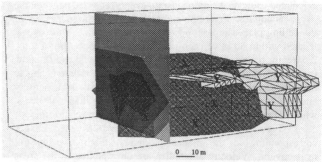

Figure 7a. Perspective view of the remnant pillar showing the active mining stope (marked "Y"), an isosurface of event density (marked "X") and the sequential velocity image on a vertical north-south cutting plane.

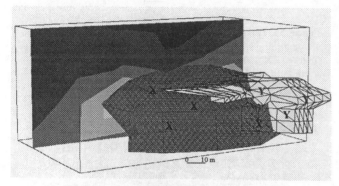

Figure 6a. Perspective view of the remnant pillar showing the active mining stope (marked "Y"), an isosurface of event density (marked "X") and the static velocity image on a vertical east-west cutting plane.

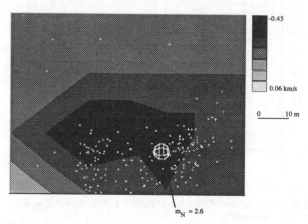

Figure 7b. Orthogonal view of the north-south sequential velocity cutting plane, superimposed with the location of events located within 30 m of the plane.

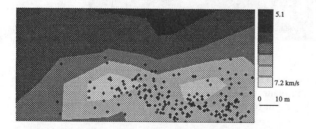

Figure 6b. Orthogonal view of the east-west static velocity cutting plane, superimposed with the location of events located within 30 m of the plane.

Figure 6c. Velocity image error estimates.

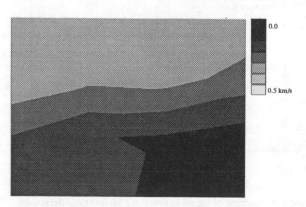

Figure 7c. Velocity image error estimates.

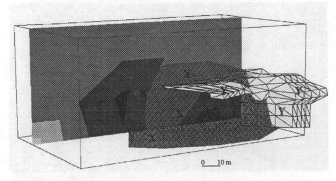

Figure 8a. Perspective view of the remnant pillar showing the active mining stope (marked "Y"), an isosurface of event density (marked "X") and the sequential velocity image on a vertical east-west cutting plane.

Figure 8b. Orthogonal view of the east-west sequential velocity cutting plane, superimposed with the location of events located within 30 m of the plane.

Figure 8c. Velocity image error estimates.

ACKNOWLEDGEMENTS

We wish to thank the Natural Science and Engineering Research Council of Canada, the Mining Research Directorate, Queen's University and Falconbridge Limited for funding of the work described here. We also wish to thank members of the Engineering Seismology Laboratory for assisting various aspects of the work.

REFERENCES

Amato, A., Alessandrini, B., Cimini, G.B., and Selvaggi, G.. 1991. Three-dimensional P-velocity structure and seismicity in the upper mantle of Italy. EOS Trans. Am. Geophys. Union, 72, 349.

Evernden, J.F.. 1969. Identification of earthquakes and explosions by use of teleseismic data, J. Geophys. Res., 74, 3828-3856.

Ishida, M., and Hasemi, A.K.. 1988. Three-Dimensional Fine Velocity Structure and Hypocentral Distributions of Earthquakes Beneath the Kanto-Tokai District, Japan. J. Geophys. Res., 93, 2076-2094.

Ito, A. 1990. Earthquake swarm activity revealed from high resolution relative hypocenters-clustering of microearthquakes, Tecton., 175, 47-66.

Kormendi, A., Bodosky, T., Hermann, L., Dianiska, L., and Kalman, T.. 1986. Seismic measurements for safety in mines. Geophysical Prospecting, 34, 1022-1037.

Lee, W.H.K., and Stewart, S.W. 1981. Principles and applications of microearthquake networks, special issue of Advances in Geophysics, Academic Press, New York.

Lees, J.M.. 1990. Tomographic P-wave velocity images of the Loma Prieta earthquake asperity. Geophys. Res. Lett., 17, 1433-1436.

Lees, J.M., and Malin, P.E.. 1990. Tomographic images of P-wave velocity variation at Parkfield, California. J. Geophys. Res., 95, 21793-21804.

Mason, I.M. 1981. Algebraic reconstruction of a two-dimensional velocity inhomogeneity in the High Hazles seam of the Thoresby colliery, Geophys., 46, 298-308.

Maxwell, S.C., and Young, R.P.. 1992a in press. Sequential velocity imaging and microseismic monitoring of mining-induced stress change, submitted to a special issue of J. Pure and Applied Geophys. on Induced Seismicity.

Maxwell, S.C., and Young, R.P.. 1992b in press. A comparison between active source and passive source images. submitted to Bull. Seismo. Soc. Am..

McGaughey, W.J.. 1990. Mining Applications of Crosshole Seismic Tomography. Ph.D. dissertation, Queen's University, Kingston, Ontario, Canada.

Michael, A.J., and Eberhart-Phillips, D.. 1991. Relations among fault behaviour, subsurface geology, and three-dimensional velocity models. Science, 253, 651-654.

Nicholson, C, and Lees, J.M.. 1992. Travel time tomography in the Northern Coachella Valley using the aftershocks of the 1986 M_L 5.9 Paula Springs earthquake. Geophys. Res. Lett., 19, 1-4.

Ogata, Y., Imoto, M., and Katsura, K.. 1991. 3D Spatial variation of b-values of magnitude-frequency distribution beneath the Kanto district, Japan, Geophys. J. Int., 104, 135-146.

Popandopulo, G.A.. 1990. Three-dimensional velocity model using small earthquakes data. (in Russian) Regime Geophysical Observations, Acad. Sci. USSR, Moscow-Garm, 104-125.

Young, R.P., Talebi, S., Hutchins, D.A., and Urbancic, T.I.. 1989. Source mechanism studies of mining-induced microseismic events at Strathcona Mine, Sudbury, Canada. Special Issue of Pure and Appl. Geophys. on Mining Induced Seismicity, 129, 455-474.

Young, R.P., Hutchins, D.A., McGaughey, W.J.. 1990. Seismic imaging ahead of mining in rockburst prone ground, in *Rockburst and Seismicity in Mines*. (Ed. Charles Fairhurst, Balkema, Rotterdam, 231-236).

Young, R.P., and Maxwell, S.C.. 1992. Seismic characterization of a highly stressed rock mass using tomographic imaging and induced seismicity. J. Geophys. Res, 97, 12361-12373.

Zhao, D.. 1990. A tomographic study of seismic velocity structure in the Japan Islands, Ph.D.dissertation, Tohoku University.

Seismicity in the Sudbury Area mines

D. M. Morrison
Inco Limited, Mines Research Department, Copper Cliff, Ont., Canada

ABSTRACT: The problem of rockbursts and seismicity in a few of the mines of the Sudbury Basin was recognized some 50 years ago and since the early 1980's has become more widespread and the severity has increased. Today, about half of the deep mines in the Basin are monitored continuously by microseismic systems providing source locations and some provide full wave-form analysis of events. However, not all deep mines in the area are seismically active and while there are some variations in the depth, geology and mining conditions in these mines, several display little or no seismicity. This is most clearly illustrated by groups of two mines which are coupled together and which appear to have very similar mining configurations but which display dramatically different levels of seismicity. The implications of this phenomena for the criticality of some of the factors which control the occurrence and onset of seismicity in mines are discussed.

1 INTRODUCTION

From the perspective of the mining industry, the most important aspect of mining-induced seismicity is the occurrence of the largest events as they have the potential to cause severe damage. The smaller events typically cluster, in time and space, around the creation of an excavation and their effects are therefore more easily managed. The larger events tend to be much less predictable, occurring with little obvious correlation to the timing of parts of the mining cycle and in locations which can only be loosely correlated with major excavations or major geological structures.

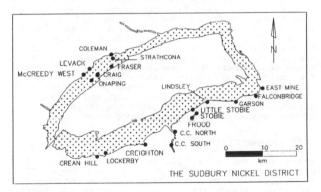

Figure 1. Area of the Sudbury Basin showing the Operating Mines.

Of the 19 hardrock mines operating over the last few years in the Sudbury Basin (Figure 1), only about 8 are subject to significant levels of seismicity. From the mining perspective, it is important to be able to identify which of the current and future mines in the Basin are likely to become seismically active and which are not. Most of the current research into mining-induced seismicity in the area is focused on the seismic activity itself, in order to gain an understanding of the source parameters and focal mechanisms of the events. Although detailed analysis of seismic events can reveal an understanding of the mechanisms and processes involved in the violent release of energy in rock, it can reveal nothing about the processes occurring within aseismic mines. The Sudbury Area mines exist in a relatively well constrained regional geological setting and appear to be generally very similar but display a range of seismic responses to mining. Consequently, a comparison of the characteristics of both seismic and aseismic mines, which has so far been ignored, could provide a valuable contribution to the study of mining-induced seismicity.

In the case of the mines which do become seismically active, there are other important questions to be addressed. It is reasonable to suppose that those characteristics which influence the occurrence of seismicity may also influence the timing, location and severity of the activity. Questions which can be raised in this regard are firstly, when in the life of an active mine will the activity become significant, and are there conditions under which such activity will cease or become insignificant before complete extraction is achieved? Secondly, how rapid is the transition from the initial aseismic phase to the seismic phase and, similarly, is it possible to identify or characterise the transition to a final aseismic phase? Thirdly, during the seismic phase, what is the distribution and severity of the largest events likely to be, with respect to mining activity, to the excavation of the mine as a whole and to the stability of mine-wide geological structures?

The mines in the Sudbury Area display a diversity of response to mining which may make it possible to discern those attributes of a mining environment which make a violent response to mining on a large scale more or less likely. The breadth of scope of a project to study the aseismicity/seismicity relationships in the Sudbury Basin makes it very difficult to arrive at definitive conclusions early in the project. Presented here is an initial report on an on-going project whose principal objective is to promote discussion rather than to present definitive conclusions.

2 SUDBURY AREA MINES

The seismically active mines are distributed in both the North and South Ranges of the Basin, areas of quite different lithological and structural complexity. The Sudbury Area mines have been divided into four types (Naldrett 1984) as shown in Table 1 and the more recently operating mines have been categorized according to the degree of seismicity they have displayed. All three categories of activity are represented in all four types of deposits. In each of the four types there is a pair of mines which are located very close together or are even connected underground and which show sharply contrasting responses to mining. Because these 'mine couples' are similar in almost everything except the level of seismicity, they offer a unique insight into the factors which most strongly influence the occurrence of seismicity. The seismic and aseismic members of these couples are Falconbridge and East Mines; Frood and Stobie Mines; Copper Cliff North and Copper Cliff South Mines; Lockerby and Crean Hill Mines; and Strathcona and Fraser Mines. Three of the South Range couples are briefly described.

TABLE 1. Four Types of Ore Deposits in the Sudbury Basin, showing mines with a high level of seismic activity (**), a moderate level of seismic activity (*) or no seismicity (). [(I) Inco Limited, (F) Falconbridge Limited]

NORTH RANGE DEPOSIT			SOUTH RANGE DEPOSIT		
	Levack	(I)	**	Creighton	(I)
	McCreedy	(I)			
				Little Stobie	(I)
*	Coleman	(I)			
**	Strathcona	(F)		Crean Hill	(I)
*	Fraser	(F)	**	Lockerby	(F)
*	Onaping	(F)			
*	Craig	(F)			

DEFORMED MARGINAL DEPOSIT			OFFSET DEPOSIT		
**	Falconbridge	(F)	**	C.C. North	(I)
	East	(F)		C.C. South	(I)
*	Garson	(I)	**	Frood	(I)
				Stobie	(I)

2.1 Falconbridge and East Mines

Falconbridge Mine and East Mines (Fig. 2) are located on the south-east end of the Basin and are Deformed Marginal deposits of the South Range. The orebodies are set in identical geological environments, along the faulted contact between the South Range Norite in the footwall and very siliceous metasediments or ('greenstones') in the hangingwall. The emplacement of the ore at both Falconbridge and East Mines is dominated by the Main Fault, with up to 0.7m of gouge, which forms the footwall contact. The Falconbridge #5 Shaft Orebody extended some 1200 to 1500m along strike, the East Mine Orebody extended 650m along strike and the two were separated by a waste zone of about 650m. The Falconbridge Mine is dissected by several other major faults, the No.1 Flat Fault, the 14E and 78E Vertical Faults and the No.1 Orepass Fault. Of these only the No.1 Flat Fault affects the East Mine, intersecting it at the 6050 level, 2000m below surface. In all other respects the geology of the two orebodies is very similar.

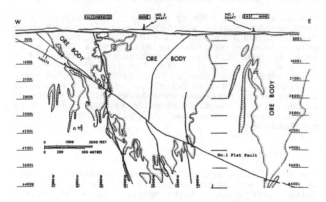

Figure 2. Longitudinal Section of Falconbridge and East Mines.

The Falconbridge Mine had experienced a few isolated events since the early 1950's but by the early 1980's was becoming continuously active and a monitoring system was installed. By 1983 some large events (M>2.0) had occurred. In 1984 a series of very large events (M>3.0) occurred within a few hours, causing severe damage from 4025 level to 4375 level resulting in four fatalities and the subsequent closure of the mine. East mine which was being mined at the same time was virtually aseismic with only one event recorded during the 1980's.

The series of large events at Falconbridge Mine was attributed to fault-slip movement on the major faults in and around the hoistroom pillar area on the 4025 level (West 1985). The significant faults in this area were the Main Fault, the No.1 Flat Fault and the No.1 and No.2 Orepass Faults. Another contributing factor could have been that the mine was largely mined out except for the block of ore in the middle of the

orebody around the #9 Shaft Hoistroom. The No.1 Flat Fault was suspected of being responsible for the one significant seismic event recorded at the East Mine 6050 level in 1983.

A comparison of these two mines suggests several factors which dominate the marked difference in seismic response to mining. Firstly, the difference in the size or lateral extent of these two mined-out orebodies relative to the mechanical strength of the host rocks. Perhaps East Mine is simply too small to cause convergence large enough to generate large releases in energy. Secondly, the structural complexity caused by the existence of the major faults could be a significant factor. Finally, it may be that the proximity of the large excavation of Falconbridge Mine is sufficient to cause East Mine to react more passively.

2.2 Frood and Stobie Mines

The lithological setting of Frood and Stobie Mines is identical with moderately jointed fine-grained metasediments in both the footwall and hangingwall. The footwall is only distinguished from the hangingwall by the presence of a prominent shear zone (fault) about 7m into the footwall which runs more or less parallel to the footwall ore contact.

The Frood Mine was one of the most seismically active mines in Canada in the 1940's and was a major part of a previous investigation (Morrison, 1942). The mining sequence was identified as a major contributing factor to the problem of rockbursts within the orebody and while the occurrence of rockbursts was significantly reduced by making major strategic changes to the mining sequence, very large events continued to be recorded until the 1960's. These large events were felt throughout the mine and on surface but usually caused little or no damage to the mine workings and would today be recognised as fault-slip type events in the host rocks. However, the mine is now in the final stages of pillar recovery, and appears to have made a transition to a completely aseismic phase, displaying very large, gradual displacements along major mine-wide discontinuities.

Stobie Mine is very similar to the Frood Mine in depth and shape although the present extent of extraction at Stobie Mine is significantly less. However, bulk mining has occurred at depths at which bursting was recorded in the Frood Mine. The higher grade, eastern end of Stobie Mine is being mined by VRM mining and is roughly about 1000 ft along strike. Because of the geological similarity to Frood Mine, only two of the Falconbridge/East Mine explanations can be employed for Stobie's relatively passive response. Firstly, the limited extent of the currently mined-out zone, which is very similar to East Mine, and secondly the effect of the close proximity of the large excavation of the extensively mined-out Frood Mine.

2.3 Copper Cliff Mines

There are two mines located on the Copper Cliff Offset; Copper Cliff North Mine and Copper Cliff South Mine. The Offset is transected by the Creighton Fault and straddling it are a series of distinct orebodies. At the south and north ends of the Offset respectively, there are the largest orebodies of the two mines, the South Mine 810 orebody and the North Mine 120 Orebody. The 120 Orebody sulphides are entirely contained within an envelope of quartz diabase which is located at the boundary of the Creighton granites to the west and the Stobie metavolcanics to the east. South Mine is located within the matasediments and the 810 Orebody is emplaced with quartz diabase in the footwall and metasediments in the hangingwall. While the stress regime of the South Mine 810 orebody is consistent with the Basin's regional stress model, the field stresses of North Mine's 120 Orebody do not conform to this model.

At North Mine the maximum stress is roughly horizontal and perpendicular to the strike of the ore and increases with depth. However, the intermediate and minor principal stresses are anomalously low, only 60% of those predicted by the regional stress model. Numerical modelling studies of a series of major ground failures and collapses which occurred in 1986 (Morrison and Galbraith 1990) confirmed that these could not have occurred under the conditions of the regional stress model but could have occurred under this anomalous stress regime. There are no major faults through the orebody but there are two principal joint sets which are mutually perpendicular and sub-vertical. In plan, the orebody also has a slightly arcuate shape with the azimuth of the north and south limbs rotating through about 25 degrees. The strike of the southern limb of the orebody is roughly perpendicular to the maximum principal stress and is oriented about 45 degrees to the two joint sets. The intermediate stress is vertical and the minimum stress is along the strike of the southern limb.

In contrast to South Mine, North Mine has proven to be one of the most seismically active in the Basin and although the confining stresses are anomalously low, mining has induced several very large events (M>3.0) over the past three years. The mine was re-opened as a bulk mining operation in 1983, mining stopes 130m high, which is about twice the normal height. The onset of seismicity in 1986 was very rapid with some large events occurring within a few days of the first indication of any kind of seismicity. The occurrence of seismicity in such a small mined-out zone at such modest depth was completely unexpected.

Once the microseismic monitoring system was installed, it revealed that the majority of events were occurring in the hangingwall of the orebody with many fewer events occurring in the northern end of the orebody. At the lower levels in the mine, between 3400 level and 3935 level, the two limbs are separated by a waste pillar just where the strike changes azimuth. Numerical modelling indicated that this pillar is large enough to isolate the effect of mining the two limbs of the orebody. While it is clear that a significant degree of failure occurred in the walls of the entire orebody, the more violent response and the most severe damage, was generally restricted to the southern limb. At higher levels where there is no waste pillar and the arcuate shape is less pronounced, there is little difference in the severity of seismicity in the north and south ends of the orebody.

The response to mining of the southern limb of North Mine's 120 Orebody as compared to the northern limb of the 120 Orebody or to the 810 Orebody suggests three contributing factors. Firstly, the extent of extraction at South Mine is much smaller than at North Mine and the stress regime at South Mine conforms to the regional model. Secondly, the quartz diabase which envelops the 120 Orebody at North Mine is inherently more brittle than the host rock at the 810 Orebody at South Mine. Thirdly, since the quartz diabase envelops the entire 120 Orebody even when the northern limb changes orientation relative to the joint sets, it may be that the size, shape and relative orientation of the rock blocks in the walls influences the nature of the response.

3   NUMERICAL EXAMPLE

The behaviour of a very simple system of rock blocks was examined using a 2-dimensional Udec model representing a rectangular excavation (40m x 10m) in a system of 5m square blocks. The blocks were oriented at 0,15,30 and 45 degrees to the long axis of the excavation (0 indicating perpendicular to strike of excavation). The maximum stress (60 MPa) was perpendicular to the long axis of the excavation with minimum stress of 20 MPa. The joint behaviour was simple Mohr-Coulomb, with both sets with identical properties or one set slightly weaker.

Figure 3 shows that the total energy released by the systems is greatest for joints oriented at 30 degrees as would be expected. But it also reveals that the rate of release is significantly different and given the sensitivity of the activity in some Sudbury mines (Morrison et al. 1992) the relative change in shape of these curves is easily sufficient to describe either passive or active response to mining.

4   RESPONSE OF A ROCKMASS TO MINING

Seismicity in mines is the result of the manner in which the walls of an excavation converge into the opening in response to the applied loads and how regional pillars resist that convergence. The scale of the excavation, i.e. the magnitude of the deviation from equilibrium of the system, determines the magnitude of the energy which has somehow to be dissipated in order for the system to regain equilibrium. How the system does this, how the energy in the system is dissipated, determines whether a mine becomes seismically active.

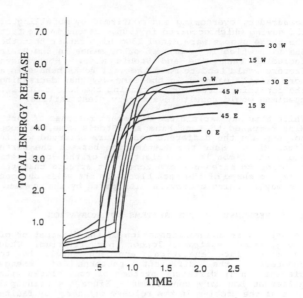

Figure 3.   Total Energy released by system of 5m square rock blocks. Angle gives orientation of blocks (0 = perpendicular to strike of the excavation). Curves denoted by 'E' indicate joints of equal properties, those denoted by 'W' indicate the orientation of the weaker joint set.

Laboratory testing has shown how different rock lithologies or different joint orientations and different loading rates combine to cause either a passive or a violent failure of a rock specimen. The behaviour of the specimen is also influenced by its relative dimensions and boundary conditions. Short specimens exhibit a more ductile response than do tall, thin specimens of identical material. Increasing confining stress produces a more ductile response and the kind of shear applied by the plattens also influences its response. Similar factors apply to in situ behaviour and for a given excavation size, material properties and loading conditions, the total energy to be released by the system should be the same whether the response is violent or passive. Clearly the rate of energy release and the pattern of energy dissipation as the system equilibrates, indicates the nature of the response. A uniform pattern of energy release indicates a passive response and a non-uniform release rate indicates a more violent, seismic response.

Within the mines of the Sudbury Basin the orebody host rocks show some geological diversity but there is a fairly limited range of geomechanical behaviour. While there is certainly a range of material properties, all of the rock types can be classified as relatively strong and the different lithologies have, at one time or place, displayed both passive and seismically active responses to mining. The behaviour of a host rockmass around an orebody, like any other material, is neither intrinsically brittle nor intrinsically ductile. Of course, the inherent properties of the material do play a role in the behaviour of the material under load, but only in limiting how brittle or ductile the material may become. The relative ductility of the behaviour is determined also by the boundary conditions (field stresses and excavation geometry and sequence). In the Sudbury Basin at least, it appears to be the boundary conditions rather than the inherent material properties which determine the nature of the response to mining and how that response may change as mining proceeds.

The seismic activity at the two larger mines, Frood and Falconbridge, could be explained simply because of the scale of the mine relative to the material properties of the surrounding rockmass. The scale of excavation is so extensive and the amount of energy to be dissipated so large that the system cannot accomplish this passively or gradually. The seismic activity at North Mine cannot be explained in this way since the scale of excavation is similar to Stobie and East Mines which responded passively. There are two conditions at North Mine which could explain the violent response of the host rocks. One is the lack of confining stress in the plane of the orebody as measured by overcoring and confirmed by modelling of the caving which occurred in 1986. Second is the fact the these stopes were mined 130m high rather than the 65m high which is common at other mines in the Basin including East Mine and Stobie Mine. Both these factors would tend to make the wall rocks behave in a more brittle manner, in the same way that decreasing the confining stress or increasing the height of test specimen would result in a more violent failure.

While this may explain the different response of North Mine compared to East Mine and Stobie Mine, it does not explain the different behaviours observed within North Mine. Here the distinction between the north and south limbs is the orientation of the joint sets relative to mined-out ore zone. It appears that the relative shape of the rock blocks in the walls changes the way in which energy is dissipated by the system.

5 RESPONSE OF ROCK SYSTEMS TO EXCAVATION

Seismicity is one manifestation of the dissipation of energy from a system no longer in equilibrium. Thus an appropriate analytical approach should be to understand the different patterns of energy dissipation in different systems of rock blocks with different boundary conditions. Since the principal part of the problem is the release of energy by fault-slip events, then the patterns of energy released by (or stored in) shear should also be of particular interest. These different patterns can then be interpreted as the response of a relatively more brittle (seismic) or ductile (aseismic) behaviour.

The boundary conditions are the field stress regime, the size and shape of the overall mined-out zone, the size and shape of regional pillars within the mined-out zone and the presence of major regional faults. The major regional faults are included as part of the boundary conditions because of the role they play in the mines in the Sudbury Basin.

In the South African goldfield the location, magnitude and frequency of the largest events is linked directly to the major faults which transect the reef (Brummer and Roarke 1988). A comparison of the Klerksdorp, Carletonville and Wilkomm areas, for example, suggests that the magnitude and frequency of events is related to the amount of throw on the faults which intersect the reef. The more faulted the reef and the larger the throw, the larger the remnant (waste) pillar and the larger the maximum magnitude of the events (Wilkom). The area with the least faulting has the smallest maximum magnitude events but the largest number of events (Carletonville).

The largest events in the Sudbury Basin mines are also fault-slip type events but here the major faults are typically geomechanically weak structures. They are often at least 10-100mm wide usually in-filled with mud gouge or a highly fractured or friable material and consequently they have very low shear strength. They are not likely to be the source of the largest events except where a section is locked-up as in a solid bridge (Morrison 1989). Very often the distribution of the most severe damage in these mines does not correlate directly with the largest events which may indicate that the larger events triggered other smaller but more damaging events.

At the Falconbridge Mine some of the M>3.0 events which occurred in June 1984 have been attributed to movement on the No.1 Flat Fault. However, at the location where the movement on this fault was measured there was no damage to the mine workings - the damage was located several tens of meters away. At Strathcona where much of the early activity was located in or around the Main Dyke and Small Dyke the most severe damage was not located at these dykes but again tens of meters away. At Creighton where several prominent faults have been identified either by mapping or from the small scale activity, the large damaging events seldom occur directly on the faults. Thus, while the major faults are playing a role in controlling the occurrence of seismicity, they do not directly cause the most severe damage as in South Africa. However, the largest events are fault-slip events and are almost certainly caused by slip along one of the high strength joints which define the system of rock blocks which are often (but not always) driven by aseismic slip on the major faults.

6 CONCLUSION

The way in which energy is dissipated in the Sudbury Mines can be investigated by systems of rock blocks defined by high shear strength discontinuities which may or may not be dominated by larger, more mobile faults. The geological similarity of the mines may suggest that the boundary conditions, rather than material properties, are dominant factors controlling the distribution of seismicity in the mines in the Sudbury Area. Rather than using numerical models to identify the very specific conditions and material properties which can simulate the occurrences of particular case histories, what is important at this stage is a thorough understanding of how comparatively simple systems respond to different mining configurations. This can be accomplished by examining the seismic (and aseismic) response of the area mines and by using numerical models to investigate the generic characteristics of energy dissipation of different block systems as they equilibrate.

Recently, we have begun to recognise seismicity as yet another of those non-linear dynamic systems, such as the weather, which can be studied and understood by the new science of chaos. But when they began to use choas theory, the meteorologists already had a fairly rigorous, mathematical understanding of how individual phenomena operated and chaos theory was the result of the interaction of these relatively simple systems.

In the study of mining-induced seismicity, the analogy with meteorology is a valid one since very minor variations in initial conditions can have dramatic effects on developing storm systems. However, while millions of detailed measurements are essential to provide local confirmation of how weather systems behave, time-lapse satellite photography of world-wide cloud patterns conveys far more understanding of the meteorological phenomena. Now that we have the data gathering systems in place in so many mines, it is appropriate to take a broader view of the phenomena and to develop a generic understanding of the way simple systems store and dissipate energy.

This approach to the aseismicity/seismicity relationships for so many mines is difficult because of the breadth and nature of the information which has to be synthesised. In the short term such a study will do little to alleviate the specific problems at individual mines. But this approach also helps to address a fundamental problem in the way the mining industry has approached rockbursting in the past, as a problem which has to be solved - rather than as a phenomena which has to be understood. The ability to develop solutions to the problem of rockbursts will only become possible when the behaviour of both seismic and aseismic mines is fully understood.

REFERENCES

Brummer, R.K. and A.J. Rorke 1988. Case Studies on large rockbursts in South African gold mines. 2nd Int. Symp. on Rockbursts and Seismicity in Mines, Minneapolis, pp 323-330. Rotterdam, Balkema.

Hedley, D.G.F., S. Bharti, D. West & W. Blake 1985. Fault-slip Rockbursts at Falconbridge Mine. Proc. 4th Conf. Acoustic Emissions and Microseismic Activity in Geological Structures and Materials, Penn. State USA.

Morrison, D.M. 1989. Rockburst Research at Falconbridge's Strathcona Mine. Pageoph, vol.129, Nos.3/4, pp.619-645.

Morrison, D.M. and J.E. Galbraith, 1990. A Case History of Inco's Copper Cliff North Mine. Proc. 31st U.S. Symposium on Rock Mechanics, Golden Colorado, pp.51-58. Rotterdam, Balkema.

Morrison, D.M., T. Villeneuve & A. Punkkinen 1991. Factors Influencing Seismicity in Creighton Mine. Proc. 5th Conf. on Acoustic Emissions and Microseismic Activity, Penn. State, USA.

Morrison, R.G.K. 1942. Report on the Rockburst Situation in Ontario Mines. Trans. CIM vol. 45, pp. 225-272.

Naldrett, A.J. 1984. Ni-Cu Ores of the Sudbury Igneous Complex. The Geology and Ore Deposits of the Sudbury Structure, Ontario Geological Survey Special Publication, pp.302-308.

Rockbursts and Seismicity in Mines, Young (ed.) © 1993 Balkema, Rotterdam, ISBN 90 5410 320 5

A laboratory experiment for development of acoustic methods to investigate condition changes induced by excavation around a chamber

Yoshiki Nakayama, Akira Inoue & Masahiro Tanaka
Nihon Public Co, Inc., Japan

Tsuyoshi Ishida
Yamaguchi University, Japan

Tadashi Kanagawa
Central Research Institute of Electric Power Industry, Japan

ABSTRACT: To develop acoustic methods to investigate condition changes around a chamber induced by excavation, we conducted a laboratory experiment simulating stress distribution around a chamber. In this experiment, cracks observed on surfaces of the specimen seemed to spread following the spread of AE hypocenters. P-wave monitoring along two paths through the specimen indicated that the amplitude decreases according to accumulation of microfractures represented by clusters of AE hypocenters while the velocity decreases according to relatively large fractures represented by open cracks. These findings demonstrate that AE and P-wave monitoring can be effectively used to detect progress of fractures around a chamber.

1 INTRODUCTION

For risk assessment of nuclear waste isolation systems, it is important to develop methods to investigate changes in the conditions around a chamber caused by excavation. To understand condition changes, if stress redistribution around a circular tunnel is considered, stress redistribution is divided into two types; one is release of radial normal stress (σr) and the other is increase in tangential normal stress ($\sigma \theta$). (See the left side of Fig. 1) The release of σr mainly results in displacement expanding to the chamber due to elastic expansion of rock itself and opening of cracks. The phenomenon can be measured by conventional multi-extensometers, a sliding micrometer (Kovari and Petter, 1983) and a borehole television system (Kanaori, 1983). The increase of $\sigma \theta$ results in fractures and spread of cracks, causing a decrease in supporting capacity of rock masses. For these phenomena, acoustic methods are useful as shown by the in situ experiments of Talebi and Young (1992) and we have also developed methods through small scale in-situ experiments (Ishida et al. 1986, 1992) and preliminary monitoring in a real underground chamber (Ishida et al. in press). However, generally speaking, it is not long since such modern acoustic methods started to be used. Therefore, in order to investigate and develop suitable methods, we conducted a laboratory experiment simulating stress distribution around a chamber. In this paper, we will describe some results obtained in the experiment.

2 METHOD

2.1 Specimen and method of loading

Around a chamber, tangential normal stress increases with the approximation to the chamber wall as shown on the right side of Fig. 1. To simulate stress distribution, load was eccentrically applied to a left half of the upper end of the specimen so as to make the vertical normal stress increase with approximation to one side of the specimen as shown in Fig. 2. The specimen was a rectangular parallelepiped Ohya tuff, measuring 300 mm high, 200 mm wide and 60 mm thick. In loading, a 200 tonf servo-controlled machine was used to control the displacement between the ram and the cross-head.

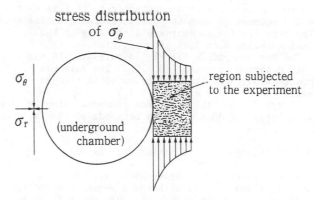

Figure 1. Illustration of stress distribution around an underground chamber.

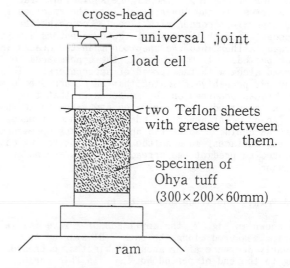

Figure 2. Specimen of the laboratory experiment.

2.2 Fracture monitoring methods

The load, the displacement, AE(Acoustic Emission), amplitude and velocity of the P-wave and strains on the surfaces were measured to observe progressive fracturing of the specimen (See Fig. 3).
The load was measured using a 50 tonf load cell

every 30 seconds before the maximum load and every 15 second thereafter. The measured values were stored on a floppy disk via a digital strainmeter controlled by a personal computer.

The displacement between the ram and the cross-head of the loading machine was measured using an LVDT.

AE events were detected by twelve disk-shape sensors, measuring 17.4 mm in diameter and 16.3 mm thick and having a resonance frequency of 150 kHz, set on the four surfaces in the upper part of the specimen. AE waveforms were recorded on floppy disks for source location and AE count rates were printed out, using the system shown in Fig. 4.

Amplitude and velocity of P-waves were measured every one minute along two paths from transmitter T to recievers R1 and R2.

3 RESULTS AND DISCUSSION

3.1 Load, displacement and AE count rate

The displacement was controlled as shown in Fig. 5; a displacement of 0.75 mm was caused for 90 seconds at the begining to shorten the time needed for the experiment and the displacement rate was then decreased and kept at a constant rate of 1 mm every 4,000 seconds to make the fracture progress slowly. The load started to decrease after showing the maximum value 9.85 tonf after 34 minutes.

The bar diagram in Fig. 5 indicates the AE count rate detected by sensor No. 10. The rate was counted only when the amplitudes of AE signals amplified by 70 dB exceeded 2.4 V.

Figure 5 shows that the load decreased step by step with bursts of AE along with progress of the failure.

3.2 AE hypocenters and surface cracks

Figure 6 shows the AE hypocenters located from the arrival times of P-waves using a measured velocity 2.11 km/s. In the figure, we can see that the AE hypocenters clustered in the upper-left corner of the specimen in period No. I and that they spread to the lower-right into the specimen with time in periods Nos. II, III and IV. The figure shows that cracks also spread to the lower-right into the specimen from the upper-left corner with time. Comparing the hypocenters with the cracks, the hypocenters always spread farther into the specimen than the cracks in each period. In other words, the cracks seem to spread along with the spread of hypocenters. This findings probably indicates that AE occurred before the cracks occurred on the surfaces or that the cracks occurred in the specimen and extended to the surfaces. Whichever it indicates, since it means that AE could be monitored ahead of crack appearance on the surfaces, we conclude that AE monitoring can be used to predict the spread of cracks around a chamber.

3.3 P-wave amplitude and velocity

As shown in Fig. 7, the amplitude of P-wave initial motion measured along the path T-R1 started to clearly decrease after about 30 minutes, corresponding to the end of period No. I. In this period, because AE hypocenters clustered at the left end of the specimen where receiver R1 was located, P-wave propagation along the path was most likely influenced by microfractures. However, P-wave velocity started to decrease after 44 minutes, and only two minutes after that it became difficult to measure owing to disappearance of the initial motion (see the waveform in Fig. 7). In period No. II including the time, because an open crack appeared just to the right of the position of the receiver R1 (see Fig. 6), the crack most likely caused a decrease in P-wave

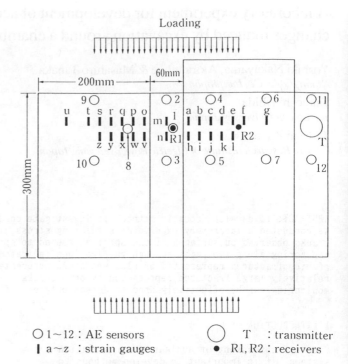

Figure 3. Locations of AE sensors, a transmitter and recievers for P-waves and strain gauges on the surfaces of the specimen.

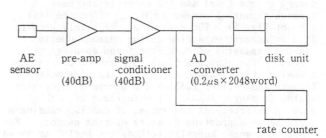

Figure 4. Block diagram of AE monitoring system.

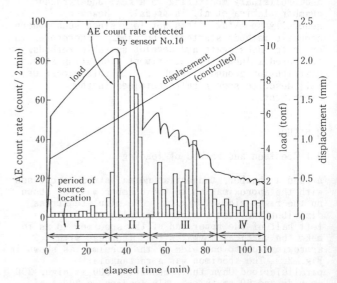

Figure 5. Load, displacement and AE count rate as a function of elapsed time.

384

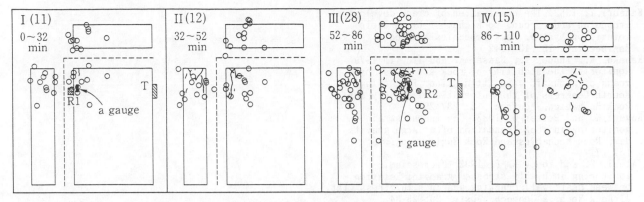

Figure 6. Located AE hypocenters (O) and cracks newly appearing on the surfaces of the specimen (solid lines) for each period from I through IV shown in Figure 5. The number of located AE hypocenters and elapsed time are shown in the upper left of each figure. T and R (R1 and R2) indicate positions of a transmitter and receivers for P-wave monitoring.

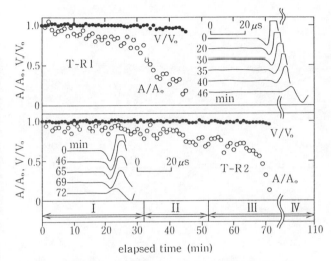

Figure 7. Changes in P-wave velocity (V) and P-wave initial motion amplitude (A) with elapsed time measured along the two paths T-R1 and T-R2. Both are shown as ratios to the initial values Vo and Ao.

velocity and disappearance of the initial motion.
Along path T-R2, a decrease of the amplitude seems to start after about 65 minutes. In the time corresponding to the middle of period No. III, AE hypocenters have spread to near the receiver R2 (see Fig. 6). Therefore, the decrease is most likely caused by the cluster of AE hypocenters as well as along path T-R1. The velocity started to decrease after 70 minutes and the initial motion disappeared only two minutes after the velocity decrease. These are similar to the results of monitoring along path T-R2.
Along both paths T-R1 and T-R2, we can see that the amplitude decrease appears ahead of the velocity decrease. Kaneko et al. (1979) showed that the amplitude is more sensitive than the velocity to rock fracture in the laboratory and in in situ experiments. The findings obtained by us are consistent with their results. In addition, our results seem to indicate that the amplitude decreases according to accumulation of microfractures represented by clusters of AE hypocenters while the velocity decreases according to relatively large fractures represented by open cracks. These findings demonstated that P-wave monitoring can be effectively used to detect progress of fractures around a chamber.

4 CONCLUSIONS

To develop acoustic methods to investigate changes in condition around a chamber caused by excavation, in particular, caused by an increase of tangential normal stress, we conducted a laboratory experiment simulating stress distribution around a chamber using a rectangular Ohya-tuff specimen. The following results are obtained:

(1) Cracks observed on surfaces of the specimen seemed to spread following the spread of AE hypocenters. The finding suggests that AE monitoring can be used to predict the spread of cracks around a chamber.

(2) P-wave monitoring along two paths through the specimen showed that an amplitude decrease is more sensitive than a velocity decrease to fracture. This coincides with the results obtained by other researchers in laboratory and in in situ experiments. In addition to this, our monitoring seemed to indicate that the amplitude decreases according to accumulation of microfractures represented by clusters of AE hypocenters while the velocity decreases according to relatively large fractures represented by open cracks. The findings demonstate that P-wave monitoring can be effectively used to detect the progress of fractures around a chamber.

As mentioned above, from laboratory experiments it was found that AE and P-wave monitoring are good methods to detect the progress of fractures around a chaber. To apply these methods to risk assessment of nuclear waste isolation systems, we would like to improve and develop monitoring equipment and techniques that we have now (Ishida et al. 1992, in press).

REFERENCES

Ishida, T., Kanagawa, T., Sasaki, S. and Urasawa, Y. 1986. AE monitoring during the in-situ direct shear test applied to an underground cavern. Proceedings of the Japan Society of Civil Engineers, 376(III-6): 141-149 (in Japanese)
Ishida, T., Kitano, K., Kinoshita, N. and Wakabayashi, N. 1992. Acoustic emission monitoring during in-situ heater test of granite, Journal of Acoustic Emission. 10: S42-S48
Ishida, T., Kanagawa, T., Tsuchiyama, S and Momose, Y. (in press). High frequency AE monitoring with excavation of a large chamber, Proceedings of The Fifth Conference on Acoustic Emission/Microseismic Activity in Geologic Structures and Materials

Kanaori,Y. 1983. The observation of crack development around an underground rock chamber by borehole television system. Rock Mechanics and Rock Engineering. 16: 133-142

Kaneko, K., Inoue, I., Sassa, K. and Ito, I. 1979. Monitoring the stability of rock structures by means of acoustic wave attenuation. Proceedings of Fourth Congress of the International Society for Rock Mechanics, Montreal, 2: 287-292

Kovari,K. and Petter,G. 1983. Continuous strain monitoring in rock foundation of a large gravity dam, Rock Mechanics and Rock Engineering. 16: 157-171

Talebi, S. and Young R. P. 1992. Microseismic monitoring in highly stressed granite: Relation between shaft-wall cracking and in situ stress, Int. J. Rock Mech. & Geomech. Abstr. 29: 25-34

Rockbursts and Seismicity in Mines, Young (ed.) © 1993 Balkema, Rotterdam, ISBN 90 5410 320 5

The application of an energy approach in fault models for support design

Larry K.W. Ng & Graham Swan
Falconbridge Ltd, Sudbury Operations, Ont., Canada

Mark Board
Itasca Consulting Group, Minn., USA

ABSTRACT: Since rockbursts are the result of a violent release of energy, it is appropriate to use an energy approach to explain the mechanics of fault-slip behaviour and the risk of its occurrence. In this paper the release of seismic energy for simple mine excavation sequences and fault geometries is examined through the use of the 2-D Distinct Element code, UDEC. A variety of factors, thought to influence the release of energy, have been considered in the study, including the fault geometry, orientation and mechanical properties. Both a static and a dynamic modelling technique is presented. Where appropriate, the two cases of excavating towards and retreating from a fault are examined.

For the case of the simple geometry used, results from the static sensitivity analyses provide a good indication of the propensity for large slip to occur. This propensity has been expressed in the paper as a function of mining step or sequence and through comparison with a base case. Results generated from as many as 50 variations are presented conceptually in the form of a risk map, which may provide the basis for the assessment of ground support design.

An example of a dynamic analysis using UDEC is described by means of an asperity shear simulation, given the same simple geometry of the static case. In this case the calculation provides results in terms of energy released and, perhaps more importantly for support design, particle velocities at excavation boundaries.

1 INTRODUCTION

Typically, the layout and planning of excavations in rockburst-prone ground has been based on the practical experience of the mine engineer, occasionally using seismic instrumentation and broad rock mechanics principles for assistance. Some efforts have been made to apply quantitative methods for the prediction of the potential for rockbursting and the approximate magnitude which may be expected. The Chamber of Mines Research Organization (COMRO) of South Africa pioneered the use of boundary element methods and the calculation of energy release rate (ERR) as a means of quantifying the rockburst potential for deep, thin reef excavations, Cook et al (1966). Currently the mine engineer needs design procedures which allow a fairly simple means of quantifying the rockburst potential or risk for various mining geometries or sequences. By necessity, these procedures must take into account geologic structures which are important for the initiation of seismicity.

It is recognised that a majority of the large rockbursts in Canadian mines are a result of unstable slip on pre-existing fracture or fault surfaces. For this reason a portion of the work within the Canadian Rockburst Research Program (CRRP) of the Mining Research Directorate (MRD) is aimed at developing a quantitative procedure for risk and potential magnitude assessment of slip-induced rockbursting.

In this paper we discuss the development of an approach presented earlier (Ng et al, 1992) and based upon the numerical analysis of fault-slip using the Distinct Element method. With support from the MRD Program, logic has now been added to the UDEC code (Christianson and Board, 1992) which allows determination of the various stored and released energy components associated with the intact rock blocks and discontinuities. The change in the released energy resulting from fault-slip can be used as an indicator of rockburst potential and the magnitude of the resulting events. Problems may be run in either static or dynamic modes, the latter allowing an interaction of the slip-induced seismic waves with the adjacent excavation to be simulated. This is considered important since the damage from slip events is often the result of the seismic wave as it passes through the surrounding rock mass. Peak particle velocities at excavation surfaces can be used to assess damage levels.

2 ENERGY FORMULATION

Salomon (1984) presented the basic equations for energy storage and release in elastic ground as a result of excavation. His equations can be modified slightly to account for the energy storage occurring in elastic compression of discontinuities and the energy dissipation due to inelastic slip, plastic deformation of the rock mass and/or backfill. The energy balance is then given by:

$$W_r = W - (U_c + U_b + W_j + W_p) \qquad (1)$$

where W_r = released energy,
W = total boundary loading work supplied to the system,
U_c = total stored strain energy in material,
U_b = total change in potential energy of the system,
W_j = total dissipated energy in discontinuity shear,
W_R = total dissipated work in plastic deformation of intact rock.

The above energy components can all be determined from the stresses and deformations within the blocks, discontinuities and along the boundary of a UDEC model.

UDEC uses a form of dynamic relaxation for solution of the equation of motion for each block. For static problems, mass damping is used to extract kinetic energy from the system and to allow the system to come to equilibrium. Alternatively, for dynamic analysis, Raleigh damping is used with non-reflecting boundaries on the outer surfaces of the model. These boundaries allow the waves generated within the model to pass through them without generating reflections, thus simulating an infinite surrounding rock mass. This arrangement allows the released energy to be calculated in terms of the damped energy and that absorbed by the non-reflecting boundaries:

$$W_r = U_k + W_k + W_v + U_m \qquad (2)$$

where U_k = current value of kinetic energy in the system,
W_k = total work dissipated by mass damping,
W_v = work done by viscous (non-reflecting) boundaries, and
U_m = total strain energy in the excavated material.

For dynamic analyses, the amount of damped energy is generally a small proportion of the total released energy; that absorbed by the non-reflecting boundaries provides a convenient measure of the seismic energy released in slip.

In the UDEC code, the various energy components are determined at each time-step, thus giving an incremental and total change in energy as a function of time (in the dynamic case) or as a function of mining step (in the static case).

3 MINING APPLICATIONS

The following three examples, Figure 1, are selected from Falconbridge Ltd., Sudbury Operations, to illustrate conditions where an energy approach might usefully assist in the mine planning and design process. The first, Figure 1a, is taken from a rockburst incident in the back/face of a drift development, 46L Fraser Mine (Swan and Semadeni, 1992). Next, we show a convergence-driven stick-slip burst problem complicated by a weak Diabase dyke at Lockerby Mine. The resulting failure zone extends beyond the dyke to eventually compromise the ore pass. Finally, Figure 1c depicts an asperity/fault geometry containing a mining sequence which has been considered in a recent review of the Falconbridge Mine, #5 Shaft rockburst incident of 1984 (Board, 1992).

3.1 Static modelling technique

As discussed earlier, UDEC can now calculate changes in the energy dissipated due to slip in the vicinity of an excavation using either a static or a dynamic technique. In this section we will describe the static case, using the simple example of an excavation passing through a fault surface in a series of discrete steps. An application of the dynamic simulation technique will be discussed briefly in Section 5.

In an earlier study, Ng et al (1992), UDEC was used to examine the fault-slip induced seismic energy of the simple geometry shown in Figure 2. The excavation was carried out in ten mining steps of equal volume, each step being 5m high by 4m long. The two cases of mining towards and retreating from the fault were investigated. A continuously-yielding joint model was used with input parameters shown in Table 1. The code determined where fault-slip occurred and calculated the corresponding stress drop for the location. These data were then used in available analytic expressions for a circular fault model to determine the released seismic energy. Finally, an "event" magnitude was estimated from the empirical expression given by Hedley (1992):

$$\log W_k = 1.3 \, Mn - 1.75. \qquad (3)$$

Results from the earlier base case study indicated that the total release of seismic energy was about 20% greater in the case of advancing towards the fault compared to retreating from the fault. As well, the release of seismic energy was more evenly distributed over the mining steps. This result was later confirmed

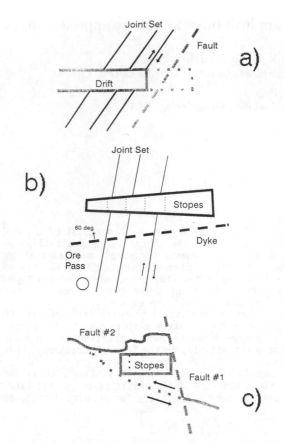

Figure 1 Simplified fault slip geometries in the Sudbury district for which rockbursts have been observed:
 a) Fraser Mine, 46 L, side view;
 b) Lockerby Mine, 31L, plan view;
 c) Falconbridge Mine, 40L, plan view.

by Christianson and Board, 1992, expressed directly by UDEC in terms of seismic energy, Figure 3.

In the present work, use has been made of UDEC's new explicit capability of calculating the energy components due to slip, while again using equation (3) to calculate magnitude. Under these circumstances it is a relatively easy task to perform a large number of variational calculations, with the original model as the base case. In each variation UDEC was run to achieve an equilibrium condition before a determination of the slip event, energy release, etc was made.

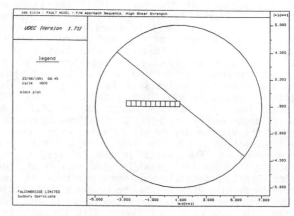

Figure 2 UDEC model geometry.

Table 1 Input parameters, base case model

PROPERTY	VALUE
E (MPa)	40.e3
υ	0.2
k_n (MPa/m)	1.e4
k_s (MPa/m)	1.e4
ϕ_{peak} (°)	55
ϕ_{res} (°)	20
R Factor	0.0005

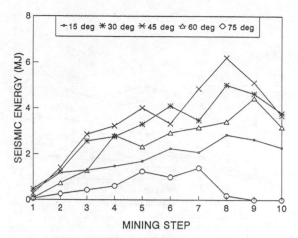

Figure 4 Seismic energy released as a function of mining step and fault orientation.

4.3 Residual friction angle

Three residual frictional angles of 5°, 10° and 20° were examined again for the 30° fault orientation case. Figure 5 indicates that there is a significant increase in seismic energy in the first two mining steps when the residual frictional angle was reduced to 5° from 20° (2 MJ/m to 15-20MJ/m). This was caused by a sudden large stress drop and excessive shear displacement in the initial two mining steps.

4.4 Rock mass stiffness

The elastic modulus of the rock mass was doubled to 80 GPa for both the 30° and 45° fault orientation case. The results show that the liberated seismic energy is reduced by more than 50 %.

4.5 Insitu stress

The horizontal stress was increased from 60 MPa to 100 MPa, and the vertical stress was increased from 40 MPa to 60 MPa. The retreat simulation was carried for all five fault orientations. Figure 6 shows that the seismic energy more than doubled for all cases except for the 75° case. It is conceivable that a high horizontal insitu stress would provide significant clamping action on the steeply dipping fault, thus reducing the amount of slip.

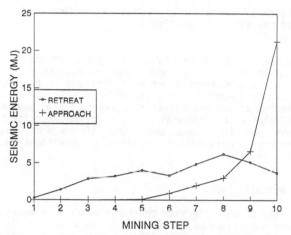

Figure 3 Comparison seismic energy released by mining step, excavation approaching and retreating.

3.2 Sensitivity analysis

The sensitivity analysis was carried out to observe the effect of certain input parameters on the amount of energy liberated due to fault-slip. For the purpose of illustrating the results of such a study, only the retreat mining sequence with UDEC's continuously-yielding joint model under static conditions was considered. The parameters and geometries which were varied in the model were: fault orientation, fault surface peak and residual friction angles, rock mass stiffness, insitu stress condition and the presence of a pillar in the second panel under the fault.

4 RESULTS

A summary of the findings from the sensitivity analysis are discussed in the following sections.

4.1 Fault orientation

Figure 4 shows that the amount of liberated seismic energy increases in the following order of fault orientation : 75°, 15°, 60°, 30° and 45°.

4.2 Peak friction angle

Peak friction angles of 30°, 55°, and 70° were used for the 30° fault orientation case. Results suggest that the released seismic energy is relatively insensitive to the change of peak friction angle within the range considered.

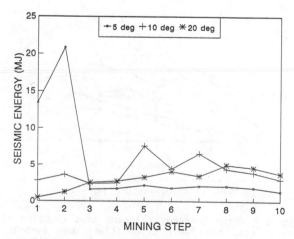

Figure 5 Seismic energy released as a function of mining step and residual friction angle.

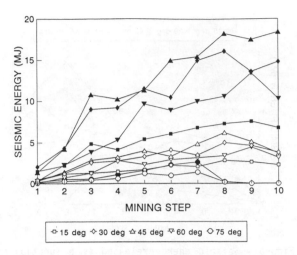

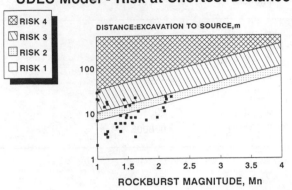

Figure 6 Seismic energy released as a function of mining step and joint orientation. Open symbols, base case; filled symbols, increased insitu stress.

Figure 8 Representation of risk at shortest distance for retreating excavation.

4.6 Remnant pillar, panel 2

Except for the 15° case, the calculated seismic energy reduced by more than 50 % for each subsequent panel excavation beyond the pillar location. The results are summarized in Figure 7. Note that panel 2 was not excavated, thus there was no seismic energy liberated in mining step 2.

4.7 Discussion

The results presented above describe the broad behaviour and system sensitivity of a simple fault configuration. On this scale of analysis, such results can be obtained relatively quickly and can provide an insight of the probabilistic rather than the deterministic propensity for slip to occur. Thus the approach can quickly identify areas of concern where uncertainty exists. For example, our results suggest that the residual friction angle and the insitu stress conditions have a strong influence on the magnitude of the seismic energy released.

A drawback with this approach is its inability to determine whether the energy is released violently or non-violently. Obviously this leaves the important issue of risk and likelihood of a seismic event open

if use is made of these results for support design. An example of such a use was presented elsewhere (Ng et al, 1992 and Morrison et al, 1993) and is presented again in Figure 8. Here the UDEC slip locations are plotted as seismic events relative to the excavation. The plot area represents an empirical damage criterion developed in Canada and South Africa for ground support requirements under bursting conditions (Hedley, 1992).

5 ASPERITY SHEAR

In the previous Section the fault material properties were assumed to be uniform across the entire length of the fault and the behaviour of each contact surface was described by the same constitutive relation. In reality, the topography and physical characteristics of the fault will vary across its surface. There currently is much debate over the mechanisms for earthquakes and fault-slip rockbursts. Stick-slip (velocity-weakening) models have been suggested by some (eg. Hobbs, 1990), whereas others have argued an asperity model as the primary source mechanism.

In the asperity model, topographic features along the surface act as stress concentrations as the fault undergoes shear deformation, eg. Figure 1c. Little is documented regarding fault topography, but the asperity may be undulations in the fault surface, splays, offsets or intersections with dykes or other geologic complications. In any case it appears possible to have essentially stable Mohr-Coulomb slip along portions of the fault surface which then tend to load and shear off asperities, leading to the generation of seismic waves which are radiated from the surrounding rock mass.

5.1 Dynamic modelling technique

Using the same UDEC model as above, the technique involves a calculation of the radiated seismic energy from a simulated asperity failure, rather than, as previously, predicting when slip occurs under static conditions (ie. by mining step). Thus the model establishes a locked asperity condition by assigning a segment of the fault a high (65° in this case) friction angle. Next, the entire excavation is created in one step, allowing the model to achieve equilibrium under the new induced stress state. Finally the friction angle of the asperity segment is instantaneously reduced to a low residual value (25° for a 25 m long segment in this case). This results in a shear dislocation with a given stress drop and associated energy release. Non-reflecting boundaries are used to absorb the radiated wave and to prevent reflections back into the model. The change in released seismic energy is

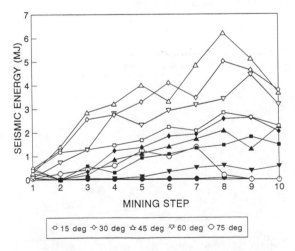

Figure 7 Seismic energy released as a function of mining step and fault orientation. Open symbols, base case; filled symbols, with remnant pillar in panel 2.

given by the sum of the viscous boundary energy and the damping work.

5.2 Results

With the simulated asperity shear, a maximum of about 3.5 cm of relative shear displacement occurs on the fault. The change in released energy due to slip is approximately 7.8 MJ/m of fault strike length. This is equivalent to an event of Mn=3.1 using equation (3) with a slip length of 25 m in the strike direction (ie. the slip occurs over a 25 m x 25 m area of the fault). Since the simulation is dynamic in nature, the particle velocity at various positions in the model may be sampled. Figure 9 shows the particle velocity in the roof of the excavation at the fault intersection, see Figure 2. We observe a positive first motion with a magnitude of 1.6 m/s. Such information can be very useful for the development of an empirical risk map/damage criteria, as was illustrated for the static case, Figure 8.

Additional results have been obtained with this model (Board, 1993) which show that a large energy release can be produced even with the assumption of a relatively small asperity radius. Such events are observed to produce very large local shear stresses while the remainder of the fault slips in a non-violent fashion. This inertial effect is interesting since it demonstrates that the seismic event may be independent of the size of the ruptured asperity.

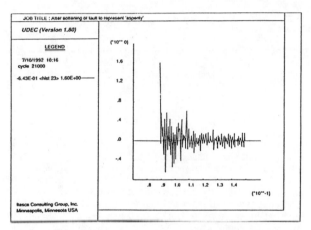

Figure 9 Velocity history after asperity shear, in the back of excavation at fault intersection.

6 CONCLUSIONS

From this work with a UDEC representation of a simple fault geometry we have shown that the model can be used in two basic ways. First, static analyses can be used to explore the general effect of mine geometry and geology on rockburst potential. At this stage nothing can be said about the nature of the released seismic energy in terms of unstable or stable slip. Results can be presented however which show the risk of slip relative to mine openings and structures.

Second, the model can be used to explicitly simulate rockburst mechanisms, leading to the determination of induced velocity in surrounding excavations. This second method can be used to explore the influence of mine geometry on the damage risk for excavations surrounding the expected hypocentre. Since it is known that damage does not necessarily follow a simple distance-magnitude relationship, this type of forward modelling would be useful for developing a mine-specific damage criteria, forming the basis of a support design.

REFERENCES

Board, M. 1992. Review of mining conditions and seismic data for the June 20-22, 1984 rockbursting at Falconbridge Mine. Research Report, MRD/CRRP, November 1992.
Board, M. 1993. Personal communication.
Christianson, M. and Board, M. 1992. Energy calculations in UDEC. Research Report, MRD/CRRP, July 1992.
Cook, N.G.W., Hoek, E., Pretorius, J.P.G., Ortlepp, W.D. and Salamon, M.D.G. 1966. Rock mechanics applied to the study of rockbursts. J. S. Afr. Inst. Min. Metall., 66, pp. 425-528.
Hedley, D.G.F. 1992. Rockburst Handbook for Ontario Hardrock Mines. CANMET Special Report, SP92-1E, Energy, Mines and Resources, Canada.
Hobbs, B.E. 1990. Chaotic behaviour of frictional shear instabilities. Proc. 2nd Int. Symp. Rockbursts and Seismicity in Mines, Minneapolis, June 1988, Ed. C. Fairhurst, Rotterdam: A.A. Balkema.
Morrison, D.M., Swan, G. and Scholz, C.H. 1993. Chaotic behaviour and mining-induced seismicity. In Proc. 3rd. Int. Symp. Rockbursts and Seismicity in Mines, Kingston, August 1993, Ed.: R.P. Young.
Ng, L.K.W., Swan, G. and Hedley, D.G.F. 1992. An examination of seismic energy in fault slip. Submitted to Proc. Induced Seismicity Workshop, Santa Fe, June 1992, Ed.: C. Ramseyer.
Salomon, M.D.G. 1984. Energy considerations in rock mechanics: fundamental results. J. S. Afr. Inst. Min. Metall., 84(8), pp. 233-246.
Swan, G. and Semadeni, T. 1992. Rockbursts in a development drift: field observations. Submitted to Proc. Induced Seismicity Workshop, Santa Fe, June 1992, Ed.: C. Ramseyer.

Rockbursts and Seismicity in Mines, Young (ed.) © 1993 Balkema, Rotterdam, ISBN 90 5410 320 5

The Agnico-Eagle Mine seismic survey

M. Plouffe
*Experimental Mine, Mining Research Laboratories, CANMET,
Energy, Mines and Resources Canada, Val d' Or, Que., Canada*

I. Asudeh
*Geological Survey of Canada, Energy, Mines and Resources
Canada, Ottawa, Ont., Canada*

D. V. Lachance
Agnico-Eagle Mines Limited, Joutel Division, Que., Canada

R. Aguila & L. Turgeon
Mining Engineering, Ecole Polytechnique, Montréal, Que., Canada

ABSTRACT: Following a magnitude 3.4 rockburst that hit the Agnico-Eagle Mine, Joutel, Québec, a seismic survey was conducted between December 1991 to October 1992 in order to locate and evaluate the seismic activity. Two types of portable seismographs were used in a network to record the seismic waveforms, both using triaxial transducers. The instruments performed well monitoring mining induced seismic events under conditions markedly different from a typical aftershock survey. About 350 events were recorded during the survey and about half of these were located. It was found that many seismic events were located in areas near the perimeter of the main area mined out. The size of these seismic events were compared to size of the production blasts, characterized by powder factors. No major mining induced seismic events were recorded during the survey.

1 INTRODUCTION

On November 7, 1991, at 08:45 EST (13:45 U.T.), a rockburst of magnitude 3.4 hit the Agnico-Eagle mine, Joutel, Québec. This rockburst was the largest seismic event originating from a Québec mine in recent mining history. This event was felt in the community of Joutel, 7 km from the mine.

Only minor damage was observed. According to the damaged sites, location of the event could only be estimated to be within the upper part of the main area mined. Surprisingly, however, an uncommonly shallow depth for this seismic event could be extrapolated.

A seismic survey was conducted by CANMET/MRL in order to locate and evaluate the seismic activity originating from the mine. The survey was conducted between December 1991 to October 1992. Seismically active areas were identified during this time period.

Two types of portable seismographs were used to record seismic waveforms with triaxial sensors. A Teledyne-Geotech PDAS-100, operated by CANMET, was used to record waveforms in order to conduct a complete waveform analysis. Three portable seismographs, Scintrex PRS-4, belonging to the Geological Survey of Canada (GSC) were used to complete the network. The data recorded by these instruments were used for event location and size determination. GSC used this opportunity to test the PRS-4 seismographs in a mining environment for the first time. The instruments performed well monitoring rockbursts under conditions markedly different from a typical aftershock survey.

About 350 events were recorded during the survey and half of these were located. Many seismic events were located in areas surrounding the main area mined out. A few events were located in the other parts of the mine.

The size of these seismic events were compared to the size of the production blasts which were characterized by powder factors.

2 THE AGNICO-EAGLE MINING COMPLEX

Agnico-Eagle Mines Limited operate three gold mines at their Joutel division located 200 km north of Val d'Or. The Agnico-Eagle deposit, located in the Joutel Township, was discovered and delineated in 1962-1964 during a diamond drilling survey of an electromagnetic anomaly. Production first began in 1974 at the Eagle Mine. In 1984, the Telbel Mine, 1.3 km from the Eagle Mine, came on-stream, and the Eagle West open-pit started in 1991 (figure 1). In June 1990, the millionth ounce of gold was poured in Joutel.

The orebody lies at the contact of the Joutel volcanic complex with overlying Harricana River sedimentary sequence. These two regional units have, at the mine scale, a NW-SE strike, subvertical dips and face towards the NE. The Harricana fault, straddles this contact. Gold mineralisation is 2 to 5 m thick and is related to argillites and chemical sediments (footwall - Rock Quality Designation (RQD): 0 to 25%) and carbonatized argillites (hangingwall - RQD: 10 to 40%). The ore contains 20% disseminated cataclastic pyrite and gold occurs as fine particles directly associated with the pyrite.

The main ore zone has been mined downward with sub-level retreat and shrinkage methods. The voids created by mining are filled with waste rock sloughing off the walls.

Ground stability has always been a concern at the Telbel Mine since operations commenced when the shaft was sunk to a depth of 1215 m. Stress levels increased considerably over the years, mainly in the lower portion of the orebody, where stresses three to four times higher than the natural stress levels were observed (Emond 1990). The walls are not very competent, showing strong convergence in drifts and raises. This situation forced the company to use cable lacing in the backs and the walls, as well as cable bolting.

Although ground support helped mining activities, several mining induced seismic events (i.e. air blasts) occurred and compelled the company to abandon some mining areas and to leave ore pillars.

3 SEISMIC HISTORY

Not many large seismic events had been recorded at the Agnico-Eagle Mine, prior to the magnitude 3.4 rockburst.

Only three other large events originating from Agnico-Eagle Mine occurred on June 06 and September 23, 1990 and had magnitude ranging between 1.8 and 2.0, as measured with the National Seismograph Network (NSN). Some large production blasts in the initial phase of the open-pit were also recorded by the NSN.

4 SEISMOGRAPH SETTINGS

A network of four portable seismographs was installed in the mine in order to locate and size of the seismic events. This seismic survey was conducted from December 1991 to October 1992. Three seismographs were located in the Telbel Mine surrounding the main mine-out area and one in the Eagle Mine. When possible, quiet sites were selected far from on-going mining operations. The network finally covered a volume 1500

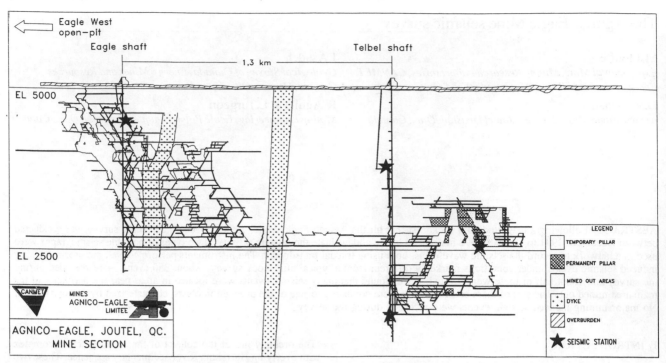

Figure 1. Longitudinal section of the Agnico-Eagle Mining Complex in Joutel, Québec.

m long, 150 m wide and 900 m high of the mines.

Two types of portable seismographs were used in the network to record the seismic waveforms using triaxial sensors.

4.1 The PDAS-100 portable seismograph

The Teledyne-Geotech PDAS-100 portable seismograph, operated by CANMET, is made by Teledyne-Geotech. This seismograph has been previously used by CANMET/MRL in some mining-induced seismicity surveys (Laverdure & Plouffe 1991, Plouffe 1990a, Plouffe 1990b) in several mines in northern Ontario. The instrument houses a microcomputer and enables the recording of seismic events on its 4.5 Mb of ROM. Data acquisition parameters are downloaded and recorded data files are uploaded using Teledyne-Geotech software.

A full scale, low-gain input of 2.0 V was maintained during the survey. Thus, according to calibration, the maximum peak particle velocity that could be recorded is around 250 mm/s. A trigger level was set to have a lower threshold equivalent to 15% of the input. The triggered events are sampled at a frequency of 1000 Hz. The sensors house three orthogonal Geotech Model S-100 accelerometers with a frequency response ranging from 0.20 to 2000 Hz.

4.2 The PRS-4 portable seismographs

Three PRS-4 portable seismographs operated by the Geological Survey of Canada (GSC), were used to complete the network. These instruments are made by Scintrex Ltd (Markle 1992) and use solid-state memory as recording medium. The PRS-4s were initially designed for rapid deployment in aftershock studies (North et al. 1989, Lamontagne et al. 1993), including longer term seismicity surveys. They can also be used on refraction surveys in timed mode.

The PRS-4 seismograph has four channels, the first three being the sensor channels, each with a differential input amplifier, an anti-aliasing filter and a high-pass filter section to eliminate DC offset. The fourth channel is used to digitize time marks. At the Agnico-Eagle Mine, the PRS-4s were connected to sets of triaxial geophones.

Communications with the PRS-4 are done via a computer running the LithoSEIS software. This software, developed by the

GSC, helps to set up the initial data acquisition parameters, to collect the event data recorded, to monitor the instrument status and to optimize the event trigger for the location.

Both large and small events can be recorded through a 132 dB dynamic range, consisting of 60 dB autogain ranging and 13 bit A/D conversion. The preamplifier has selectable gains. A selectable threshold, varying accordingly to the ambient noise surrounding the seismograph site, was a function of the short-term average/long-term average (STA/LTA) ratio. The seismic data was recorded at the maximum sampling rate of 200 Hz.

The GSC used this opportunity to test the PRS-4 seismographs for the first time in a mining environment. The data recorded by these latter instruments were used for event location and size determination.

5 SEISMIC ACTIVITY RECORDED

At the request of the Agnico-Eagle mine management, a seismic survey was conducted by CANMET/MRL scientists in order to locate and evaluate the importance of the seismic activity originating from the mine. Major waveform analysis tasks were divided in waveform discrimination, seismic velocity determination, seismic event location and size determination.

This paper summarizes results from December 1991 to May 1992. This 6-month period allowed the identification of seismically active areas.

5.1 Seismic velocities

The seismic wave velocities were determined using signals from blasts with known locations recorded by several portable seismographs.

The average velocities for the P- and S-waves were evaluated to be approximately 5500 m/s and 2900 m/s respectively. These velocities are less than the characteristic velocities, 6200 m/s and 3650 m/s respectively, used by the GSC in the Canadian Shield (Drysdale et al. 1991). This difference could be explained by the effects of the mined-out areas between the source and the sensors, which impacts more on wave attenuation within short travel distances.

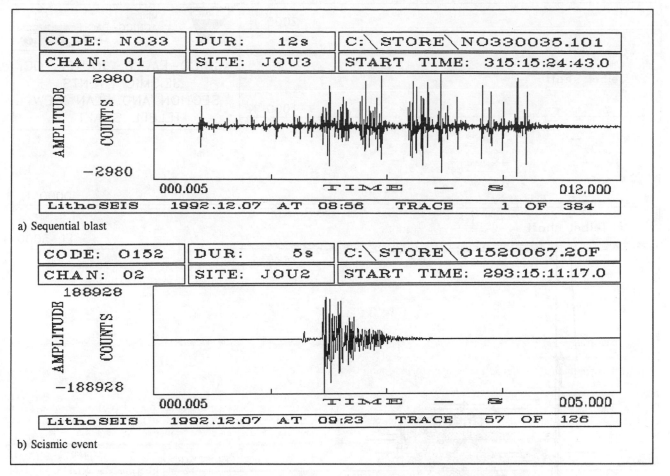

CODE: NO33 | DUR: 12s | C:\STORE\NO330035.101
CHAN: 01 | SITE: JOU3 | START TIME: 315:15:24:43.0

AMPLITUDE COUNTS 2980

-2980

000.005   TIME — s   012.000

LithoSEIS   1992.12.07   AT   08:56   TRACE   1   OF   384

a) Sequential blast

CODE: O152 | DUR: 5s | C:\STORE\O1520067.2OF
CHAN: 02 | SITE: JOU2 | START TIME: 293:15:11:17.0

AMPLITUDE COUNTS 188928

-188928

000.005   TIME — s   005.000

LithoSEIS   1992.12.07   AT   09:23   TRACE   57   OF   126

b) Scismic cvent

Figure 2. Distinctive signatures of a sequential blast and a seismic events, recorded by a PRS-4 seismograph.

5.2 Waveform classification

The waveforms which were recorded were classified in three categories: seismic events, sequential blasts and mining noises.

The mining noises originate from the mining machinery, the electrical current and the mining personnel. These events were easily distinguished and subsequently discarded, using the LithoSEIS software, because of their distinctive waveform signatures.

Waveforms produced by sequential blasting in development headings, as illustrated in figure 2a, could similarly be distinguished and eliminated from the analysis.

The seismic events left were the mining induced seismic events, or rockbursts, and the borehole blasts. Their differentiation was more difficult because both of them appear as single event (see figure 2b) in the recorded time windows. Times of blasts were obtained from blasting logs; unconfirmed events were classified as mining induced seismic events.

Of the 2000 triggers recorded by the four portable seismographs in the 6-month period survey, only 495 were classified as blasts (mostly borehole, 149) and mining induced seismic events (346).

5.3 Determination of the event location

Determination of seismic locations was possible using the difference between the wave arrival times. The alternate procedure, whereby all absolute waveforms arrival times are used in the computation of the location, could not be applied because of the difficulty to synchronize the two types of seismographs.

Of the 346 seismic events being classified as mining-induced, computation of the location of 144 events was possible. The majority of the events were located around the main area mined out (figure 3), mainly in the upper portion and in the permanent pillar. In plan view, the straight line seems to indicate the position of the permanent pillar. The other events seem to be located at the perimeter of the mined out area.

Due to the very wide grid, the confidence range of the event location could not be specified to be less than 10 to 15% of the distance between the event location and the closest seismographs, which can be compared to the error in location as defined by Hedley (1992) for macroseismic systems (± 50 metres). Also, the precision of time difference was about 0.005 second, which created a potential error of about 30 meters in this analysis.

5.4 Quantification of size

The size of the seismic events was determined using an empirical powder factor relationship.

The vector sum of the maximum particle velocities of 75 borehole blasts with known quantities of explosives, ranging from 44 to 1300 kg was calculated. Scale factors, based on the cubic root of the weight (W) of explosives, in kg, and the hypocentral distances (R, in metres) between the source and the sensors, were then related to these amplitudes (A_{max}) in a relationship as described by Dowding (1982):

$$A_{max} = k \left[\frac{R}{W^{1/3}} \right]^n$$

Mathematical relationships were developed according to the spatial distribution of the borehole blasts, divided in three main blasting zones, and the transducers. These relationships were used to quantify the size of the located seismic events according

395

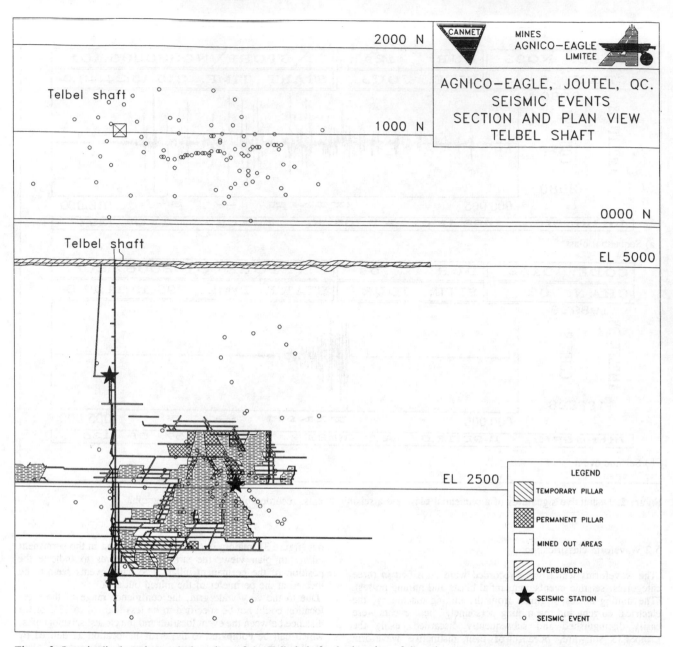

Figure 3. Longitudinal section and plan view of the Telbel shaft: the location of the seismographs and the seismic events.

to the closest blasting zones; in the case of locations in the main mined-out area, average values from several relationships were determined. For the unlocated events, an average mathematical relationship was derived using the distance derived from the difference between the wave arrival times.

Figure 4 shows the size distribution for the 279 located and unlocated seismic events, for which a magnitude was estimated. Eighty-eight per cent of the events were equivalent to blasting less than 10 kg of explosives. Consequently, no major events and/or blasts during the survey period could be recorded, and for which magnitude could be determined by the NSN.

6 CONCLUSION

This survey made it possible to compare the performance of a portable seismograph network to a permanent one. The instruments performed well while monitoring mining-induced seismic events under conditions markedly different from a typical aftershock survey.

During a relatively short time, seismically active zones were defined where stresses are potentially at their highest level, namely the upper area of the main mined-out section of the Agnico-Eagle Mine. No major mining-induced seismic events were recorded during the survey.

The study has demonstrated the benefits which can be achieved with a seismic network, which can be rapidly installed and efficient to identify potentially dangerous areas for on-going mining operations. A larger number of seismographs and a greater range of blast types and blast locations would have undoubtedly improved the location and size analysis of the mining-induced seismic events recorded at the Agnico-Eagle Mine.

An on-going survey, from October 1992 to June 1993, should strengthen the initial conclusions of this analysis.

ACKNOWLEDGEMENTS

The full cooperation and assistance provided by the mining engineering personnel of Agnico-Eagle Mines, Joutel Division, is gratefully acknowledged. The authors wish to specially thank Mr. B. Bédard, Agnico-Eagle Mines Ltd, Messrs J. Thomas and T. Carthwrigth, Instrumentation Laboratory, GSC, Ottawa, for their help in the conduct of the seismic survey and Mr. Y. Lizotte, CANMET/MRL, for proof-reading of this paper.

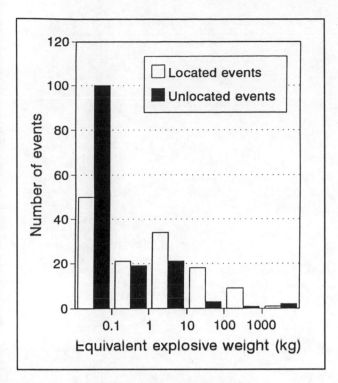

Figure 4. Size distribution of the seismic events.

REFERENCES

Dowding, C.H. 1982. *Blast Vibration Monitoring and Control*: Northwestern University.

Drysdale, J.A., R.B. Horner, R. Kolinsky & M. Lamontagne 1991. Canadian Earthquakes - National Summary: January-March 1990. Geological Survey of Canada, Energy, Mines and Resources Canada.

Emond, R. 1990. Sécurité et contrôle de terrain, Mines Agnico-Eagle, Division Joutel. Internal Report, Agnico-Eagle Mines.

Hedley, D.G.F. 1992. Rockburst Handbook for Ontario Hard rock Mines. CANMET, Energy, Mines and Resources Canada, Special Report SP92-1E.

Lamontagne, M., H.S. Hasegawa, D.A. Forsyth, G.G.R. Buchbinder & M.G. Cajka 1993. The Mont-Laurier, Québec, earthquake of 19 October 1990 and its seismotectonic environment (in preparation, to be submitted to the *BSSA*).

Laverdure, L. & M. Plouffe 1991. Full-waveform analysis at Kidd Creek Mine - Progress report. CANMET, Energy, Mines and Resources Canada, Division Report MRL 91-104.

Markle, B. 1992. PRS-4 Portable Recording Seismograph/Accelerograph. Second LithoSEIS Workshop, Ottawa.

North, R.G., R.J. Wetmiller, J. Adams, F.M. Anglin, H.S. Hasegawa, M. Lamontagne, R. du Berger, L. Seeber & J. Armbruster 1989. Preliminary results from the November 25, 1988 Saguenay (Québec) earthquake. *Seismological Research Letters* 60:89-93.

Plouffe, M. 1990. A local seismic survey at Creighton Mine. CANMET, Energy, Mines and Resources Canada, Division Report MRL 90-076.

Plouffe, M. 1990. A local seismic survey at Denison Mine. CANMET, Energy, Mines and Resources Canada, Division Report MRL 91-018.

Simard, J.-M. & R. Genest 1990. Géologie de la Mine Agnico-Eagle, Joutel (Québec). *La ceinture polymétallique du Nord-Ouest québécois: Synthèse de 60 ans d'exploration minière*: 373-381. CIM, Special volume 43.

Figure 5 Size distribution of the seismic events.

REFERENCES

Rockbursts and Seismicity in Mines, Young (ed.) © 1993 Balkema, Rotterdam, ISBN 90 5410 320 5

Monitoring of large mining induced seismic events – CANMET/MRL's contribution

M. Plouffe, P. Mottahed, D. Lebel & M. Côté
Rockburst Group, Mining Research Laboratories, CANMET, Energy, Mines and Resources Canada, Canada

ABSTRACT: CANMET/MRL's contribution to the Canadian Rockburst Research Program is the study of the mining induced seismic events located in the upper range of magnitude, i.e., magnitude 1.0 and above. To achieve this objective, a remote-controlled macroseismic system was designed and developed which is now operating in four mines across northern Ontario. A menu-driven waveform analysis software was also developed in-house to analyze the waveforms recorded. To determine magnitude from local mine camps, a network of eight regional seismograph stations is being operated by MRL. The most important part of this network is the Sudbury Local Telemetered Network that includes three uniaxial and one triaxial outstations. Empirical local magnitude scales were developed in order to extend national magnitude values to smaller seismic events.

1 INTRODUCTION

Rockburst activity in Canadian mines increased significantly during the 1980's, particularly in Ontario. A particularly active year was 1984, that culminated with the largest seismic event in modern Canadian mining history; a magnitude 4.0 event recorded at INCO's Creighton Mine in Sudbury. The high level of rockbursting raised concern from the industry. In response, the Canada-Ontario-Industry Rockburst Project was initiated in 1985 (Hedley & Udd 1987). The objectives of this project were to develop state-of-the-art monitoring systems capable of capturing complete seismic waveforms and investigate the causes and mechanisms of rockbursts. Field trials would also be used to evaluate methods to reduce magnitude and frequency of rockbursts. The final output of this first phase was published in 1992 (Hedley 1992).

Since April 1990, an expanded second phase of the Rockburst Project, the Canadian Rockburst Research Program (CRRP), is in progress. This second, five-year research project involves government agencies, universities and the Canadian & Chilean mining industries. On the basis of the knowledge established from the first phase, the research program will cover five aspects of mining seismicity: the reduction of mining seismicity using stiff backfill, the determination of guidelines in ground support and control, the improvement of mine layout through full waveform analysis, the automation of waveform data analysis, and a better understanding of fault-slip rockburst mechanisms.

CANMET's contribution to the CRRP, through the Mining Research Laboratories (MRL), is the study of the mining induced seismic events located in the upper range of magnitude, i.e., magnitude 1.0 and above. To achieve this objective, a remote-controlled macroseismic system was developed which operates at a lower sampling rate and with the sensor grid being much wider than a microseismic system. This system is now operating in four mines in northern Ontario. A menu-driven waveform analysis software was also developed in-house.

In order to determine the magnitude of small mining induced seismic events originating from some Canadian mine camps, a network of eight regional seismograph stations is now operating. The most important part of this network is the Sudbury Local Telemetered Network (SLTN) that includes three uniaxial and one triaxial outstations. This network was recently upgraded to increase data acquisition of scientifically meaningful data. Empirical local magnitude scales were developed in order to extend national magnitude values for the analysis of these smaller seismic events.

2 THE MACROSEISMIC NETWORKS

Macroseismic systems are being used to investigate the seismic source parameters of rockbursts, which are defined below. These systems have the capability of recording the seismic waveforms of events greater than magnitude 1.0 using triaxial sensors.

MRL developed its own in-house remote-controlled macroseismic system, now operating in four mines in northern Ontario. These systems are located at Falconbridge's Strathcona Mine, INCO's Creighton Mine at Sudbury, Placer Dome's Campbell Mine at Red Lake and Lac Mineral's Macassa Mine at Kirkland Lake. Each system consists of five strong-motion triaxial sensors installed in boreholes underground and/or on surface, usually within a kilometre of the mine workings. They are specifically designed to capture the complete seismic waveforms to allow source location determination as well as seismic parameters. These systems also benefit the mine operators by allowing them to monitor the seismic activity originating from their mine. These in-house systems were commissioned between August 1990 (Strathcona Mine) and February 1992 (Creighton Mine). The macroseismic networks are permanently installed in these mines. Such an installation requires months of labour, software, etc. A Mark I portable version of this macroseismic system is currently being completed and will shortly be undergoing *in situ* trials.

All four networks have basically the same hardware components. All the installations use geophones, except for the Strathcona Mine system which use accelerometers. The differential signals from the transducers are initially amplified and routed via cables in the mine shaft to the data acquisition computer system.

All processing units are located on surface, where the seismic signals are conditioned and passed on to the data acquisition system. After initial signal conditioning and filtering, the signals are then routed to both the trigger detection controller and the data acquisition board. The signals from each channel are digitized by the data acquisition board at a rate varying from 1024 to 4500 samples per second.

2.1 The data acquisition system

Traditionally, high speed data acquisition was only possible using expensive, high powerful computers. However, with the technological advances made by the computer industry, it is now more cost-effective to use microcomputers to perform this task. Although high speed data acquisition hardware is now available, most software packages available operate under the operating system DOS, which lacks the flexibility and power of multi-

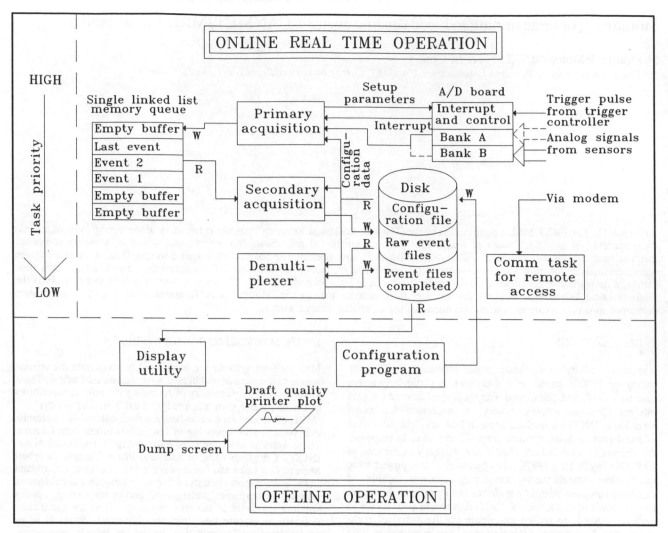

Figure 1. Flow chart of the data acquisition system used by CANMET's macroseismic networks.

tasking operating systems. This dictated the use of another system providing this convenience. Hence, CANMET opted for the QNX™ (© Quantum Software Systems Ltd.) operating system. This allows the data acquisition system to be made up of several smaller tasks, which can run concurrently on a priority basis. Tasks not requiring immediate processing can be put on hold and not compete for processor time. Remote access to the system is also possible, simply by having a communication task run in the background, which requires minimal processor time.

The data acquisition software consists of three main modules: primary and secondary acquisition programs and a demultiplexer program. A configuration program, a start batch file and a stop batch program complete the package. Optional program modules are also available to display waveforms and simulate events under software control. The dataflow instructions and interactions which occur between the various program modules is shown on figure 1. Each program runs as a stand alone task on a priority basis, the higher priority tasks being at the top of the flow chart.

The start batch file initiates the acquisition process by starting the primary acquisition task. This primary acquisition task performs all executive functions including creating the secondary acquisition task and the demultiplexer task, which run at lower priorities. The primary acquisition module reads in parameters from the configuration file on disk and then performs a complete hardware and software initialization. It is also responsible for detecting, storing and queuing events in memory. The secondary acquisition module is responsible for transferring the acquired events from the memory buffers to hard disk storage.

A sophisticated triggering controller has also been developed to eliminate the recording of very small seismic events, machine noise or electrical spikes which could saturate the system. The controller has the following features: selectable trigger windows, selectable number of channels for a valid trigger, short-term integration for spike suppression, selectable trigger threshold, individual channel enable/disable switches, event simulator to test the trigger controller, software enable/disable of the triggering, optional hardware selectable retrigger timer and optional sensor amplifier gain control. If the trigger controller determines that a valid event has occurred, then two seconds of seismic data are stored for each channel, including half a second of pre-trigger data. Except for the Creighton Mine system, the acquisition is interrupted for about 200 milliseconds while the data are downloaded to the hard disk.

All macroseismic systems are connected via modems to MRL's Elliot Lake Laboratory. At the mines, the data files can be edited and transferred to Elliot Lake or some functions of the trigger controller can be altered.

2.2 Waveform analysis

An in-house menu-driven waveform analysis software was also developed to analyze the recorded waveforms. The schematic of this software system is illustrated in figure 2.

The waveform analysis software is used to calculate seismic parameters of mining induced seismic events, and to estimate their size in the four Ontario mines, or for other research purposes.

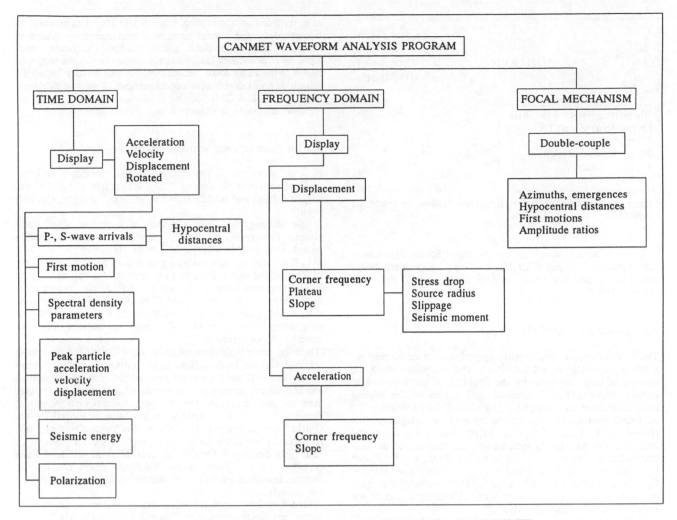

Figure 2. Schematic of the waveform analysis software developed by CANMET.

The signature of the waveforms can be displayed in acceleration, velocity and displacement modes. Thus, peak particle acceleration and velocity can be determined and used to assess damage. Waveforms can be rotated in order to obtain transverse and radial waveforms for polarization studies and, for source parameter studies, P and S waves are also available. Mine operators can be provided with the seismic energy and equivalent Nuttli (1973) magnitude values in order to quantify the size of the seismic events. Double-couple focal mechanisms can also be analyzed.

In the frequency domain, corner frequencies and low-frequency spectral density levels (plateaus) allow the determination of the source radius, seismic moment and stress drop values.

Currently, new additions to the software is being developed which include automation of seismic parameters computations, polarization analysis, considering surface and mined zone effects, raypath analysis, attenuation, non-double-couple focal mechanisms and near-field/far-field analysis.

This software enables the analysis of more than 250 large mining induced seismic events originating from the four mentioned mines. Of these, 180 were originating from Creighton Mine.

3 REGIONAL SEISMOGRAPH STATION NETWORK

CANMET is responsible for the operation of eight regional seismograph stations. These stations are located in the vicinities of the Northern Ontario and North-West Québec (figure 3) mining camps, in order to provide more detailed information about the source properties of the mining-induced seismic activity in the local mines. Four of these stations are part of the Sudbury Local Telemetered Network (SLTN).

3.1 The Sudbury Local Telemetered Network

This network is one of the achievements of the Canada-Ontario-Industry Rockburst Project (Hedley and Udd 1987) in order "to increase the range for recorded rockbursts and to improve the source location of previously unlocated rockbursts" (Brehaut and Hedley 1986).

The SLTN was initially a three-station network located around the Sudbury Basin (figure 4). It was established as an initiative of the Canada-Ontario-Industry Rockburst Project in 1984 and has been fully operational since May 1987 (Plouffe 1988).

The three initial outstations, located at Joe Lake, Long Lake and Chicago Mine Road, consisted of a vertical single-component short-period seismometer and a 60 Hz digitizing package. These outstations were linked via dedicated phone lines to a central processing facility at the Science North natural sciences museum in Sudbury. The processor would save triggered events using a simple short-term/long-term average algorithm.

All triggered event files were automatically transmitted, via a dedicated phone line, to the National Seismological Laboratory in Ottawa where they were analyzed along with regional earthquake data.

Using the waveforms recorded by the three original stations, local magnitude scales were developed in order to determine magnitude values for small seismic events, which had not been detected by the National Seismograph Network (NSN) as defined by Munro et al (1988). The scales were developed for three

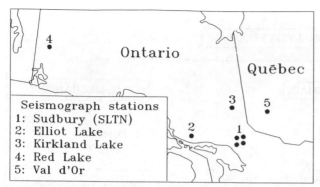

Figure 3. Location of the seismograph stations operated by CANMET.

seismically active mines in the Sudbury Basin: Strathcona, Creighton and Copper Cliff North Mines. Equivalent Nuttli magnitude values could now be determined down to about 1.0 (Plouffe 1992).

3.2 Upgrading of the SLTN

The SLTN network was recently upgraded in order to: provide a better automatic data transmission and analysis; reduce the volume of data generated by the doubling of the system size; rapidly implement new stations, and centralize the mining induced seismic data analysis at the Elliot Lake Laboratory.

A fourth station has been located close to the Creighton Mine (figure 4). By virtue of the installation of extensive micro and macroseismic monitoring systems and the presence of the new seismograph station above the mine workings, the Creighton Mine will be the best equipped mine for seismic monitoring in Canada. Events could now be recorded from magnitudes ranging between -4 to 4 without interruption, this being a part of the CRRP Source Mechanism Project.

The digital signal, processed at the sites of the four stations, is

now recorded at a sampling rate of 100 Hz. It is transmitted through dedicated phones lines to a host computer located at Science North. From there, a communication software, developed by Nanometrics Inc., transmits automatically the triggered events to the Elliot Lake Laboratory. This laboratory includes a seismic analysis centre which could retrieve all seismic data from the SLTN stations. It is planned to transmit the data from all the regional seismograph stations to this central unit.

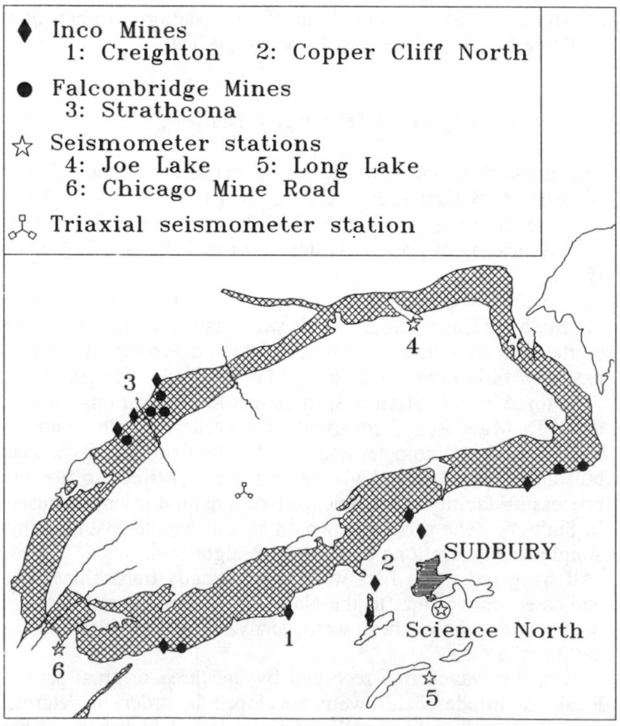

3.3 The other regional seismograph stations

In addition to the SLTN stations, MRL operates four other regional seismograph stations. They are located at Elliot Lake, Kirkland Lake and Red Lake, Ontario, and Val d'Or, Québec, (figure 3).

These seismograph stations consist of a vertical single-component Teledyne-Geotech S13 seismometer, which usually records in a frequency range of 1 to 16 Hz.

At all the stations, the analog signal was initially recorded on a helicorder; the paper chart was changed once a day and sent to Elliot Lake once a month. Presently, at Red Lake, the signal is now added to the data recorded by the local macroseismic system. For Elliot Lake and Val d'Or, the signal is now digital only, using a new generation of seismograph stations, manufactured by Nanometrics Inc.

Duration magnitude relationships were developed for these regional stations located close to the mines. These local magnitude scales allow a quantification of the size of the mining induced seismic events for these mines, at a much lower level than the one given by the National Seismograph Network. Presently, a local magnitude scale has been developed for the Quirke/Denison Mines seismic events (Rochon and Hedley 1987) recorded by the Elliot Lake seismograph; a second preliminary scale was developed for the Macassa Mine at Kirkland Lake (Plouffe 1992). These local magnitude scales enables the magnitude determination for seismic events as low as 0.7 and 1.1 respectively.

These duration magnitude values are equivalent to the national Nuttli magnitude values given by the Eastern Canada Telemetered Network (ECTN) (Munro et al. 1988).

4 CONCLUSION

Mining Research Laboratories of CANMET's involvement with the rockburst research in Canada, first in the phase I (COIRRP) and later as a partner in the Canadian Rockburst Research Program, which is to last until 1995, has contributed to the better understanding of the problems in rockburst engineering. This has been achieved through the co-ordination of the research efforts of the various mining camps, project management, research in hardware and software technology for monitoring and analyzing of rockbursts/micro/macroseismic data and technology transfer seminars.

These efforts and endeavour, augmented by the financial contribution from the industry, the provincial government and their agencies, and the research efforts from the Canadian institutions of learning have brought the Canadian mining industry to the edge of the state-of-the-art of rockburst technology. Today, it could be claimed that the Canadian mining industry, after a quantum leap since 1984, has become the leader in the forefront of rockburst research in the world.

5 ACKNOWLEDGEMENTS

The full cooperation and assistance provided by the mining engineering personnel of the mining companies contributing to the Canadian Rockburst Research Program is gratefully acknowledged. The authors also wish to thank the Canadian Rockburst Research Program Management Committee for their continuing support and Mr. Y. Lizotte, CANMET/MRL, for proof-reading of this paper.

◆ Inco Mines
  1: Creighton   2: Copper Cliff North

● Falconbridge Mines
  3: Strathcona

☆ Seismometer stations
  4: Joe Lake   5: Long Lake
  6: Chicago Mine Road

⚝ Triaxial seismometer station

Figure 4. Map of the Sudbury Basin showing the location of the active mines and the SLTN seismograph stations.

REFERENCES

Brehaut, D.H. & D.G.F. Hedley 1986. 1985-1986 Annual Report of the Canada-Ontario-Industry Rockburst Project. CANMET, Energy, Mines and Resources Canada. Special Report SP86-3E.

Hedley, D.G.F. 1992. Rockburst Handbook for Ontario Hard rock Mines. CANMET, Energy, Mines and Resources Canada. Special Report SP92-1E.

Hedley, D.G.F. & J.E. Udd 1987. The Canada-Ontario-Industry Rockburst Project. *Proc. 6th Int. Rock Mech. Congr., ISRM*: 115-128. Montreal.

Munro, P.S., R.J. Halliday, W.E. Shannon & D.R.J, Schieman 1988. Canadian Seismograph Operations - 1986. Geological Survey of Canada, Energy, Mines and Resources Canada. Paper 88-16.

Nuttli, O.W. 1973. Seismic wave attenuation and magnitude relations for eastern North America. *J. Geophys. Res.* 78:876-885.

Plouffe, M., M.G. Cajka, R.J. Wetmiller & M.D. Andrew 1988. The Sudbury Local Telemetered Seismograph Network. *Proc. 2nd Int. Symp. on Rockbursts and Seismicity in Mines*: 221-226. Minneapolis, Minnesota.

Plouffe, M. 1992. Preliminary local magnitude scales for mining-induced seismicity at some mines in the Sudbury Basin. CANMET, Energy, Mines and Resources Canada. Technical Report MRL92-109.

Plouffe, M. 1992. Preliminary report on scaling of mining induced seismicity at Kirkland Lake, Ontario. CANMET, Energy, Mines and Resources Canada. Technical Report MRL92-113.

Rochon, P. & D.G.F. Hedley 1987. Magnitude scaling of rockbursts at Elliot Lake. CANMET, Energy, Mines and Resources Canada. Division Report MRP87-156.

Rockbursts and Seismicity in Mines, Young (ed.) © 1993 Balkema, Rotterdam, ISBN 90 5410 320 5

Acoustic emission activities in hydraulic fracturing of coal measure rock

M. Seto & K. Katsuyama
National Institute for Resources and Environment, Japan

ABSTRACT: During the hydraulic fracturing experiments at Ashibetsu coal mine, Japan, observation of acoustic emission (AE) were carried out. The experimental site was located at the depth of about 900 m from the surface. Six piezoelectric accelerometers were used for the observation. Hypocenters of 632 events were located during the experiment. The results of the observation revealed the following characteristics of the AE, (1) the spatial distribution of hypocenters of AE is a fractal, and the fractal dimension is 2.08, (2) the plane of AE hypocenters' distribution is approximately perpendicular to the direction of the minimum principal stress in rock of the experimental site.

1 INTRODUCTION

The first theoretical study on hydrofracturing was by Hubbert and Wills(1957) that pointed out the tensile nature of the fractures at the borehole wall. Laboratory experiments by Lockner and Byerlee(1977) have characterized the fracturing process in terms of tensile and shear failure depending on the differential stress, rock permeability, viscosity and injection rate of the fracturing fluid. In addition to these studies, numerous computer programs have been written to the model the hydrofracturing process. The aim of most of those model studies has been to understand the pressure-transient records.

The models are complex enough to include well-bore storage, fracture storage, fracture damage in the form of a fracture skin or restricted fracture flow capacity, asymmetric fracture geometry and deformable proppant-fracture systems. Until now, the hydrofracturing industry has relied upon analytical solutions and numerical models to relate the pressure time-history records to infer crack geometries and extent. But the utility of these models has been limited by a lack of knowledge of actual mechanisms involved in fracture extension as a function of in situ rock and stress properties. Direct observations of hydrofractures by mine-back and mapping their surfaces have provided such information , however, this method is not practical for most routine production related simulated work. The acoustic emission (AE) method offers considerable potential as an effective monitoring tool. One of the purpose of the work described in this paper has been to evaluate the AE method of fracture detection through the field experiments. The other purpose is to asses the effect of the rock's discontinuity on the extension of hydrofractures experimentally.

With the increase of mining depth, the danger of coal and gas outburst has increased. The drainage of methane gas from seam is indispensable for preventing the burst. We applied the hydraulic fracturing technique to the methane gas drainage from seam. And, we investigated the fracture extension in the coal measure rock using the AE technique. We described here the result of the AE activity during the hydraulic fracturing.

In the experiment we used the AE monitoring to detect AE activity and relate the AE to the hydrofractures properties. The approach in the experiment was to use 3-D arrays of AE sensors to surround the hydrofractures zone to locate and characterize the AE activity associated with the hydrofracturing process.

2 EXPERIMENTAL METHOD

Hydraulic fracturing experiment was carried out at Mitsui Ashibetsu coal mine, Hokkaido, Japan. The site is located at the deep gallery. The depth of the site is about 900 m from the surface. Figure 1 shows the experimental site (plan view).

The hydraulic fracturing experiments was performed in a bore hole drilled from the gallery (-695 L oku-NT2). The diameter

and the length of the borehole are 80 mm and 17 m, respectively.

Figure 2 shows the method of hydraulic fracturing. The procedure was to insert the high pressure-resistant rod into the borehole. The water for fracturing was then sent to the bottom of the borehole through rods. Coal and rock were fractured at a pumping rate of 20 l/min.

Six boreholes were drilled for the instrumentation of AE transducers. The acoustic emissions were monitored using piezoelectric transducers. These transducers were set up at the bottom of the boreholes. Figure 3 shows the mounting method of AE transducer. The signals from transducers were

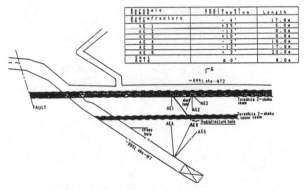

Fig. 1 Experimental site (plan view).

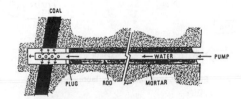

Fig. 2 Method of hydraulic fracturing.

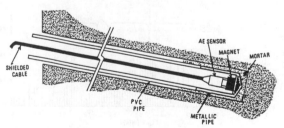

Fig. 3 Mounting method of AE transducer.

amplified 20dB by preamplifiers between 300 Hz and 10 kHz. These signals were recorded on a data recorder. AE recording sysytem used in the experiment is shown in Figure 4. The transducers and preamplifiers are frameproof structure. The analysis and the source location of recorded events were conducted in the laboratory. AE analysing system used in this study is shown in Figure 5.

3 EXPERIMENTAL RESULTS

Figure 6 is the rate of AE activity versus pumping pressure. The AE events shown in this figure were detected by AE3 transducer shown in Figure 1.

AE activity increased with pressure and there was significant activity in harmony with the drastic pressure decrease. However, AE activity continued after "shut-in" (about 80 min). The ongoing activity after shut-in may be related to the degree to the amount of well-bore and fracture storage built up during the pumping. The ongoing AE activity in this case may also be due to a fracture storage effect from a zone of fractures, possibly pre-existing, surrounding the hydrofracture zone that have been influenced by the pressure. The increase of AE activity was consistent with the rapid pressure decrease (about 45 and 60 min). This AE increase may be caused by the formation of larger fracture due to the combination of small cracks.

Figures 7 shows the AE hypocenters' distribution in the hydraulic fracturing. 632 AE events were plotted in this figure.

The events' locations distributed downward from the hydrofracture hole and went across "Torashita 2-shaku seam" almost perpendicularly. The field experimental site was located under the active mining face and the rock in the experimental site was densely fractured. The downward migration may be due to the pre-existing fracture system.

The correlation integral $C(r)$ for the hypocenter distribution $(p1, p2, ---, pN)$ shown in Figure 7 was calculated

in three dimensions. They are given by
$$C(r) = 2 \cdot Nr(R \langle r) / N(N1)$$
where $Nr(R \langle r)$ is the number of pairs (pi, pj) with a distance smaller than r, and N is the number of hypocenters. If the distribution has a fractal structure, $C(r)$ is expressed by
$$\log(C(r)) = D \cdot \log(r) + k$$
where D is a kind of fractal dimension called the correlation exponent that gives the lower limit of Hausdorff dimension. The correlation integral versus the distance for the hypocenters' distribution shown in Figure 7 are plotted on a double logarithmic scale in Figure 8. The actual data fall on the straight line, which indicates that the spatial distribution of AE during the field hydraulic fracturing has a

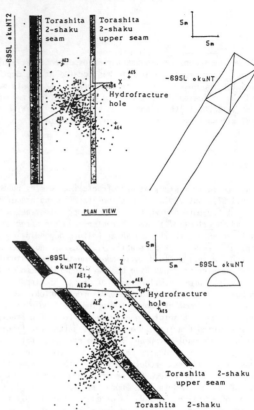

Fig.7 AE hypocenters' distribution in hydrofracturing. 632 AE events are plotted in this graph.

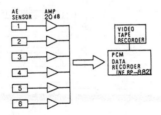

Fig. 4 AE recording system (underground).

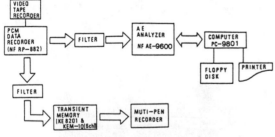

Fig. 5 AE analysing system (laboratory).

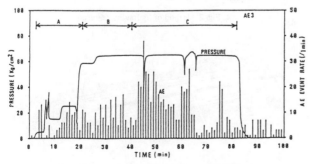

Fig. 6 AE event rate (AE3) and pressure versus elapsed time.

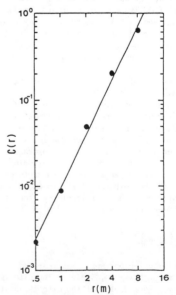

Fig.8 Correlation integral versus distance for 632 AE hypocenters determined in hydrofracturing. Fractal dimension is 2.08.

406

fractal structure. The fractal dimension estimated from the slope was 2.08. The fractal dimension obtained in the present study indicates that the distribution of AE events during the hydraulic fracturing was planer.

In this experiment we also conducted the stress measurement using the stress relief method. The stress measurement was performed in the stress hole shown in Figure 1.

Figure 9 shows the pole plot of principal stresses on the Schmitt net. The maximum, S1, intermediate, S2, and minimum, S3, principal stresses are 20.4 MPa, 8.0 MPa and 3.9 MPa, respectively. These values are shown as closed circles. We approximated the AE events' distribution as a plane using the least square method and calculated the normal line of the approximate plane. The direction of the normal line is also illustrated as a star mark on the Schmitt net shown in Figure 9. The direction of the normal line is exactly the same as that of minimum principal stress. This result indicates that the approximate plane of AE events' distribution is perpendicular to the direction of minimum principal stress. In other words, hydrofractures in this field experiment was formed as a plane which is perpendicular to the direction of

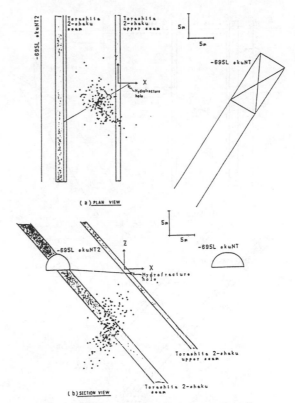

(a) PLAN VIEW

(b) SECTION VIEW

Fig.11 AE hypocenters' distribution in B stage shown in Fig 6

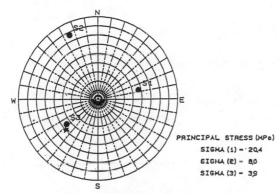

PRINCIPAL STRESS (MPa)

SIGMA (1) = 20.4

SIGMA (2) = 8.0

SIGMA (3) = 3.9

Fig.9 Pole plot of principal stress (●) and the normal line to AE hypocenters' distribution (★). (upper reference hemisphere)

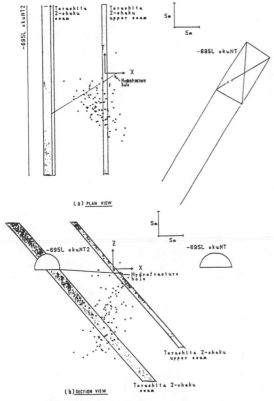

(a) PLAN VIEW

(b) SECTION VIEW

Fig.10 AE hypocenters' distribution in A stage shown in Fig 6

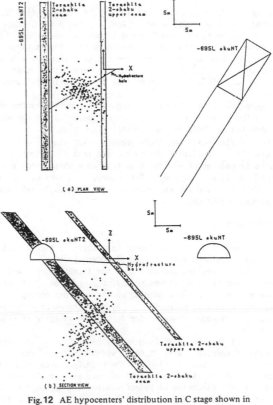

(a) PLAN VIEW

(b) SECTION VIEW

Fig.12 AE hypocenters' distribution in C stage shown in Fig 6

minimum principal stress.

The change of AE hypocenters' distribution in this experiment is shown from Figures 10 to 12.

The results shown in Figures 10, 11 and 12 are the distribution of AE measured in the stage of A, B and C shown

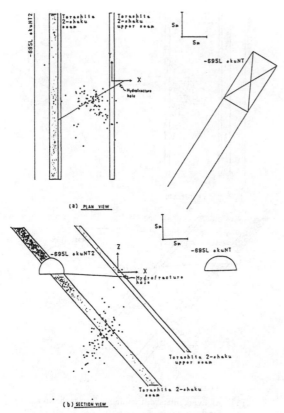

REFERENCES

Hubbert, M.K. and Willis, D.G. 1957. Mechanics of hydraulic fracturing. Trans. Am. Inst. min. Engrs 210: 153-163.
Lockner, D. and Byerlee, J. 1977. Hydrofracture in Weber sand stone at high confining pressure and differential stress. J. Geophys. Res 82: 2018-2026

(a) PLAN VIEW

(b) SECTION VIEW

Fig.13 AE hypocenters' distribution after shut-in.

in Figure 6, respectively. In the A stage AE distibuted randomly between two seams. And, in the B and C stage AE events concentrated on and around Torashita 2shaku seam. So, It seems to us that the extension of hydrofractures and surrounding zone did not start from the hydrofracture hole but from Torashita 2-shaku seam. This fracture extension process may be the effect of pre-exsiting fractures in coal measure rock.

Figure 13 shows the distribution of AE occurred after shut in. In the stage after shut in AE occurred in the region where AE occurred actively during the water injection.

The results described above indicate that the hydrofracture process is a complicated phenomenon and that the hydrofracture system is made up of a network of fractures which forms a zone around a main fracture. The overall direction of the hydrofracture and the surrounding zone is strongly influenced by stress conditions.

4 CONCLUSIONS

It is obvious from the experiment that there is significant AE activity associated with the hydrofracture process.

The AE activities indicate that the formation of fractures were caused by the stress condition , and that the fracture changed its direction depending on the rock anisotropy in the stage of fracture extension. Hydrofractures in this field experiment extended downward due to the pre-existing fractures in the rock.

The distribution of AE hypocenters has a fractal structure. Fractal dimension was 2.08, which indicate that the hydrofractures was a plane. The hydrofractures' plane was perpendicular to the direction of minimum principal stress.

The overall direction of hydrofractures and surrounding zone is strongly influenced by rock inhomogeneities and anisotropy as well as the stress direction.

These results will be applicable to the prevention of outbursts and the development of coalbed methane gas.

Rockbursts and Seismicity in Mines, Young (ed.) © 1993 Balkema, Rotterdam, ISBN 90 5410 320 5

Seismic attenuation in deep-level mines

S. M. Spottiswoode
Chamber of Mines Research Organisation of South Africa, Johannesburg, South Africa

ABSTRACT: P- and S-wave spectra from three deep-level gold mines and one platinum mine in South Africa were found to be well described by the f^{-2} model with attenuation (Q). Q was found to vary widely, from 20 for ray-paths through highly fractured ground to 1000 through solid rock, necessitating independent determination of Q for each phase. I was able to determine spectral fits for moment, corner frequency and f_{max} (Hanks 1982), even for low values of f_{max} less than the corner frequency. Under these conditions of "negative bandwidth", in excess of 98 percent of the radiated energy was lost to attenuation and peak velocity and acceleration were reduced by an order of magnitude. Stress drops ranged between 0.5 and 5 MPa and were independent of moment for moment values in the range of 10^8 to 10^{13} N-m.

1 INTRODUCTION

One of the requirements of modern mine seismic systems is the ability to provide accurate and reliable estimates of source parameters, typically moment and stress. Seismic moment estimates are important for quantitative studies of mine seismicity and relate to the problem of mining sequences and layouts and the resulting seismic deformations (Spottiswoode 1990). Stress drops drive the near-field ground velocities and are therefore directly relevant to the rockburst problem. In addition, Stewart and Spottiswoode (1993) found that times during which small events have higher stress drops than normal are more likely to be followed by seismic events with magnitude M>0. This report addresses the seismological problems of rock-mass attenuation on amplitudes and on the interpretation of body-wave spectra.

A number of authors, including Spottiswoode (1984), have characterized the high-frequency body-wave displacement spectral fall-off by assuming a value for Q and modelling the frequencies above the corner frequency ($f>f_0$) as $f^{-\gamma}$ where $\gamma \geq 2$. However, Hanks (1982), Hanks and McGuire (1981) and many other authors have argued for a flat acceleration spectrum above the corner frequency, and therefore for $\gamma = 2$, out to some maximum value, f_{max}, beyond which the spectrum crashes, or falls off more rapidly. More recently, a number of authors (e.g. Rebollar et al 1990) have described the high-frequency spectra by f^{-2} with frequency-independent Q. In this paper, this behaviour is written as:

$$\tilde{A}(f) = \frac{\tilde{A}(0)\exp(-\kappa f)}{1 + (f/f_0)^2} \qquad (1)$$

with kappa (κ) described by:

$$\kappa = \kappa_0 + \frac{\pi R}{Q V_c} \qquad (2)$$

where

$\tilde{A}(f)$ is the displacement spectral density,

$\tilde{A}(0)$ is the spectral plateau,

f_0 is the corner frequency,

R is the hypocentral distance,

Q is the quality (attenuation) factor,

κ_0 is attenuation attributed to near-source or near-geophone effects, in addition to the the effect of Q in the unfractured rock-mass, and

V_c is the phase velocity (for P or S waves).

Hank's (1982) f_{max} can be defined as:

$$f_{max} = 1/\kappa \qquad (3)$$

Churcher (1990) showed that ray-paths that avoid the stope and its associated region of fractured rock result in impulsive waveforms with little attenuation ($Q \approx 200$). Ray-paths that traversed the stope region were much more attenuated ($Q \approx 20$) and had a more complex appearance. Clearly, we cannot use a single value of Q for all ray-paths within a mine seismic network.

Analysis of small earthquakes is hampered by attenuation, and a wide bandwidth ($f_{max} - f_0 \gg 0$) is usually required for source mechanism studies (Hanks 1982 and McGarr 1984). This can be achieved either by studying only the larger events or by installing geophones closer to the source. As we do not wish to restrict mechanism studies only to the larger events, there is a need to work with bandwidths as small as possible.

In this paper, I report on the results of applying equation (1) to data from networks at several deep-level gold mines and one platinum mine, covering a range of mining conditions. The resulting attenuation has a profound effect on radiated energy and peak ground motion parameters. I will show that it is possible to invert for source parameters even for small events, even for phases having a "negative" bandwidth, defined as

$f_{max} - f_0 < 0$. Under these conditions, we record very little of the total radiated energy.

2 METHOD

I developed an algorithm (SOURCEQ) that automatically selected time windows, calculated spectra and performed the three-parameter fit for $\tilde{A}(0)$, f_0 and κ to each spectrum, according to equation 1. SOURCEQ was used either in graphics mode to display the interpretation for each individual seismogram or in batch mode to process any number of events at a time, as follows.

Windows were automatically selected to cover both the P and S pulses, based on previously picked arrival times, as well as noise preceding the P-wave arrival. A cosine taper was applied at the beginning and end of each window. Windows that were too short or had poor signal-to-noise performance were rejected. Three-point smoothing was applied to the power spectra of the body waves and further smoothing to the noise spectra so that the range of frequencies with good signal-to-noise ratio could be identified automatically.

Initial values of $\tilde{A}(0)$ and f_0 were obtained from the peak in the velocity spectrum, corrected by a user-selected value for κ. My algorithm for fitting equation (1) to the observed spectra improved on these initial values as follows:

1. A least-squares procedure was used to find $\tilde{A}(0)$ and f_0 by minimizing the sum of the squares of the difference between the theoretical velocity spectrum and the observed velocity spectrum, corrected by the current value of κ. The least-squares procedure ensured that the radiated energy of the observed spectra was equal to the energy of the fitted spectra.

2. A new value of κ was then found from the slope of log {observed spectrum / theoretical spectrum} as a function of frequency.

These two steps were repeated until satisfactory convergence occurred. Calculations for each phase took about 0.2s on a 50MHz 486 PC.

As the effect of both κ and f_0 is a rapid fall-off in the spectral energy at high frequencies, inversion is fundamentally difficult for $\kappa f_0 \gtrsim 1$. In addition, the shape of the spectrum at $f \approx f_0$ in equation 1 does not have a strong physical basis. These effects resulted in the inversion often failing for $\kappa f_0 \gtrsim 1$, despite many attempts to stabilize the interaction between stages 1 and 2 by changing the inversion method. As seismologists usually select for source mechanism or ground motion studies only those seismograms for which $\kappa f_0 \ll 1$ (e.g. McGarr 1984), any analysis that relaxes this condition provides additional opportunities.

3 DATA SETS

Data for detailed analysis were selected from networks at three deep-level gold mines and one platinum mine, covering a range of mining conditions. The gold mine networks used COMRO's Portable Seismic System (Pattrick et al 1990), with geophones grouted at the ends of holes drilled some 3 to 10m out of access tunnels. The platinum data were recorded by a GENTEL system developed by General Mining (More O'Ferral et al, pers comm 1992).

Sampling rates varied from 1000 to 20 000 samples per second, depending on the network extent. Seismograms from SENSOR 4.5Hz or 14Hz geophones were amplified with a gain factor of 100 and digitized with a 12-bit 10V analogue-to-digital converter (Pattrick et al 1990). Peak velocity values were greater than 0.02 mm/s for all events: smaller events were usually rejected at the mine. Only a small sample of the recorded data for mines A, B and C were studied, obtained at random on previous visits to the mines.

Here follows a brief description of each network.

3.1 *Mine A*

The seismic network was installed to monitor seismicity around longwalls and stabilizing pillars in a mine in the East Rand. Eight geophones were placed in follow-behind off-reef drives, usually 25m in the foot-wall, across a region some 3km in extent at depths of 2~3km below surface.

3.2 *Mine B*

This network was installed to study scattered mining and remnant removal in a region with seismogenic faults and dykes in the Orange Free State mining district at depths of 2~3km below surface. 16 geophones covered a region 800m across and were placed in haulages some 80m below reef.

3.3 *Mine C*

This project was a test site within a shaft pillar and was aimed at investigating the fracturing around two 3m by 3m experimental drifts which were spaced 7m apart skin-to-skin, mined in a field stress of about 100MPa. Geophones were placed at the ends of 10m holes some 40m behind the advancing drifts. Most of the recorded events were associated with the development blasts and care was taken to exclude these events from this study.

3.4 *Mine D*

This system consisted of one set of triaxial geophones installed at the bottom of a 40m vertical borehole drilled from surface. The data presented here originate from platinum mining where the tabular excavations were supported with a system of regional and crush pillars at a depth of 750m to 1km below surface.

As surface reflections closely followed the direct wave, it was necessary to adjust the theoretical spectral shape by the reflected waves, P, SV and SH, allowing for their appropriate amplitudes (Aki and Richards 1980). SV waves outside the shear-wave window were excluded from the analysis to avoid the associated surface waves.

In addition to the four primary data sets, I used some results from an off-reef site some 100m behind a highly stressed longwall in the Carletonville area (Hemp and Goldbach 1993) and from a study of mechanisms and ground motions by McGarr (1984).

4 ANALYSIS

In this section I will firstly justify the use of equation (1) using two events: figure 1 for a seismogram from mine A and figure 2 for mine B. As the conventional amplitude spectrum simply falls off sharply above either f_0 or f_{max}, the spectra were plotted as acceleration spectra. The acceleration spectral values were plotted as a function of log (frequency) to show the flat spectrum between f_0 and f_{max}, particularly for figure 2b, followed by a curved spectral fall-off for $f \geq f_{max}$. This curved region transforms to a linear fall-off in the log-linear plots in figures 2c and 2c, in clear agreement with

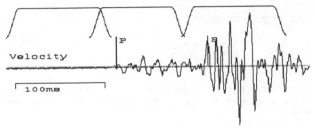

Figure 1a. A velocity seismogram for mine A, with windows marked for noise, P and S waves.

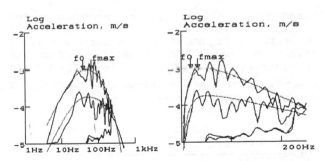

Figures 1b & c. From top to bottom: S, P and noise acceleration spectra from figure 1a. The fitted theoretical spectra as derived from equation 1 are the dotted, smooth curves. The derived values of f_0 and f_{max} are marked for the S-wave spectrum.

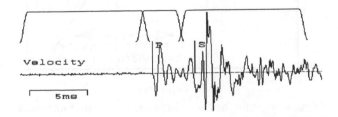

Figure 2a. Seismogram for mine B, as for figure 1a.

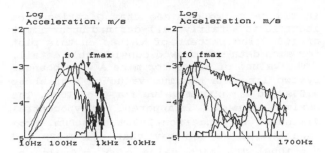

Figures 2b & 2c. As for 1b & 1c, derived from the seismogram in figure 2a.

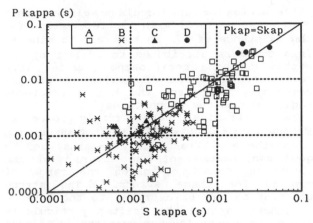

Figure 3. Attenuation as measured from P waves at different geophone sites compared to attenuation as measured from S waves. Data from the four mine sites are identified by the symbols in the box, as with the following figures.

frequency-independent attenuation (Q) from equations 1 and 2.

In figure 3 we can see that κ is approximately the same for P and S waves, and therefore $Q_P/Q_S \approx V_S/V_P$. The increased scatter for small values of κ for mines A and B is an effect generated by the log scale from errors in measurement.

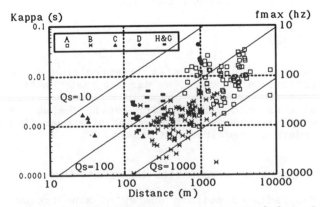

Figure 4. Attenuation (κ) as measured from S waves at different geophone sites for each mine as a function of hypocentral distance, R. Lines of constant Q are drawn for reference. In addition, data studied by Hemp and Goldbach (1993) were included, marked "H&G".

411

In figure 4, we can see that Q varies from about 20 to a fairly well-defined upper limit of 1000. The pattern of values on this plot for each data set showed considerable detail.

The values of kappa for mine A were bounded by two lines: a maximum value of κ of about 0.03s and a maximum value for Q of 1000. The maximum value of κ_0 is apparent for hypocentral distances varying from 1000m to 4000m. From equation 2, we can say that $\kappa_0 \leq 0.02$s and Q = 1000. The high value of $\kappa_0 = 0.02$s, or $f_{max} = 50$Hz, could be ascribed to the extensive fracturing around the large stope spans that were generated by the longwall mining.

Attenuation for mine B could be well described by Q = 200 to 1000, with little evidence for the existence of κ_0, in contrast to mine A. This could be attributed to smaller spans, better geophone positioning (further from reef) and the role of faults in drawing seismicity away from the highly fractured region around the stope faces.

Events for mine C were strongly grouped at $R \approx 40$m and $\kappa \approx 0.001$s. This was not surprising given the small range in ray-paths, and is good confirmation of the accuracy of the inversion procedure, given the range of moments for these events. The strong attenuation can be ascribed to the intense fracturing around the drifts, particularly since the events located near the drift ends and the rays travelled back along paths in the vicinity of the drifts.

The value of κ for mine D, 0.02s to 0.04s, or $f_{max} = 25$Hz to 50Hz, is in good agreement with values of f_{max} reported by Hanks (1982) for seismograms written at surface stations.

The low values of Q (Q<20) for the events studied by Hemp and Goldbach (1993) again illustrated high attenuation in the region around the stopes.

Static stress drops (τ) were calculated from Brune (1970):

$$\tau = \frac{7 M_0}{16 r_0^3} \qquad (4)$$

where:
M_0 is the seismic moment and

$$r_0 = \frac{2.34 V_c}{2 \pi f_0}$$

is the source radius as inferred from Brune (1970). In figure 5, we can see that the data studied here were compatible with previous mine seismic data (McGarr 1984). The stress drop was independent of event size over five orders of magnitude in moment.

5 EFFECT OF ATTENUATION ON SEISMOGRAMS

The radiated seismic energy of mine tremors is usually estimated by integrating the square of velocity over time, averaged over a number of geophones, assuming geometric spreading:

$$E = k \int v^2 dt \qquad (5)$$

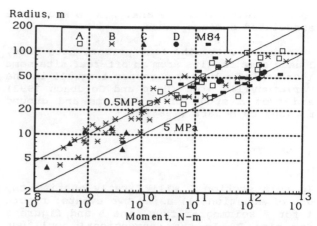

Radius, m

Figure 5. Source dimension as a function of seismic moment. Data marked as "M84" are mining events from McGarr (1984).

where k is a constant. From Parceval's theorem, the integration can be performed in the frequency domain, giving:

$$E = k \int \hat{v}^2 df \qquad (6)$$

and from equation 1,

$$E = k \hat{A}(0)^2 \int \frac{f^2 \exp(-2 \cdot \kappa \cdot f)}{(1 + (f/f_0^2))^2} df \qquad (7)$$

For integration from zero to infinity, and substituting f for f/f_0, equation 7 reduces to:

$$E = k f_0^2 \hat{A}(0)^2 \int \frac{f^2 \exp(-2 \cdot \kappa \cdot f_0 \cdot f)}{(1 + f^2)^2} df \qquad (8)$$

Numerical integration is necessary for evaluating equation 8. The results are presented in figure 6, which shows the radical effect of small or "negative" bandwidth provided by $f_{max} < f_0$. Even for $\kappa f_0 = 0.2$ (or $f_{max} = 5 f_0$), more than 50 percent of the radiated energy is lost to attenuation.

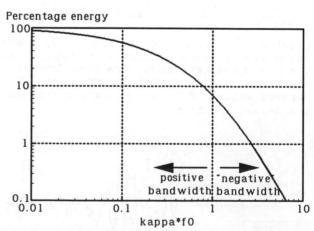

Percentage energy

kappa*f0

Figure 6. The effect of f0*kappa on recorded energy.

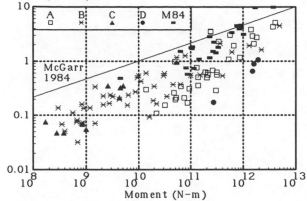

Figure 7. Hypocentral distance times peak velocity (Rv$_{max}$) as a function of seismic moment. An empirical relationship from McGarr(1984) for a depth of 2.5km is drawn for comparison.

We can see in figure 7 that the relationship of McGarr (1984) is a poor fit to the data. The fit for each data set is even worse for smaller values of M$_0$ than for larger values, for reasons that will become apparent.

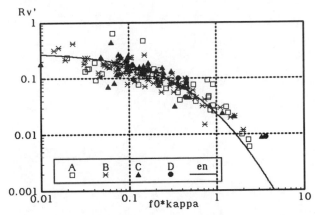

Figure 8. Rv′ = $Rv_{max}/(\tau r_0)$, in arbitrary units, as a function of $f_0\kappa$. The energy attenuation curve from equation 8 and figure 6 is also plotted for comparison.

The peak velocity should be proportional to the product of stress drop and source radius (e.g. McGarr 1984). In figure 8, Rv′ = $Rv_{max}/(\tau r_0)$ is compared to $f_0\kappa$ for individual seismograms. There is a very strong decrease in peak velocity with increasing values of $f_0\kappa$, in close agreement with the theoretical decrease in energy. Figure 8 shows that the peak velocity can generally be estimated within a factor of two, given M$_0$, τ and $f_0\kappa$.

Similar analysis of the ground motion parameter $\rho R a_{max}$, where ρ is the density and a_{max} is the peak ground acceleration, showed a very similar trend to that shown for peak velocity in figure 8.

6 CONCLUSIONS

I have found that Q varies widely, from 20 for highly fractured ground to 1000 for long ray-paths through solid ground. These extreme variations make corrections for an assumed constant Q subject to large errors.

1. Source mechanism studies of small mine events are feasible only if spectra are simultaneously inverted for attenuation, κ.

2. Simultaneous inversion for M_0, f_0 and κ is possible, even for $f_0\kappa < 1$.

3. Attenuation has a profound effect on energy, peak velocity and peak acceleration.

Acknowledgement

This work forms part of the research program of the Chamber of Mines. Thanks to Dr W. Rymon-Lipinski, Ms D. Hemp, and Messrs P. Chetty, N. Cook, M. Grave, R. Kersten, D. Minney, R. More O'Ferral and M. Spengler for providing me with seismic data. Thanks also to Dr N.C. Gay and Mr R.D. Stewart for reviews and comments.

REFERENCES

Aki, K. & P.G. Richards 1980. *Quantitative seismology*. San Francisco : W.H. Freeman.

Brune, J.N. 1970. Tectonic stress and the spectra of seismic shear waves from earthquakes. *J. Geophys. Res.* 75: 4997-5009. (Correction. 1971, *J. Geophys. Res.* 76: 5002).

Churcher, J.M. 1990. The effect of propagation path on the measurement of seismic parameters. *Proceedings of the 2nd International Symposium on Rockbursts and Seismicity in Mines*: 205-209. Minnesota: Balkema.

Hemp, D.A. & O.D. Goldbach 1993. The effect of backfill on ground motion in a stope during seismic events. *Proceedings of the 3rd International Symposium on Rockbursts and Seismicity in Mines*: this issue. Kingston: Balkema.

Hanks, T.C. 1982. f_{max}, *Bull. Seism. Soc. Am.*, 72, 1867-1879.

Hanks, T.C. & R.K. McGuire 1981. The character of high-frequency ground motion. *Bull. Seism. Soc. Am.* 71: 2071-2095.

McGarr, A. 1984. Scaling of ground motion parameters, state of stress, and focal depth. *J. Geophys. Res.* 89: 6969-6979.

Pattrick, K.W., A.M. Kelly & S.M. Spottiswoode 1990. A portable seismic system for rockburst applications. *International Deep Mining Conference : Technical challenges in deep level mining*: 1133-1146. Johannesburg: SAIMM.

Rebollar, C.J., L. Munguía, A. Reyes, A. Uribe & O. Juménaz 1991. Estimates of shallow attenuation and apparent stresses from aftershocks of the Oaxaca earthquake of 1978. *Bull. Seism. Soc. Am.* 81: 99-108.

Spottiswoode, S.M. 1984. Source mechanisms of mine tremors at Blyvooruitzicht Gold Mine. *Proceedings of the First International Symposium on Rockbursts and Seismicity in Mines*: 29-37. Johannesburg, S. Afr. Inst. of Min. Metall.

Spottiswoode, S.M. 1990. Volume excess shear stress and cumulative seismic moments. *Proceedings of the Second International Symposium on Rockbursts and Seismicity in Mines*: 39-43. Rotterdam: Balkema.

Stewart, R.D. & S.M. Spottiswoode 1993. A technique for determining the seismic risk in deep-level mining. *Proceedings of the 3rd International Symposium on Rockbursts and Seismicity in Mines*: this issue. Kingston: Balkema.

414

Rockbursts and Seismicity in Mines, Young (ed.) © 1993 Balkema, Rotterdam, ISBN 90 5410 320 5

Implications of harmonic-tremor precursory events for the mechanism of coal and gas outbursts

Peter Styles
Department of Earth Sciences, University of Liverpool, UK

ABSTRACT: Harmonic, monochromatic seismic events showing only a single, emergent seismic phase have been observed as precursors to coal and methane outbursts and at times of abnormal gas emission. These are considered to be close analogues to slow volcanic tremor observed before major eruptions, to long period, harmonic events recorded during hydrofracture and also to icequakes. They appear to be due to premonitory, dilatant microfracturing during which adsorbed gas suffers a rapid change of phase from capillary condensation to free gas. Resonant oscillations of slow, dispersive crack-wave modes give rise to the harmonic-tremor like events. An outburst appears to be a catastrophic coalescence of these events in conjunction with body forces generated by the abnormal gas pressure gradients.

1 INTRODUCTION

Outbursts are instantaneous catastrophic failures of the coal mine structure, characterised by emissions of large quantities of finely divided coal-dust and gas (methane, carbon dioxide or a mixture of the two) from a coal face. Outbursts are a global problem with occurrences in Australia, Canada, China, Europe, Japan, the USA and Russia and many incidents which are clearly Outburst initiated are not reported as such particularly from Eastern Europe and China. Although outbursts are generally associated with coal, they have occurred in other rock types, for example in potash at Boulby in the United Kingdom and salt in Louisiana (Belle Isle Mine , New Iberia, Molinda 1988) and in East Germany (Werra and South Hartz Districts). Outbursts can have devastating consequences, for instance 93 fatalities occurred during and subsequent to an outburst at the Yubari Shin mine near Hokkaido in 1981. Outbursts are particularly frequent in coal as a consequence of the complex nature of the interaction between coal gases and the coal matrix and appear to be particularly violent where the gas involved is carbon dioxide although methane outbursts are obviously associated with extreme ignition risks. Coal gas is generally a mixture of carbon dioxide and methane, with one component predominating. However, other gases are sometimes found in coal seams, such as nitrogen and hydrogen sulphide. In the Bowen and Sydney Basins in Australia for example, the seam gas can range from pure methane to pure carbon dioxide. Carbon dioxide is found in the seam gas in France, Poland and Czechoslovakia. Quantities of coal in excess of 10,000 Tonnes and gas volumes of 100,000 m^3 have been reported.

The coal and gas system constitutes a three-phase medium as coal has the capacity to adsorb enormous quantities of methane, carbon dioxide or both onto the coal matrix surfaces, in a liquid-like mono-molecular layer by capillary condensation. Free gas also exists within the system within the pores and a small quantity of gas will exist in solution in any water present (Litwiniszyn 1985, Paterson 1990). As much as 28 m^3 per Tonne of methane has been reported from outburst coal from Cynheidre, South Wales (Richards pers. comm.)

Outbursts involving carbon dioxide tend to be more violent than those involving methane. The sorptive capacity of coal for carbon dioxide is two or three times greater than for methane. In addition, the desorption of carbon dioxide is much faster than

methane, which implies that the pressure gradients are higher when carbon dioxide is present, which would explain the greater violence of carbon dioxide outbursts. In addition laboratory work on coal saturated with CO_2 has shown a reduction in elastic moduli of as much as 35% (Ketslakh and Vinokurova 1983)

Outbursts are often associated with structural features such as faults or folds or even microscopic irregularities in the coal micro-structure. They may also be associated with sedimentary features, for example palaeo-fluviatile channels within the coal seam, or with the presence of a thick competent sandstone roof although it is the stress-containment aspect of this which is critical to outburst occurrence. The threshold depth for the occurrence shows considerable variation globally from c 200 metres to more than 1000 metres although at least part of this variation is because CO_2 outbursts seem to happen at much shallower depths than do Methane and many of the Australian and Nova Scotian events are of this type.

2 MODELS FOR THE OUTBURST MECHANISM

Although many thousands of outbursts have occurred and many field investigations have been carried out the mechanism by which they occur is not well understood. There are two principal models for the process.

2.1 Structural failure of coal due to anomalous pressure gradients

Paterson (1986, 1990) has suggested that the excess stress caused by gas flow within the coal can exceed gravitational body forces by ten times and may lead to structural failure due to the formation of tensile fracturing. Evidence for the weakening effect of high pressure gas on the structural integrity of porous materials has been shown by Ujihira et al. (1984) and the reduction in the mechanical strength of coal when saturated by CO_2 has already been mentioned (Ketslakh and Vinokurova 1983).

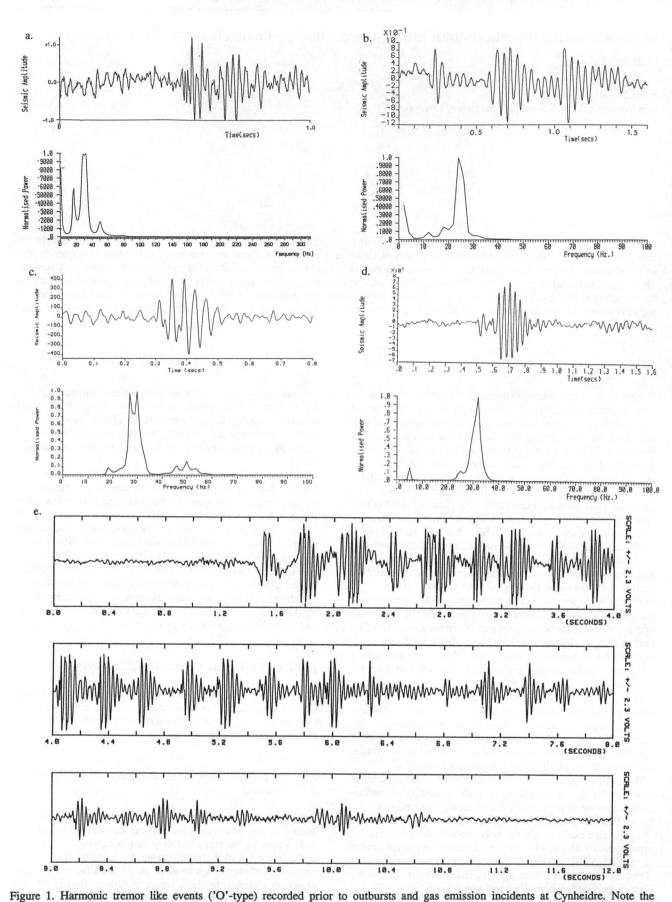

Figure 1. Harmonic tremor like events ('O'-type) recorded prior to outbursts and gas emission incidents at Cynheidre. Note the consistent waveform and distinctive 30 Hz frequency.

a. Event detected at a distance of 6 km. before an outburst on 15 November 1982, b. Event detected before an outburst on 24 June 1985. c. Event detected before a gas emission incident on 24 June 1988, d. Event detected during a period of high methane on 20 August 1987, e. Sequence of c 30 harmonic events occurring in less than 12 seconds and associated with audible 'pouncing' and emergency evacuation of a heading on 16 December 1988.

416

2.2 Destruction of the coal integrity due to shock-induced phase transition.

Litwiniszyn (1985) has suggested that a rarefactional shock wave can precipitate a near-instantaneous phase change from the condensed phase to the gaseous phase with the associated 'crushing wave' causing the disintegration of the coal structure.

Paterson is sceptical of Litwiniszyn's model because of the occurrence of outbursts in salt mines where capillary condensation cannot occur. However, Molinda (1988) does point out that methane bubbles at an estimated 500 to 1000 atmospheres can occur entrained on halite crystal interfaces and within individual crystals. Both of these models invoke a very rapid progression to failure and associated Outburst. It is not the purpose of this paper to present another detailed model for the outburst mechanism but rather to attempt to show how microseismic observations carried out prior to, during and subsequent to outbursts demand that the development of an outburst is a progressive phenomenon which only becomes catastrophic in its final stages. He and Shining (1992) also make this point.

3 MICROSEISMIC/ACOUSTIC EMISSION MONITORING DURING OUTBURSTS

The first acoustic emission monitoring of outbursts was carried out by McKavanagh and Enever (1980) who recorded signals in the 100 to 5 Khz range prior to outbursts in West Cliff Colliery, N. S. Wales. noting a reduction in the rate of micro-seismic activity in the 10 minute period immediately prior to the outbursts. Grczl, Lcung and Ahmed (1984) subsequently deployed underground systems at both Leichardt and Metropolitan collieries with accelerometers in the coal rib with a system bandwidth of 2 Hz to 1.5 Khz. They clearly detected in-seam Channel waves which gave characteristic coal dispersion curves. Enhanced background seismicity occurred for 17 days before a 25 tonne outburst again with a reported reduction in activity several shifts before the outburst. Leighton (1984) reports microseismic monitoring during an outburst at an unspecified site but clearly notes that 'the outburst was comprised of several failures occurring very rapidly...'. Nakajima et al (1984,1989) have reported acoustic emission prior to an outburst during a drivage in Sunagawa Mine noting a clear correlation between methane concentrations and rate of acoustic emission and that the outburst itself was the coalescence of a sequence of individual acoustic events.

3.1 Microseismic monitoring of Outbursts at Cynheidre

In the United Kingdom, outbursts have represented a serious problem in the anthracite zone of the South Wales Coalfield since the last century, particularly in the Gwendraeth Valley where anthracite of very high quality is mined, with numerous casualties and 6 fatalities in 1971. The Outbursts generally occur in the Big Vein and Pumpquart seams although others have been involved. They are frequently associated with the major thrusting which is present in the Lower Coal Measures In the Gwendraeth Valley 236 outbursts have occurred of which 105 were spontaneous, 113 were induced by shot firing and 18 were unclassified (Davies, Styles and Jones 1986).

Miners have traditionally reported hearing rock noises or 'pounces' prior to these outbursts. These have been described as the sound of a 'two-stroke motorbike engine'. The pounces are the audible expression of acoustic emission/microseismic (AE/MS) activity occurring as failure of the seam occurs.

Tapes recorded using an underground accelerometer and a portable tape recorder showed in-seam micro-seismic activity which increased at the time of roof collapse, as the powered supports were advanced. Subsequently, a spontaneous outburst, during March 1981 was recorded and although the signal was contaminated by crosstalk from the colliery telephone system it did show the presence of frequent 'pouncing' for many minutes prior to the outburst.

Using a network of eight vertical component seismometers deployed in a 6 km by 6 km array at the surface of the colliery it was possible to detect, identify and locate microseismic activity associated with normal mining activity, ie extraction of coal and the subsequent collapse of the waste and to make preliminary correlations with the rate of advance and hence the volume of coal won.

Microseismic activity on the surface seismometer network was monitored for a very extensive period (1982 to 1989). Full descriptions of this work can be found in Styles (1983), Kusznir et al (1985), Styles et al (1987) and Styles et al (1991).

Two distinct styles of microseismic activity originating from within the colliery and generated by the extraction of coal were recognised.

3.1.1 Normal Activity

This is characterised by impulsive, short-duration events which propagate across the colliery at velocities of about 3.5 km/s, and show clearly distinguishable P and S phases particularly at more distant stations. At this colliery they are of very low magnitude, generally between -1 and 1 but can be very frequent with as many as 800 in an 8-hour period. They show normal attenuation and are severely reduced in amplitude by the time they have travelled 2 km or so. The most telling piece of evidence that this type of activity is generated directly by extraction of coal is the correlations between the rate of extraction and the number of microseismic events per shift. Hypocentral locations show that this activity is localised around of the face, presumably close to the front abutment zone. The very impulsive nature of the events and the ubiquitous, downwards first motions suggest they are generated by an almost instantaneous brittle failure, probably fracturing of the sandstone roof together with failure in the floor. This is confirmed by detailed monitoring of coal faces in Staffordshire and Warwickshire from three-component borehole sensors (Toon and Styles 1992, Styles et al 1992)

3.1.2 Outburst Activity

More importantly a completely new type of seismic activity ('O'-type activity) was recognised which appears to be associated with the microfracturing of the coal and emission of gas during periods of abnormal face conditions. They have the following distinctive characteristics:

i   These events appear to be rare during the working of a normal face but to increase rapidly in frequency of occurrence immediately prior to outbursts or outburst-prone conditions.They often occur in closely spaced sets with several events within a few seconds. They begin to occur at least ten days prior to an outburst and gradually increase in rate of occurrence until they form a coalescing continuum of acoustic activity at the time of an outburst

ii   They have emergent onsets, durations of about 600 millisecs, a symmetrical 'beaded' envelope and appear to be mono-phasic with a monochromatic spectrum with a principal frequency component at 32Hz.

iii   They show little attenuation as they propagate across the

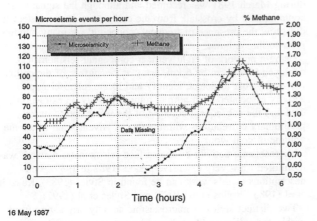

Comparison of Surface Microseismicity
with Methane on the coal-face

16 May 1987

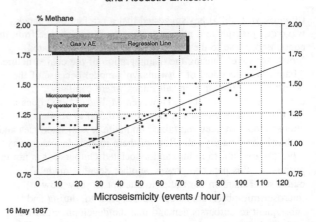

Correlation between Gas Emission
and Acoustic Emission

16 May 1987

Figure 2. Microseismic activity detected at the colliery surface by the automated monitoring system compared with the methane measured on the face and the correlation between these two activity indices. The excellent agreement suggests that the methane has been emitted as a consequence of the dilatant microfracturing associated with the microseismicity.

colliery and have been detected up to 6 kilometres away from an outburst in comparison with normal events which attenuate rapidly with distance.

iv   The velocity of propagation of these events is often very slow and variable with velocities between 300 m/s to 1 Km/s.

v   There is very little variation in magnitude of these events in contrast to normal mining-induced activity which can vary over several orders of magnitude (Styles et al 1992)

Although it was initially considered that these characteristics were likely to be determined by propagation path effects, analysis of their nature over a very long period leads to the conclusion that they are source dependent. These events have been recognised during outbursts from both the Pumpquart and Big Vein seams at all seismometers within the network, from pulsed infusion firings (detonation of explosives under water pressure in boreholes) and from numerous gas emission incidents which were controlled without progressing to an outburst. Figure 1 shows a selection of this type of event recorded during the monitoring of this colliery. As Figure 2 shows the events show a remarkable correlation with the emission of methane gas from the face As events rose to over 100 per hour the percentage of methane rose to 1.65%. The microprocessor monitoring the network gave an alarm which required that cutting be stopped and activity and methane emission dropped to a safe level without progressing to an outburst.

4 VOLCANIC TREMOR, HARMONIC TREMOR, ICE-QUAKES AND 'O'-TYPE EVENTS

Although seismic events with the above characteristics have rarely (Wong et al 1989) been reported to be associated with mining, events with remarkably similar signatures have been recognised for many years from a broad range of seismic regimes.

Seismic monitoring of volcanoes has revealed the presence of puzzling event types which included volcanic tremor, an emergent, low-frequency (c 1Hz), monochromatic oscillation which can persist for several minutes or more, (Aki et al 1977,

Fehler 1983, Hofstetter and Malone 1986), 'B'-type earthquakes which have 'emergent P-wave onsets, no clear S-phase' (McNutt 1986). Figure 3 (from McNutt 1986) shows an example of a B-type event from Pavlov Volcano (after McNutt 1986). McNutt points out the low range of magnitudes (-0.5 to 1.3) observed for these events and suggests that both the harmonic tremor and the B-type events are related to degassing of the magma during eruption.

Seismic monitoring of glaciers in Prince William Sound reveals seismic events which are characterised by '1) an emersive (sic) onset lacking a distinct first arrival; 2) a weakly developed P phase; 3) an obscured S arrival....4) a monochromatic, low-frequency (1-2 Hz) signature...' (Wolf and Davies 1986).

St Lawrence and Qamar (1979) noted the similarities between these icequakes and volcanic tremor and attributed them both to hydraulic transients; in water for glaciers and for magma in a conduit for volcanoes. We have also recorded events of this type during monitoring of the Bakananbreen glacier on Svalbard.

Dobecki and Romig (1983) and Dobecki (1985) describe events which were recorded during a hydrofracture experiment carried out by Sandia National Laboratories at the Nevada Test site. They detected events with a single polarisation which they associated with resonance within the fracture due to pressure-transients within the hydrofrac fluid. A subsequent experiment where a previously fractured well was remonitored gave considerable acoustic emission which was all of the 'single phase/ single polarisation type'. Majer and Doe (1986) report a hydrofracture experiment in granite where they observed two types of seismic event, 'Type-1', impulsive events and 'Type-2', slow or harmonic tremor events and recognised the correspondence between these and volcanic tremor.

A very large volume of work both on volcanic tremor (Aki et al 1977, Fehler and Chouet 1982, Ferrazini and Aki 1987) and hydrofracturing events, (Chouet 1988, Ferrazini et al 1990) has been carried out by Aki and co-workers leading to quantitative models of long-period resonant events observed during hydrofracturing (Chouet 1986). Figure 4 shows the remarkable correspondence between volcanic and hydrofracture long-period events even though the timescales differ by a factor of 100. Of particular note is the fact that their model shows that the period of oscillation of a fluid-filled crack may be much longer than would be expected from the normal modes of acoustic vibration of the fluid-filled crack, because of the existence of very slow

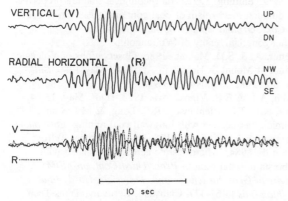

Figure 3. 'B-type' volcanic event detected at Pavlov volcano. Note the simlarity between the waveform and that observed for gas emission events during outbursts (After McNutt 1986).

waves which they call crack waves. The actual resonant frequency shows a complex dependence on the stiffness factor of the crack, its aspect ratio and the acoustic properties of the fluid but may be only 10% or less of that predicted by normal acoustical modes. The detailed signature of the radiated waveform generated by the resonance is also controlled by the position of the excitation which triggers the resonance but generally matches the character of the harmonic tremor well.

The 'O'-type events which we observe during methane outbursts appear to have all of the characteristics of these tremor-type events and with an apparent resonant frequency of 30 Hz, fall nicely between the volcanic events at c 1-2 Hz and the hydrofracture events at c 1 Khz. A normal organ-pipe resonance of this frequency would require an oscillating cavity length of c 10 metres which seems very unlikely to be present within a coal seam. However, if slow, crack-wave oscillation is a permitted mode for methane-filled cracks the work of Chouet(1986) would suggest that this length may be reduced by a factor of 10 at least and a sub-metre resonant fracture would seem plausible within a coal seam. In particular, as it is possible to stimulate the generation of the 'O'-type events by pulse-infusion firing which is carried out by detonating explosive charges in a pressurized hole in the coal, (in effect a mini-hydrofrac) it certainly seems possible that they have their origin in these 'slow' resonant oscillations of fluid (in this case methane) in micro-fractures in the coal.

5 HARMONIC TREMOR EVENTS AND THEIR RELEVANCE TO THE MECHANISMS OF OUTBURSTS

Microseismic monitoring during outbursts invariably shows that the outburst itself is the coalescence of many acoustic emission/microseismic events but that the process was initiated much earlier, perhaps several days before the final catastrophic culmination. It is also clear that structural weaknesses within the coal control the location of the outburst. The occurrence of the harmonic-tremor like events which were associated with gas emission in Cynheidre would seem likely to be due to the onset of dilatant microcrack development (Brace et al 1966) within the coal probably initiating in the weakened disturbed zones well in advance of the face position . As the microfracture opens this may trigger the quasi-instantaneous phase change from capillary condensation to free gas in the manner described by Litwiniszyn (1985) and the gas-filled crack will oscillate generating 'slow', crack-waves which will radiate harmonic-tremor events with a characteristic frequency. As the face then approaches the potential outburst zone the degree of dilatancy will increase and the associated microfracturing will release more gas. The permeability will also rise as the degree of dilatancy increases as demonstrated by Majewska and Marcak (1989). Shining and He (1991) recognise the importance of microcrack development during the fracture of coal in the laboratory under tri-axial conditions and note that the strength of coal is considerably reduced in the presence of high gas content. The remarkable correlation between event rate and methane on the face (Figure 2) certainly seems to indicate that the excess methane is being released by the same process which is generating the microseismic activity. The fact that is possible to control the rate of methane emission by ceasing the cutting and allowing microseismic activity to die away seems to indicate that the process need not always become catastrophic. However, if mining activity proceeds sufficiently rapidly it then seems certain that eventually the dilatant microfracturing will release sufficient quantities of gas that the pressure gradients established may participate in the disintegration of the coal and an eventual outburst. It is clear that gas must play some dynamic role in the eventual catastrophic failure because if cutting ceases and the gas allowed to escape either naturally or by stimulation of the coal through pulse-infusion firing then it is possible to resume cutting the coal and although it has suffered severe structural disintegration it appears to have a reduced outburst risk.

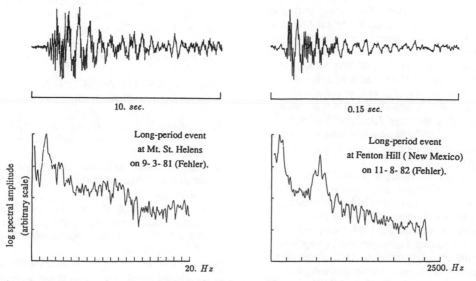

Figure 4. Comparison between volcanic tremor at Mount St Helens and long-period harmonic tremor during hydrofracture at Fenton Hill (After Ferrazzini et al 1990)

6 CONCLUSIONS

The seismic characteristics of the emergent, monophasic, monochromatic, microseismic events observed during methane outbursts and gas emission incidents at Cynheidre Colliery, S. Wales suggest that they should be classified as Harmonic Tremor associated with slow crack-wave oscillations and as such fall into the same category as volcanic tremor events (eg Mount St Helens), hydrofracture long-period events (eg Fenton Hill) and icequakes. It is suggested that they occur during the formation of dilatant, tensile microfracturing of the coal in a potential outburst zone with adsorbed gas instantaneously changing from the capillary condensed phase to free gas and triggering the crack-wave resonances. The detection of this type of activity many days before an outburst and their gradual increase in frequency of occurrence until they coalesce into continuous acoustic emission at the time of the outburst suggests that gas outbursts are caused by a progressive disintegration of the coal structure from a combination of dilatancy and abnormal gas pressure gradients.

Recognition of the onset of occurrence of this type of activity may indicate that dilatancy has begun and that there is some risk of an imminent outburst. Experience from Cynheidre suggests that if this is recognised and cutting ceases and gas is allowed to dissipate the outburst may be prevented.

REFERENCES

Aki, K., M. Fehler & S. Das. 1977. Source mechanism of volcanic tremor:Fluid-driven crack models and their application to the 1963 Kilauea eruption. *J. Volcanol. Geotherm. Res.* 2:259-287.

Brace, W. F., B.W. Paulding. & C. Scholz. 1966. Dilatancy in the fracture of crystalline rocks, *J. Geophys. Res.*, 71: 3939-3954.

Chouet, B. 1986. Dynamics of a fluid-driven crack in three dimensions by the finite difference method. *J. Geophys. Res.* 91:13967-13992.

Davies, A.W., P. Styles & V. K. Jones. 1987. Developments in Outburst Prediction by Microseismic Monitoring from the surface. *Mining Engineer.* 147:486-498.

Dobecki T. L. & P.R. Romig. 1983. Acoustic signals generated by hydraulic fractures - In-situ observations. *36th Annual Meeting of the Midwest Society of Exploration Geophysicists*, March 7-9th, 1983, Denver, Colorado.

Dobecki T. L. 1989. Acoustic emissions from hydraulic fractures: Long term observations. *Proceedings of the Fourth Conference on Acoustic Emission/ Microseismic Activity in Geological Structures and Materials, Hardy and Leighton (eds.):* 403-415. Claustal, Germany:Trans-Tech Publications.

Fehler, M. C. 1983. Observations of volcanic tremor at Mount St Helens volcano. *J. Geophys. Res.* 88:3476-3484.

Fehler, M. C. & B. Chouet 1982. Operation of a digital seismograph network on Mount St Helens volcano and observations of long-period seismic events that originate under the volcano. *Geophys. Res. Lett.* 9:1017-1020.

Ferrazzini, V. & K. Aki. 1987. Slow waves trapped in a fluid-filled infinite crack:implication for volcanic tremor. *J. Geophys. Res.* 92:9215-9223.

Ferrazzini V., B. Chouet. M. Fehler & K. Aki. 1990. Quantitative analysis of long-period events recorded during hydrofracture experiments at Fenton Hill, New Mexico. *Journal of Geophysical Research*, 95:871-21,884.

Grezl, K. J, L, Leung, & M. Ahmed. 1984. Development of Seismoacoustic Monitoring Techniques for Underground Mines in Australia. *Proceedings of the Third Conference on Acoustic Emission/Microseismic Activity in Geological Structures and Materials*: 659-670. Clausthal, Germany: Trans. Tech. Publications

He, X. & Z. Shining. 1992. The rheological fracture properties and outburst mechanism of coal containing gas. *Proc. 11th Int. Conf. Ground Control in mining*:575-579. Wollongong:University of Wollongong.

Hofstetter, A. & S.D. Malone. 1986. Observations of Volcanic Tremor at Mount St Helens in April and May 1980. *Bull. Seis. Soc. Amer.* 76, No4:923-928.

Ketslakh, A.I. & E.B. Vinokurova. 1983. *Ugol'*:May. 13-14.

Kusznir, N.J., T.G. Blenkinsop, R.E. Long, & M.J. Smith. 1985. Seismicity associated with longwall coal mining. *Earthquake Engineering in Britain*:111-128. London:Thomas Telford.

Leighton, F. 1984. Microseismic activity associated with outbursts in coal mines. *Proc. Third Conf. on AE/MS activity in Geol. Structures and Materials, Hardy and Leighton (eds.)*:467-477. Clausthal, Germany: Trans-Tech Publications.

Litwiniszyn, J. 1985. A model for the initiation of Coal-Gas Outbursts. *Int. J. Rock Mech. Min. Sci. & Geomech. Abstr.* 22:39-46.

Majer E. L. & T. W. Doe. 1986. Studying hydrofractures by high frequency seismic monitoring. *Int. J. Rock Mech. Min. Sci. & Geomech. Abstr.*, 23: 185-199.

Majewska, Z & H. Marcak. 1989. The relationship between acoustic emission and permeability of rock under stress, *Min. Sci and Tech.*, 9: 169-179.

McKavanagh, B.M. & J.R. Enever. 1980. Developing a microseismic outburst warning system. *Proceedings of the Second Conference on Acoustic Emission/Microseismic Activity in Geological Structures and Materials*: 211-225. Clausthal, Germany: Trans. Tech. Publications.

McNutt, S.R. 1986. Observations and analysis of B-type earthquakes, explosions and volcanic tremors at Pavlov volcano, Alaska. *Bull. Seis. Soc. Amer.* 76:153-175.

Molinda, G.M. 1988. Gas outbursts studied at domal salt mines. *Geotimes*, January:12-14.

Nakajima, I.Y. Watanabe & T. Fukai. 1984. Acoustic Emission during advance boring associated with the prevention of coal and gas outbursts. *Proc. of the Third Conf. Acoustic Emission/Microseismic Activity in Geological Structures and Materials, Hardy and Leighton (eds)*:529-548.Clausthal, Germany: Trans-Tech Publications.

Nakajima, I.Y., K. Itakura & M. Ujihira. 1989. Consideration for acoustic emission activity from caol and gas outbursts. *Proc. of the Fourth Conf. Acoustic Emission/Microseismic Activity in Geological Structures and Materials, Hardy and Leighton (eds)*:207-244. Clausthal, Germany: Trans-Tech Publications.

Paterson, L. 1986. A model for outbursts in coal. *Int. J. Rock Mech. Min. Sci. & Geomech. Abstr.* 23: 327-332.

Paterson, L. 1990. The mechanism of outbursts in coal and the prevention of outbursts by gas drainage. *Rockbursts and Seismicity in Mines*, Fairhurst (ed.). Rotterdam: Balkema. 285-287.

Shining, Z. & X. He. 1991. Rheological hypothesis of the coal and outburst mechanism. *Proc. Aus. Inst. Min. Mett.* no 2:19-23.

St. Lawrence W.S. & A. Qamar. 1979. Hydraulic transients: a seismic source in volcanoes and glaciers, *Science*, February 16:654-656.

Styles, P. 1983, Microseismic precursors to a spontaneous outburst in Cynheidre Colliery, Dyfed, W. Wales; *Geophys. J. R. Astr. Soc.*:299.

Styles, P., T. Jowitt & E. Browning. 1987. Surface Microseismic Monitoring for the prediction of Outbursts, *22 Int Conf.Safety in Mine Research, Dai Guoquang (ed.)*: 767-780.Beijing, China.

Styles, P., Emsley, S.,J., & Jowitt, T., 1988, Microseismic monitoring for the prediction of outbursts at Cynheidre Colliery, Dyfed, S. Wales. *Engineering Geology of*

Underground Movements, Geological Society Engineering Geology Special Publication No.5: 423-433. London: Geol. Soc. London.

Styles, P., S. J. Emsley & E.A. McInairnie. 1991. Microseismic prediction and control of coal outbursts in Cynheidre Colliery, South Wales, United Kingdom. *Proc. of the Fifth Conf. Acoustic Emission/Microseismic Activity in Geological Structures and Materials, Hardy (ed):* Clausthal, Germany: Trans-Tech Publications.

Styles. P., I. Bishop, S. Toon & R. Trueman. Surface and borehole microseismic monitoring of longwall faces; their potential for three-dimensional fracture imaging and the geomechanical implications. *Proc. 11th Int. Conf. Ground Control in mining*:177-192. Wollongong:University of Wollongong.

Toon S. & P. Styles. 1993. Microseismic imaging of fractures around longwall coal faces using down-hole three component geophones. *Geophysical Prospecting* (in press).

Ujihira, M., T. Isobe & K. Higuchi. 1984. On the flaking destructive phenomena of porous material induced by high pressure gas. Min. Metall. Inst. Japan. 100:225-232.

Wolf, L.W. & J.N. Davies. 1986. Glacier-generated earthquakes from Prince William Sound, Alaska. *Bull. Seis. Soc. Amer.* 76,no. 2:367-379.

Wong, I. G., J.R. Humphrey & W.J. Silva. 1989. Microseismicity and subsidence associated with a potash solution mine, South-Eastern Utah. *Proc. of the Fourth Conf. Acoustic Emission/Microseismic Activity in Geological Structures and Materials:287-306. Hardy and Leighton (eds)*,Clausthal, Germany: Trans-Tech Publications.

Rockbursts and Seismicity in Mines, Young (ed.) © 1993 Balkema, Rotterdam, ISBN 90 5410 320 5

Design of a microseismic monitoring system for the investigation of tunnel excavation damage

Shahriar Talebi
Mining Research Laboratories, CANMET, Elliot Lake, Ont., Canada

R. Paul Young
Engineering Seismology Laboratory, Department of Geological Sciences, Queen's University, Kingston, Ont., Canada

ABSTRACT: A microseismic monitoring system was designed in order to monitor the activity associated with the excavation of a circular tunnel, 46 m long and 3.5 m in diameter, at the 420 m level of AECL's Underground Research Laboratory. The process of array optimization and design of the full-waveform microseismic monitoring system was based on the examination of parameters affecting the quality of source location and source parameter determinations. Sixteen triaxial sensors were installed around the "Mine-by" tunnel. Several hundred events were recorded after the extraction of each 1-meter round of the tunnel which was performed by careful line drilling and reaming, followed by mechanical breaking. A clear clustering of event source locations was observed at two opposite sides of the tunnel along a subvertical axis, in agreement with the in-situ stress field and observed overbreaks.

1 INTRODUCTION

Atomic Energy of Canada Limited has undertaken fundamental research at the Underground Research Laboratory (URL) in order to assess the concept of safe and permanent disposal of nuclear fuel waste in the plutonic rock of the Canadian Shield. Localized failure of rock leading to the formation of overbreaks around the walls of deep underground excavations, such as those observed at the URL, is of considerable importance to the above concept since cracks could form pathways for radionuclide leakage. The present paper describes the design optimization and some results from microseismic monitoring in the case of the "Mine-by experiment" undertaken at the 420 m level of the URL. Emphasis will be placed on the methodology used for the design of the data acquisition system and preliminary source location results.

2 DESCRIPTION OF THE SITE

The Underground Research Laboratory (URL) in southeastern Manitoba is a major research facility in the Canadian Nuclear fuel waste management program operated by Atomic Energy of Canada Limited (AECL). A multi-disciplinary research program has been undertaken at the URL to develop the technology needed for the safe and permanent disposal of Canada's nuclear waste. The URL vertical shaft has been excavated to a depth of 443 m into an undisturbed part of a plutonic rock mass and major research activities take place in the 240 m and 420 m levels (Martin and Simmons, 1992).

2.1 Geology

The URL is located in the Lac du Bonnet granite batholith where several granite phases are present. Everitt and Brown (1986) and Brown et al (1989) have described the geology of this site. Three major low-dipping fracture zones identified during site evaluation control the ground water movements within the rock mass (Figure 1). The in-situ observations have shown that the grey granite (below a depth of 220 m) is much less fractured than the pink granite (between surface and the Fracture Zone 2). Except for a few minor closed fractures of a very limited extent, natural fractures are rare below this fracture zone and the rock is essentially massive unfractured grey granite.

2.2 Stress field

Data from different stress measurement techniques have been compiled in order to gain insight about the in situ stress field at the URL . Martin (1990) distinguished two different stress domains at the URL based on the data from the orientation of the maximum in situ horizontal stress. These domains extend respectively above and below the Fracture Zone 2 (see Figure 1). In the top zone, the maximum horizontal stress strikes about 040°, parallel to the major subvertical joint sets present in the pink granite. The maximum horizontal stress rotates 90° below the fracture zone 2 and is oriented 130°.

Read and Martin (1991) discuss the in situ stress state in the area of the mine-by experiment. The magnitude of the maximum and the intermediate principal stresses which are subhorizontal are respectively 55 MPa and 48 MPa. The minimum principal stress is subvertical and about 14 MPa.

2.3 Failure observations

The excavation of the shaft below 255m into the highly-stressed rock above Fracture Zone 2 was associated with spalling of the walls and sometimes buckling of the shaft bottom (Figure 2). The failure process observed also in the galleries of the 420 m level often involves small thin slabs of rock popping off the walls of the excavation, resulting in the formation of V- shaped notches. The appearance of well-bore breakouts are closely related to the magnitude and the orientations of the in situ stress at the excavation boundary.

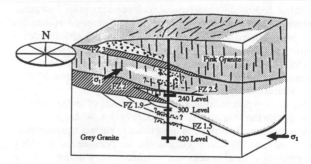

Figure 1. Generalized geological setting of the URL (after Read and Martin, 1991).

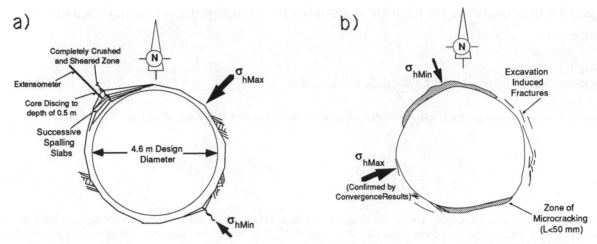

Figure 2. (a) Shaft failure at 277.5 m of depth (b) Zones of microcracking at the 300 level station (after Martin, 1990)

3 DESCRIPTION OF THE EXPERIMENT

3.1 *Background*

The excavation of the shaft at URL between 324 m and 443 m of depth was monitored using a microseismic system (Talebi and Young, 1992). The microseismic activity following each blast clustered mainly around the walls and the bottom of the shaft near the newly created faces. A very interesting feature observed was the clustering of the events in a preferential NE-SW orientation, in agreement with the available data on stress field and observations of overbreaks. The accuracy of the source location procedure was better than 1 meter.

Spectral analysis of the signals resulted in corner frequencies of P-and S-waves in the range 1 kHz to 4 kHz. The intensity of the events, estimated from their moment magnitude, was in the range -3.6 to -1.9 (Gibowicz et al, 1991). An increase was observed in the seismic moments versus depth, particularly during the excavation of the 420 m level, suggesting that the response of the rock mass depends on the direction of the excavation (Talebi et al 1990).

3.2 *Procedures*

Figure 3 shows a 3-D view of the final arrangement of the mine-by experiment. The circular test tunnel (Room 415) is 46 m long, 3.5 m in diameter and was excavated towards azimuth 225° so as to maximize the in-plane stress ratio to promote the occurrence of excavation damage development (Read and Martin, 1991). An incline (Room 409) and a decline (Room 413) provided access to instrument galleries and areas around the test tunnel for sensor installation prior to and during excavation.

The excavation method consisted of perimeter line drilling and reaming, followed by mechanical rock breaking using hydraulic splitters. This consists of a series of 0.1 m diameter 1 m long boreholes around the perimeter of the tunnel. A hydraulic rock splitter was then used to break the rock. This method was preferred, over the drill-and- blast technique used for the shaft excavation, in order to have better control over the induced excavation damage and experiment monitoring.

Previous experiments had shown the effects of temperature changes on the instruments and rock behavior. Although corrections are performed in most cases to correct measurement results affected by changes in ambient temperature and humidity an enthalpy control system was installed to keep the air temperature at 10° C (the ambient rock temperature) and the humidity at 90%.

4 DESIGN OF A MICROSEISMIC SYSTEM

The main objective of microseismic monitoring in the present project was to contribute in the evaluation of excavation damage around the mine-by tunnel through improved accuracy in the determination of location and mechanism of microseismic events. In order to achieve this objective the design of the array of sensors and the data acquisition system was optimized. This section describes the parameters taken into account in and the results of the optimization process. Emphasis will be placed on the parameters affecting source location accuracy.

4.1 *Array of sensors*

The basic parameter to take into account in the selection of sensor type (accelerometer or geophone) is the frequency band of interest. In the present case accelerometers were chosen since the expected activity is in the kiloHertz range. The ideal procedure for the sensor installation is cementing them inside boreholes since this allows a perfect coupling between the sensor and the borehole and the elimination of a number of complexities related to the presence of a high frequency three component sensor in a fluid-filled borehole.

The number of sensors to be used depends on the requirements of source location and source mechanism procedures. An absolute minimum of 4 arrival times is required for a source location determination and a minimum of 6 determinations of seismic moment is required for the inversion of the seismic

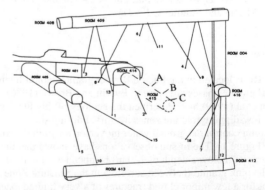

Figure 3. A three-dimensional view of the mine-by tunnel and the location of the 16 triaxial sensors (after Read and Martin, 1991).

moment tensor. Since the absolute minimum is insufficient to provide a reliable estimate, a total of 16 sensors was used in order to achieve the required accuracy in source location and source mechanism determinations.

The fundamental requirement in the determination of the geometry of the array is a satisfactory coverage of the active area in order to obtain reasonable accuracy in source location and source mechanism determinations. Figure 3 shows a 3 dimensional view of the designed array of sensors. Indeed, a perfect symmetry in the location of the sensors relative to the excavation axis would not have been ideal. The resolution in source location determinations and the spatial coverage of some potential source areas could be improved by introducing slight changes in sensor location relative to the symmetrical arrangement.

4.2 Parameters affecting source location accuracy

The most reliable source location methods use first arrival times of P- and S-waves radiated by a microseismic event. The quality of a source location determination depends on the quality of the arrival times picked which in turn depends on the quality of the signals. The parameters affecting the accuracy of source location determinations could be divided into two categories: those related to the recording system and those affecting the waves during their propagation in the rock mass (Talebi and Young, 1990).

4.2.1 Path effects

Propagation of P- and S-waves through a rock mass could distort the signals in a number of ways. Rock masses are often heterogeneous and anisotropic, properties which affect the travel time and frequency content of body waves. The monitoring area

at URL could be considered as a reasonably homogeneous rock mass, as indicated by the results obtained during the shaft extension period. Similarly, anisotropy is weak to absent at URL: a maximum of 4% of P-wave anisotropy was observed by Talebi and Young (1992). Even such a weak anisotropy could be successfully taken into account in the source location procedure.

The other factor affecting the waves during their propagation is their attenuation: the selective filtering of higher frequencies leading to a lower frequency content of the signal. This phenomenon is unlikely to affect the signals in any significant manner in this project due to the very short distances involved (a few tens of meters) and the competent nature of the URL granite: no evidence of any significant attenuation of the signals was discerned in the shaft excavation data (Talebi et al, 1990).

4.2.2 Recording parameters

The monitoring system could, on the other hand, affect the properties of the signals and cause, ultimately, inaccuracies in source locations. Three factors are noteworthy: the array geometry, the signal to noise ratio and the sampling frequency.

The array geometry does not in itself induce errors but can amplify them. An adequate geometry of the array of sensors will ensure good source location results. A high signal to noise ratio is required for accurate arrival time picking. The error in the measurement of first arrival time decreases inversely in proportion to the frequency bandwidth. The other parameter affecting source location accuracies is the sampling frequency. A high sampling frequency ensures a high accuracy in source location results. The acquisition system allows a maximum sampling frequency of up to 62.5 kHz per channel, more than 1 order of magnitude higher than the maximum frequency already observed at URL.

4.2.3 Calibration tests

In order to test the ability of the present sensor array for accurate source location and source mechanism determinations, three tests were performed.

The first test consisted of source location error-space determination for the area around the test tunnel (see Figure 4).

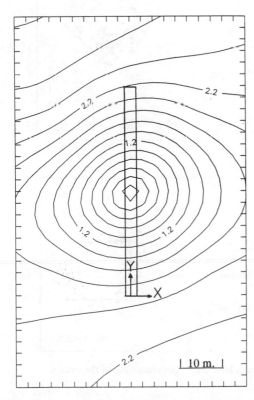

Figure 4. Error-space analysis for an event located in the middle and on the top edge of the tunnel. The contours show iso-values of error in milliseconds in the horizontal plane.

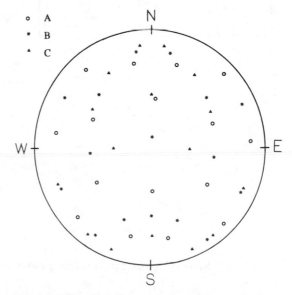

Figure 5. Equal-area projection of source-sensor vectors for the three event locations A, B, and C (see Figure 3) located along the top edge of the tunnel.

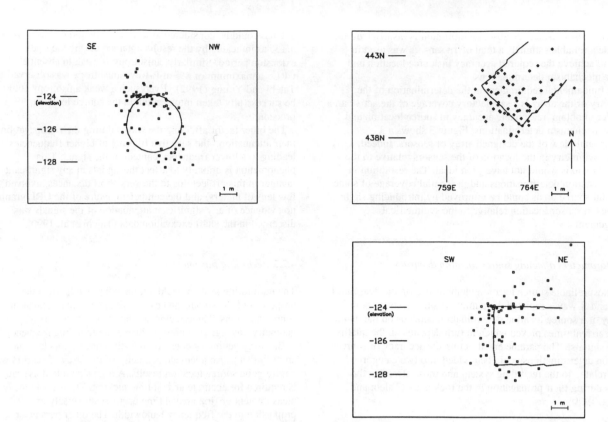

Figure 6. Source locations from round 415- 13, calculated from automatically computer- picked P-wave arrival times, shown in three two dimensional views of the tunnel.

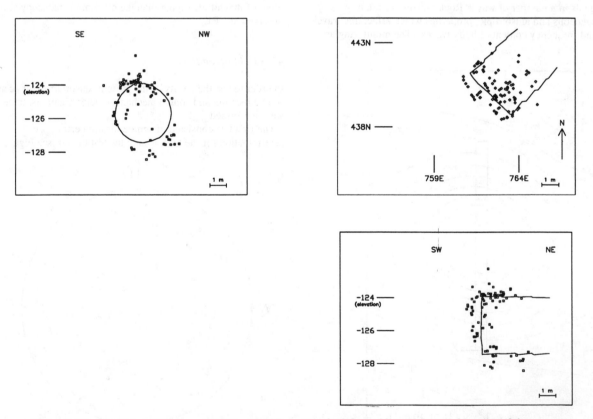

Figure 7. Same as Figure 6 but for manually picked P-wave arrival times. Note the better clustering of the events.

The final result of such a test for a convenient array of sensors should show a minimum close to the actual source location and a rather smooth transition path from different areas to that location. Similar results were obtained for two other trial source locations along the tunnel.

The second test consisted of generating artificial P-wave arrival times for 24 assumed sources, located on the tunnel wall along a spiral pattern and using these arrival times to find the source locations. This was based on the realistic hypothesis of a homogeneous medium where raypaths are straight lines. An error in percent of the travel time is then added to each arrival time. These results indicated that source location accuracy could be much better than half a meter (Talebi and Young 1990). A general observation was an even higher accuracy for the central area of the tunnel.

In a final test, the spatial coverage of three potential sources located along the tunnel were examined (figure 5). The results were satisfactory, with a similarly higher degree of coverage in the central area of the proposed tunnel.

4.3 *The acquisition system*

Signals from 16 triaxial sensors pass through preamplifiers installed in junction boxes on top of each sensor hole at the 420 m level. The borehole inclination was surveyed prior to sensor installation using a laser system in order to achieve the required accuracy in sensor coordinate determinations. The amplification levels varied between 3 and 30 for the most part. The signals were then transmitted to surface through a 50-pair, 19 gauge self-supporting shaft cable with overall shield. A differential amplifier and anti-aliasing filter board on surface was used to provide an additional 10 dB of amplification and band-pass filtering of 50- 10,000 Hz with a 72 dB per octave roll-off rate. The output then passed through a trigger box before being recorded. The sampling frequency was 50 kHz for each channel resulting in an 80 milli-seconds recording window(Collins and Young, 1992).

The calibration test of the acquisition system showed an approximately flat response in the 1-6 kHz frequency range (Feustel and Young, 1992). The acquisition computer was linked to two other computers: one provided real-time processing ability and the other was used for more in-depth analysis. The data was then backed up on optical disks for archival purposes.

5 RESULTS

The tunnel was excavated in 1-meter rounds for the most part using perimeter line drilling and reaming. Each step was followed by a 12 hour monitoring period when several hundred events were recorded. The total number of events over the course of the experiment was approximately 25,000. The recording parameters such as the channels and amplitude levels to trigger and the amplification level for each channel were closely monitored and set throughout the duration of the experiment.

On-site real-time processing was performed using a computer package automatically picking P-wave arrival times and calculating source location estimates (Figure 6). An increase in source location accuracy was achieved using the sum of the squares of the 3 waveforms from each sensor as the input signal (Collins and Young, 1992). Although this provided a reasonable initial estimate of the areas of seismicity, best results were obtained from manually-picked data (Figure 7). The general pattern of seismicity, however, did not change significantly. The quantitative comparison of the "error" defined as the distance represented by the product of the P-wave velocity and the RMS value of residuals of P-wave arrival times after source location was as following:

automatic picking : 0.95 ± 0.45 m
Manual picking : 0.35 ± 0.17 m

Although these values do not represent the true error in source location determinations, they explain the better clustering of the results for the second case. The microseismic activity clustered mainly around the newly-created faces and migrated forward with the excavation front as it proceeded. A clear tendency was observed for more events to cluster above rather than below the face. This phenomenon is currently under investigation and could be related to confinement and/or gravity. The event location pattern for the majority of the extracted rounds shows a clear clustering of source locations, in the plane perpendicular to the tunnel axis, at two opposing sides of the tunnel along a subvertical axis. This is consistent with the fact that the tunnel axis is perpendicular to the maximum principal stress which is sub-horizontal and the orientation of observed overbreaks in the tunnel.

6 CONCLUSION

Several hundred microseismic events were recorded following the extraction of each 1- meter round of the test tunnel, excavated using careful line drilling and mechanical rock breaking. The optimization of the array and the design of the data acquisition system revealed to be successful in that the data collected showed the required quality to fulfill the objectives of the design. An improved source location accuracy was obtained using manually-picked P-wave arrival times rather than those picked automatically. The clear clustering of the events at two opposite sides of the test tunnel along a subvertical axis is compatible with the orientations of the principal components of the in-situ stress field and overbreaks observed in the tunnel.

ACKNOWLEDGEMENTS

This work was supported by the Canadian Nuclear Fuel Waste Management Program with Joint Funding from AECL Research and Ontario Hydro under the auspices of the CANDU Owners Group. The authors wish to thank the above organizations, as well as the Natural Sciences and Engineering Research Council of Canada for additional funding. We are grateful to D.S. Collins for his help in the management and processing of the data and other members of the Engineering Seismology Laboratory for their contribution to different aspects of this work.

REFERENCES

Brown, A., Soonawala, N.M., Everitt, R.A. and Kamineni, D.C. 1989. Geology and geophysics of the Underground Research Laboratory site, Lac du Bonnet Batholith, Manitoba. *Can. J. Earth Sci.* 26: 404-425.

Everitt, R. and Brown, A. 1986. Subsurface geology of the Underground Research Laboratory: an overview of recent developments. *Proc. 20th Information Meeting of the Canadian Nuclear Fuel Waste Management Program*, Winnipeg, Canada, 1: 146- 181.

Collins, D.S. and Young, R.P. 1992. Monitoring and source location of microseismicity induced by excavation of the Mine-by tunnel : Preliminary analysis. Report to Atomic Energy of Canada Limited.

Feustel, A.J. and Young, R.P., 1992. Calibration of the Queen's microseismic system at the Underground Research Laboratory. Report to Atomic Energy of Canada Limited.

Gibowicz, S.J., Young, R.P., Talebi, S. and Rawlence, D.J. 1991.Source parameters of seismic events at the Underground Research Laboratory in Manitoba, Canada: scaling relations for events with moment magnitude smaller than -2. *Bull. Seism. Soc. Am.* 81: 1157-1182.

Read, R.S. and Martin, C.D., 1991. Mine-by experiment final design report. Report #AECL-10430, Atomic Energy of Canada Limited.

Martin, C.D., 1990. Characterizing in situ stress domains at AECL's Underground Research Laboratory. *Can. Geotech. J.* 27: 631-646.

Martin, C.D. and Simmons, G.R., 1992. The Underground Research Laboratory - An opportunity for basic rock mechanics. *ISRM News Journal*, 1: 5-12.

Talebi, S. and Young, R.P. 1990. Design of a Microseismic system for the URL Mine-by experiment. Report to Atomic Energy of Canada Limited.

Talebi, S. and Young, R.P. 1992. Microseismic monitoring in highly-stressed granite: Relation between shaft-wall cracking and in situ stress. *Int. J. Rock Mech. Min. Sci. & Geomech. Abstr.* 29: 25-34.

Talebi, S., Young, R. P., Rawlence, D. J. and Gibowicz, S. J. 1990. Source mechanism characterization of microseismic events induced by shaft excavation at the Underground Research Laboratory. Report #RP005AECL to Atomic Energy of Canada Ltd.

Rockbursts and Seismicity in Mines, Young (ed.) © 1993 Balkema, Rotterdam, ISBN 90 5410 320 5

Improved seismic ground conditions with the double-cut mining method in wide tabular reef extraction

J. P. Tinucci
Itasca Consulting Group, Minneapolis, Minn., USA

A. R. Leach
Western Deep Levels South Division, Carletonville, South Africa

A. J. S. Spearing
Anglo American Corporation, Gold Division, Johannesburg, South Africa

ABSTRACT: In wide reef areas, experience has shown that fewer problems occur if the face is advanced in two cuts rather than in one full-face cut. Severe face-bursting and hangingwall stability problems can occur if the face is advanced in a single cut. This paper describes data obtained to evaluate the difference in ground conditions between double-cut and full-face single-cut mining in wide reef tabular stopes at Western Deep Levels South Mine in South Africa. In practice, the double-cut method produces more favorable ground conditions than the single-cut method, and a reduction in accidents, groundfalls, lost blasting shifts, and seismicity have been observed without reducing production. This paper provides some insight as to why better conditions are observed with the double-cut mining method as a result of using cemented fill and more efficient support.

1 INTRODUCTION

Deep-level gold mines in the West Rand Region of South Africa have typical stoping widths of 0.9 m to 1.6 m. In some areas, the reef width is greater due to the manner in which the formation was deposited. High channel width areas were first intersected in a few panels at Western Deep Levels South Mine in early 1989, and, by the end of that year, all panels within two major long-walls (78-49 and 80-49) on the east side of 2400 m level had channel widths in excess of 2.5 m, with some over 4 m. This area of high channel width represents a remnant of a high river terrace in the Ventersdorp Contact Reef (VCR) landscape.

Initially the full channel width was mined as one single cut, with support comprised of very long camlock props as temporary support, plus 2.2 by 1.1 packs as permanent support. With the introduction of double-cut mining and stiff fill, ground conditions were noticeably improved. This paper presents the data gathered before, during and after double-cut mining was used in wide channel areas. Data on production, seismicity, lost production blasts, injuries, hangingwall overbreak, and depth of rock ejected from the face are discussed. Energy release rates and conditions of face stability are evaluated to examine the effect of increasing stope width. The information presented is based on work by Leech (1991) and Tinucci and Dawson (1992).

1.1 Experience with wide reef mining

Several mining problems soon became evident when mining the wide channel in a single cut: (1) the stope hangingwall was often seen to be up to 1 m into the lava hangingwall formation, and (2) the top part of the face frequently overhung the working area, creating dangerous rock fall conditions.

These problems arose for several reasons. First, drillers would drill the top row of blast holes upward, often penetrating the lavas. Top holes were not as far forward in the face as bottom holes. The blast would then free a greater depth of rock at the bottom than the top, leaving an overhang.

The second reason for hangingwall falls was the relative ineffectiveness of timber packs. Closure rates in the VCR were low (i.e., 5 to 10 mm/m face advance) and, given the height of the packs, the loads generated were very low. Pre-stressing bags were not effective because the rise required to get the same pre-load was greater than at a low stope width. Even with Vetter air bags and wedging, it would take a long time before the packs became effective at these low closure rates. Between packs there

was essentially no support to prevent hangingwall falls, particularly during seismic-induced events. In some cases extensive umbrellas were constructed at the top of packs to try to control falls.

A further problem that was noted was that minor seismic events (i.e., strain bursts) appeared to occur more frequently on the high single-cut faces than previously at narrower stoping widths. Face burst events appeared to cause greater damage, as more rock was ejected from faces in high stope width areas compared to narrower areas.

1.2 Motivation for making assessments

In an attempt to solve these problems, a double-cut mining method was introduced during 1990. The top cut was mined at 1.5 m stope width and supported with rows of hydraulic props. The bottom cut of the remaining reef was then mined in the low stress environment. Initially, packs were installed behind the bottom cut and split sets, or Delkor bolts were installed to control the span between props and packs.

At about the same time, classified tailings fill was being placed in the back area between rows of packs in some single-cut areas. Early attempts to go with classified tailings only and no packs resulted in severe shrinkage such that almost no fill contacted the stope hangingwall. After several months, extensive fractures were noted in the hangingwall near the face inclining over the stope, suggesting that large blocks were being mobilized in a virtually unsupported hangingwall. Fill shrinkage had to be eliminated.

Experiments with cemented fill, using Fosroc's two-part (binder and accelerator) Fillset silicate material added to classified tailings showed that shrinkage problems appeared to be controlled, as good hangingwall contact was generally maintained even at 4 m stoping widths.

Following these trials, a combined cemented fill and double-cut layout was adopted for all stoping width exceeding 2.5 m. The final combined layout is shown in Fig. 1.

The double-cut method was used at Western Deep Levels South Mine until channel widths fell to normal widths again in mid-1991. However, as the method was known to be effective in controlling hangingwall stability and face bursting problems, the reasons it worked needed to be better understood, especially since wide channel areas might be encountered in the future. Wide reef mining is anticipated in the future because of the high recoverable tonnage.

2 DATA COLLECTED

It is difficult to compare data from both single- and double-cut methods, since there are many factors that might control apparent trends. Factors such as limited experience in wide channel areas, variable ground conditions, seismic equipment changes over time, and experience with a new mining method will influence the data and must be taken into account.

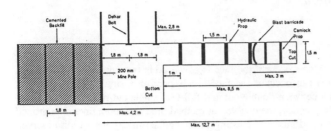

Fig. 1 Double-cut mining layout with cemented fill

Figure 2 shows a plan view layout of the 78-49 and 80-49 mini longwalls where channel widths over 2 m have been mined. Panels mined by the double-cut method are indicated. Data over a three-year period were collected to assess the effectiveness of the double-cut method.

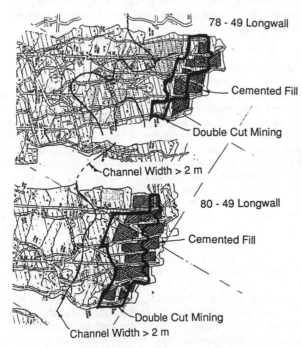

Fig. 2 High channel width areas mined during 1989 to 1991

2.1 Production considered unaffected by new method

Production is measured in monthly tonnage and square meters of face advance, and there is a relation between the two if stoping widths are constant. However, as the stoping width increases, the rate of face advance decreases, but there is more tonnage per meter advance. Production levels shown in Fig. 3 indicate a sharp decrease in the 79-49 longwall during October/November 1990, coinciding with the introduction in the double-cut method. The reduction is mainly due to reduction in face length with pillar cutting along the Wuddles Dyke. On the 80-49 longwall, double-cut mining has had no effect on production. Production drops in the new year and peaks in midyear are considered typical.

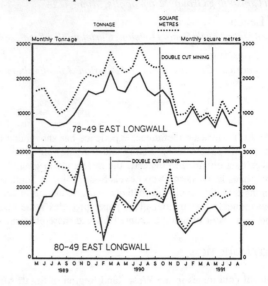

Fig. 3 Monthly production for both longwalls from May 1989 to August 1990

2.2 Inconclusive evidence of seismic activity changes

Evaluating seismicity is meaningless without knowing something about the accuracy of the seismic system. Of primary interest is the frequency of events of different magnitudes. Under normal mining conditions, proportionally more small magnitude events should occur compared to large ones. On a lognormal plot this relation should be approximately a straight line. The graph for the years 1989, 1990 and 1991 in Fig. 4 shows that detection accuracy appears to have been constant during 1989 and 1990, but dropped off in 1991, particularly in the 80-49 area as indicated by the flattening of the line in the lower magnitude ranges. This means that probably not all small events were detected. It is likely that some events less than 0.5 on the 78-49 longwall are missing and events less than 1.0 on the 80-49 longwall are missing. Therefore, assessments of the seismic data will not focus on small events, the occurrence of which may be affected by face height.

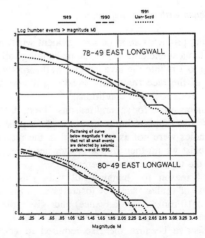

Fig. 4 Distribution of seismic events for 1989, 1990 and 1991

Plots of number of seismic events per month (Fig. 5) indicate that the use of double-cut mining appears to have had negligible effect on, at least, the larger seismic events. In both longwalls, levels remained nearly constant, except for the increase in late 1989 on the 78-49 longwall, which was attributed to the presence of a zone of minor faulting that was mined through by 1990. The higher levels in the 78-49 longwall compared to 80-49 are attributed to higher stress levels (see section 3.3).

Sorting the seismic data by magnitude of event (Fig. 6) shows only slight differences in seismicity before, during and after the double-cut mining. Small magnitude event data below the threshold limits of each longwall are not reliable. Any differences attributable to change in mining method should be indicated by similar trends in both longwalls, which is not the case.

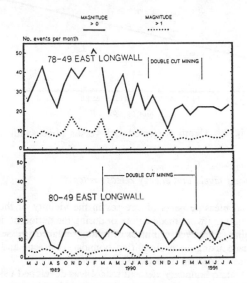

Fig. 5 Frequency of seismic events versus time

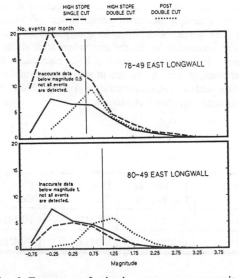

Fig. 6 Frequency of seismic events versus magnitude

3 GROUND CONDITIONS

Though production and seismicity did not show a substantial difference with the double-cut method, ground conditions appeared to improve. Unfortunately, quantifying ground conditions is difficult because of the large number of factors influencing seismically induced damage. Factors such as lost blast shifts due to rockbursts and rock falls, and rock-related injuries from

production records are qualitative measures since decreases might be attributed to improving conditions. More tangible factors such as depth of hangingwall overbreak and volume of rock ejected from the face during facebursts can be related back to the mechanics of rock behavior. Both types of data have been collected.

3.1 Reduction in lost blasts and injuries

After a panel has been damaged by a seismic event, the face has to be cleaned and re-supported before production blasting can commence. Records of lost blast shifts provide an indication of the amount of time required to reestablish the face and that time is usually proportional to the severity of the damage. On both longwalls the number of lost blasts (Fig. 7) appears to have fallen a dramatic 80% when double-cut mining was introduced, indicating that more effective support was introduced with the method.

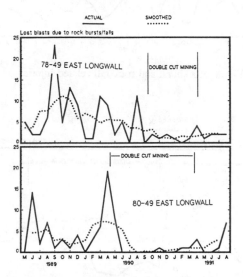

Fig. 7 Lost blast shifts due to rockbursts and rock falls

From damage report descriptions, only a few instances involving rock being kicked from the stope face occurred during double-cut mining, while such rockburst damage was frequently seen while mining high channel widths with a single cut. However, instances of props punching into the footwall, occasionally causing the bottom cut face to punch out, were recorded several times during double-cut mining.

Another measure of the severity of seismic events is the number of workers injured during such events. Overall injury rates (Fig. 8) did fall during the double-cut mining period. A smoothing curve is shown as large variations in injury data occur from month to month. These data were taken from reports of rock related injuries in Section 121 East, which included the 78-49 and 80-49 longwalls, involving all fatal, disabling and reportable injuries.

3.2 Reduction in hangingwall overbreak and volume of ejected rock

As a routine part of the mining sequence, the channel width and stoping width are measured. Assuming good footwall control, the monthly overbreak in the hangingwall can be estimated from this data (Fig. 9). On a longwall basis, only a slight reduction in the overbreak height is apparent, although more noticeable in 78-49

431

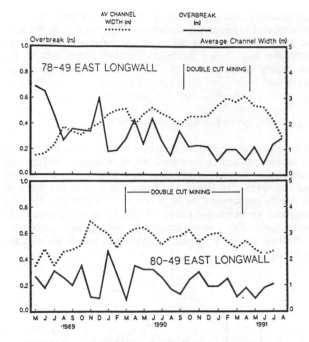

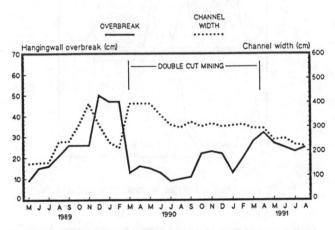

Fig. 8 Rockburst and rock fall related injuries

Fig. 9 Hangingwall overbreak history on a longwall basis

longwall. However, there were troublesome panels, such as 80-49 E2, where hangingwall overbreak substantially improved with the introduction of double-cut mining (Fig. 10).

The severity of face bursting was reported to have decreased with

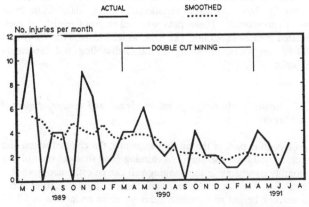

Fig. 10 Hangingwall overbreak history for panel 80-49 E2

the introduction of double-cut mining. One way to quantify this is by the volume of rock ejected from the face as a result of face bursting. This data is recorded along with stoping widths as a routine part of fatal injury reports. Plotting the depth of ejected rock versus stope width for 19 faceburst cases at the mine (Fig. 11) generally indicates an increasing ejected volume with increasing stope width.

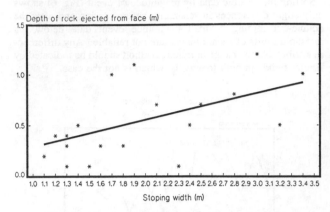

Fig. 11 Depth of ejected rock versus stope width

3.3 Stress levels and energy release rates

No actual measurements of stresses in the vicinity of the wide channel mining were made before or during the period of double-cut mining. However, estimates of mining-induced stresses have been made using both simplified boundary element and finite difference numerical models.

A 3-D static boundary element model was developed using the MINSIM-D program (Napier and Stephansen 1987) where a close approximation of the actual mini-longwall panel layout was simulated. The rock mass was assumed to behave as an elastic continuum, although the nonlinear behavior of backfill was simulated. Properties of the stiffer hangingwall lavas and Wuddle dyke were accounted for in the equivalent rock deformability properties. Energy release rates (ERR), calculated from the elastic stresses, provide a measure of stress concentrations ahead of the face and in the hangingwall. Average ERRs for both longwalls were calculated as follows.

Table 1.

Longwall	Pre-Double Cut Period	Double Cut Period	Post-Double Cut Period
78-49	15 MJ/m^2	12 MJ/m^2	10 MJ/m^2
80-49	9 MJ/m^2	9 MJ/m^2	8 MJ/m^2

These results indicate that the 78-49 longwall is the higher stressed of the two, and as mining progresses in time beyond the double-cut period, the ERR decreases.

A simplified 2-D static finite difference model of the geometry in Fig. 1 was used to evaluate the difference in released kinetic energy between double-cut and single-cut mining. Released kinetic energy is the kinetic energy of motion that leaves the model due to damping. The model was developed using the UDEC code (Itasca, 1992) with effective continuum deformability and Mohr-Coulomb strength properties used to represent rock mass behavior. Stoping widths were varied from 1 m to 3 m and compressible material used to represent cemented backfill behavior. For each mining increment, the kinetic energy released through motion, as determined by the accumulation mass damping work, was recorded.

The rate that kinetic energy is released for a 3 m channel width was estimated as 8 MJ/m² (Fig. 12). Very similar values are estimated for other channel widths. These 2-D results indicate that non-elastic energy release rates are more-or-less the same for both methods. The total kinetic energy released is slightly higher in double-cut mining, although not much energy is stored in the bench cut. It cannot be determined from these analyses what portion of the kinetic energy is released in an unstable, or seismic, manner.

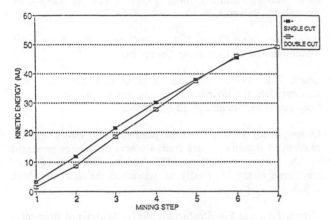

Fig. 12 Cumulative kinetic energy release for a 3 m stoping width

4 DISCUSSION

The main reason for shifting from single-cut mining to double-cut mining was to try to decrease hangingwall stability problems and face bursting problems. Analyses of the data presented above seem to show that:

(1) production levels were not significantly affected by introduction of the double-cut method, considering normal production decrease in December-January and the reduction in active face length in 78-49 longwall;

(2) the incidences of seismic events over $M_n=1$ were probably not affected by double-cut mining, but very small events (less than $M_n=0.5$) may have decreased in frequency, although the accuracy of the seismic location makes this difficult to confirm;

(3) the incidence of lost production blasts due to rockbursts a rock falls decreased as much as 80%, which is proportio to face and hangingwall volumes requiring rehabilitation;

(4) injuries decreased by approximately 30%, which might directly be the result of support being more effective;

(5) hangingwall overbreak was reduced in at least one panel by 10%, which is considered to be due to improvements in blast hole orientation control and support effectiveness;

(6) rockburst damage in the form of rock being kicked from the stope face was rarely seen in double-cut mining, while it was frequently observed when a single high cut was mined; and

(7) Calculated ERRs from simplified 3-D elastic models and 2-D plasticity models did not indicate sufficiently different conditions to explain the improvements in ground conditions.

Improvements in ground conditions can be separated into two categories: improved hangingwall support and improved face stability. Improved hangingwall conditions are in part attributed to the fact that cemented fills were introduced during the double-cut mining period, and they have a higher early stiffness than conventional classified tailings. Data from instruments in 3 m wide stopes in the 80-49 E3 panel are summarized as follows.

Table 2.

Face Advance since Backfilled	Stope Closure (Wo=3 m)	Strain	Vertical Load
1 m	5 mm	0.001	0.1 MPa
25 m	30 mm	0.012	1.0 MPa
45 m	90 mm	0.037	2.0 MPa

These closure rates are low, which is characteristic of the VCR hangingwall lavas, but the rate decreases to 1.2 mm/m at 25 m face advance and then increases to 2 mm/m. Also, the backfill stiffness is high at these low strains, compared to other measurements in cemented fills (e.g., Clark, 1991). This suggests that the cemented fill has a high early stiffness (that may not be sustained at higher strains) which serves to provide confinement to the hangingwall, hence increasing stability of the hangingwall. The fill is definitely stiffer than timber packs placed the full stoping width.

The hangingwall is further stabilized by the improvements in temporary supports. The support capacity of props is primarily a function of the length-over-radius ratio squared. The long camlock props used in single-cut mining would have to be of a substantially larger diameter to maintain the capacity of similarly spaced hydraulic props used in double-cut mining. Therefore, the hydraulic props used in double-cut mining provide greater hangingwall support, and the hangingwall is less disturbed when backfill is placed.

Improved face stability conditions are evident from the observation that rock being kicked from the stope face was rarely seen when mining with two cuts. The mechanism involved in face bursting is thought to be surface buckling instability of the face and may be influenced by inhomogeneities in rock mass strength brought about by mining-induced fracturing. It is interesting to note that in analyses of surface instability conditions by Vardoulakis (1984), the layer thickness, or stoping width in our case, does not enter into the equation for stability when using a Mohr-Coulomb constitutive relation. He suggests adding the dimension of length through the constitutive relation (e.g., microcrack surface density) to compare with the stoping width.

Linkov (1992) has developed the following expression for stability conditions at the face for rocks that soften after peak strength is attained. A condition of face instability exists when

$$\frac{(1 - v^2)\, k_I^2}{E} \geq \frac{UCS^2}{2M}\, h \tag{1}$$

where v is Poisson's ratio, E is Young's modulus, M is the post-peak softening modulus, h is the stope width, UCS is the uniaxial compressive strength, and k_I is the mode I stress intensity factor given by $k_I^2 = \sigma_y^2 \pi c$, where c is the half span of the longwall. The left-hand side of equation (1) is the ERR, and the right-hand side is maximum energy dissipation rate. This instability condition is based on a thin slit assumption to represent the stope.

Assuming $\nu=0.20$, E=70 GPa, M=100 MPa, and UCS=200 MPa for typical VCR quartzites at a depth of 2500 m, an ERR of 12 MJ/m^2 is estimated from eq. (1) for the 140 m span in the 78-49 longwall. This is the same as estimated from the 3-D boundary element model. In the other longwall (i.e., 180 m span in which one side is unmined), the ERR ranges from 6 to 12 MJ/m^2 which, on the average, is the same as the 3-D boundary element model. The maximum energy dissipation rate is estimated from eq. (1) as (0.2 h) MJ/m^2, which is less than the ERR. This means the face is theoretically predicted to be unstable for both thin and wide stoping widths.

When the onset of instability occurs (i.e., both sides of eq. 1 are equal), the size of the failure zone at the face can be expressed as (Linkov, 1992)

$$a = \frac{\pi}{4(1-\nu^2)} \left(\frac{E}{M}\right) (h)\,\mu(f_o) \qquad (2)$$

where f_o is the ratio of ERR to the maximum energy dissipation rate and the function $\mu(f_o)$ accounts for the equivalent loading stiffness of the face. The f_o energy ratio reaches a critical value of 1.0 at the stresses at this depth, whereby the limiting value of μ is 0.4655. The size of the failure zone at the face for M=100 GPa is estimated from equation (2) as a = 0.27 h. The 0.27 is very close to the slope of the least squares line in Fig. 11.

5 CONCLUSIONS

Data has been presented illustrating the difference between double-cut and single-cut mining. Double-cut mining is known to produce more favorable ground conditions. Improvements in hangingwall conditions are, at least in part, attributed to better temporary hydraulic prop support and permanent cemented fill support.

The difference in face conditions between the double-cut and single-cut methods are not explained by stability conditions, as even a 1.5 m top cut is expected to have face instability. However, estimates of the depth of failed zone appear to closely compare with the measured data. It is possible that the explanation for the observation that ejected rock was rarely seen when mining with two cuts lies in the manner in which mining-induced fracturing occurs in the rock ahead of the face. This factor was not evaluated but should be in the future via numerical modeling, microseismic monitoring of small seismic events, and fracture mapping of the face.

Results from this work are directly applicable to the deep gold mining environment in South Africa when reef channel widths exceed about 2.0 m. Such conditions occur in the West Rand Region near Carletonville. As mining progresses deeper and costs increase, there will be pressure to mine less favorable areas, including high channel areas. Hence, understanding the conditions controlling stability is important for safety.

Future studies will focus on verifying whether it is possible for fractures created during the top bench of the double-cut to dissipate energy in a more stable manner than fractures created during single-cut mining as a result of post-peak softening conditions. Also, the relation between unstable energy release and damage will be evaluated using dynamic models in which peak particle velocities can be quantified.

6 ACKNOWLEDGMENTS

The authors wish to thank Anglo American Corp., Gold Division, and the management at Western Deep Levels South Mine for support of this work and whose data are extremely valuable for supporting this type of research. Also, valuable discussions with Itasca staff members are greatly appreciated.

7 REFERENCES

Clark, I.H. 1991. The cap model for stress path analysis of mine backfill compaction processes. *Computer Methods and Advances in Geomechanics (Proceedings of the 7th International Conference, Cairns, Australia, May 1991)*: Vol. 2, 1293-1298. Rotterdam: Balkema.

Itasca Consulting Group 1992 UDEC manual. Minneapolis, Minnesota: Itasca Consulting Group, Inc.

Leech, A.R. 1991. Analysis of production, seismicity, and accident data for double cut mining areas. Anglo American Corporation memorandum, 13 September.

Linkov, A.M. 1992. Dynamic phenomena in mines and the problem of stability. Notes from a course of lectures presented by Dr. Linkov in 1992 as MTS Visiting Professor of Geomechanics at the University of Minnesota, Minneapolis, MN, U.S.A.

Napier, J.A.L. & S.J. Stephansen 1987. Analysis of deep-level mine design problems using the MINSIM-D boundary element program. *APCOM 87. Proceedings of the Twentieth International Symposium on the Application of Computers and Mathematics in the Mineral Industries*: Vol. 1, Mining, 3-19. Johannesburg: SAIMM.

Tinucci J.P. & E. Dawson 1992. An analysis of released energy in wide reef stopes. Itasca Consulting Group report to Anglo American Corporation, June. Minneapolis, MN.

Vardoulakis, I. 1984. Rock bursting as a surface instability phenomenon. *Int. J. Rock Mech. Min. Sci. & Geomech. Abstr.* 21(3): 137-144.

Rockbursts and Seismicity in Mines, Young (ed.) © 1993 Balkema, Rotterdam, ISBN 90 5410 320 5

Strategies for clamping faults and dykes in high seismicity tabular mining conditions

John P. Tinucci
Itasca Consulting Group, Inc., Minneapolis, Minn., USA

A. J. S. Spearing
Anglo American Corporation, Gold Division, Johannesburg, South Africa

ABSTRACT: This paper describes the results of a study to evaluate the dynamic effects of fault slip on the stability of tabular stopes in high stress conditions. A seismic event was simulated using a dynamic numerical model, and possible rockburst damage locations evaluated. A comparison of the calculated and actual geophone record for the Brand Fault Event of January 1989 is made to demonstrate the accuracy with which such events can be simulated. The recorded event was a Mn = 4.7 fault-slip event which resulted in extensive rockburst damage. The model was used to examine the distribution of peak particle velocities and induced failure resulting from wave attenuations. From an understanding of the dynamic effects on rock mass stability, strategies toward clamping faults and dykes are developed.

1. INTRODUCTION

In deep hard-rock tabular ore body mining, mining-induced seismic events are the most formidable source of ground control problems because of problems in determining where, when and how large events will occur and the resulting damage. As seismic energy is released, the induced ground motion will shake the rock around stopes. Under certain conditions, this shaking causes the rock mass in highly-stressed ground to fail violently, thereby releasing stored strain energy in an unstable manner (i.e., rockbursts). Under other conditions, the shaking simply redistributes stresses in marginally stable ground, allowing blocky ground to collapse (i.e., rockfall).

These large events are known to produce the greatest amount of damage, and studies of such events have shown that they are most frequently associated with shear failure of geologic structures (Gibowicz, 1990). The breakdown of the location of major events at the Western Deep Level mines is: 14% strain bursting of pillars, 4% seismic slip of geologic structures in pillars, 6% strain bursting of abutments, 19% seismic slip of structures in abutments, and 57% slip bursting of structures within mining panels. The need for adequately designed bracket pillars along structures is therefore very evident.

Strategies have been developed toward mining around seismically active geologic structures (e.g., COMRO, 1988) and generally consist of either mining through or leaving bracket pillars adjacent to such structures. However, much controversy surrounds the issue of whether it is better to:

(1) mine along strike of the fault early in the mining sequence and run the risk of suddenly unlocking on asperity storing large amounts of excess energy over a small area; or

(2) leave bracket pillars adjacent to structures that serve to clamp against slippage and run the risk of storing moderate excess energy over a large area.

This paper presents the results of a study (Tinucci and Damjanac, 1992) in which a specific seismic event was simulated using a dynamic numerical model. Both rock and fault properties were calibrated to known ground conditions and a particular seismic event on the Brand Fault at the President Brand Gold Mine in the Orange Free State of South Africa. The calibrated model was then modified for various bracket pillar widths to evaluate stress conditions and develop strategies for clamping bracket pillars.

1.1 Brand Fault event description

The following descriptions of the Brand Fault and main seismic event in late January 1989 are intended to provide sufficient background for understanding the objectives of the study and the assumptions made in the models. This is not intended as a complete description of all events associated with the Brand Fault event. The reason this event was selected was that this large event resulted in significant damage in nearby drifts, the mining geometry was uniform, a good seismic velocity record was obtained, and fault displacements were measured at five locations.

The Brand Fault is located in the western portion of the President Brand Mine, approximately 300 m west of the No. 2 and No. 3 shafts, striking north-south (Fig. 1). The fault is bracketed by the Arrarat Fault on the north and the Dagbreek Fault on the south, both of which are considered major geologic structures. The Brand fault intersects the main ore-bearing reef at about 1830 m below ground, where the reef has been down-thrown on the

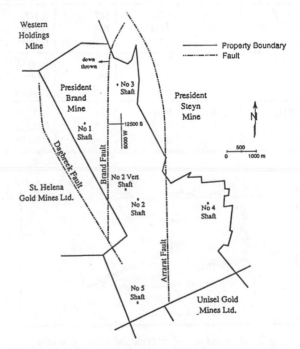

Fig. 1 Location of Brand Fault at President Brand Mine

west side by about 400 m.

The main reef dips about 17.5° toward the southeast and has been extensively mined at 1.65 m stoping width. There is another reef 18 m above the main reef, but only limited mining has occurred in close proximity to the Brand Fault. The lithographic units above the main reef are considered typical for the Orange Free State gold mines; however, there is a distinct 1 m thick shale bed, known as Khaki shale, located immediately above the main reef, which is known for producing hangingwall stability problems because of its low strength. In the 400 m fault-loss area between the down-thrown portion of the main reef, the rocks in the fault zone are characterized as pyrophyllite-rich rocks of low strength whose formation probably is related to hydrothermal alteration of the fault footwall quartzite and the smeared Khaki shales (ISS, 1992). Bracket pillars approximately 20 m wide were left on either side of the fault.

In late January 1989, there was a series of seismic events along the Brand Fault. The main shock was located on the fault approximately 700 m from the No. 1 shaft, but well below the main reef. A magnitude of $M_n = 4.7$ was estimated, and extensive damage occurred in drifts in close proximity to the bracket pillar area. Fourteen (14) geophones in the Welkom regional network recorded the event, of which the closest phone was 1000 m from the source in a direction normal to the fault (Table 1). The point of rupture initiation was located approximately 309 m beneath the intersection of the Brand Fault and the down-thrown portion of the main reef. Figure 2 shows the location of the main event relative to the faults and mined-out reef.

It has been estimated, based on observed displacements in drifts intersecting the fault (Table 2) and the condition of the remnant pillars, that the main source of seismic energy, coinciding with the maximum amplitude on the wave forms, would be at least 280 m up-dip from the point of rupture initiation (van Aswegen, 1992). The event locations were positioned in the Welkom seismic network such that the accuracy of the location is estimat-

Table 1. Geophone locations and maximum recorded velocities (ISS, 1992).

Station Name	Hypocenter Distance	Hypocenter Unit Direction	Motion Component	Max. Velocities (m/s)
PB1	982 m	0.06 0.69 0.73	NS EW Vert.	0.166 0.178 0.123
WH4	6041 m	0.98 0.20 0.04	NS EW Vert.	0.0044 0.0056 0.0166
SH8	3379 m	0.99 0.15 0.02	NS EW Vert.	0.0166 0.0230 0.0249

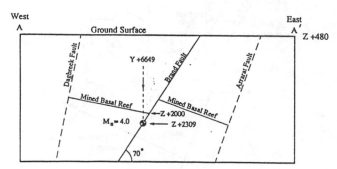

Fig. 2 Idealized model of fault and stope geometry

Table 2. Post-event measured displacements (ISS, 1992).

Measurement Number	Hypocenter Distance	Hypocenter Unit Direction	Displacement
1	1306 m	0.90 NS 0.21 EW 0.38 Vert.	2.5 cm Vert.
2	357 m	0.22 NS 0.14 EW 0.96 Vert.	22.0 cm Vert.
3	748 m	0.53 NS 0.22 EW 0.82 Vert.	37.0 cm Vert.
4	1021 m	0.92 NS 0.01 EW 0.39 Vert.	32.0 cm Vert.
5	1228 m	0.97 NS 0.04 EW 0.24 Vert.	8.0 cm Vert.

ed to be within ± 100 m (ISS, 1992). Analysis of the records (ISS, 1992) indicated a corner frequency of about 5.2 Hz and a dominant frequency of about 8.6 Hz. A radius of rupture of 297 m was estimated assuming a circular source radius and static stress drop was estimated as 5.9 MPa.

1.2 Motivation for using dynamic numerical models

In a longwall mining environment, the bracket pillar must not only effectively clamp the feature but also not cause over-stressing of any nearby off-reef footwall excavations. In the design of bracket pillars presently, elastic 3-D continuum models are used to estimate the excess shear stress (ESS) on geologic features given as:

$$ESS = \tau - \sigma_n \tan \phi \quad (1)$$

where τ is the shear stress, σ_n is the normal stress, and ϕ is the friction angle of the feature. For a given mining layout, three categories of features have been defined based on ESS levels: (1) the feature is considered stable if ESS<0 MPa; (2) there is a potential for slip if 0<ESS< 10 MPa; and (3) the feature will definitely slip if ESS>10 MPa.

This design technique has an inherent problem in that ESS values based on elasticity do not consider stress redistributions due to rock mass failure or fault slip. Finite difference discontinuum codes are well suited for this type of analysis because: (1) rock mass failure and fault slip can be directly accounted for; (2) static analyses can be used to estimate effectiveness of clamping and footwall stress concentrations; (3) spatial variations in properties can be directly accounted for; and (3) dynamic analyses can be used to quantify PPVs for estimating potential damage. Numerical models have been shown to be reliable when used to qualitatively compare different mining layouts and define the possible range of conditions.

2. DESCRIPTION OF MODEL

Simple 2-D discontinuum models were developed using the FLAC finite difference and UDEC distinct element codes to evaluate the redistribution of ESS along a fault due to slip. The models were calibrated to recorded velocities at the nearest geophone and measured fault displacements from the Brand Fault event.

2.1 Asperity behavior

The key to estimating stresses in bracket pillars and along the fault is simulating the strength behavior of the fault. Using a Coulomb slip model would not store ESS since slip occurs when the shear stress equals the strength. However, an asperity model of high initial strength could store the ESS and could be suddenly unlocked by reducing its strength. The unlocking would induce a stress wave to propagate out from the point of slip initiation. Using such a simple asperity behavior would provide the necessary information on pillar stresses and PPVs while not attempting to simulate details of the asperity (i.e., en echelon faulting, undulating roughness, hydrothermal annealing, etc.).

Mining-induced stresses were allowed to come to static equilibrium with a portion of the fault locked at high strengths. Non-asperity portions of the fault would slip as necessary. The asperity was then "unlocked" by reducing the friction and the model run dynamically. Two asperity locations were evaluated: (1) a "near-stope" asperity extended 250 m immediately below the lower stope, and (2) a "deep" asperity which started 200 m below the lower stope extending 700 m downward.

2.2 Calibrating strength behavior of the asperity

Velocities from the model were calibrated to the nearest geophone record and the residual strength of the asperity adjusted appropriately. Simplified elastic scoping models indicated PPVs to be insensitive to damping frequencies over a 5 to 15 Hz range but sensitive to stress drop and fault slip. Remnant pillars from the scatter mining method were represented by 40 m pillars uniformly spaced 280 m apart, matching the 85% extraction ratio. Bracket pillars on both sides of the fault were evaluated for 20 m and 30 m widths.

Limited fracturing was simulated in the discontinuum UDEC model by including 5 bedding planes parallel to the reef spaced at twice the stoping width. Two sets of opposite dipping shear fractures spaced at 4 times the stoping width were also simulated. The Khaki shale was represented by a 1 m thick bed of low strength material in the immediate hangingwall. Rock mass properties in the FLAC continuum models were adjusted to account for fracturing.

The section of the fault between the upper and lower stopes was assumed to have a low friction of 5°. In-situ stresses were simulated by lithostatic vertical stresses and a stress ratio of 0.5. These initial stresses were sufficient to induce slip even before mining was simulated (i.e., 1.5 cm over 400 m). This initial slip resulted in stress concentrations along the fault in the bracket pillar area (Fig. 3). If this stress and strength non-uniformity is true in reality, this suggests that portions of the fault could be locked before mining, and mining-induced stresses could continue to lock the fault at higher stress levels.

The mining-induced stress state with the asperity locked indicated average elastic vertical stress concentrations in the lower bracket pillar of about 500 MPa but only about 250 MPa in the upper bracket pillar. Mining was simulated by simultaneously removing all stopes and the asperity was locked with an 80° friction angle. The simulated pre-event mining-induced fault slip is about 9 cm between the upper and lower stopes (Fig. 4). This "static" slip results in a redistribution of shear stress localized at the top of the asperity (Fig. 5).

3. COMPARISON TO SEISMIC RECORDS

When the asperity was unlocked, sudden slip occurred and a

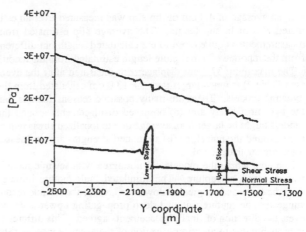

Fig. 3 Shear and normal stresses along the fault for pre-mining conditions from continuum base case mode

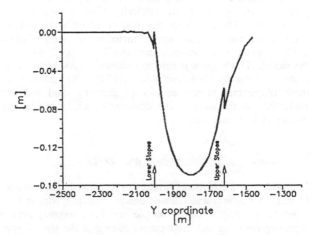

Fig. 4 Mining-induced shear displacements along fault from continuum near-stope asperity model with asperity locked

dynamic stress wave propagated away from the fault. Non-reflecting boundaries were used to prevent waves from reflecting off boundaries. An average shear stress drop of 33 MPa occurred over 64 m (Fig. 5). The shear displacement along the fault averaged 4.2 cm over a 456 m radius, centered at the lower reef contact with the fault. Averaging this stress drop over the slip length gives an average stress drop of 2.2 MPa, which is less than the 5.6 MPa reported from analysis of the record. By compari-

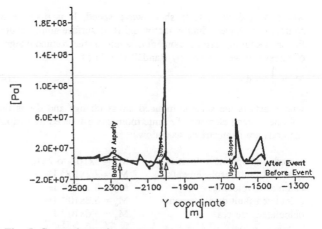

Fig. 5 Comparison of pre-event to post-event static shear stresses along the fault for the discontinuum near-stope asperity model

son, an average of 4.0 cm of dip slip was measured over an estimated 890 m in slip length. The average slip estimated from measurements is quite close to the calculated length but different from the reported 600 m rupture length estimated from the record.

The maximum 37.0 cm displacement measured after the event (see Table 2) is much larger than the 13.0 cm calculated from the dynamic models. There are many possible reasons for this, but the two most likely are: (a) observed displacements around the tunnels might be larger than average due to localized stress reductions around the tunnel, or (b) simulated strengths and stiffnesses were higher than actual.

One area where the models did not agree with seismic records was direction of rupture. The simulated fault rupture initiated near the stope and progressed downward, while the seismic record suggested the rupture initiated deep propagating upward. However, the direction of final displacements agreed. This difference is probably due to oversimplification of the asperity model or the true 3-D nature of the onset of slip instability.

Spectral density plots for the discontinuum near-stope asperity model indicate a close agreement in frequency to the recorded peak frequencies of 3.6 Hz vertically (Fig. 6) and 5.5 Hz horizontally. The low frequency plateau of s-waves from the displacement power spectra was estimated in the range of $|\Omega_o| = 2 \times 10^{-5}$ to 8×10^{-5} m-s from the record and $|\Omega_o| = 2 \times 10^{-4}$ m-s from the model. This value was used to estimate the seismic moment using a relation by Hanks and Wyss (1972). The sensitivity of these frequencies to the amount of damping used was not evaluated. In the model, mass damping of 2% of critical frequency of 8 Hz was used.

3.1 Estimates of seismic moment and energy

The other area where model results can be compared is estimates of seismic moments and released energy. Starting with seismic moments, an analytic double-couple model is commonly used to represent fault-slip and the moment quantifies the size of each couple. There are at least four ways to estimate seismic moments from the models: (1) velocity histories; (2) fault displacements; (3) stress changes; and (4) unbalanced forces.

Seismic moments estimated from displacements are given by Aki and Richards (1980) as

$$M_o = GDA \qquad (2)$$

where G is shear modulus, D is average slip and A is slip area. Moments estimated from integrating the velocity records are given by Hanks and Wyss (1972) as

$$M_o = 4\pi\rho C^3 R|\Omega_o| / F^c R^c \qquad (3)$$

where ρ is density, C is shear wave speed, R is distance to source, F^c is wave radiation factor and R^c is surface amplification factor. Defining seismic moment as a tensor, the volume integral of stress change is given by Randall (1971) as

$$M_o = \int \Delta\sigma \, dV \qquad (4)$$

where $\Delta\sigma$ is the seismic-induced stress change and dV is the volume of each element. Seismic moments estimated by various methods are summarized as follows:

recorded geophone velocities	$M_o = 5 \times 10^{13}$ to 2×10^{14} J
recorded fault displacements	$M_o = 2.2 \times 10^{15}$ J
calculated velocities	$M_o = 5.1 \times 10^{14}$ J
calculated fault displacements	$M_o = 8.2 \times 10^{14}$ J
calculated stresses	$M_o = 5.6 \times 10^{14}$ J

For each of these methods, the out-of-plane dimension was assumed circular or spherical, depending on the method and

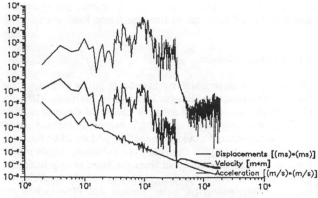

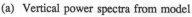

(a) Vertical power spectra from model

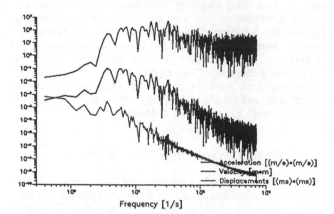

Frequency [1/s]

(b) Vertical power spectra from record

Fig. 6 Comparison of calculated power spectra at geophone location to seismic records for the discontinuum, near-stope asperity model

calculated values are from the continuum near-stope asperity model.

Energy components were directly calculated in the UDEC code (Christianson and Board, 1992) and converted to 3-D by assuming a slip radius of 425 m and a ratio of spherical volume to circular area of (4/3)r. The energy components include kinetic energy, stored strain energy in rock and fractures and changes in potential energy, while the work components include mass damping work viscous boundary work, frictional sliding work and boundary loading work. Subtracting the energy in from energy out of the problem gives a released energy of 55.4×10^9 J for the continuum model. Approximately 2.4 times more energy is released from the discontinuum model.

These estimates of seismic moments and energies compare well to those estimated from the record and provide confidence that the asperity model is capable of producing reliable results.

3.2 Comparison of Asperity Models and Pillar Widths

Unlocking the asperity was necessary to reproduce the seismic event. The best fit asperity model was the near-stope asperity in which the asperity extended from the footwall of the lower stope to a depth of 250 m below. Judging the best-fit model was based on PPVs, dominant frequencies, shear stress drop, fault displacement, slip length, and predicted damage. Although, from runs with lower stiffnesses and strengths, only slight adjustments are necessary in rock mass properties to have the other models compare as well with the actual records, due to the sensitivity of stability conditions.

The best fit model was run with 20 m and 30 m bracket pillars.

The results were quite similar in the magnitude of average stress drop and average fault slip. However, the maximum stress drop and fault slip were less with wider bracket pillars. Footwall average stress concentrations were about 30% lower along the upper stope bracket pillar but only 11% lower in the lower stope bracket pillar. The case of no bracket pillar was not evaluated.

The main factors controlling whether a seismic event is generated appear to be: (1) proximity of the locked asperity to mining-induced stress concentrations (or ESS lobes), (2) the magnitude of pre-mining and mining-induced stresses, (3) pre-event fault slip, and (4) the post-peak deformability of the surrounding rock mass. The stiff testing machine analogy can be applied here, where the "sample" is the fault releasing (or supplying) the strain energy and the "machine" is the rock mass taking on strain energy according to its stiffness. However, more research into defining stability conditions is required.

4. COMPARISON TO DAMAGE OBSERVATIONS

Results indicate that stope damage is not only a function of event magnitude and distance from the source, but also stope orientation relative to the direction of wave propagation and the geometry of remnant pillars. Contour plots of peak velocities (Fig. 7), induced shear stresses (Fig. 8), and peak tensile stresses, as well as results from the discontinuum model (Fig. 9), suggest that the upper stopes are as susceptible to hangingwall failure as the lower stopes, even though they are farther from the source. This agrees with actual damage being located in both upper and lower stopes nearest the fault even though upper stopes were more than 400 m farther away. The upper stopes experience primarily vertical waves and the footwall free-surface interferes with wave transmis-

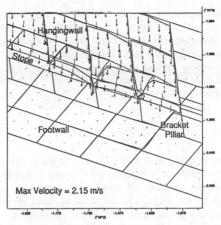

Fig. 9 Post-Event Block Displacements from Discontinuum Near-Stope Asperity Model

sion to the hangingwall by focusing waves through pillars. These contour plots were developed using FISH routines in FLAC.

It is noted that all models have limitations as to their ability to "predict" areas of damage, but one must be a little more careful with results from the discontinuum models because of the larger number of assumptions inherent in assembling fracture geometries. Nonetheless, the discontinuum models demonstrate that a statically stable hangingwall can be seismically unstable under certain conditions.

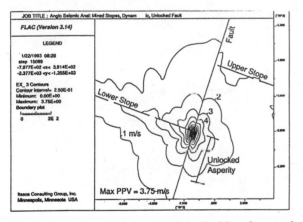

Fig. 7 Contours of peak particle velocities for continuum near-stope asperity model

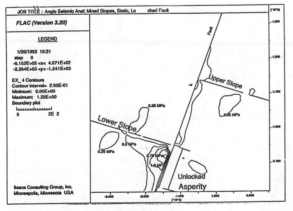

Fig. 8 Contours of induced shear stresses for continuum near-stope asperity model

5. STRATEGIES FOR CLAMPING ACTIVE GEOLOGIC STRUCTURES

This study alone will not resolve the pro and con issues of fault and dyke clamping. However, from the results showing small ESS lengths, large slip lengths, nonuniform PPVs and minor sensitivity to bracket pillar widths, the following strategies for minimizing negative effects can be inferred:

(1) Mining away from geologic structures serves to reduce clamping stresses early in mining and prevent the buildup of stress concentrations later in mining. However, as seen in these models, large amounts of ESS can accumulate on small asperities. This strategy is appropriate for brittle features in highly jointed ground where sudden energy releases can be transferred in a stable manner.

(2) Continuous wide bracket pillars may be more effective with soft geologic features in brittle unfractured ground because ESS transferred from the feature will have a larger area to distribute over. Mining should progress parallel to the bracket pillars and not toward them so as not to compound ESS on mining-induced fractures.

(3) Mining bracket pillars first and replacing them with stiff fill should be experimented with. In this way 100% of the ore reserves are recovered and the fill (which is softer than the reef material) will provide long-term clamping to the features. This might work better with mining methods that use pre-development to locate faults and dykes.

(4) Hybrid systems of periodic bracket pillars and stiff fill abutments or ribs would serve to partially relax stresses on the structure without allowing unrestrained slip and yet provide a softer system of support to prevent an unstable release of energy. The intact pillars will minimize sudden unlocking over long distances. Backfilling against bracket pillars allows a gradual transition of footwall stresses.

These strategies, of course, should be supplemented with strategies for regional support (i.e., stabilizing pillars, backfilling, etc.) and local support (i.e., yielding props and packs, systematic bolting and lacing, etc.)(e.g., COMRO, 1988). The developments of a mine-worthy ground penetrating radar system and a dense sensitive digital seismic system are receiving high priority to aid in locating such features.

6. CONCLUSIONS

It has been demonstrated that: (1) dynamic asperity models of fault-slip can reasonably reproduce PPVs and dominant frequencies of seismic events; (2) rockburst damage is not only a function of source magnitude and distance from the source, but also the mining geometry relative to the propagating waves and the relative deformability of the fault and nearby rock mass; and (3) mining strategies should be based on partial relief of ESS along the feature without permitting unrestrained slip. The use of clamping pillars whose deformability is similar to the feature to allow stable energy transfer to the deforming rock mass.

The most significant results presented in this study are contour plots of PPVs and induced shear stresses. Such plots offer the ability to identify possible locations where rockburst damage might be expected for a feature of given strength and stiffness. Future analyses should evaluate the importance of stress-dependent stiffness for intact rock and discontinuities, displacement- or velocity-dependent fault slip behavior, and post-peak stability conditions. Also an evaluation of 3-D dynamic fault-slip effects on the velocity response of remnant and isolated pillars should be made.

Results from this work are directly applicable to the deep gold mining environment in South Africa. Strategies for mining in high seismicity areas will become more important as mining depths progress to 4000 m and below. As it is expected that seismic conditions will persist at greater depths, these strategies must focus on identifying seismically unstable conditions in areas near active mining.

7. ACKNOWLEDGEMENTS

The authors wish to thank Anglo American Corp., Gold Division, and Integrated Seismic Systems International for support of this work and whose case history data are extremely valuable for supporting this type of research. Also, valuable discussions with Itasca staff members are greatly appreciated.

8. REFERENCES

Aki, K., & P.G. Richards 1980. *Quantitative seismology: theory and methods*. San Francisco: W. H. Freeman.

Chamber of Mines Research Organization (COMRO) 1988. *An industry guide to methods of ameliorating the hazards of rockfalls and rockbursts (user guide 12)*. Johannesburg: COMRO.

Christianson, M., & M. Board 1992. Energy calculations in UDEC (draft), Itasca Consulting Group Report to Mining Research Directorate, Laurentian University, July.

Gibowicz, S. J. 1990. The mechanism of seismic events induced by mining - a review. *Proceedings of the 2nd international symposium on rockbursts and seismicity in mines: 3-27* C. Fairhurst, Ed. Rotterdam: A. A. Balkema.

Hanks, T.C., & M. Wyss 1972. The use of body-wave spectra in the determination of seismic source parameters. *Bull. Seis. Soc. Am.* 62: 561-589.

Integrated Seismic System International, Ltd. (ISS) 1992. Memo on fault geometry, seismic records map of displacements for modeling the Brand Fault event, Integrated Seismic System Report, Welkom, RSA.

Randall, M.J. 1971. Shear invariant and seismic moment for deep-focus earthquakes. *J. Geophys. Res.* 76: 4991-4992.

Tinucci, J.P., & B. Damjanac 1992. Analysis of mining-induced fault slip at the President Brand Mine: Part I: Phase I study (draft), Itasca Consulting Group Report to Anglo American Corporation, August.

van Aswegen, G. 1992. Letter to John Tinucci concerning comments on assumptions for modeling the Brand Fault Event, Integrated Seismic System International, Ltd., Welkom, RSA, 22 April.

Rockbursts and Seismicity in Mines, Young (ed.) © 1993 Balkema, Rotterdam, ISBN 90 5410 320 5

Microseismic event location around longwall coal faces using borehole in-seam seismology

S. M. Toon & P. Styles
Microseismology Research Group, Department of Earth Sciences, University of Liverpool, UK

ABSTRACT: During mining using the longwall method of extraction, as the waste collapses behind the advancing face fractures propagate upwards from the edges of the face. These fractures are a potential hazard if an aquifer is present above the seam, because they can breach the aquifer causing flooding of the mine. We are developing a method to image fracture development around the face by passively monitoring microseismic activity using three component downhole geophones. We use spectral matrix analysis to give a precise estimate of the source direction for events detected at the geophones. The precise location of the events can be found from the difference between arrival times of 2 phases or by using two or more boreholes to locate the event by triangulation. It is hoped to be able to develop a fully automatic system for detection and location of events to enable results to be obtained in real time.

INTRODUCTION

A number of relationships have been postulated in order to predict the extent and nature of fracture zones around total extraction panels. The majority of these link the seam thickness to the extent of rock mass failure. Much of this work is based upon physical and numerical modelling studies that have largely not been completely validated from measurements. Currently the extent of caving and fracturing above the seam is considered to be in the region of 50 times the seam thickness for a 200 m wide longwall panel, with an approximately linear decrease with decreasing panel width; a figure of 10 times seam thickness has been suggested for a 40 m wide panel for example (Choi and McCain, 1982, Bieniawski, 1987, Peng and Chiang, 1984, Follington, 1988). There is no general consensus as to the depth of fracturing in the floor, other than it is significantly less than for the roof. Although seam thickness will undoubtedly influence the extent of fracturing, the complex interaction between seam thickness, depth, magnitude and direction of in-situ stresses, panel width and geomechanical properties of the strata will be the controlling factor. Although numerical models have increased in sophistication and accuracy, they are not as yet a reliable guide in this complex geomechanical environment. More measurements of the extent of fracturing are needed. Microseismic monitoring holds the potential of being able to determine and delineate the full extent of the fracture zone, in real time and at relatively low cost. The data provided by such a system will significantly increase our understanding of the mechanisms leading to strata failure. There are a number of conditions where a knowledge of fracturing would significantly aid design.

Knowledge of the extent of fracturing above a longwall or pillar extraction panel is essential if the working seam lies in close proximity to an aquifer or surface water source, as these fractures can act as a conduit to the workings. Likewise gas can migrate from below longwalls if fractures extend to other seams. Where multi-seam mining is practised or planned, the extent of fracturing of seams above and below total extraction panels could influence the way in which they are to be mined. The extent and timing of caving where seams are extracted under strong roofs influences the mining process, as periodic weightings and air blasts can occur. All of these conditions represent problems currently faced by the underground mining industry.

EVENT LOCATION

Two experiments have been carried out in the English Midlands at Littleton and Coventry collieries (Toon, 1990 and Toon *et al.*, 1992) These involved cementing sondes in exploration boreholes,

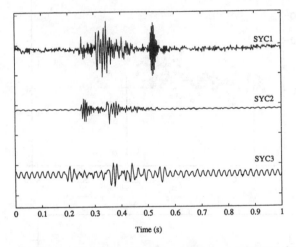

Figure 1. Examples of data recorded at Littleton for each borehole. A guided wave can clearly be seen at SYC1.

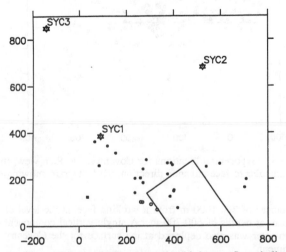

Figure 2. Hypocentral locations for Littleton. Stars indicate the boreholes, open circles are locations for SYC1, closed circles for SYC2, squares for SYC3.

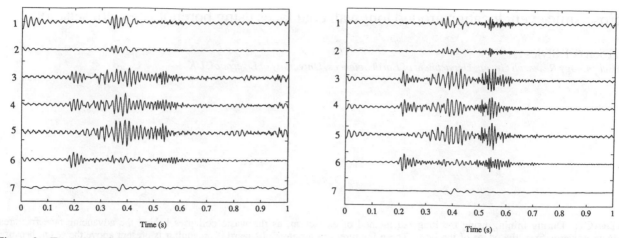

Figure 3. Examples of data recorded at Coventry. Channels 1 and 2 are vertical, 3-6 are horizontal components and 7 is the surface seismometer.

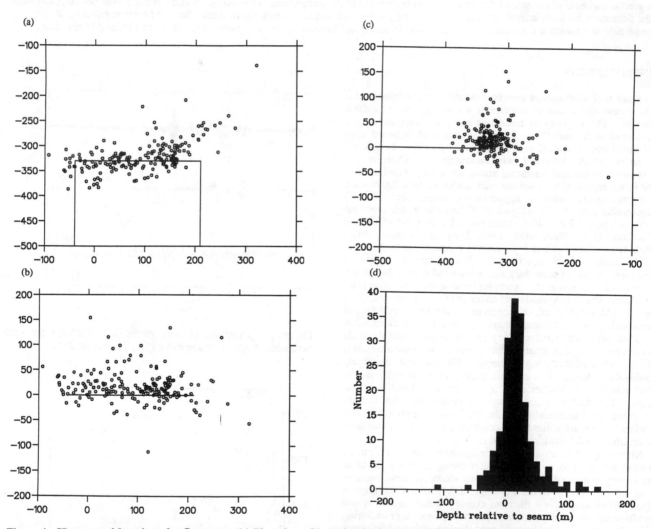

(a)

(b)

(c)

(d)

Figure 4. Hypocentral locations for Coventry. (a) Plan view, (b) section looking parallel to face advance direction, (c) section looking perpendicular to face advance direction, (d) histogram showing depth distribution of events.

at distances of 300-1000 m from a working face, at the level of the seam at depths of 600-700 m. A single vertical component seismometer was also deployed at the surface for the Coventry experiment.

For the Littleton experiment three boreholes, Sycamore 1, 2 and 3 (SYC1, SYC2, SYC3) were used and data was recorded over two days (25th-26th August 1988). The data obtained was of very high quality and guided waves can clearly be seen, Fig. 1, with

the dispersive characteristics associated with seam waves. It was assumed that the events occurred within or very close to the seam and only the horizontal components of motion were considered. The azimuth of the event is determined from P-wave hodograms, assuming the low frequency compressional arrival is polarised along the source-detector axis (Jackson, 1989). The distance to the event is determined from the time difference between P-wave and S-wave arrivals, using the velocities $V_p=3.5$ kms^{-1} and $V_s=2.0$

kms[-1]. The hypocentral locations obtained are shown in Fig. 2. Naturally occurring microseismic events have been successfully detected at SYC3 which is more than 1 km away from the active longwall face at Littleton Colliery and it has proved possible to obtain satisfactory results using these data.

For the Coventry experiment only one borehole was used and a single vertical component seismometer was also deployed at the surface, to see if any events could be detected at the surface. Fifteen hours of data were recorded over a period of three days (24th-26th September 1991). A typical event is shown in Fig. 3. The source direction was determined using the spectral matrix method. This method was first used by Samson(1973) in the analysis of ultra-low frequency magnetic fields. It has since been successfully used in the analysis of local network seismic data and Acoustic Emission data. The distance to the event is obtained from the time difference between P-wave and Airy phase arrivals, because of problems in picking the S-waves. Examples of the distribution of the hypocentres, relative to the geophones, are shown in Fig. 4. It can be seen that most of the hypocentres concentrate around the working face, with a few up to 200 m away which may be due to bed separation. In the sectional views the hypocentres appear to delineate inclined surfaces hading away from the longwall corners, as predicted by numerical models of the mode of failure for this type of mine opening. The events are generated above and below the seam but because of the spatial spread of the Green's Function (the displacement response of the ground to a point impulse situated at the event location) seam waves will be stimulated even if the source is some considerable distance from the seam, with the decay of the amplitude of the Airy phase in particular being very sensitive to vertical distance from the seam. It may be possible to use the ratio of the amplitudes of the various phases to establish the vertical location of the events with considerable precision, probably to better than 5 m, particularly for events within 50 m of the seam. Additionally, induced events are found to be much more efficient stimulators of guided waves than are explosive shots in the seam, probably because of the inherent asymmetry of induced events and the broad spectrum of frequencies they generate. Some of the events were also detected at the surface (Fig. 3) and it is thought that it may be possible to integrate this data with the downhole data to further improve the accuracy of hypocentral locations.

SOURCE PARAMETERS

The data collected during the Coventry experiment has also been used to determine seismic source parameters of the events. The parameters calculated are source radius, stress drop and seismic moment using P wave spectra. The parameters were calculated using the source models of Brune (1970) and Madariaga (1976).

A contour plot of stress drop around the longwall face is shown in Fig. 5. The values of stress drop fall in the range 1 to 35 bars and regions of high stress drop occur just ahead and behind the centre of the face. Figure 6 shows a plot of seismic moment versus source radius. The source radii fall in the range 6 to 15 m and seismic moment in the range 0.1 to 5 GJ.

CONCLUSIONS

Borehole microseismic monitoring permits delineation of the spatial and temporal development of the fractures around an active longwall face. Ideally, detectors in boreholes would be deployed around a colliery with multi-component geophones grouted at 100 m intervals from seam level to the surface and recorded for subsequent off-line discrimination and analysis, with eventual progression to real-time monitoring. The combination of multi-component detectors in-seam and throughout the overlying rock mass would permit high precision hypocentral locations (± 5 m) and the determination of the fracture characteristics including approximate dimensions, orientations and stress drop. This could be carried out in existing exploration boreholes or in piezometric boreholes drilled for hydro-geological investigations.

Three-dimensional imaging of the dynamic development of

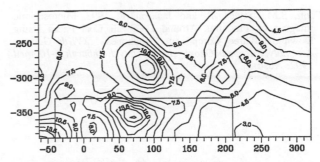

Figure 5. Contour plot of stress drop for events around the longwall face at Coventry. Values of stress drop are in bars.

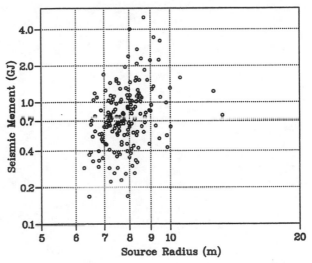

Figure 6. Plot of seismic moment versus source radius for events at Coventry.

fracturing may eventually lead to the demarcation of the intersection of propagating fractures and sub-level caving with aquifers and mapping of floor fractures controlling excess gas migration from underlying seams. It may allow modification of the mining parameters to enhance strata control and provide 'ground-truth' input to finite and distinct-element simulations of the mine opening to improve numerical modelling techniques.

REFERENCES

Bieniawski, Z.T. 1987. *Strata Control in Mineral Engineering.* AA Balkema, Rotterdam.

Brune, J.N. 1970. Tectonic stress and the spectra of seismic shear waves from earthquakes, *J.Geophys.Res.* 75:4997-5009.

Choi, D.S. and McCain, D.L. 1982. Design of Longwall Systems. *Trans. SME-AIME*, 268

Follington, I.L. 1988. *Geotechnical Influences upon Longwall Mining.* Ph.D. Thesis (Unpublished), University of Wales.

Jackson, P.J. 1989. Measuring seismic wave polarisations. Paper for EAEG meeting and exposition, Berlin 1989.

Madariaga, R. 1976. Dynamics of an expanding circular fault. *Bull. Seism. Soc. Am.* 66:639-666.

Peng, S.S. and Chiang, H.S. 1984. *Longwall Mining.* Wiley, New York.

Samson, J.C. 1973. Descriptions of the polarisation states of vector processes: Applications to ULF magnetic fields, *Geophysical Journal of the Royal Astronomical Society.* 34:403-419.

Toon, S.M. 1990. *The Location of Faults in Coal Seams Using Microseismic Activity.* B.Sc. Dissertation (Unpublished), University of Liverpool.

Toon, S.M., Styles, P. and Jackson, P. 1992. Microseismic imaging of Fractures Around Longwall Coal Faces Using Downhole Three Component Geophones. *XVIIth General Assembly, European Geophysical Society, Edinburgh, 6-10 April 1992*. Abstract.

Rockbursts and Seismicity in Mines, Young (ed.) © 1993 Balkema, Rotterdam, ISBN 90 5410 320 5

Post-closure seismicity at a hard-rock mine

R.J.Wetmiller
Geophysics Division, Geological Survey of Canada, Ottawa, Ont., Canada

C.A.Galley
Mining Research Laboratories, CANMET, Ottawa, Ont., Canada

M.Plouffe
Mining Research Laboratories, CANMET, Val d'Or, Que., Canada

ABSTRACT: The Falconbridge Mine (Sudbury, Ontario, Canada) was closed in 1984 after a serious rockburst incident on June 20 seriously damaged the mine workings and killed four miners. Significant seismic activity continued to occur at the mine for a few years after closure but the rate of activity dropped to minimal levels after July, 1987. In November, 1990 the mine began to flood. It remained seismically quiescent until May, 1991 when a magnitude 1.8 event occurred which has been followed by a sequence of strong events including a magnitude 2.7 event on April 25, 1992 and a magnitude 2.5 event on June 17, 1992. This report documents the seismic activity that has occurred at this mine since its closure and shows the relationship of the rate of seismic activity to the flooding within the mine.

1 INTRODUCTION

The Falconbridge Mine is located on the southeast rim of the Sudbury Basin (Figure 1) and has been in operation since 1929. It has a tabular copper–nickel sulphide orebody which strikes east–west for about 2000 m on surface and extends to about 1800 m at depth. The orebody has been mined by cut-and-fill methods from the ground surface to a depth of 6,050 feet (1840 m). Several major fault structures intersect the orebody, including the Main Fault, the Flat Faults and the Ore Pass Faults (see Figure 2 and 3 of Hedley et al., 1985) with generally a north-to-northeast strike and intermediate-to-steep dips. The mine was equipped with a microseismic monitoring system in 1983 which operated until 1989 (G. Swan, Falconbridge, pers. comm., 1992).

Rockbursts have occurred in the Falconbridge Mine since at least 1955 (Bharti and West, 1984). On June 20, 1984, a series of three exceptionally strong rockbursts (magnitude 3.4 at 14:12 UT, magnitude 3.5 at 16:10 and magnitude 3.3 at 16:18) caused damage between the 4025 and 4375 levels and resulted in four casualties on the 4025 level. Most of this seismic activity has been associated with the No. 1 Flat Fault, the No. 1 and 2 Ore Pass Faults and the Ropeway Dyke (Hedley et al., 1985; Board, 1992). Location of the rockbursts from the microseismic system and visual inspection of the mine workings following the rockbursts confirmed fault–

slip failure mechanisms for these events. Striation marks were observed on the gouge of the No. 1 Flat Fault (strike 330, dip 45 E) indicating that the west side of that fault had moved 1 or 2 cm laterally (SE) towards the orebody during a M 2.2 rockburst on July 5, 1984 (Hedley et al., 1985).

These rockbursts provided the first clear evidence of fault–slip mechanisms in Ontario mines and were one of the main incidents that spurred the formation of the Canada/Ontario/Industry Rockburst Project in 1984 (Hansen et al., 1987).

2 MONITORING SEISMIC ACTIVITY IN THE SUDBURY BASIN SINCE 1984

One of the aims of the Rockburst Project was to improve the level of seismic monitoring in the Sudbury Basin. One way this was achieved was by establishing a regional three–station local seismograph network, the Sudbury Local Telemetered Network (SLTN), around the rim of the Sudbury Basin, which has monitored seismic activity throughout the Basin since 1987 (Plouffe et al., 1989). SLTN is operated jointly by the Geological Survey of Canada (GSC) and the Canada Centre for Mineral and Energy Technology (CANMET). In its routine operation, SLTN has proved to be a consistent and reliable monitor of seismic activity throughout the Basin

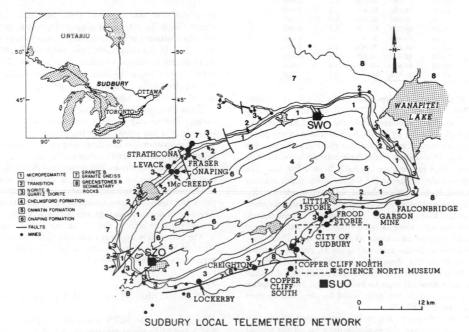

Figure 1. Distribution of the sensors of the Sudbury Local Telemetered Network with respect to the mining operations in the Sudbury Basin, Ontario, Canada. Large dots indicate mines active in the 1980's.

and has regularly detected seismic events at various mines that have not been detected by other means (Wetmiller et al. 1989).

Prior to 1987, the seismic activity in the Sudbury Basin had been monitored by the National Seismograph Network (located entirely outside the Sudbury Basin) supplemented by one local seismograph station within the Basin. SUD, a conventional analogue seismograph station was installed in the Sudbury Basin in 1967 and operated until 1984 when it was replaced by SUO, a digital telemetered seismograph station now part of SLTN. The location capabilities of the supplemented National Network were poor for seismic events in the Sudbury Basin but the exact location of seismic events in the Basin could often be confirmed by the various mining operators from felt reports, rockfall within their mines or data from their local seismic arrays. As a result, a number of events with magnitudes 1.5 or greater are known at the Falconbridge Mine but it is clear that many of the smaller events during this period would probably have gone undetected. Since 1987 however, SLTN data have enabled reliable detection and location of seismic events at the Falconbridge Mine with magnitudes as low as magnitude 0. Table 1 lists all the known seismic events (not including blasts) located at the mine since 1985.

Table 1. Post–closure seismicity at the Falconbridge mine

No.	Date y-m-d	Time UT h:m	Lat. °N	Long. °W	Mag.	$\sum$ Moment N–M x10^{13}
1	85-04-17	06:19	46.580	80.800	2.0	0.159
2	85-04-17	06:22	46.580	80.800	1.7	0.242
3	86-03-29	10:50	46.580	80.800	1.6	0.309
4	86-10-12	17:19	46.580	80.800	2.2	0.555
5	87-04-16	03:46	46.580	80.800	1.5	0.609
6	87-07-25	01:14	46.580	80.800	1.6	0.676
7	88-02-04	04:35	46.588	80.753	0.1	0.678
8	88-07-01	02:34	46.580	80.800	0.3	0.682
9	88-11-01	19:56	46.583	80.801	0.4	0.687
10	88-11-22	19:58	46.580	80.804	0.6	0.695
11	89-01-05	19:55	46.582	80.810	0.6	0.703
12	89-07-20	18:51	46.583	80.702	0.4	0.708
13	89-08-02	18:58	46.579	80.721	0.5	0.714
14	89-11-28	15:00	46.593	80.760	0.6	0.721
15	90-03-27	19:39	46.580	80.774	0.5	0.728
16	91-02-10	19:16	46.580	80.800	0.3	0.732
17	91-02-16	19:06	46.580	80.800	0.4	0.737
18	91-03-04	19:26	46.580	80.800	0.1	0.739
19	91-03-25	19:23	46.580	80.800	0.6	0.747
20	91-05-05	17:18	46.580	80.800	1.8	0.850
21	91-05-21	07:02	46.580	80.800	1.5	0.904
22	91-07-17	01:57	46.580	80.800	0.3	0.908
23	91-07-19	01:54	46.580	80.800	0.6	0.916
24	91-07-30	01:04	46.601	80.625	0.4	0.921
25	91-07-31	01:54	46.580	80.800	0.5	0.927
26	91-08-06	18:14	46.580	80.800	0.8	0.939
27	91-10-09	08:25	46.580	80.800	1.9	1.067
28	91-12-16	01:29	46.580	80.800	1.6	1.134
29	92-04-18	14:14	46.580	80.800	0.7	1.144
30	92-04-23	01:50	46.580	80.800	0.2	1.147
31	92-04-25	00:40	46.580	80.800	1.8	1.250
32	92-04-25	23:05	46.580	80.800	2.7	1.975
33	92-05-28	06:47	46.580	80.800	0.1	1.978
34	92-06-17	15:38	46.580	80.800	2.5	2.448
35	92-06-17	21:48	46.580	80.800	1.2	2.476
36	92-08-11	17:16	46.580	80.800	1.4	2.520
37	92-09-02	06:02	46.580	80.781	1.7	2.603
38	92-09-23	14:56	46.580	80.800	1.4	2.646
39	92-09-23	15:00	46.580	80.800	2.0	2.806
40	92-09-23	15:01	46.580	80.800	2.4	3.184
41	92-09-29	05:02	46.580	80.800	0.4	3.189
42	92-11-21	15:32	46.578	80.798	0.3	3.193

The Falconbridge Mine technically lies outside the SLTN, 23 km from the closest station SWO (Figure 1), but SLTN still provides reasonably accurate locations. A number of blasts confirmed at the Falconbridge Mine (Figure 2) and well recorded by SLTN have established the phase order and time delays for seismic events from this site and confirmed the location accuracy as well. Mislocation errors for confirmed seismic events at mines in the Falconbridge region have typically been a few km. The closest operating mine to Falconbridge is the Garson Mine which is about 4 km to the west. As well, there are a number of old inactive mines within a few km of Falconbridge and two town sites, Conniston and Capreol, are close by. Historically, there has been little mining–induced seismic activity at the Garson Mine so there is little chance that activity from that mine has been misidentified as activity at Falconbridge. There have been a few surface explosions in the region (Figure 3), probably carried out for construction projects, but these have been easily identifiable by their distinctive waveforms. We can not rule out the possibility of seismic events at abandoned mine sites adjacent to Falconbridge, but we assume such events have been rare and, in any case, could be identified by their slightly different delay times on SLTN.

Figure 4 shows the waveforms of a typical seismic event at the Falconbridge Mine. We have reviewed the available data for all the seismic events recorded by SLTN and located at or near the Falconbridge Mine since 1985 and feel confident that Table 1 represents as complete and reliable a record as possible of the seismic activity associated with this mine since its closure.

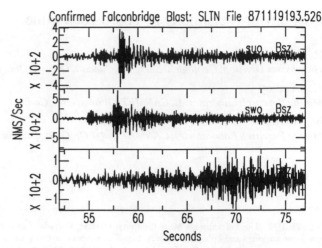

Figure 2. SLTN traces for a confirmed blast in the Falconbridge Mine.

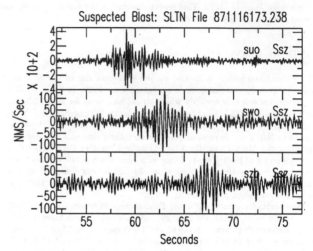

Figure 3. SLTN traces for a suspected surface blast near the Falconbridge Mine.

3 HISTORY OF POST–CLOSURE SEISMICITY

Since the mine was closed in 1984 there have been at least 42 seismic events (Table 1) ranging in magnitude from 0 to 2.7. Significant seismic activity continued at the mine after closure for a period of 2 to 3 years until mid–1987. In the subsequent three years, 1988, 1989 and 1990, a very few small events continued to occur, then in May of 1991 significant seismic activity again began to occur at the mine and has continued to date of writing (December 1992). In April, 1992 the largest event, magnitude 2.7, in the post–closure history occurred.

Figure 5 shows the cumulative moment release for the Falconbridge events. The seismic moment of the events has been calculated from the magnitude of the events using the moment–magnitude relation for earthquakes in eastern Canada developed by Hasegawa (1983). Since fault slip is proportional to seismic moment, the cumulative moment release gives some indication of the overall strain or deformation that has been occurring within the mine in association with this seismicity. Three phases of the post–closure seismicity are clearly evident on the moment–release curve and the anomalous nature of the events since 1991 is also apparent. The rate of seismic moment release since April, 1991 has been approximately twice as high as the rate in the period immediately after mining.

The cumulative moment release from the mine (approximately 3×10^{13} Newton–Meters) since closure in late 1984 has been equivalent to one magnitude 3.3 seismic event.

4 CHARACTERISTICS OF THE FALCONBRIDGE EVENTS

Figure 4 shows the SLTN traces for a typical seismic event at the mine. SWO and SUO are approximately equidistant from the mine (23 and 25 km respectively); SZO is approximately twice the distance of the other 2 stations (53 km). Three phases of ground motion are visible: the direct P phase, Pg, the direct S phase, Sg, and the Rayleigh surface wave, Rg. The peak ground motion on all the traces occurs in the Sg phase. Each of

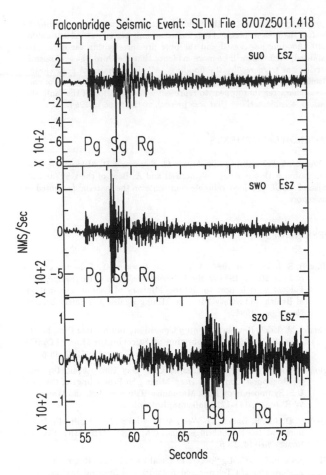

Figure 4. Typical SLTN traces for an induced seismic event in the Falconbridge Mine.

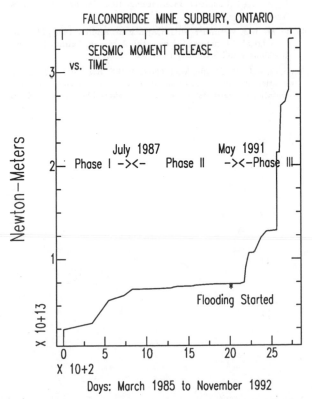

Figure 5. Cumulative seismic moment release at the Falconbridge Mine after 1984.

these phases is radiated from a slip–event on a fault surface with different radiation patterns which vary with the strike and dip of the fault surface, the direction of slip on the fault surface and the take–off angle for the ray going to a particular seismograph station. The relative amplitudes of the various local phases recorded on a seismograph station thus each vary in a different manner with changes in the fault–slip parameters. Consequently, if some of the Falconbridge events were produced by slip on faults with significantly different strike, dip or slip–directions, we should expect to see significant differences in the relative local phase amplitudes.

Figure 6 shows the SWO traces for three different Falconbridge events. The first occurred in July, 1987 and represents an example of the activity which continued after the major rockburst sequence in 1984 and the subsequent cessation of mining. The second trace shows an event in May, 1991 immediately after the resumption of significant seismic activity and the third trace gives an example of one of the most recent events (September, 1992). The events chosen all have approximately the same magnitude, M 1.5–1.7 and are typical of the events recorded in their particular time periods. The equivalent traces from SUO are shown in Figure 7.

From these Figures, it is appears that the overall character of the seismicity has not changed much over the five–year period represented by these plots. The first and last events are almost identical while the middle event has some extra cycles in its phases, which suggest a multiple slip event. However, there is little in the waveforms to suggest major changes in the mode of seismic deformation in the mine, from fault–slip to pillar–burst events for instance, nor significant changes in the fault–slip parameters. Instead, the similarities in the waveforms suggests a continuation of the dominant mode of seismic slip that was present at the end of active mining in the Falconbridge Mine in 1984 throughout the post–closure regime.

5 HISTORY OF FLOODING IN THE FALCONBRIDGE MINE

The pumps in the Falconbridge Mine were stopped in November, 1990 (G. Swan, Falconbridge, pers. comm., 1992). Thereafter, the mine began to flood. Detailed records of the rate and extent of flooding are lacking, but a reliable measurement of the water level within the mine on December 4, 1992 showed that the water had risen to within 1205 feet of the surface (B.

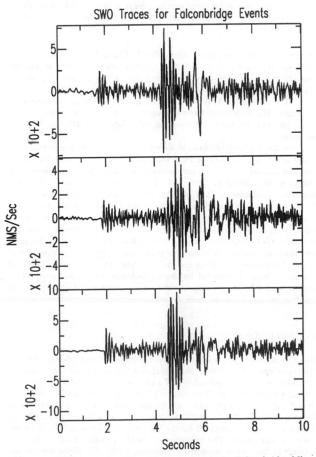

Figure 6. SWO traces for three seismic events at the Falconbridge Mine: upper - July 25, 1987; middle - May 21, 1991; lower - September 2, 1992.

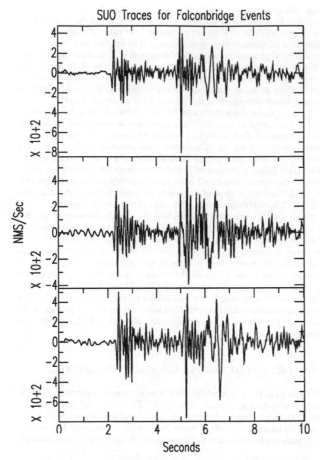

SUO Traces for Falconbridge Events

Figure 7. SUO traces for three seismic events at the Falconbridge Mine: upper - July 25, 1987; middle - May 21, 1991; lower - September 2, 1992.

Mikkila, Falconbridge, pers. comm., 1992). This indicated that the water had risen approximately 4800 feet in approximately 760 days, an average flooding rate of 6.3 feet/day. Assuming this rate was constant from the start of flooding, the water level within the mine should have reached the 4800-foot level in May, 1991 when the seismic activity reappeared in the mine. The major rockburst sequence of June, 1984 is known to have taken place in the range between the 4025-foot and 4375-foot levels in the mine.

There are many studies in the literature that show that increasing the water levels in reservoirs or increasing pore pressure by pumping fluids into the ground during oil/gas recovery or waste disposal schemes can trigger seismic activity. Board et al. 1992 discuss some of the more important examples of fluid-induced seismic activity. They also report on a test study in a deep-level South African gold mine where small scale seismicity was induced by injecting water along a fault surface. In the mine study they found that the amount of pressure needed to trigger seismic activity depended on the in situ state of stress along the fault and that the magnitude of the induced events depended on the area of elevated pore pressure along the fault surface. The most recent phase of seismic activity experienced by the Falconbridge Mine could easily have been triggered by the increased pore pressures that would accompany the flooding acting on one or more critical faults.

6 CONCLUSIONS

The Falconbridge Mine has experienced three distinct phases of post-closure seismicity. The first phase included the period from immediately after closure in 1984 until mid-1987 and was characterised by a moderate rate of seismic activity. The second phase began in mid-1987 and continued until April, 1991 and was characterised by a much lower rate of seismic activity. The third phase, beginning in 1991 and continuing to date, has been characterised by the highest level of seismic activity since the mine was closed. The initiation of the third phase of seismic activity appears to have been triggered by flooding within the mine. Seismic activity had subsided to negligible levels prior to the flooding and restarted at significant levels after the start of flooding in November, 1990. However, detailed knowledge of the flooding history is not available to allow

a precise comparison. At the time of writing, the mine is continuing to flood. If the present phase of heightened seismic activity at the mine stops after flooding has ceased and the pore pressures within the mine have stabilised, then that will be more evidence that the third phase of seismic activity was triggered by the flooding. All three phases of post-closure seismicity at the Falconbridge Mine have produced seismic events of consistent character that appear to represent a continuation of the fault–slip phase of seismic activity that was present at the mine in 1984.

7 ACKNOWLEDGEMENTS

We thank Falconbridge employees G. Swan and B. Mikkila for their cooperation with our study. W. McNeil and A. Bent of the GSC and M. Cajka of AECL provided valuable comments on the material presented in this report.

REFERENCES

Bharti, S. & D. West 1984. A Technical Assessment of the Rockbursts of June 20–23, 1984 at the Falconbridge Mine, Falconbridge, Ltd., Falconbridge Report to Ontario Ministry of Labour, Occupational Health and Safety Division, Mining Health and Safety Branch, November, 1984.

Board, M. 1992. Review of Mining Conditions and Seismic Data for the June 20–22, 1984 Rockbursting at Falconbridge Mine. ITASCA Report to Mining Research Directorate September 1992, 45 p.

Board, M., T. Rorke, G. Williams, & N. Gay 1992. Fluid Injection for Rockburst Control in Deep Mining, in Proceedings of the 33rd U.S. Symposium on Rock Mechanics, Tillerson, J.R. and Wawersik, W.R. (eds.), 111-121, Balkema, Rotterdam

Hedley, D.G.F, S. Bharti, D. West, & W. Blake 1985. Fault-slip Rockbursts at the Falconbridge Mine, CANMET MRP/MRL Division Report 85-114, 13 p.

Hasegawa, H.S. 1983. Lg Spectra of Local Earthquakes Recorded by the Eastern Canada Telemetered Network and Spectral Scaling, Bull. Seism. Soc. Am., 54, 1041-1061.

Hansen, D., P. Rochon, & T. Semadeni 1989. Seismic Monitoring Systems Being Used in the Canada/Ontario/Industry Rockburst Project; in Proceedings of the Fred Leighton Memorial Workshop on Mining Induced Seismicity, International Rock Mechanics Congress, Montreal, 59–168.

Plouffe, M., R.J. Wetmiller, M.G. Cajka & M. Andrew 1990. The Sudbury Local Telemetered Seismograph Array. In: Fairhurst, C. (ed) Rockbursts and Seismicity in Mines, 221-226.

Wetmiller, R.J., M. Plouffe, M.G. Cajka & H.S. Hasegawa 1990. Investigation of mining-related and natural seismic activity in northern Ontario. In: Fairhurst, C. (ed) Rockbursts and Seismicity in Mines.

Author index